BIRD ATLAS OF NATAL

BIRD ATLAS OF NATAL

DIGBY CYRUS

NIGEL ROBSON

illustrated by

TONY CLARKSON

UNIVERSITY OF NATAL PRESS Pietermaritzburg 1980

University of Natal Press
Box 375, Pietermaritzburg
3200, South Africa.

ISBN 0 86980 215 1

Printed by:
L. Backhouse
Pietermaritzburg

To

HILARY ROBSON and ROSE THOMPSON

The publication of the *Bird Atlas of Natal* is timed to coincide with the Fiftieth Anniversary of the Southern African Ornithological Society which was founded in Natal.

Contents

Maps, Figures and Tables

Foreword

The ornithology of Natal has been linked with the names of the Woodward brothers (*Natal birds*, 1899) and P. A. Clancey (*The birds of Natal and Zululand*, 1964). Now it will also be linked with the names of Digby Cyrus and Nigel Robson whose *Atlas* is the first to be produced for the region, and the first to be published for any part of southern Africa. It is a fitting contribution to the Jubilee Celebrations of the Southern African Ornithological Society at Hluhluwe in Natal this year.

Until now, the only atlas coverage that any part of Africa has received has been in the magnificent *Atlases of speciation in African birds*, the *Passerines* by Hall & Moreau (1970) and the *Non-passerines* edited by Snow (1978). But the scale of these maps and the nature of records have been useful only for the grossest distribution patterns. Natal forms less than 1% of the Afrotropical Region (Africa south of the Sahara), yet its economy must be almost the largest per unit area, and its rate of progress close to the highest on the whole continent.

A measure of the natural resources of Natal has become more critical than ever, therefore. As with any other part of the world, its wildlife is irreplaceable, yet threatened to a greater degree than at any time in the past. Baselines for future comparisons become more necessary every day, so an atlas of distribution in time and space becomes an invaluable tool in the hands of planners, agriculturalists, conservationists and biologists, because it is in their hands that the future of the land and the quality of human life may ultimately depend.

Cyrus and Robson embarked on a labour of love, and have come up with a major achievement. Milestones and monuments are well worn words endlessly applied to important works of any kind, but I am going to wear them a little further by saying that the *Bird Atlas of Natal* is a milestone in the ornithology of Africa and a monument to the dedication of two men who spent their own time and money to produce one of the finest documentaries of its kind in the world.

Perhaps modesty should curb the scope of my eulogy, since I have been closely associated with the *Atlas* in an advisory capacity since its early stages, but the basic ideas, the planning, execution and production of this work have been those of Cyrus and Robson – not of mine. The *Atlas* is their baby – my role may be likened more to that of godfather.

The greatest apparent shortcoming of any atlas is that it is out of date even as it comes off the press. This is as it should be, because it illustrates the dynamic nature of biological systems (including man). New records were being added up to the very moment of going to press; an Angola Pitta arrived in Pietermaritzburg on 29 November 1979 just in time for inclusion – the first record of the species for Natal. But the new and unusual do not diminish the value of a biological atlas; rather does the atlas become increasingly valuable as it highlights the changes that are constantly occurring. And that, after all, is what it's for.

Gordon Maclean
Pietermaritzburg, January 1980

Preface

The idea of a Natal bird distribution project was put forward to the Pietermaritzburg branch of the University of Natal Wildlife Society by its Chairman, Digby Cyrus, towards the end of 1975. A project was subsequently initiated to determine the distribution of birds in Natal, using methods similar to those of the atlassing project in the Transvaal, for which fieldwork was in progress at the time. Early in 1976 the organisation of fieldcards in Natal was taken over by Nigel Robson, Recorder of the Natal Bird Club.

It was decided to combine efforts towards the *Bird Atlas of Natal* and the project gained momentum. We decided to include records for the decade 1970-9, to allow time for detailed coverage of the area. The Natal Bird Club under the chairmanship of Hardy Wilson assisted by providing the fieldcards, covering the cost of all correspondence and setting aside space in its newsletter, *Albatross*, for regular progress reports. As a result, we were able to keep members informed of our progress and to encourage them to fill in fieldcards. The Natal Parks Board, through Dr John Hanks, then Chief Professional Officer, distributed fieldcards to staff in all its Reserves. By publicising the project and making personal approaches, assistance was obtained from other birdwatchers living in outlying districts. Before long, information was being gathered on a large scale from many parts of Natal. A sense of purpose was given to even the most casual birdwatching. The project was registered with the Southern African Ornithological Society, the Natal Parks Board, the Department of Forestry and, latterly, with the Department of Zoology, University of Natal, Pietermaritzburg, under the supervision of Professor Gordon Maclean, who was already acting in an advisory capacity.

By 1978, atlassing in Natal was well established. We decided that, apart from providing up-to-date distributional data for use by other researchers, the *Atlas* should be expanded to include illustrations and short informative texts to complement the maps. Tony Clarkson of Ladysmith, who was contributing fieldcards, readily agreed to illustrate the book.

Up to the end of 1978, funding of the project was on a small scale and running expenses were borne mainly by us. In 1979, sufficient funds were forthcoming to cover basic costs of the monthly atlassing excursions undertaken by us to the poorly covered parts of Natal.

The completion of this project concludes what is only a stage in the investigation of bird distribution in Natal. It is to be hoped that the gathering of information will continue and that, with the use of computers, updated distribution maps will be produced in the future. This *Atlas* could then be used as a yardstick for determining changes in the avifauna which may have occurred in the interim.

Acknowledgements

During the preparation of the *Atlas*, we have been helped and encouraged by a large number of people and organizations. Many gave us practical assistance with the collection of data, some gave us advice on compiling and arranging the material, and others gave financial assistance. Our sincere thanks go to them all. In the lists that follow, we hope that we have included everyone.

Publication of the results of our work was made possible by generous sponsorship which is most warmly acknowledged:

Unilever South Africa lent us a large sum without interest.
Natal Bird Club also helped with an interest-free loan.

Substantial donations were received from:

John Voelcker Bird Book Fund and **the Natal Parks Board.**

Financial support also came from:

Natal Bird Club
Natal Bird Club — individual members
SAPPI Kraft (Pty) Ltd
Wildlife Society — Mandini Centre

We received much help in compiling the material for publication. Hilary Robson, throughout the duration of the project, and Rose Thompson, in the latter stages, were totally involved. Professor Gordon Maclean was always ready to discuss and advise on *Atlas* matters, and, with Professor John Hanks, helped to make our work known to radio listeners all over South Africa. The Committee of the Natal Bird Club gave us whole-hearted support and allowed us space in the N.B.C. Newsletter *"Albatross"* as a communication medium. The Natal Parks Board helped in every way that they could, including giving us access to their gamebird data; John Line assisted with the extraction of these records. Mr P. A. Clancey advised on matters related to species distributions and, as Director of the Durban Museum, allowed us to consult its records of specimens. The Southern African Ornithological Society permitted us to use data on species distributions contained in the new *Checklist of southern African birds.* The Centre for the Rehabilitation of Wildlife supplied us with records of avian casualties. Dave White, Steve and Tessa Blaber, and Geoff Nichols all helped at some time with plotting the information on fieldcards on to maps. Peter Frost gave guidance on the interpretation section of the *Atlas*, and Bruno Martin aided us with advice and practical assistance with the cartography; Dr R. E. Shulze supplied the data for the temperature maps. Drafts and the final manuscript were typed by Marjory Hill and thoroughly checked by Gordon Bennett. Mr H. C. Cyrus made constructive comment on the text of the introductory sections. For finally putting the book together, we are indebted to our publishers, the University of Natal Press, and in particular to Margery Moberly and Jean Cockcroft.

Fieldwork was the most important single aspect on which this *Atlas* is based. Many people played an active part in this, and their practical assistance in the collection of the data illustrates how much amateurs can contribute to the science of ornithology. We would like to single out a few stalwarts. Bob, Trish and John Turner undertook monthly trips through western Zululand from Babanango and Tony Clarkson helped to cover western and north-western Natal from Ladysmith. Geoff Nichols and Steve Blaber accompanied us on several fieldtrips and regularly contributed fieldcards from a wide range of localities. Hamish Campbell co-ordinated *Atlas* work in the Durban area and assisted with covering southern Natal. Marlene Sach provided regular information from KwaZulu areas in northern and north-eastern Zululand, and Hugh Chittenden did the same from the Nkandla-Eshowe district and Cape Vidal. Les Nutting and David Manry provided comprehensive data from the Estcourt and Mooi River environs, Ray and Sheila Hewitt covered much of the Estcourt, Winterton and Bergville areas whilst Harry Nevill thoroughly investigated the Richmond district. Hardy Wilson travelled widely throughout the study area returning a large number of fieldcards, and Brian Dennis accompanied us on many day trips to specific parts of Natal. The Natal Parks Board arranged for their field staff to return fieldcards from many of their Reserves.

The following contributed fieldcards during the *Atlas* period; those whose names are marked with an asterisk covered particular squares on a monthly basis:-

C. Adam*, B. E. S. Agar, C. J. Albertyn, M. H. & W. Astrup, S. B. Baines, G. A. Baird, S. & G. Baldo*, V. A. Ball, A. Barclay, B. Barnes, J. Barnes, J. Barrie, P. & R. Bennet, G. Bennett, I. & A. Bensimon, R. M. F. Berard*, R. T. F. Bernard, A. Berruti, C. Bingham, S. J. M. & T. Blaber, R. Blaker, C. Bleksley, A. Boland, M. L. Borrageiro, M. & P. Bosman, A. O. Bourquin, H. Brassey, H. E. D. Briscoe, R. K. Brooke, N. Brooks, C. Brown, L. Brunjes, M. N. Bruton, C. A. Bruzas, J. Burman, J. Burn, J. E. Butler, D. Calder, H. A. Campbell*, J. & P. Campbell*, R. Cassidy, D. Chalton, H. Chittenden*, J. Clarke, A. C. V. Clarkson*, C. Coetser*, D. Coghlan, B. D. Colahan, E. A. Cole, K. M. Collins, D. Cook, K. H. Cooper, J. A. Copelyn, L. Cowan, A. V. A. Cowen, R. Cowgill, A. J. F. K. Craig*, R. A. Crook*, J. Crooks, J. Crowson, N. Cunningham*, A. R. Currie, H. C. & I. D. Cyrus, M. Damp, M. Dark*, D. H. Day, S. de Jager*, J. de Lange, B. Dennis*, D. Densham*, N. J. Devenish, P. W. de Vos, A. Dixon, R. G. & J. R. Dove, R. J. Dowsett, F. Dowsett-Lemaire, W. Dreboldt, T. Dugmore, D. C. Dunton, W. Duthie, R. A. Earle*, I. Espie, R. E. Evans, J. Fenwick, J. Field*, A. T. & J. Forbes, O. C. Foreman, A. S. & P. D. Formo, P. G. H. & S. Frost*, J. & D. Galatis, I. F. Garland*, D. Geiger, J. Gerber, C. A. Gerrans, S. M. Gillatt, R. Goetz, P. S. Goodman*, N. Greeff*, E. Green*, B. C. Grobler, N. J. Grobler, T. Gunton, B. Gwala, D. Hall, P. Hancock*, D. Hanmer, B. Harris, E. W. Harris*, W. R. Harris*, C. Hawkins, E. Helm, T. G. Henderson*, S. & R. Hewett*, L. Hilkovitz*, C. Hilton-Taylor, C. Hines, P. M. Hitchins*, R. & S. Hitchner*, N. J. Hori, R. Horner, H. B. Huntley, J. Hutt, D. Irwin, P. Irwin, B. Jamison, R. A. C. Jensen, C. H. Jerome, S. Jex*, B. & T. Johnson, J. Johnson, G. W. & K. A. Johnstone, W. Juta, A. G. Kee, D. M. Kee, A. & M. Kemp, F. J. Kleyn, R. Kleyn, S. Large, A. Lawson, L. Lawson, H. T. Laycock, T. D. Le Brasseur*, P. le Hausse, D. Lee, K. Leo-Smith, R. W, Leslie, M. L. Leslie, M. Lewis, B. Liggitt, J. Line, I. G. B. Little, D. Lloyd, G. Lowe, P. B. Lowry, R. G. Lucey, D. R. MacDevette, I. A. W. MacDonald*, F. M. MacDonald, C. Mack, I. Macfadyen, G. L. Maclean, L. G. C. Maclear, M. B. Madonsela, A. P. Malherbe, Q. V. Mann, D. E. Manry*, A. Marchant, J. Massingham, D. McCarthy, A. G. McClean, L. McCormack, D. J. McCulloch, L. J. McDonald, J. D. M. McLaren*, A. &. S. Meinesz, H. P. Mendelsohn*, J. & C. Mendelsohn, B. Meredith*, K. L. Milne*, P. Milstein, D. Mitchell*, J. Morphew, C. L. Mowat, V. Muller, R. & A. Murray, R. Narraway, M. Neethling, W. M. Neethling, H. Nevill*, B. A. Newman, P. J. Newman, G. H. Nicholls, G. Nichols, M. Norris-Rogers, L. Nutting*, T. B. Oatley*, C. Oberholster, M. Oliver, D. Olsen, T. Orpwood, M. Page, N. M. Page, R. Parker, G. Parton, D. D. Penn, P. Perlstein, P. & J. Phelan, R. Physick, S. E. Piper, R. N. Porter, A. Rabie, C. Rattray*, C. M. Rautenbach, J. P. Reed, C. Reim, R. Reinertsen, I. F. Rennie*, J. Rennie*, S. Rennie, D. S. Reynolds, J. H. Robson, N. Rochfort, O. C. Rogers, A. E. A. Root, G. Root*, K. B. Rosewall, D. Rossiter, P. Rowsell, A. L. & G. S. Ruddock*, M. Sach*, K. Saunderson, B. Schmidt, A. M. Schofield, D. Schultz, I. Schumann, P. Scogings, J. Scotcher, L. Scott*, I. S. Scott-Barnes, A. Sharp*, J. B. Shephard, N. Short, G. Simmons, W. Small, W. P. Small, G. R. Smith, J. Stannard*, P. & E. Steineger, B. A. Stewart*, R. Stubbs, T. J. Sutcliffe, G. Tatham, M. G. Taylor*, R. Taylor, T. Teague, S. Tedder, P. Thompson, R. Thompson, R. P. Thorpe, A. C. Tingle, A. J. Tomkinson*, S. Trathen, R. G. & P. Turner*, J. Turner*, R. A. &

A. Turner, A. Twiss, H. Valentine, G. van Heerden*, G. van Rensberg, J. van Rensberg, C. Vernon, R. Wakeford, E. R. Walker, C. Walley, R. Way, C. Wex*, D. F. & B. White, J. Whitecross, P. & S. Whitehouse, A. K. Whitfield, H. W. Whitfield*, K. & J. Willan, A. P. Williams, A. Wilson, H. E. Wilson, A. R. Wright, J. Wyatt, D. J. Young, G. A. Zaloumis*.

The cost of special *Atlas* fieldwork in poorly-covered areas was largely borne by ourselves. However, we are most grateful to the following for financial assistance towards this aspect of the *Atlas* project:

Natal Bird Club
Natal Bird Club — individual members
Natal Parks Board
Percy Fitzpatrick Institute of African Ornithology
Southern African Ornithological Society
University of Natal Wildlife Society
Wildlife Society of Southern Africa

Gorgeous Bush Shrike.

And, finally, a special word of thanks to Tony Clarkson. We were indeed fortunate to have him as artist for this book. His achievement in illustrating every bird in the map section and a selection of the rarer species is considerable and a measure of his special talent.

DIGBY CYRUS
NIGEL ROBSON

References

Brooke, R. K. and Sinclair, J. C. 1978. A preliminary list of southern African seabirds. *Cormorant* 4:10-17.

Clancey, P. A. 1955. Results of the Durban Museum ornithological expedition to Pondoland in August 1954. *Durban Mus. Novit.* 4(13):187-214.

Clancey, P. A. 1964. *The birds of Natal and Zululand.* Edinburgh: Oliver and Boyd.

Dean, W. R. J. 1971. Breeding data for the birds of Natal and Zululand. *Durban Mus. Novit.* 9(6):61-91.

de Jager, J. M. and Schulze, R. E. 1977. The broad geographic distribution in Natal of climatological factors important to agricultural planning. *Agrochemophysica* 9:81-91.

Dowsett, R. J. (in prep.). *An atlas of Zambian birds.*

Dyer, T. G. J. and Tyson, P. D. 1977. Estimating above- and below-normal rainfall periods over South Africa, 1972-2000. *J. Appl. Met.* 16:145-147.

Hall, B. P. and Moreau, R. E. 1970. *An atlas of speciation in African passerine birds.* London: British Museum (Natural History).

Kemp, M. I. (in press). *An atlas of Transvaal bird distribution.* Pretoria: Transvaal Museum.

Liversidge, R. 1959. Tropical mountain birds south of the Zambesi. *Ostrich Suppl.* 3:68-78.

Liversidge, R. 1962. Distribution of birds in relation to vegetation. *Ann. Cape Prov. Mus.* 2:143-151.

McLachlan, G. R. and Liversidge, R. 1978. *Roberts birds of South Africa.* Cape Town: Trustees of the John Voelcker Bird Book Fund.

Pentz, J. A. 1945. An agro-ecological survey of Natal. *Dept. Agric. Forestry, Soil and Veld Conservation Series* 7(250):1-10.

Perring, F. H. and Walters, S. M. 1962. *Atlas of the British flora.* London: Nelson.

Rowan, M. K. 1971. On temperature and disease as limiting factors in distribution. *Ostrich Suppl.* 9:147-151.

Sharrock, J. T. R. 1976. *The atlas of breeding birds in Britain and Ireland.* Berkhamsted: T. and A. D. Poyser.

Siegfried, W. R., Frost, P. G. H., Cooper, J. and Kemp, A. C. 1976. *South African Red Data Book – Aves.* Pretoria: CSIR.

Snow, D. W. Ed. 1979. *An atlas of speciation in African non-passerine birds.* London: British Museum (Natural History).

Southern African Ornithological Society. 1980. *S.A.O.S. checklist of southern African birds.* Johannesburg: S.A.O.S.

Tyson, P. D. and Dyer, T. G. J. 1975. Mean fluctuations of precipitation in the summer rainfall region of South Africa. *S.Afr. Geog. J.* 57:104-110.

Winterbottom, J. M. 1959. Some zoogeographical remarks on the South African avifauna. *Ostrich Suppl.* 3:40-46.

Winterbottom, J. M. 1962a. The east coast elements in the South African avifauna. *Proc. 1st Fed. Sci. Congr.*: 189-194.

Winterbottom, J. M. 1962b. A note on zoogeographical limits in south-east Africa as suggested by the avifauna. *Ann. Cape Prov. Mus.* 2:152-154.

Woodward, R. B. and Woodward, J. D. S. 1899. *Natal birds.* Pietermaritzburg: Davis.

Yates, D. W. 1966. *The pattern and structure of fresh milk dairy farming in the Durban milkshed.* Pietermaritzburg: Unpublished B.A. thesis, University of Natal.

Abbreviations

G. R. Game Reserve
N. R. Nature Reserve
S. W. Sewage Works

Introduction

BACKGROUND TO ATLASSING

Biological atlases had their precedent nearly twenty years ago when in 1962 The Botanical Society of the British Isles published the *Atlas of the British Flora* by F. H. Perring and S. M. Walters. Using a 10-km square gridded map, plant distributions were plotted on a presence or absence basis. This pointed the way for a similarly comprehensive and equally objective mapping of the distributions of breeding birds, and *The Atlas of Breeding Birds in Britain and Ireland* by J. T. R. Sharrock, was published in 1976. Bird distribution atlases are currently in preparation in many parts of the world, including Europe, North America, Africa and Australia. In Southern Africa, *An Atlas of Transvaal Bird Distribution* by M. I. Kemp, is in the press, and fieldwork is under way towards producing atlases for Zambia (Dowsett in prep.) and Malawi.

AIM AND OBJECTIVES

The aim of the project was to map the distributions, by presence or absence per month, of all bird species occurring in Natal during the decade 1970-9, using the National Quarter-Degree Grid as the square unit. Each of these 'squares' represents one of the maps in the 1:50 000 Topocadastral map series produced by the Trigonometrical Survey Office. The objectives were to present baseline data on the current distributions and seasonal occurrences of birds in Natal, against which future changes in the avifauna could be measured. The area covered by this work is shown on Map 1 and includes that portion of the Republic of Transkei encompassed by Natal, and the historical areas of Natal, East Griqualand and Zululand, which together form the Province of Natal in the Republic of South Africa.

DATA SOURCES, COLLECTION AND STORAGE

All known sources of information were exploited. Data were extracted from records kept by the Natal Bird Club, Durban Museum and the Natal Parks Board, and to a lesser extent from other organizations, individuals' notes and published records. A programme of fieldwork was undertaken which developed on a widespread and co-operative basis, involving members of the Natal Bird Club, field staff of the Natal Parks Board, individuals of the Department of Forestry and others, such as landowners, with a knowledge of birds.

A fieldcard was designed for listing birds observed in the field, and to facilitate the plotting of the data on to maps. Supplies of fieldcards were distributed to all participants for regular monthly completion for particular squares, or in respect of casual visits to any other squares. Where comprehensive lists were available in field notebooks, backdated fieldcards were completed. Many cards were filled in during atlassing trips organized by ourselves, usually of several days' duration each month, to squares in little-known parts of the study area. With limited time available, we would cover four squares a day by car, having first consulted maps and planned routes to include as many habitats as possible. The number of species recorded per square varied, but it was not unusual for our lists each to contain over 80 species. The bulk of the data used in the *Atlas* were collected during the active fieldwork period (see Maps 2 and 3).

Species cards were used to catalogue individual records, such as those extracted from notebooks, files and publications. These included hard-to-find species (for example nocturnal birds), single observations and records of specimens lodged with the Durban Museum. All map entries stem from data on either a fieldcard or a species card. This information is filed and stored with the Natal Bird Club, enabling all records to be traced back to the locality, date and observer.

DATA PRESENTATION

The data are presented in two sections. The first comprises information on the 530 most commonly recorded species occurring in Natal, and the second summarises records of the rare species. Those species recorded in Natal before 1970 but not during the *Atlas* period are listed in Table 4 on page 12, thus completing the list of birds known to have occurred in Natal (see checklist, page 307).

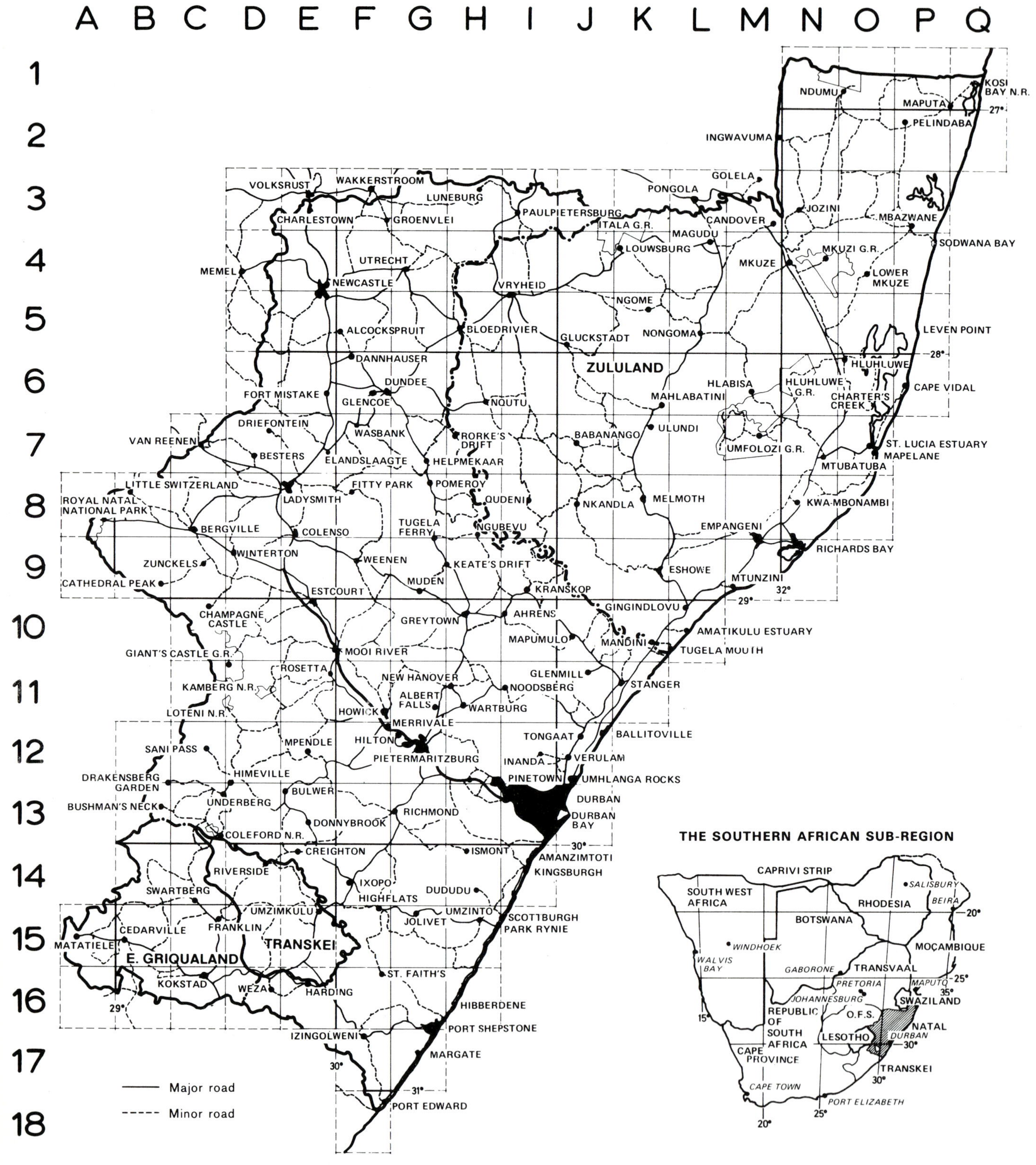

Map 1: Natal showing quarter-degree squares with their major place-names.

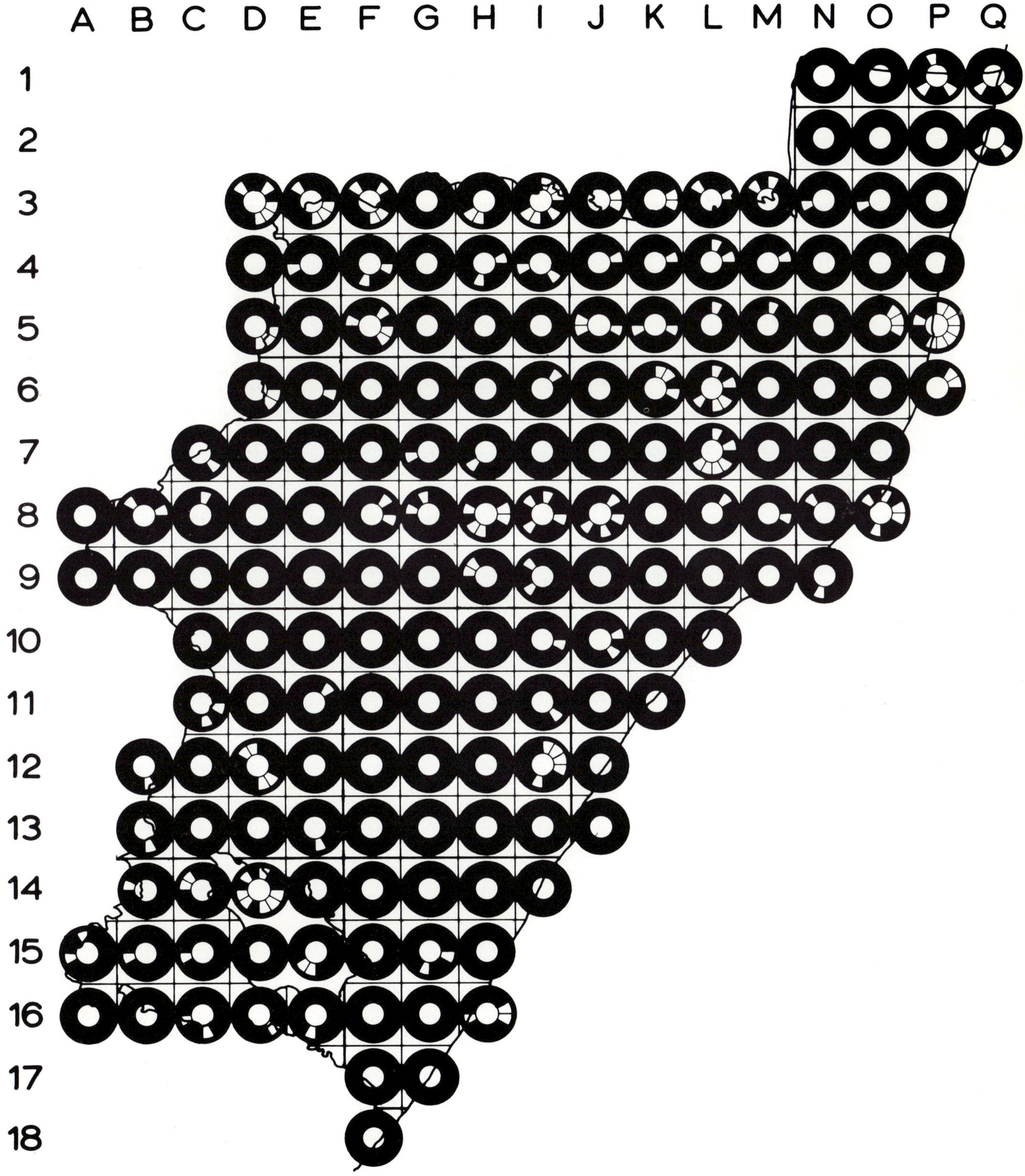

Map 2: Coverage attained during the study period, January 1970 – December 1979 (months per quarter-degree square).

Figure 1 shows an enlarged version of the symbol used on the maps of the main section. A circle, placed within each quarter-degree square on the map, is divided into twelve segments. Each segment represents a month of the year, clockwise from January to December, for the 10-year period. A bold outline applied to the circumference indicates that the species was recorded in that square, and the blacked-in segments indicate the months during which it was recorded.

Figure 1: Symbol used to illustrate monthly information recorded per quarter degree square.

Figure 2 illustrates three typical seasonal patterns which could occur within a quarter-degree square: resident (present all year), summer visitor (present from September to March) and winter visitor (present from April to August). The maps illustrate distributional patterns at a glance, and seasonal patterns in detail.

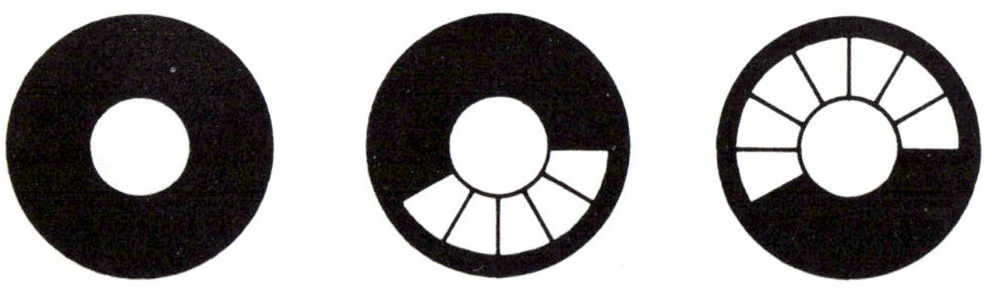

Resident   Summer Visitor   Winter Visitor

Figure 2: Symbols showing typical seasonal patterns occurring within a quarter-degree square.

In three cases, two quarter-degree squares were coupled because of the inaccessibility of one of them. All the information from the accessible square was used for the other. Decisions to couple squares were taken after we had visited such areas and found them to contain habitats similar to those of the adjacent squares. These were: A 9 linked to A 8, B 12 to B 13, and A 16 to B 16.

In seven cases Natal Parks Board reserves straddle the boundaries between squares and, because of the impracticability of determining precisely in which square a species was seen, the record was mapped for both squares. The Reserves in question were the Kosi System (Q 1 and Q 2), Ndumu G.R. (N 1 and O 1), Mkuzi G.R. (N 4 and O 4), Itala G.R. north (J 3 and K 3), Itala G.R. south (J 4 and K 4), Giant's Castle G.R. west (C 10 and C 11) and Giant's Castle G.R. east (D 10 and D 11). Where squares border neighbouring territories, the whole square has been incorporated in the study area. Species observed outside Natal in border squares, which would have been new to the Natal list, were omitted.

Because only actual records were plotted and no assumptions were made in presenting the data, it follows that the maps show all squares in which a species was observed and not necessarily all in which it occurs.

In the species write-ups, the number and English name of each species follows *Roberts birds of South Africa* (1978), while the scientific nomenclature follows the *S.A.O.S. checklist of southern African birds* (1980). The world distribution, followed by the distribution within the Southern African Sub-region, is given for each species, mainly using data from the S.A.O.S. checklist. The habitats in which the bird is found in Natal are based on our experience.

Mention is made of the breeding season in Natal for each species, based largely on Dean (1971). Where records of breeding during a period of consecutive months were published, breeding is stated as 'recorded'. When data were incomplete, the season is given from the earliest to the latest date, or simply as during the summer or winter. Finally, the number of squares in which the bird was recorded from 1970-9 is given, as well as the percentage that it represents of the 166 squares covering Natal.

Changes made to the nomenclature during the *Atlas* period resulted in four species each being split into two. They were Cuckoo (340), Blue-cheeked Bee-eater (406), Rockjumper (540) and Boubou Shrike (709), in the cases of the first two, all four 'new' species occur in Natal.It was impossible at that stage to separate records and therefore the original nomenclature was retained and the species mapped accordingly. The latter two species presented simpler cases, as only one of each of the 'new' species occurs in Natal and the English name was therefore retained and the scientific nomenclature changed.

SUMMARY MAPS

The maps and figures in this section are included to assist in orientation and to give details pertaining to coverage of the study area and the number of species recorded per square.

To enable the reader to relate records in squares on the species map to localities, an orientation map is included (Map 1). This will also be useful to birdwatchers who wish to establish whether birds they observe in a certain locality have been previously recorded in that particular square and for that month.

The historical boundaries of East Griqualand, Natal and Zululand, which together form the Province of Natal, are included on Map 1 as in many instances in the text distributional data are referred to by these regions.

The coverage attained, in terms of at least one fieldcard with a minimum of 25 species listed, per month per quarter-degree square, is summarised on Map 2. Map 3

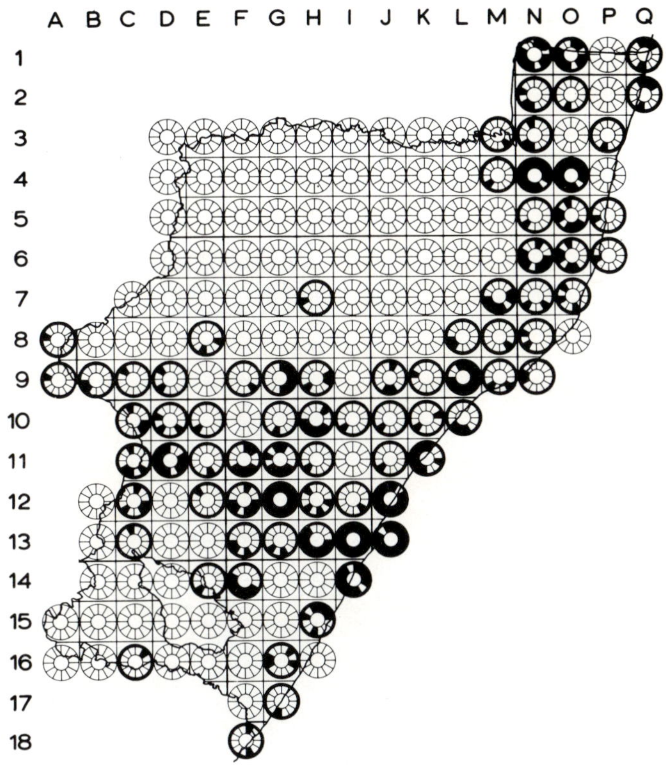

Map 3: Coverage, January 1970 — October 1976 (months per quarter-degree square).

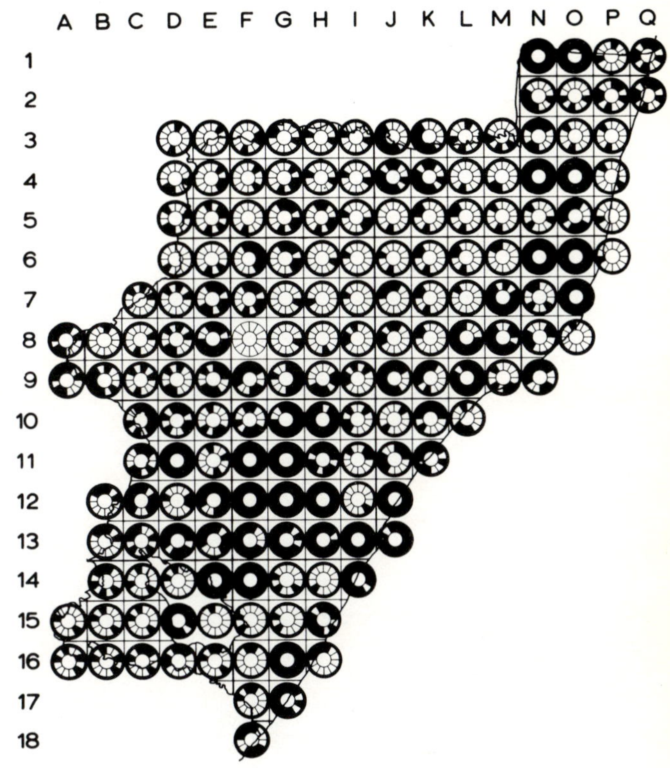

Map 4: Coverage, January 1970 — March 1978 (months per quarter-degree square).

illustrates the coverage position shortly after the start of the project, for the period January 1970 — October 1976 and reveals observer bias and the limited extent of existing available information. The coverage approximately half-way through the active fieldwork period is illustrated on Map 4 and shows the progress made towards covering outlying areas and reducing observer bias. Our field trips were aimed mainly at eliminating this bias and attaining as wide a knowledge as possible of the birds and their habitat preferences throughout the study area. Our coverage during these trips is summarised on Map 5.

The numbers of returned fieldcards, totalling 4855, used for this project are shown in Figure 3. Map 6 illustrates species diversity during the *Atlas* period, and the numbers of species per square are shown in groups of 50 to enable comparisons to be made between squares. The actual number of squares having similar species totals is given in Figure 4. Squares with similar numbers of returned fieldcards are shown on Map 7. When related to Map 6, it will be seen that squares with a low number of returned fieldcards do not necessarily have a low number of species recorded. This indicates the extent to which observer bias was eliminated. Actual species and fieldcard numbers are included in Table 1. This table provides a cross-reference between the grid codes in the *Atlas* and the official map references and names used by the Trigonometrical Survey Office.

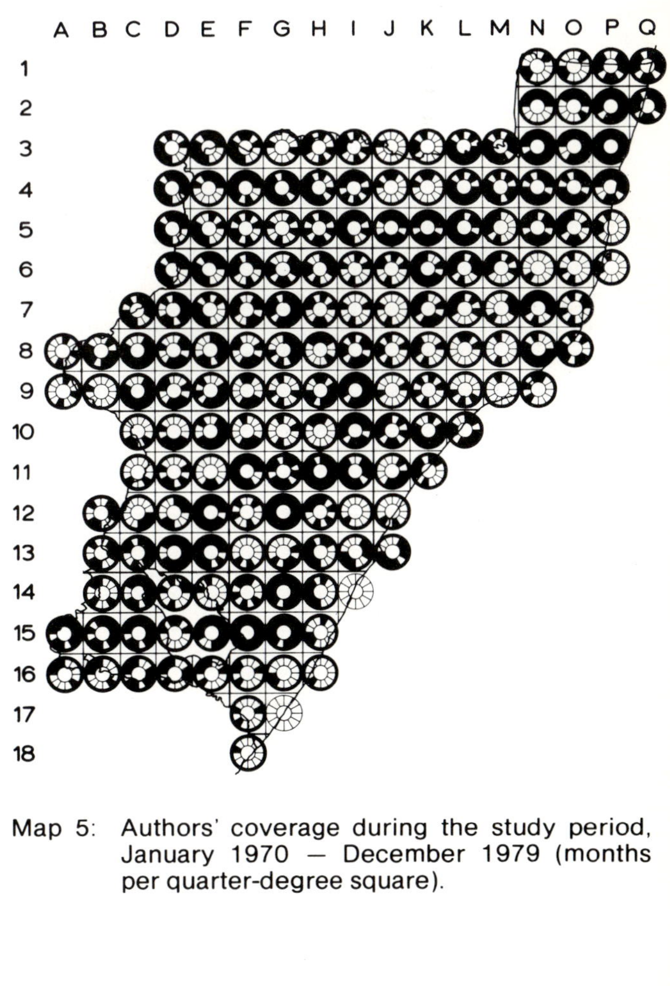

Map 5: Authors' coverage during the study period, January 1970 — December 1979 (months per quarter-degree square).

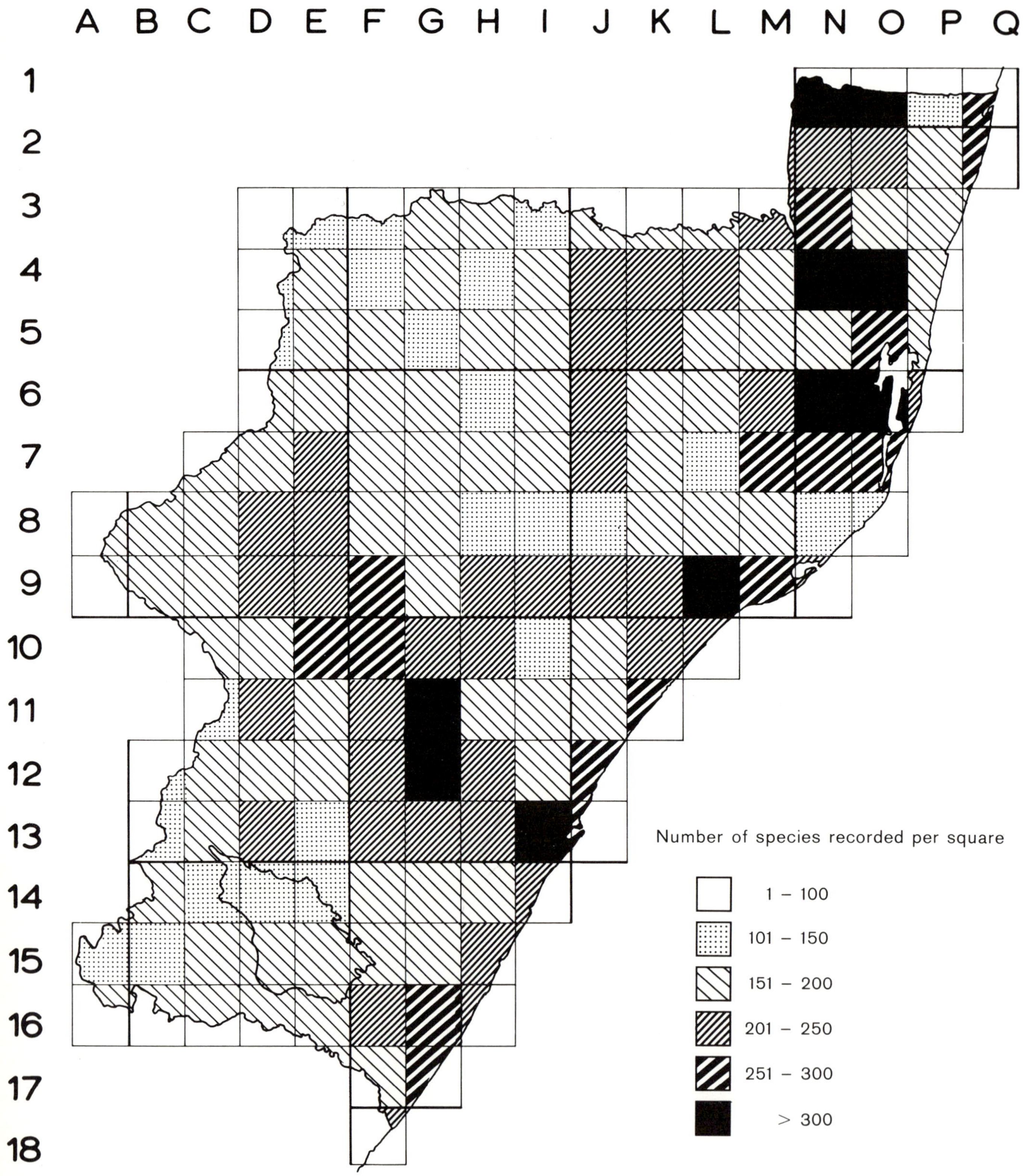

Map 6: Species diversity.

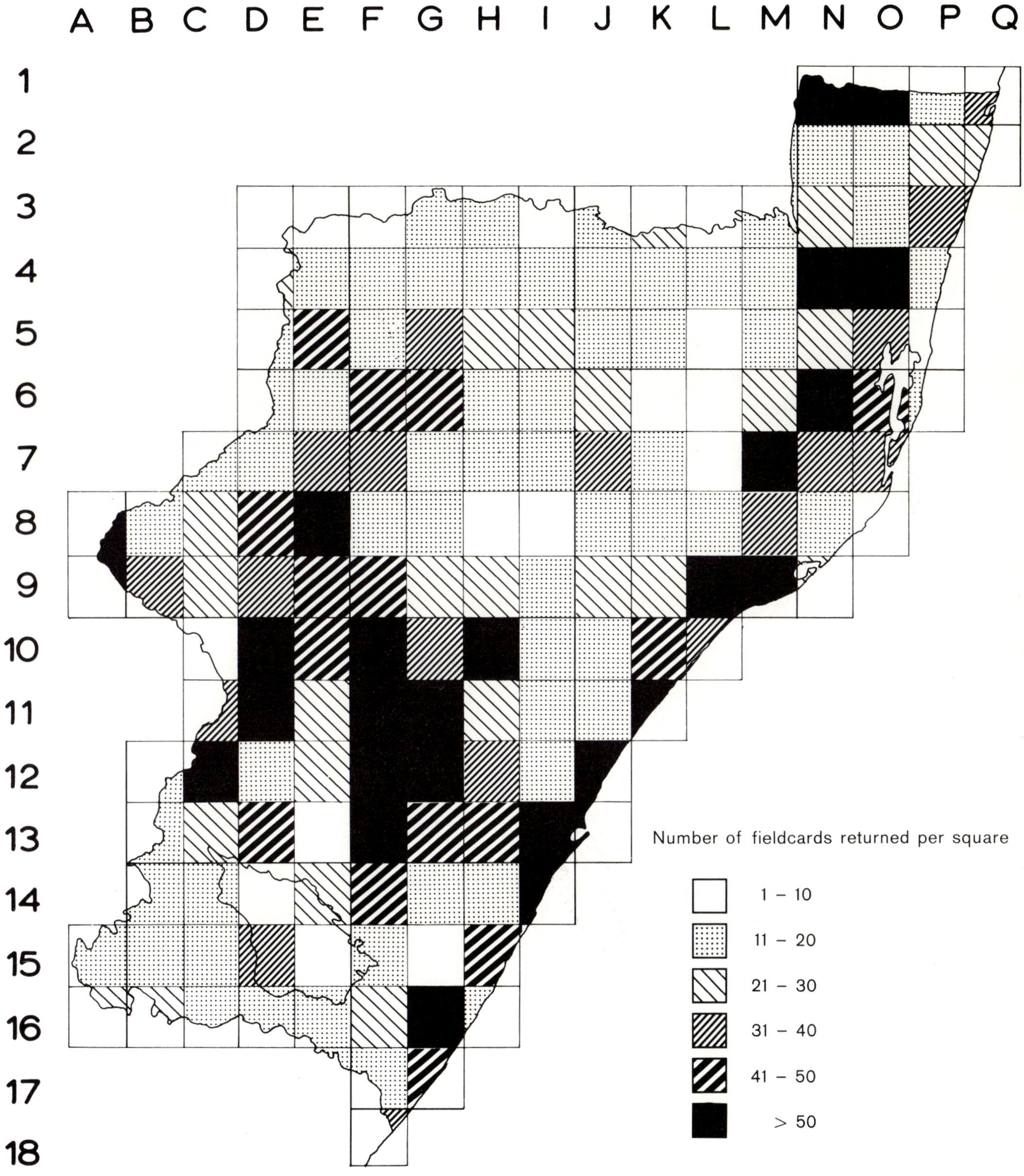

Map 7: Distribution of returned fieldcards.

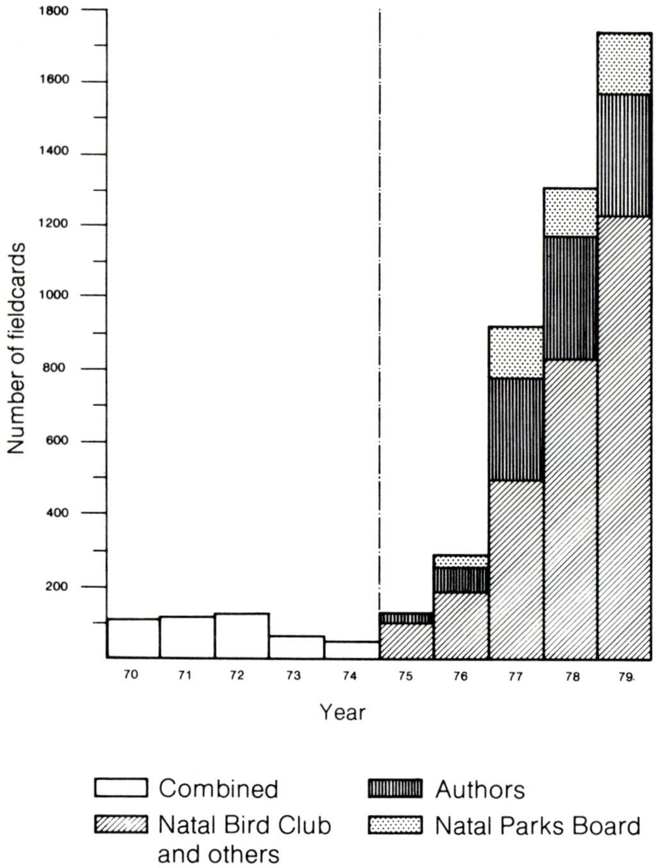

Figure 3: Fieldcard returns 1970-79.

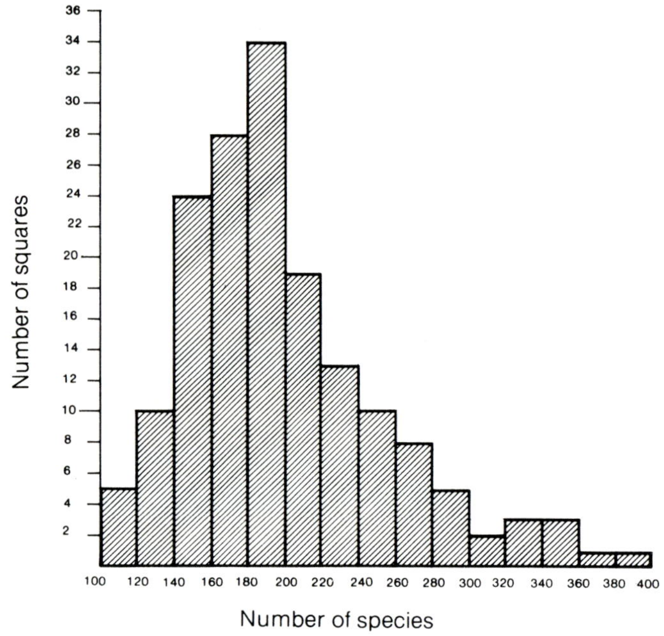

Figure 4: The number of squares with similar species totals.

Table 1: Quarter-degree squares in Natal, with actual numbers of fieldcards returned and species recorded.

Atlas Reference	1:50 000 Reference	1:50 000 Title	Fieldcards	Species
A 8	2828 DB	Witsieshoek	69	193
A 9	2828 DD	Mont-Aux-Sources	69	193
A 15	3028 BD	Matatiele	11	145
A 16	3028 DB	Luyengweni	23	156
B 8	2829 CA	Oliviershoek	16	196
B 9	2829 CC	Cathedral Peak	38	165
B 12	2929 CA	Sani Pass (West)	16	129
B 13	2929 CC	Bushman's Neck	16	129
B 14	3029 AA	Taylerville	19	159
B 15	3029 AC	Cedarville	12	144
B 16	3029 CA	Ntola	23	156
C 7	2829 AD	Van Reenen	16	163
C 8	2829 CB	Bergville	24	181
C 9	2829 CD	Zunckels	21	194
C 10	2929 AB	Champagne Castle	50	173
C 11	2929 AD	Giant's Castle	34	136
C 12	2929 CB	Sani Pass	64	188
C 13	2929 CD	Underberg	24	171
C 14	3029 AB	Swartberg	14	131
C 15	3029 AD	Franklin	13	151
C 16	3029 CB	Kokstad	11	166

Atlas Reference	1:500 000 Reference	1:50 000 Title	Fieldcards	Species
D 3	2729 BC	Streepfontein	8	114
D 4	2729 DA	Memel	21	128
D 5	2729 DC	Mont Pelaan	12	154
D 6	2829 BA	Santiago	13	156
D 7	2829 BC	Besters	20	188
D 8	2829 DA	Spioenkop	49	240
D 9	2829 DC	Winterton	32	215
D 10	2929 BA	Ntabamhlope	51	169
D 11	2929 BC	Kamberg	141	213
D 12	2929 DA	Himeville	15	179
D 13	2929 DC	Pevensey	43	201
D 14	3029 BA	Riverside	6	106
D 15	3029 BC	Glengarry	31	152
D 16	3029 DA	Weza	20	195
E 3	2729 BD	Volksrust	10	123
E 4	2729 DB	Ingogo	21	166
E 5	2729 DD	Newcastle	44	196
E 6	2829 BB	Fort Mistake	23	172
E 7	2829 BD	Elandslaagte	38	214
E 8	2829 DB	Ladysmith	54	238
E 9	2829 DD	Frere	43	221
E 10	2929 BB	Estcourt	43	296
E 11	2929 BD	Nottingham Road	26	193
E 12	2929 DB	Impendle	24	192
E 13	2929 DD	Donnybrook	18	148
E 14	3029 BB	Creighton	24	147
E 15	3029 BD	Umzimkulu	10	155
E 16	3029 DB	Harding	16	175
F 3	2730 AC	Wakkerstroom	10	139
F 4	2730 CA	Balelesberg	12	132
F 5	2730 CC	Alcockspruit	13	169
F 6	2830 AA	Dundee	47	195
F 7	2830 AC	Wasbank	32	186
F 8	2830 CA	Jobskop	14	174
F 9	2830 CC	Weenen	48	266
F 10	2930 AA	Weston	61	270
F 11	2930 AC	Howick	64	241
F 12	2930 CA	Merrivale	55	229
F 13	2930 CC	Byrne	56	235
F 14	3030 AA	Ixopo	43	200
F 15	3030 AC	Highflats	13	168
F 16	3030 CA	St. Faiths	24	211
F 17	3030 CC	Izingolweni	15	192
F 18	3130 AA	Port Edward	35	236
G 3	2730 AD	Vredehof	15	170
G 4	2730 CB	Utrecht	15	176
G 5	2730 CD	Doringberg	35	138
G 6	2830 AB	Tayside	43	187
G 7	2830 AD	Helpmekaar	16	189
G 8	2830 CB	Pomeroy	18	181
G 9	2830 CD	Muden	24	190
G 10	2930 AB	Mount Alida	34	244
G 11	2930 AD	Albert Falls	81	309
G 12	2930 CB	Pietermaritzburg	196	326
G 13	2930 CD	Richmond	47	219
G 14	3030 AB	Nhlavini	13	153
G 15	3030 AD	Jolivet	10	180
G 16	3030 CB	Port Shepstone	52	270
G 17	3030 CD	Margate	49	251
H 3	2730 BC	Luneburg	14	152
H 4	2730 DA	Nkambule	12	146

Table 1 (Cont.): Quarter-degree squares in Natal with actual numbers of fieldcards returned and species recorded.

Atlas Reference	1:50 000 Reference	1:50 000 Title	Fieldcards	Species
H 5	2730 DC	Kingsley	29	169
H 6	2830 BA	Nqutu	15	149
H 7	2830 BC	Rorke's Drift	17	176
H 8	2830 DA	Collessie	10	118
H 9	2830 DC	Nadi	21	208
H 10	2930 BA	Greytown	55	215
H 11	2930 BC	New Hanover	29	198
H 12	2930 DA	Cato Ridge	32	203
H 13	2930 DC	Hammarsdale	47	231
H 14	3030 BA	Dududu	13	189
H 15	3030 BC	Scottburgh	49	247
H 16	3030 DA	Hibberdene	12	214
I 3	2730 BD	Paulpietersburg	6	125
I 4	2730 DB	Hlobane	14	180
I 5	2730 DD	Vryheid	24	187
I 6	2830 BB	Nondweni	15	180
I 7	2830 BD	Fort Louis	19	184
I 8	2830 DB	Qudeni	9	141
I 9	2830 DD	Kranskop	19	237
I 10	2930 BB	Ahrens	16	146
I 11	2930 BD	Noodsberg	12	151
I 12	2930 DB	Inanda	12	158
I 13	2930 DD	Durban	315	358
I 14	3030 BB	Umkomaas	65	226
J 3	2731 AC	Hartland	17	186
J 4	2731 CA	Coronation	20	249
J 5	2731 CC	Gluckstadt	10	212
J 6	2831 AA	Nhlazatshe	24	212
J 7	2831 AC	Babanango	32	242
J 8	2831 CA	Nkandhla	15	129
J 9	2831 CC	Mbongolwane	24	210
J 10	2931 AA	Mapumulo	16	176
J 11	2931 AC	Glendale	22	190
J 12	2931 CA	Verulam	123	287
J 13	2931 CC	Durban	115	300
K 3	2731 AD	Klipwal	24	195
K 4	2731 CB	Louwsburg	20	250
K 5	2731 CD	Ngome	11	214
K 6	2831 AB	Mahlabatini	10	164
K 7	2831 AD	Dingaanstat	18	171
K 8	2831 CB	Melmoth	20	188
K 9	2831 CD	Eshowe	27	237
K 10	2931 AB	Tugela	41	250
K 11	2931 AD	Stanger	58	259
L 3	2731 BC	Pongola	10	170
L 4	2731 DA	Magudu	15	213
L 5	2731 DC	Nongoma	9	166
L 6	2831 BA	Maphophoma	9	145
L 7	2831 BC	Ntiningwe	8	143
L 8	2831 DA	Nkwaleni	16	180
L 9	2831 DC	Blackburn	78	317
L 10	2931 BA	Gingindhlovu	33	220
M 3	2731 BD	Golela	14	237
M 4	2731 DB	Nkonkoni	20	198
M 5	2731 DD	Ngxongwane	15	162
M 6	2831 BB	Hlabisa	27	219
M 7	2831 BD	Mfolozi Reserve	105	292
M 8	2831 DB	Empangeni	32	193
M 9	2831 DD	Felixton	51	270

Atlas Reference		1:50 000 Reference	1:50 000 Title	Fieldcards	Species
N	1	2632 CC	Abercorn Drift	68	359
N	2	2732 AA	Ingwavuma	16	239
N	3	2732 AC	Jozini	27	273
N	4	2732 CA	Umbombo	105	381
N	5	2732 CC	Mhlosinga	25	198
N	6	2832 AA	Hluhluwe Reserve	100	321
N	7	2832 AC	Mtubatuba	33	252
N	8	2832 CA	Kwa-Mbonambi	19	142
N	9	2832 CC	Richards Bay	21	216
O	1	2632 CD	Ndumu	71	355
O	2	2732 AB	Sihangwane	17	225
O	3	2732 AD	Tshongwe	18	195
O	4	2732 CB	Lower Mkuze	104	378
O	5	2732 CD	Kuleni	35	286
O	6	2832 AB	Hluhluwe	48	328
O	7	2832 AD	St. Lucia Estuary	39	263
O	8	2832 CB	Cape St. Lucia	7	118
P	1	2632 DC	Muzi	11	146
P	2	2732 BA	Ngutshana	23	189
P	3	2732 BC	Sibayi	32	198
P	4	2732 DA	Sordwana Bay	12	190
P	5	2732 DC	St. Mary's Hill	5	162
P	6	2832 BA	Hluhluwe	15	208
Q	1	2632 DD	Kosibaai	33	270
Q	2	2732 BB	Black Rock	24	261

BIAS AND EXTENT OF COVERAGE

Over the study period, January 1970—December 1979, over 80% of the 166 quarter-degree squares within the study area were covered by at least one fieldcard for each of the 12 months of the year (Table 2) with an overall coverage of 91%. The average number of species recorded per square was 201 with a range of 106 to 381.

Table 2: **Percentage of quarter-degree squares covered per calendar month during the period 1970-79.**

Month	J	F	M	A	M	J	J	A	S	O	N	D
%	92	92	89	84	88	91	92	92	92	95	91	97

North-western Natal, parts of KwaZulu and East Griqualand were found to have lower species numbers than elsewhere, mainly as a result of low habitat diversity, but also because of the difficulty of access or the absence of local observers. Observer bias was reflected in the higher numbers shown from the Durban and Pietermaritzburg areas, the Drakensberg resorts and the Zululand Game Reserves. The result of efforts to reduce this bias is shown by comparing the number of species (over 350) recorded from the Ndumu and Mkuzi Game Reserves (from where thorough lists were returned each month over several years), with the average number of species (over 200) recorded from the intervening squares with a fraction of the coverage of the Reserves (c.f. numbers of fieldcards returned). Some species which appeared to be sparsely distributed are more difficult to locate than others. These include rails, crakes and some warblers which are secretive; night herons, owls and nightjars which are nocturnal; and albatrosses, shearwaters and petrels which are pelagic and, as a result of there being few suitable land-based observation points, were under-recorded.

The tape recorder proved to be a useful tool for establishing the presence of certain species by calling them up, for example Water Rail (197), Natal Nightjar (375), Stierling's Barred Warbler (618), Rudd's Apalis (624) and Woodwards' Batis (676).

SIGHT RECORDS OF RARE OR SIMILAR-LOOKING SPECIES

Sight records of birds rare in Natal or from unlikely localities were accepted or rejected on the basis of our own experience of birdwatching. Criteria for acceptability of identification included, where necessary, a conclusive field description, a knowledge of the observer's experience with the species, and details of the conditions pertaining to the observation. In cases of similar-looking species possibly being confused, for example Greater Double-collared Sunbird (758) and Lesser Double-collared Sunbird (760), the habitat in which the bird was recorded was taken into account. Although certain birds, such as some raptors, larks and warblers, are difficult to identify, their distribution patterns appear to be well defined. Other species, such as the bishops and weavers, are difficult to identify when not in breeding plumage and, as a result, their maps appear to indicate (incorrectly) that they are summer visitors to certain areas.

NATAL LIST AND THE ATLAS

The total number of species recorded during the *Atlas* period was 630 of which 56 were not listed by Clancey (1964) (Table 3). A further 28 species were recorded prior to, but not during, the *Atlas* period (Table 4). The total number of species recorded in Natal at 31 December 1979 is 658 (see Checklist, page 307).

Table 3: Species recorded in Natal during the *Atlas* period, 1970-79, not listed by Clancey (1964).

	Macaroni Penguin		Franklin's Gull
9	Grey-headed Albatross	289X	Sabine's Gull
	Northern Giant Petrel	290X	Gull-billed Tern
15	Silver-grey Fulmar		Black-naped Tern
18	Schlegel's Petrel		White-cheeked Tern
	Kerguelen Petrel	292	Antarctic Tern
21X	Medium-billed Prion	293	Roseate Tern
26	Cory's Shearwater	295X	Bridled Tern
26X	Flesh-footed Shearwater	301	White-capped Noddy
26Z	Wedge-tailed Shearwater	305X	Black Tern
36	Black-bellied Storm Petrel	329X	Ring-necked Parakeet
37	White-bellied Storm Petrel	401	Grey-hooded Kingfisher
116Y	Sooty Falcon	455	Angola Pitta
120	Western Red-footed Kestrel	456	Singing Bush Lark
121	Dickinson's Kestrel	463	Thick-billed Lark
136	Lesser Spotted Eagle	523X	House Crow
163	Dark Chanting Goshawk	596	Icterine Warbler
225	Black Korhaan	597	Olive Tree Warbler
231X	European Oystercatcher	615	Stierling's Barred Warbler
234	Mongolian Sandplover		Grey Wagtail
240X	Eastern Golden Plover	730	White-crowned Shrike
251X	Pectoral Sandpiper	733	European Starling
	Buff-breasted Sandpiper	769	Blue-throated Sunbird
251Z	Broad-billed Sandpiper	821	Cut-throat Finch
254	Knot	850	Purple Widowfinch
271	Red-necked Phalarope	854	Cuckoo Finch
285	Pomarine Skua	860X	Lemon-breasted Canary
287X	Lesser Black-backed Gull	865	White-throated Seedeater

Table 4: Species recorded in Natal prior to, but not during, the *Atlas* Period 1970-79

3	Rockhopper Penguin	*Eudyptes chrysocome*
28	Baillon's Shearwater	*Puffinus lherminieri*
39	Red-tailed Tropicbird	*Phaeton rubricauda*
117	Red-necked Falcon	*Falco chicquera*
	Long-legged Buzzard	*Buteo rufinus*
157	Ovambo Sparrowhawk	*Accipiter ovampensis*
200	Striped Crake	*Aenigmatolimnas marginalis*
201	Spotted Crake	*Porzana porzana*
204	White-winged Flufftail	*Sarothrura ayresi*
218	Ludwig's Bustard	*Neotis ludwigii*
249	Great Snipe	*Gallinago media*
265	Black-tailed Godwit	*Limosa limosa*
271	Grey Phalarope	*Phalaropus fulicarius*
303	Common Noddy	*Anous stolidus*
306	African Skimmer	*Rynchops flavirostris*
307	Namaqua Sandgrouse	*Pterocles namaqua*
342	Lesser Cuckoo	*Cuculus poliocephalus*
366	Barred Owl	*Glaucidium capense*
446	Bennett's Woodpecker	*Campethera bennettii*
448	Knysna Woodpecker	*Campethera notata*
473	Rudd's Lark	*Heteromirafra ruddi*
525	Grey Tit	*Parus afer*
568	Capped Wheatear	*Oenanthe pileata*
604X	European Reed Warbler	*Acrocephalus scirpaceus*
611	Knysna Scrub Warbler	*Bradypterus sylvaticus*
738	Blue-eared Glossy Starling	*Lamprotornis chalybaeus*
743	Burchell's Glossy Starling	*Lamprotornis australis*
747	Yellow-billed Oxpecker	*Buphagus africanus*

ADDITIONAL RECORDS

Although we have included as complete a list as possible of birds recorded in Natal, some records may have eluded us. The following species were reported in Natal during the *Atlas* period but, in our opinion, the records require confirmation.

Table 5: Species recorded during the *Atlas* period which require confirmation.

24 Great Grey Shearwater

Procellaria cinerea

Occurrence indicated by Brooke and Sinclair (1978) — 'rarely to Natal'. No records submitted.

25 Great Shearwater

Puffinus gravis

Occurrence indicated by Brooke and Sinclair (1978) — 'to all waters of Namibia and South Africa'. No records submitted.

28 Baillon's Shearwater

Puffinus lherminieri

Occurrence indicated by Brooke and Sinclair (1978) — 'scarce winter visitor to Natal'. No records submitted.

31 Leach's Storm Petrel

Oceanodroma leucorhoa

Occurrence indicated by Brooke and Sinclair (1978) — 'all South African waters'. No records submitted.

153 Augur Buzzard

Buteo augur

Reported from Steilrand (J 5) in August 1978.

249 Great Snipe

Gallinago media

Reported from Durban (I 13) in September 1970.

271 Grey Phalarope

Phalaropus fulicarius

Occurrence indicated by Brook and Sinclair (1978) — 'rare off Natal'. No records submitted.

Herring Gull

Larus argentatus

Occurrence indicated by Brook and Sinclair (1978) — 'two visual records from Natal'. No records submitted.

345 Thick-billed Cuckoo

Pachycoccyx audeberti

Reported from Ndumu G.R. (N/O 1) in April 1977.

446 Bennett's Woodpecker

Campethera bennettii

Reported from the Pongola flood plain (O 2) in June 1978.

592 Thrush Nightingale

Luscinia luscinia

Reported from the Bluff, Durban (J 13) and Umhlanga S.W. (J 12) in February, and from KwaMashu S.W. (I 13) in March 1974.

623 Black-headed Apalis

Apalis melanocephala

Reported from Cape Vidal (P 6) in December 1979.

655X White-collared Flycatcher

Ficedula albicollis

Reported from the Umgeni Mouth, Durban (J 13) in October 1979.

762 Yellow-bellied Sunbird

Nectarinia venusta

Reported from Nhlolo pan (0 2) in July 1979.

USING THE ATLAS IN THE FIELD

It is hoped that the *Atlas* will be useful to birdwatchers in many ways. Should they wish to determine where a particular species may be located, they can check the distribution maps for localities and refer to the text for habitat preferences. If any species are recorded in new localities, observers are asked to submit full details to The Recorder, Natal Bird Club, c/o Durban Museum, P.O. Box 4085, Durban 4000. New distributional data will be published regularly in *Albatross,* the newsletter of the Natal Bird Club.

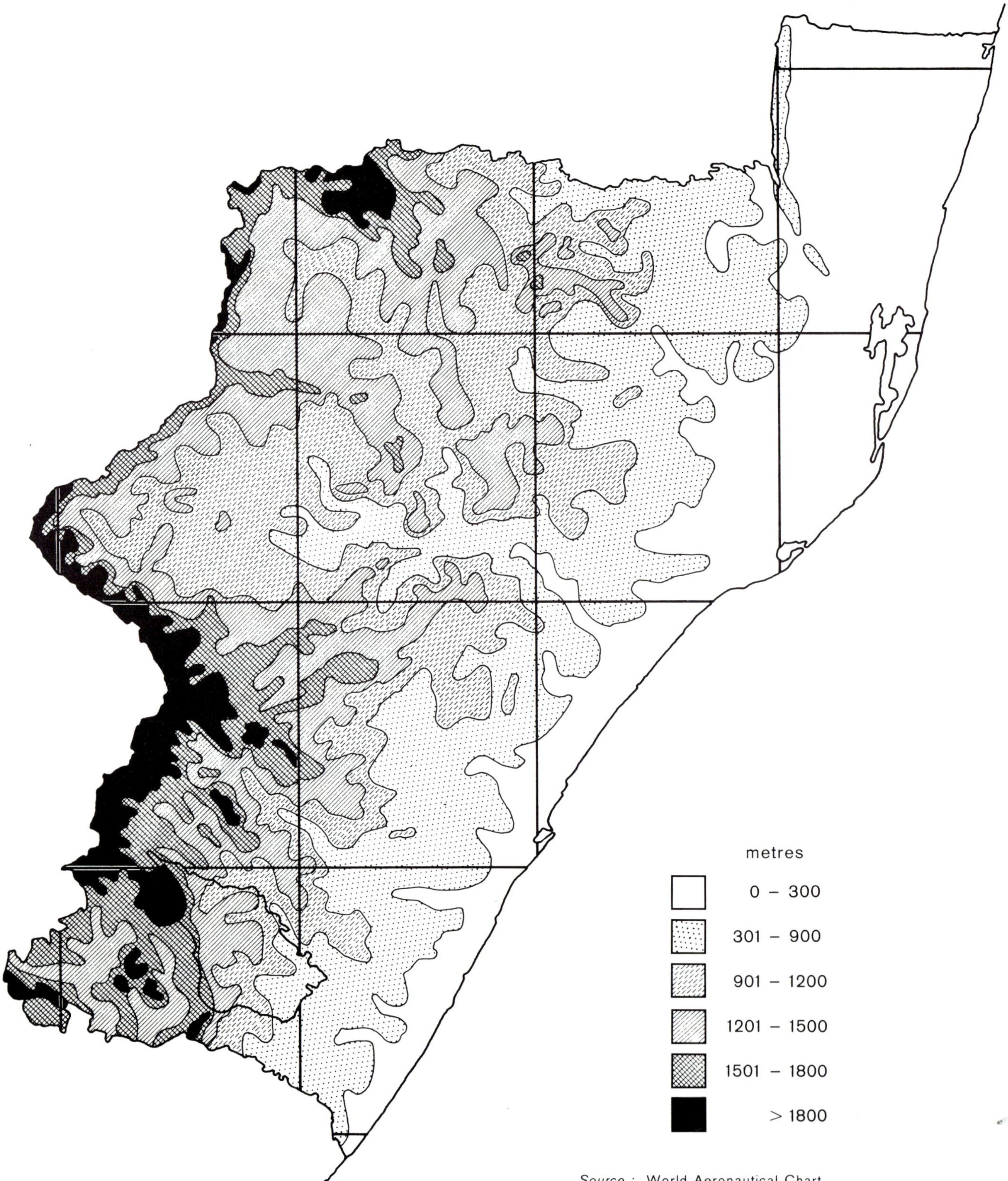

Map 8: Altitude

Atlas Interpretation

INTERPRETATION MAPS

Bird community patterns result from the interactions between species over long (evolutionary) and short (ecological) time scales. The *Atlas* provides a basis for those wishing to make assessments of changes in bird distribution patterns over short time-scales. Maps showing ecological factors which may affect the distribution of species are included to enable the reader to draw his own conclusions. These are: Altitude (Map 8), Mean Daily January and July Temperatures in °C (Maps 9 and 10), Mean Annual Rainfall in mm (Map 11), Mean Monthly December and January Rainfall in mm (Maps 12 and 13) and Vegetation Types (Map 14). Conservation and KwaZulu areas (Map 15) are shown because of the effect they have on the habitat in terms of protection and land use. The Altitude and Vegetation maps are also provided in overlay form for use with the species maps.

The importance of ecological factors in bird distribution has been stated by several authors. Bird distribution patterns result from a combination of factors rather than just one. For example, Liversidge (1962) states that 'dominant climatic factors in a region influence the birds of that area', indicating that several, but not necessarily all, climatic factors may combine to affect the distribution of a species. Rowan (1971) has discussed the effects altitude and temperature may have on the Cape Robin (581). Liversidge (1959) found that zones of highest mean temperature in January and July coincided with the limits of distribution of many species in the southern parts of their ranges in southern Africa. Rainfall, apart from having short-term effects on aquatic species, has little direct influence, but exerts major influences on vegetation growth and distribution. During the *Atlas* period, rainfall was above average and sustained droughts occurred only in northern Natal and Zululand during the summer 1978-9. This above-average rainfall may account for the fact that several species previously recorded in Natal were found to be rare or absent, for example Namaqua Sandgrouse (307) (unrecorded) and Chestnut-backed Finchlark (484) (very local). Some climatologists (Tyson and Dyer 1975) hold the view that annual rainfall varies above or below the mean in approximately 10-year cycles. This is shown by Dyer and Tyson (1977) who plotted the mean yearly percentage deviation of rainfall for South Africa (summer rainfall region), as well as the estimated wet and dry spells to the year 2000 (Fig. 5). The *Atlas* period coincided with a wet cycle and it is possible therefore that over the next ten years some species from the dry interior of Southern Africa may move into Natal. On the other hand, an influx of certain species, such as White-bellied Stork (78) and European Roller (412) into many parts of Natal during the summer 1978-9 occurred simultaneously with drought conditions elsewhere.

Figure 5: Rainfall deviation above and below the mean annual (1910-2000), showing wet and dry cycles. (Based on Dyer and Tyson 1977).

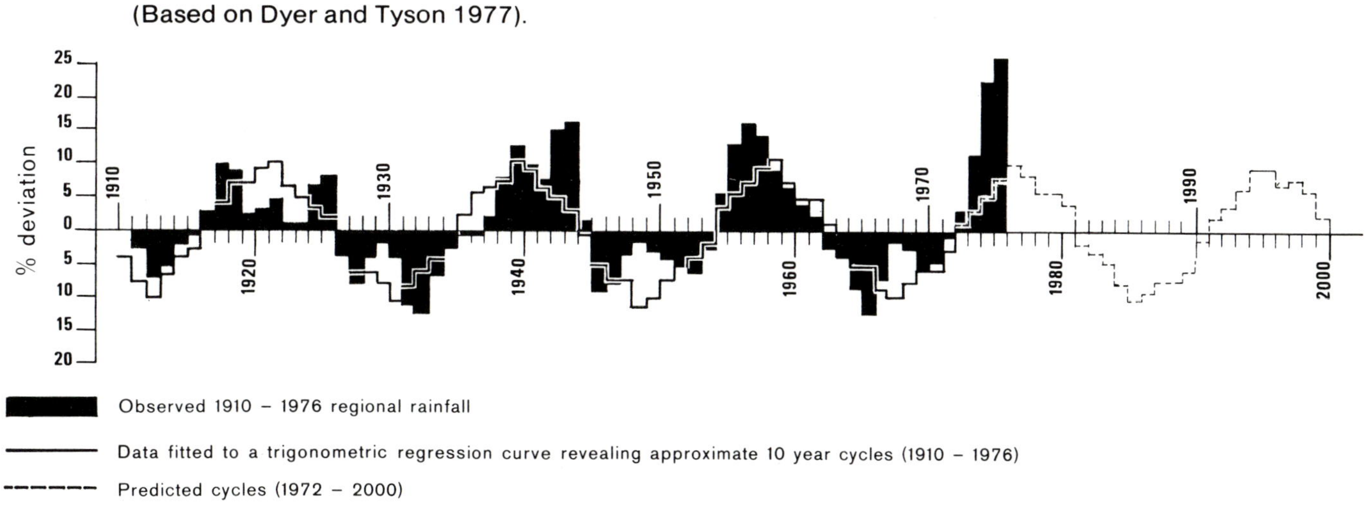

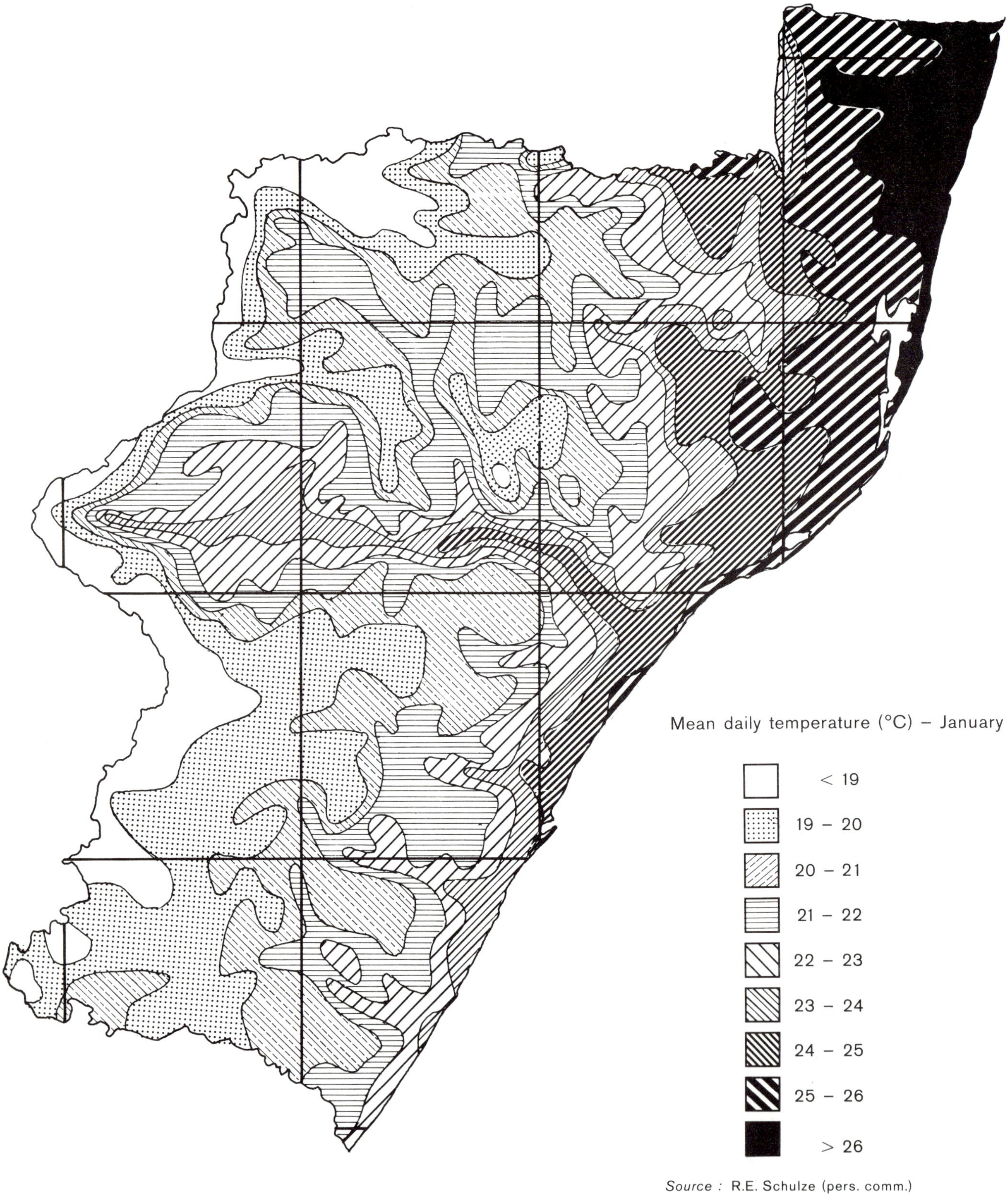

Map 9: Mean daily temperature (°C) — January

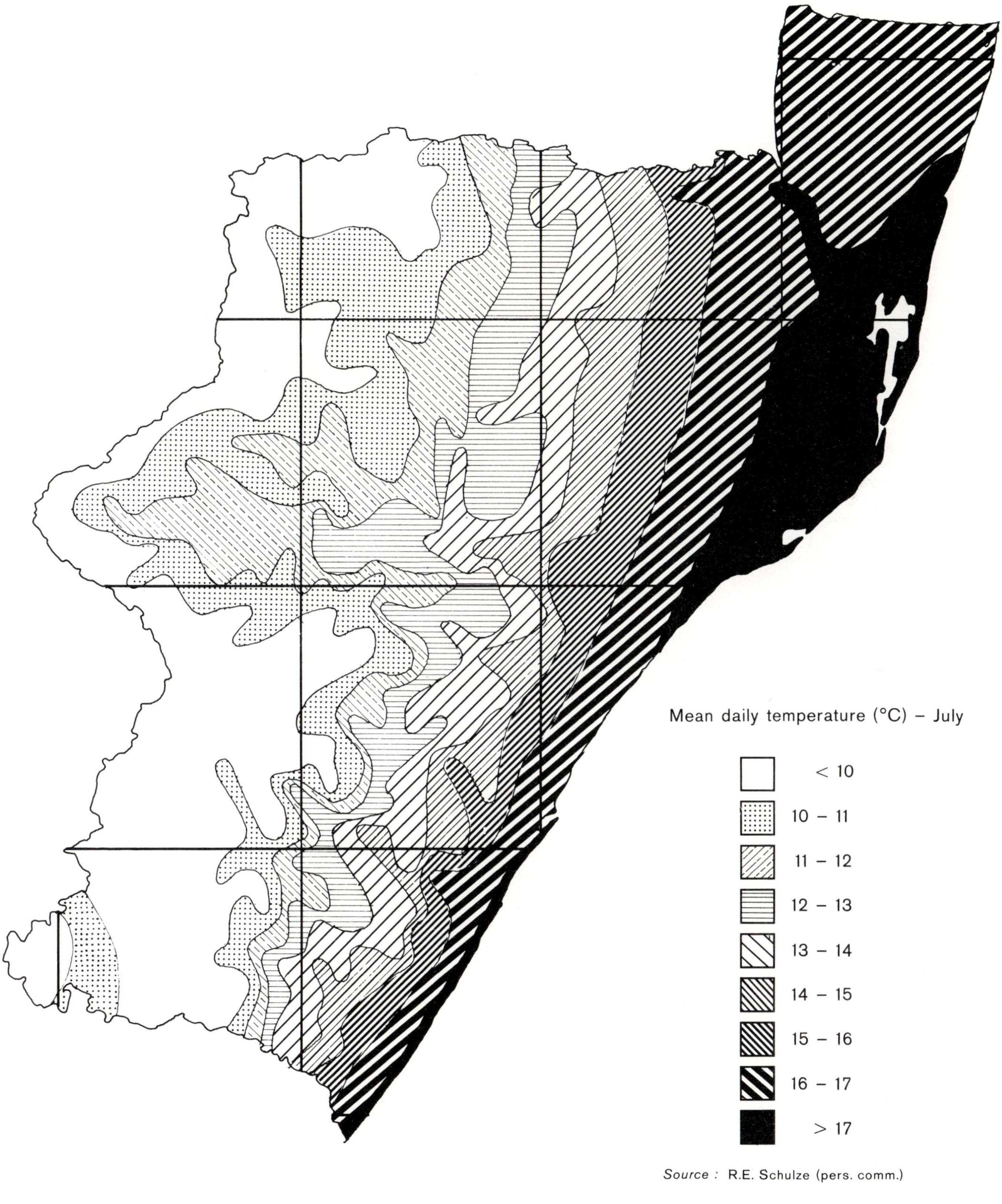

Map 10: Mean daily temperature (°C) — July

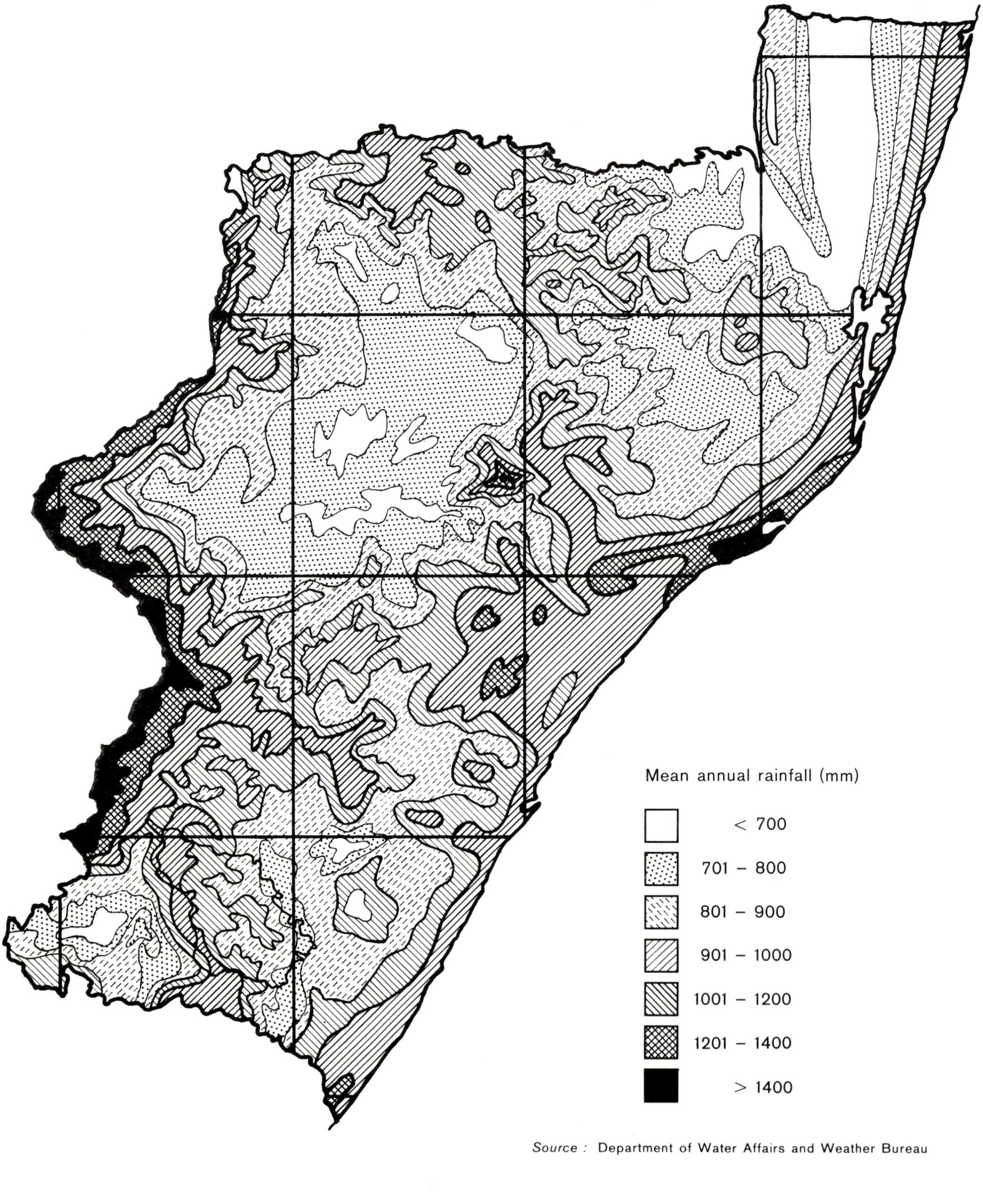

Source : Department of Water Affairs and Weather Bureau

Map 11: Mean annual rainfall (mm)

Map 12: Mean monthly rainfall (mm) — December.

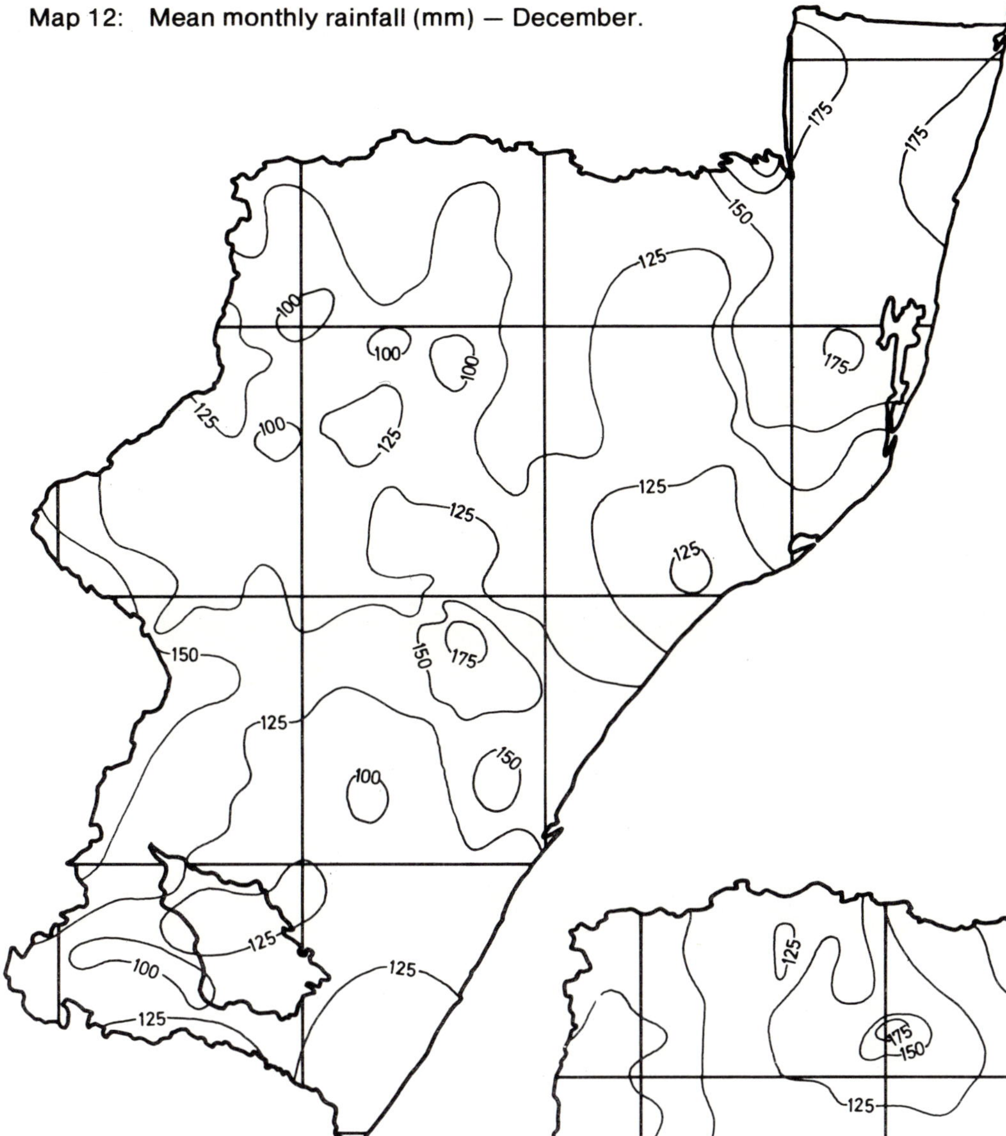

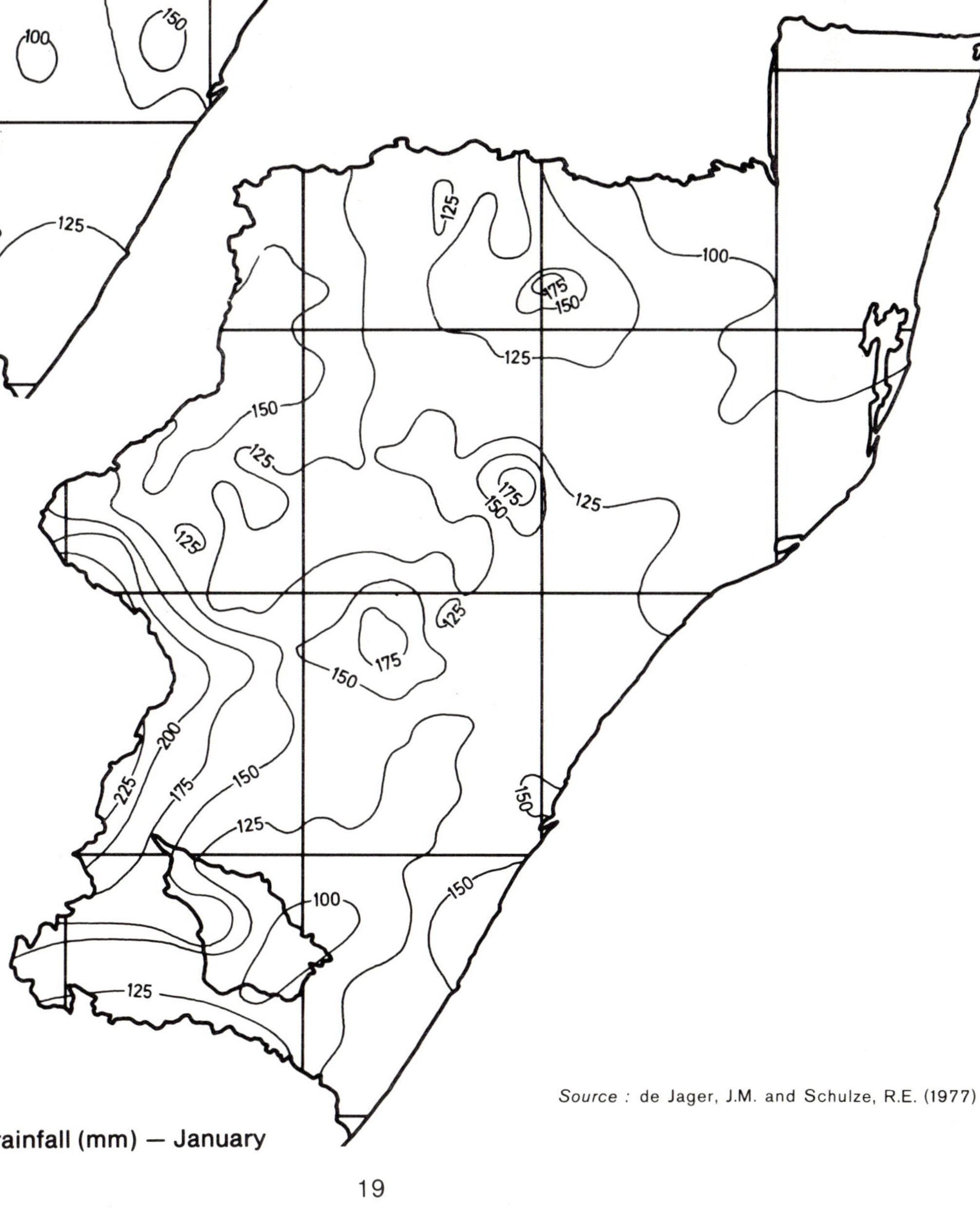

Source : de Jager, J.M. and Schulze, R.E. (1977)

Map 13: Mean monthly rainfall (mm) — January

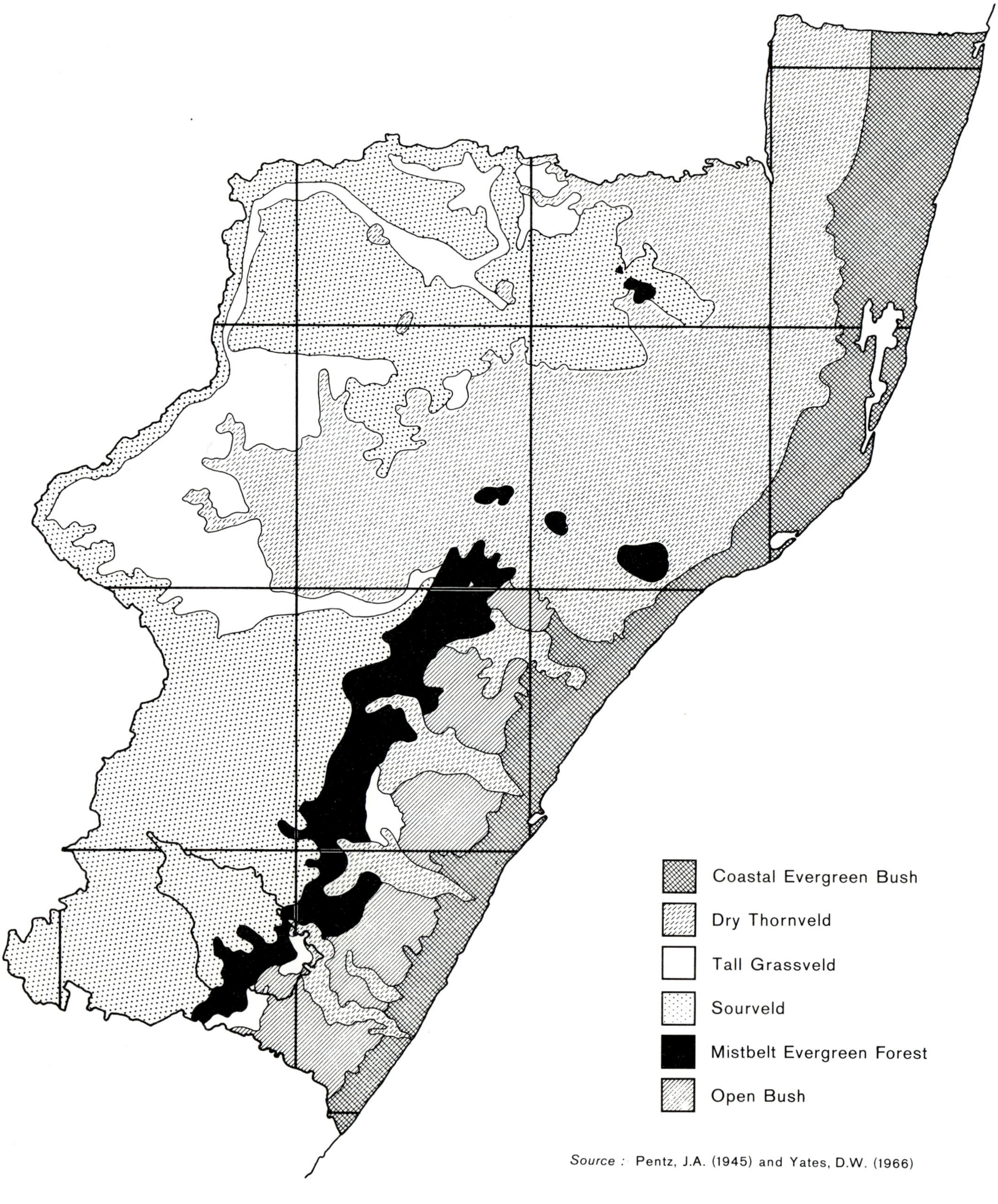

Source : Pentz, J.A. (1945) and Yates, D.W. (1966)

Map 14: Vegetation

VEGETATION TYPES

Liversidge (1962) found that individual plant species or types are often associated with bird distributions. Broadly speaking, it may be said that vegetation types can be used in assessing bird distributions. The vegetation types (Map 13) and the following descriptions are based on Pentz (1945) and Yates (1966) with modifications.

(i) Coastal Evergreen Bush

This vegetation type is typical of the coastal area from sea level to an altitude varying from 300 to 450 m a.s.l. Where it is intersected by river valleys, there is a change to coastal thornveld. The annual rainfall is well distributed throughout, increasing from south to north. The vegetation of the Coastal Evergreen Bush consists of open thornveld with patches of forest, particularly along streams and on steep slopes. The forest is low and dense towards the coast and in more exposed places, but becomes closed high forest in valleys and farther inland against mountainsides. The low forest is very varied and rich in plant species, of which the following are typical: *Millettia caffra, Celtis kraussinana, Canthium* spp., *Scutia* sp., *Combretum* spp., *Pavetta lanceolata, Syzygium cordatum, Trimeria* spp., *Cordia* spp. and *Fagara* spp. The high forest includes: *Celtis kraussiana, Alberta magna, Millettia caffra, Combretum* spp., *Cussonia umbellifera, Albizia* sp., *Syzygium cordatum, Trichilia emetica, Protorhus longifolia, Rauwolfia natalensis, Croton* spp., together with a dense ground flora. The coastal thornveld includes *Acacia benthami* and other *Acacia* spp., *Dichrostachys* sp., with small clumps of the short forest species in a dense, tall grass flora; the latter includes *Aristida junciformis*, species of *Hyparrhenia, Sporobolus, Setaria, Bothriochloa, Eragrostis* and *Chloris gayana*.

(ii) Dry Thornveld

This vegetation type is distinct from the coastal thornveld. It follows the rivers from the coast to an altitude of just over 900 m a.s.l. In southern Natal, it is restricted mainly to the river valleys, but in the midlands, northern Natal and Zululand, after following the rivers from the coast, it embraces the whole of the district of Weenen, part of the Estcourt, Ladysmith and Nqutu districts, and a large part of Zululand. In it, a number of different veld types are grouped together, ranging from the *Acacia-Boscia-Euphorbia-Aloe* bush of Weenen to the *Combretum-Dombeya-Spirostachys* veld in Zululand. The Dry Thornveld is very rich in species, the following being the most common: *Vitex rehmanii, Euphorbia ingens, Euphorbia tirucalli* and other *Euphorbia* spp., *Spirostachys africana, Panicum* spp., *Dombeya* spp., *Euclea* spp., *Gymnosporia* spp., *Olea verrucosa, Cymbopogon plurinodis, Themeda triandra, Bothriochloa insculpta, Digitaria* spp., *Schotia brachypetala, Acacia* spp., *Urochloa pullulans, Eragrostis* spp., *Sporobolus fimbriatus*, and *Pappea capensis*. On the whole, the rainfall is low and unevenly distributed, occurring in heavy storms.

(iii) Tall Grassveld

This vegetation type occurs at an altitude of 1050 to 1375 m a.s.l. and extends from Estcourt through Winterton, Bergville and portions of the districts of Ladysmith and Dundee, and then along a narrow strip below the foothills of the Drakensberg Range to Utrecht and Vryheid. The country is undulating, consisting of large plains with outcrops of dolerite forming small koppies. The rainfall averages about 750 mm per annum, but is unevenly distributed and hail is common in summer. The only trees associated with this vegetation type are *Acacia lasiopetala* and *Acacia caffra*, usually as isolated trees or in small clumps. The main grasses of the area are *Hyparrhenia* spp., *Sporobolus indicus, Setaria nigrirostris, Tristachya hispida, Themeda triandra* and *Setaria flabellata*.

(iv) Sourveld

This combines Highland, Sandy and East Griqualand Sourveld types.

a) Highland Sourveld occurs at an altitude of 1375 to 1825 m a.s.l. In southern Natal, it forms a broad belt at the foot of the Drakensberg Range and includes parts of the districts of Underberg, Ixopo, Donnybrook, Bulwer, Impendhle, Howick, Nottingham Road and Estcourt. From Estcourt, it extends in a narrow belt below the Drakensberg Range to Volksrust and Vryheid. There are also a few isolated highlying areas with this type of veld in the districts of Helpmekaar, Louwsburg, Nongoma, Babanango and Qudeni. The country is undulating. The annual rainfall is high (875-1250 mm) and well distributed, although most of it occurs in the summer. Mists are frequent. The following species are typical: *Tristachya hispida, Themeda triandra, Monocymbium cerisaeforme, Andropogon appendiculatus, A. filifolius, A. amplectens, Harpechloa falx, Eragrostis plana, Trachypogon plumosus, Elionurus argenteus, Alloteropsis semialata, Digitaria trichloaenoides*, and *D. diagonalis. Podocarpus*, temperate forest and *Leucosidea* scrub-forest occur occasionally in kloofs.

b) Sandy Sourveld occurs between 900 and 1225 m a.s.l. and occupies a large area in the districts of Newcastle, Dundee, Utrecht and Vryheid. This area is fairly flat. The annual rainfall ranges from 625 to 875 mm, with poor distribution. Hail is frequent. The vegetation consists of grassveld of which the typical species are: *Andropogon eucomus, eragrostis* spp., *Digitaris trichloaenoides, Monocymbium cerisiforme, Cymbopogon excavatus, Elyonurus argenteus, Eragrostis gammiflua, Hyparrhenia* spp., *Themeda triandra, Pogonarthria squarrosa*, and sedges.

c) East Griqualand Sourveld occurs at an altitude of 1225 to 1525 m a.s.l. and extends in a belt from near Swartberg to Matatiele, with a small patch occurring around Kokstad. The grass cover is fairly dense with the principal grasses of undisturbed veld being *Themeda triandra, Cymbopogon* spp., *Tristachya hispida, Brachiaria serrata, Setaria nigriostris, Andropogon amplecteris, Digitaria trichloaenoides* and *Trachypogon capensis*. Where veld has been disturbed or ploughed and left fallow, *Cynodon dactylon, Eragrostis curvula* and *Sporobolus pyramidalis* become dominant.

(v) Mistbelt Evergreen Forest

This vegetation type occurs at an altitude of 900 to 1225 m a.s.l. It runs in a narrow belt from Weza in the south to Kranskop in the north, extending in isolated patches farther north as far as Ngome. The country is undulating. The annual rainfall ranges from 750 to 1125 mm and is well distributed; heavy mists are also a feature of this region. At one time this area probably carried much natural bush or forest. To-day, where it is not cultivated, there is still a great deal of relict bush occurring in deep kloofs or on the sides of the hills. All the undulating land is

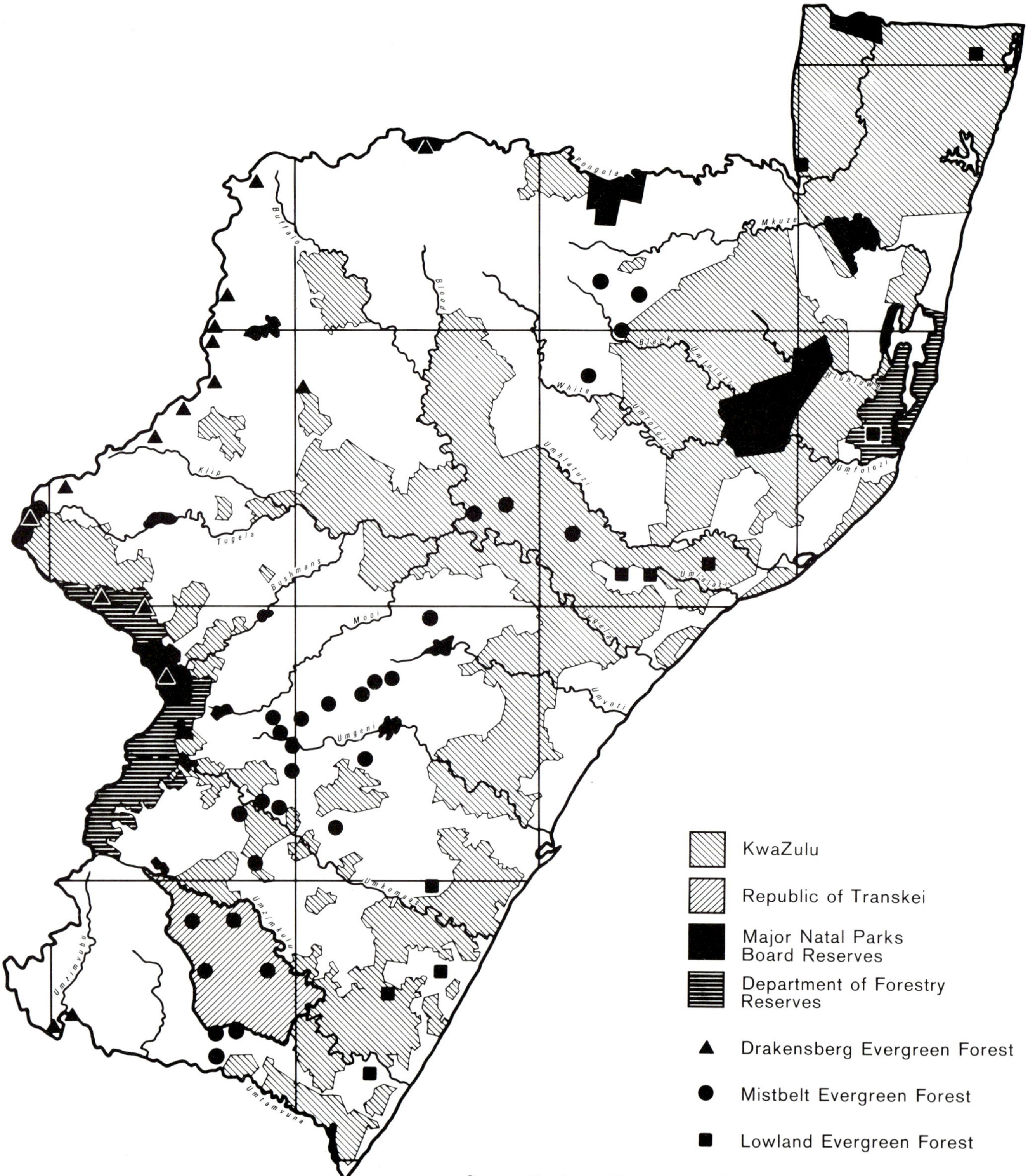

Source : KwaZulu – Department of Co-operation and Development
Conservation – Natal Parks Board and Department of Forestry

Map 15: KwaZulu and major conservation areas, forests and rivers

covered with grassveld of a poor type. The main tree species found in the area are: *Podocarpus* spp., *Scolopia mundtii, Cussonia spicata, Heteromorpha arborescens, Bowkeria* spp., *Ilex mitis, Apodytes dimidiata, Olinia cymosa, Rapanea melanophloeos, Pittosporum viridiflorum, Olea foveolata, Trimeria grandifolia* and *Leucosidea sericea.*

(vi) Open Bush

This vegetation type occurs between 600 and 900 m a.s.l., lying between the Coastal Evergreen Bush and Mistbelt Evergreen Forest areas and stretching in patches from the Umzimkulu River in the south to the Tugela River in the north. The country is undulating. The rainfall is moderate (500 to 750 mm per annum), with a fair distribution throughout the summer months. There are fairly considerable variations in climatic conditions from east to west. In this vegetation type, as in the Mistbelt Evergreen Forest region, kloofs carry a closed forest with similar main species, but on the undulating land bush is mixed with the grass. Thornbush predominates. Among the commonest grass species are: *Aristida junciformis, Sporobolus* spp., *Aristida* spp. and *Digitaria tricholaenoides.*

BIRD DISTRIBUTION PATTERNS

The species maps show various patterns of bird distribution. The following well-defined patterns are selected as examples.

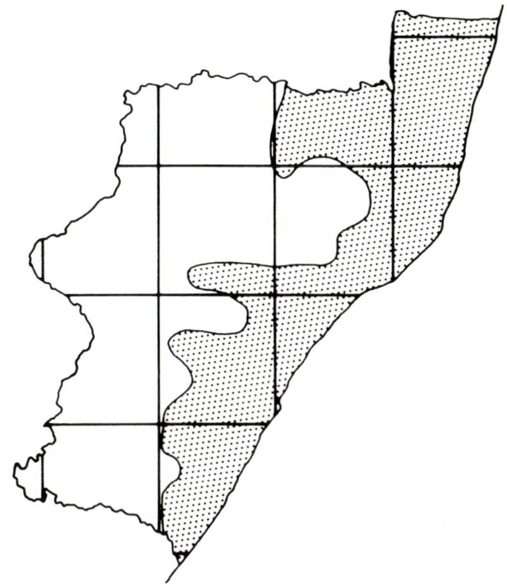

Map 17: Natal Robin (579).
A distribution along the littoral and coastal plain, extending up river valleys. There are two broad groups; species inhabiting damp grassland, e.g. White-browed Coucal (356) and Thick-billed Weaver (804), and others inhabiting riverine woodland, e.g. Gorgeous Bush Shrike (721), with a number reaching mistbelt forests, e.g. Emerald Cuckoo (350) and Trumpeter Hornbill (422).

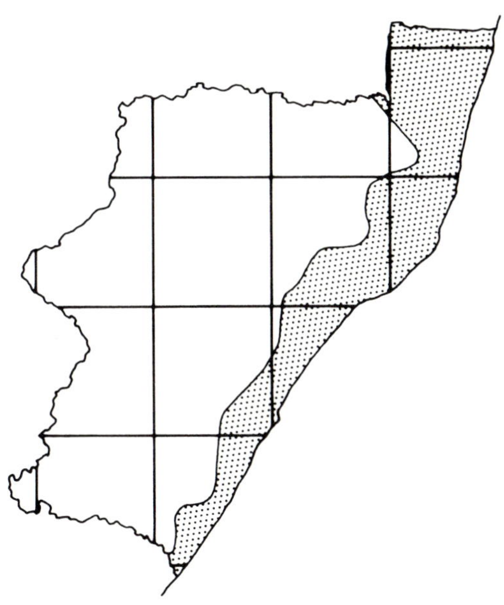

Map 16: Golden-rumped Tinker Barbet (438).
A distribution confined to the littoral and coastal plain. Some species, e.g. Woodwards' Batis (676) and Yellow White-eye (777) are restricted to the littoral plain; others extend southwards in varying degrees down the coastal plain, e.g. Grey Titbabbler (657), White-eared Barbet (433) and Yellow-bellied Bulbul (550).

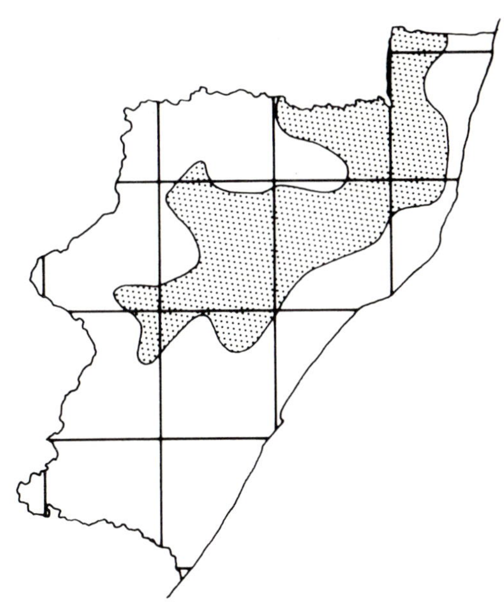

Map 18: Tawny Eagle (134).
A distribution along the northern dry thornveld region, from north-eastern Zululand west to the Bergville district. Three grades of this pattern are apparent: species confined to north-eastern Zululand, reaching Louwsburg in the west and Babanango in the south, e.g. White-backed Vulture (107), Grey Loerie (339) and Yellow-billed Hornbill (426); species with a strong presence in north-eastern Zululand and isolated pockets as far west as Ladysmith, e.g. Lizard Buzzard (144), Red-crested Korhaan (224) and Grey Penduline Tit (530); and species occurring throughout, e.g. Sabota Lark (460) and White-throated Robin (582).

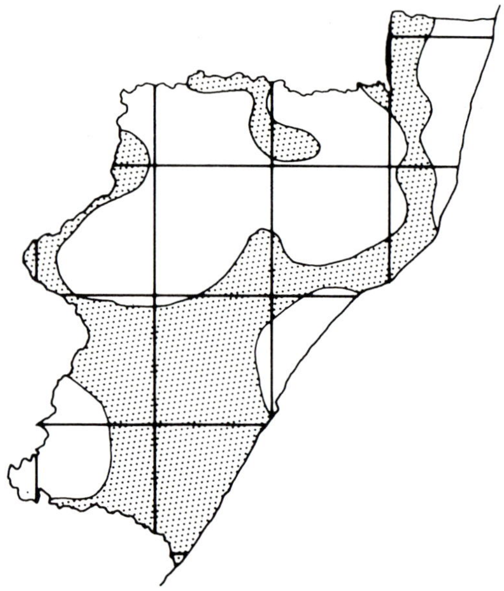

Map 19: Rameron Pigeon (312).
A distribution throughout evergreen forest. Three types of forest occur; coastal, interior mistbelt and high altitude on the slopes of the Drakensberg Range. Some species occur throughout, e.g. Starred Robin (589) and Scrub Warbler (610); others are confined to coastal and mistbelt forest, e.g. Buff-spotted Flufftail (206), and a few are restricted to mistbelt forest, e.g. Brown-necked Parrot (326) and Orange Thrush (556). Forest Canary (858) is found only in mistbelt and high altitude forest.

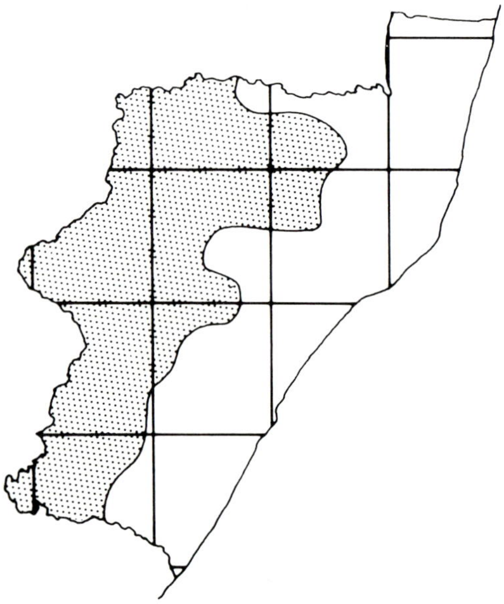

Map 20: Mountain Chat (564).
A distribution throughout montane and upland grassland. Three groups fall within this pattern: species restricted to north-western Natal, e.g. Blue Korhaan (223) and Black-throated Canary (860); species mainly confined to the high Drakensberg Range, e.g. Bearded Vulture (150) and Drakensberg Siskin (856), and species occurring throughout, e.g. Blue Crane (216) and Pied Starling (746).

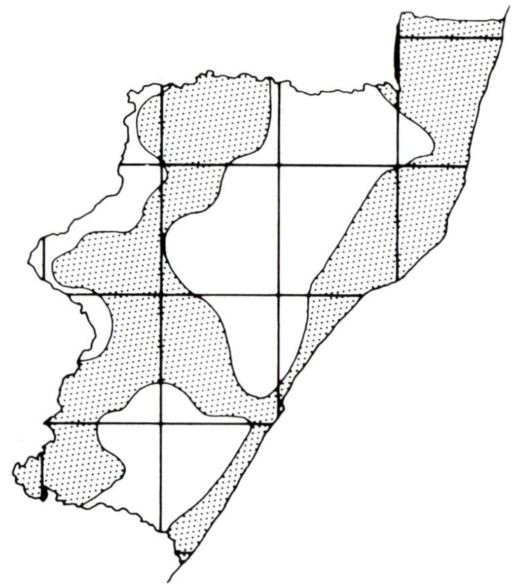

Map 21: White-breasted Cormorant (47).
A distribution resulting from the presence of standing water. A wide range of species is included, e.g. Yellow-billed Egret (60), White-faced Duck (100), Fish Eagle (149), African Jacana (228), Pied Kingfisher (394) and Cape Reed Warbler (604).

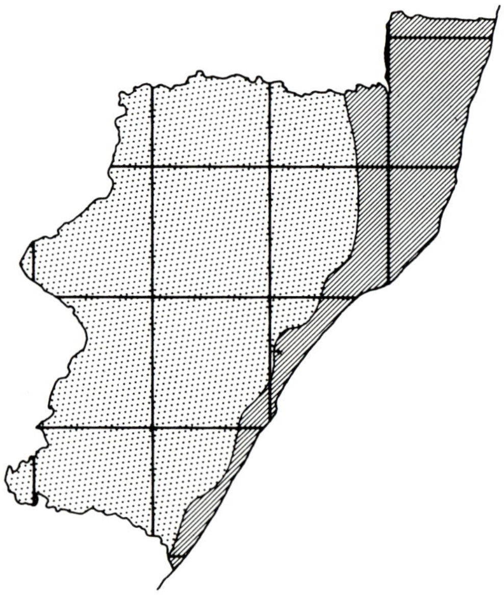

Map 22: Black-backed Cisticola (645) and Le Vaillant's Cisticola (646).
An example of replacement species which are closely related, and have only marginal overlap of their distributions. Other examples include Heuglin's Robin (580) and Cape Robin (581), and Bar-throated Apalis (622) and Rudd's Apalis (624).

Seasonal distributional patterns are shown by migrant species, the majority being summer visitors, e.g. Yellow-billed Kite (129), Red-chested Cuckoo (343) and European Swallow (493). The Fairy Flycatcher (678) is a winter visitor from the west, and in the case of the Fiscal Flycatcher (665), some birds are resident in the interior, while others move down to the coastal lowlands in winter.

When consulting the species maps, cognizance should be taken of the coverage map (Map 2) and the table listing numbers of returned fieldcards (Table 1), which may indicate the possibility that certain species, though present, have not been recorded in a particular square. The extent of habitat diversity should also be borne in mind, as this may account for the absence in a square of a species represented in adjoining squares.

ZOOGEOGRAPHIC REGIONS

A number of papers have been written on the boundaries of Avifauna Zoogeographic Regions in southern Africa. These proposed limits are largely based on the distribution of certain groups of birds, which are taken as being representative of the various regions. Although this is not the place to go into a detailed discussion regarding the boundaries, it should be mentioned that the distributional information for the period 1970-9 as shown in this *Atlas,* may well shed new light on the boundaries of those regions falling within the limits of Natal. Further information will be found in Winterbottom (1959, 1962a, b) and Clancey (1955).

RED DATA SPECIES

The known status of rare and endangered species in South Africa was summarised by Siegfried *et al.* (1976). The *Atlas* provides up-to-date information regarding the distribution of these species in Natal. Some, such as Long-crested Eagle (138), Green Coucal (358) and Rudd's Apalis (624) were found to be well distributed. Others, such as Bronze-naped Pigeon (313), Blue Swallow (497) and Orange Thrush (556) were confirmed as having restricted distributions within Natal.

SOME DISTRIBUTIONAL CHANGES

The distributions of birds in Natal were summarised by Clancey (1964). Although in many cases the distributions given were similar to those mapped here, there were noteworthy instances where changes appear to have taken place. The following are some species found to be more widely distributed than previously stated: Singing Bush Lark (456)('found sparingly in the Matatiele district of East Griqualand in the summer months') was found to be widely distributed but local in north-western Natal. Pink-billed Lark (490) ('only recently determined as occurring in Natal, a specimen being collected in the Newcastle district') was found to be widely distributed in north-western Natal reaching as far east as Rorke's Drift. Cuckoo Finch (854), unrecorded by Clancey, was discovered in a number of localities. Black-throated Canary (860) ('records from the high country on the Natal/Orange Free State border') was also found to be widely distributed in north-western Natal. Lemon-breasted Canary (860X), unrecorded by Clancey though described as 'possibly occurring', was found in a number of localities in north-eastern Zululand. Several species were found to be less widely distributed than previously stated, for example, Bittern (71), Kori Bustard (217) (one record), Ludwig's Bustard (218) (no records), Great Snipe (249) (no records), Black Coucal (353) and Fan-tailed Warbler (616). As the occurrence of seabirds and waders had not previously been well documented, a considerable volume of new information was collected.

Map 23: Zoogeographic regions of the world.

Species Maps

This section comprises information on the 530 most commonly recorded species occurring in Natal during the *Atlas* period, January 1970 – December 1979. A circle, placed within each quarter-degree square on the map, is divided into twelve segments. Each segment represents a month of the year, clockwise from January to December. A bold outline applied to the circumference indicates that the species was recorded in that square, and blacked-in segments indicate the months during which it was recorded. The maps illustrate distributional patterns at a glance, and seasonal patterns in detail. The number and English name of each species follows *Roberts birds of South Africa* (1978), while the scientific nomenclature follows the *S.A.O.S. Checklist of southern African birds* (1980). For each species the world distribution is given first. (See Map 23 for the zoogeographic regions). This information and the distribution within the Southern African sub-region is based on data from the S.A.O.S. checklist. The habitats in which the bird is found in Natal are based on the authors' experience. Mention is made of the breeding season in Natal of each species, based largely on Dean (1971). The number of squares in which the bird was recorded is given, as well as the percentage that it represents of the 166 squares covering Natal. Transparent overlay maps of vegetation types and altitude are provided to assist the reader in interpreting the effect these factors may have on bird distribution patterns.

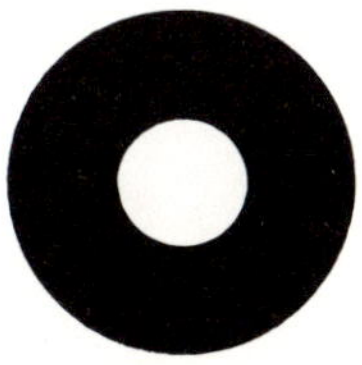

Resident

Summer Visitor

Winter Visitor

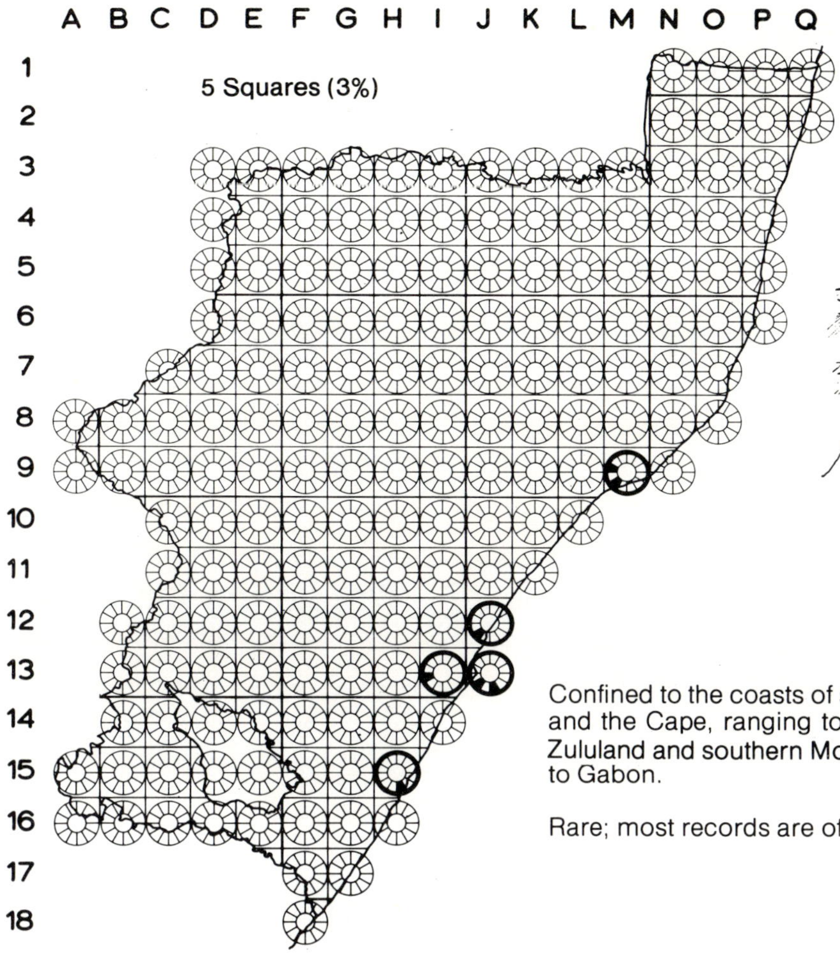

2 Jackass Penguin

Spheniscus demersus

Confined to the coasts of south-western Africa. Breeds on islets off S.W.A. and the Cape, ranging to the coasts of Angola, the eastern Cape, Natal, Zululand and southern Moçambique in the non-breeding season. Wanders to Gabon.

Rare; most records are of individuals stranded on the beach.

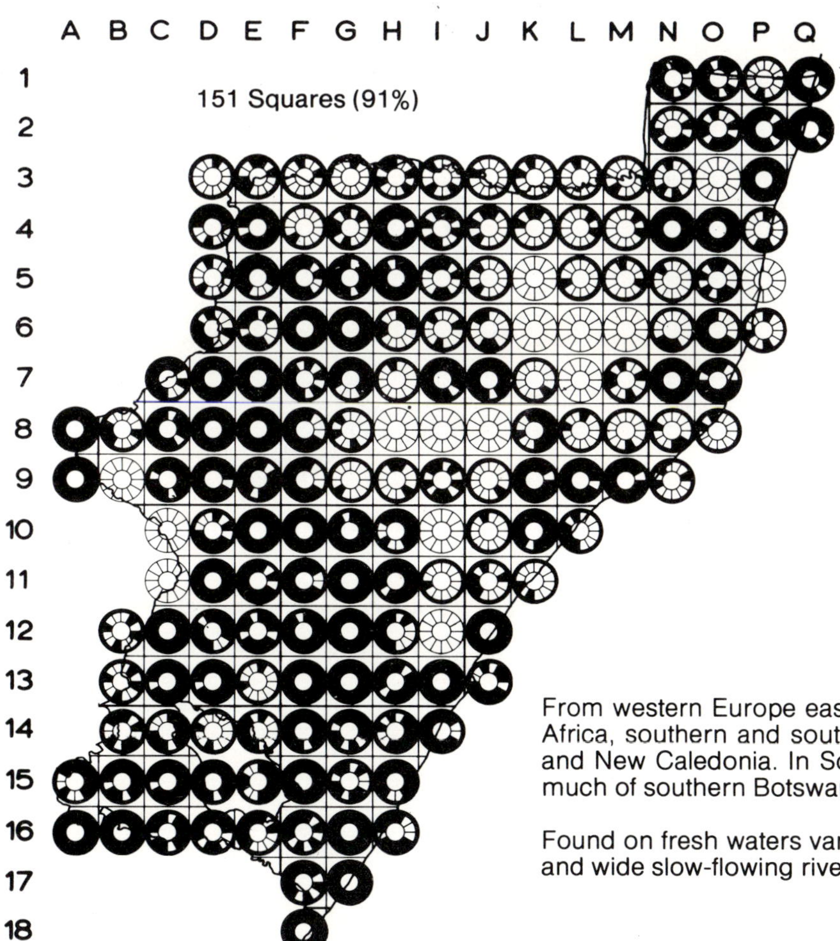

6 Little Grebe

Tachybaptus ruficollis

From western Europe east to eastern Asia and Japan, south throughout Africa, southern and south-eastern Asia and the Philippines to Australia and New Caledonia. In Southern Africa, breeds sparingly in S.W.A. and much of southern Botswana, but is locally common elsewhere.

Found on fresh waters varying from small ponds to large open stretches, and wide slow-flowing rivers. Breeds mainly from October to February.

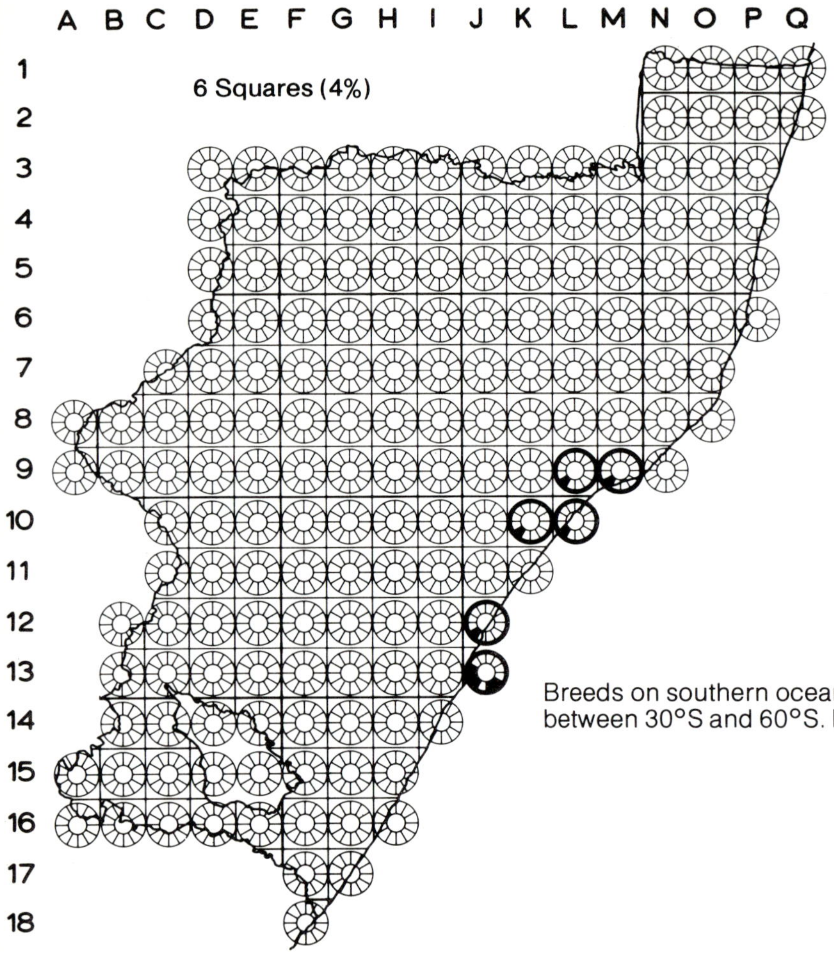

7 Wandering Albatross

Diomedea exulans

Breeds on southern oceanic islands and is pelagic over the southern seas between 30°S and 60°S. In Southern Africa, a regular visitor to all coasts.

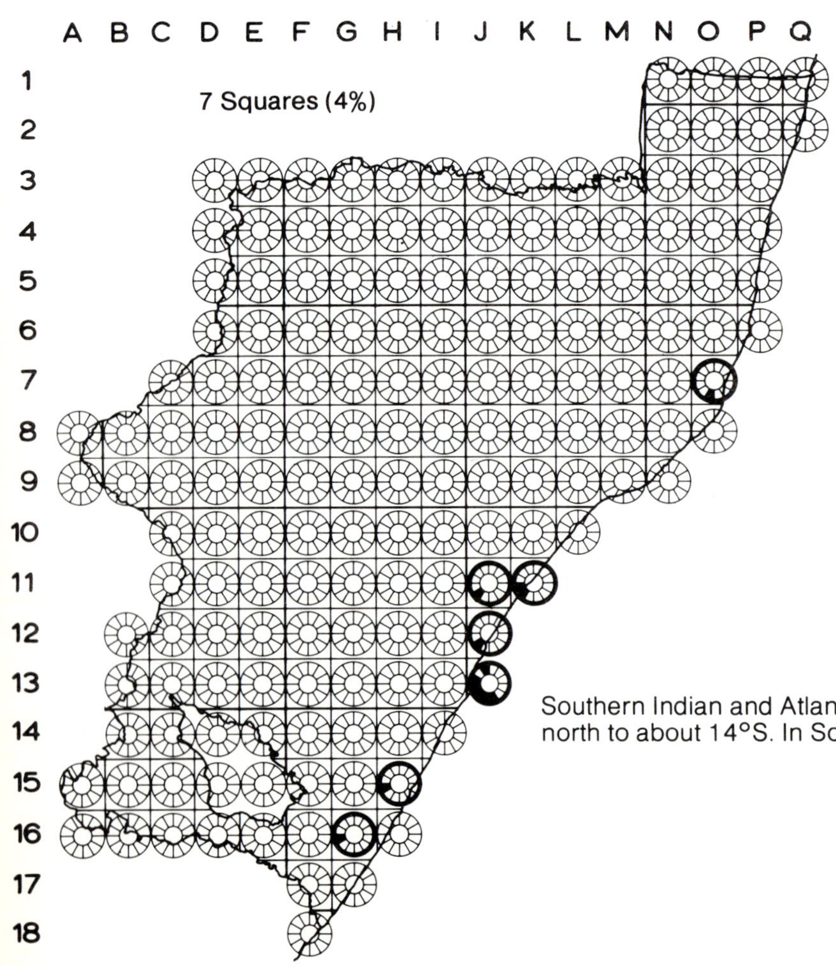

10 Yellow-nosed Albatross

Diomedea chlororhynchos

Southern Indian and Atlantic Oceans; pelagic when not breeding, ranging north to about 14°S. In Southern Africa, a common visitor to all coasts.

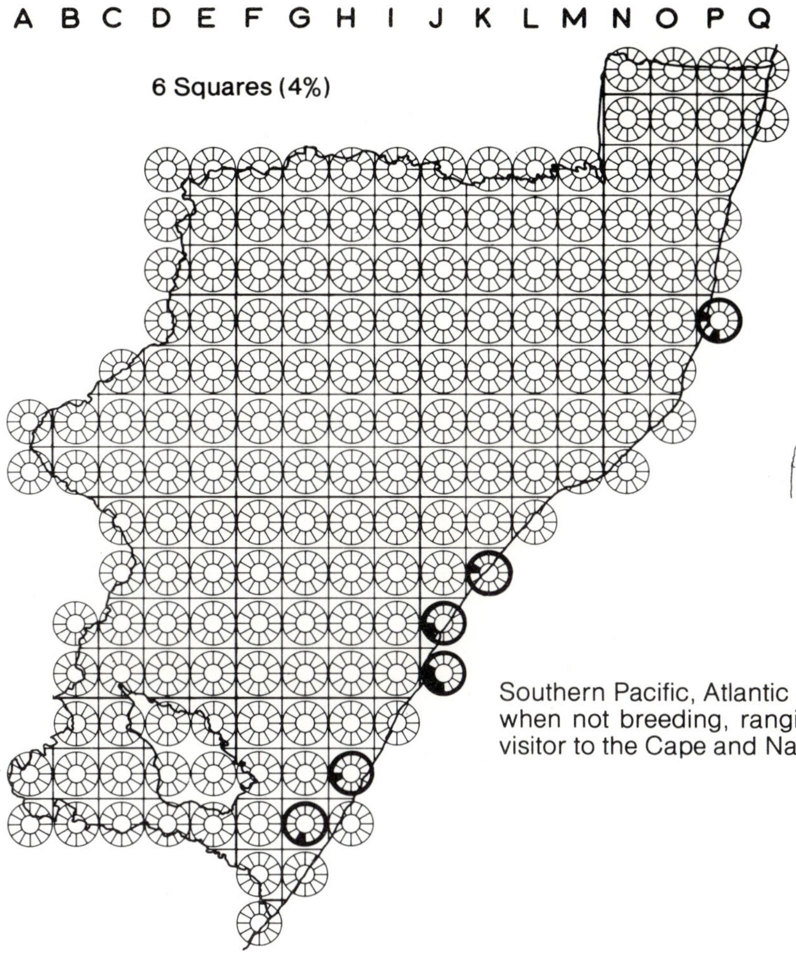

11 Shy Albatross

Diomedea cauta

Southern Pacific, Atlantic and Indian Oceans and Australian seas, pelagic when not breeding, ranging north to about 30°S. In Southern Africa, a visitor to the Cape and Natal coasts.

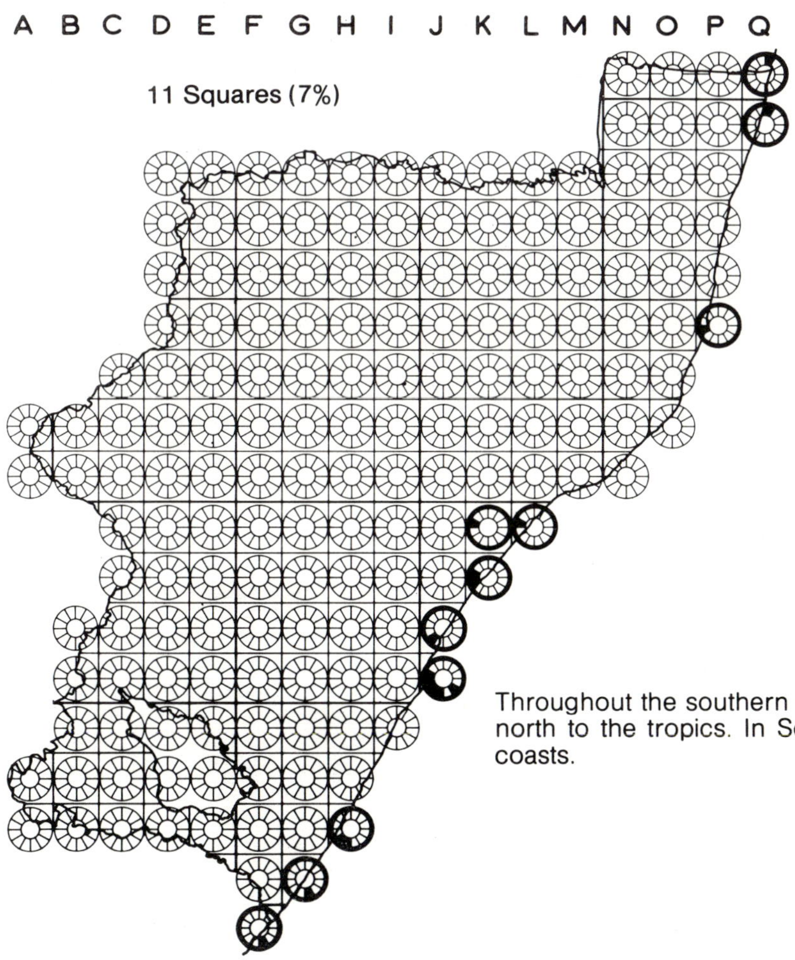

13 Southern Giant Petrel

Macronectes giganteus

Throughout the southern oceans; pelagic when not breeding, extending north to the tropics. In Southern Africa, a common winter visitor to all coasts.

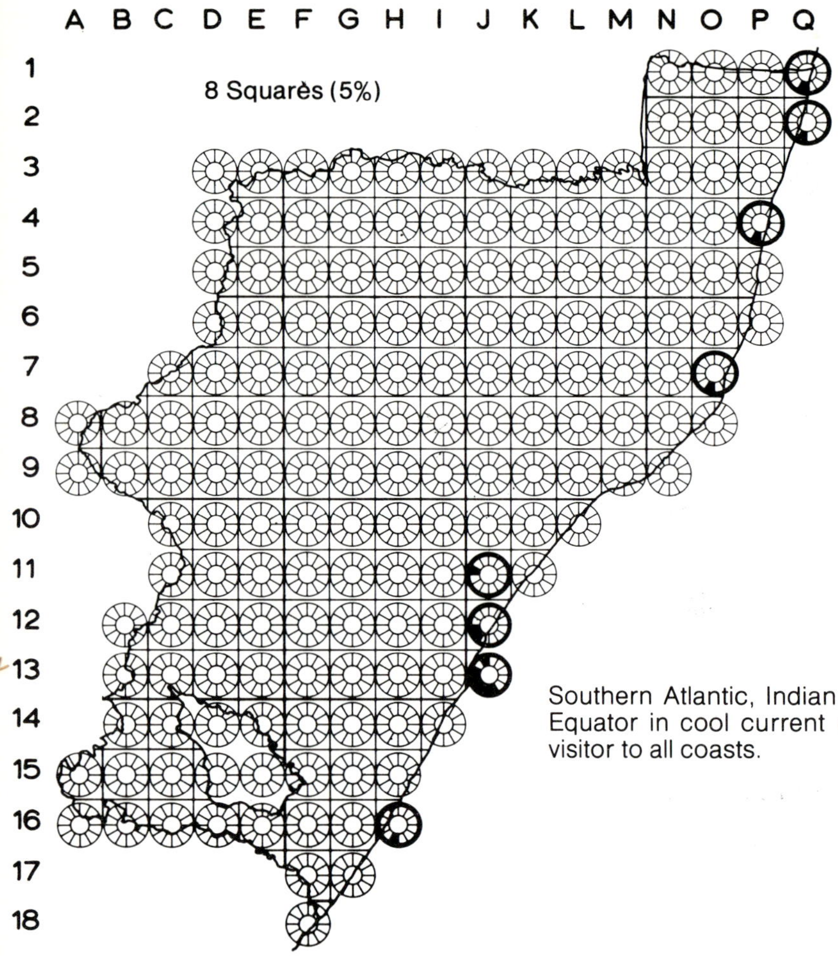

14 Pintado Petrel

Daption capense

Southern Atlantic, Indian and Pacific Oceans, ranging seasonally to the Equator in cool current regions. In Southern Africa, a common winter visitor to all coasts.

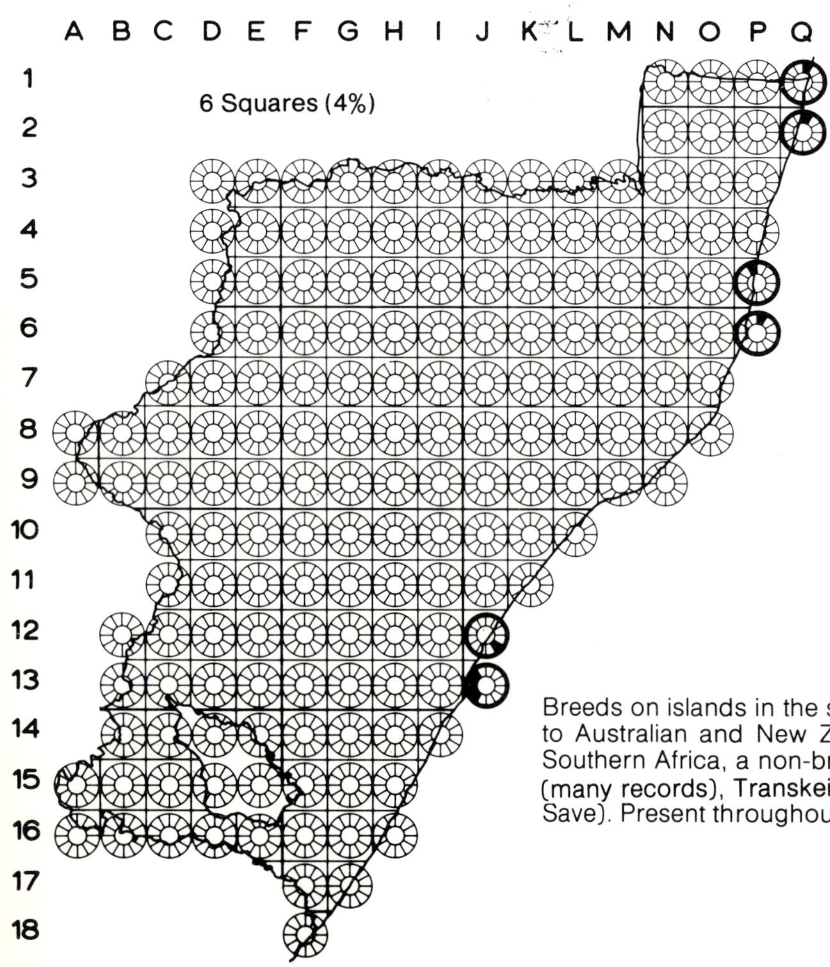

16 Great-winged Petrel

Pterodroma macroptera

Breeds on islands in the southern Atlantic and Indian Oceans, eastwards to Australian and New Zealand seas, ranging north to about 10°S. In Southern Africa, a non-breeding visitor to the coasts of S.W.A., the Cape (many records), Transkei, Natal and Zululand, and Moçambique (Sul do Save). Present throughout the year.

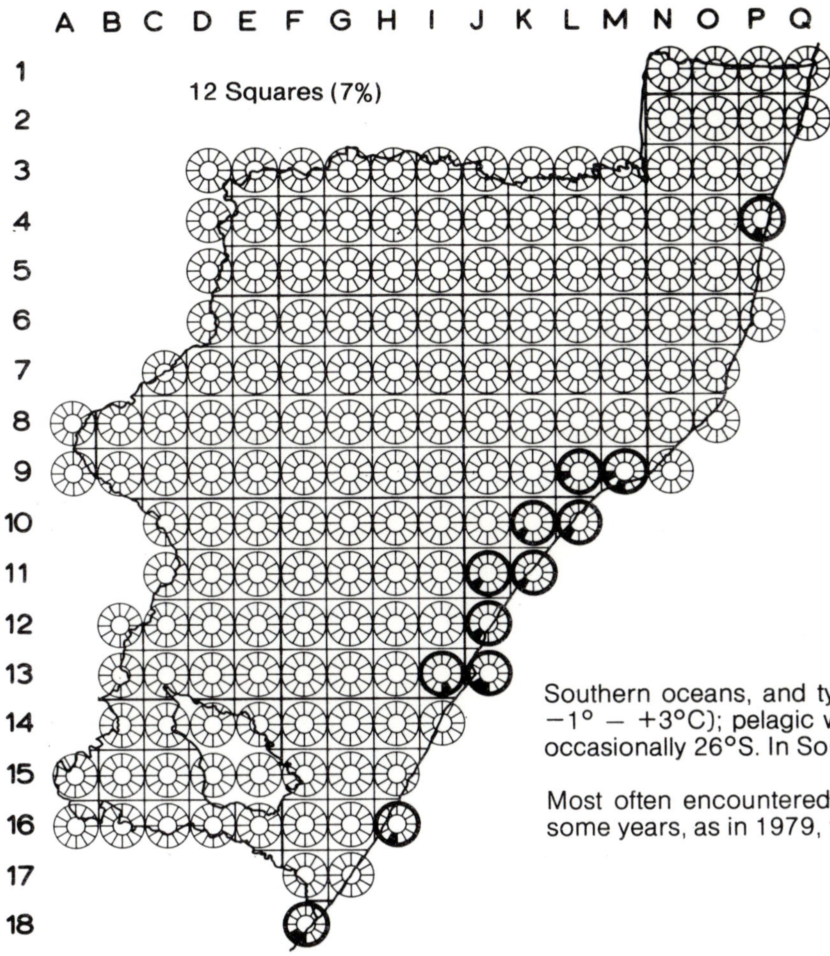

22 Dove Prion

Pachyptila desolata

Southern oceans, and typical of the Antarctic Zone (sea-water at about −1° – +3°C); pelagic when not breeding, ranging north to about 30°S occasionally 26°S. In Southern Africa, a visitor to all coastal waters.

Most often encountered from July to September, with large influxes in some years, as in 1979, when birds regularly noted close inshore.

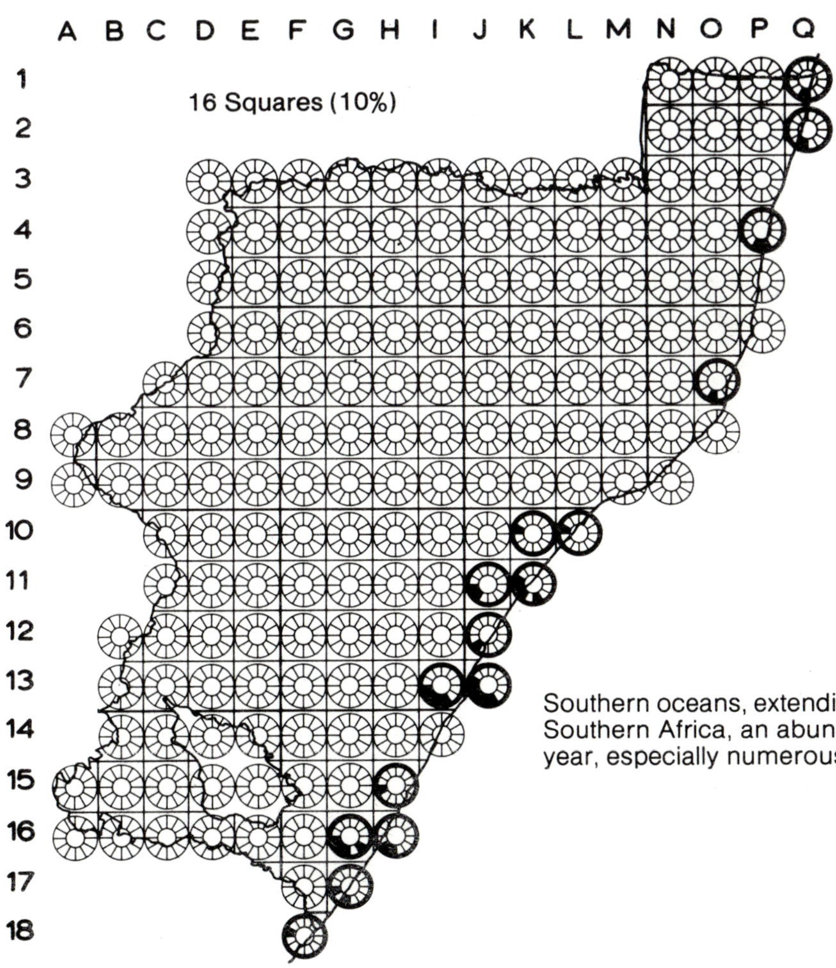

23 White-chinned Petrel

Procellaria aequinoctialis

Southern oceans, extending north to the tropics in cold current regions. In Southern Africa, an abundant visitor to all coasts, present throughout the year, especially numerous from June to August.

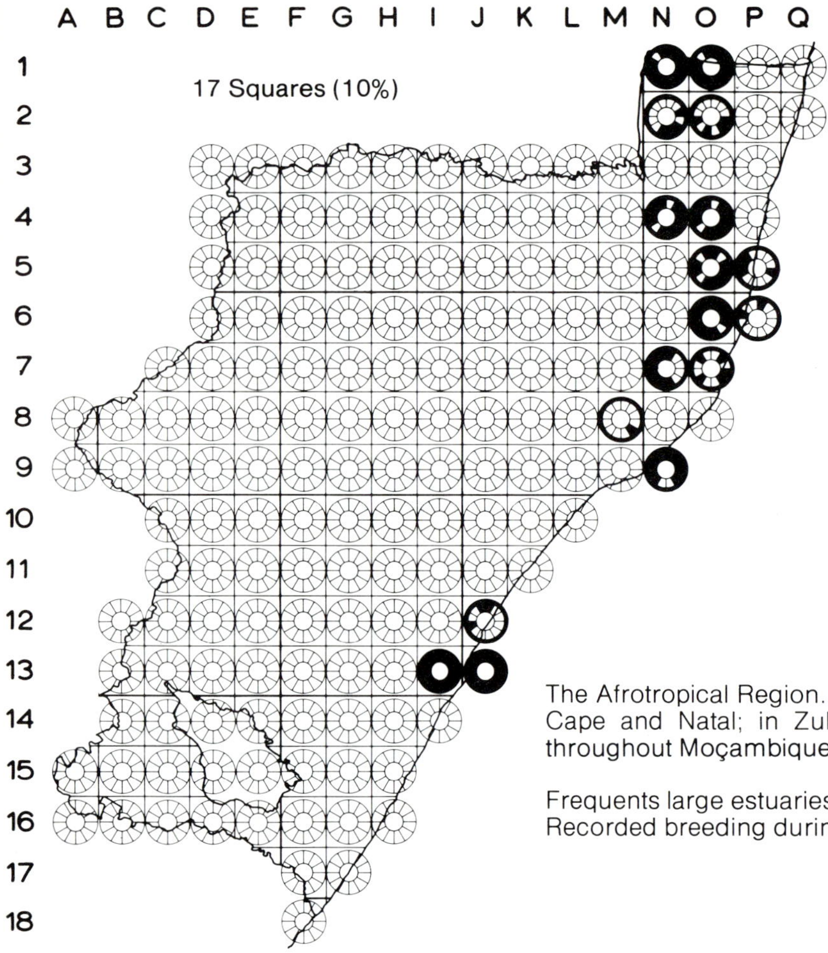

41 Pink-backed Pelican

Pelecanus rufescens

The Afrotropical Region. In Southern Africa, a non-breeding visitor to the Cape and Natal; in Zululand, breeds at Lake St Lucia, and present throughout Moçambique. A straggler to S.W.A., Botswana and Rhodesia.

Frequents large estuaries, shallow coastal lakes and pans on flood plains. Recorded breeding during the summer and winter.

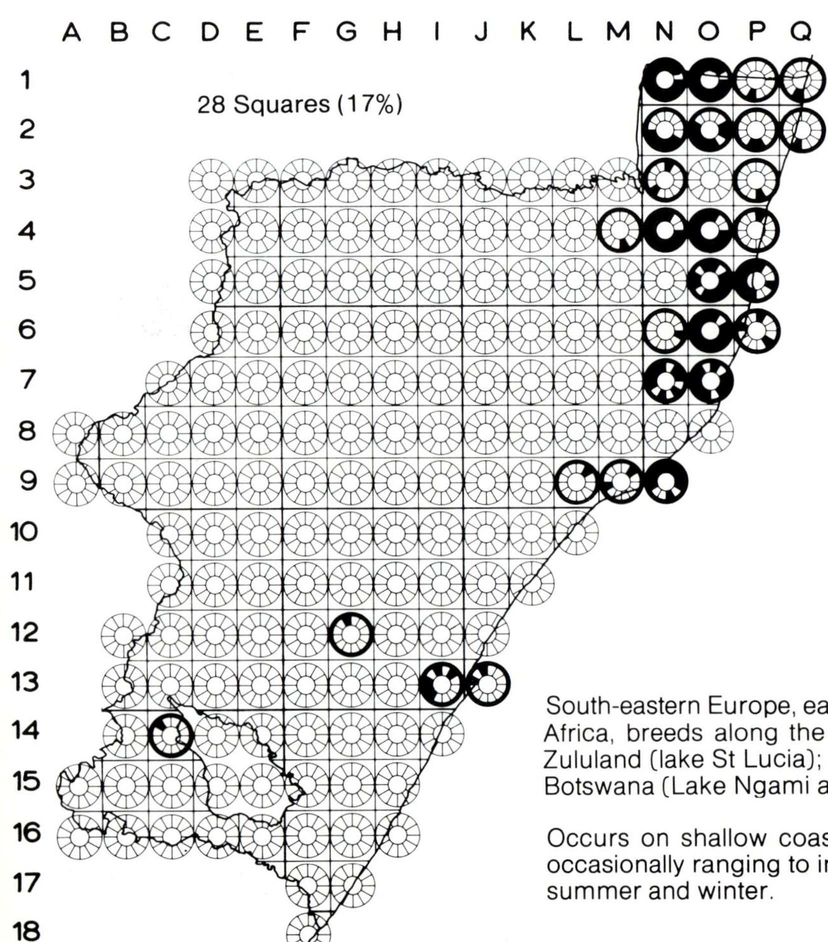

42 White Pelican

Pelecanus onocrotalus

South-eastern Europe, east to central Asia, and locally in Africa. In Southern Africa, breeds along the coasts of S.W.A., the south-western Cape and Zululand (lake St Lucia); also Moçambique (formerly the Zambesi River), Botswana (Lake Ngami and Nata) and occasionally locally elsewhere.

Occurs on shallow coastal lakes, flood plain pans and large estuaries, occasionally ranging to inland stretches of open water. Breeds during the summer and winter.

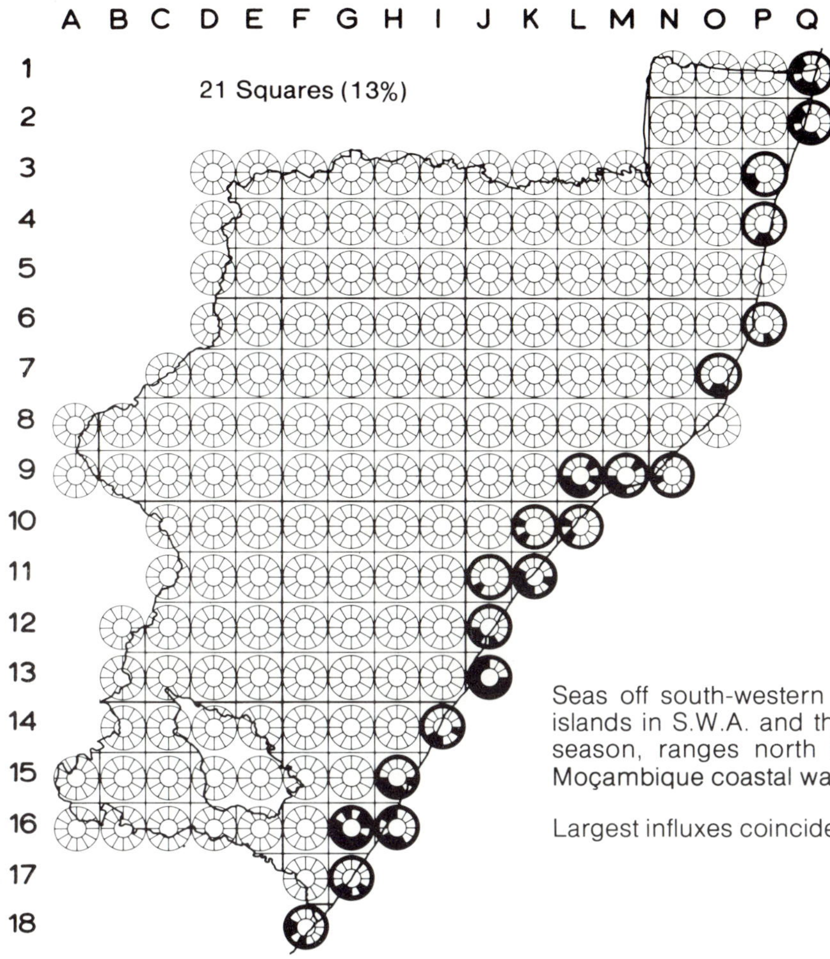

44 Cape Gannet

Morus capensis

Seas off south-western Africa. In Southern Africa, breeds on off-shore islands in S.W.A. and the Cape, east to Algoa Bay. In the non-breeding season, ranges north to the Gulf of Guinea and east to Natal and Moçambique coastal waters.

Largest influxes coincide with the annual sardine run.

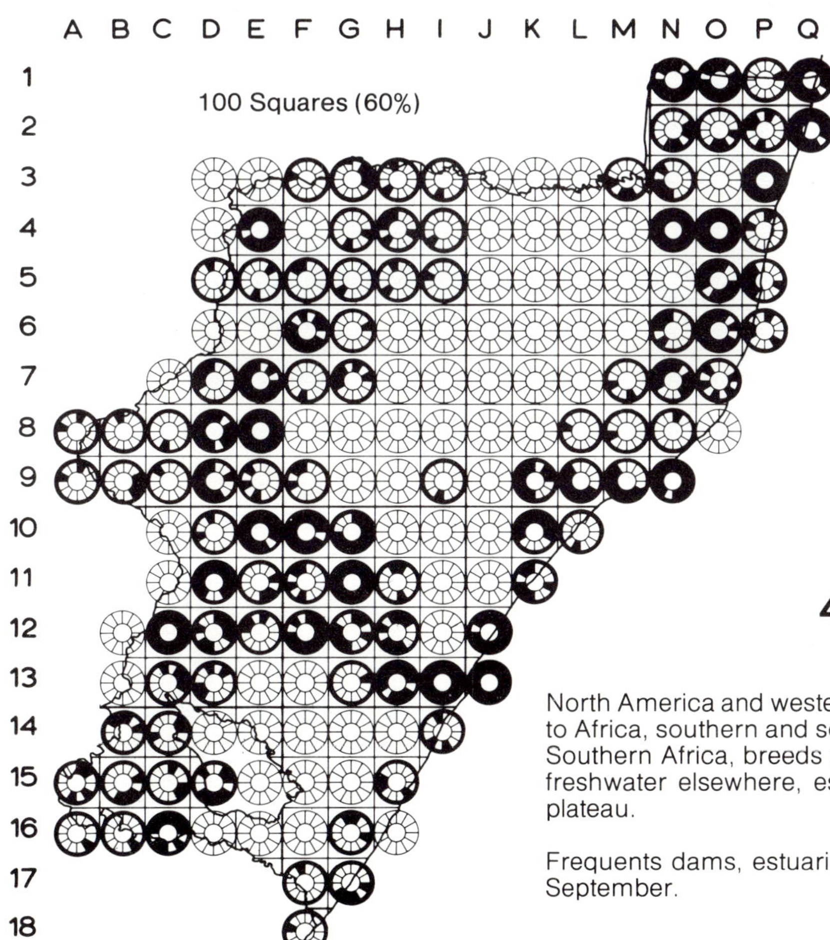

47 White-breasted Cormorant

Phalacrocorax carbo

North America and western Europe, east to eastern Asia and Japan, south to Africa, southern and south-eastern Asia, Australia and New Zealand. In Southern Africa, breeds on islets off S.W.A. and the Cape, and locally on freshwater elsewhere, especially in the southeast and over the eastern plateau.

Frequents dams, estuaries, lakes and pans. Breeds mainly from May to September.

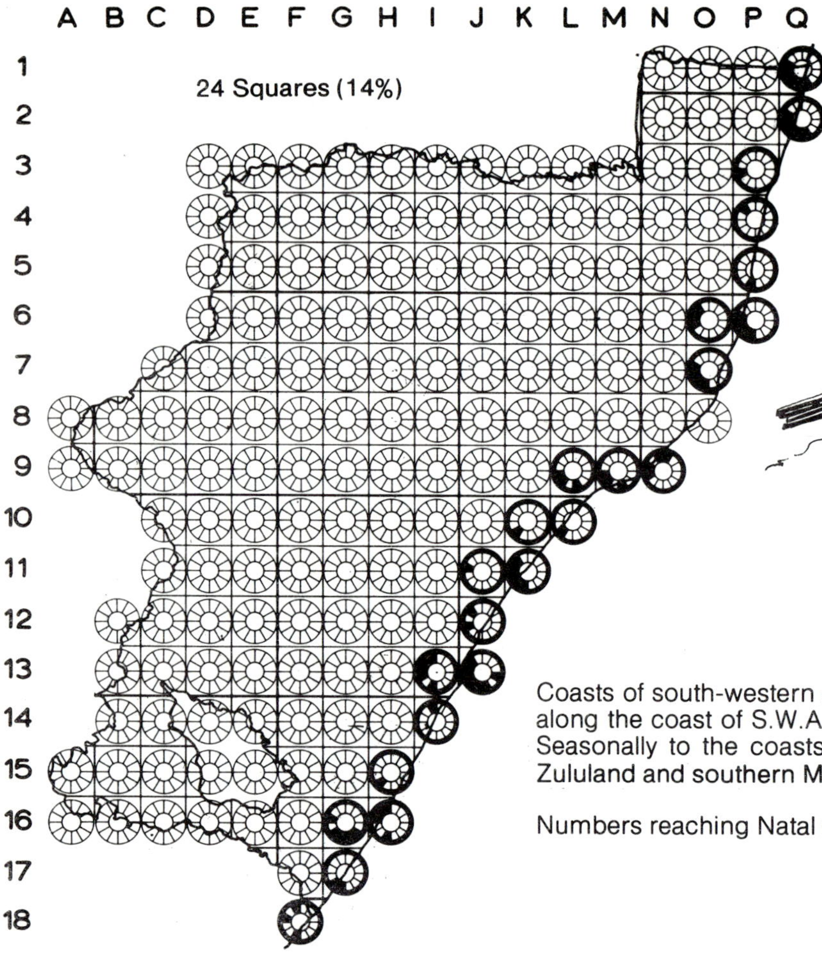

48 Cape Cormorant

Phalacrocorax capensis

Coasts of south-western and southern Africa. In Southern Africa, breeds along the coast of S.W.A. and the Cape, east to Algoa Bay (Bird Island). Seasonally to the coasts of Angola and Cabinda, and Transkei, Natal, Zululand and southern Moçambique.

Numbers reaching Natal vary from year to year.

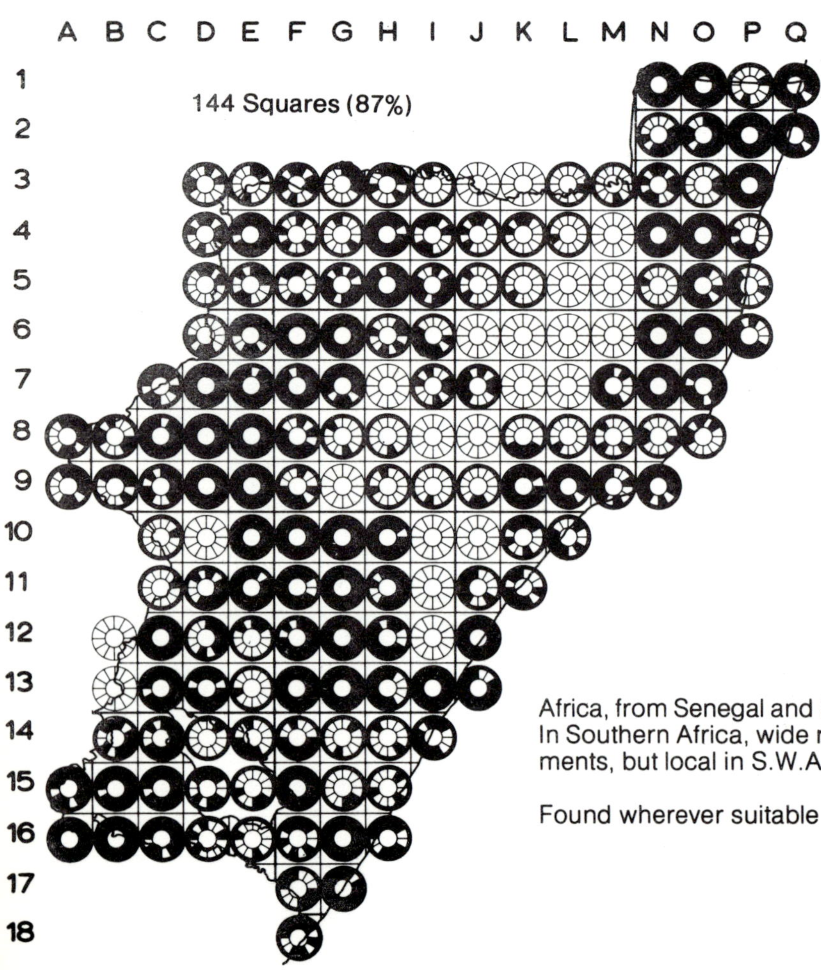

50 Reed Cormorant

Phalacrocorax africanus

Africa, from Senegal and Upper Egypt south to the Cape; also Madagascar. In Southern Africa, wide ranging on rivers, lakes and freshwater impoundments, but local in S.W.A., Botswana and the northern Cape.

Found wherever suitable habitat exists. Breeds throughout the year.

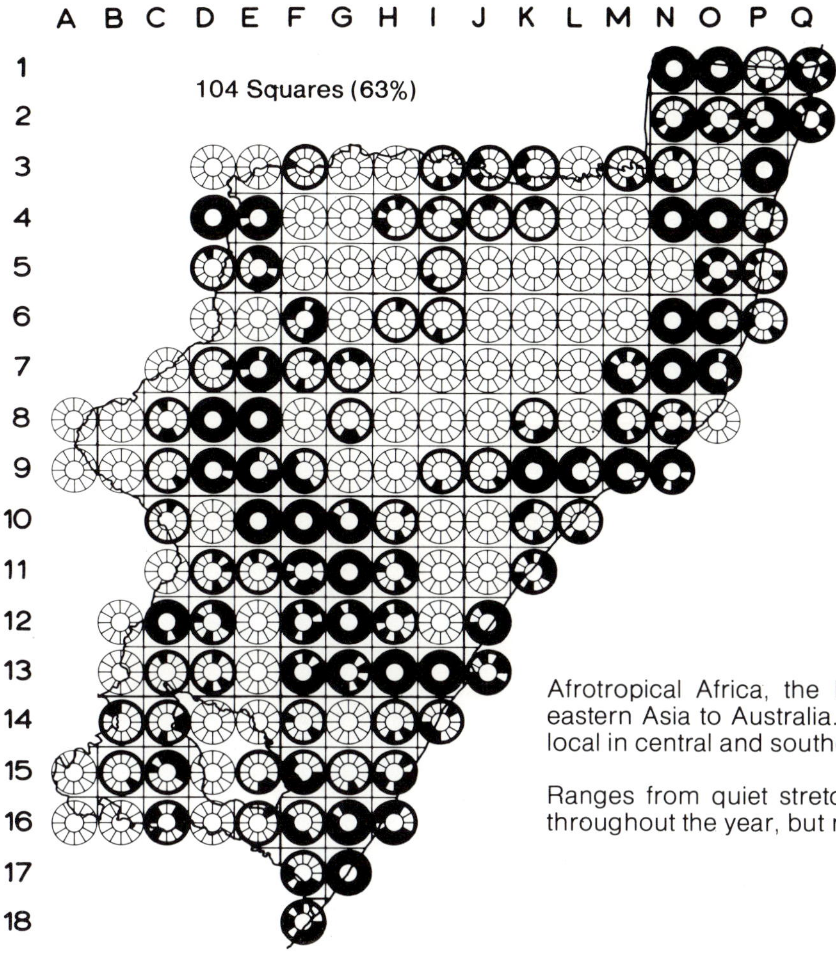

52 Darter

Anhinga melanogaster

Afrotropical Africa, the Middle East, India and Sri Lanka, and south-eastern Asia to Australia. In Southern Africa, distributed throughout, but local in central and southern S.W.A., Botswana and the northern Cape.

Ranges from quiet stretches of rivers to open sheets of water. Breeds throughout the year, but mainly during the winter.

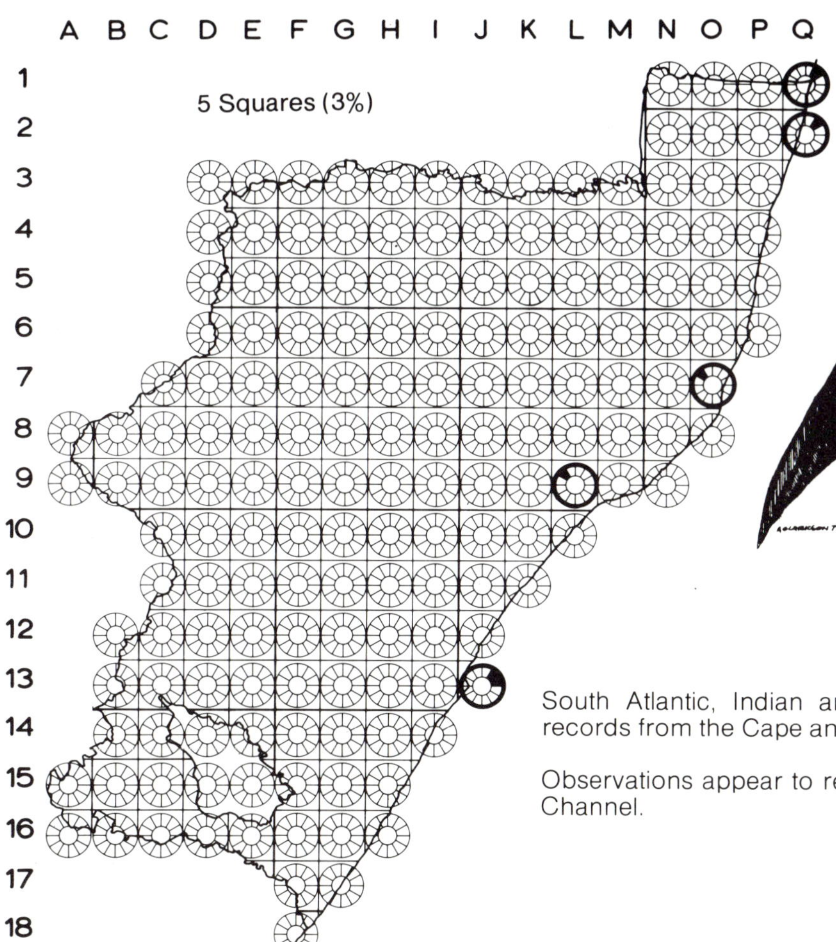

53 Greater Frigatebird

Fregata minor

South Atlantic, Indian and Pacific Oceans. In Southern Africa, sight records from the Cape and Natal.

Observations appear to relate to cyclonic conditions in the Moçambique Channel.

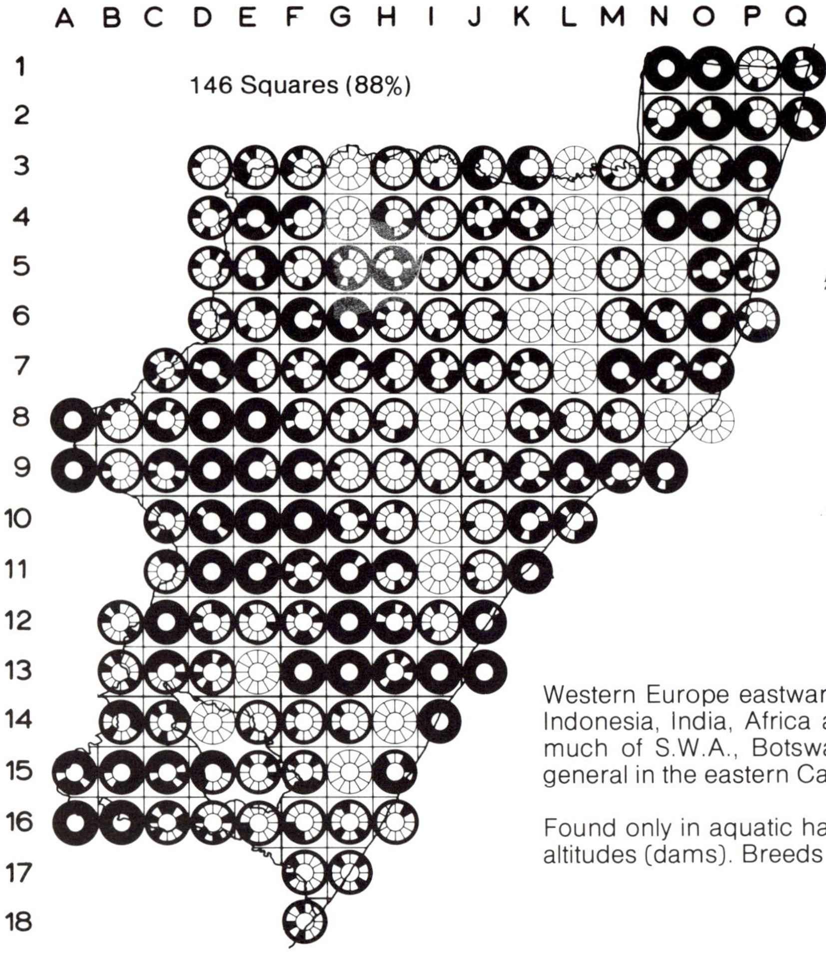

54 Grey Heron

Ardea cinerea

Western Europe eastwards to eastern Asia, Japan and Taiwan, south to Indonesia, India, Africa and Madagascar. In Southern Africa, local over much of S.W.A., Botswana and the interior and northern Cape; more general in the eastern Cape, Natal and Zululand, northwards in the east.

Found only in aquatic habitats, ranging from sea level (estuaries) to high altitudes (dams). Breeds mainly from September to February.

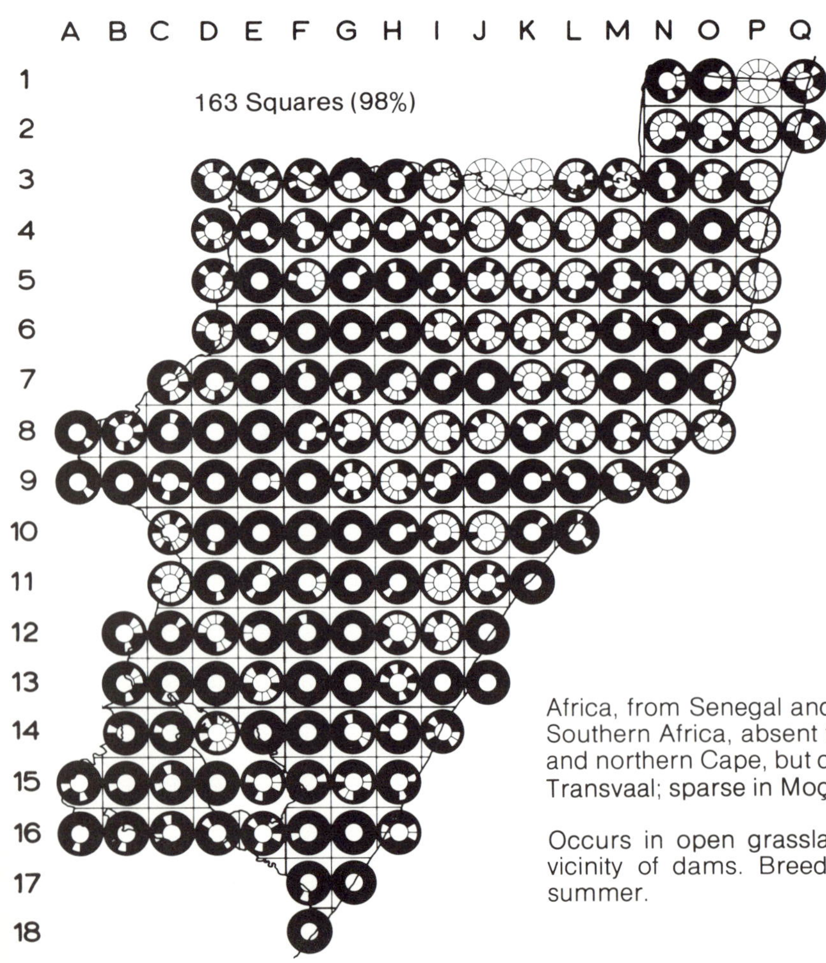

55 Black-headed Heron

Ardea melanocephala

Africa, from Senegal and Sudan south to the Cape; also Madagascar. In Southern Africa, absent from much of S.W.A., Botswana and the interior and northern Cape, but common in the eastern Cape, Natal, Zululand and Transvaal; sparse in Moçambique and Rhodesia.

Occurs in open grassland, cultivated fields and marshes, often in the vicinity of dams. Breeds throughout the year, but mainly during the summer.

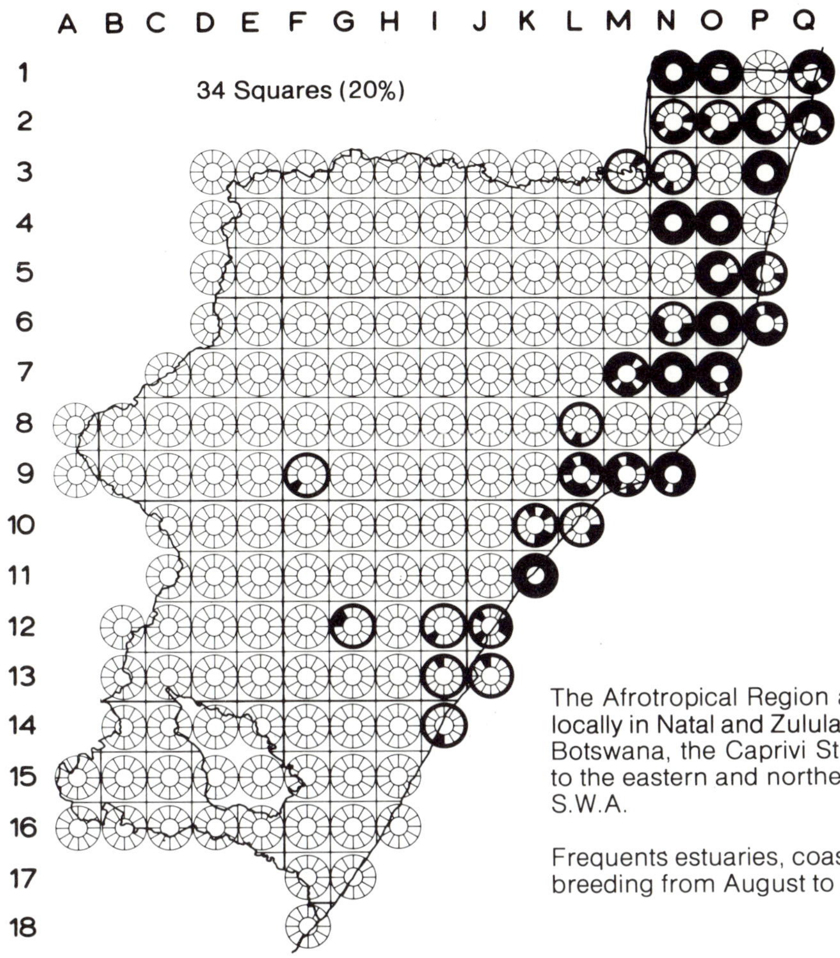

56 Goliath Heron

Ardea goliath

The Afrotropical Region and the Middle East. In Southern Africa, breeds locally in Natal and Zululand, Moçambique, Transvaal, Rhodesia, northern Botswana, the Caprivi Strip and north-eastern S.W.A. Ranges irregularly to the eastern and northern Cape, O.F.S., Lesotho (lowlands) and central S.W.A.

Frequents estuaries, coastal lakes and pans on the littoral plain. Recorded breeding from August to December.

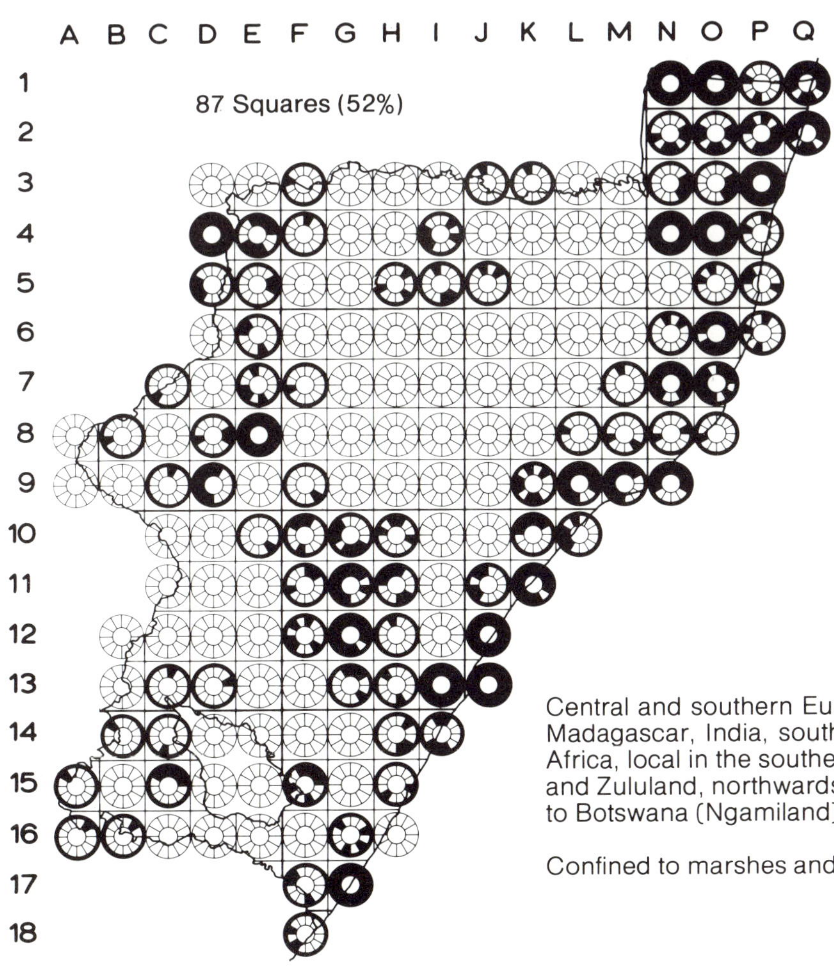

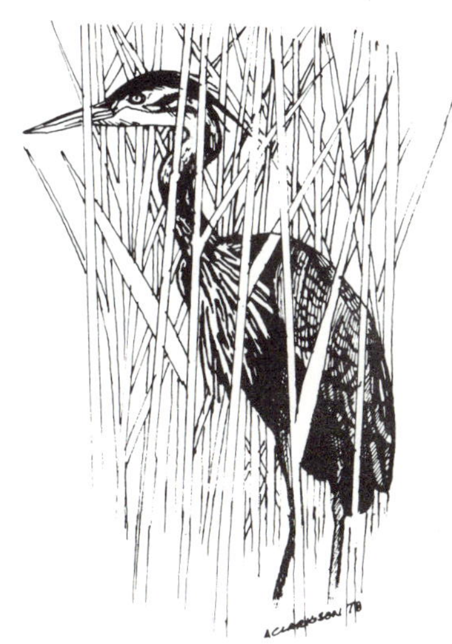

57 Purple Heron

Ardea purpurea

Central and southern Europe eastwards to eastern Asia, south to Africa, Madagascar, India, south-eastern Asia and the East Indies. In Southern Africa, local in the southern and eastern Cape, and more plentiful in Natal and Zululand, northwards in the east. In the north, west through Rhodesia to Botswana (Ngamiland), the Caprivi Strip and northern S.W.A.

Confined to marshes and reedbeds. Breeds mainly during the summer.

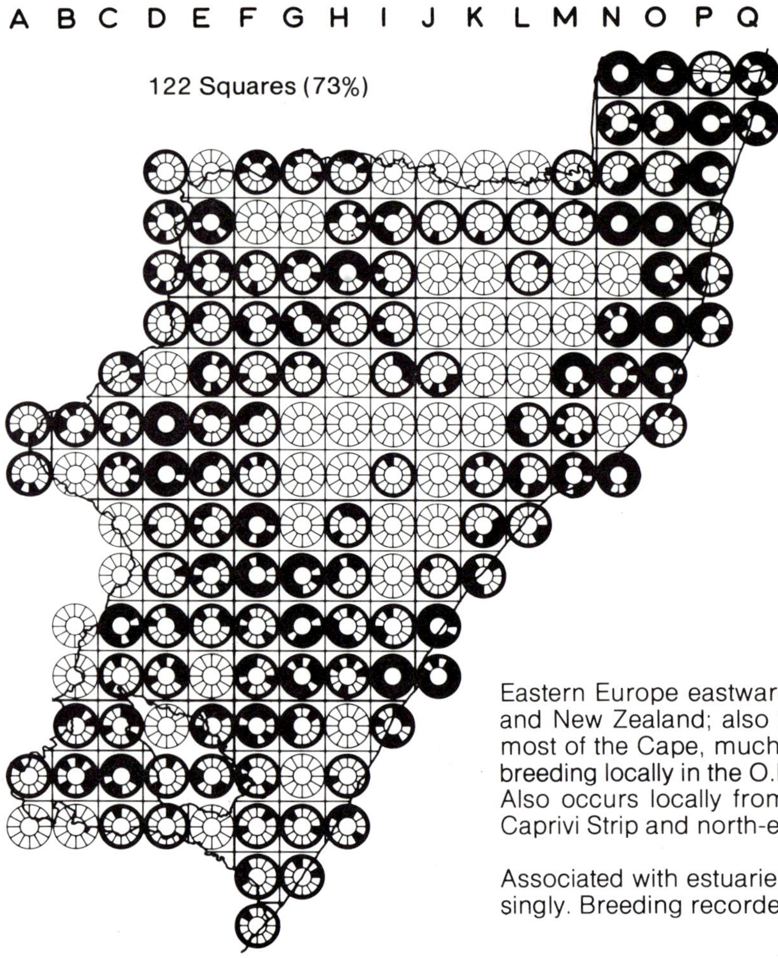

58 Great White Egret

Egretta alba

Eastern Europe eastwards to Siberia, south to Africa and Asia, Australia and New Zealand; also the Americas. In Southern Africa, a straggler to most of the Cape, much of S.W.A. and western and southern Botswana, breeding locally in the O.F.S., Transvaal, Natal, Zululand and Moçambique. Also occurs locally from Rhodesia west to Botswana (Ngamiland), the Caprivi Strip and north-eastern S.W.A.

Associated with estuaries, fringes of open waters and vleis; usually found singly. Breeding recorded from January to March.

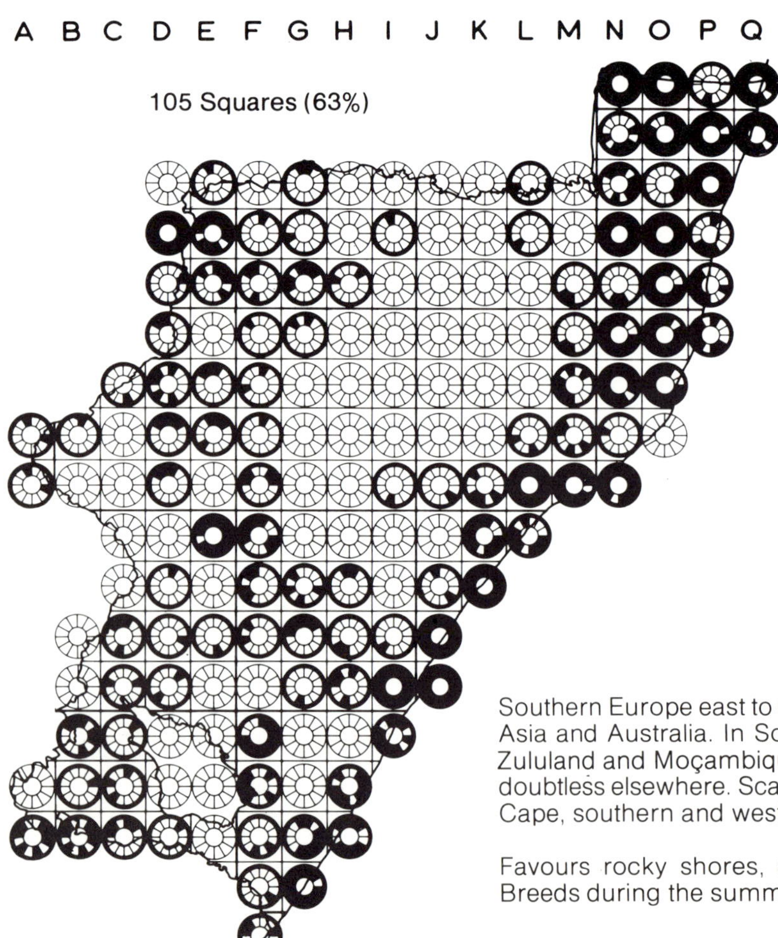

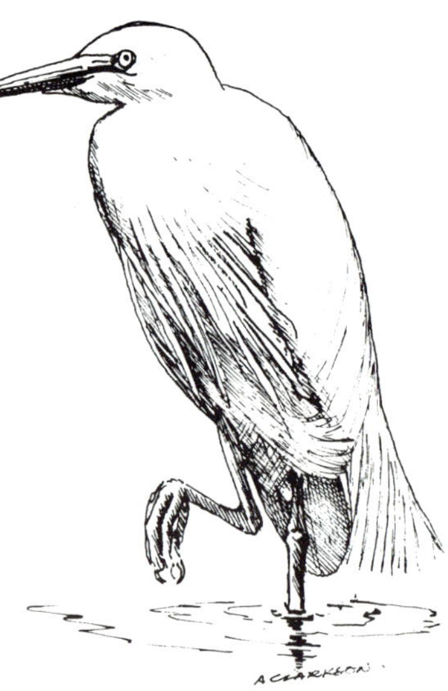

59 Little Egret

Egretta garzetta

Southern Europe east to China and south to Africa, Madagascar, southern Asia and Australia. In Southern Africa, breeds locally in the Cape, Natal, Zululand and Moçambique (where largely coastal), O.F.S. and Transvaal; doubtless elsewhere. Scarce over much of S.W.A., the interior and northern Cape, southern and western Botswana and Rhodesia.

Favours rocky shores, estuaries, rivers and shallow margins of pans. Breeds during the summer.

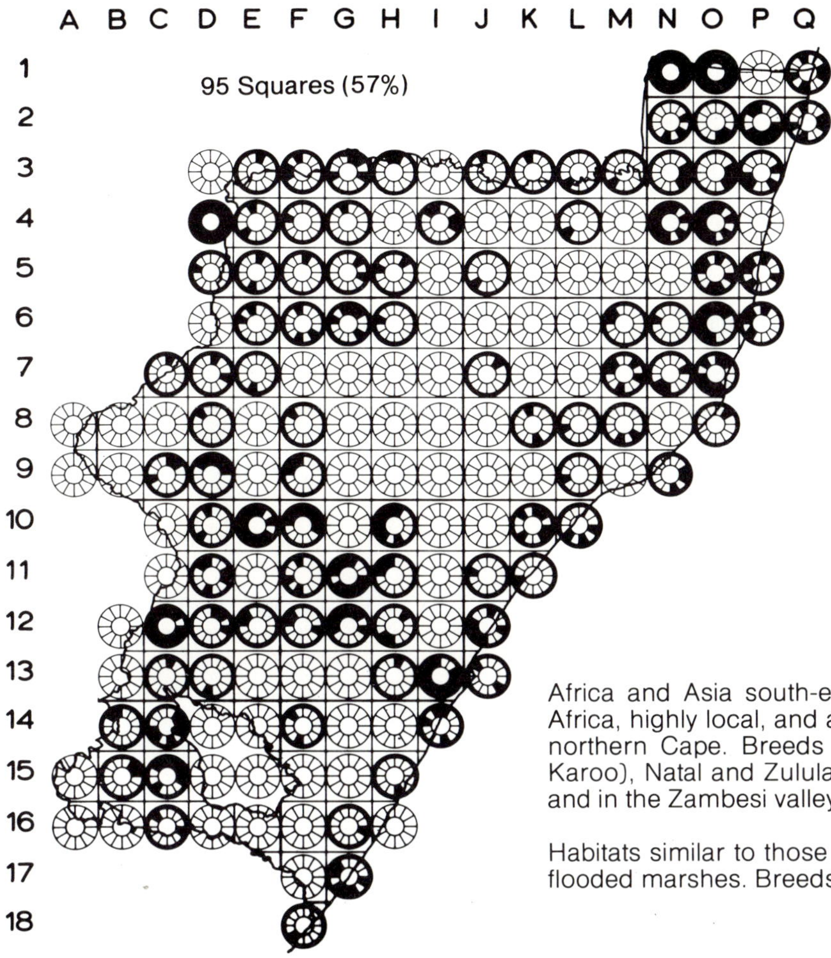

60 Yellow-billed Egret

Egretta intermedia

Africa and Asia south-east to New Guinea and Australia. In Southern Africa, highly local, and absent from much of S.W.A. and the interior and northern Cape. Breeds in the Cape (Winter Rainfall Region and Little Karoo), Natal and Zululand, O.F.S., Transvaal, Rhodesia (on the plateau and in the Zambesi valley), northern Botswana and the Caprivi Strip.

Habitats similar to those of Great White Egret (58), with a preference for flooded marshes. Breeds mainly during the summer.

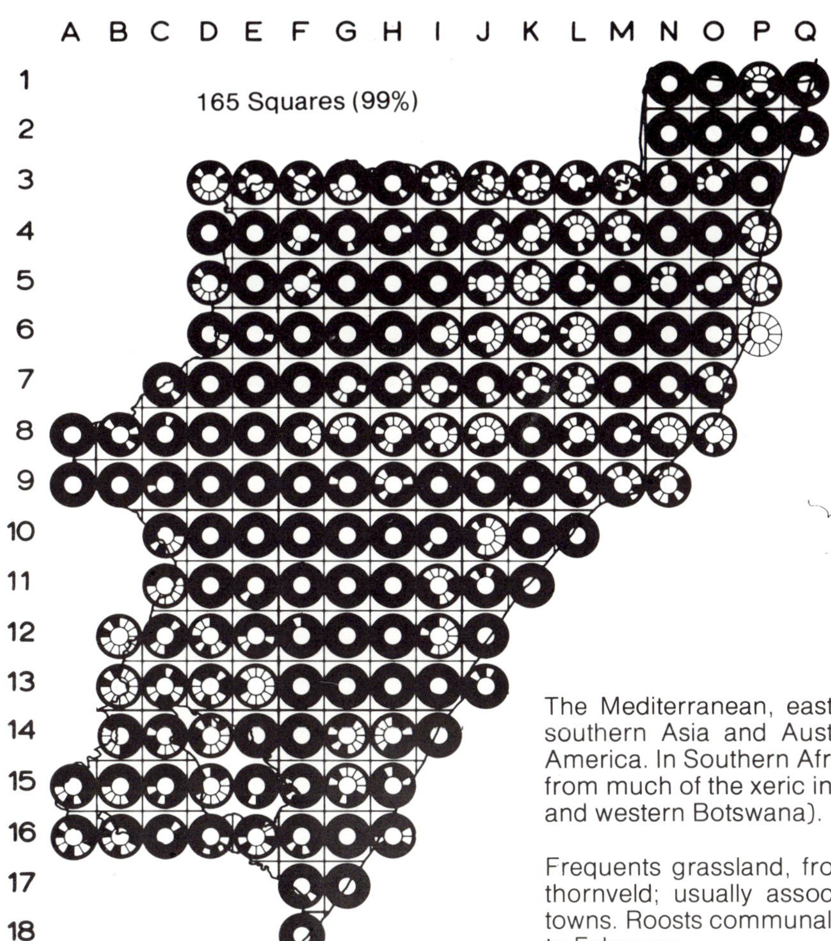

61 Cattle Egret

Bubulcus ibis

The Mediterranean, eastwards to Japan, south to Africa, Madagascar, southern Asia and Australia; also North, Central and northern South America. In Southern Africa, widespread, though sparse or largely absent from much of the xeric interior and far west (in the northern Cape, S.W.A. and western Botswana).

Frequents grassland, from large open areas to the smallest patches in thornveld; usually associated with cattle and large game. Also enters towns. Roosts communally close to water. Breeding recorded from August to February.

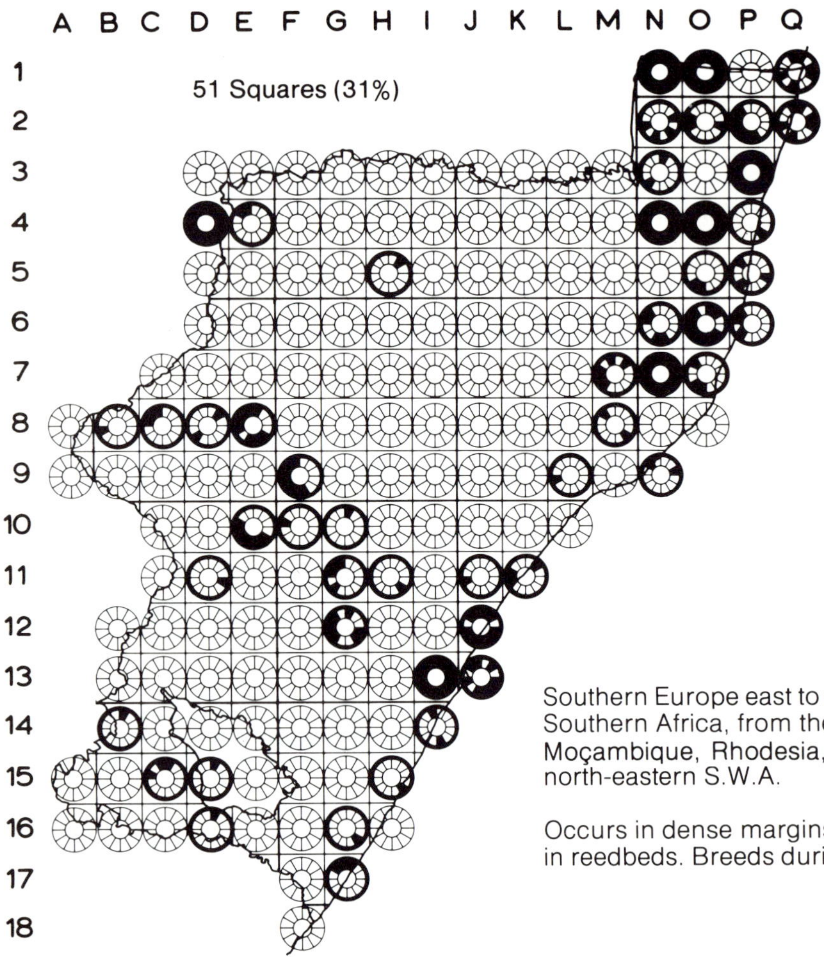

62 Squacco Heron

Ardeola ralloides

Southern Europe east to western Asia, south to Africa and Madagascar. In Southern Africa, from the eastern Cape to Natal and Zululand, Transvaal, Moçambique, Rhodesia, north-western Botswana, the Caprivi Strip and north-eastern S.W.A.

Occurs in dense margins of ponds and slow flowing streams, particularly in reedbeds. Breeds during the summer and winter.

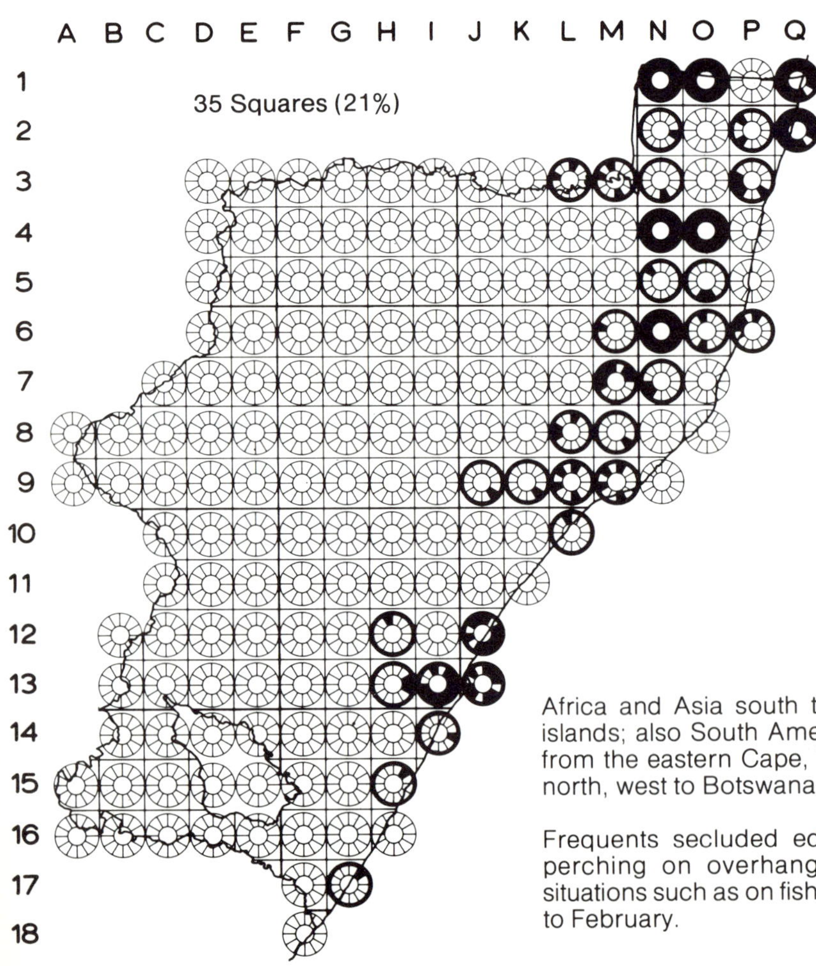

63 Green-backed Heron

Butorides striatus

Africa and Asia south to Australia, Indian and southern Pacific Ocean islands; also South America. In Southern Africa, mainly the eastern half, from the eastern Cape, Transkei, Natal and Zululand, northwards. In the north, west to Botswana (Ngamiland) and northern S.W.A.

Frequents secluded edges of rivers, pans and small ponds, usually perching on overhanging branches. Occasionally seen in exposed situations such as on fish kraals in the Kosi system. Breeds from September to February.

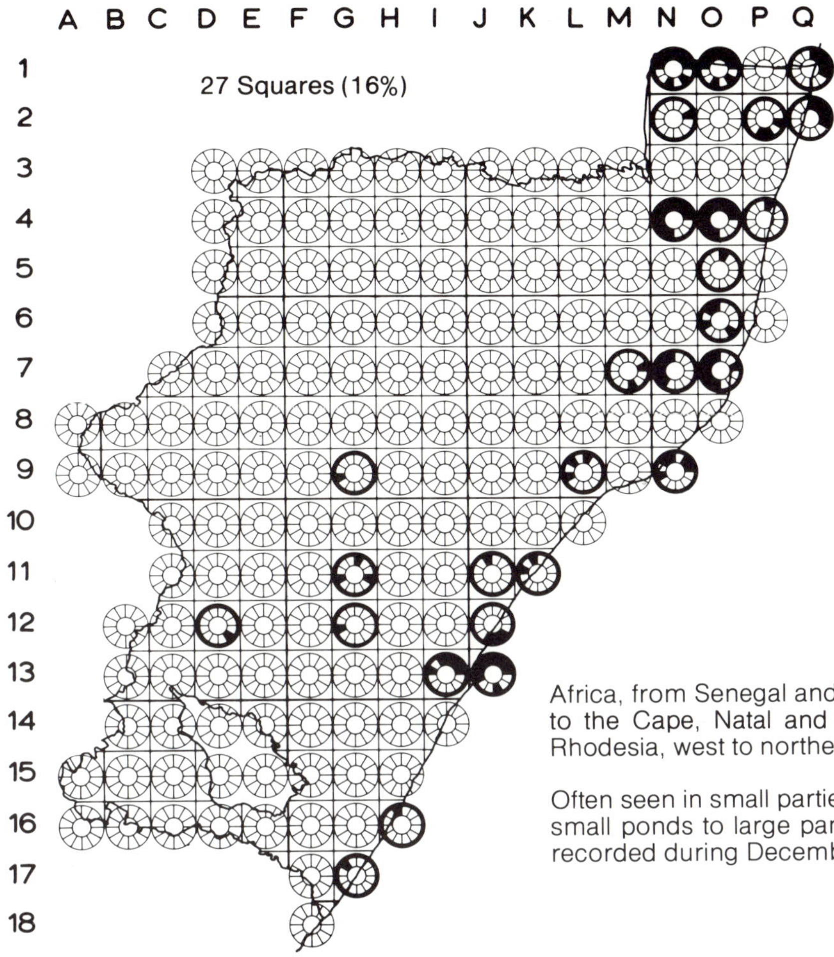

64 Black Egret

Egretta ardesiaca

Africa, from Senegal and Sudan to the Cape. In Southern Africa, a visitor to the Cape, Natal and Zululand, Moçambique, O.F.S., Transvaal and Rhodesia, west to northern Botswana and northern S.W.A.

Often seen in small parties along shallow margins of standing water, from small ponds to large pans; also at estuaries and coastal lakes. Breeding recorded during December and January.

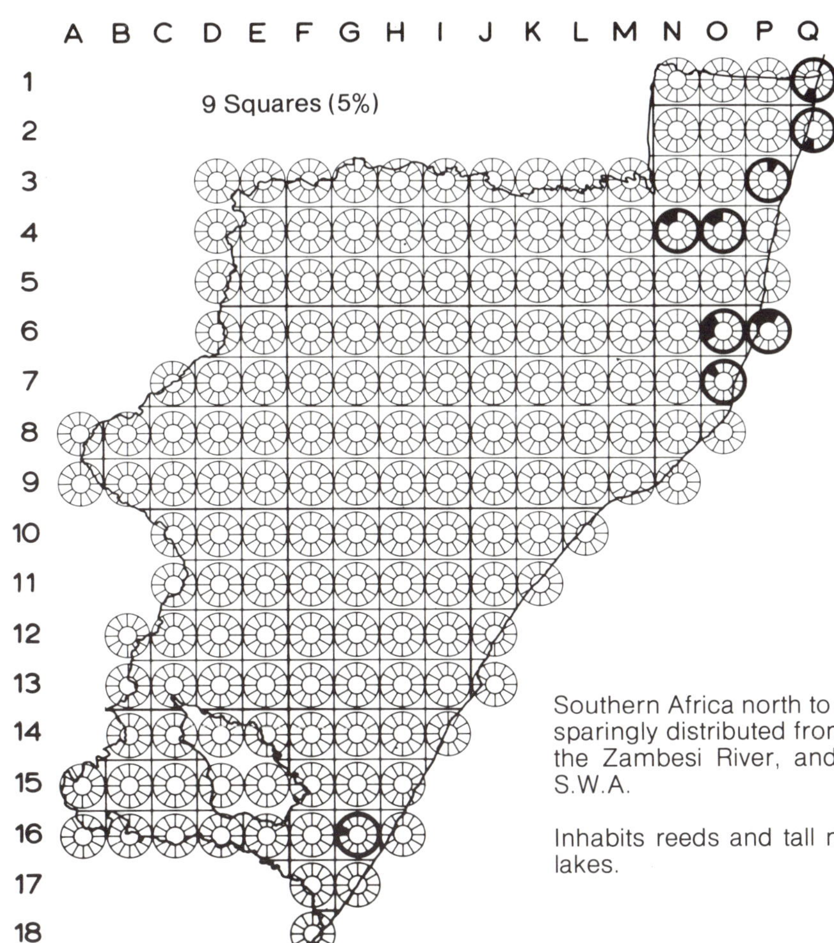

65 Rufous-bellied Heron

Butorides rufiventris

Southern Africa north to Zaïre, Uganda and Tanzania. In Southern Africa, sparingly distributed from the eastern Cape, Natal and Zululand, north to the Zambesi River, and west to Botswana (Ngamiland) and northern S.W.A.

Inhabits reeds and tall marsh grasses around ponds, pans and coastal lakes.

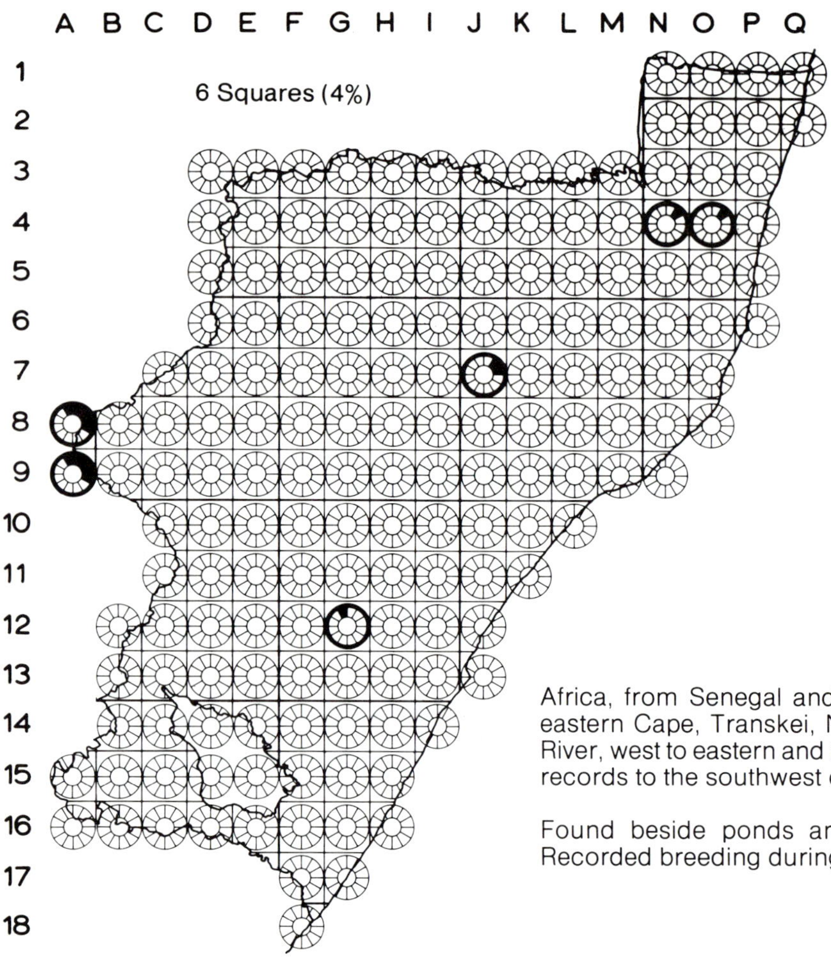

66 Dwarf Bittern

Ixobrychus sturmii

Africa, from Senegal and Ethiopia to the Cape. In Southern Africa, the eastern Cape, Transkei, Natal and Zululand, northwards to the Zambesi River, west to eastern and northern Botswana, and S.W.A. (local). Scattered records to the southwest of the range given.

Found beside ponds and dams in thornveld and exotic vegetation. Recorded breeding during December.

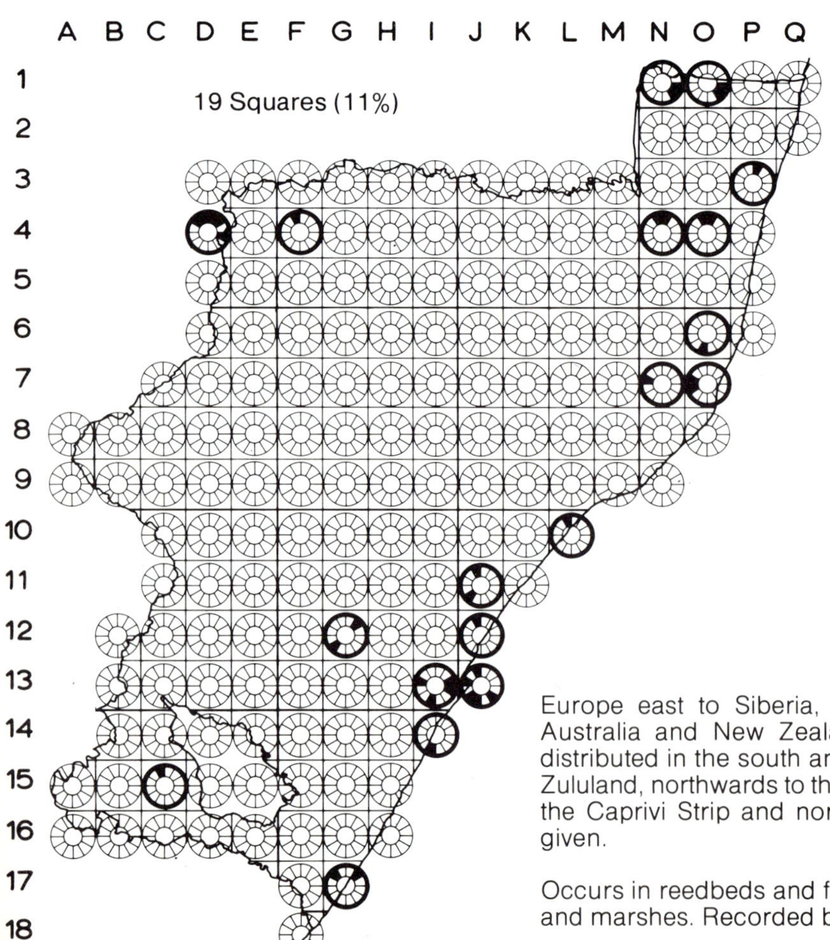

67 Little Bittern

Ixobrychus minutus

Europe east to Siberia, south to Africa, Madagascar, southern Asia, Australia and New Zealand. In Southern Africa, widely but sparsely distributed in the south and east from the south-western Cape, Natal and Zululand, northwards to the Zambesi River, west to Botswana (Ngamiland), the Caprivi Strip and northern S.W.A. Locally to the west of the range given.

Occurs in reedbeds and fringing vegetation of coastal lakes, pans, ponds and marshes. Recorded breeding from September to December.

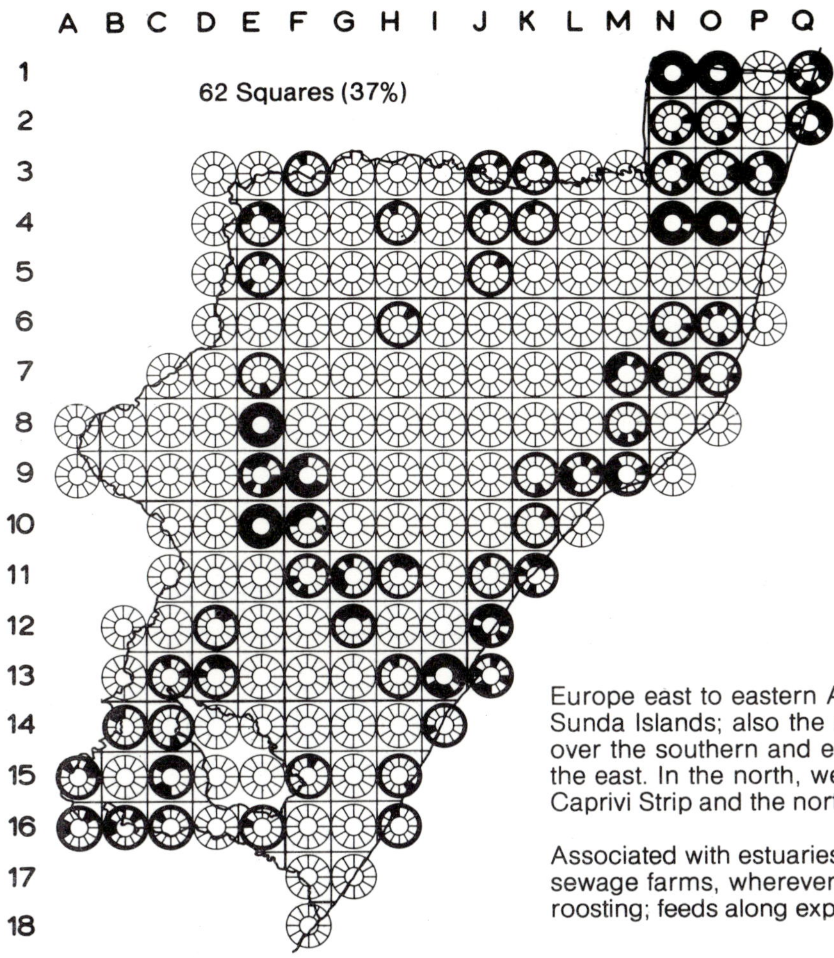

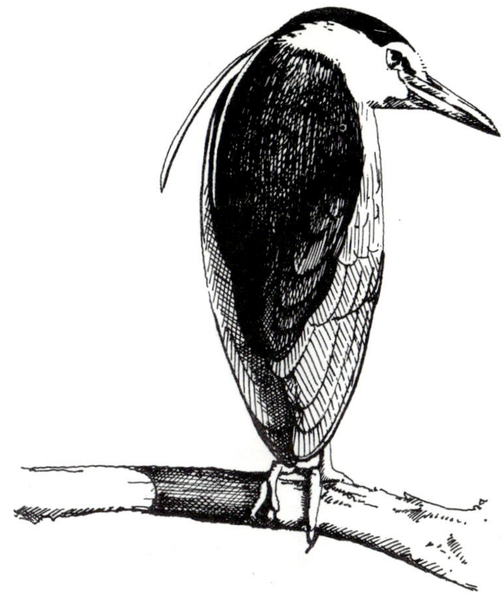

69 Night Heron

Nycticorax nycticorax

Europe east to eastern Asia, south to Africa, Madagascar, India and the Sunda Islands; also the Americas. In Southern Africa, locally distributed over the southern and eastern Cape, Natal and Zululand, northwards in the east. In the north, west through Rhodesia to northern Botswana, the Caprivi Strip and the northern half of S.W.A.

Associated with estuaries, rivers, coastal lakes, pans, dams, marshes and sewage farms, wherever isolated clumps of large trees provide cover for roosting; feeds along exposed shores at night. Breeds during the summer.

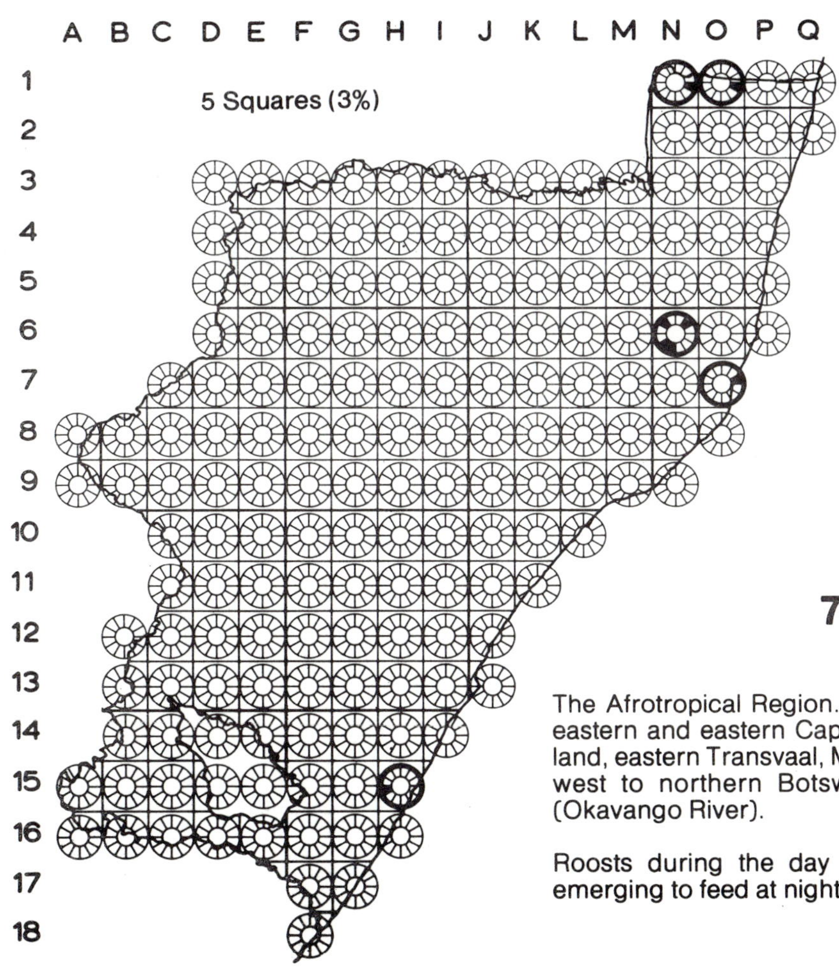

70 White-backed Night Heron

Gorsachius leuconotus

The Afrotropical Region. In Southern Africa, coastal areas of the south-eastern and eastern Cape, Transkei, Natal and Zululand, eastern Swaziland, eastern Transvaal, Moçambique and Rhodesia (except the dry west), west to northern Botswana, the Caprivi Strip and northern S.W.A. (Okavango River).

Roosts during the day in dense stands of trees overhanging rivers, emerging to feed at night. Breeds from September to January.

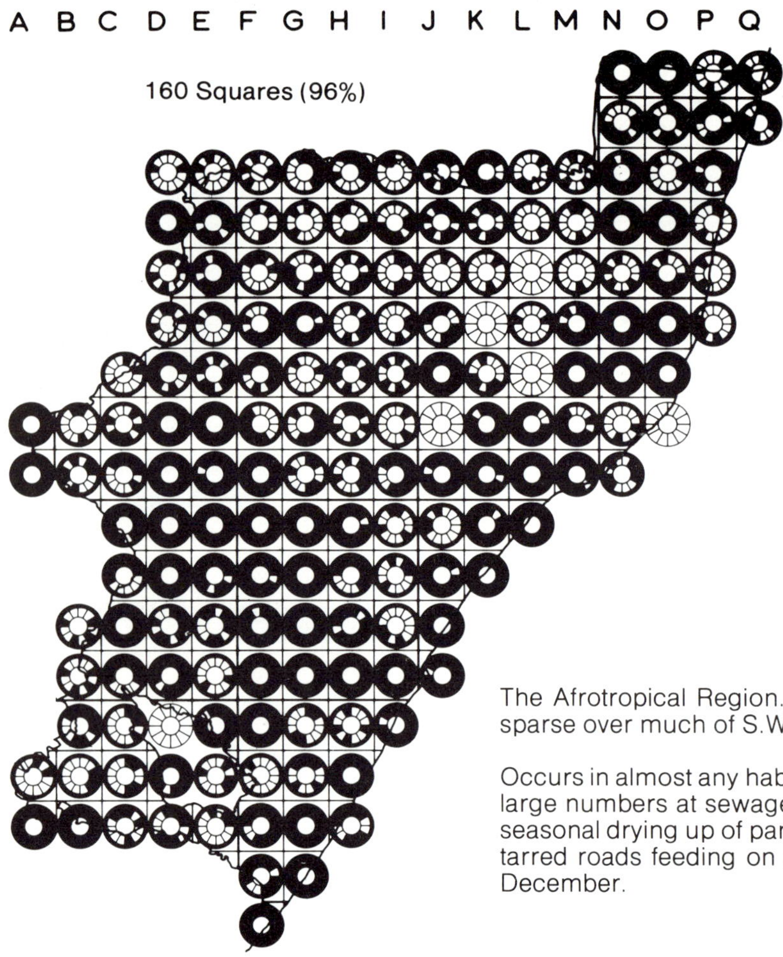

72 Hamerkop

Scopus umbretta

The Afrotropical Region. In Southern Africa, distributed throughout, but sparse over much of S.W.A., the northern Cape and most of Botswana.

Occurs in almost any habitat provided water is present; may congregate in large numbers at sewage farms and exposed margins resulting from the seasonal drying up of pans and small ponds. Seen in the early morning on tarred roads feeding on dead frogs. Recorded breeding from August to December.

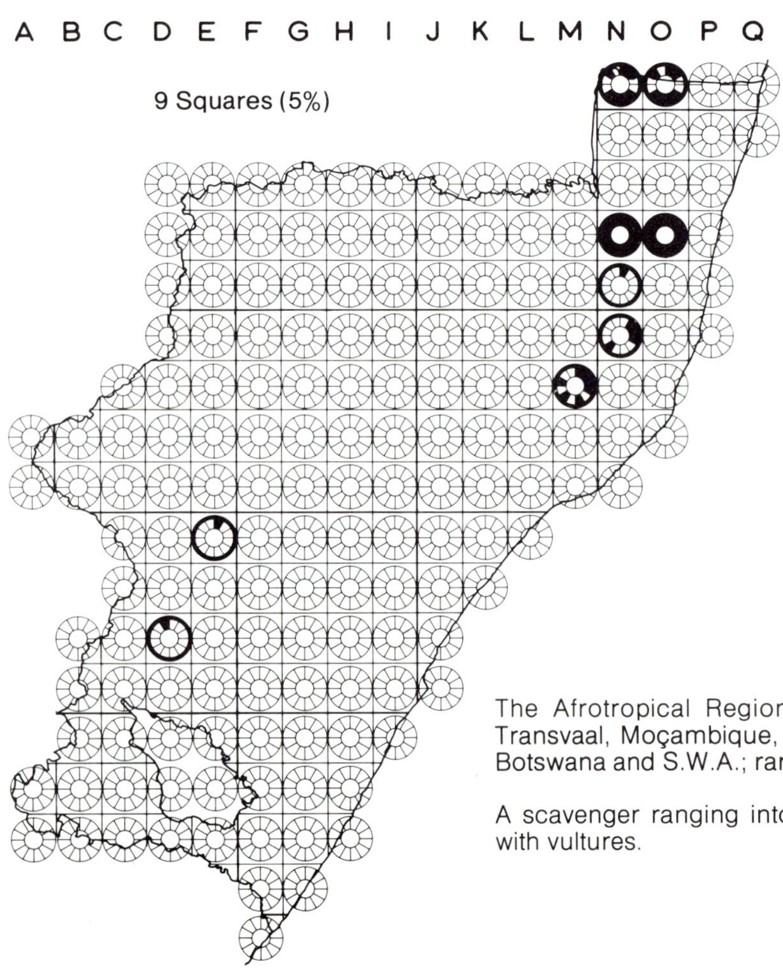

73 Marabou Stork

Leptoptilos crumeniferus

The Afrotropical Region. In Southern Africa, a visitor to Rhodesia, Transvaal, Moçambique, eastern Swaziland and Zululand, westwards to Botswana and S.W.A.; rarely to Natal and the eastern Cape.

A scavenger ranging into Zululand game reserves, where it associates with vultures.

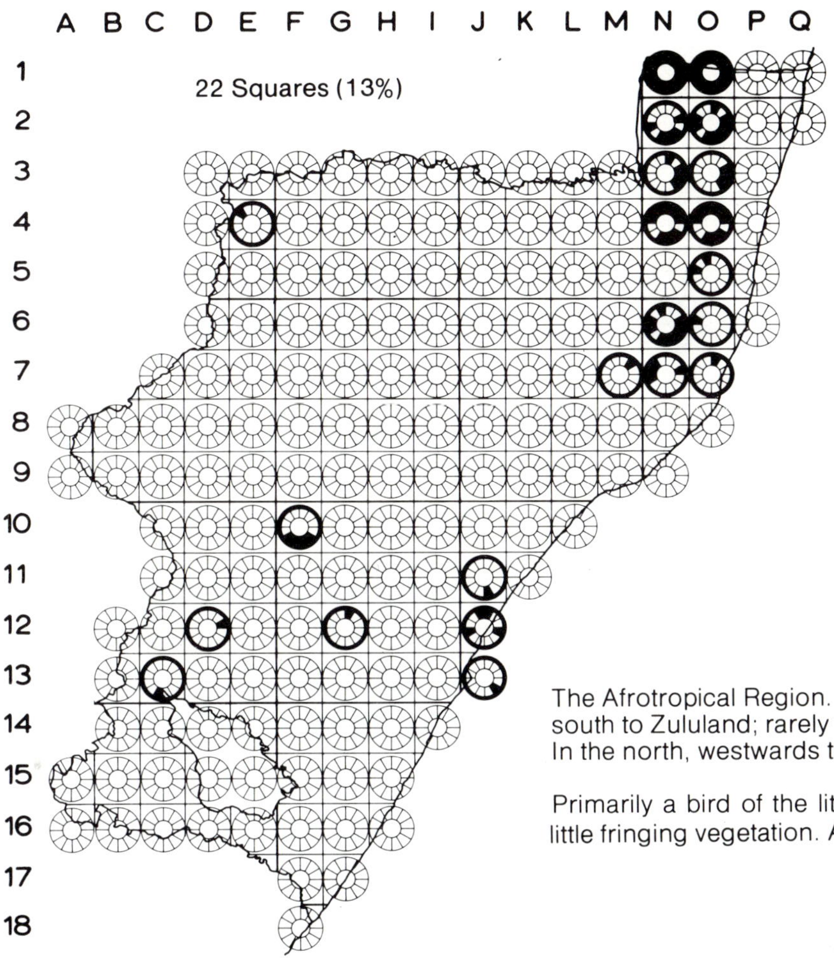

74 Openbill Stork

Anastomus lamelligerus

The Afrotropical Region. In Southern Africa, occurs on the eastern littoral, south to Zululand; rarely to coastal Natal, the eastern Cape and Transvaal. In the north, westwards to Botswana (Ngamiland) and northern S.W.A.

Primarily a bird of the littoral plain, inhabiting pans of varying sizes with little fringing vegetation. A vagrant to the midlands.

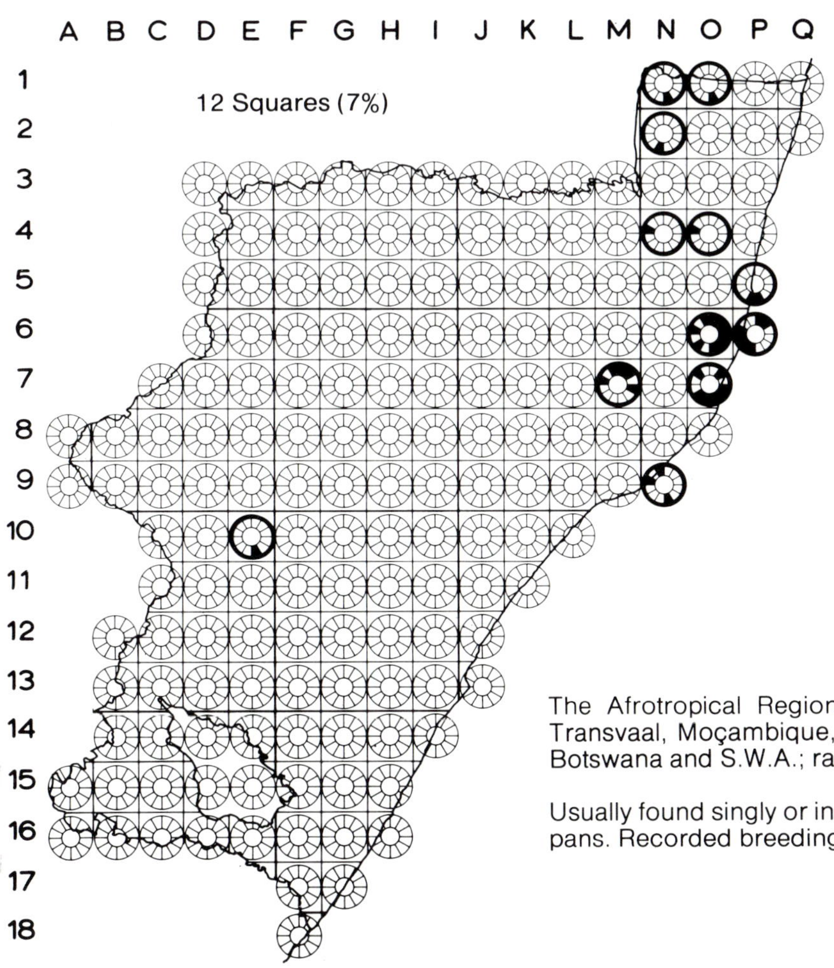

75 Saddlebill Stork

Ephippiorhynchus senegalensis

The Afrotropical Region. In Southern Africa, occurs from Rhodesia, Transvaal, Moçambique, eastern Swaziland and Zululand, westwards to Botswana and S.W.A.; rarely to Natal and the eastern Cape.

Usually found singly or in pairs in swampy grassland adjacent to lakes and pans. Recorded breeding during July at Lake St Lucia.

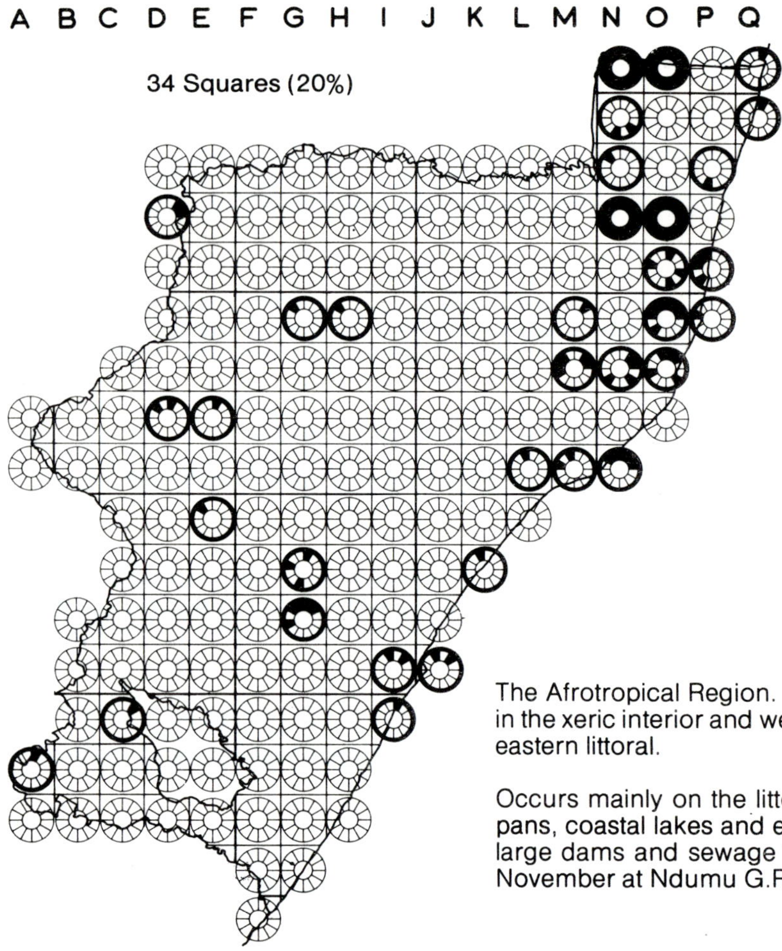

76 Yellow-billed Stork

Mycteria ibis

The Afrotropical Region. In Southern Africa, a widespread visitor. Sparse in the xeric interior and west, and most numerous in the southwest and the eastern littoral.

Occurs mainly on the littoral plain, where it inhabits exposed margins of pans, coastal lakes and estuaries. Uncommon inland, where observed at large dams and sewage farms. Recorded breeding during October and November at Ndumu G.R.

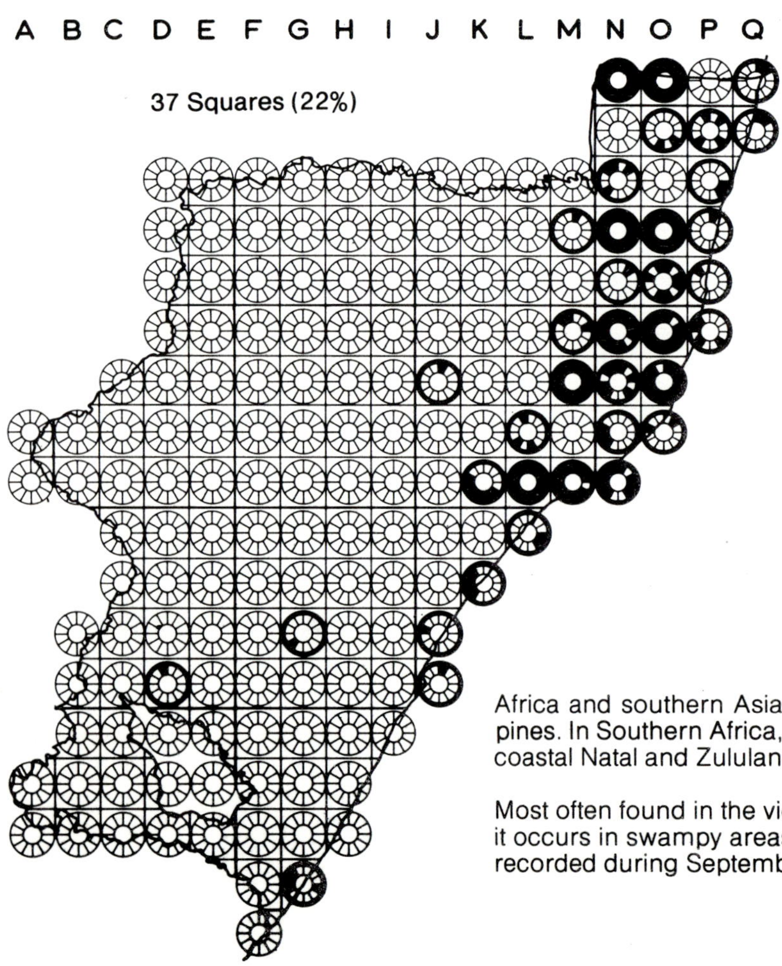

77 Woolly-necked Stork

Ciconia episcopus

Africa and southern Asia eastwards to Borneo, Sulawesi and the Philippines. In Southern Africa, a visitor to Rhodesia and Moçambique, south to coastal Natal and Zululand, and the eastern Cape.

Most often found in the vicinity of estuaries, coastal lakes and pans where it occurs in swampy areas as well as open grassland. Breeding has been recorded during September and October.

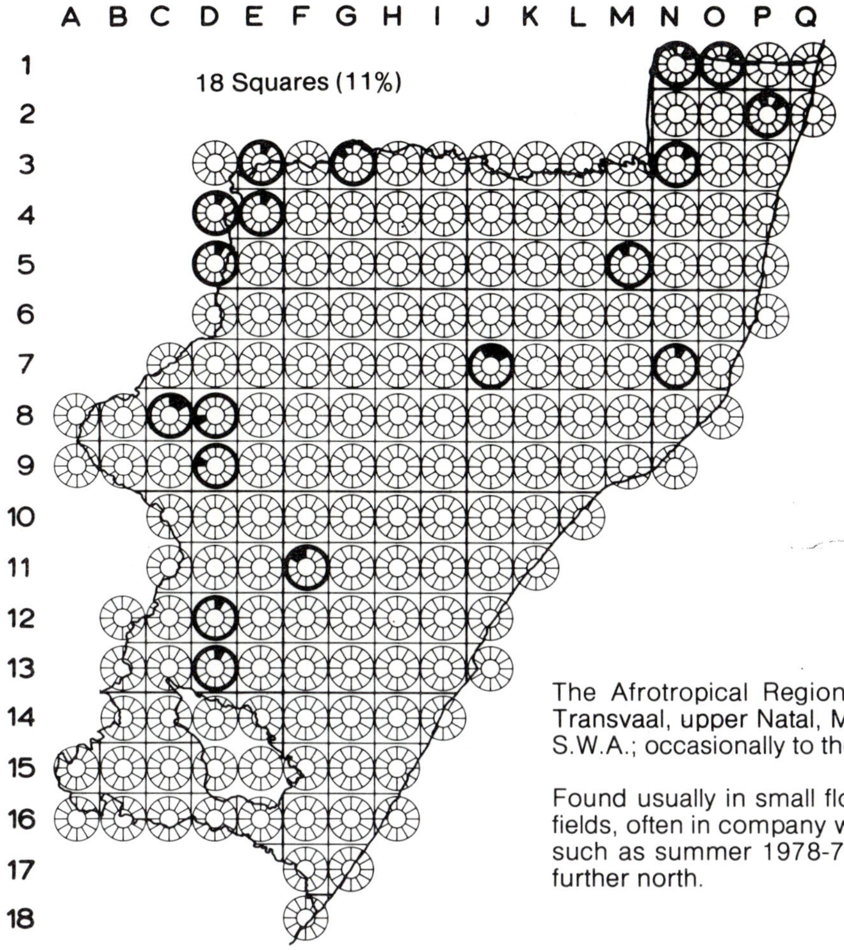

78 White-bellied Stork

Ciconia abdimii

The Afrotropical Region. In Southern Africa, a visitor to the O.F.S., Transvaal, upper Natal, Moçambique, Rhodesia, Botswana and northern S.W.A.; occasionally to the Cape. Present from November to March.

Found usually in small flocks in grassland, open savanna and cultivated fields, often in company with White Stork (80). Subject to annual influxes, such as summer 1978-79, which may be related to drought conditions further north.

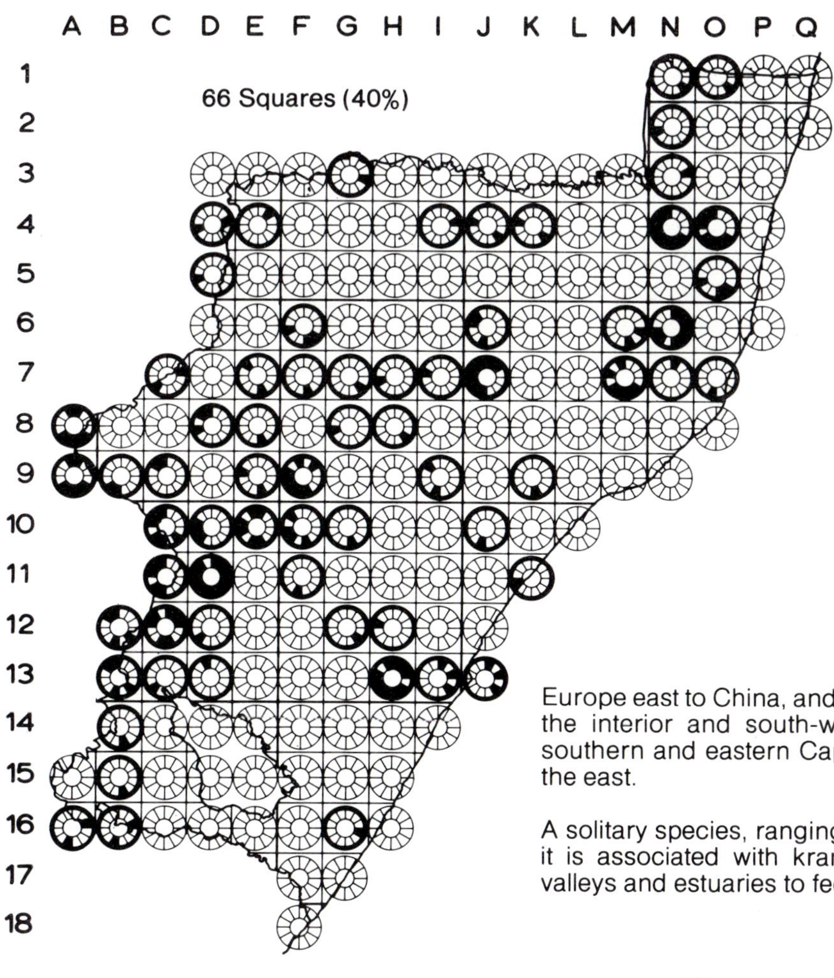

79 Black Stork

Ciconia nigra

Europe east to China, and Southern Africa. Occurs locally in S.W.A. and in the interior and south-west of the region. Recorded breeding in the southern and eastern Cape, Lesotho, Natal and Zululand, northwards in the east.

A solitary species, ranging from the Drakensberg to lowland areas where it is associated with kranses and gorges. Visits open grassland, river valleys and estuaries to feed. Breeds during the winter.

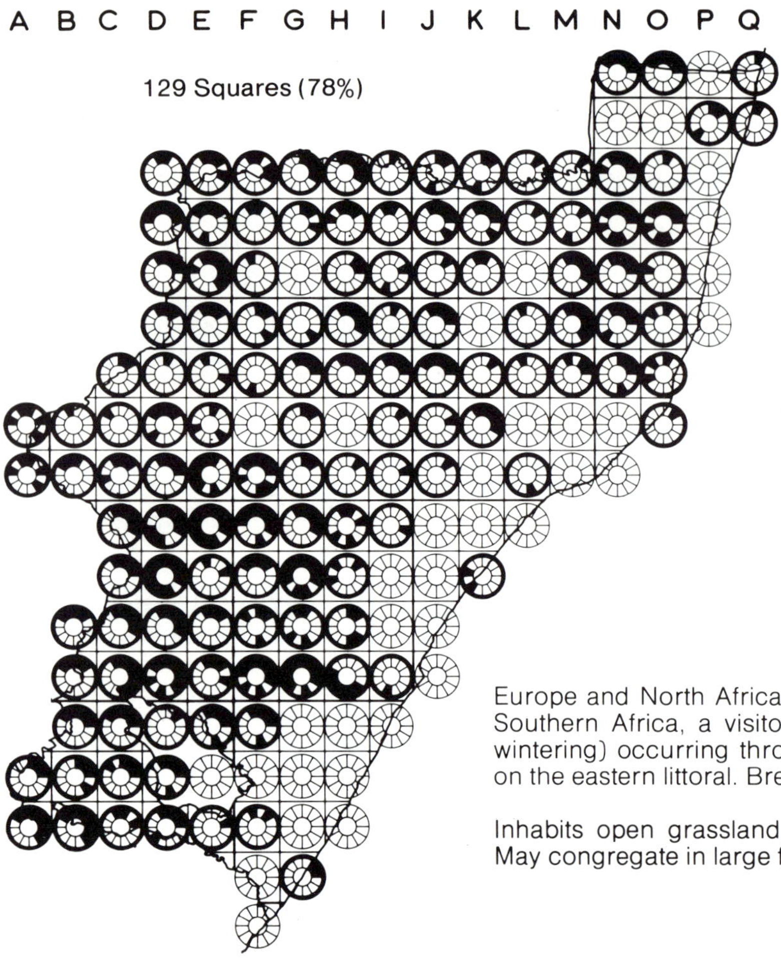

80 White Stork

Ciconia ciconia

Europe and North Africa east to eastern Asia, wintering to the south. In Southern Africa, a visitor from November to March (with some over-wintering) occurring throughout although scarce in S.W.A. and seldom on the eastern littoral. Breeds locally in the southern Cape.

Inhabits open grassland, cultivated fields and lightly-wooded savanna. May congregate in large flocks on burnt veld.

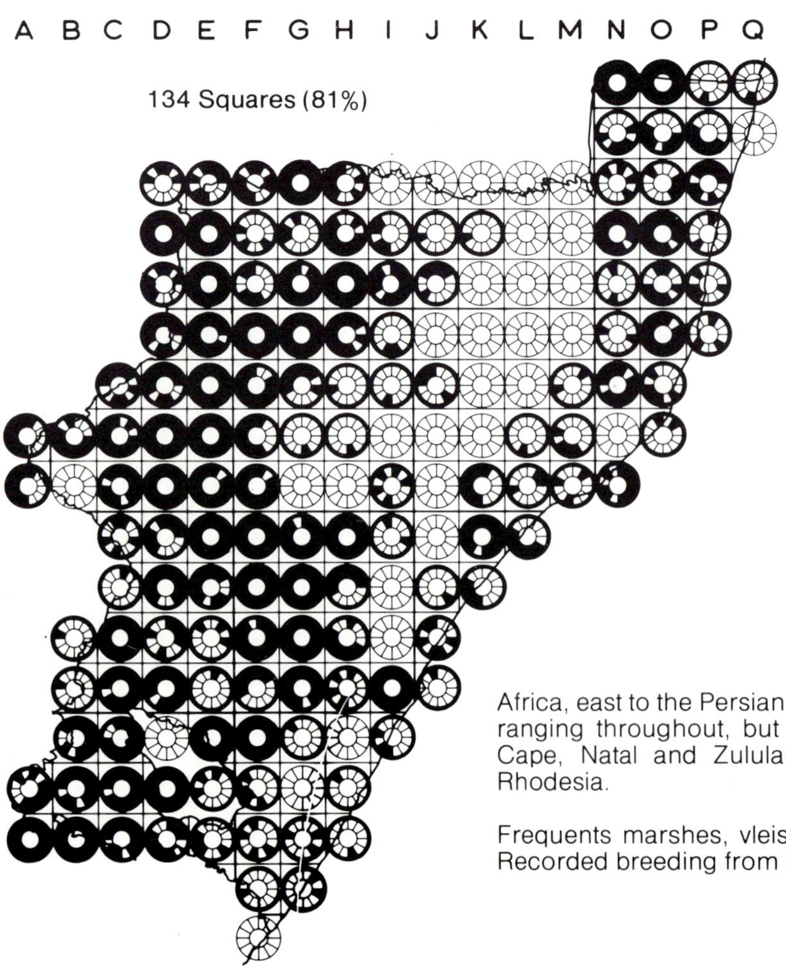

81 Sacred Ibis

Threskiornis aethiopicus

Africa, east to the Persian Gulf and Madagascar. In Southern Africa, wide-ranging throughout, but local and mainly in the southern and eastern Cape, Natal and Zululand, northwards in the east to south-western Rhodesia.

Frequents marshes, vleis, sewage farms and areas of moist grassland. Recorded breeding from October to January.

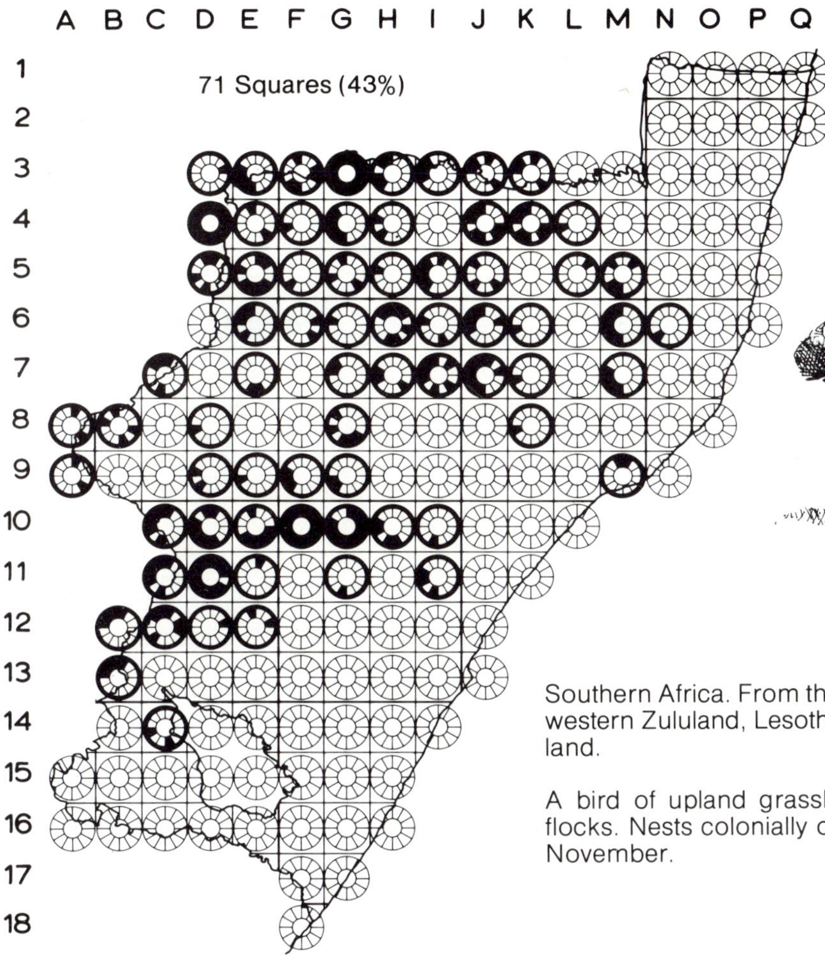

82 Bald Ibis

Geronticus calvus

Southern Africa. From the north-eastern Cape to Transkei, Natal (interior), western Zululand, Lesotho, O.F.S., southern Transvaal and western Swaziland.

A bird of upland grassland, usually in the moister parts, occurring in flocks. Nests colonially on kranses and sides of gorges, from January to November.

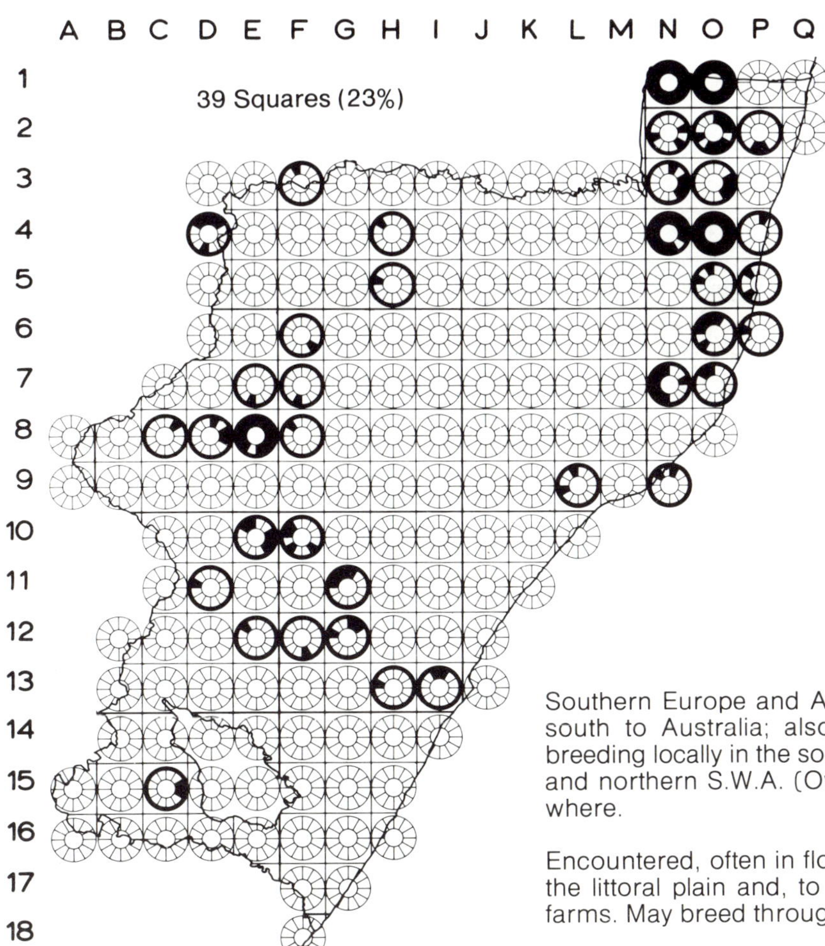

83 Glossy Ibis

Plegadis falcinellus

Southern Europe and Africa east to Asia, the Philippines and Sulawesi, south to Australia; also the Americas. In Southern Africa, recorded breeding locally in the south-western Cape, Zululand, Transvaal, Botswana and northern S.W.A. (Ovamboland), and occurs in small numbers elsewhere.

Encountered, often in flocks, on the edges of open stretches of water on the littoral plain and, to a lesser extent, inland vleis, dams and sewage farms. May breed throughout the year.

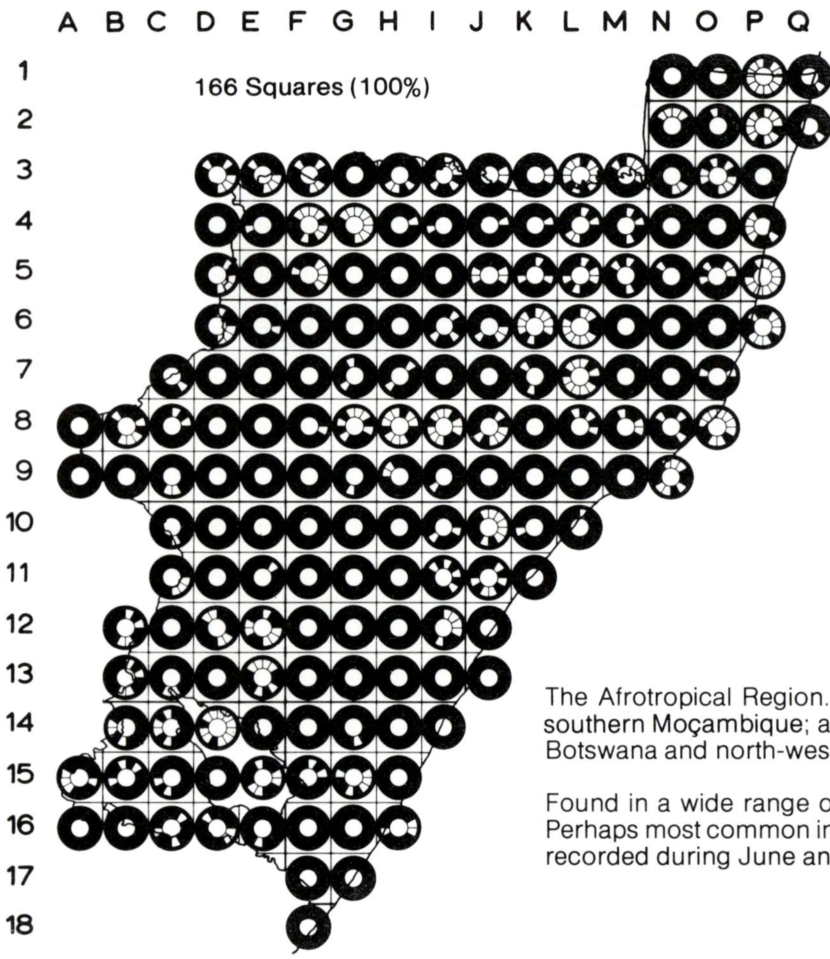

84 Hadeda

Bostrychia hagedash

The Afrotropical Region. In Southern Africa, from the southern Cape to southern Moçambique; also in northern S.W.A., the Caprivi Strip, adjacent Botswana and north-western Rhodesia.

Found in a wide range of habitats, including towns and cultivated lands. Perhaps most common in well watered areas with stands of trees. Breeding recorded during June and from September to January.

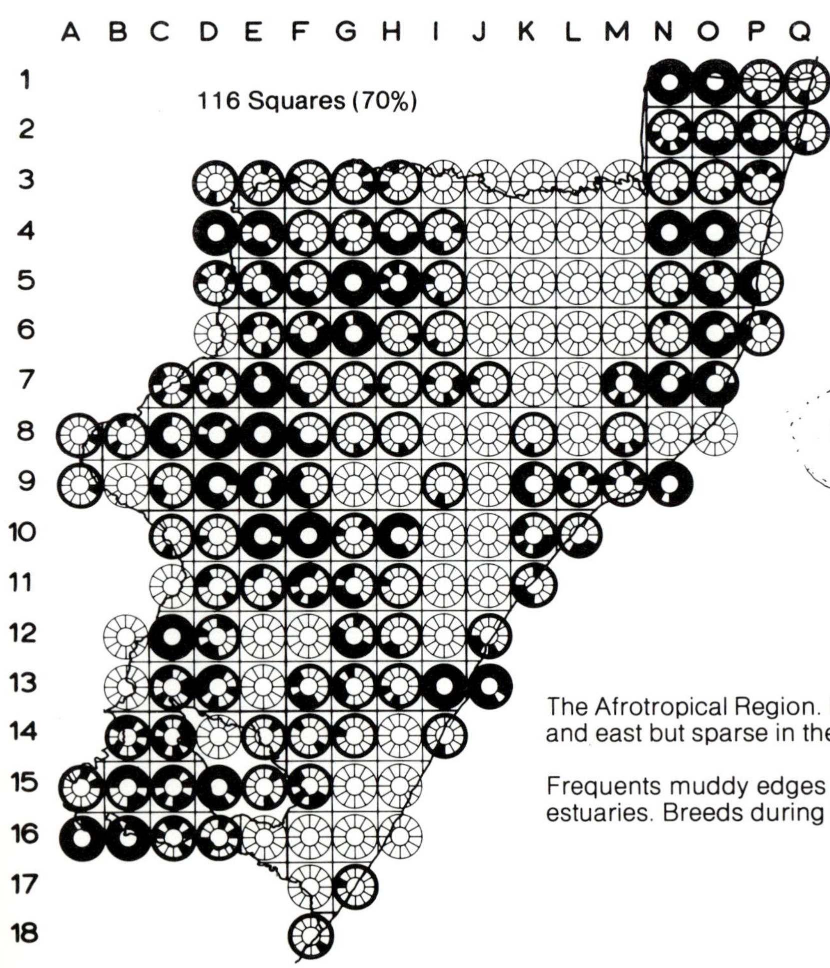

85 African Spoonbill

Platalea alba

The Afrotropical Region. In Southern Africa, widely distributed in the south and east but sparse in the xeric interior and west.

Frequents muddy edges of marshes, pans, dams and ponds; irregular at estuaries. Breeds during the winter.

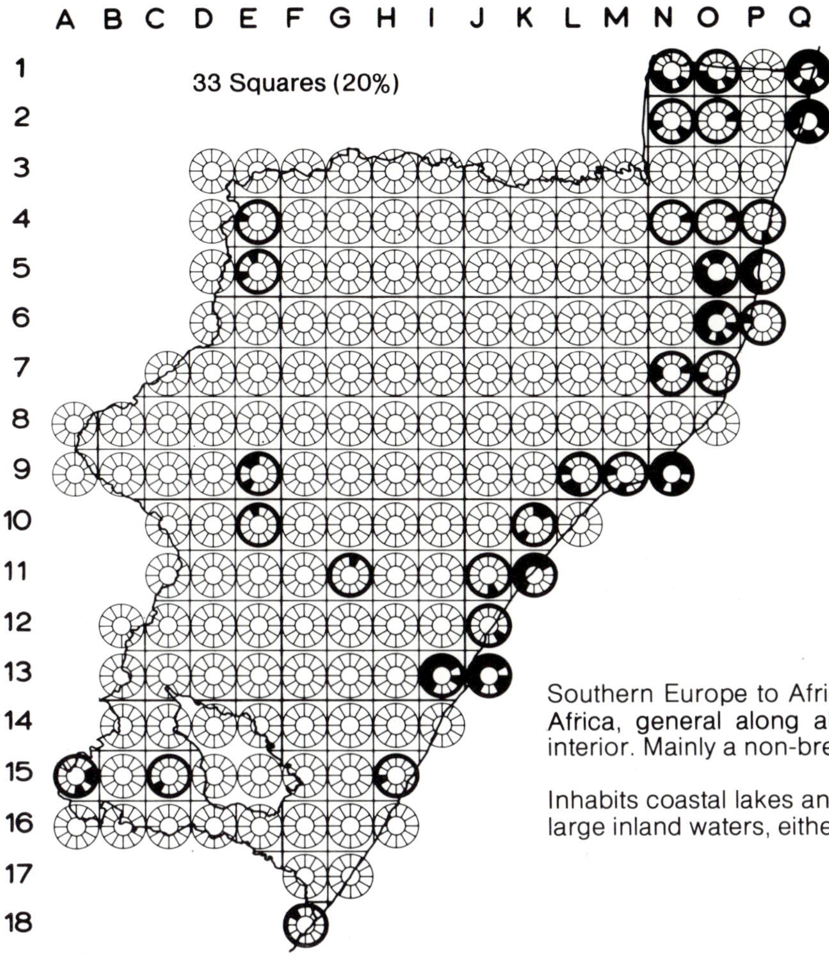

86 Greater Flamingo

Phoenicopterus ruber

Southern Europe to Africa, east to Asia; also the Americas. In Southern Africa, general along all coasts, and local on brackish waters of the interior. Mainly a non-breeding visitor.

Inhabits coastal lakes and estuaries, often in large numbers. A vagrant to large inland waters, either singly or in small parties.

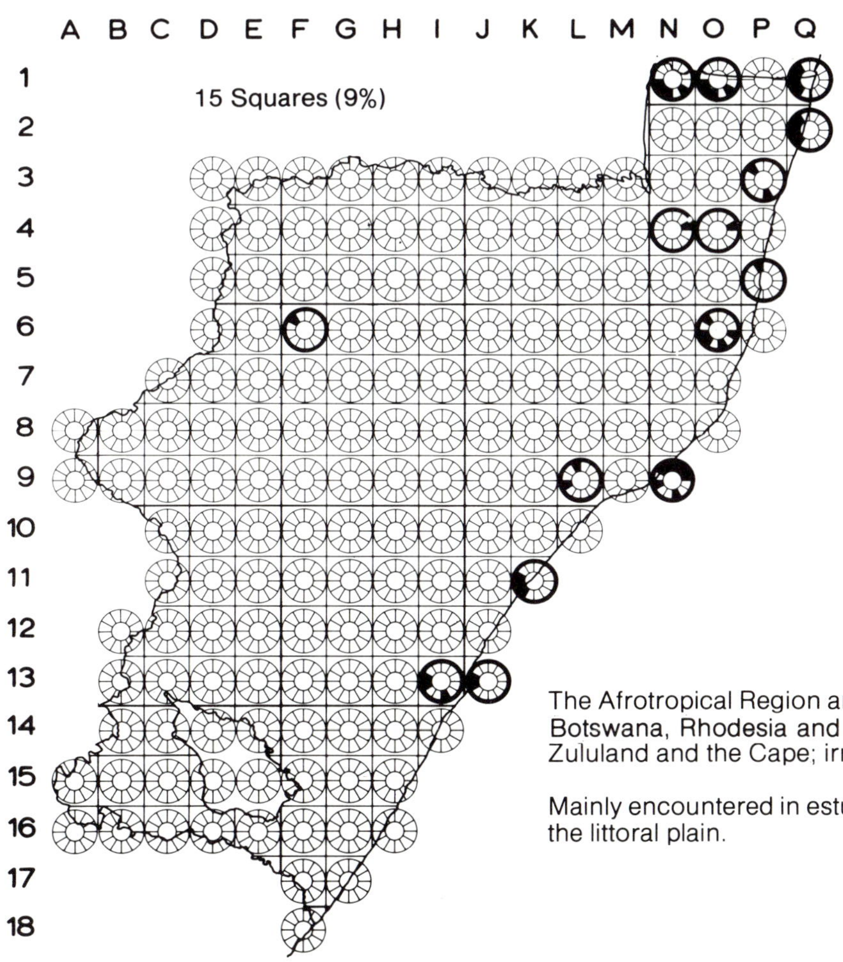

87 Lesser Flamingo

Phoenicopterus minor

The Afrotropical Region and India. In Southern Africa, a visitor to northern Botswana, Rhodesia and Moçambique, southwards in the east to Natal, Zululand and the Cape; irregular elsewhere.

Mainly encountered in estuaries; occasionally at coastal lakes and pans on the littoral plain.

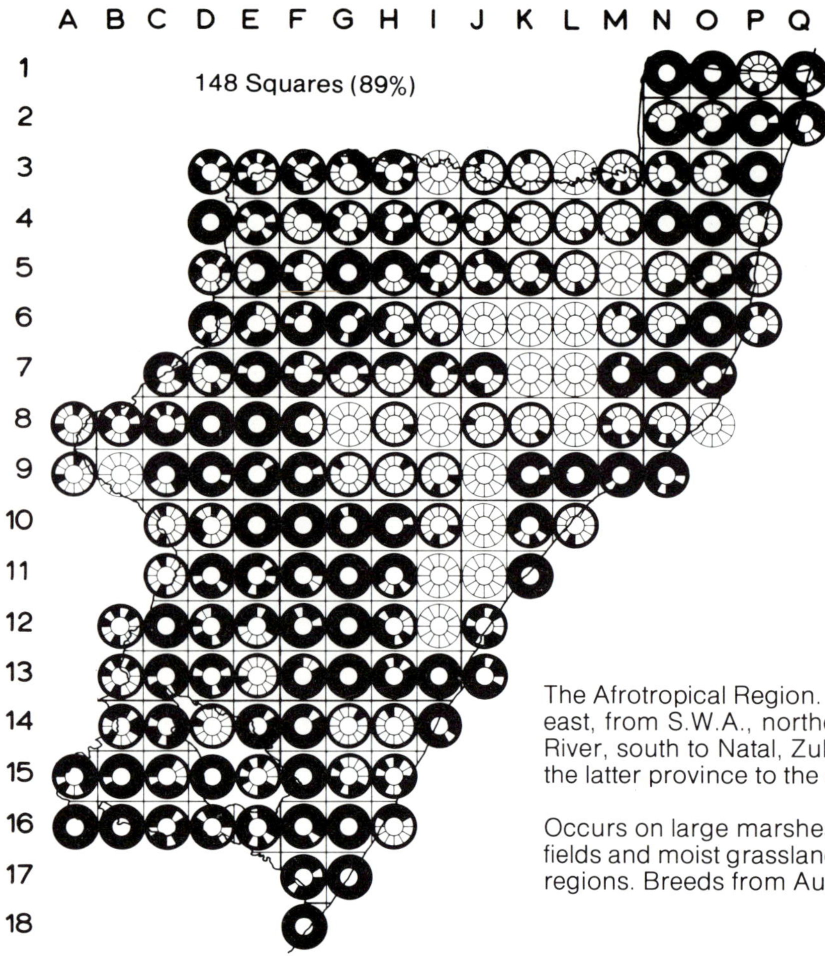

88 Spur-winged Goose

Plectropterus gambensis

The Afrotropical Region. In Southern Africa, wide-ranging in the north and east, from S.W.A., northern Botswana, Rhodesia and the lower Zambesi River, south to Natal, Zululand and the eastern Cape, and locally through the latter province to the extreme west.

Occurs on large marshes, dams, pans and estuaries; ranges to cultivated fields and moist grassland to feed. Largest concentrations found in upland regions. Breeds from August to March.

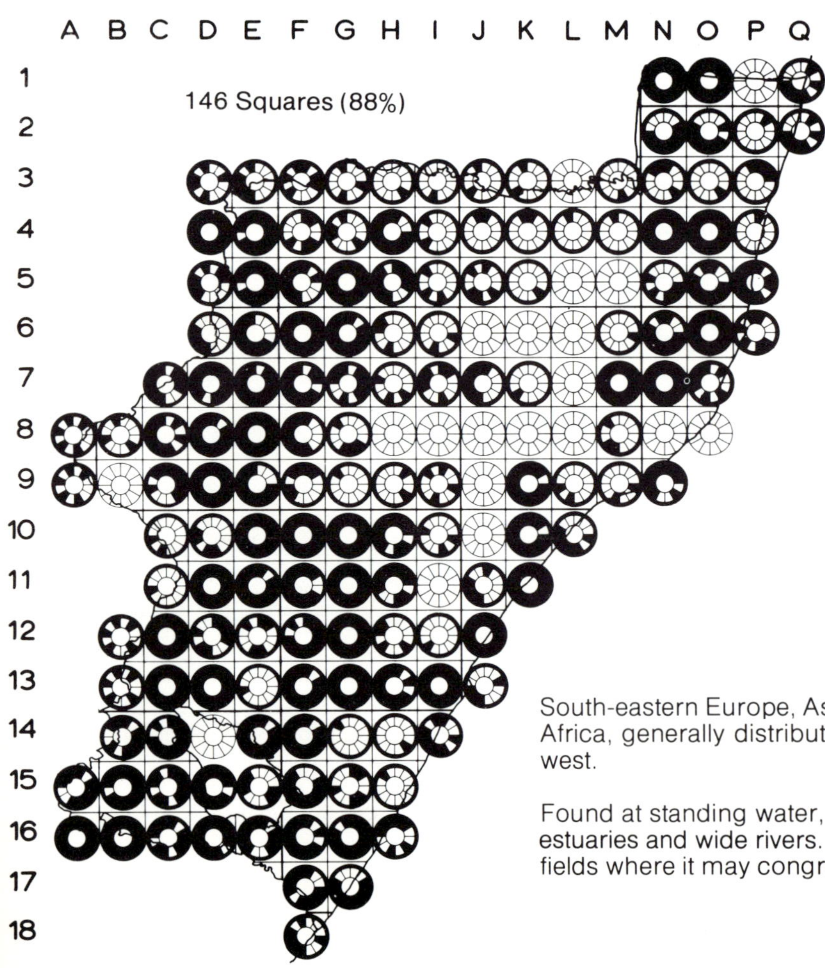

89 Egyptian Goose

Alopochen aegyptiacus

South-eastern Europe, Asia Minor, the Middle East and Africa. In Southern Africa, generally distributed, though local in the xeric interior and the far west.

Found at standing water, from open stretches to the smallest ponds; also estuaries and wide rivers. Ranges into grassland and particularly cultivated fields where it may congregate in numbers. Breeds throughout the year.

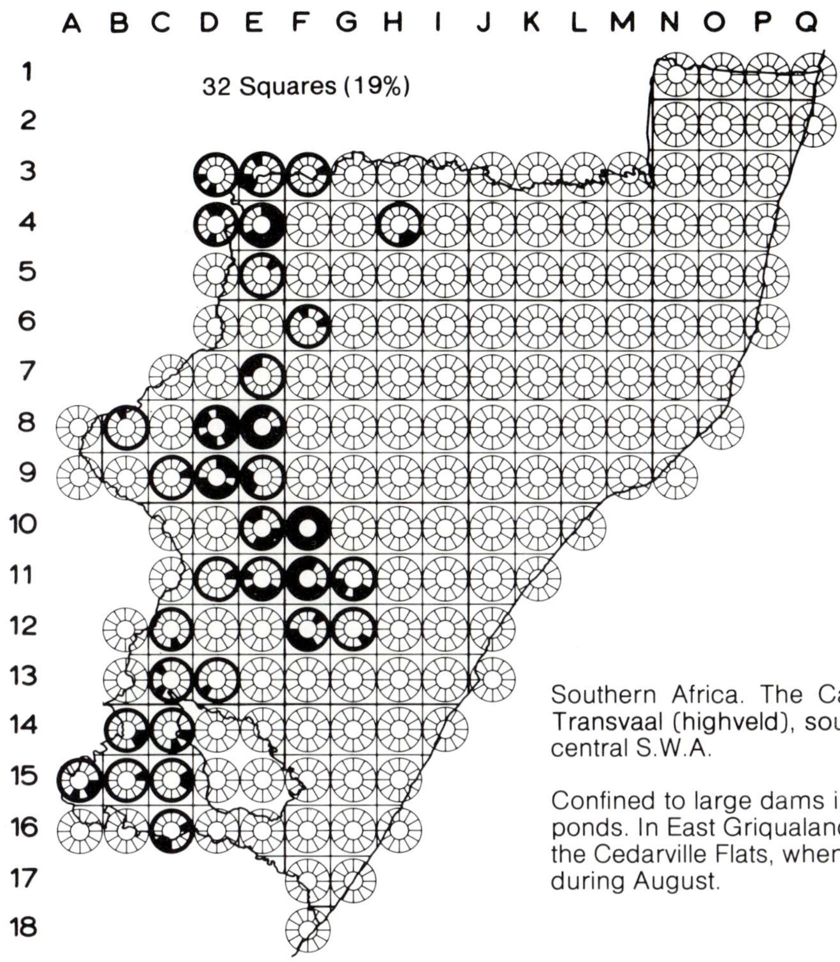

90 South African Shelduck

Tadorna cana

Southern Africa. The Cape, upper Natal, Lesotho (lowlands), O.F.S., Transvaal (highveld), southern and eastern Botswana and southern and central S.W.A.

Confined to large dams in upland areas; occasionally visits sewage farm ponds. In East Griqualand, subject to influxes during seasonal flooding of the Cedarville Flats, when numbers may exceed 100. Recorded breeding during August.

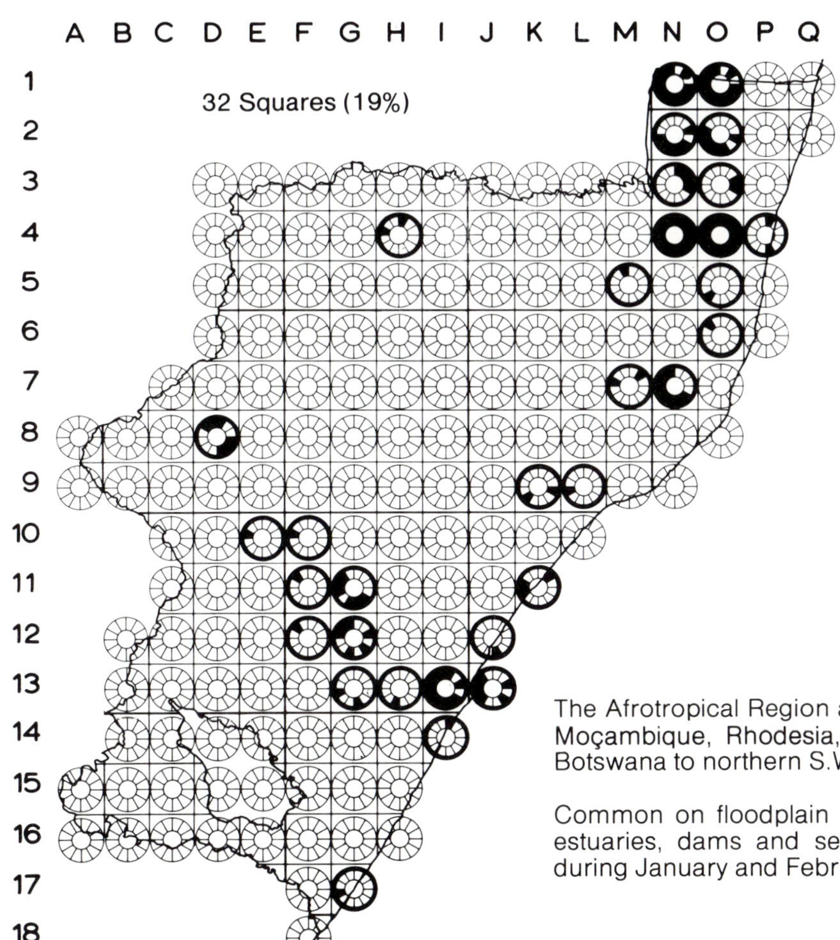

91 Knob-billed Duck

Sarkidiornis melanotos

The Afrotropical Region and southern Asia. In Southern Africa, southern Moçambique, Rhodesia, Transvaal, Natal and Zululand, west through Botswana to northern S.W.A. Irregularly farther south.

Common on floodplain pans in north-eastern Zululand; infrequently at estuaries, dams and sewage farms to the south. Breeding recorded during January and February.

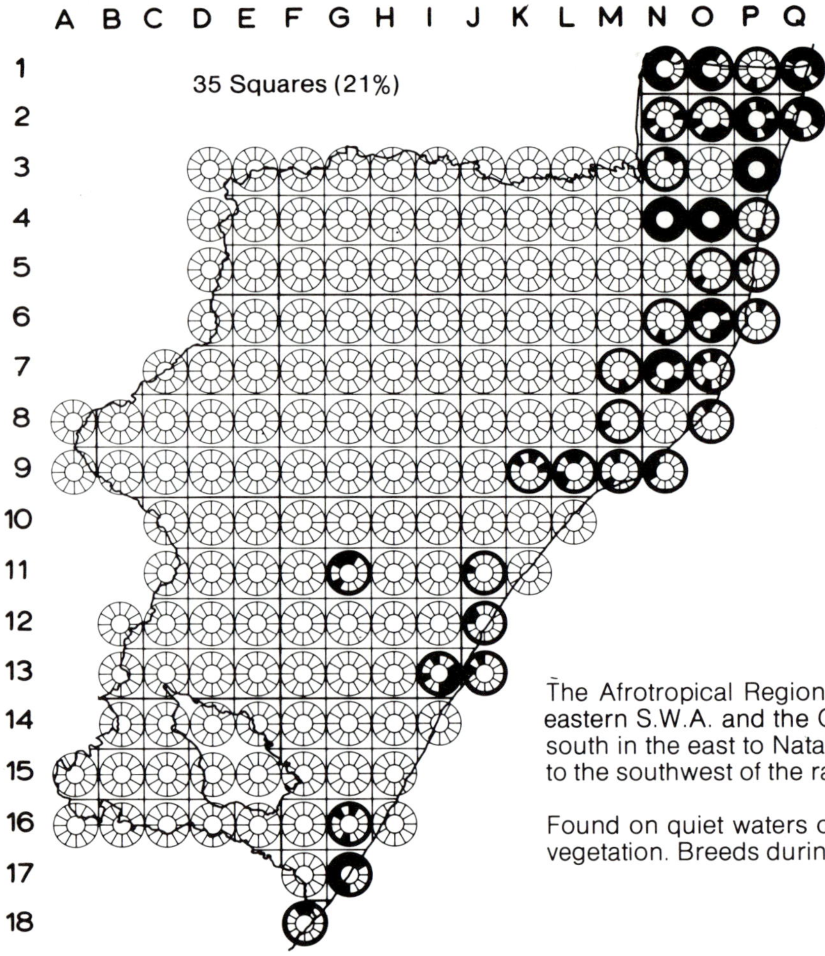

92 Pygmy Goose

Nettapus auritus

The Afrotropical Region. In Southern Africa, northern Botswana, north-eastern S.W.A. and the Caprivi Strip, east to Rhodesia and Moçambique, south in the east to Natal and Zululand, and the eastern Cape. Irregularly to the southwest of the range given.

Found on quiet waters of varying sizes, with much floating and marginal vegetation. Breeds during the summer.

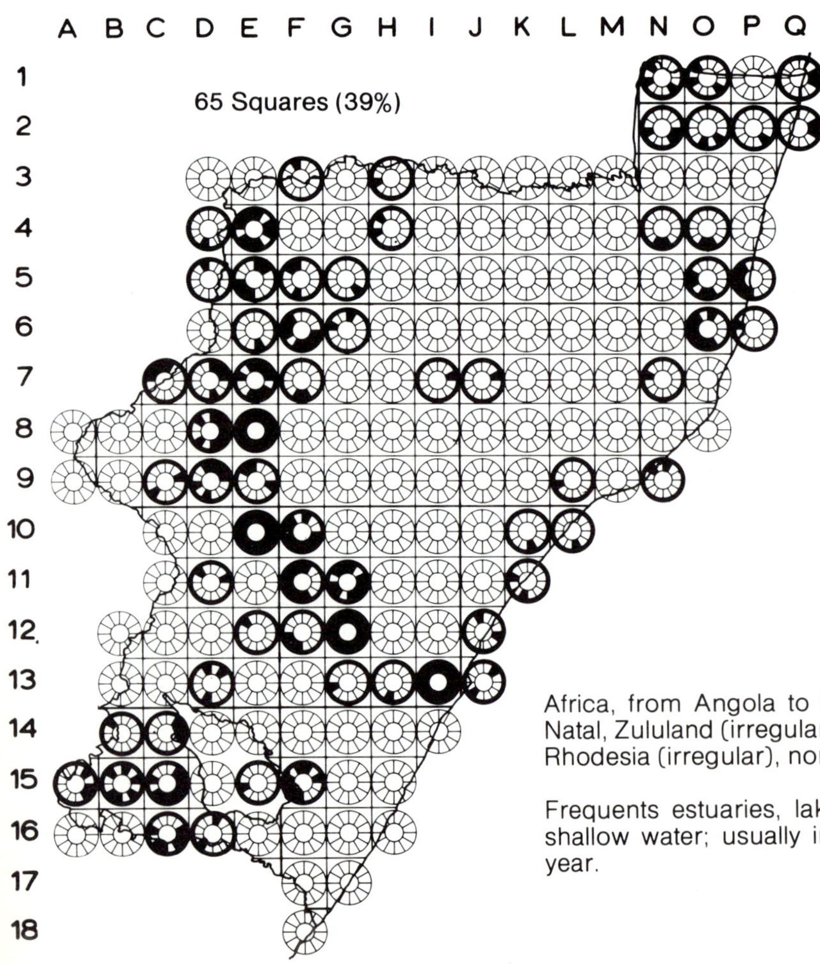

94 Cape Shoveler

Anas smithii

Africa, from Angola to the Cape. In Southern Africa, the Cape, upper Natal, Zululand (irregular), Lesotho (lowlands), O.F.S., Transvaal, western Rhodesia (irregular), northern Botswana and S.W.A.

Frequents estuaries, lakes, pans, dams and sewage farms, preferring shallow water; usually in pairs or small parties. Breeds throughout the year.

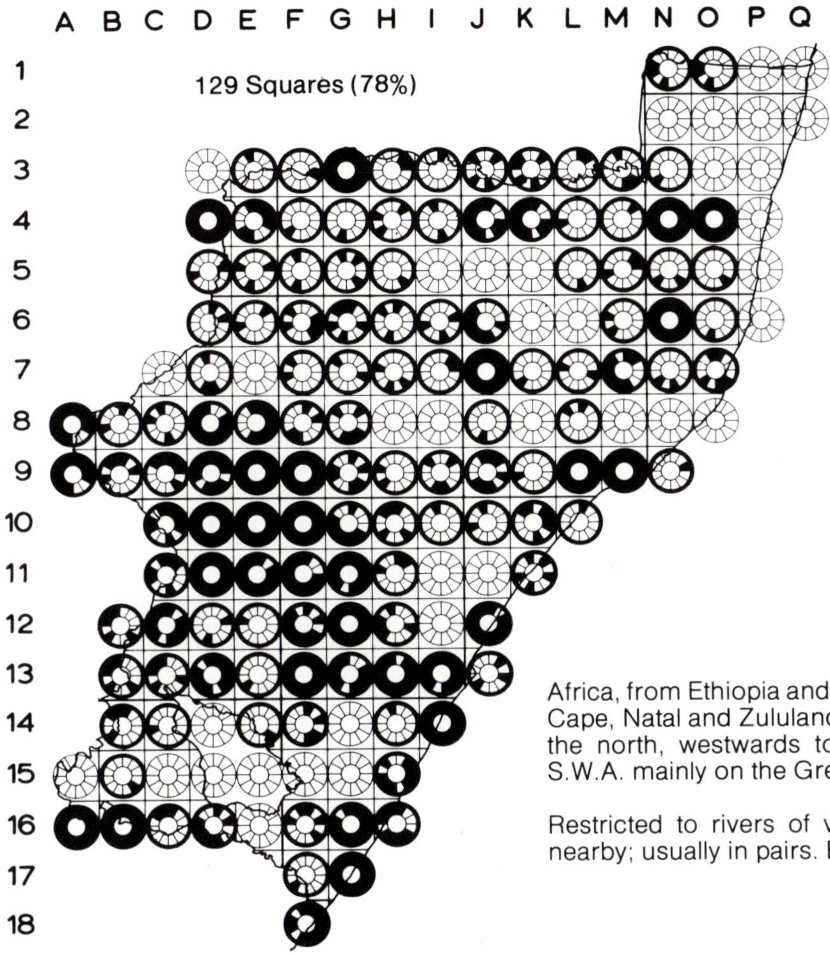

95 Black Duck

Anas sparsa

Africa, from Ethiopia and Gabon south to the Cape. In Southern Africa, the Cape, Natal and Zululand, northwards in the east to the Zambesi River. In the north, westwards to northern Botswana and the Caprivi Strip. In S.W.A. mainly on the Great Fish and Orange Rivers.

Restricted to rivers of varying sizes, occasionally visiting small ponds nearby; usually in pairs. Breeds mainly from July to February.

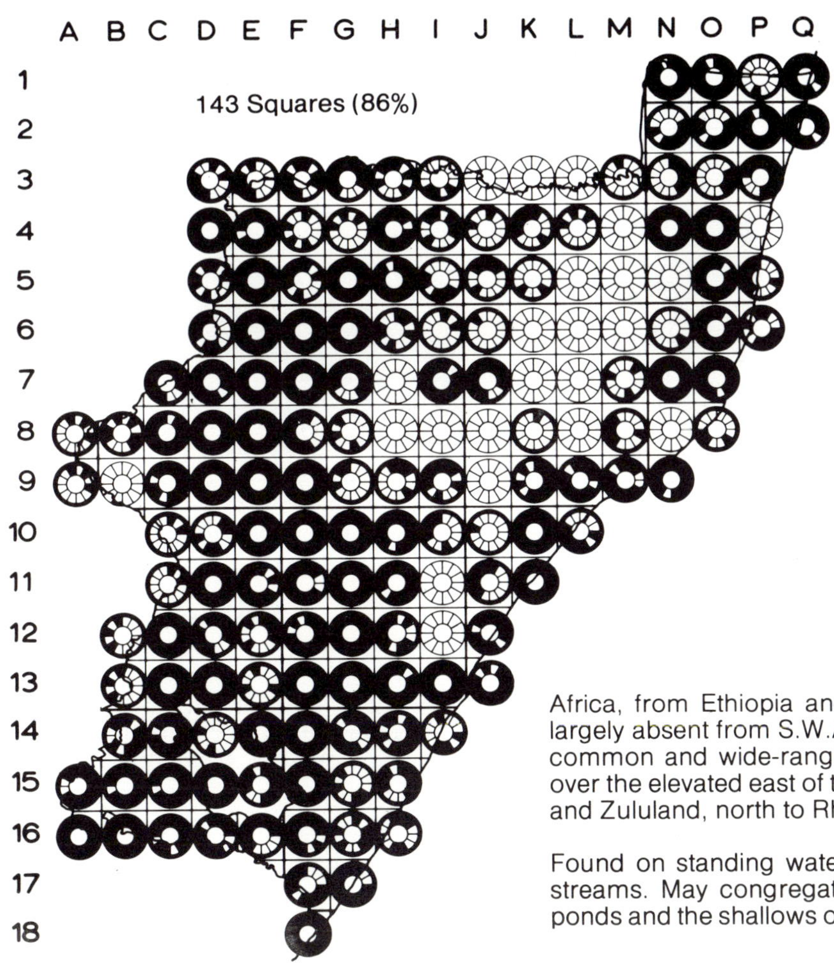

96 Yellow-billed Duck

Anas undulata

Africa, from Ethiopia and Zaïre south to the Cape. In Southern Africa, largely absent from S.W.A., Botswana and the eastern tropical littoral, but common and wide-ranging in the Cape south of the Orange River and over the elevated east of the region from the eastern Cape, Transkei, Natal and Zululand, north to Rhodesia (rare).

Found on standing water of any size and occasionally on slow flowing streams. May congregate in large numbers on marshes, sewage farm ponds and the shallows of large open waters. Breeds throughout the year.

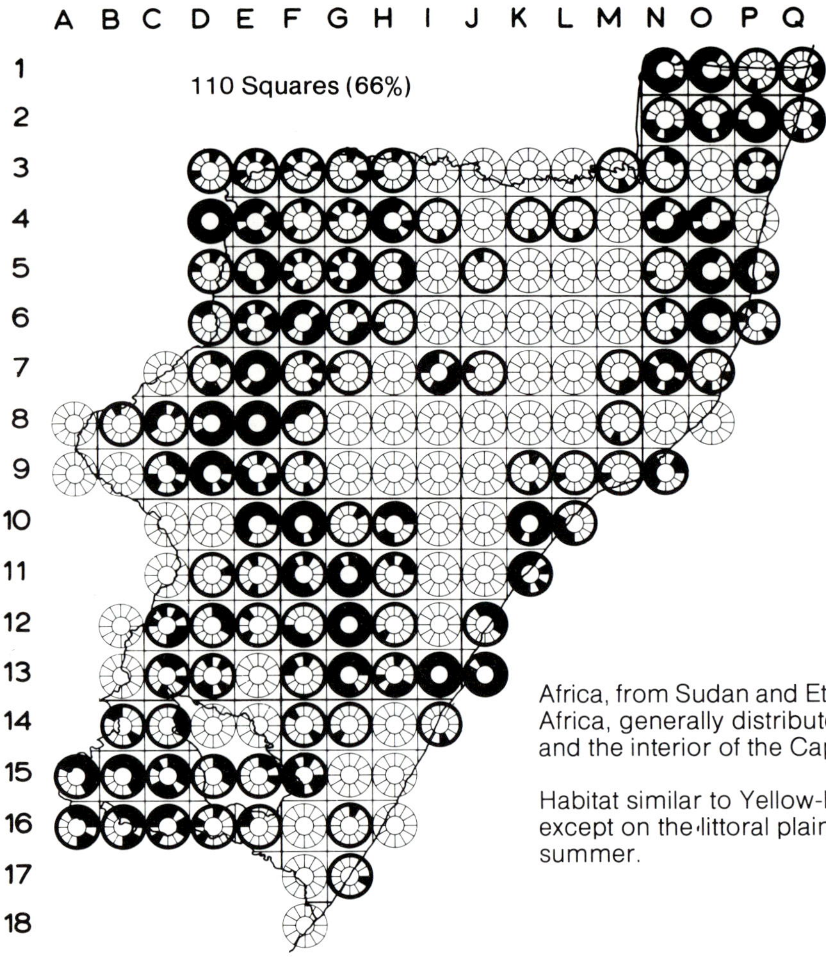

97 Red-billed Teal

Anas erythrorhyncha

Africa, from Sudan and Ethiopia to the Cape; also Madagascar. In Southern Africa, generally distributed, being the most abundant anatid in Rhodesia and the interior of the Cape, S.W.A. and Botswana.

Habitat similar to Yellow-billed Duck (96), but occurs in smaller numbers except on the littoral plain where more plentiful. Breeds mainly during the summer.

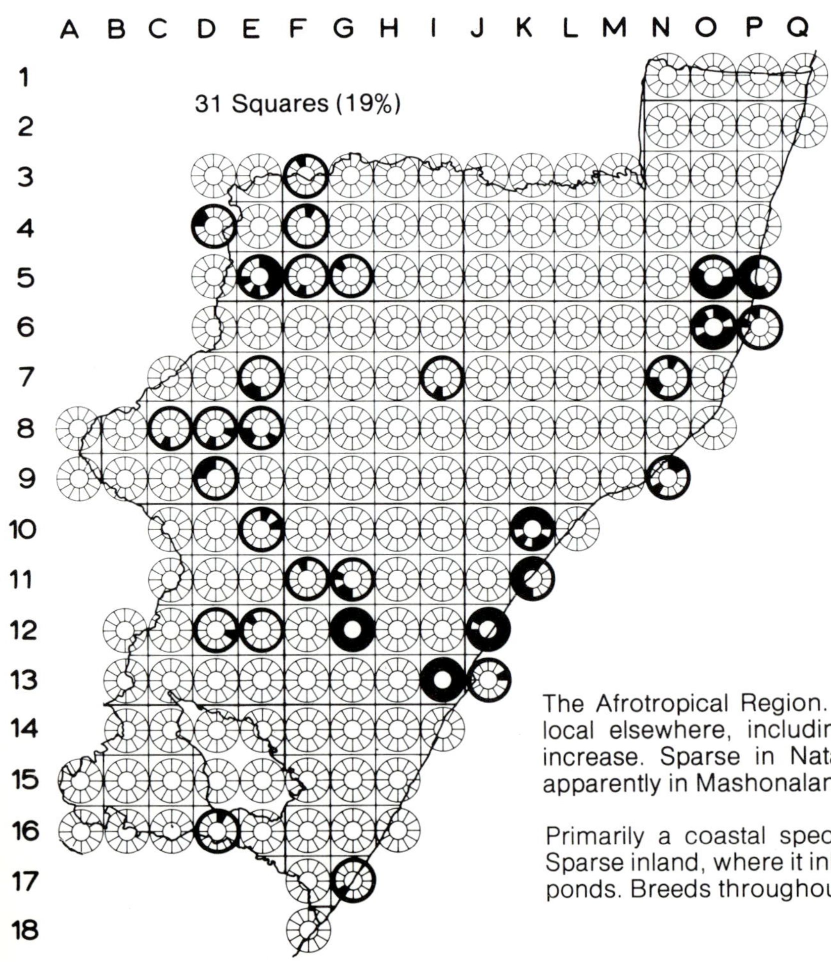

98 Cape Teal

Anas capensis

The Afrotropical Region. In Southern Africa, common in the Cape, but local elsewhere, including S.W.A., though with a recent tendency to increase. Sparse in Natal, Zululand and southern Moçambique; also apparently in Mashonaland, Rhodesia.

Primarily a coastal species frequenting estuaries; also Lake St Lucia. Sparse inland, where it inhabits larger stretches of water and sewage farm ponds. Breeds throughout the year.

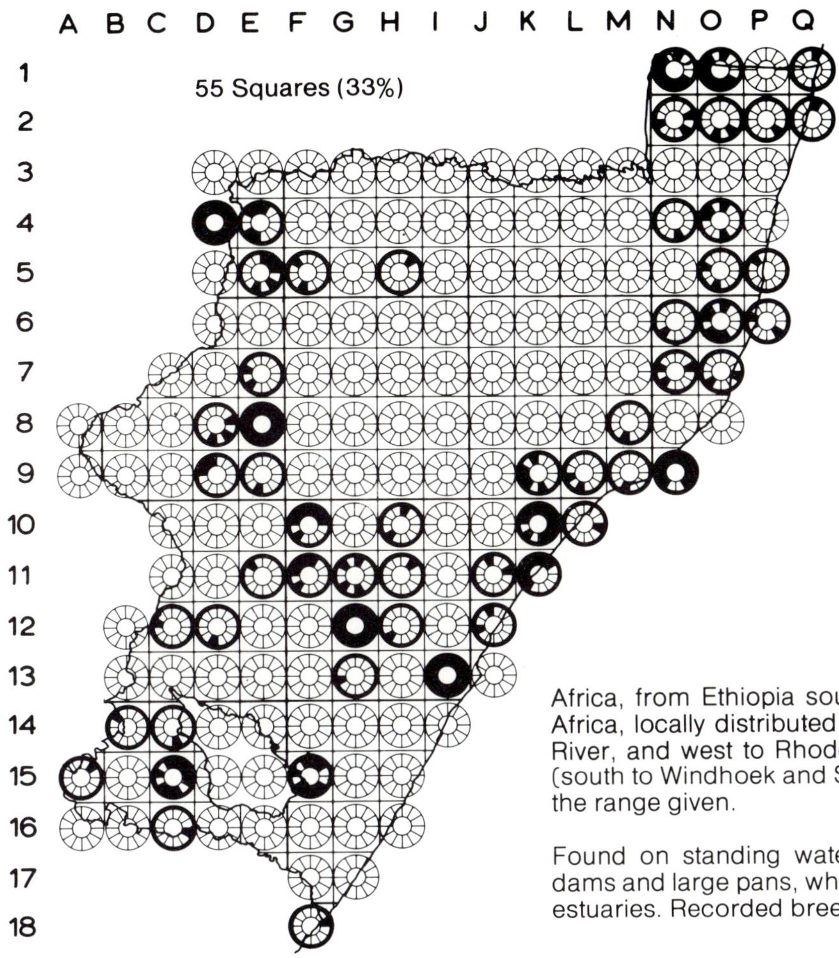

99 Hottentot Teal

Anas hottentota

Africa, from Ethiopia south to the Cape, and Madagascar. In Southern Africa, locally distributed from Natal and Zululand, north to the Zambesi River, and west to Rhodesia, Transvaal, Botswana and northern S.W.A. (south to Windhoek and Sandwich Harbour). Irregular to the southwest of the range given.

Found on standing water, from sewage farm ponds to medium-sized dams and large pans, where it keeps to the more sheltered parts; sparse at estuaries. Recorded breeding during May and June.

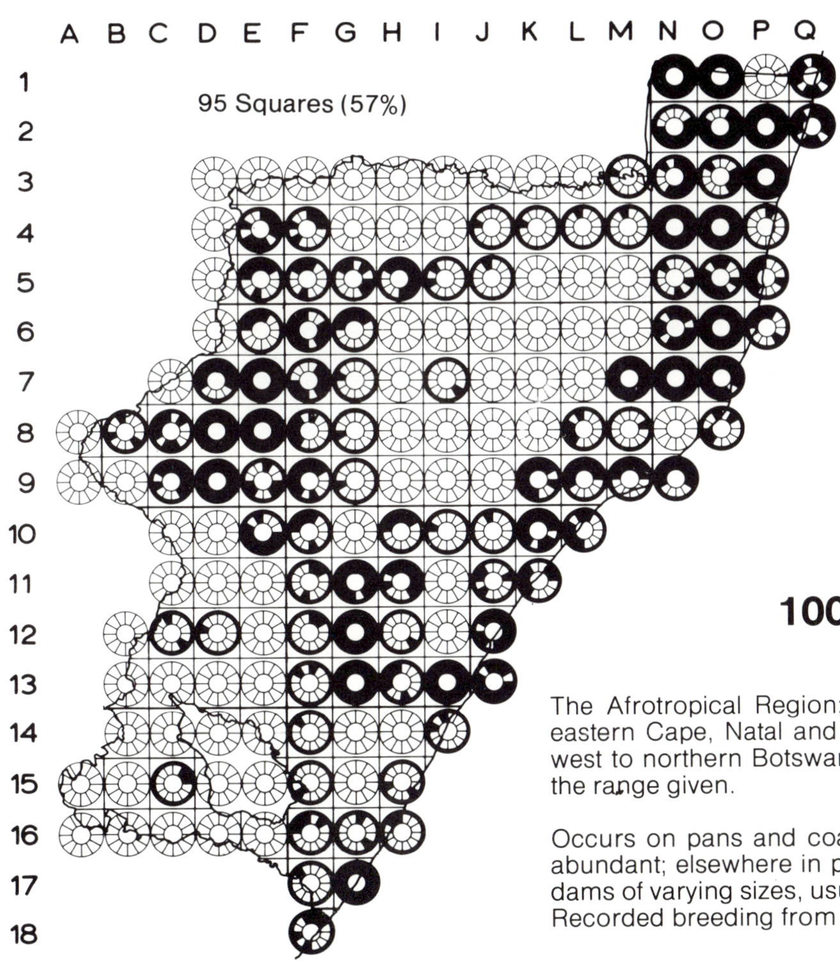

100 White-faced Whistling Duck

Dendrocygna viduata

The Afrotropical Region; also South America. In Southern Africa, the eastern Cape, Natal and Zululand, northwards in the east; in the north, west to northern Botswana and S.W.A. Occasionally to the southwest of the range given.

Occurs on pans and coastal lakes of the littoral plain where seasonally abundant; elsewhere in parties at estuaries, marshes, sewage farms and dams of varying sizes, usually fringed with overgrown vegetation or reeds. Recorded breeding from December to April.

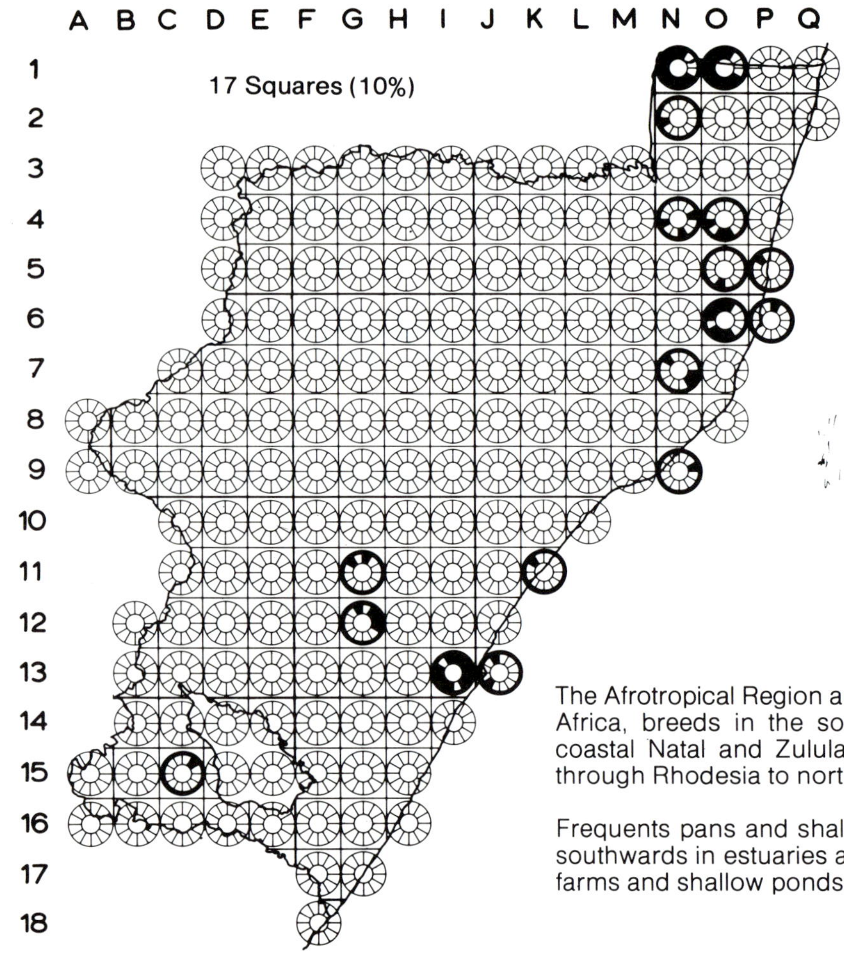

101 Fulvous Whistling Duck

Dendrocygna bicolor

The Afrotropical Region and southern Asia; also the Americas. In Southern Africa, breeds in the south-western Cape (seldom) and, locally, from coastal Natal and Zululand northwards in the east. In the north, west through Rhodesia to northern Botswana and northern S.W.A.

Frequents pans and shallow coastal lakes on the littoral plain, extending southwards in estuaries and coastal swamps. A vagrant inland to sewage farms and shallow ponds. May breed throughout the year.

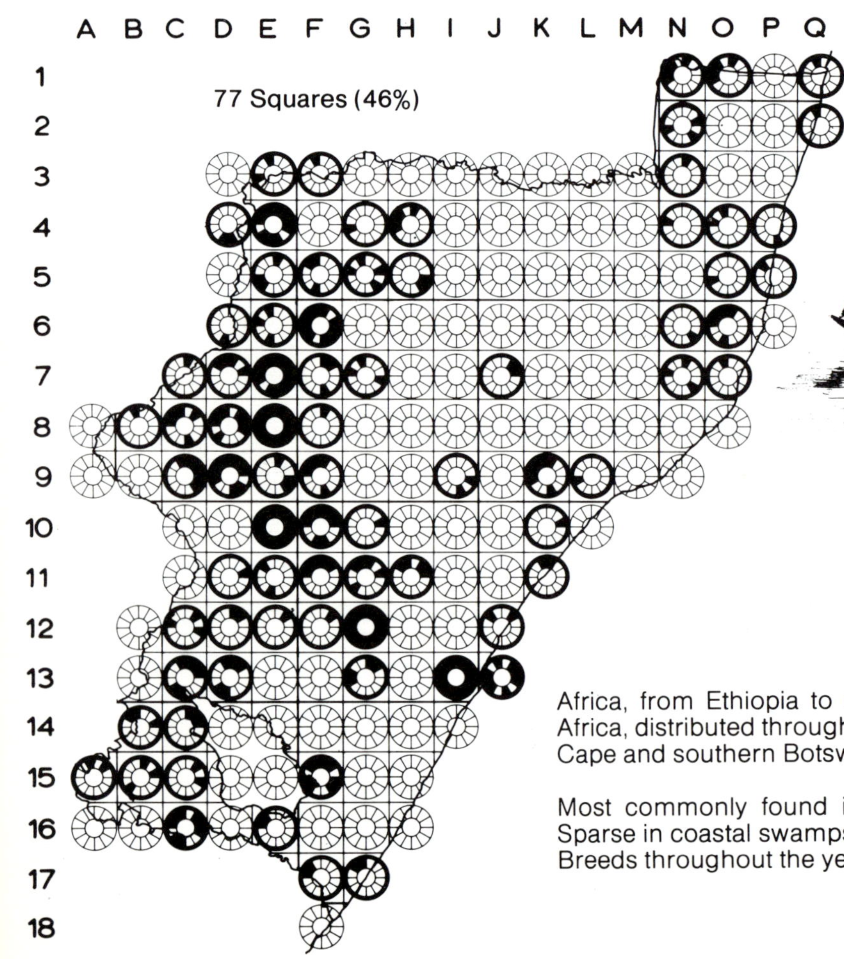

102 Red-eyed Pochard

Netta erythrophthalma

Africa, from Ethiopia to the Cape, and in South America. In Southern Africa, distributed throughout, but local over much of S.W.A., the northern Cape and southern Botswana.

Most commonly found inland on the larger stretches of open water. Sparse in coastal swamps, and pans and coastal lakes of the littoral plain. Breeds throughout the year.

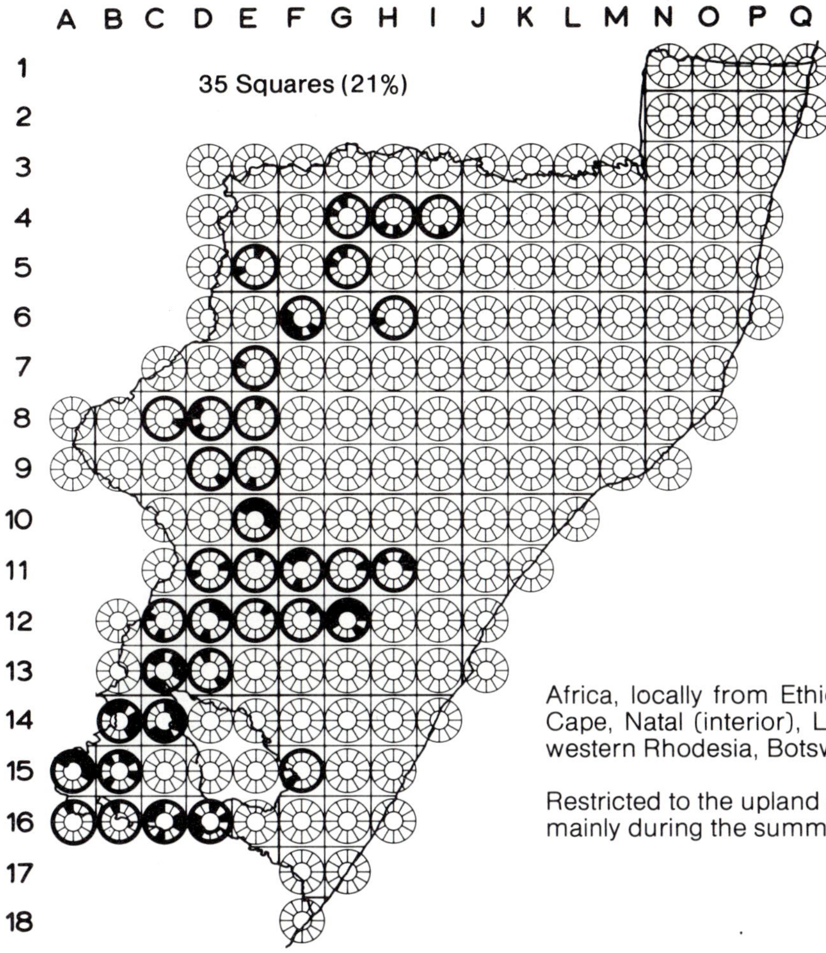

103 Maccoa Duck

Oxyura maccoa

Africa, locally from Ethiopia south to the Cape. In Southern Africa, the Cape, Natal (interior), Lesotho (lowlands), O.F.S., Transvaal (highveld), western Rhodesia, Botswana and S.W.A.

Restricted to the upland areas, where it inhabits the deeper dams. Breeds mainly during the summer.

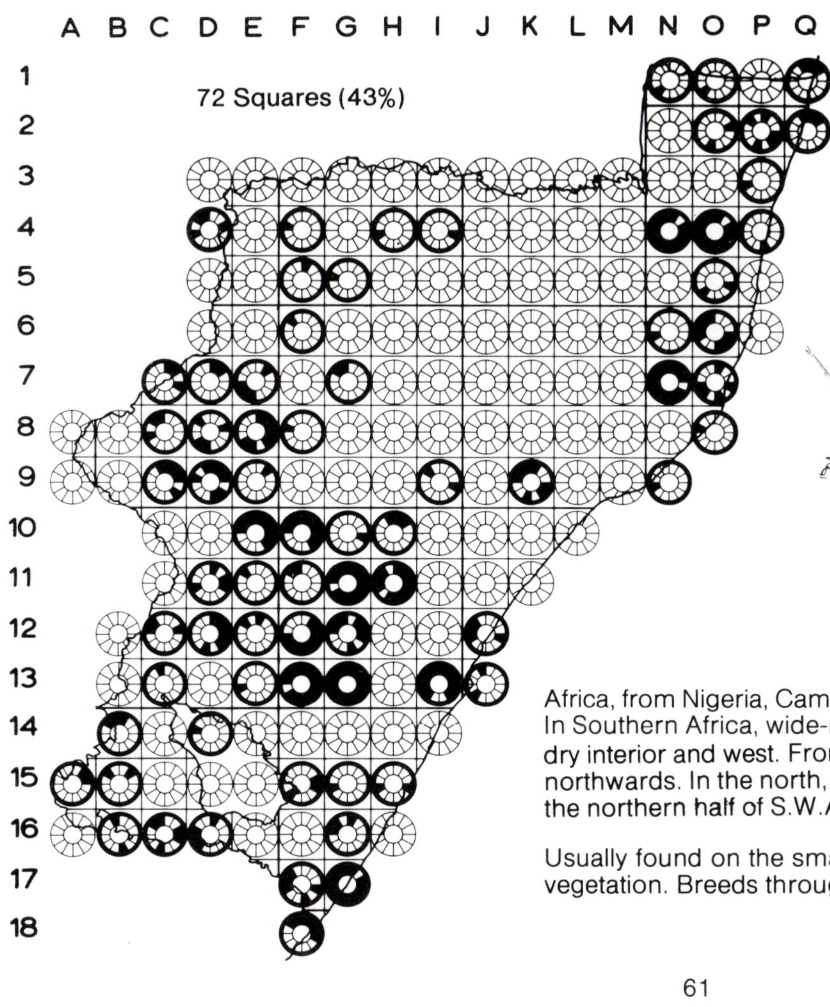

104 White-backed Duck

Thalassornis leuconotus

Africa, from Nigeria, Cameroon and Ethiopia to the Cape, and Madagascar. In Southern Africa, wide-ranging but sparse, and absent from much of the dry interior and west. From the eastern Cape, Transkei, Natal and Zululand northwards. In the north, west to northern Botswana, the Caprivi Strip and the northern half of S.W.A.

Usually found on the smaller dams and ponds with floating and marginal vegetation. Breeds throughout the year.

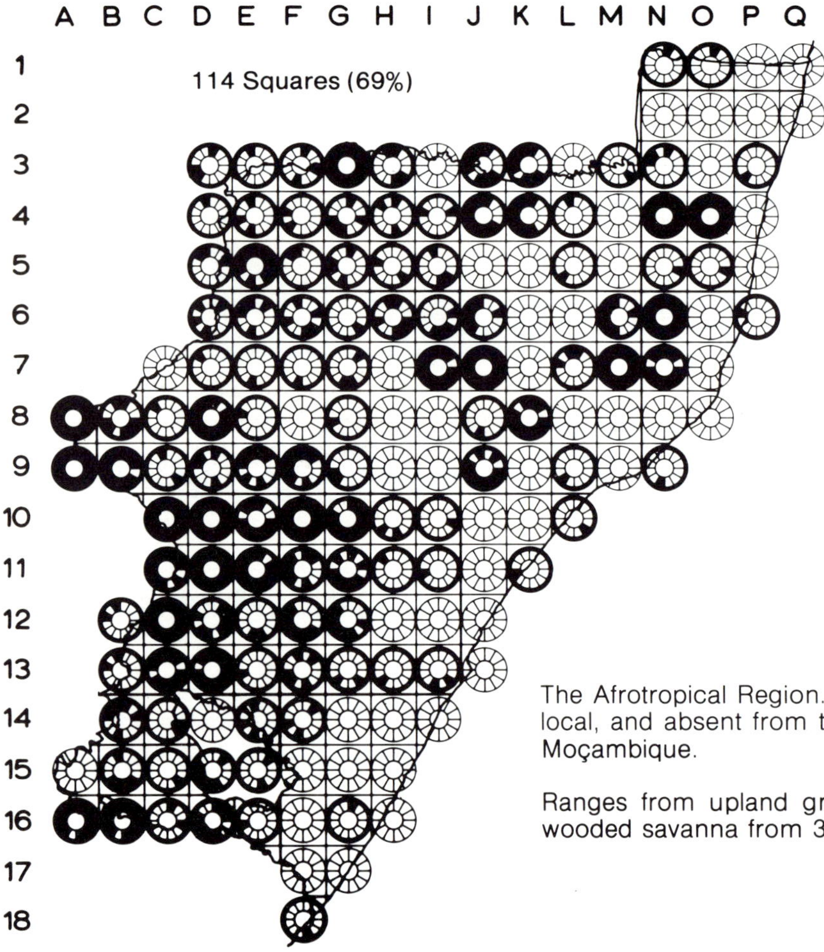

105 Secretarybird

Sagittarius serpentarius

The Afrotropical Region. In Southern Africa, distributed throughout, but local, and absent from the littoral of Transkei, Natal and Zululand, and Moçambique.

Ranges from upland grassland, including mountain slopes, to lightly-wooded savanna from 300 m a.s.l. Breeds from September to January.

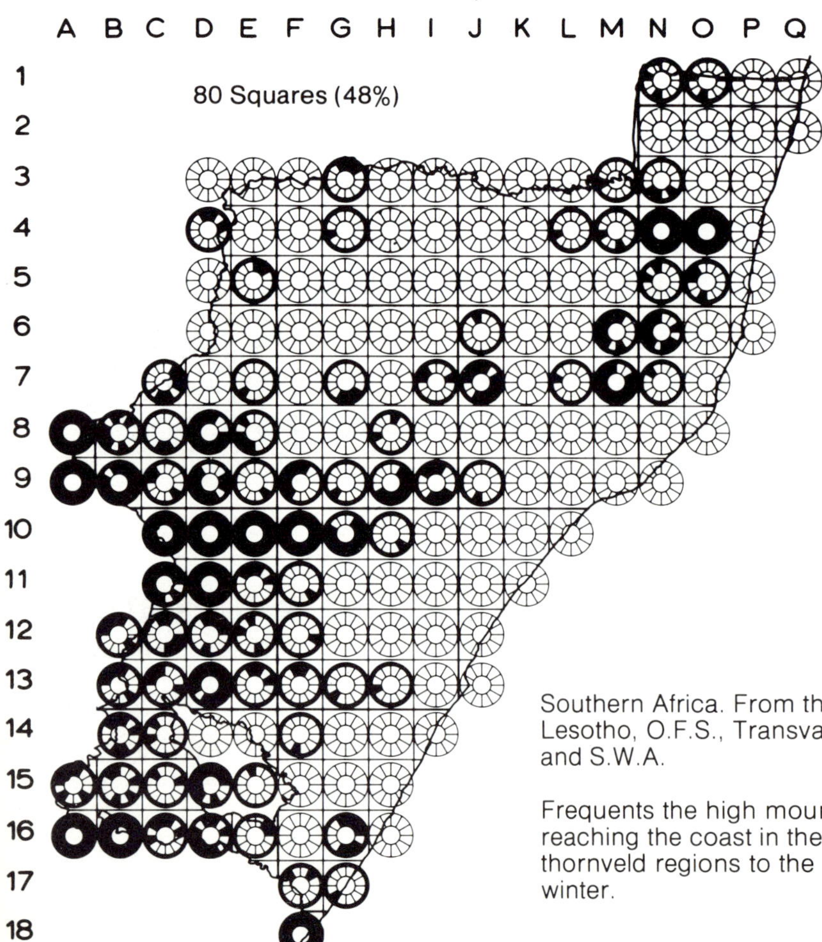

106 Cape Vulture

Gyps coprotheres

Southern Africa. From the Cape, Natal, Zululand (local) and Swaziland to Lesotho, O.F.S., Transvaal, Rhodesia (to 26°E), south-eastern Botswana and S.W.A.

Frequents the high mountains of the Drakensberg and adjacent foothills, reaching the coast in the extreme south and extending across the central thornveld regions to the Lebombo Range in Zululand. Breeds during the winter.

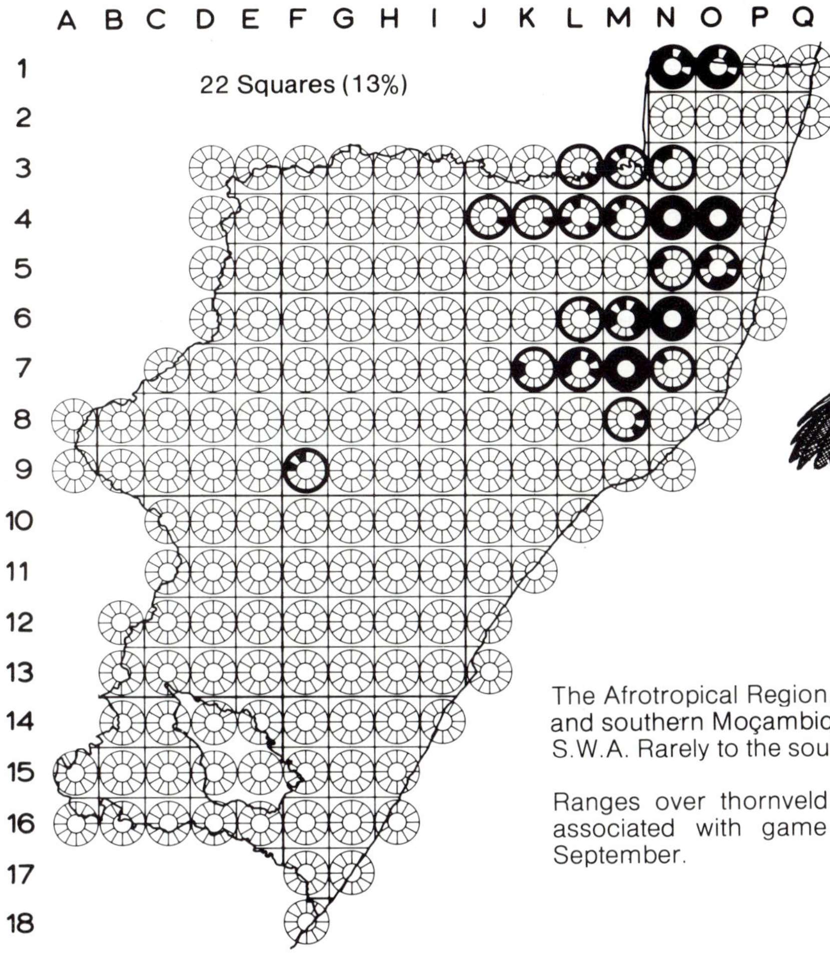

107 White-backed Vulture

Gyps africanus

The Afrotropical Region. In Southern Africa, Zululand, eastern Swaziland and southern Moçambique, westwards to O.F.S., Botswana and northern S.W.A. Rarely to the south of the range given.

Ranges over thornveld and lightly-wooded savanna; largest numbers associated with game reserves. Breeding recorded from June to September.

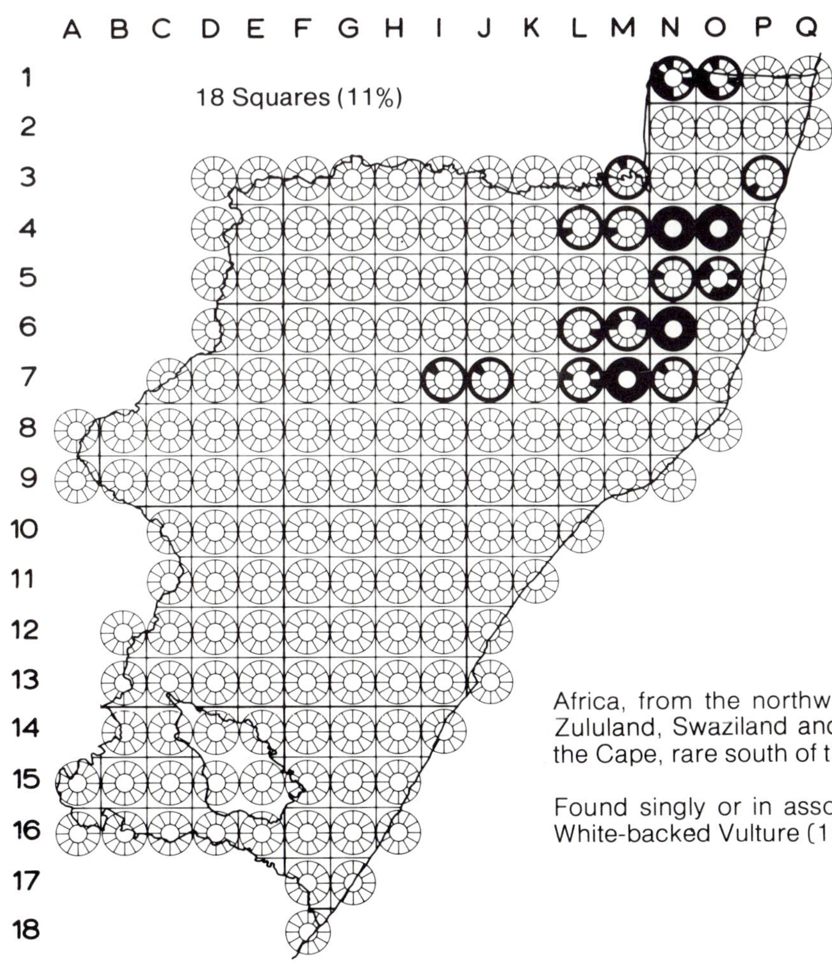

108 Lappet-faced Vulture

Torgos tracheliotus

Africa, from the northwest and Egypt to the Cape. In Southern Africa, Zululand, Swaziland and Moçambique, west to Botswana and S.W.A. In the Cape, rare south of the Orange River, and only occasionally in Natal.

Found singly or in association with other vultures in similar habitats to White-backed Vulture (107). Breeds during the winter.

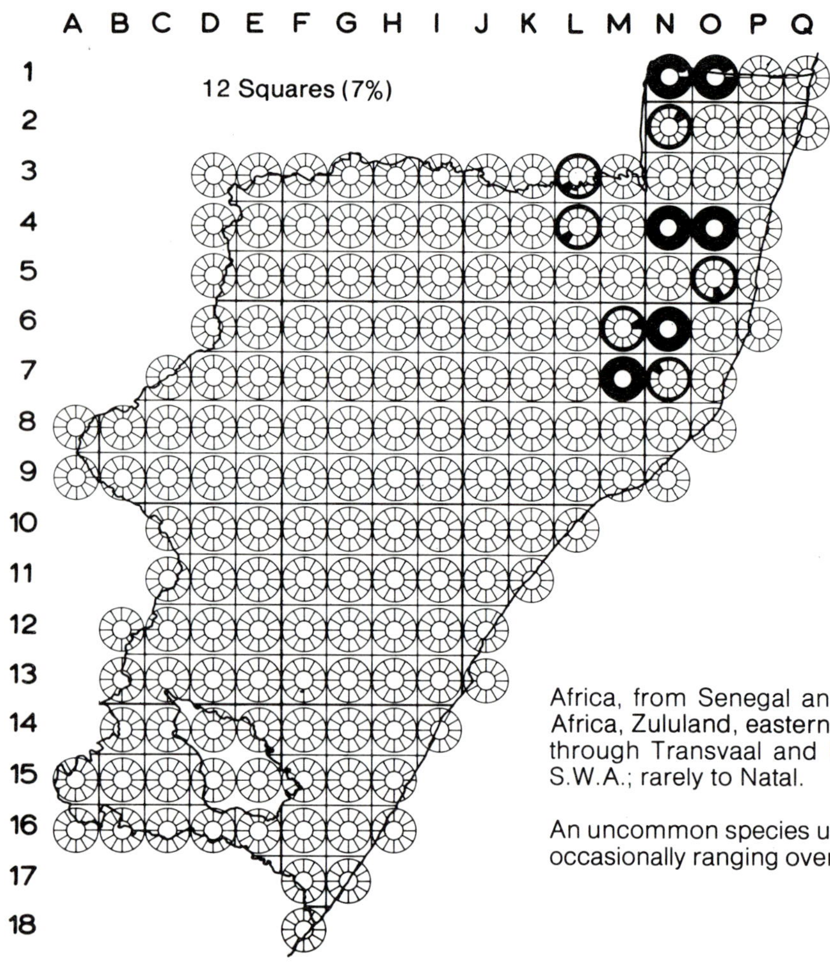

109 White-headed Vulture

Trigonoceps occipitalis

Africa, from Senegal and the Blue Nile south to the Cape. In Southern Africa, Zululand, eastern Swaziland and the interior of Moçambique, west through Transvaal and Rhodesia to Botswana, the northern Cape and S.W.A.; rarely to Natal.

An uncommon species usually seen singly, associated with game reserves, occasionally ranging over surrounding thornveld. Breeds during the winter.

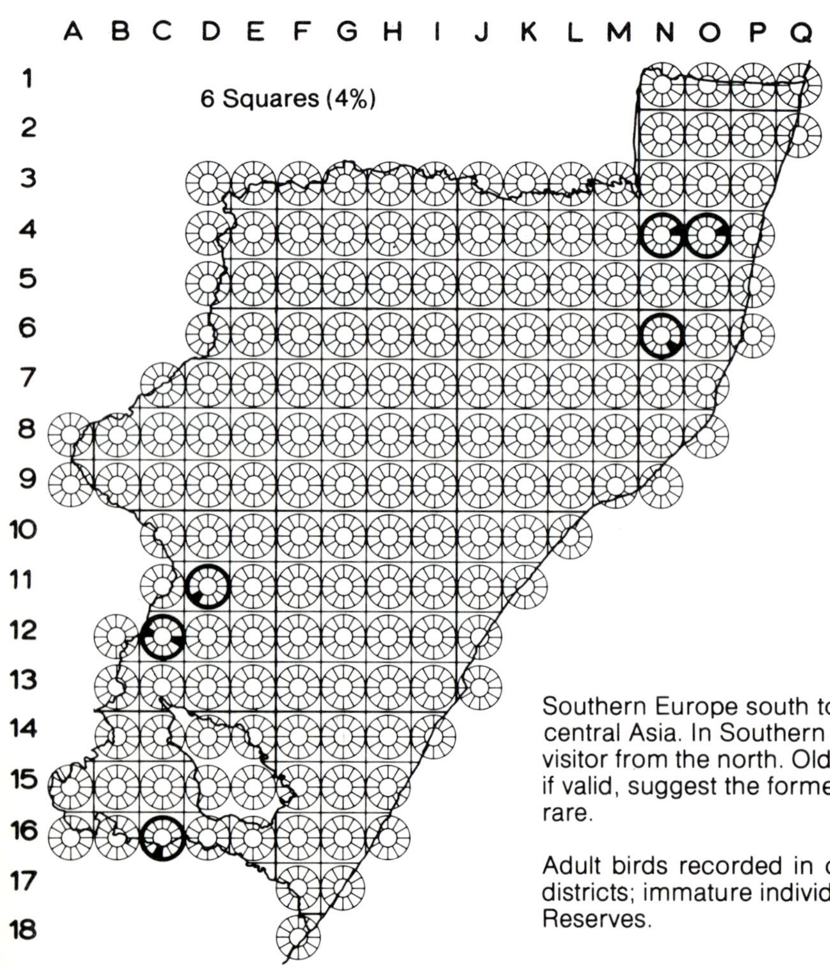

111 Egyptian Vulture

Neophron percnopterus

Southern Europe south to Africa mainly north of the Equator, and east to central Asia. In Southern Africa, formally wide-ranging as a non-breeding visitor from the north. Old records of breeding in the south-western Cape, if valid, suggest the former existence of a small endemic population; now rare.

Adult birds recorded in open grassland in the Underberg and Kokstad districts; immature individuals noted from the Mkuzi and Hluhluwe Game Reserves.

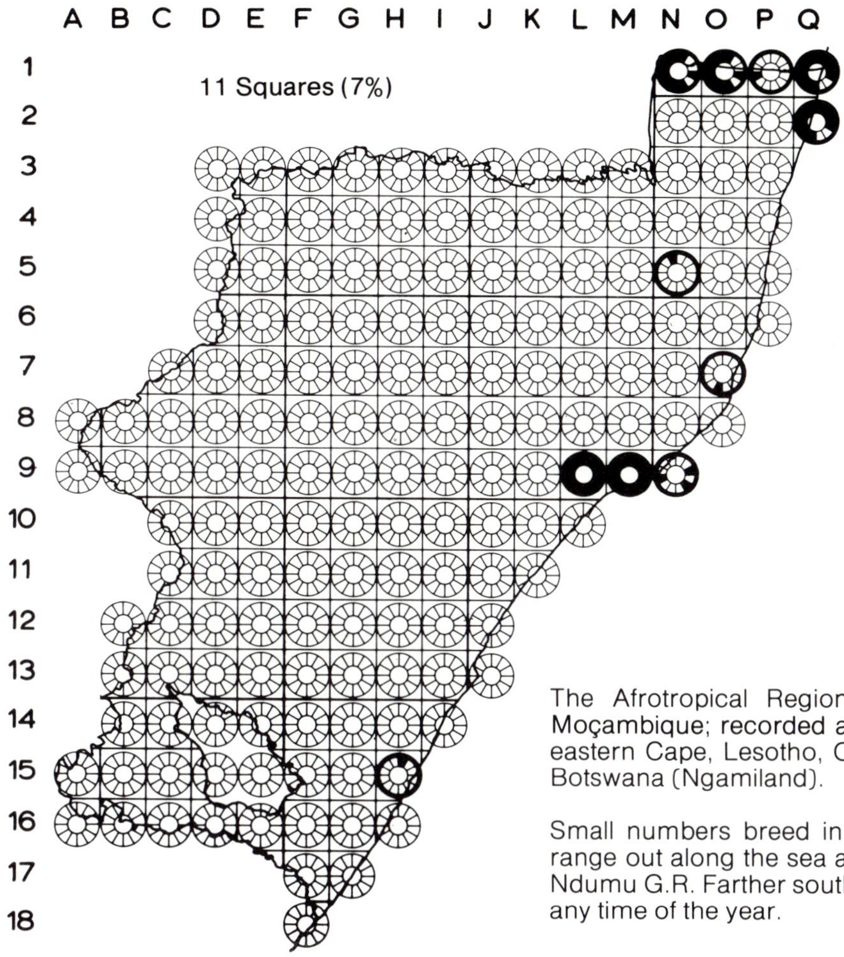

112 Palm-nut Vulture

Gypohierax angolensis

The Afrotropical Region. In Southern Africa, coastal Zululand and Moçambique; recorded as a straggler (mainly young birds) to Natal, the eastern Cape, Lesotho, O.F.S. and Transvaal. Also northern S.W.A. and Botswana (Ngamiland).

Small numbers breed in the raffia palm forest of the Kosi system and range out along the sea and lake shores, and occasionally to the pans at Ndumu G.R. Farther south, a pair nests annually at Mtunzini. May breed at any time of the year.

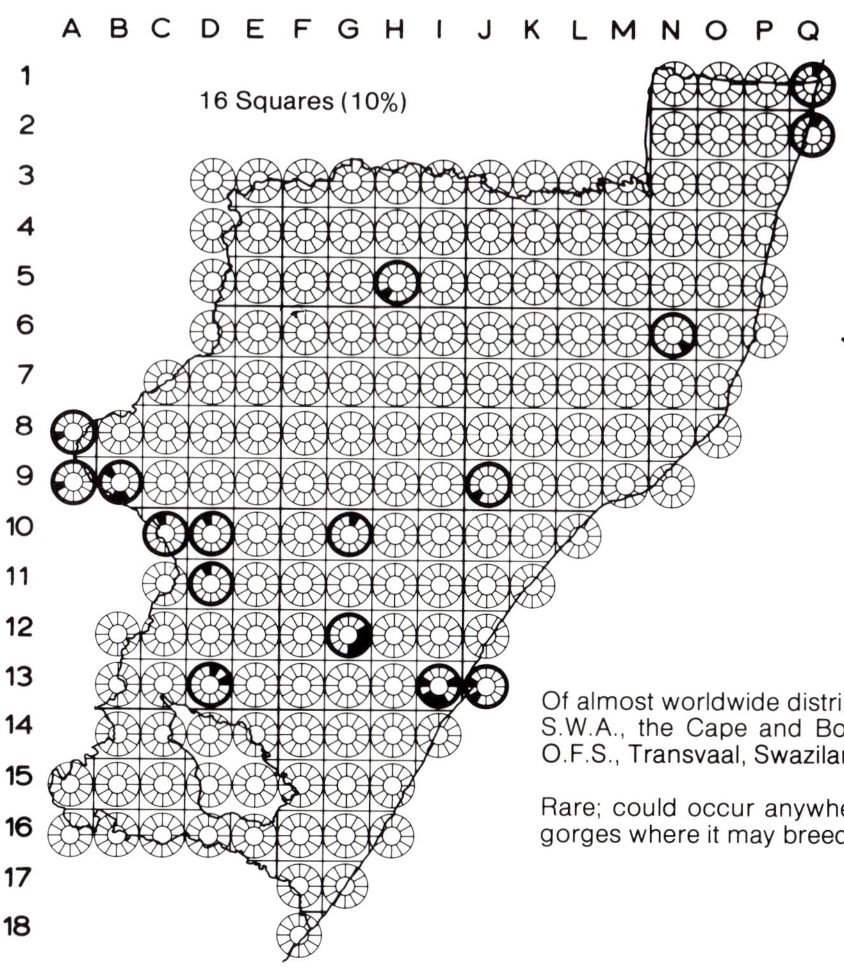

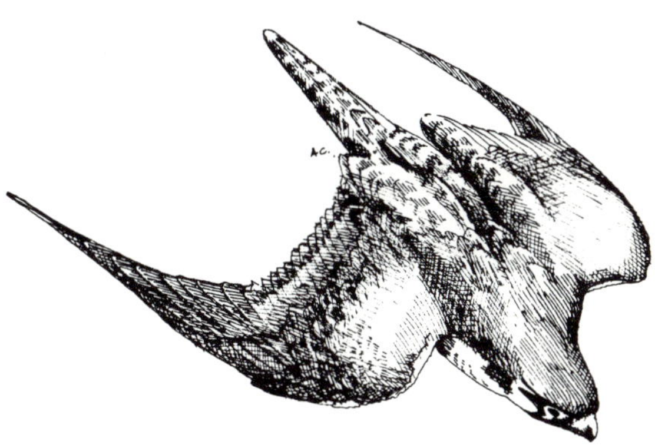

113 Peregrine Falcon

Falco peregrinus

Of almost worldwide distribution. In Southern Africa, locally distributed in S.W.A., the Cape and Botswana, east to Natal and Zululand, Lesotho, O.F.S., Transvaal, Swaziland, Rhodesia and adjacent Moçambique.

Rare; could occur anywhere but prefers kranses and the sides of deep gorges where it may breed.

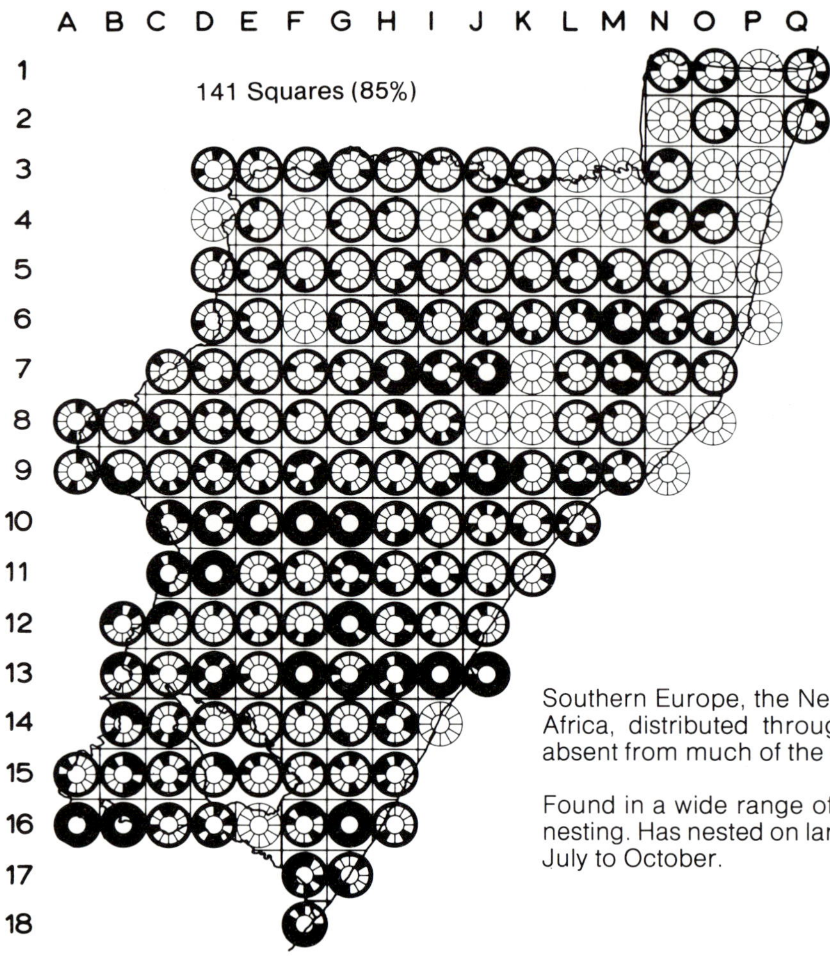

114 Lanner Falcon

Falco biarmicus

Southern Europe, the Near East, south in Africa to the Cape. In Southern Africa, distributed throughout, especially in mountainous districts, but absent from much of the eastern littoral.

Found in a wide range of habitats provided suitable rock faces occur for nesting. Has nested on large buildings in Durban. Breeding recorded from July to October.

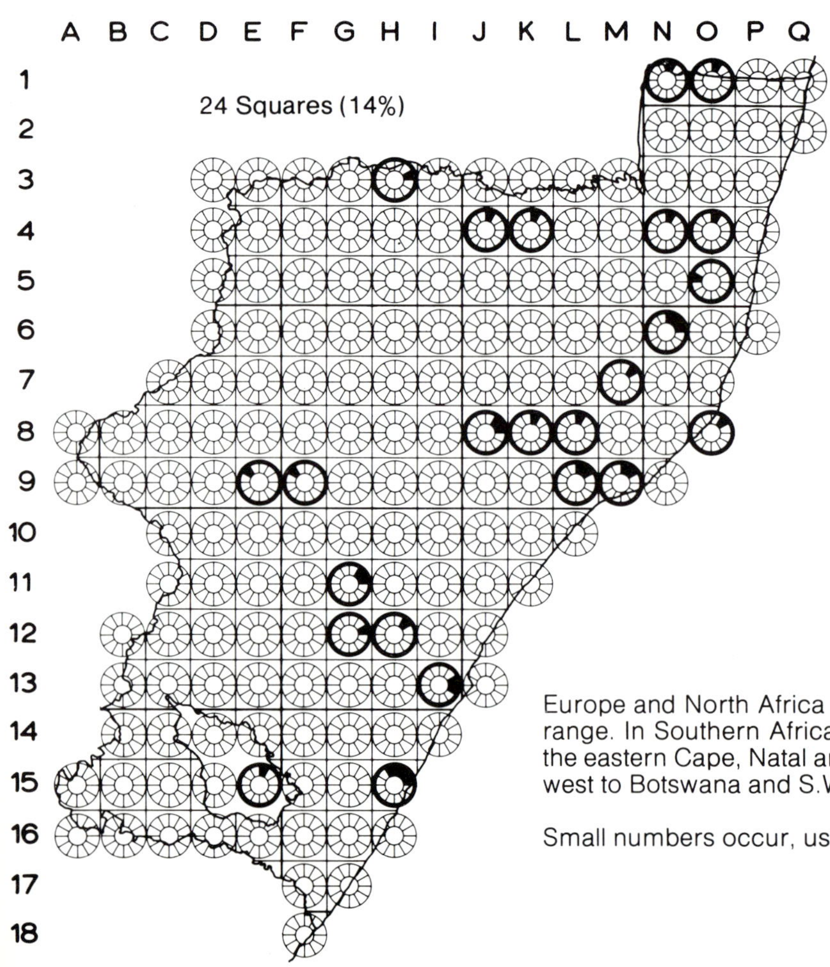

115 Hobby

Falco subbuteo

Europe and North Africa east to Asia; winters to the south of its breeding range. In Southern Africa, a visitor from October to March/early April to the eastern Cape, Natal and Zululand, northwards in the east. In the north, west to Botswana and S.W.A. Tends to be rare in the south.

Small numbers occur, usually associated with lightly-wooded habitats.

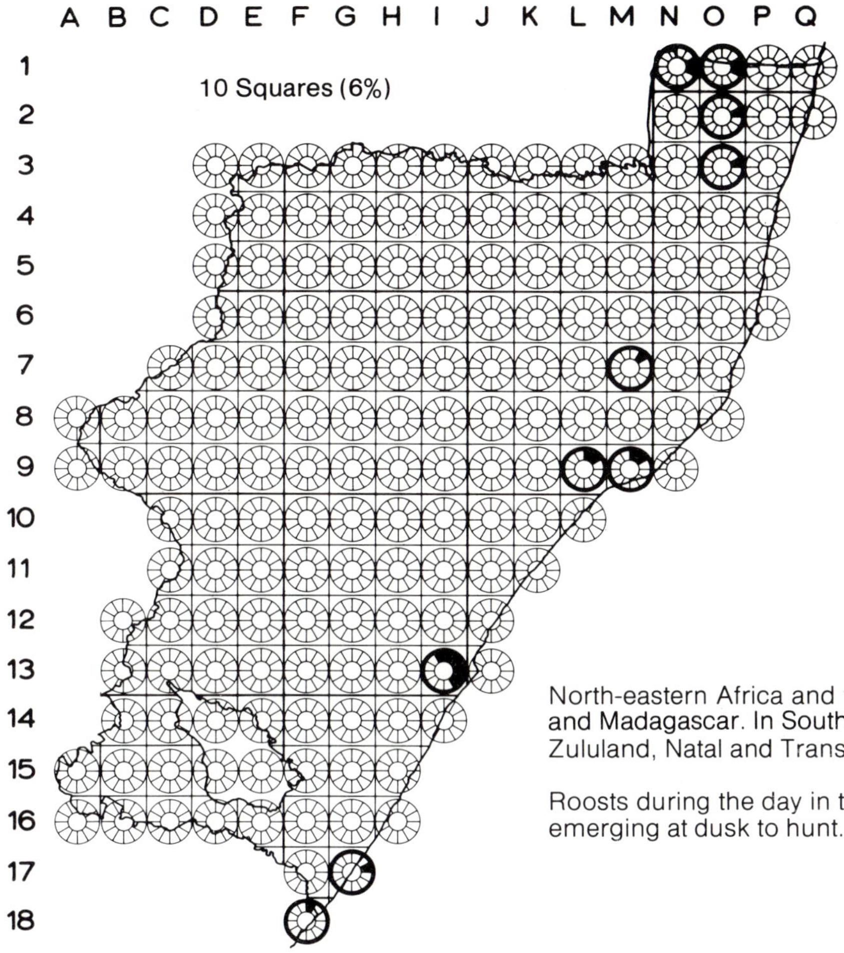

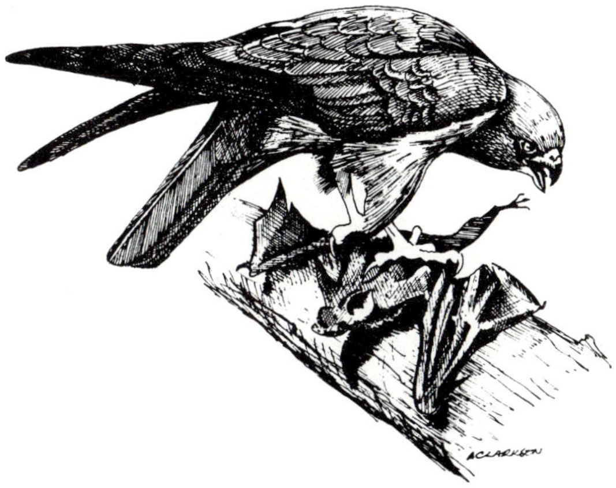

116Y Sooty Falcon

Falco concolor

North-eastern Africa and the Near East, wintering in south-eastern Africa and Madagascar. In Southern Africa, recorded from Moçambique, eastern Zululand, Natal and Transkei (coast).

Roosts during the day in tall riverine and exotic trees close to open water, emerging at dusk to hunt.

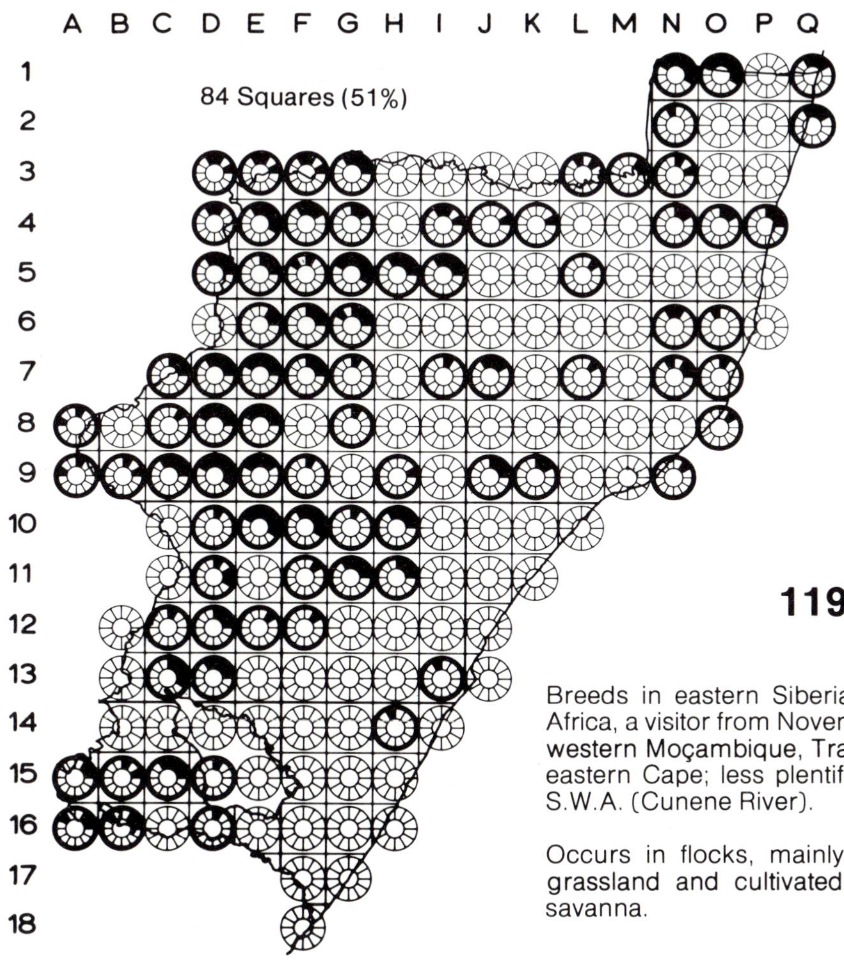

119 Eastern Red-footed Kestrel

Falco amurensis

Breeds in eastern Siberia and China, wintering in Africa. In Southern Africa, a visitor from November to March mainly to the Rhodesian highveld, western Moçambique, Transvaal, Swaziland, Natal and Zululand, and the eastern Cape; less plentiful farther west, through Botswana to northern S.W.A. (Cunene River).

Occurs in flocks, mainly in the upland areas where it favours open grassland and cultivated fields; in Zululand, frequents lightly-wooded savanna.

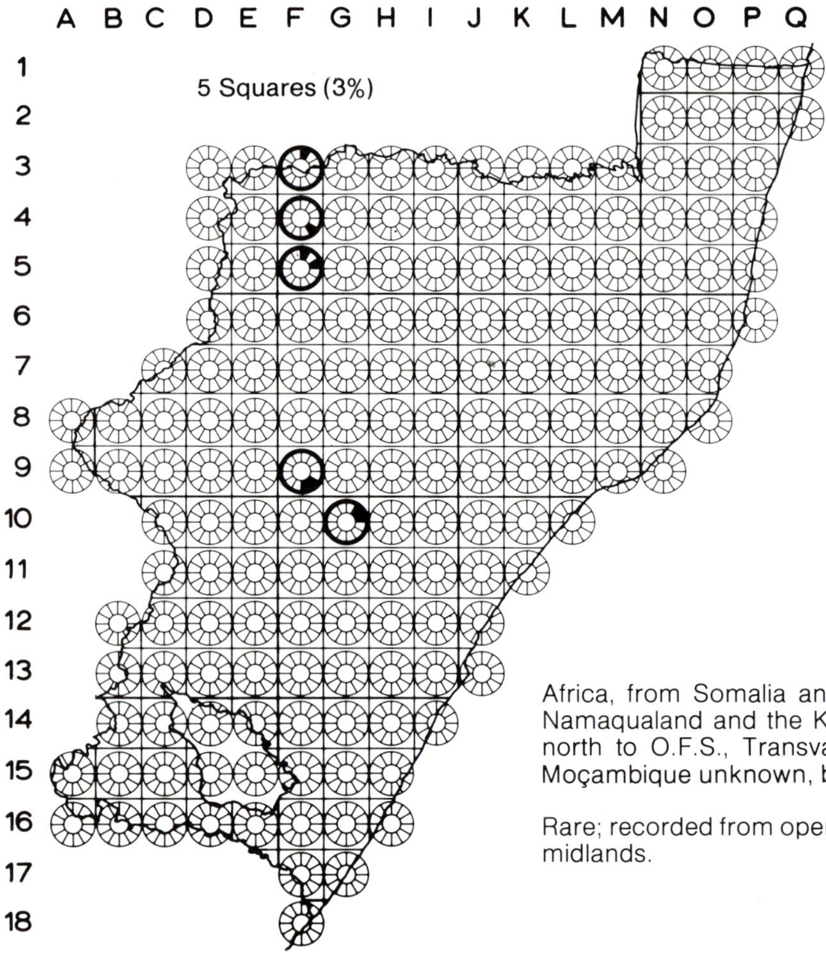

122 Greater Kestrel

Falco rupicoloides

Africa, from Somalia and Kenya to the Cape. In Southern Africa, little Namaqualand and the Karoo in the Cape, east to Natal and Swaziland, north to O.F.S., Transvaal, Rhodesia, Botswana and S.W.A. Status in Moçambique unknown, but sparse on the littoral.

Rare; recorded from open grassland in the northwest, and thornveld in the midlands.

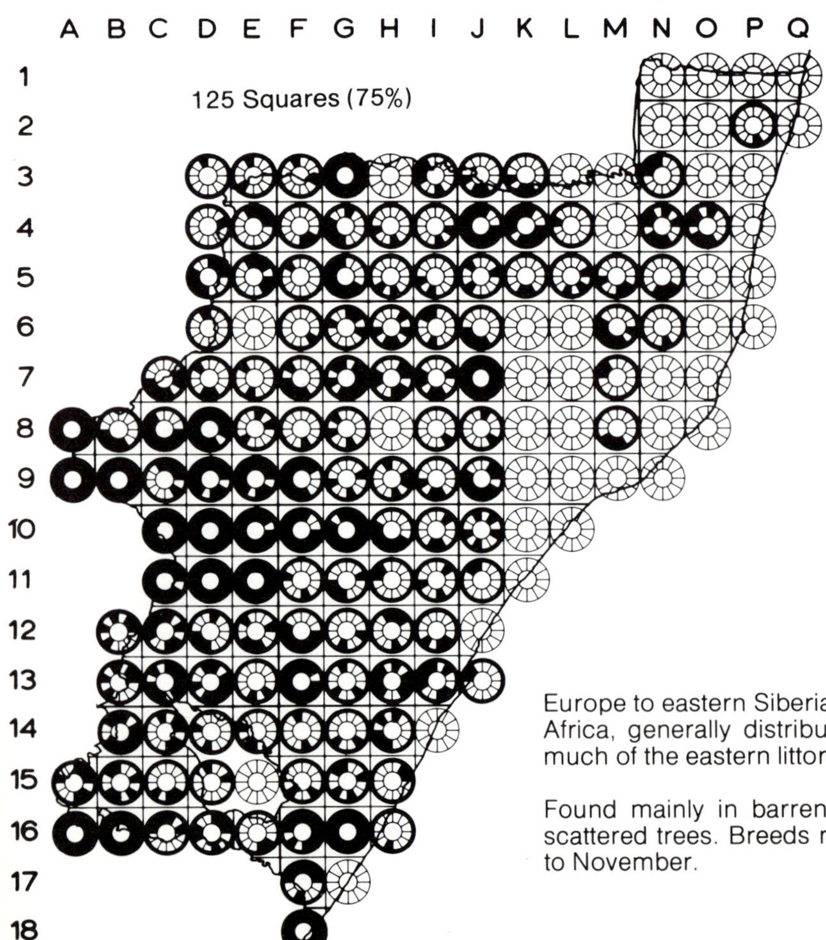

123 Rock Kestrel

Falco tinnunculus

Europe to eastern Siberia, south to Africa and southern Asia. In Southern Africa, generally distributed throughout though local, and absent from much of the eastern littoral.

Found mainly in barren hilly country, open grassland and areas with scattered trees. Breeds mainly on rock faces, recorded from September to November.

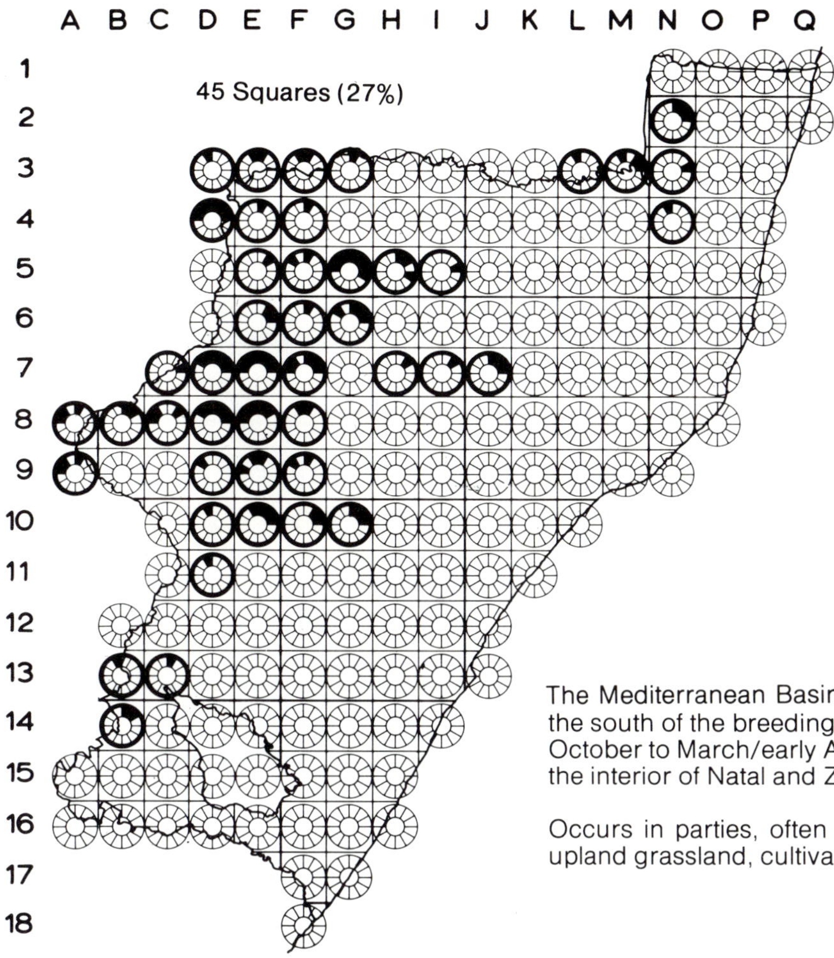

125 Lesser Kestrel

Falco naumanni

The Mediterranean Basin and southern Russia east to China; winters to the south of the breeding range. In Southern Africa, a general visitor from October to March/early April, most numerous in the eastern Cape, O.F.S., the interior of Natal and Zululand, northwards in the east.

Occurs in parties, often alongside Eastern Red-footed Kestrel (119), in upland grassland, cultivated fields and lightly-wooded savanna.

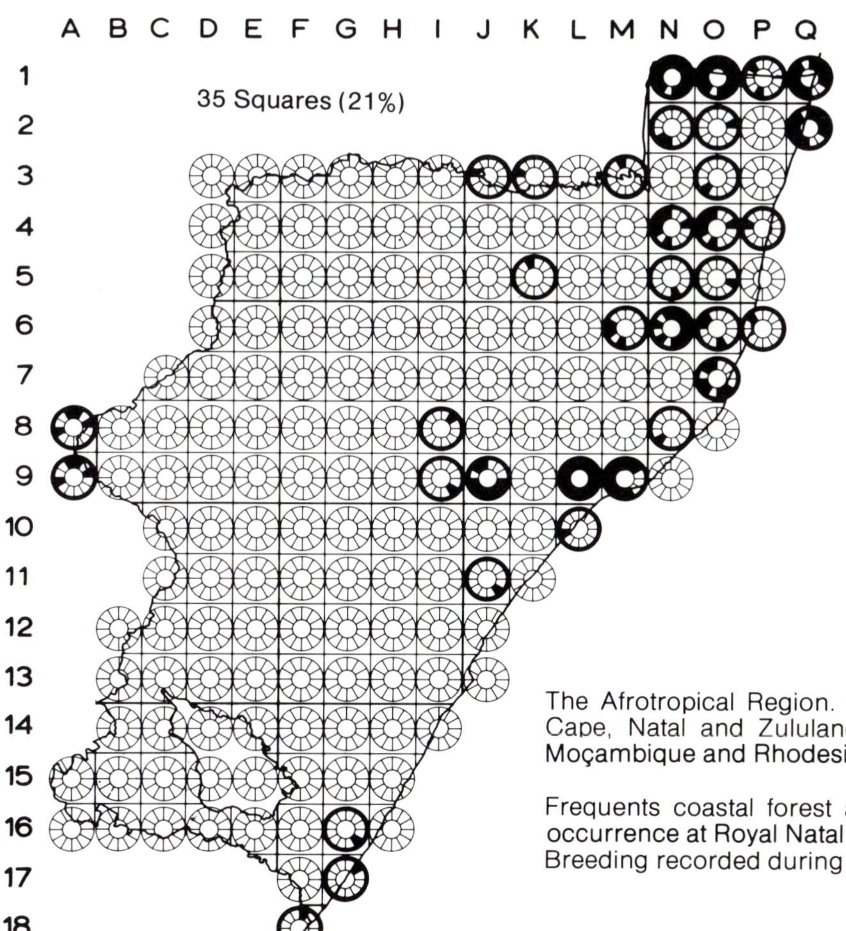

127 Cuckoo Falcon

Aviceda cuculoides

The Afrotropical Region. In Southern Africa, the southern and eastern Cape, Natal and Zululand, Swaziland, eastern and central Transvaal, Moçambique and Rhodesia (throughout); generally sparse.

Frequents coastal forest and riverine or mixed woodland. An isolated occurrence at Royal Natal National Park, where it bred in 1978 and 1979. Breeding recorded during October and November.

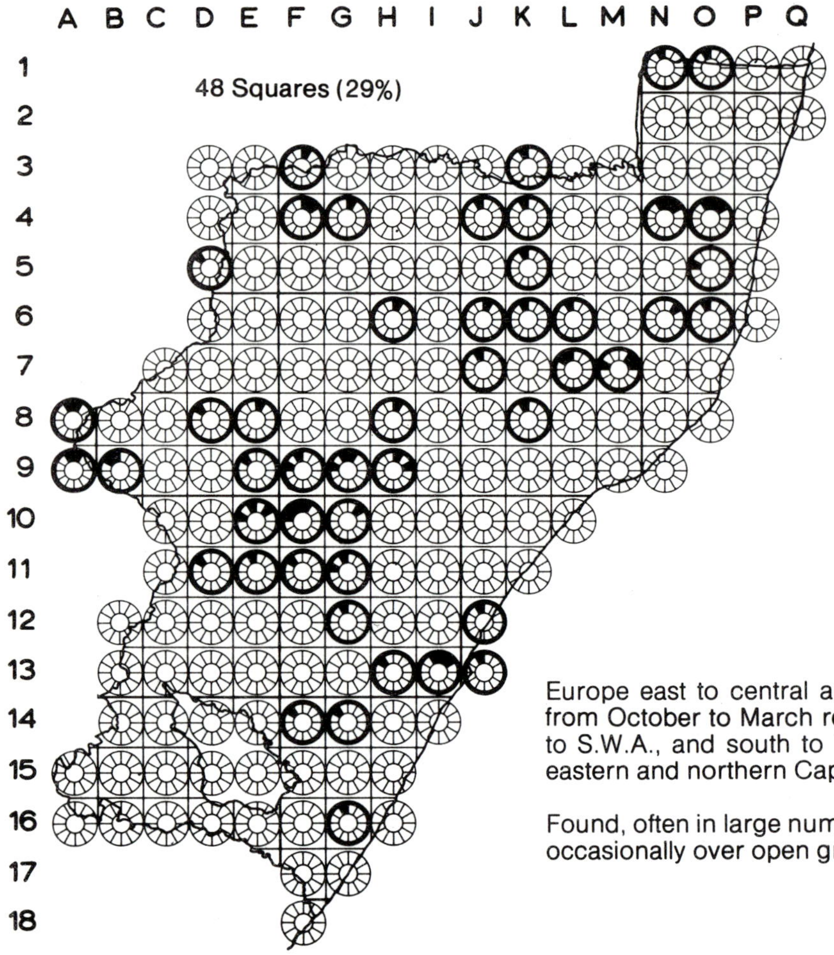

128 Black Kite

Milvus migrans migrans

Europe east to central and eastern Asia. In Southern Africa, a migrant from October to March recorded from Moçambique and Rhodesia, west to S.W.A., and south to Transvaal, O.F.S., Natal and Zululand, and the eastern and northern Cape.

Found, often in large numbers in loose association, over wooded country; occasionally over open grassland.

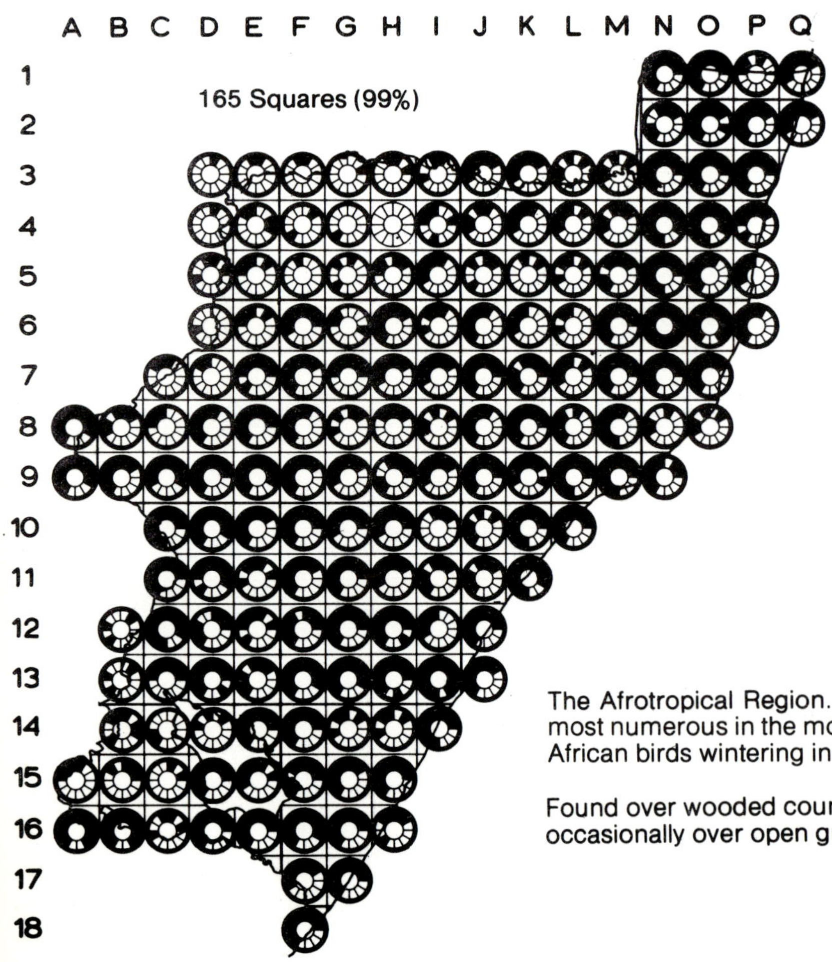

129 Yellow-billed Kite

Milvus migrans parasitus

The Afrotropical Region. In Southern Africa, generally distributed, being most numerous in the moist east. A summer breeding resident with South African birds wintering in eastern equatorial Africa.

Found over wooded country, often in large numbers in loose association; occasionally over open grassland.

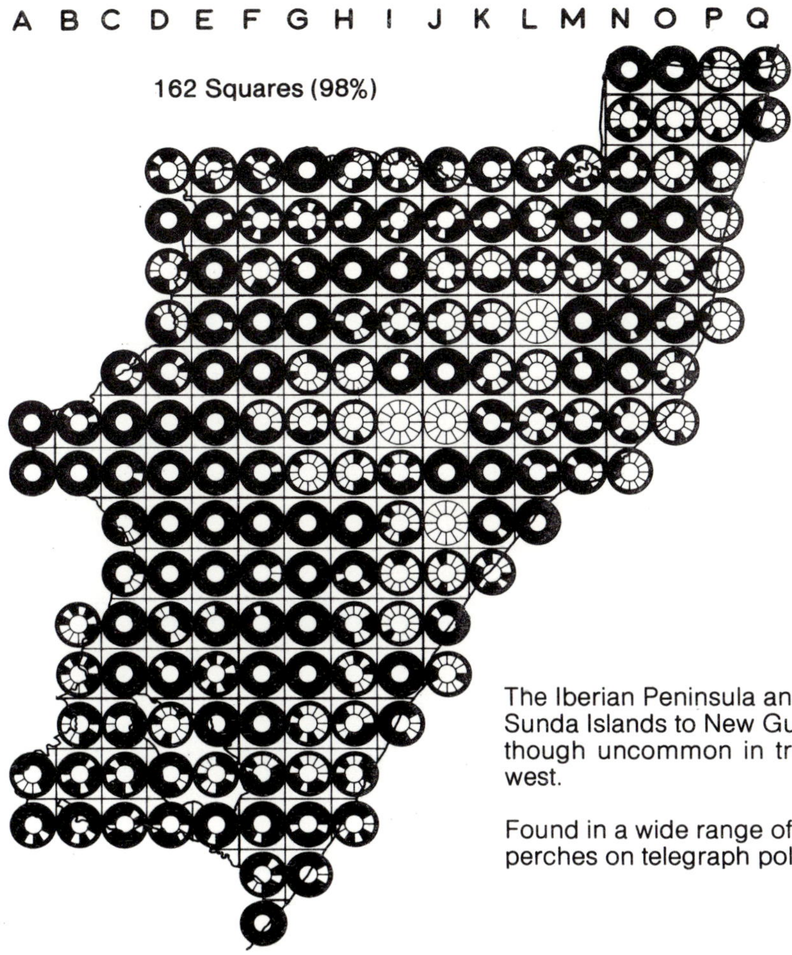

130 Black-shouldered Kite

Elanus caeruleus

The Iberian Peninsula and all Africa, Madagascar, southern Asia, and the Sunda Islands to New Guinea. In Southern Africa, distributed throughout, though uncommon in treeless and mountainous districts and the xeric west.

Found in a wide range of habitats, wherever trees are present. Frequently perches on telegraph poles. Breeds throughout the year.

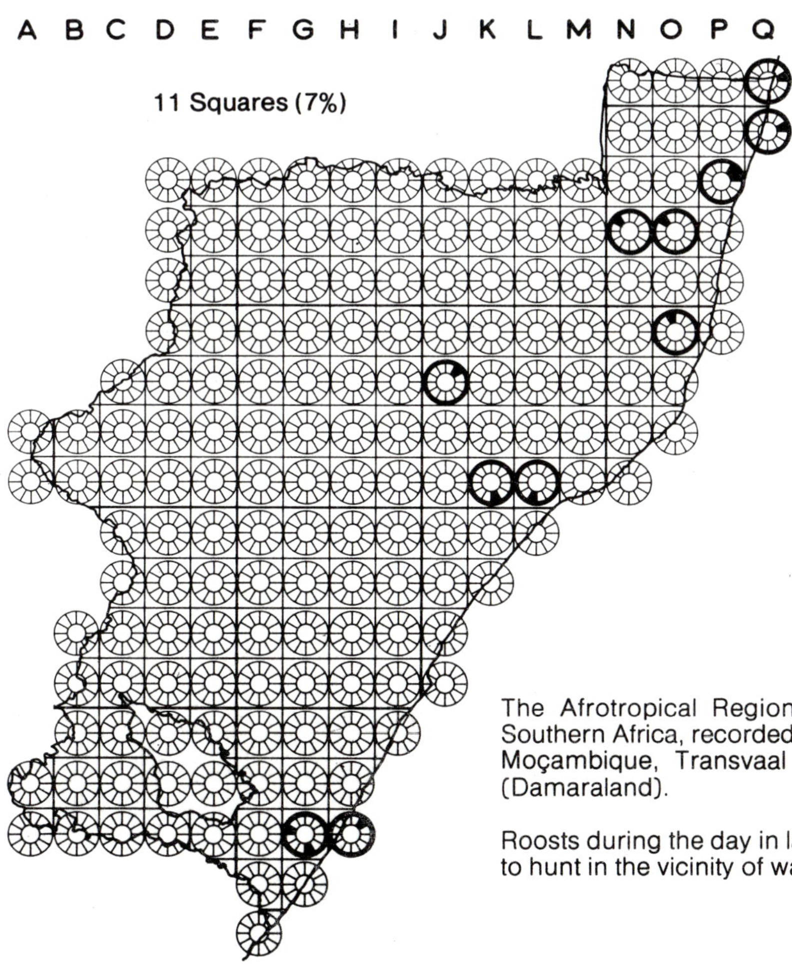

131 Bat Hawk

Macheiramphus alcinus

The Afrotropical Region and south-eastern Asia to New Guinea. In Southern Africa, recorded from coastal Natal, Zululand, eastern Swaziland, Moçambique, Transvaal and Rhodesia (widespread), west to S.W.A. (Damaraland).

Roosts during the day in large trees, emerging at dusk to feed; may prefer to hunt in the vicinity of water.

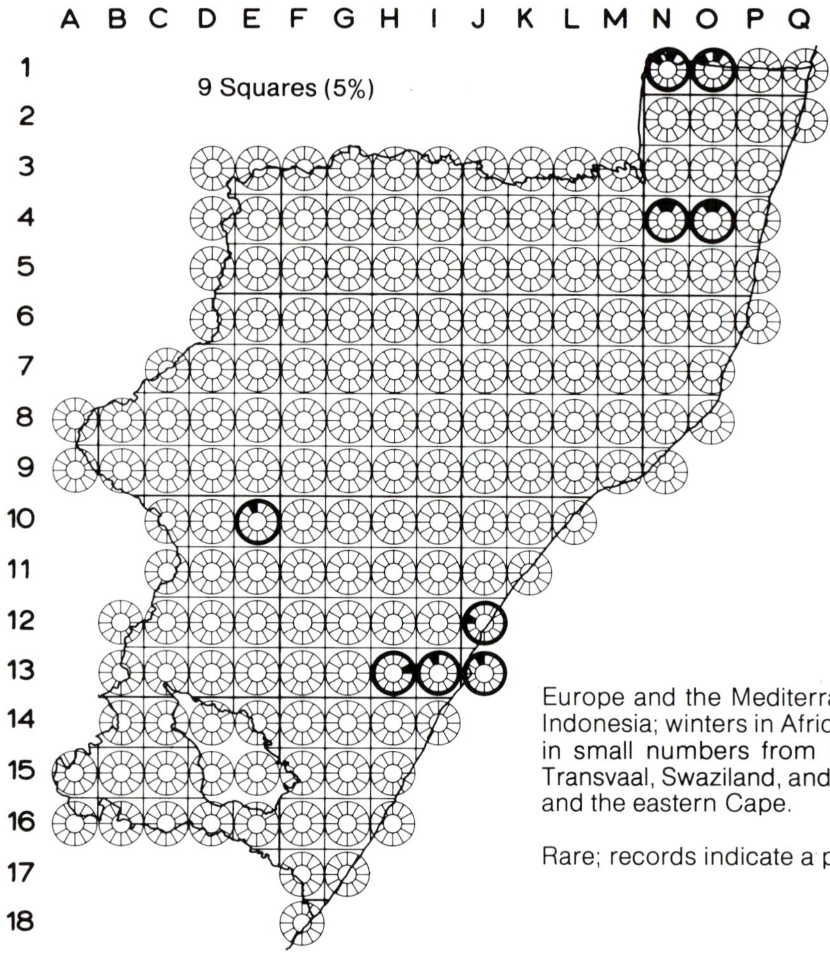

132 Honey Buzzard

Pernis apivorus

Europe and the Mediterranean Basin to Siberia and Japan, southeast to Indonesia; winters in Africa and southern Asia. In Southern Africa, a visitor in small numbers from November to April to Moçambique, Rhodesia, Transvaal, Swaziland, and Natal and Zululand. Also recorded from Transkei and the eastern Cape.

Rare; records indicate a preference for lightly-wooded areas.

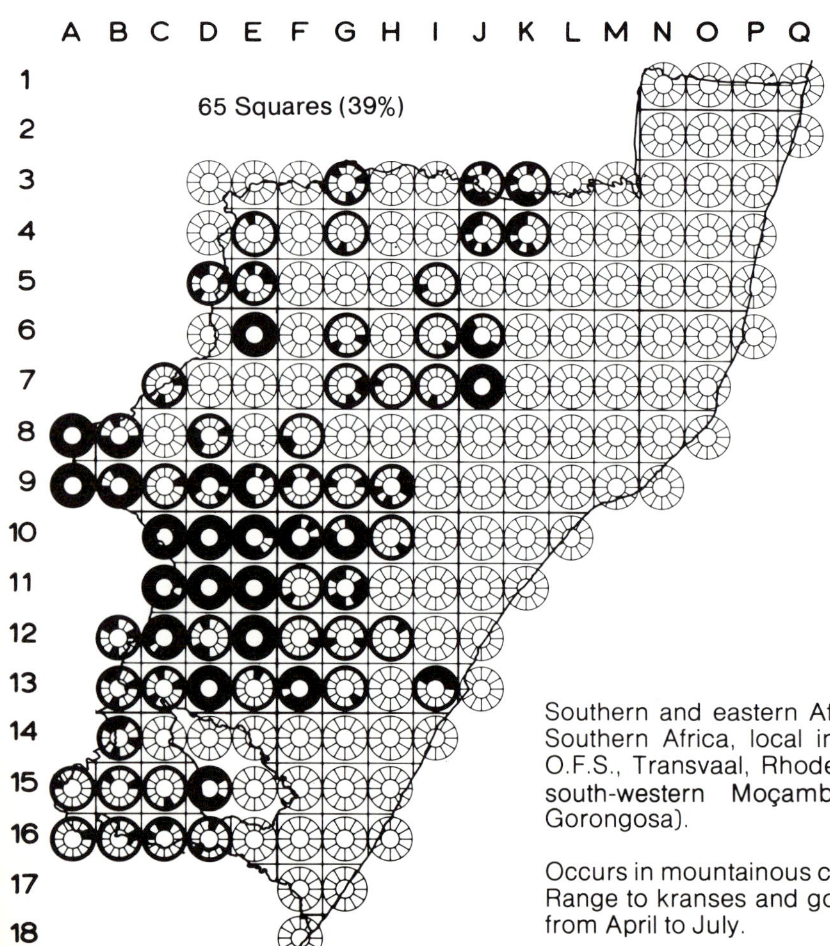

133 Black Eagle

Aquila verreauxii

Southern and eastern Africa, north to Egypt and adjacent territories. In Southern Africa, local in the Cape, Natal, western Zululand, Lesotho, O.F.S., Transvaal, Rhodesia, south-eastern Botswana, and S.W.A.; also south-western Moçambique (Chimanimani Mountains and Mount Gorongosa).

Occurs in mountainous country, from the higher parts of the Drakensberg Range to kranses and gorges down to 1200 m a.s.l. Recorded breeding from April to July.

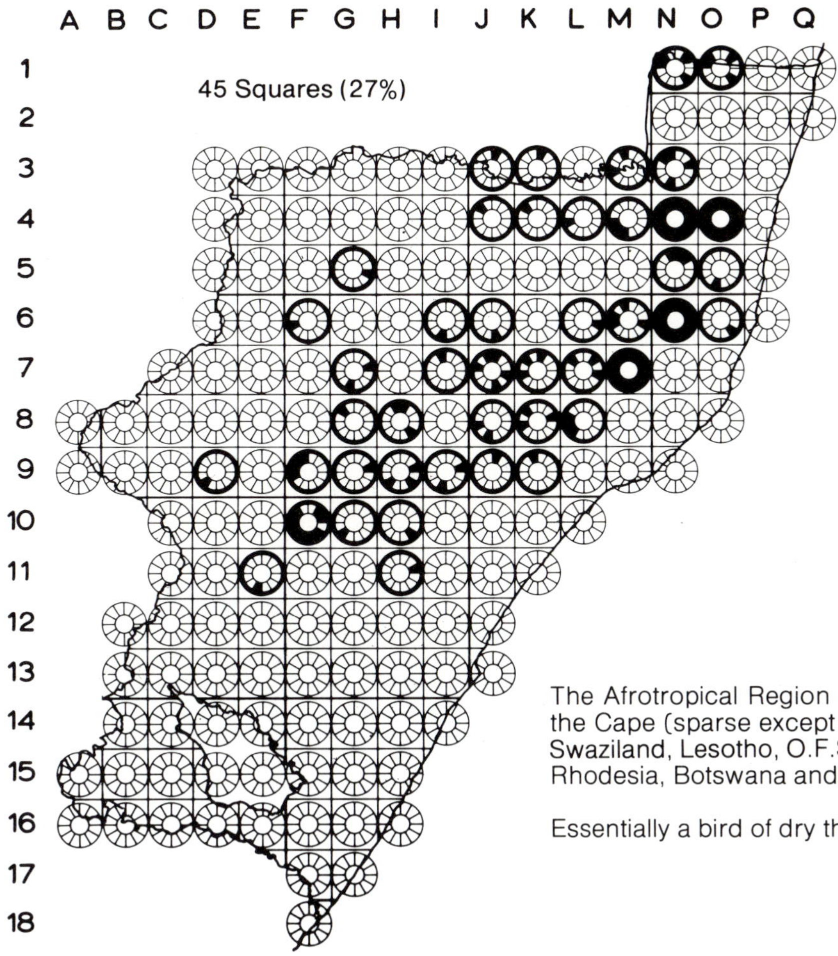

134 Tawny Eagle

Aquila rapax rapax

The Afrotropical Region east to India and Sri Lanka. In Southern Africa, the Cape (sparse except north of the Orange River), Natal and Zululand, Swaziland, Lesotho, O.F.S., Transvaal and Moçambique (interior), west to Rhodesia, Botswana and S.W.A.

Essentially a bird of dry thornveld. Breeds during the winter.

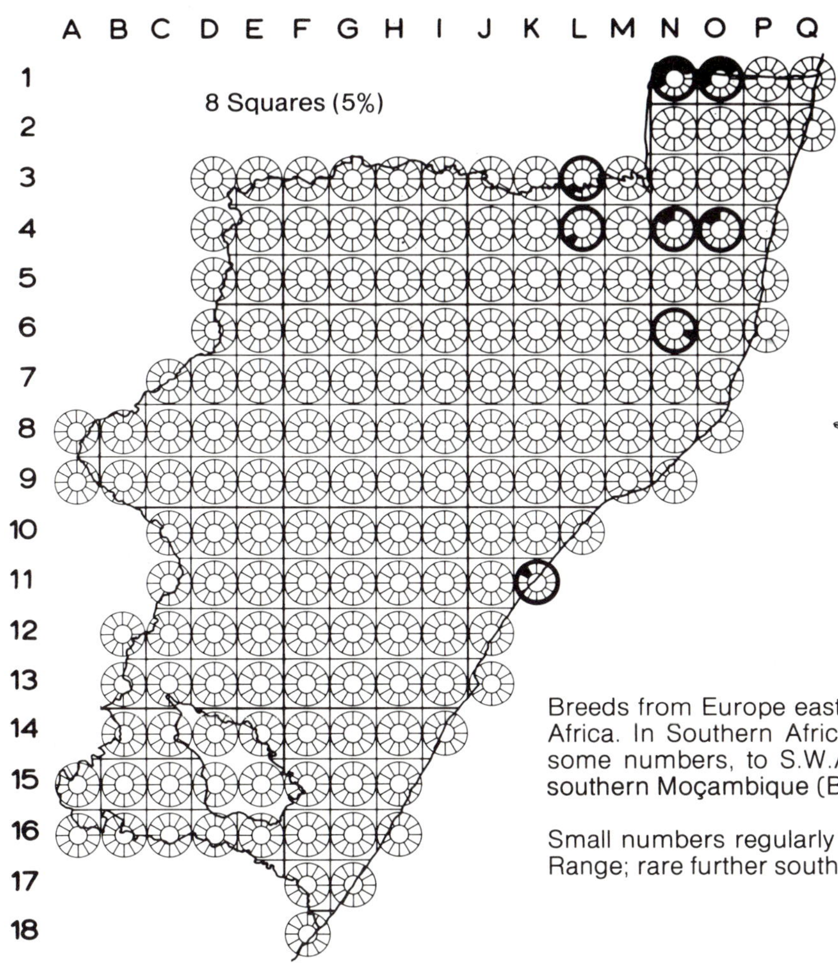

136 Lesser Spotted Eagle

Aquila pomarina

Breeds from Europe east to the Caucasus and India, wintering mainly in Africa. In Southern Africa, a visitor from November to March, often in some numbers, to S.W.A., Botswana, Rhodesia, Transvaal, Swaziland, southern Moçambique (Bella Vista), and Zululand.

Small numbers regularly occur over thornveld adjacent to the Lebombo Range; rare further south.

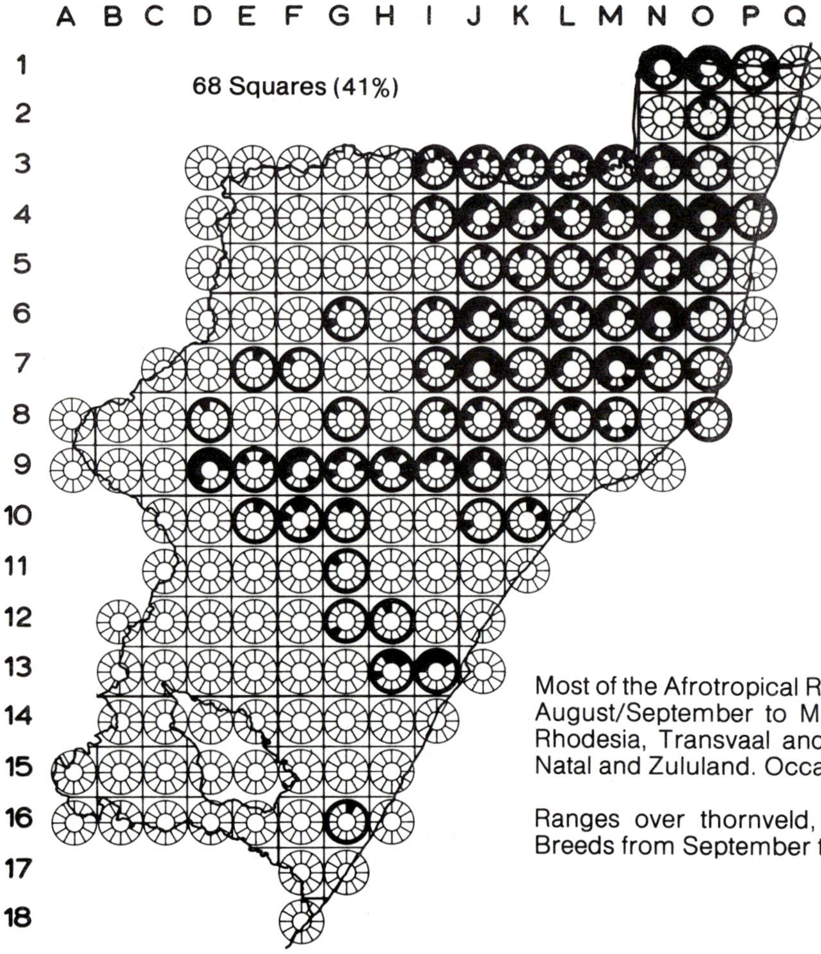

137 Wahlberg's Eagle

Aquila wahlbergi

Most of the Afrotropical Region. In Southern Africa, a breeding visitor from August/September to March to S.W.A., the northern Cape, Botswana, Rhodesia, Transvaal and Moçambique, south in the east to Swaziland, Natal and Zululand. Occasionally farther to the southwest.

Ranges over thornveld, mixed woodland and thickly-wooded valleys. Breeds from September to February.

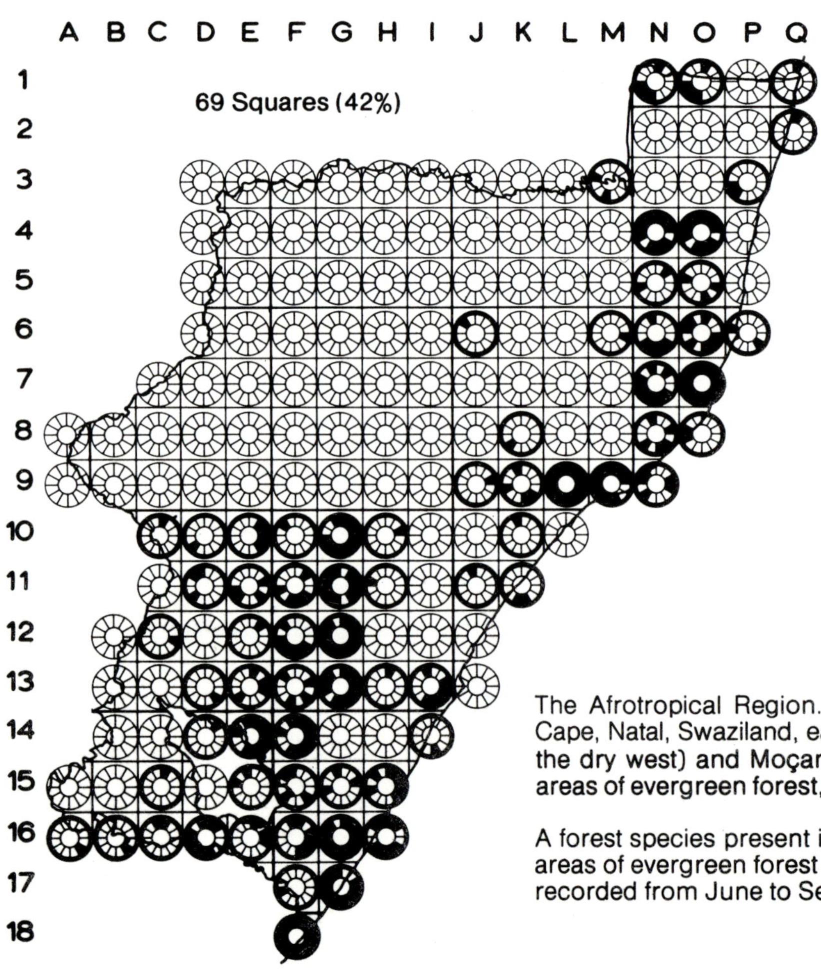

138 Long-crested Eagle

Lophaetus occipitalis

The Afrotropical Region. In Southern Africa, the southern and eastern Cape, Natal, Swaziland, eastern and northern Transvaal, Rhodesia (except the dry west) and Moçambique; also recorded from the Chobe River in areas of evergreen forest, and exotic plantations in the midlands. Breeding

A forest species present in coastal regions, extending inland to the larger areas of evergreen forest and exotic plantations in the midlands. Breeding recorded from June to September.

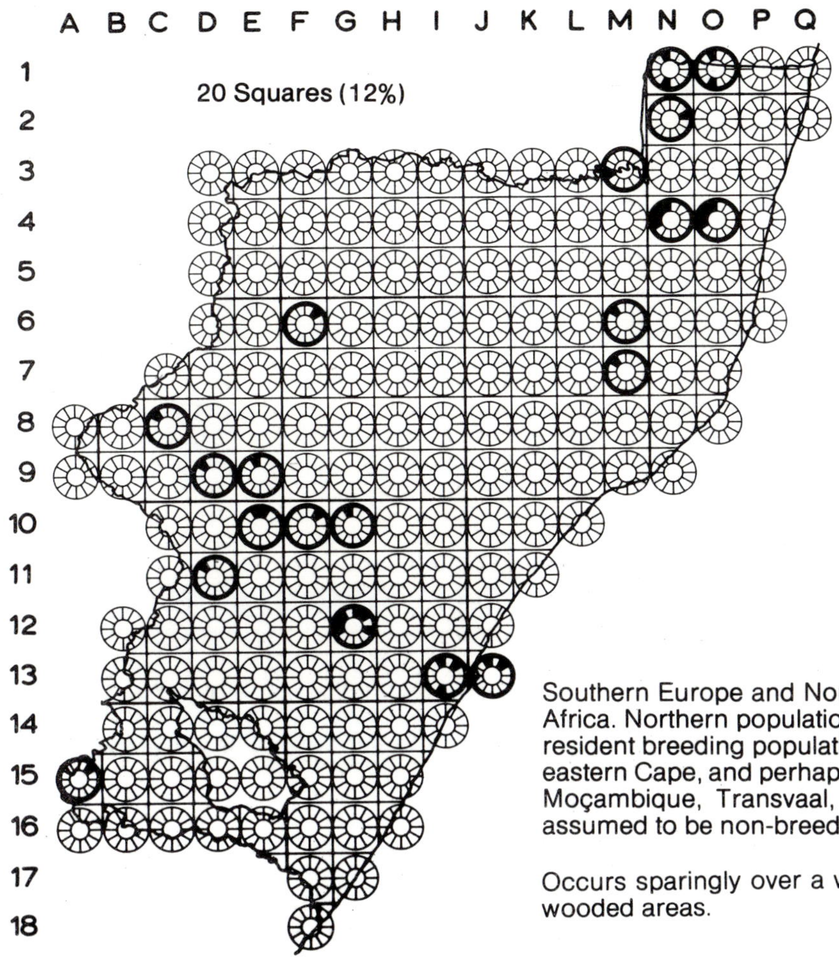

139 Booted Eagle

Hieraaetus pennatus

Southern Europe and North Africa, east to northern India; also Southern Africa. Northern populations winter to the south. In Southern Africa, small resident breeding populations present in the south-western, southern and eastern Cape, and perhaps farther east. Records from Natal and Zululand, Moçambique, Transvaal, O.F.S., Rhodesia, Botswana and S.W.A. are assumed to be non-breeding visitors from the Cape.

Occurs sparingly over a wide range of habitats, from open grassland to wooded areas.

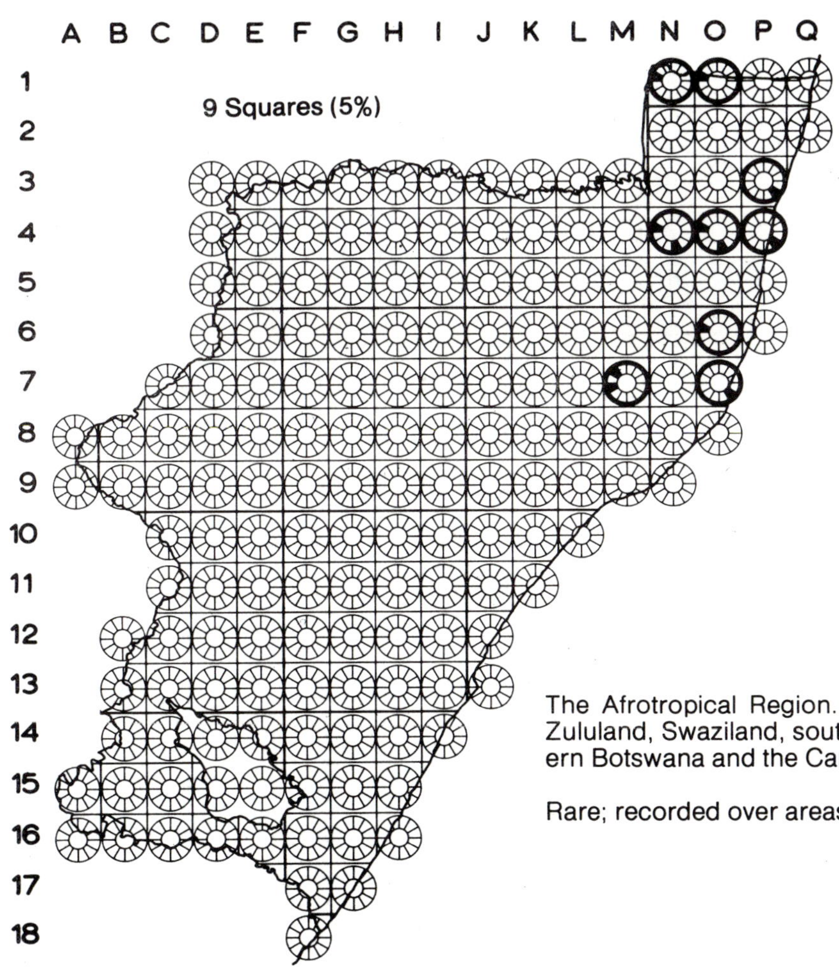

140 Ayres' Hawk Eagle

Hieraaetus ayresi

The Afrotropical Region. In Southern Africa, recorded from Natal and Zululand, Swaziland, southern Moçambique, Transvaal, Rhodesia, northern Botswana and the Caprivi Strip.

Rare; recorded over areas of dense mixed woodland and thornveld.

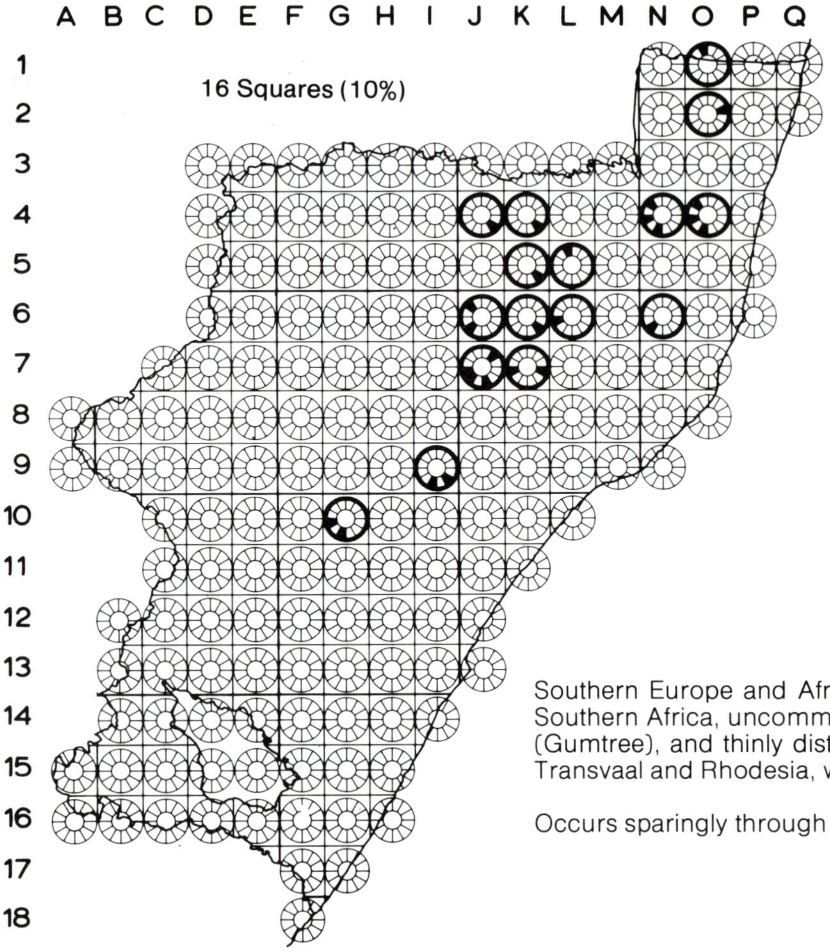

141 African Hawk Eagle

Hieraaetus fasciatus

Southern Europe and Africa, eastwards to southern Asia and China. In Southern Africa, uncommon in the Cape, Natal and Zululand, and O.F.S. (Gumtree), and thinly distributed from eastern Swaziland, Moçambique, Transvaal and Rhodesia, west to Botswana and northern S.W.A.

Occurs sparingly through the northern thornveld regions. May breed.

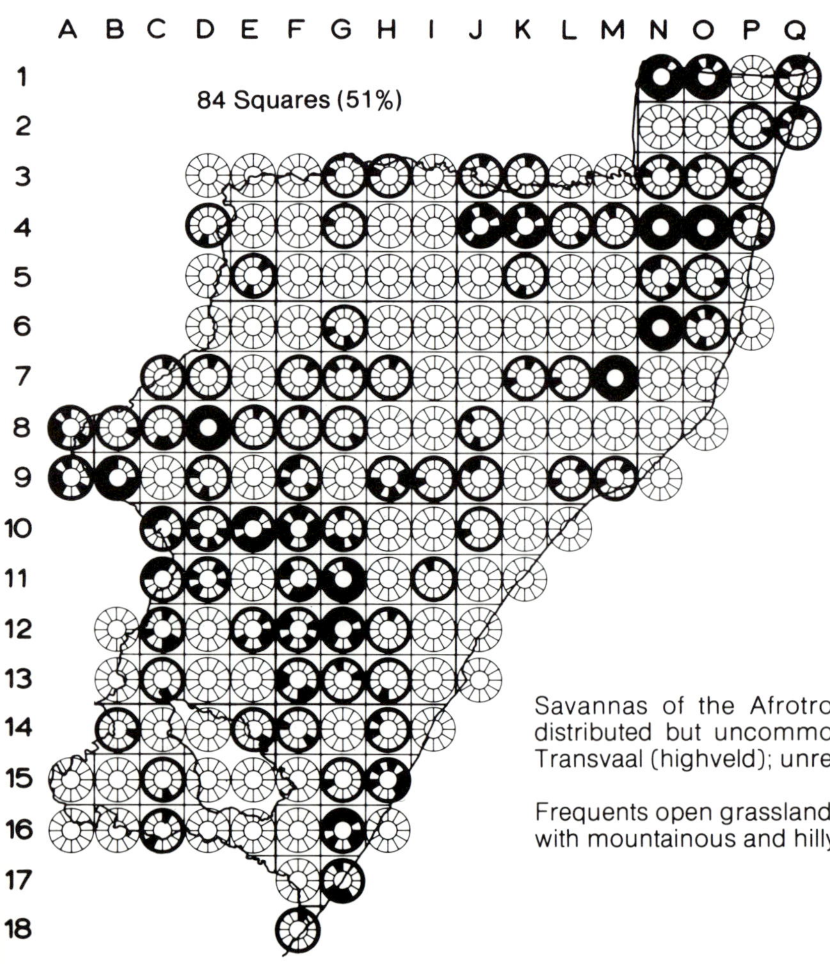

142 Martial Eagle

Polemaetus bellicosus

Savannas of the Afrotropical Region. In Southern Africa, generally distributed but uncommon over much of the Cape, Natal, O.F.S. and Transvaal (highveld); unrecorded from Lesotho.

Frequents open grassland, thornveld and woodland, generally associated with mountainous and hilly country. Breeds during the winter.

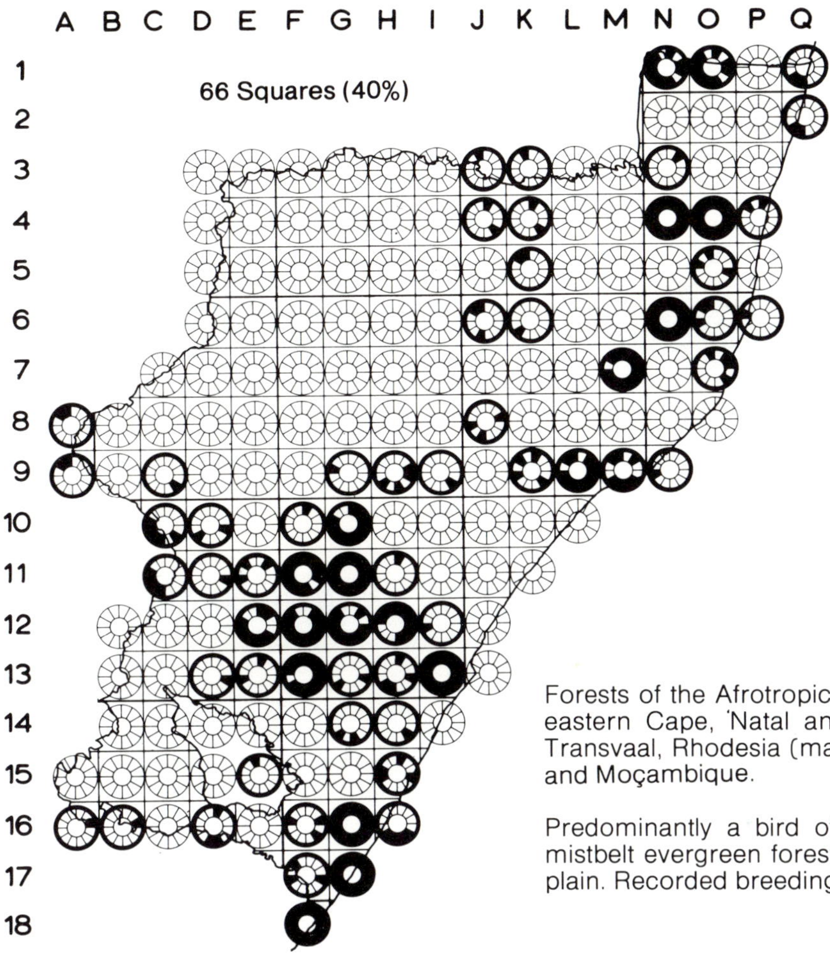

143 Crowned Eagle

Stephanoaetus coronatus

Forests of the Afrotropical Region. In Southern Africa, the southern and eastern Cape, Natal and Zululand, Swaziland, eastern and northern Transvaal, Rhodesia (mainly in the east, but locally west to the Matopos) and Moçambique.

Predominantly a bird of forested regions, most often encountered in mistbelt evergreen forest; also found in riverine woodland on the littoral plain. Recorded breeding from June to November.

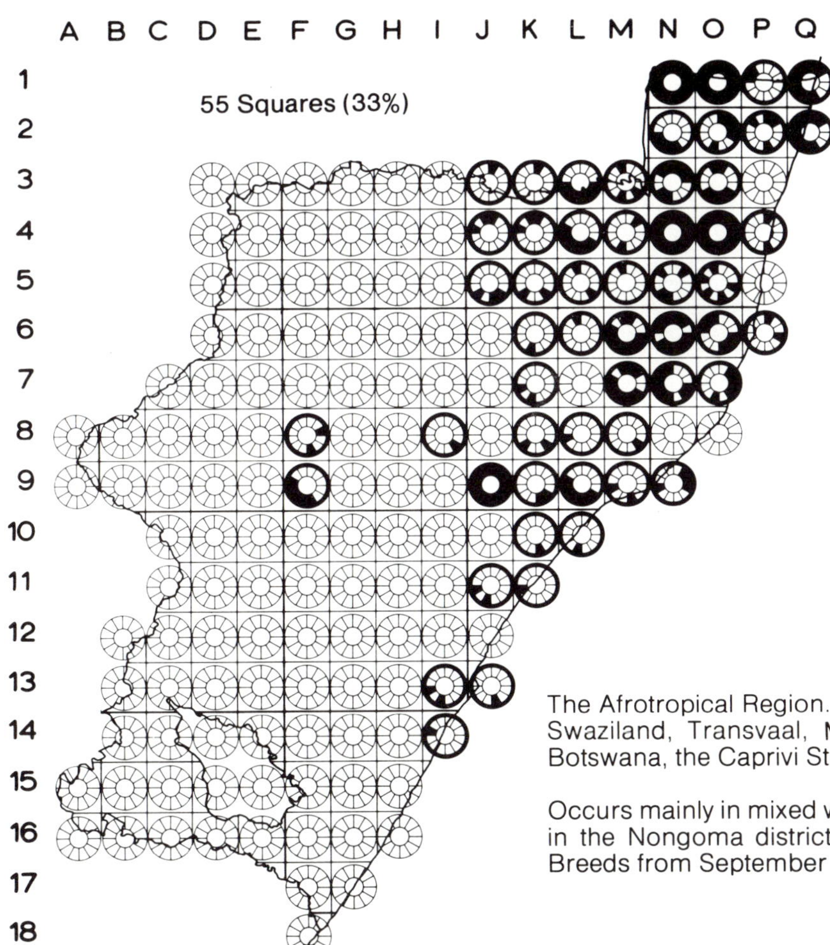

144 Lizard Buzzard

Kaupifalco monogrammicus

The Afrotropical Region. In Southern Africa, Natal and Zululand, north to Swaziland, Transvaal, Moçambique, Rhodesia, eastern and northern Botswana, the Caprivi Strip and northern S.W.A.

Occurs mainly in mixed woodland and thornveld; also noted on occasions in the Nongoma district in open grassland with stands of exotic trees. Breeds from September to November.

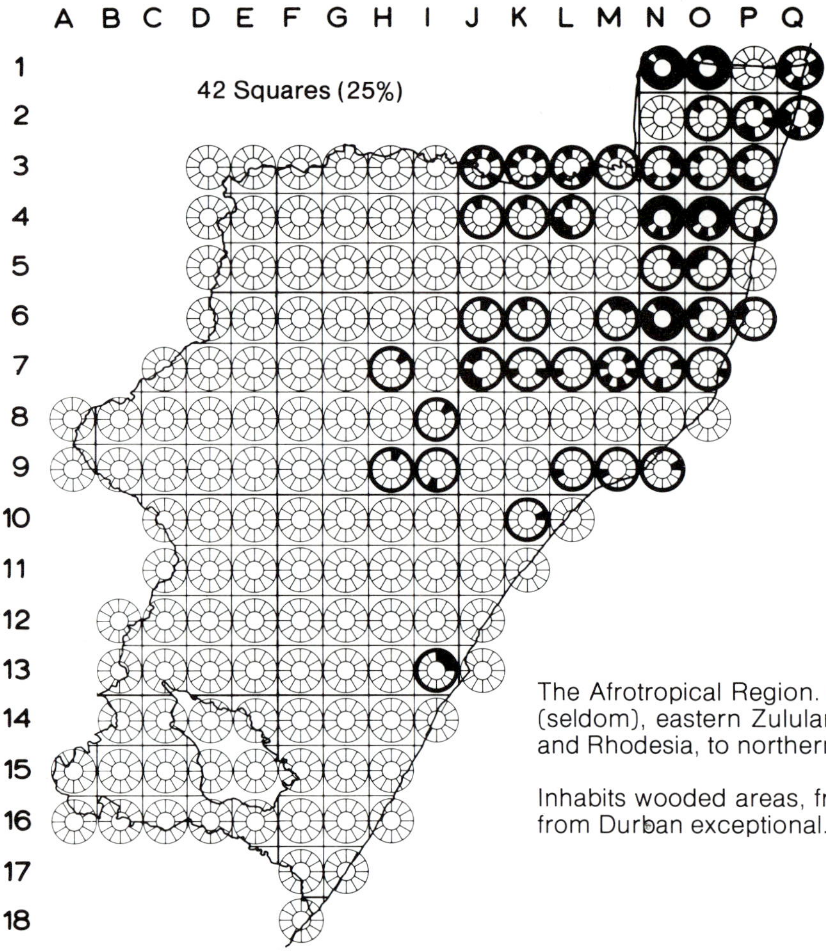

145 Brown Snake-eagle

Circaetus cinereus

The Afrotropical Region. In Southern Africa, the eastern Cape and Natal (seldom), eastern Zululand, eastern Swaziland, Moçambique, Transvaal and Rhodesia, to northern Botswana and northern S.W.A.

Inhabits wooded areas, from open savanna to dense thornveld. Records from Durban exceptional. Breeds mainly during the winter.

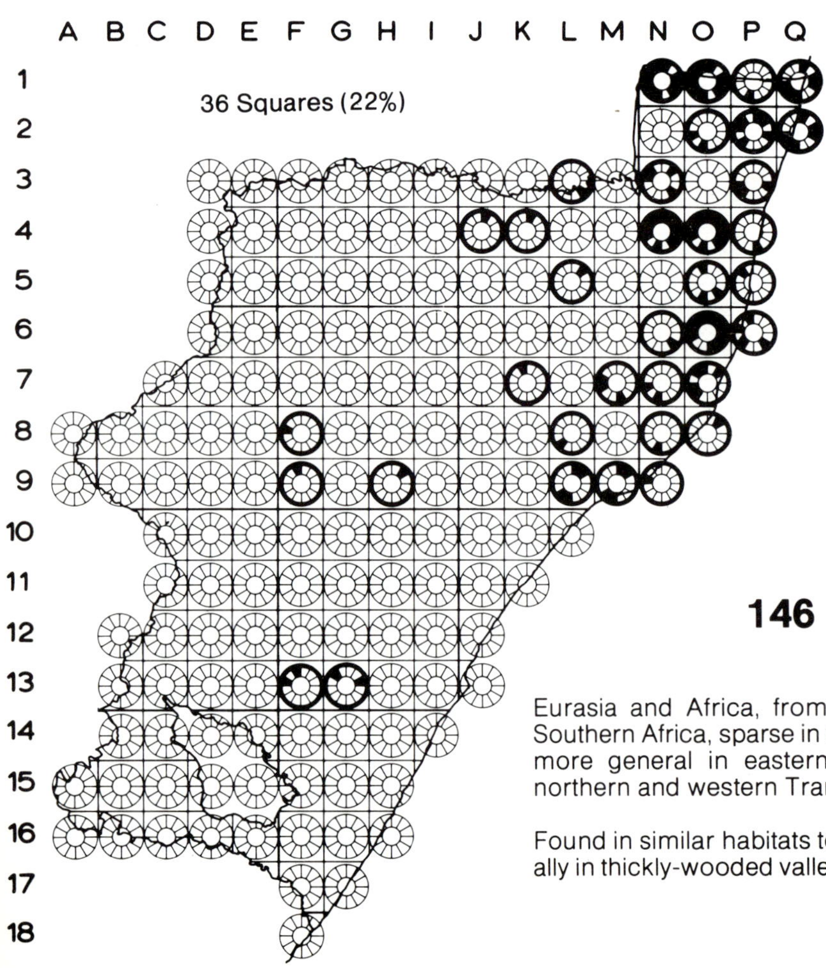

146 Black-breasted Snake-eagle

Circaetus gallicus

Eurasia and Africa, from Senegal and Ethiopia south to the Cape. In Southern Africa, sparse in the Cape, Natal, O.F.S. and Transvaal (highveld); more general in eastern Zululand, Swaziland, Moçambique, eastern, northern and western Transvaal and Rhodesia, west to S.W.A.

Found in similar habitats to those of Brown Snake-eagle (145); exceptionally in thickly-wooded valleys in the south. May breed throughout the year.

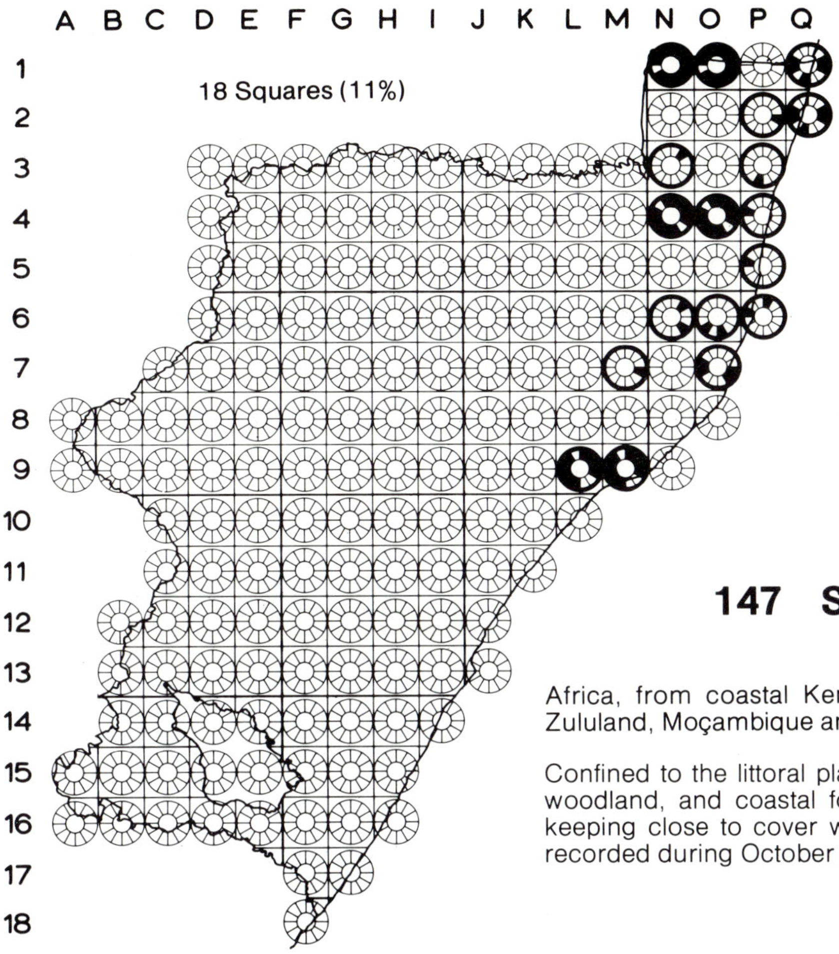

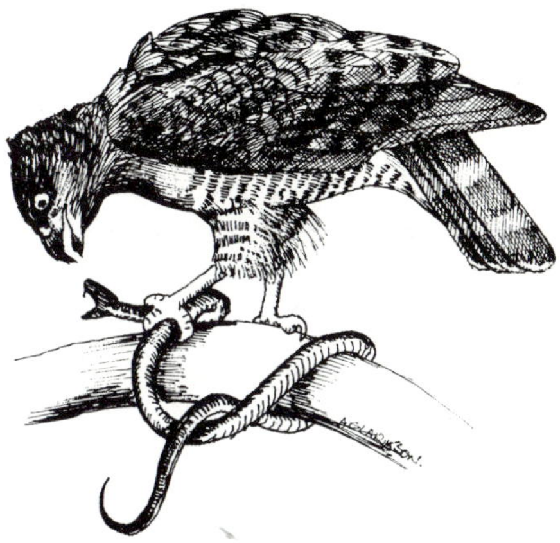

147 Southern Banded Snake-eagle

Circaetus fasciolatus

Africa, from coastal Kenya to Zululand. In Southern Africa, eastern Zululand, Moçambique and south-eastern Rhodesia.

Confined to the littoral plain, where it inhabits dense mixed and riverine woodland, and coastal forest. Usually seen perched on tops of trees, keeping close to cover when flushed; seldom noted soaring. Breeding recorded during October and November.

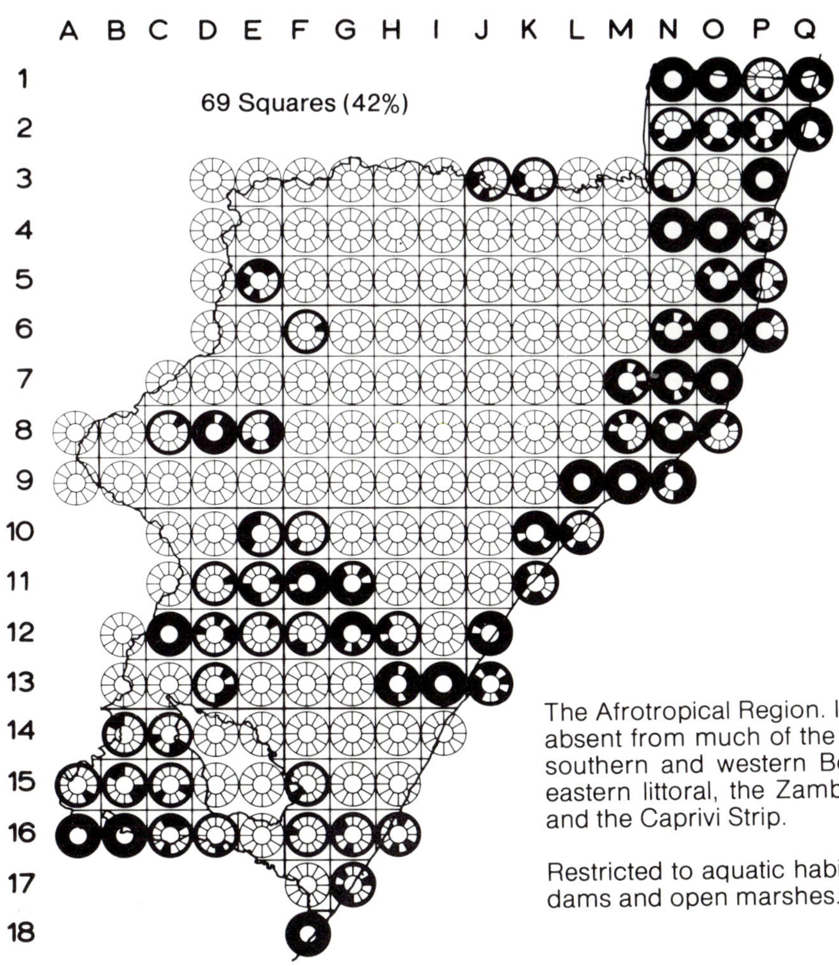

149 Fish Eagle

Haliaeetus vocifer

The Afrotropical Region. In Southern Africa, generally distributed, though absent from much of the xeric interior of the northern Cape, S.W.A. and southern and western Botswana. More common on the southern and eastern littoral, the Zambesi River, northern Botswana (swamp region) and the Caprivi Strip.

Restricted to aquatic habitats, from estuaries and coastal lakes, inland to dams and open marshes. Recorded breeding from April to August.

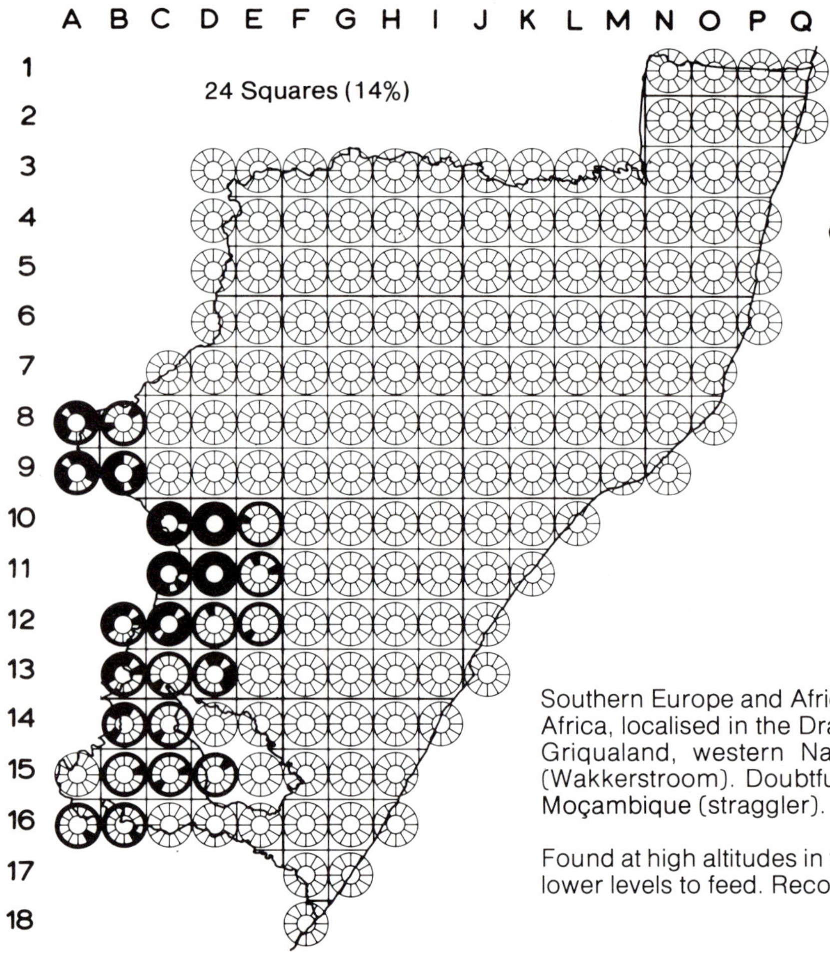

150 Bearded Vulture

Gypaetus barbatus

Southern Europe and Africa, east to northern India and China. In Southern Africa, localised in the Drakensberg Range from the eastern Cape to East Griqualand, western Natal, Lesotho, O.F.S. and southern Transvaal (Wakkerstroom). Doubtful from eastern Rhodesia, but taken in adjacent Moçambique (straggler).

Found at high altitudes in the Drakensberg Range, occasionally ranging to lower levels to feed. Recorded breeding during July and August.

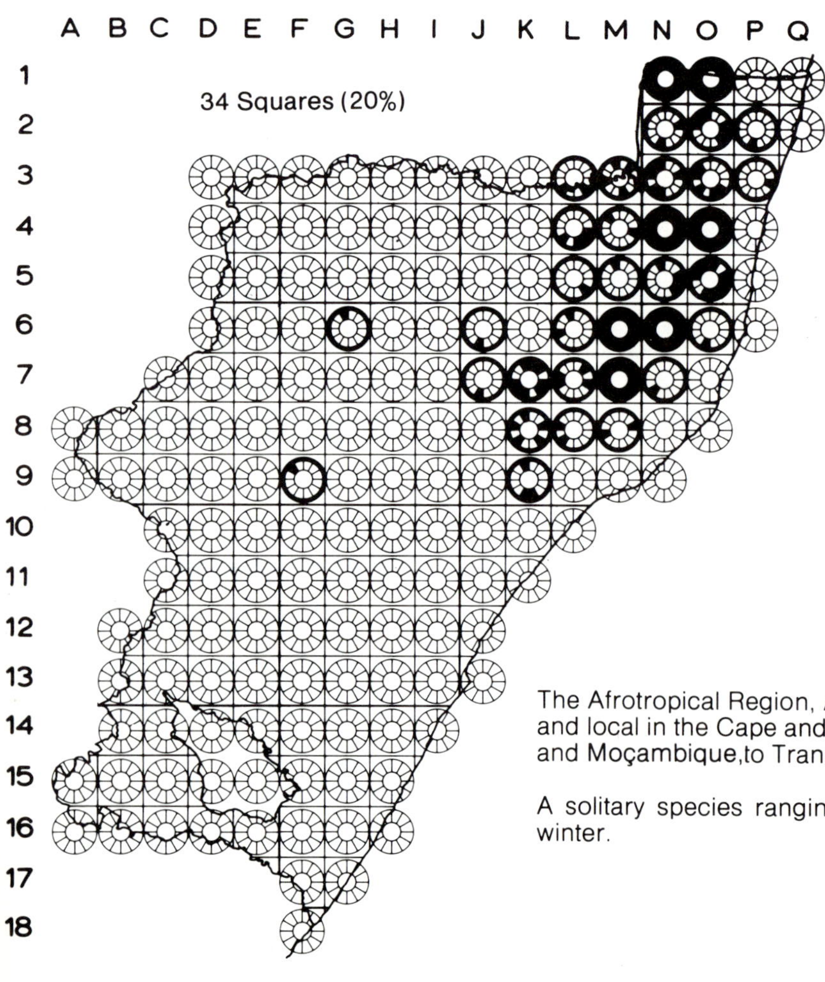

151 Bateleur

Terathopius ecaudatus

The Afrotropical Region, Arabia and Iraq. In Southern Africa, uncommon and local in the Cape and Natal; more widespread in Zululand, Swaziland and Moçambique,to Transvaal, Rhodesia, O.F.S., Botswana and S.W.A.

A solitary species ranging over savanna woodland. Breeds during the winter.

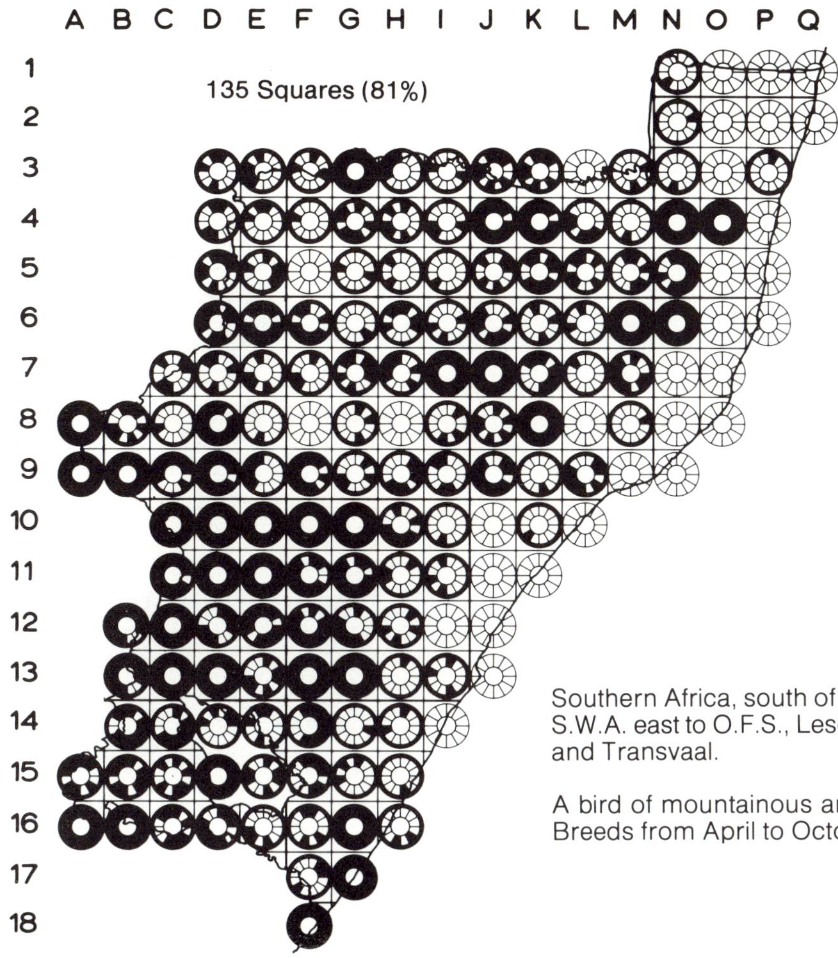

152 Jackal Buzzard

Buteo rufofuscus

Southern Africa, south of the Limpopo River. From the Cape and southern S.W.A. east to O.F.S., Lesotho, Natal, western Zululand, western Swaziland, and Transvaal.

A bird of mountainous and hilly country, reaching the coast in the south. Breeds from April to October.

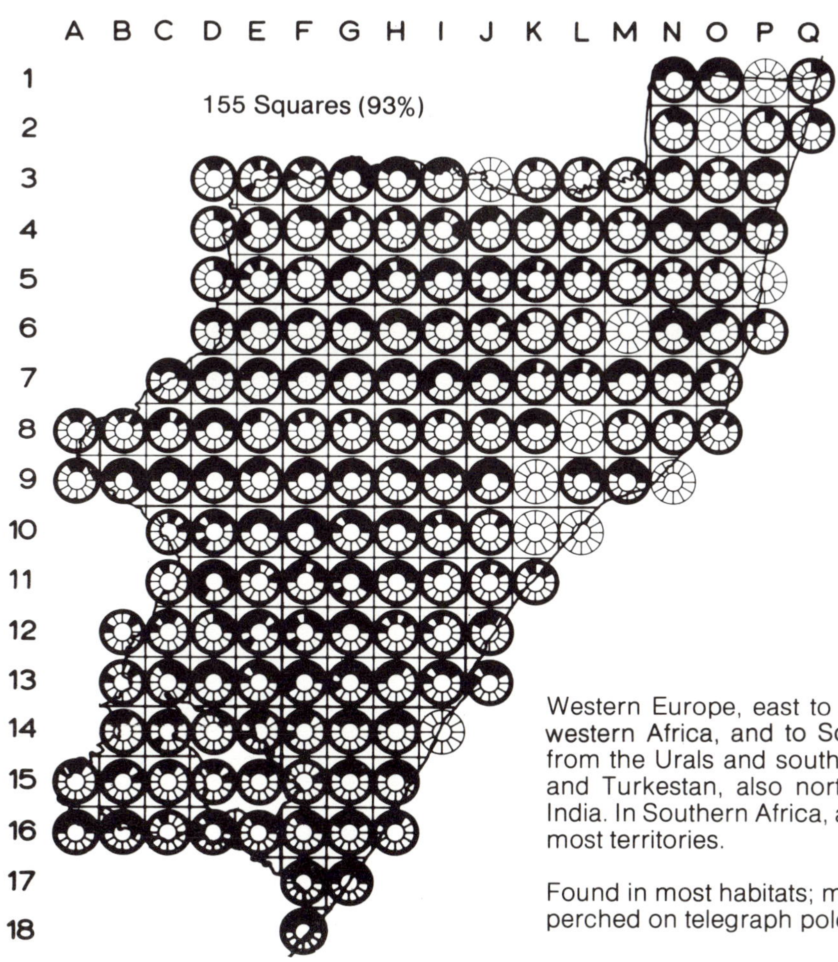

154 Steppe Buzzard

Buteo buteo

Western Europe, east to Asia, south in the west to the islands off north western Africa, and to Socotra. The subspecies *B.b. vulpinus* breeding from the Urals and south-eastern U.S.S.R., east to the Kentai Mountains and Turkestan, also north-western India; winters in Africa, Arabia and India. In Southern Africa, a visitor from October to late March/early April to most territories.

Found in most habitats; more usually in open situations and typically seen perched on telegraph poles.

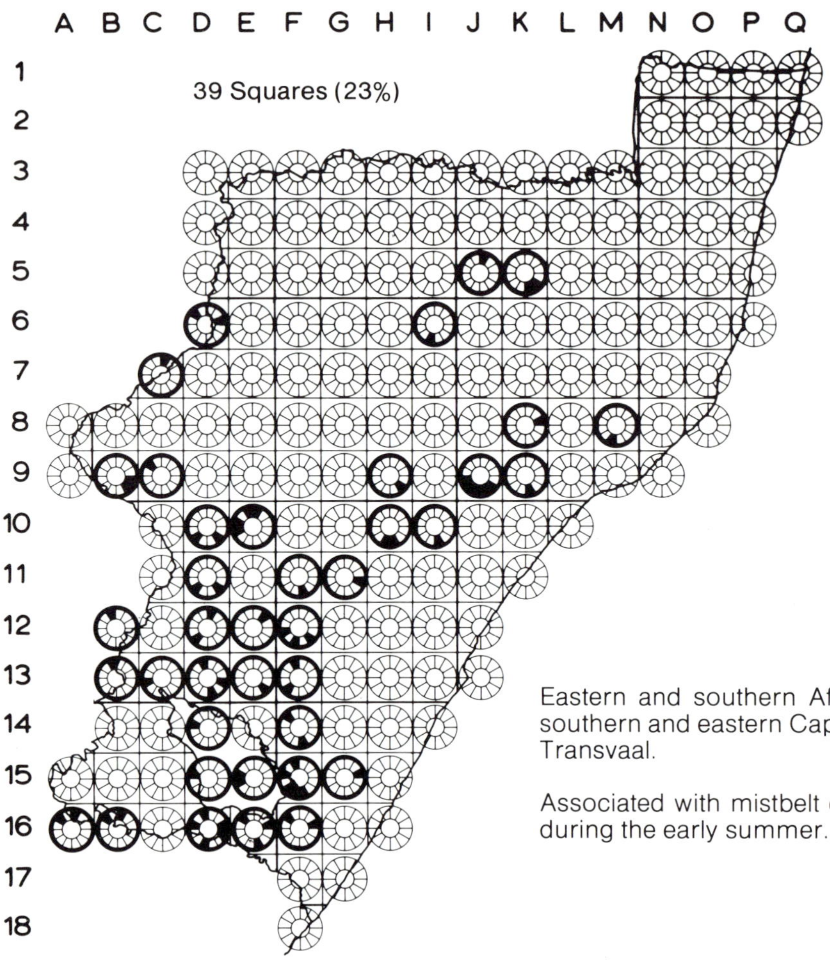

155 Mountain Buzzard

Buteo tachardus

Eastern and southern Africa. In Southern Africa, the mountains of the southern and eastern Cape, Natal (interior), western Zululand and eastern Transvaal.

Associated with mistbelt evergreen forest and exotic plantations. Breeds during the early summer.

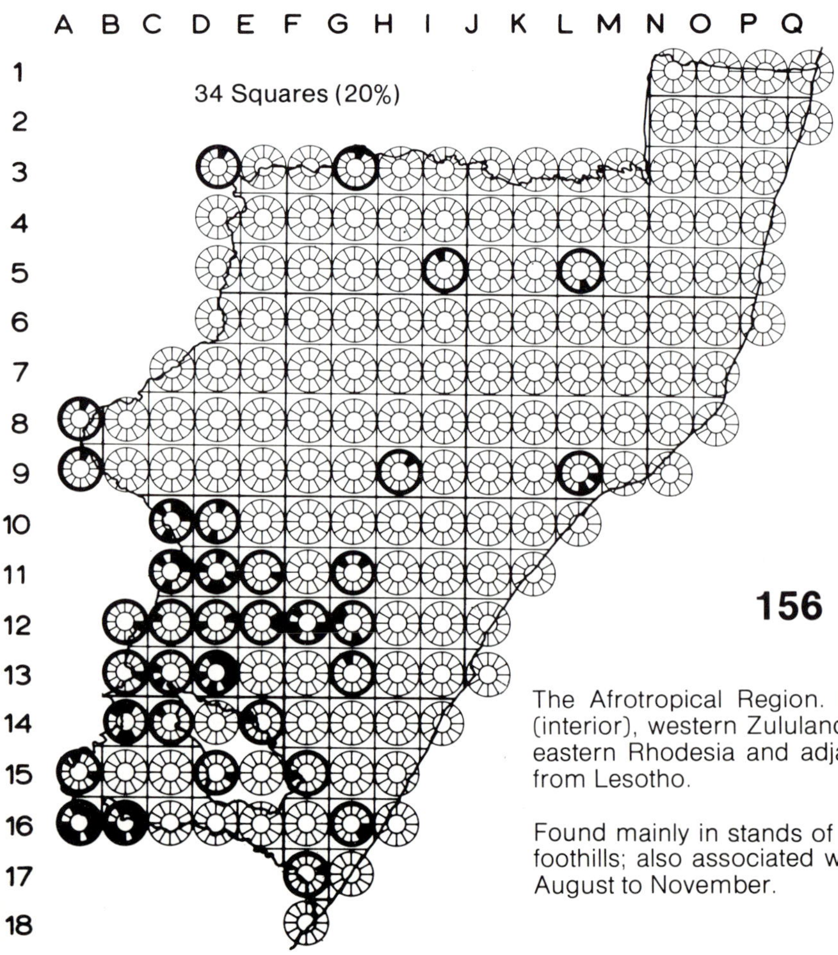

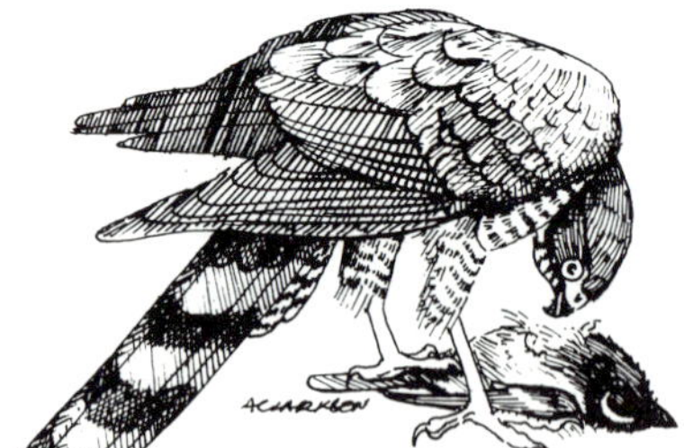

156 Red-breasted Sparrowhawk

Accipiter rufiventris

The Afrotropical Region. In Southern Africa, from the Cape to Natal (interior), western Zululand, Swaziland, eastern and northern Transvaal, eastern Rhodesia and adjacent Moçambique (highland); also recorded from Lesotho.

Found mainly in stands of exotic trees in the midlands and Drakensberg foothills; also associated with evergreen forest. Breeding recorded from August to November.

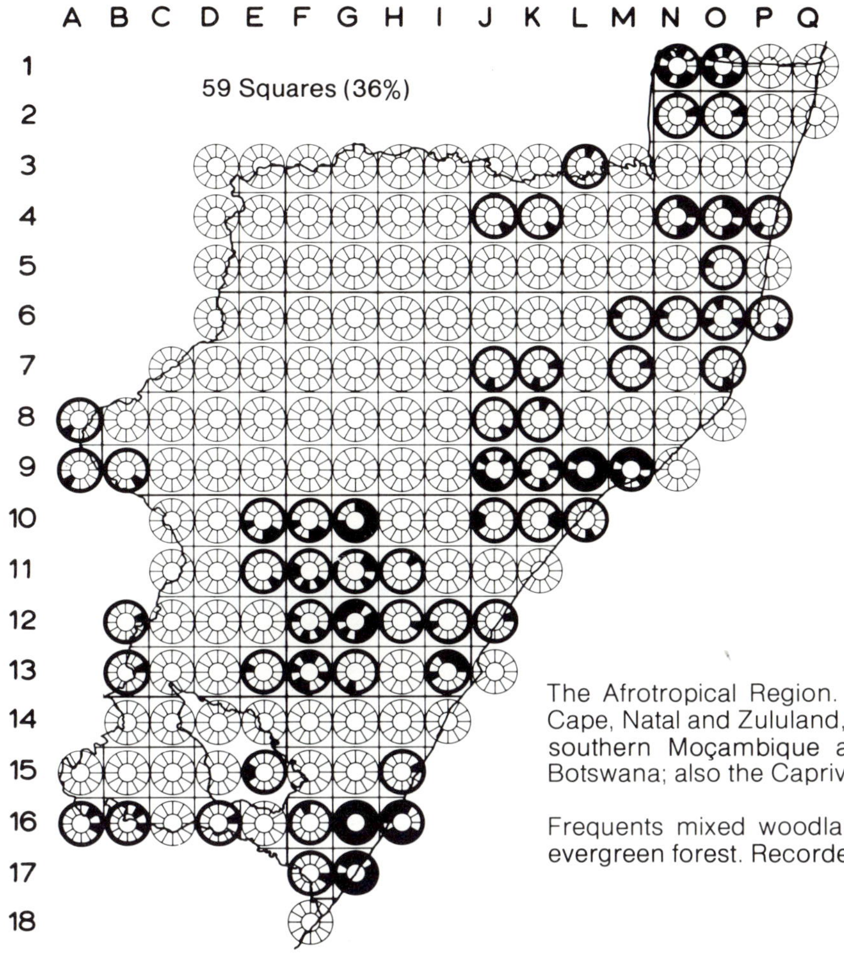

158 Little Sparrowhawk

Accipiter minullus

The Afrotropical Region. In Southern Africa, the southern and eastern Cape, Natal and Zululand, northwards in the east to Swaziland, Transvaal, southern Moçambique and Rhodesia, west to eastern and northern Botswana; also the Caprivi Strip and northern S.W.A.

Frequents mixed woodland and thickly-wooded valleys, extending into evergreen forest. Recorded breeding during October and November.

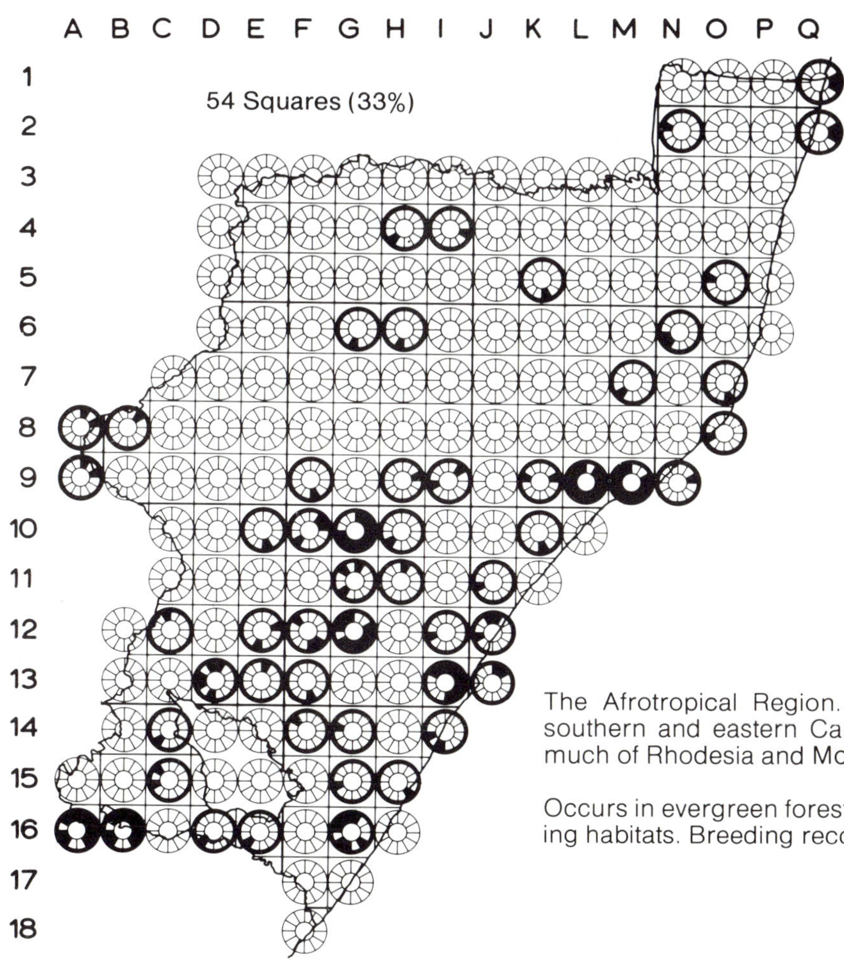

159 Black Sparrowhawk

Accipiter melanoleucus

The Afrotropical Region. In Southern Africa, wooded regions of the southern and eastern Cape, Natal and Zululand, Swaziland, Transvaal, much of Rhodesia and Moçambique; also northern S.W.A.

Occurs in evergreen forest and exotic plantations, ranging over surrounding habitats. Breeding recorded from May to November.

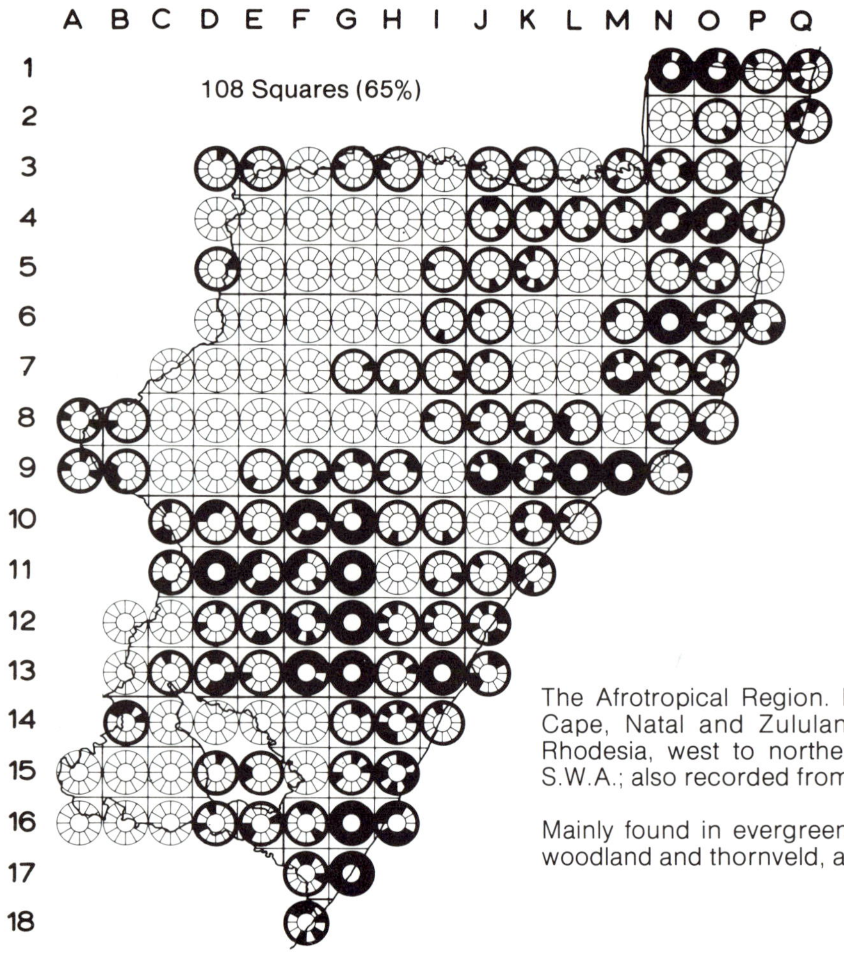

160 African Goshawk

Accipiter tachiro

The Afrotropical Region. In Southern Africa, the southern and eastern Cape, Natal and Zululand, Swaziland, Transvaal, Moçambique and Rhodesia, west to northern Botswana, the Caprivi Strip and northern S.W.A.; also recorded from Lesotho and O.F.S.

Mainly found in evergreen forest and exotic plantations; also in mature woodland and thornveld, and even gardens.

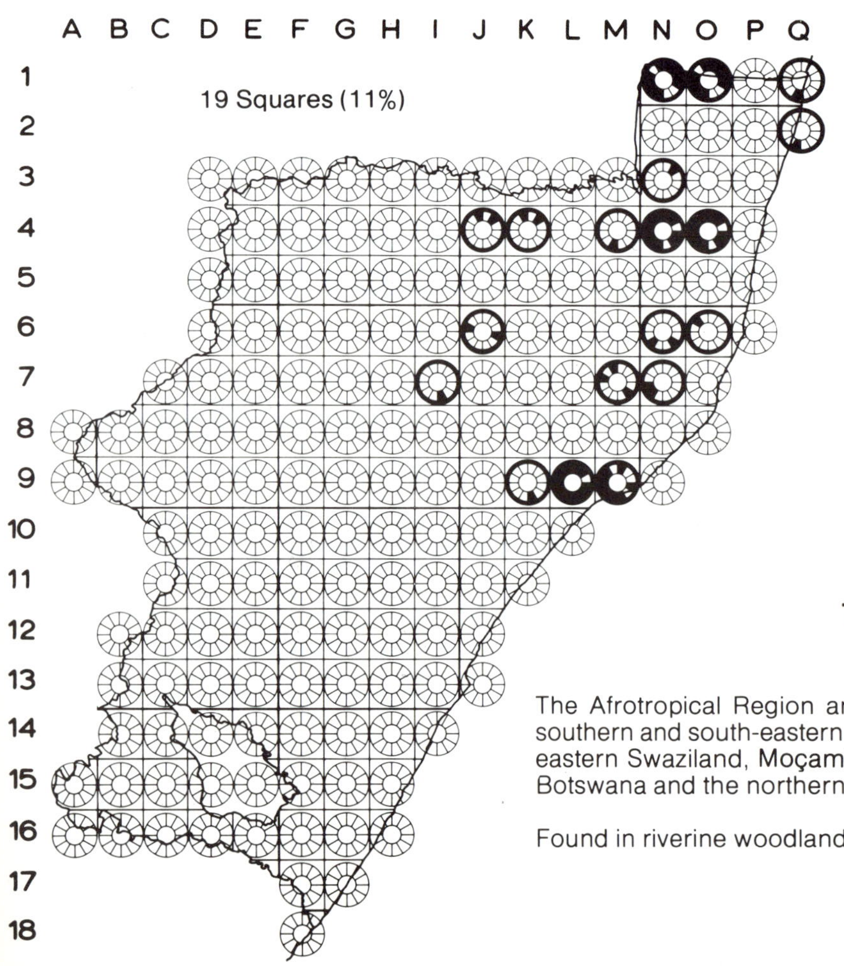

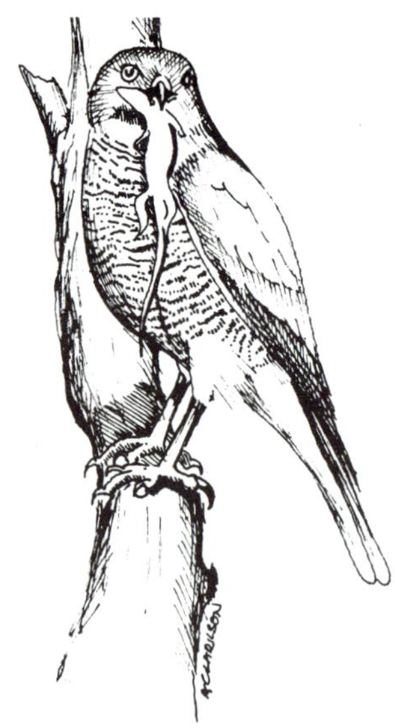

161 Little Banded Goshawk

Accipiter badius

The Afrotropical Region and the Middle East to south-central U.S.S.R., southern and south-eastern Asia. In Southern Africa, Natal (rare), Zululand, eastern Swaziland, Moçambique, Transvaal and Rhodesia, west through Botswana and the northern Cape to S.W.A.

Found in riverine woodland and thornveld. Breeds during the summer.

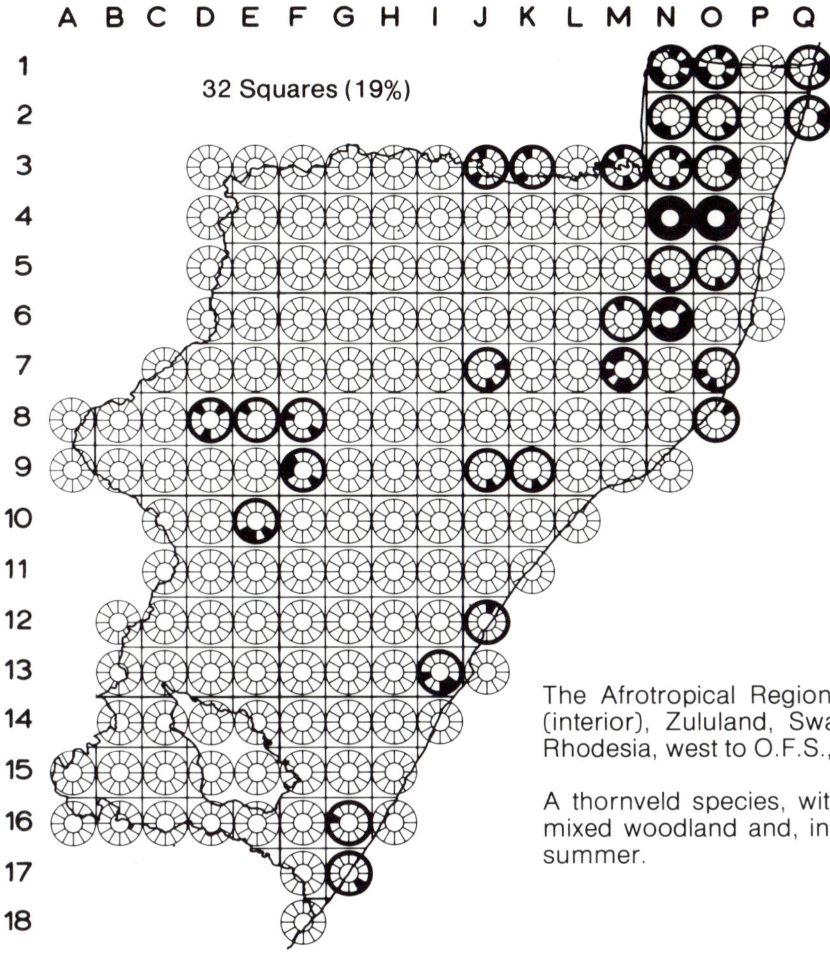

162 Gabar Goshawk

Micronisus gabar

The Afrotropical Region. In Southern Africa, the eastern Cape, Natal (interior), Zululand, Swaziland, Moçambique (interior), Transvaal and Rhodesia, west to O.F.S., the northern Cape, Botswana and S.W.A.

A thornveld species, with a preference for mature trees; also found in mixed woodland and, in the south, in coastal forest. Breeds during the summer.

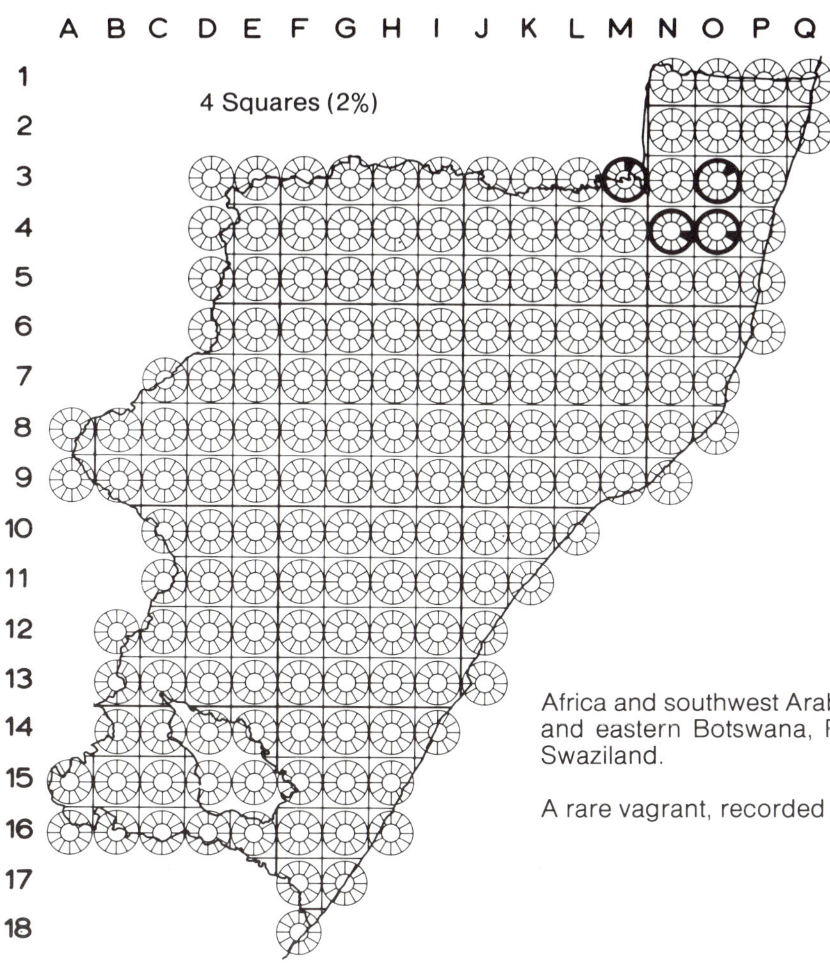

163 Dark Chanting Goshawk

Melierax metabates

Africa and southwest Arabia. In Southern Africa, northern S.W.A., northern and eastern Botswana, Rhodesia, Transvaal, Moçambique and eastern Swaziland.

A rare vagrant, recorded from mixed woodland and thornveld.

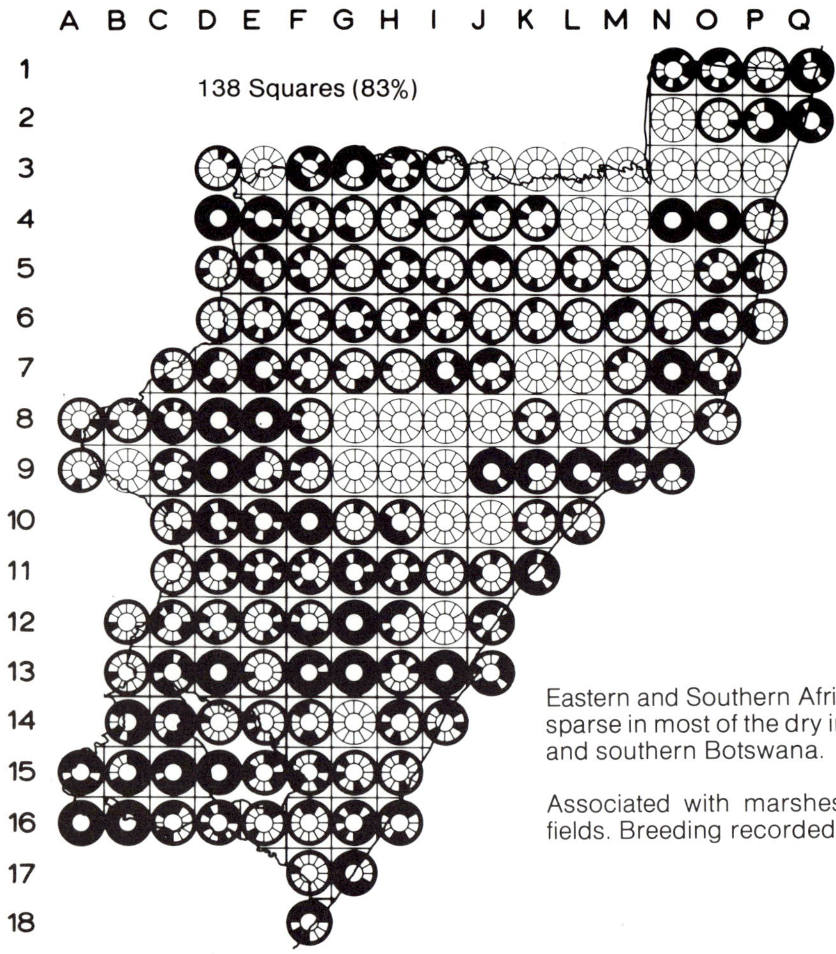

167 African Marsh Harrier

Circus ranivorus

Eastern and Southern Africa. In Southern Africa, widely distributed, though sparse in most of the dry interior and northern districts of the Cape, S.W.A. and southern Botswana.

Associated with marshes and adjacent open grassland and cultivated fields. Breeding recorded from April to November.

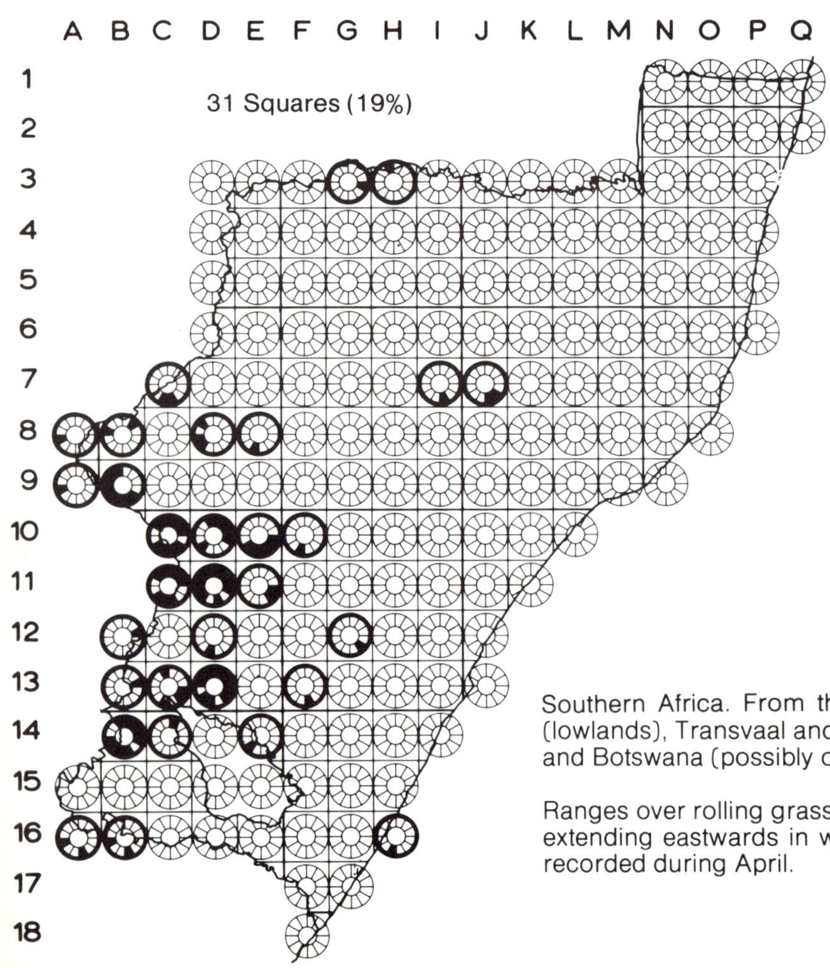

169 Black Harrier

Circus maurus

Southern Africa. From the Cape to western and upper Natal, Lesotho (lowlands), Transvaal and western O.F.S. Recently recorded from S.W.A. and Botswana (possibly off-season migrants from farther south).

Ranges over rolling grasslands in the foothills of the Drakensberg Range, extending eastwards in winter over open grassland. Breeding has been recorded during April.

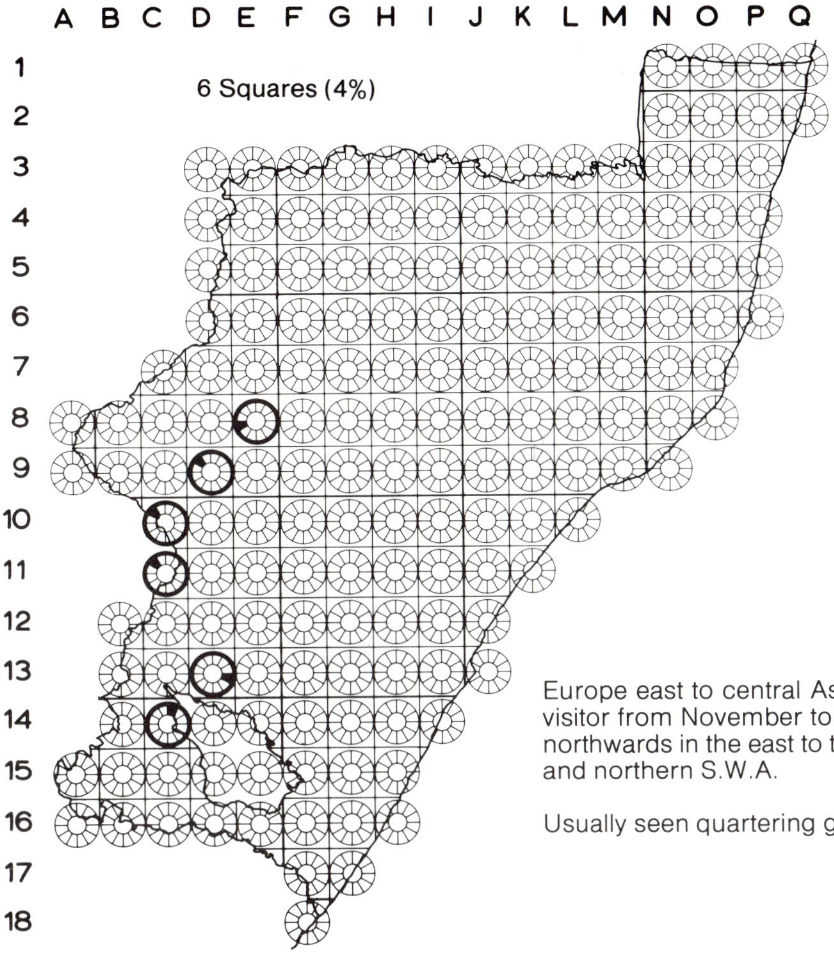

170 Montagu's Harrier

Circus pygargus

Europe east to central Asia, wintering to the south. In Southern Africa, a visitor from November to March to the eastern Cape, Natal and Zululand, northwards in the east to the Zambesi River; in the north, west to Botswana and northern S.W.A.

Usually seen quartering grassland and cultivated fields in upland areas.

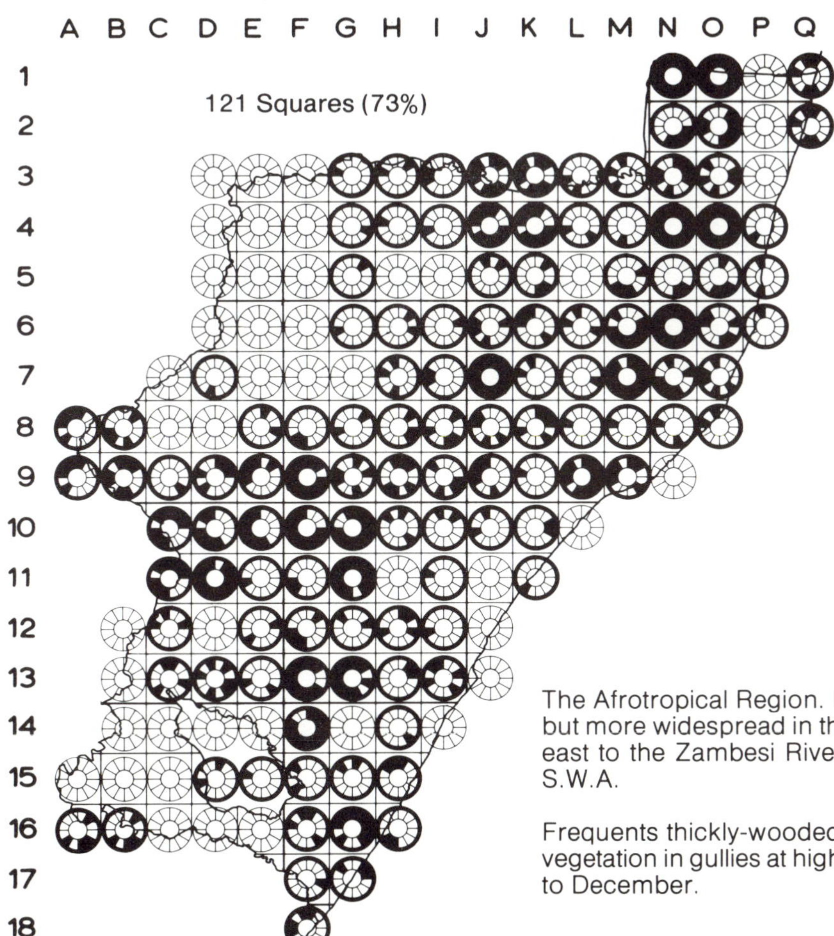

171 Gymnogene

Polyboroides typus

The Afrotropical Region. In Southern Africa, locally in the southern Cape, but more widespread in the eastern Cape, Natal and Zululand, north in the east to the Zambesi River. In the north, west to Botswana and northern S.W.A.

Frequents thickly-wooded valleys, thornveld, mixed woodland, and scrub vegetation in gullies at higher altitudes. Recorded breeding from September to December.

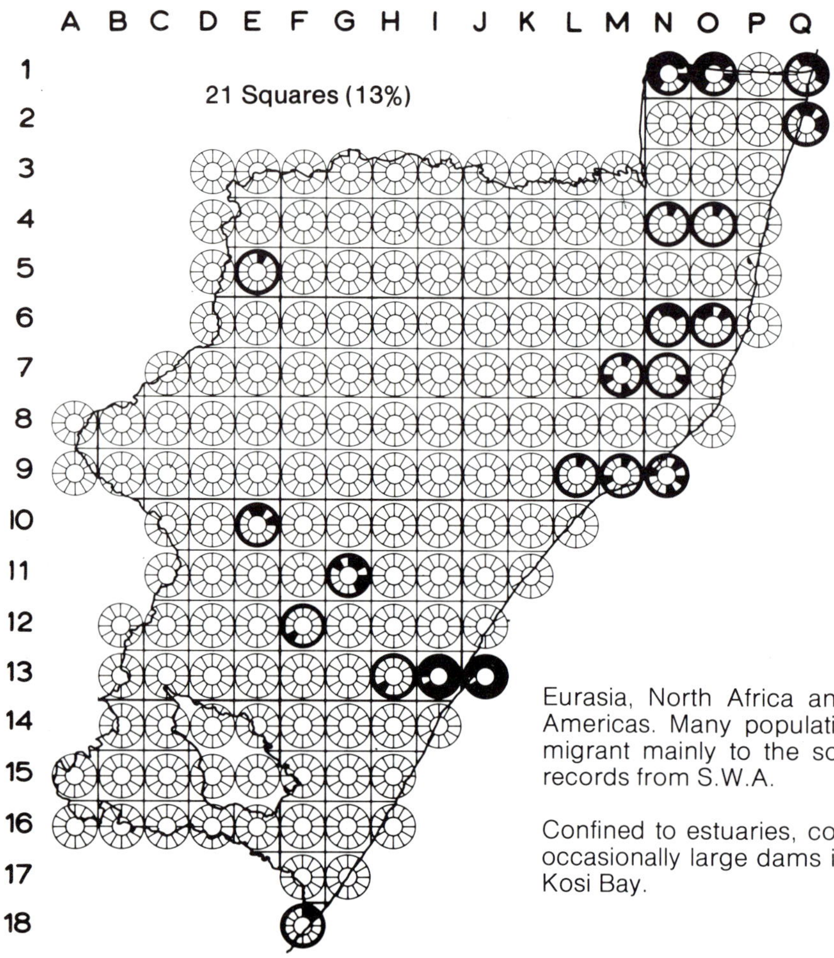

172 Osprey

Pandion haliaetus

Eurasia, North Africa and southern Asia, south to Australia; also the Americas. Many populations migratory. In Southern Africa, a summer migrant mainly to the south and east; some overwinter. A few recent records from S.W.A.

Confined to estuaries, coastal lakes, large pans on the littoral plain and occasionally large dams inland. Has been noted feeding from the sea at Kosi Bay.

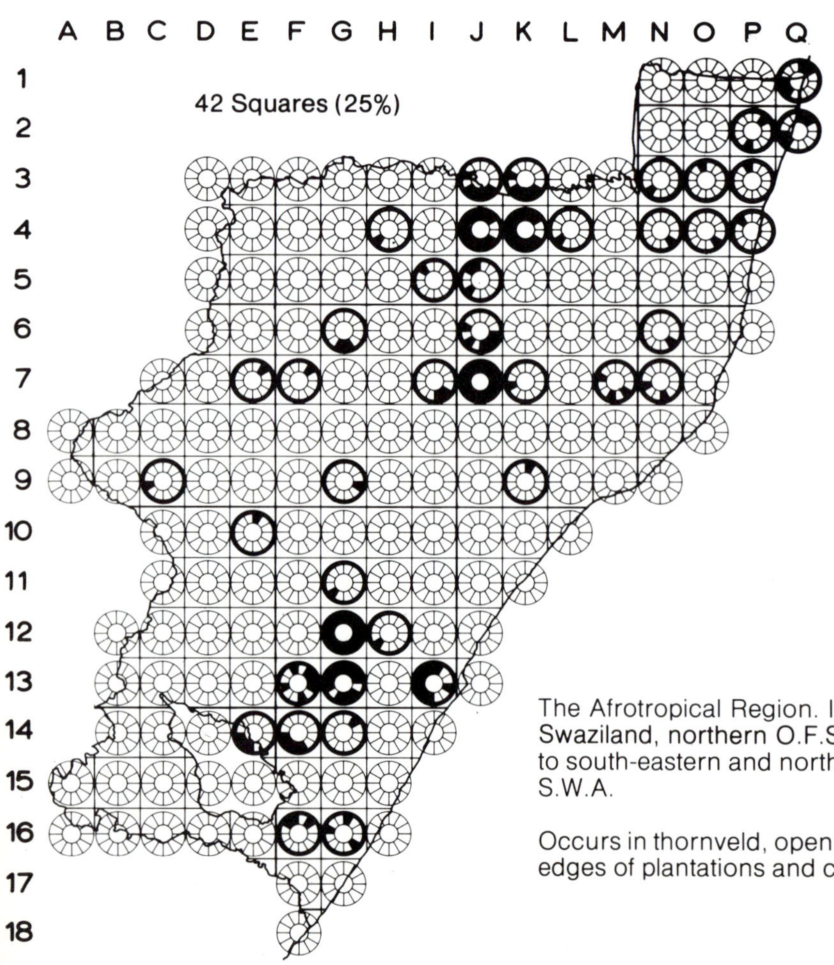

173 Coqui Francolin

Francolinus coqui

The Afrotropical Region. In Southern Africa, Natal and Zululand, eastern Swaziland, northern O.F.S., Transvaal, Moçambique and Rhodesia, west to south-eastern and northern Botswana, and northern and north-eastern S.W.A.

Occurs in thornveld, open grassland with scattered trees in lowland areas, edges of plantations and cultivated fields. Breeds from October to March.

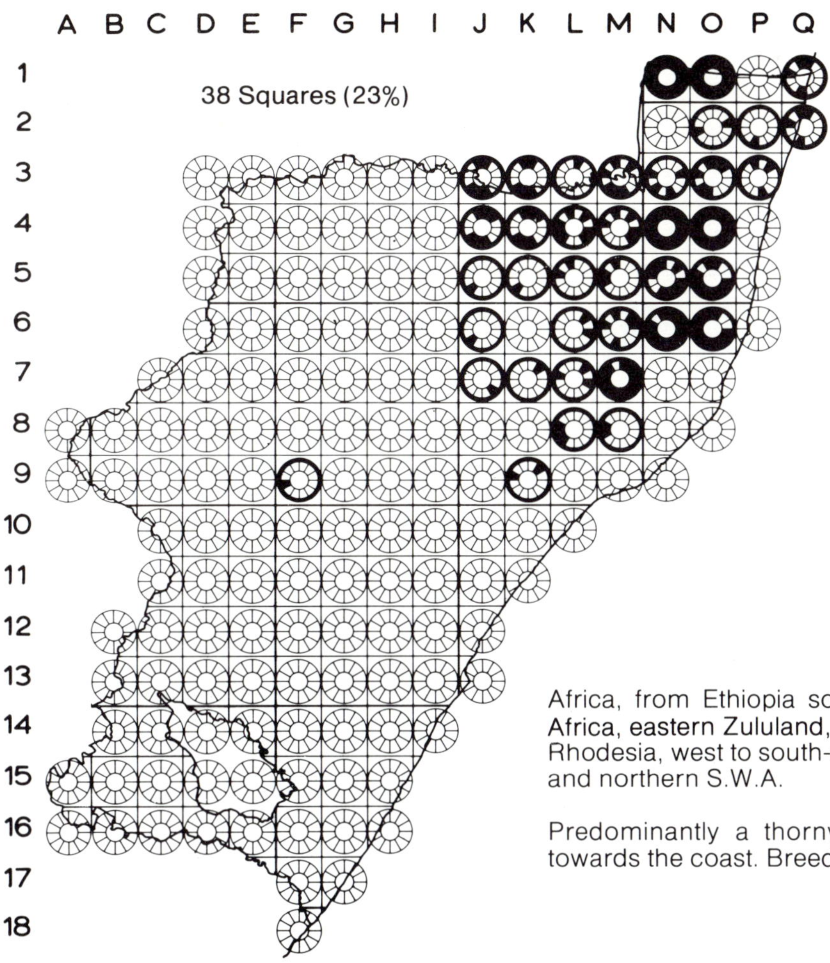

174 Crested Francolin

Francolinus sephaena

Africa, from Ethiopia south to Damaraland and Zululand. In Southern Africa, eastern Zululand, eastern Swaziland, Transvaal, Moçambique and Rhodesia, west to south-eastern and northern Botswana, the Caprivi Strip and northern S.W.A.

Predominantly a thornveld species, extending into mixed woodland towards the coast. Breeds from October to March.

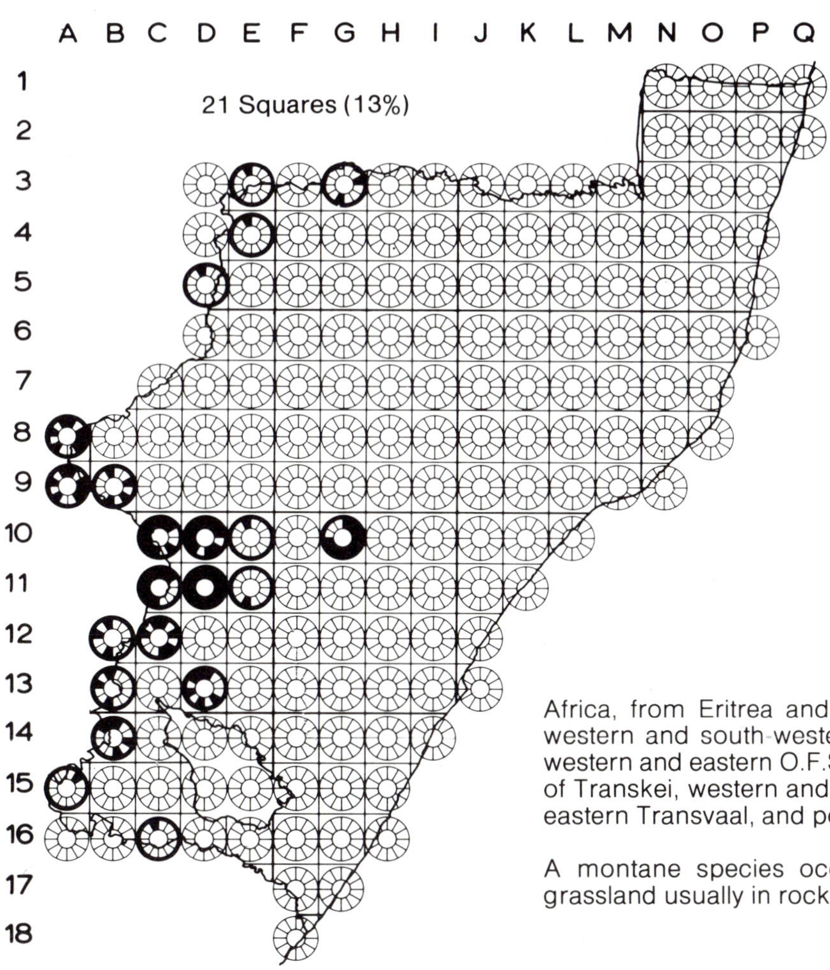

176 Greywing Francolin

Francolinus africanus

Africa, from Eritrea and Somalia to the Cape. In Southern Africa, the western and south-western Cape, east to the eastern Cape and south-western and eastern O.F.S.; also Lesotho (highlands) and adjacent districts of Transkei, western and upper Natal (including East Griqualand), south-eastern Transvaal, and possibly western Swaziland.

A montane species occurring above 1500 m a.s.l., inhabiting open grassland usually in rocky terrain. Breeds during the summer.

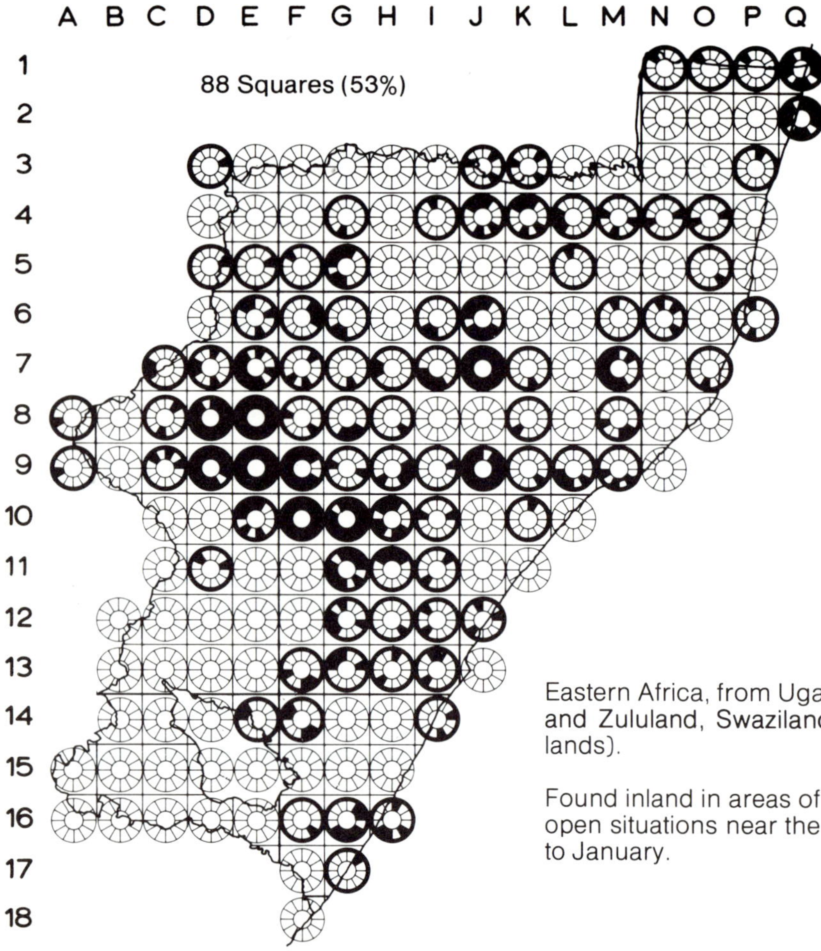

177 Shelley's Francolin

Francolinus shelleyi

Eastern Africa, from Uganda and Kenya to Natal. In Southern Africa, Natal and Zululand, Swaziland, Transvaal, Rhodesia and Moçambique (highlands).

Found inland in areas of grassland associated with thornveld, but in more open situations near the coast; also cultivated fields. Breeds from August to January.

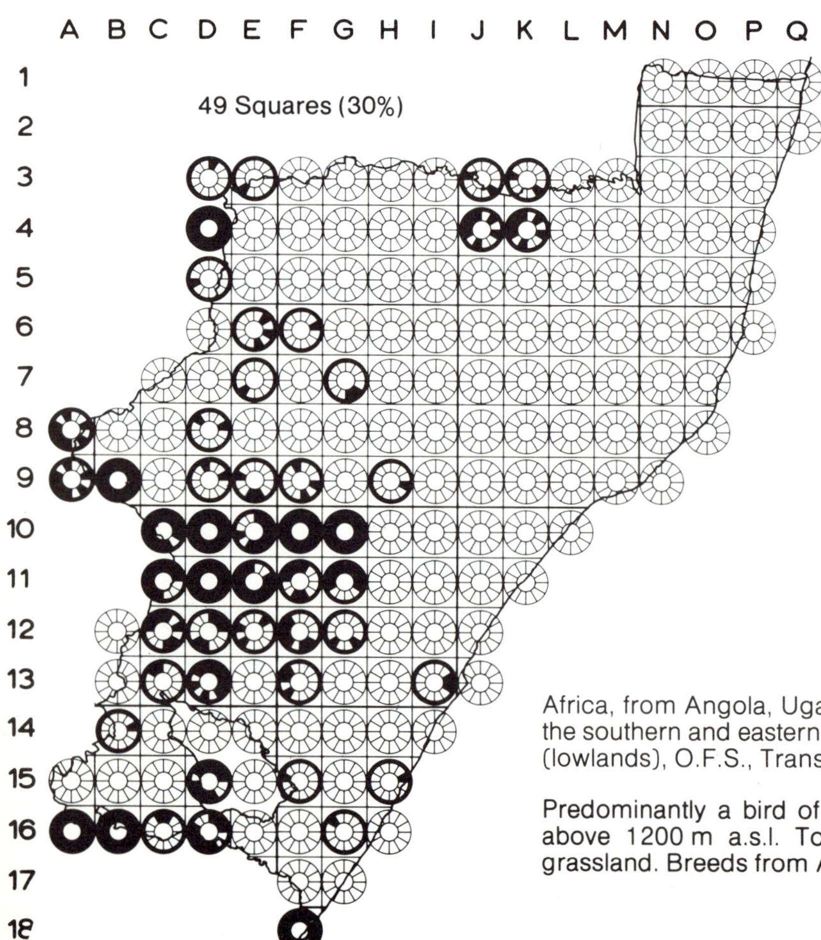

178 Redwing Francolin

Francolinus levaillantii

Africa, from Angola, Uganda and Kenya to the Cape. In Southern Africa, the southern and eastern Cape, Transkei, Natal, western Zululand, Lesotho (lowlands), O.F.S., Transvaal (highveld) and western Swaziland.

Predominantly a bird of open grassland and cultivated fields, generally above 1200 m a.s.l. Towards the coast in the south found in stony grassland. Breeds from August to February.

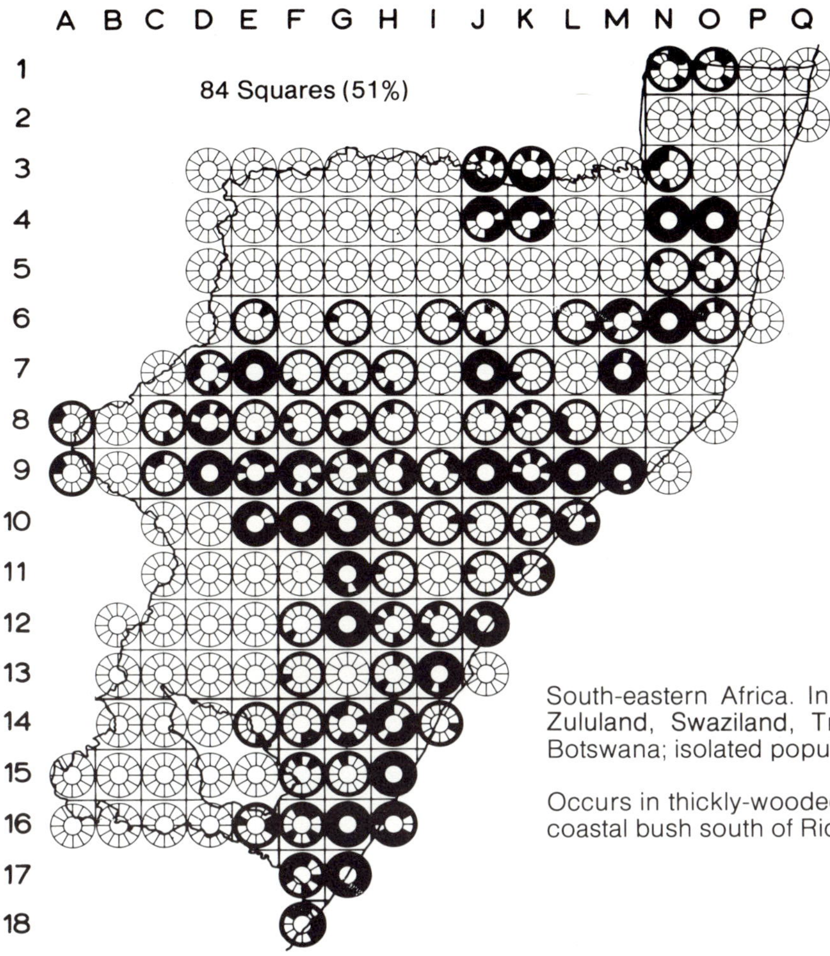

183 Natal Francolin

Francolinus natalensis

South-eastern Africa. In Southern Africa, Transkei (coastal), Natal and Zululand, Swaziland, Transvaal, Moçambique, Rhodesia and eastern Botswana; isolated population in the northern Cape and western O.F.S.

Occurs in thickly-wooded valleys, and dense thornveld on slopes; also in coastal bush south of Richards Bay. Breeds throughout the year.

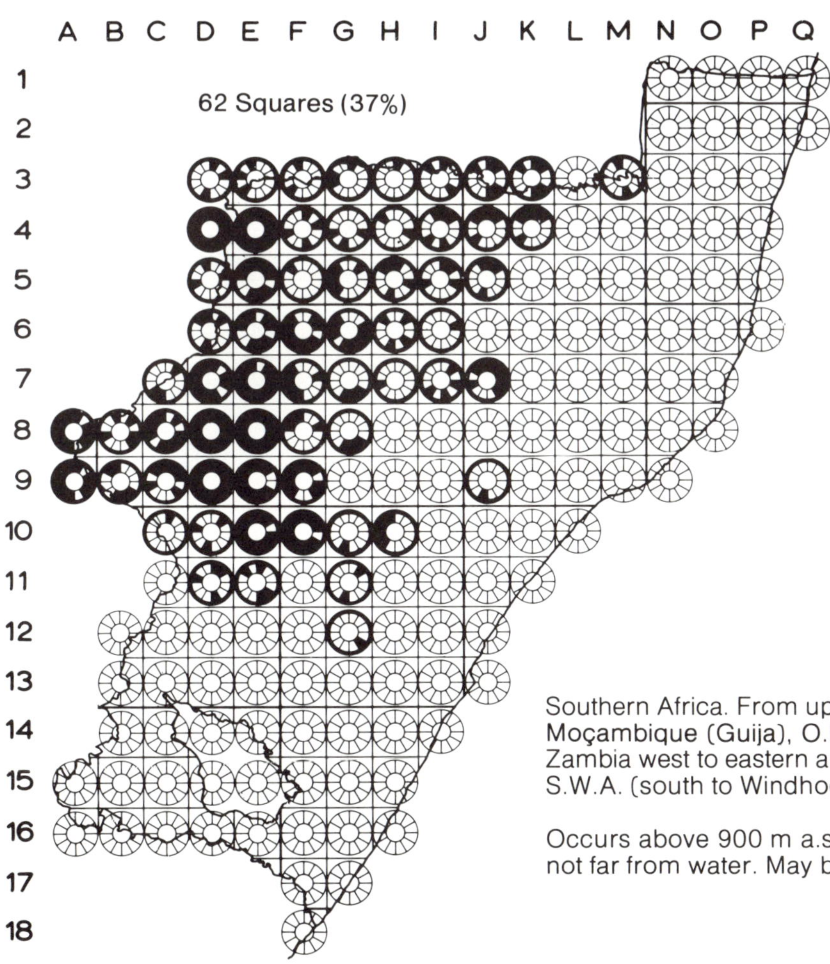

185 Swainson's Francolin

Francolinus swainsonii

Southern Africa. From upper Natal, western Zululand, Swaziland, southern Moçambique (Guija), O.F.S., Transvaal, Rhodesia, southern and eastern Zambia west to eastern and northern Botswana, the Caprivi Strip, northern S.W.A. (south to Windhoek) and adjacent Angola.

Occurs above 900 m a.s.l. in open grassland or cultivated areas, usually not far from water. May breed throughout the year.

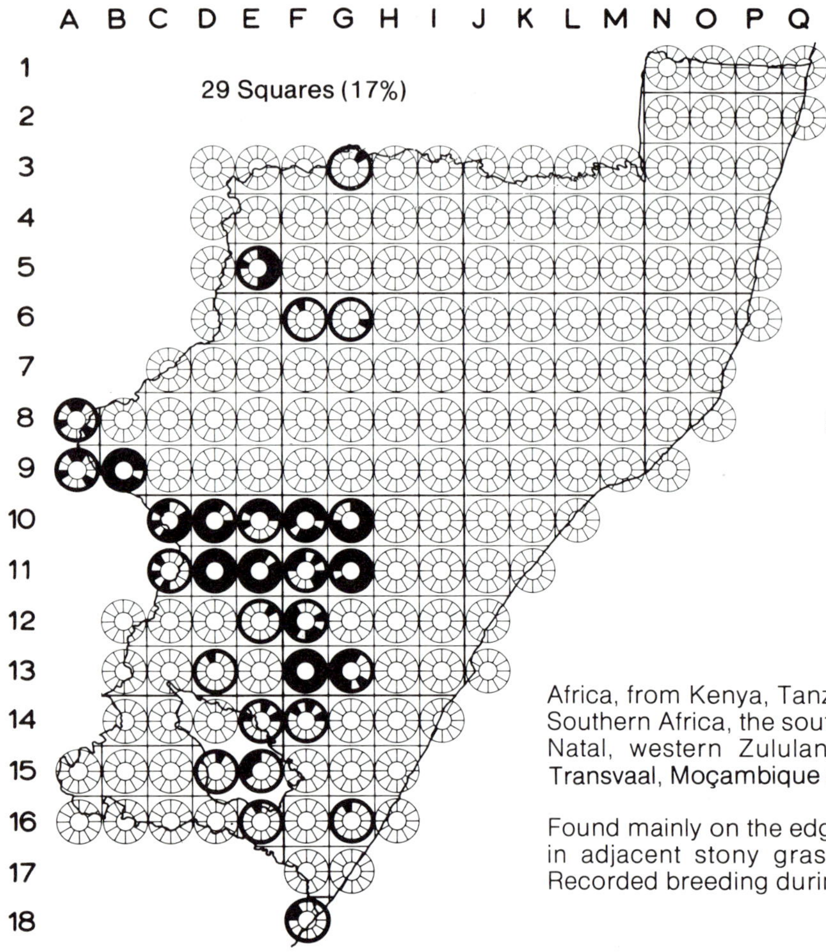

188 Red-necked Francolin

Francolinus afer

Africa, from Kenya, Tanzania, southern Zaïre and Angola to the Cape. In Southern Africa, the southern and eastern Cape, East Griqualand, interior Natal, western Zululand, western Swaziland, eastern and northern Transvaal, Moçambique and Rhodesia on the plateau also northern S.W.A.

Found mainly on the edge of indigenous forests above 1000 m a.s.l.; also in adjacent stony grassland, cultivated fields and exotic vegetation. Recorded breeding during the summer and winter.

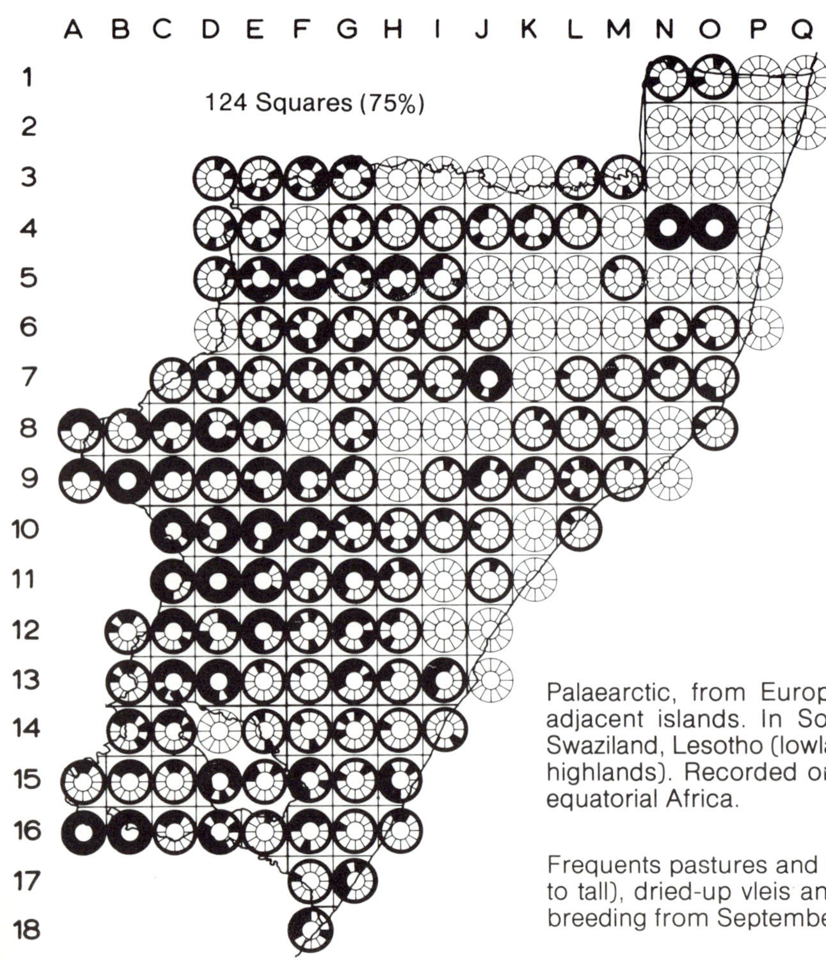

189 Common Quail

Coturnix coturnix

Palaearctic, from Europe to eastern Asia and Japan, and Africa and adjacent islands. In Southern Africa, the Cape, Natal and Zululand, Swaziland, Lesotho (lowlands), O.F.S. and Transvaal; also S.W.A. (Damara highlands). Recorded on passage in Rhodesia. Some winter in western equatorial Africa.

Frequents pastures and cultivated fields; also open grassland (from short to tall), dried-up vleis and the more open savanna woodland. Recorded breeding from September to April.

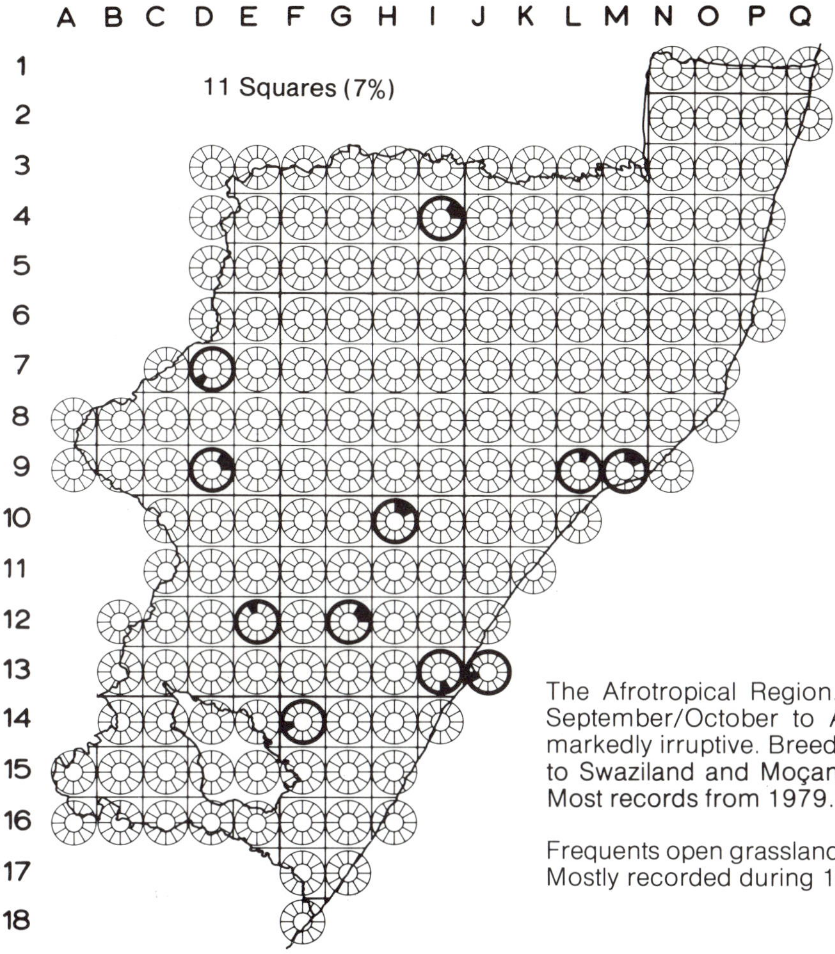

190 Harlequin Quail

Coturnix delegorguei

The Afrotropical Region. In Southern Africa, a summer resident from September/October to April; some present throughout the year, and markedly irruptive. Breeds in the eastern Cape, Natal and Zululand, north to Swaziland and Moçambique, west to Transvaal, O.F.S., the northern Most records from 1979. May breed.

Frequents open grassland, cultivated fields and occasionally open savanna. Mostly recorded during 1979. May breed.

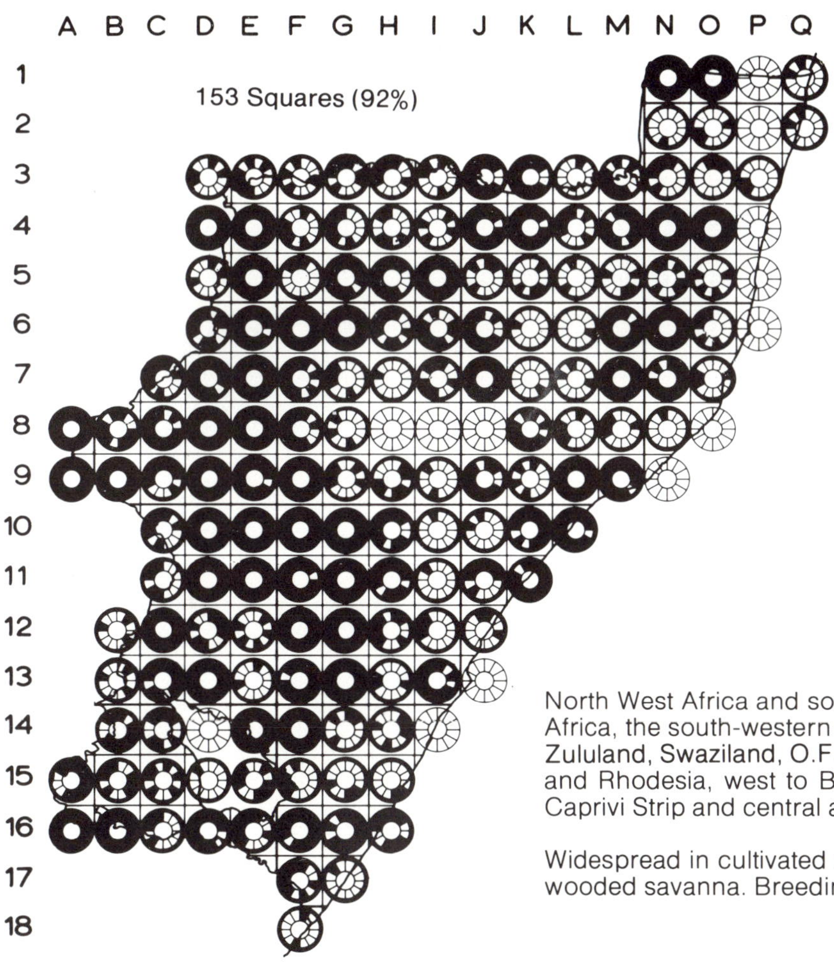

192 Crowned Guineafowl

Numida meleagris

North West Africa and south of the Sahara; southern Arabia. In Southern Africa, the south-western Cape (introduced), the eastern Cape, Natal and Zululand, Swaziland, O.F.S., Lesotho (lowlands), Transvaal, Moçambique and Rhodesia, west to Botswana (Kalahari Desert and Ngamiland), the Caprivi Strip and central and northern S.W.A.

Widespread in cultivated areas and open grassland; also found in lightly-wooded savanna. Breeding recorded from October to March.

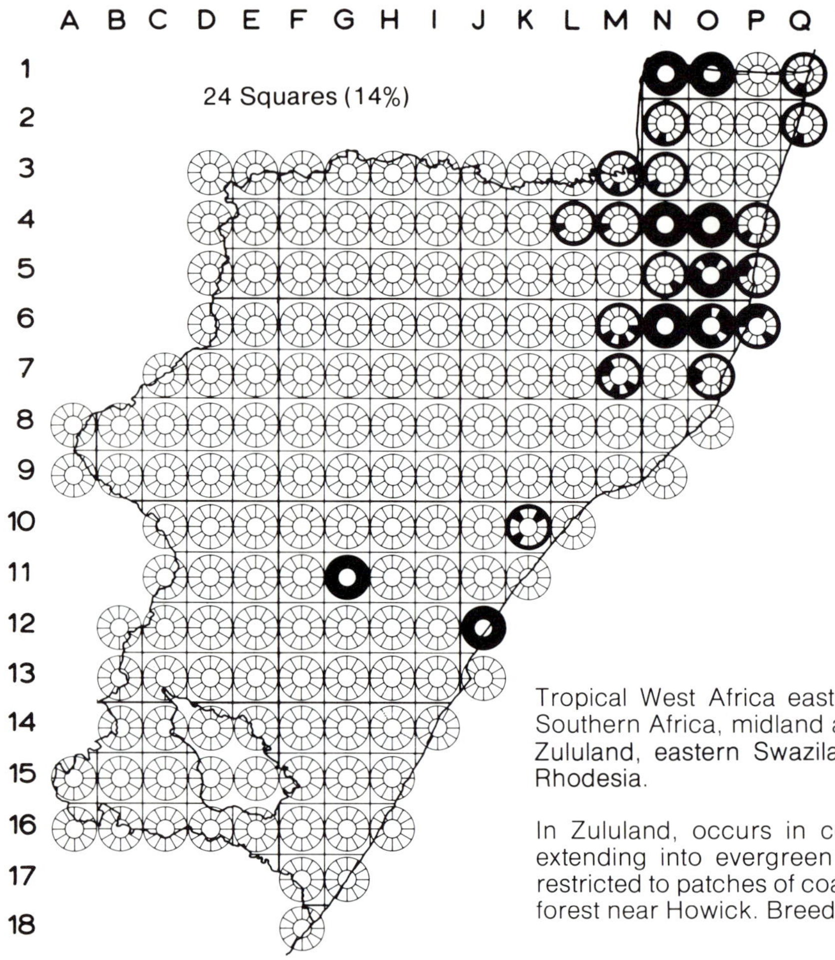

193 Crested Guineafowl

Guttera pucherani

Tropical West Africa east to East Africa, south in the east to Natal. In Southern Africa, midland and coastal Natal (north from Durban), eastern Zululand, eastern Swaziland, Moçambique, and eastern and northern Rhodesia.

In Zululand, occurs in coastal bush, dense woodland and thornveld, extending into evergreen forest in the Lebombo Range. In the south, restricted to patches of coastal forest. Isolated population found in mistbelt forest near Howick. Breeding recorded from October to January.

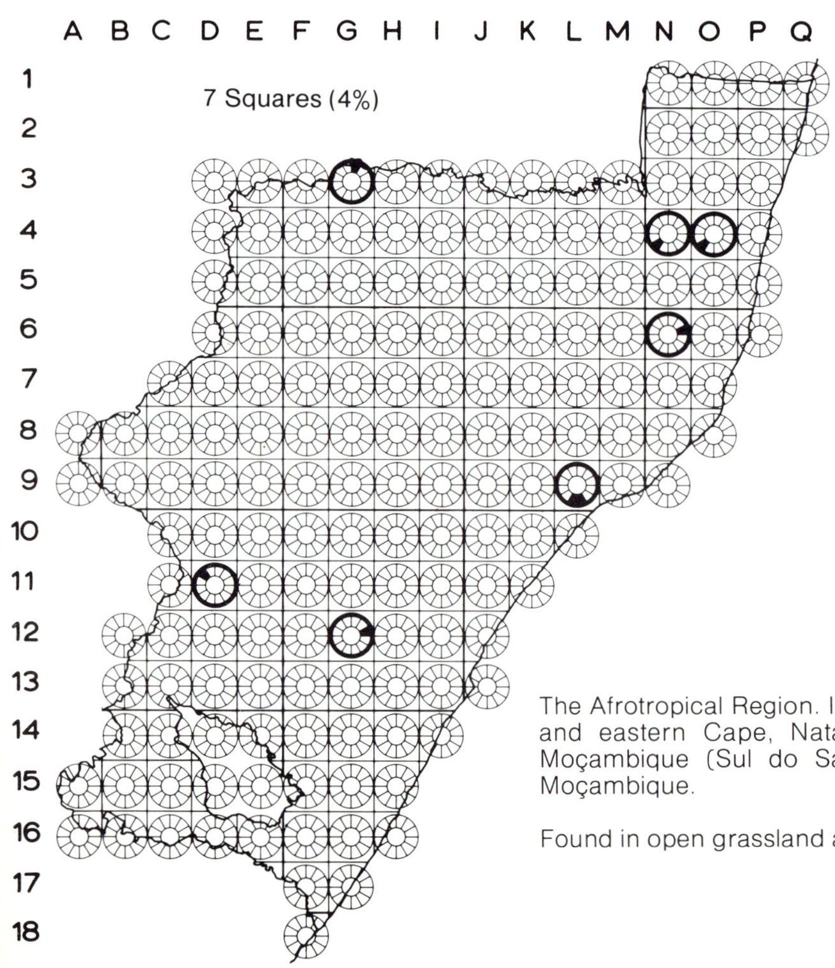

194 Hottentot Button-quail

Turnix hottentotta

The Afrotropical Region. In Southern Africa, the south-western, southern and eastern Cape, Natal and Zululand, Swaziland, Transvaal and Moçambique (Sul do Save). Uncommon in Rhodesia and adjacent Moçambique.

Found in open grassland and old lands.

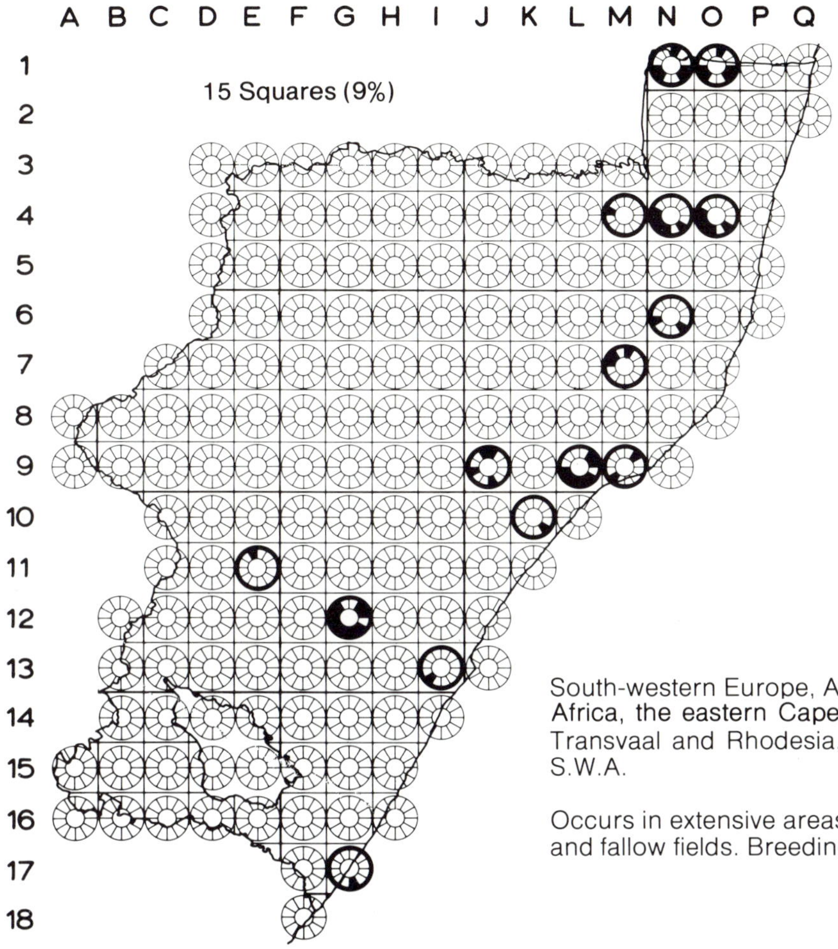

196 Kurrichane Buttonquail

Turnix sylvatica

South-western Europe, Africa and southern Asia to Australia. In Southern Africa, the eastern Cape, Natal and Zululand, Swaziland, Moçambique, Transvaal and Rhodesia, west to north-eastern Botswana and northern S.W.A.

Occurs in extensive areas of tall grass, patches of grass in open savanna, and fallow fields. Breeding recorded from October to March.

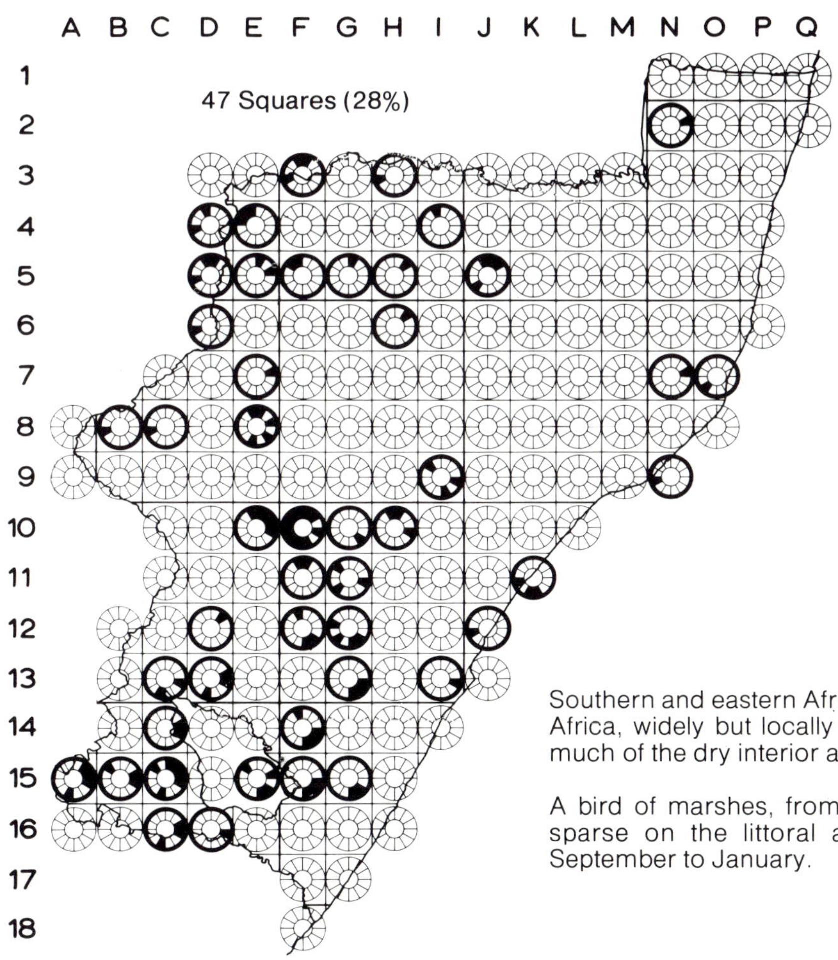

197 Water Rail

Rallus caerulescens

Southern and eastern Africa and islands in the Gulf of Guinea. In Southern Africa, widely but locally distributed in suitable habitat, and absent from much of the dry interior and west.

A bird of marshes, from large vleis to small ponds fringed with reeds; sparse on the littoral and coastal plains. Breeding recorded from September to January.

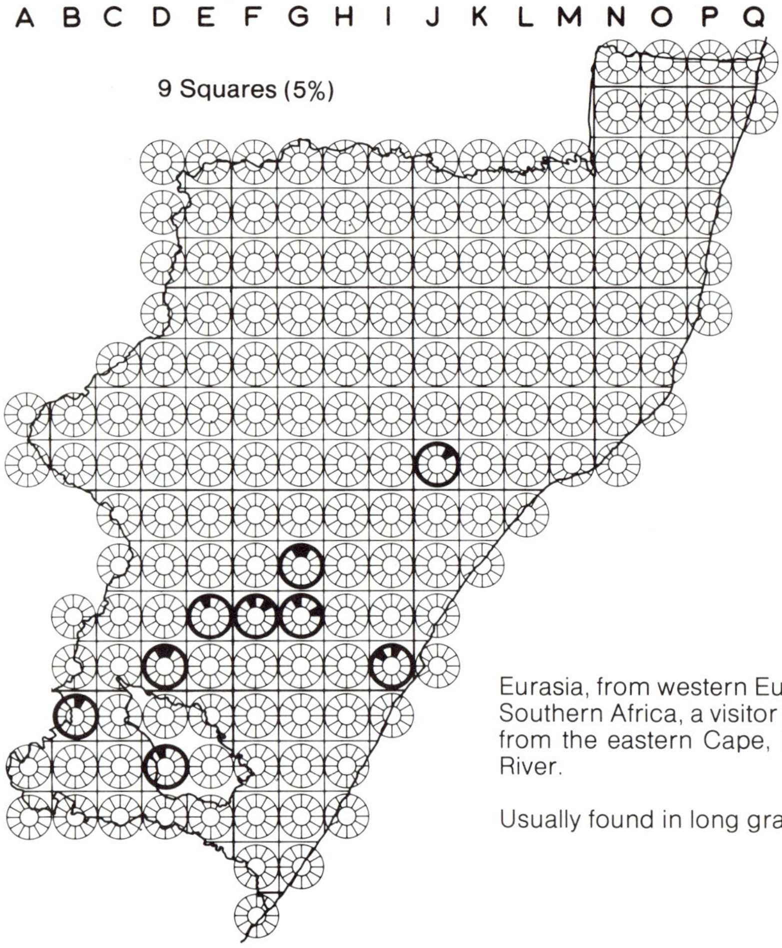

198 Corncrake

Crex crex

Eurasia, from western Europe east to central Siberia, wintering in Africa. In Southern Africa, a visitor from October to March mainly to the eastern half, from the eastern Cape, Natal and Zululand, northwards to the Zambesi River.

Usually found in long grass, lucerne and fallow fields.

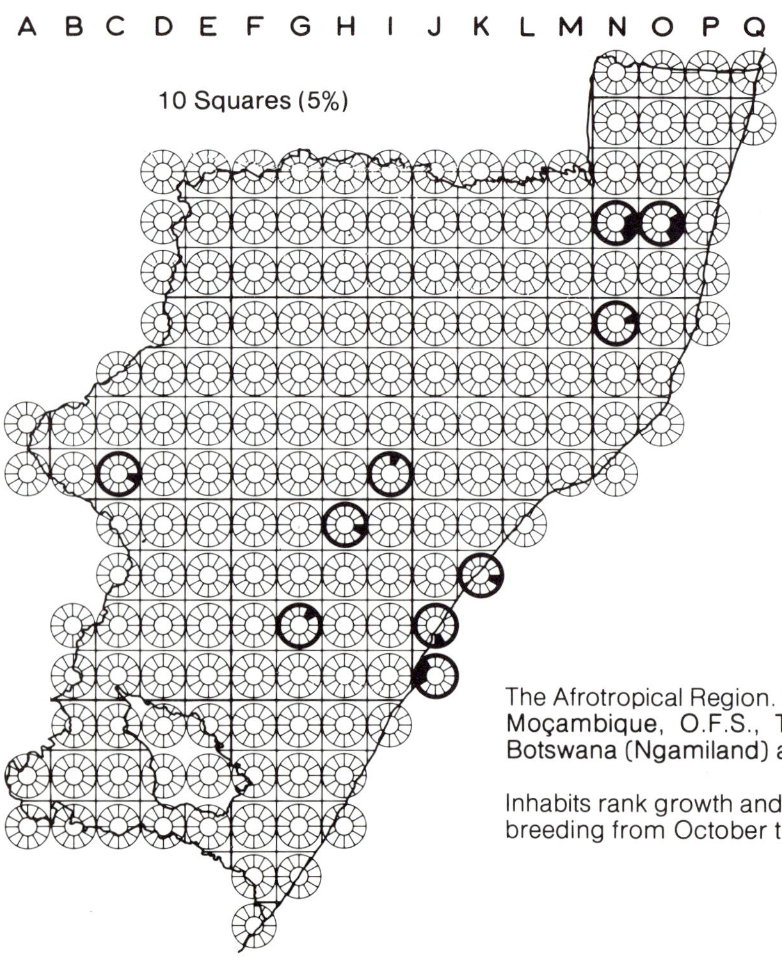

199 African Crake

Crex egregia

The Afrotropical Region. In Southern Africa, Natal and Zululand, Swaziland, Moçambique, O.F.S., Transvaal and Rhodesia, west in the north to Botswana (Ngamiland) and S.W.A.

Inhabits rank growth and reedbeds on the periphery of marshes. Recorded breeding from October to February.

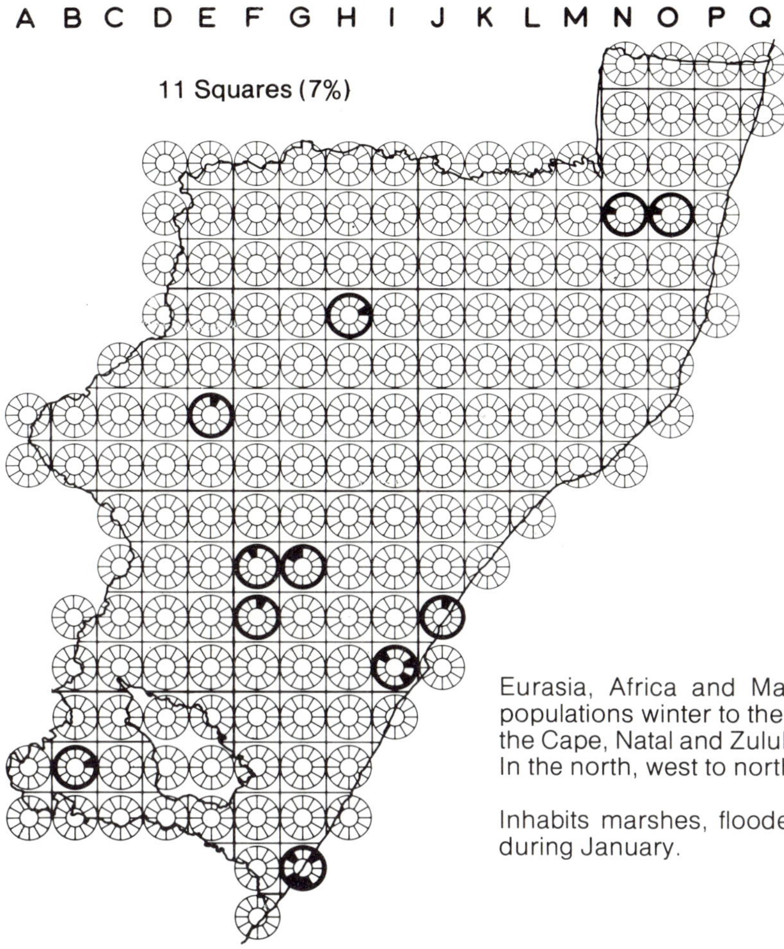

202 Baillon's Crake

Porzana pusilla

Eurasia, Africa and Madagascar, south-east to Australasia; northern populations winter to the south of the breeding range. In Southern Africa, the Cape, Natal and Zululand, northwards in the east to the Zambesi River. In the north, west to northern Botswana and S.W.A.

Inhabits marshes, flooded grassland and reedbeds. Breeding recorded during January.

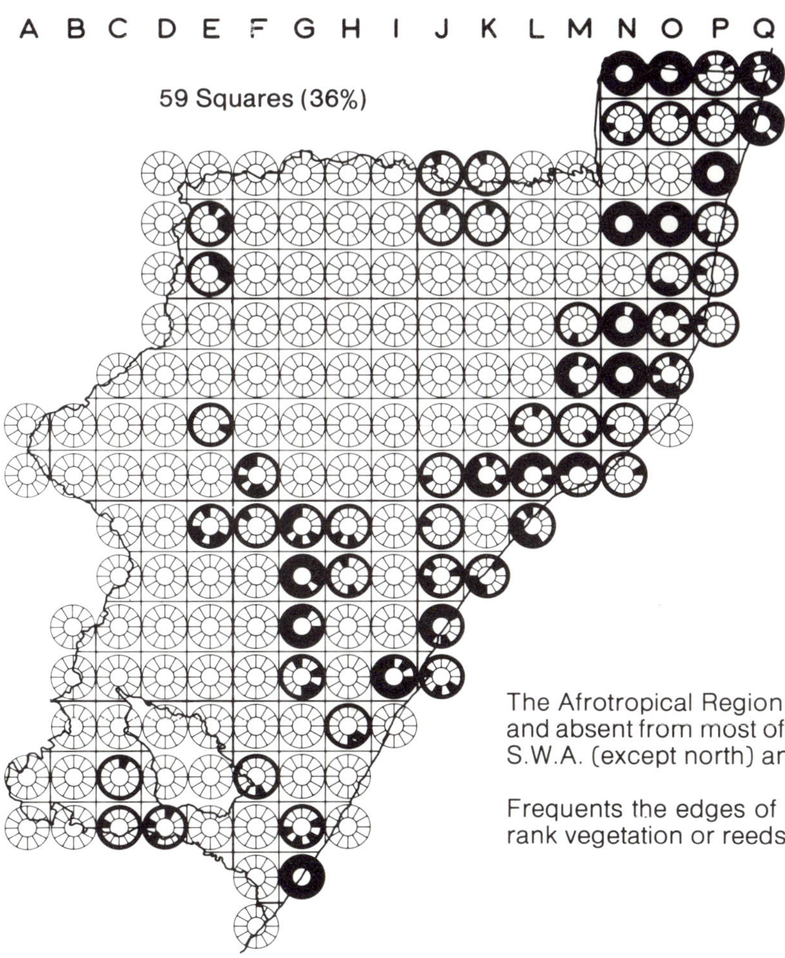

203 Black Crake

Amaurornis flavirostris

The Afrotropical Region. In Southern Africa, widely but locally distributed and absent from most of the dry interior and northern districts of the Cape, S.W.A. (except north) and southern Botswana.

Frequents the edges of ponds, dams, pans and smaller rivers, wherever rank vegetation or reeds occur. Breeds during the summer.

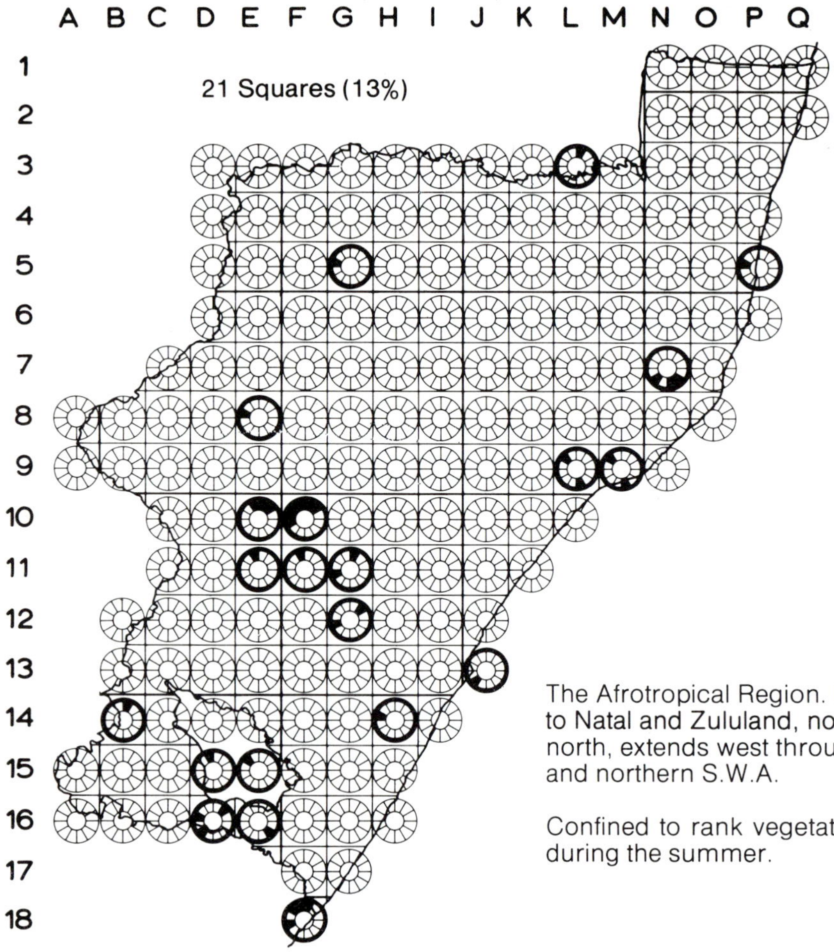

205 Red-chested Flufftail

Sarothrura rufa

The Afrotropical Region. In Southern Africa, the south-western Cape east to Natal and Zululand, northwards in the east to the Zambesi River. In the north, extends west through Rhodesia to northern Botswana (Nokaneng) and northern S.W.A.

Confined to rank vegetation and reedbeds in marshy situations. Breeds during the summer.

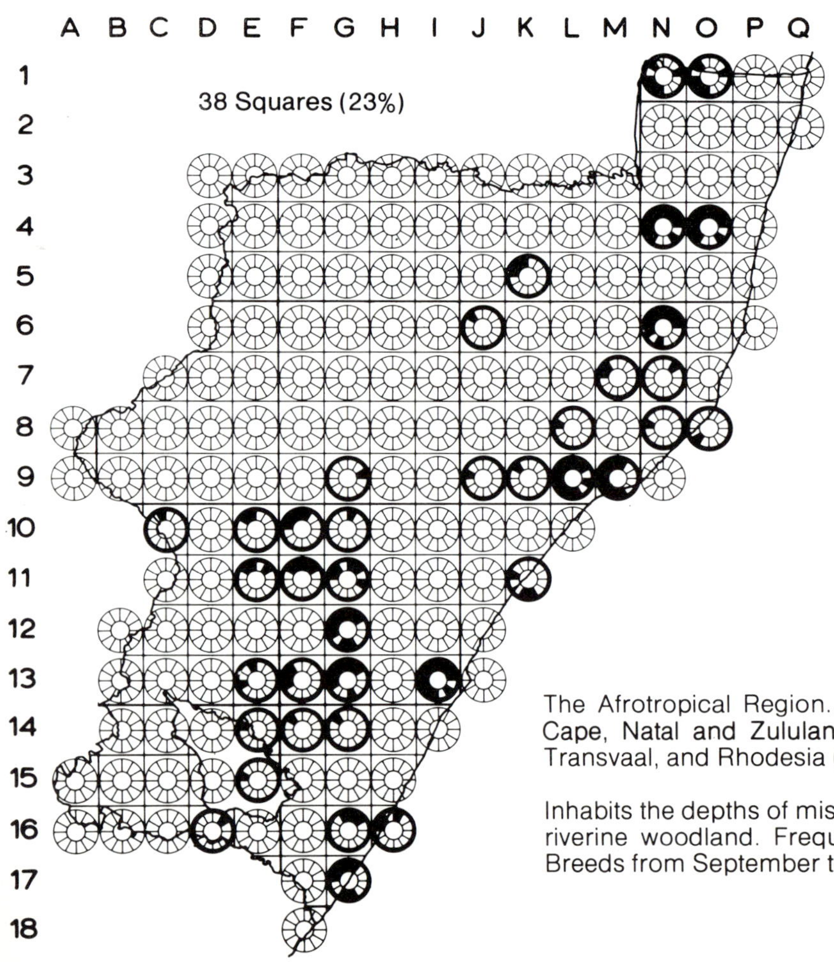

206 Buff-spotted Flufftail

Sarothrura elegans

The Afrotropical Region. In Southern Africa, the southern and eastern Cape, Natal and Zululand, Swaziland, southern Moçambique, eastern Transvaal, and Rhodesia (Mashonaland).

Inhabits the depths of mistbelt and coastal evergreen forest; also in dense riverine woodland. Frequents suburban gardens in the moister areas. Breeds from September to March.

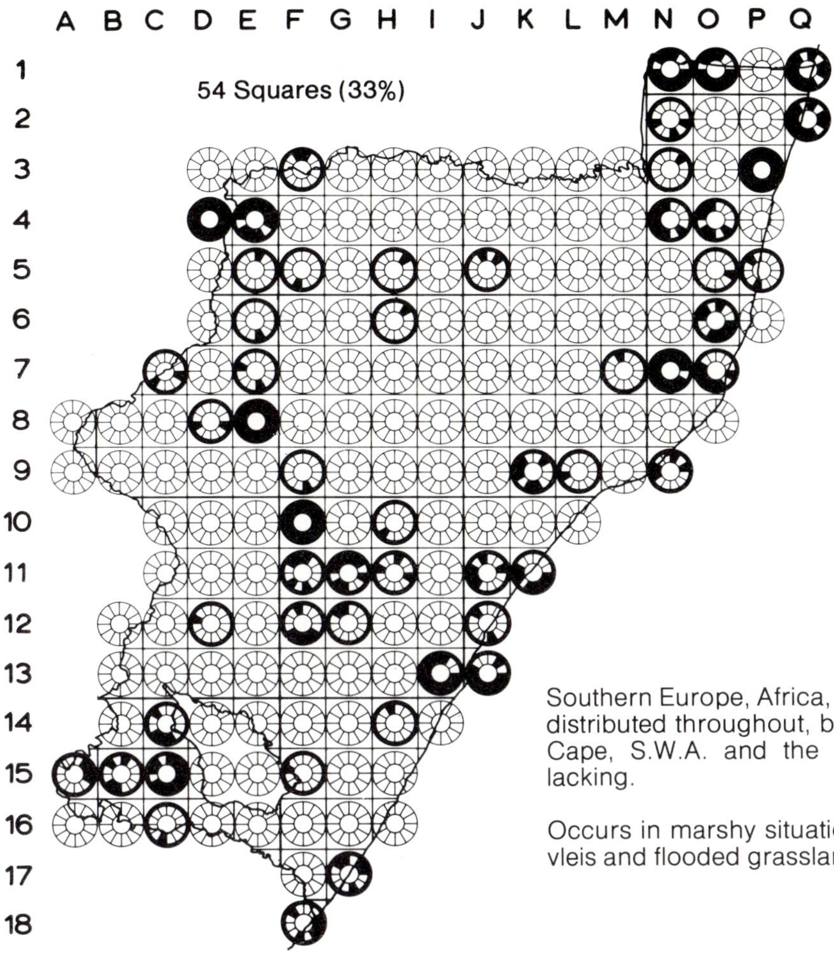

208 Purple Gallinule

Porphyrio porphyrio

Southern Europe, Africa, Madagascar and Australasia. In Southern Africa, distributed throughout, but absent from most of the interior and northern Cape, S.W.A. and the Kalahari, Botswana, where suitable habitat is lacking.

Occurs in marshy situations, from small reed-fringed ponds to extensive vleis and flooded grassland. Breeds from August to April.

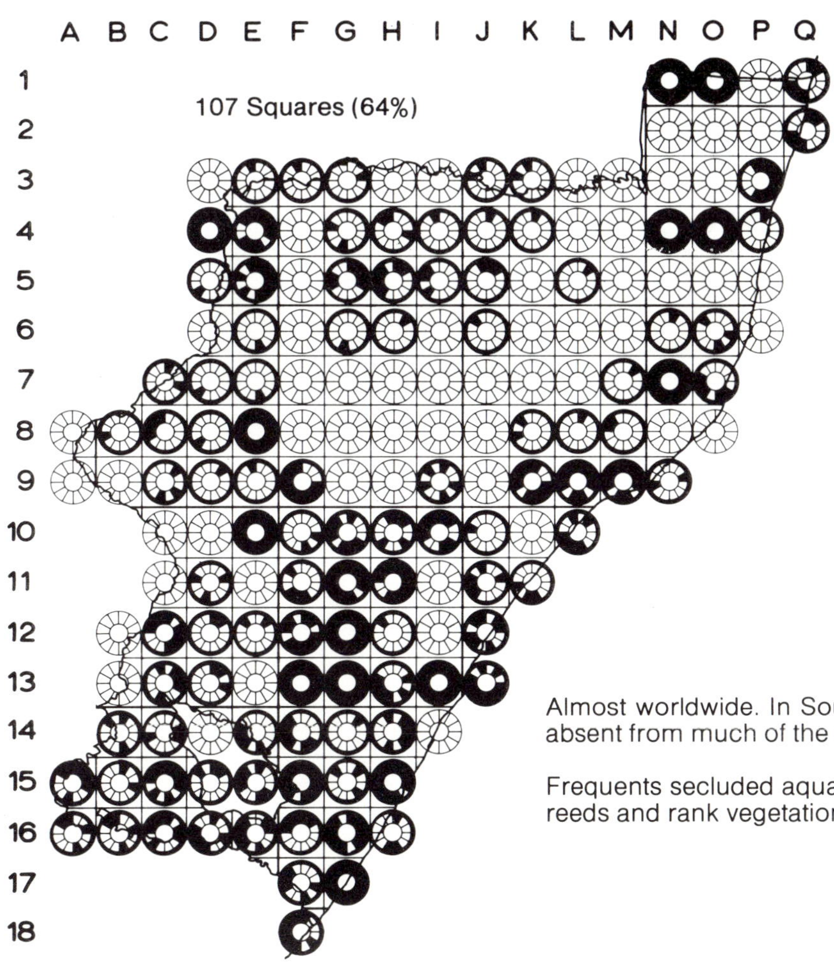

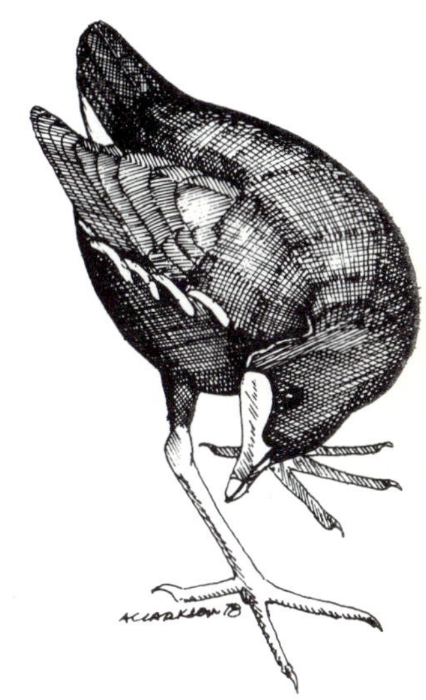

210 Moorhen

Gallinula chloropus

Almost worldwide. In Southern Africa, widely but locally distributed, and absent from much of the dry interior and west.

Frequents secluded aquatic habitats where standing water is fringed with reeds and rank vegetation. Breeds throughout the year.

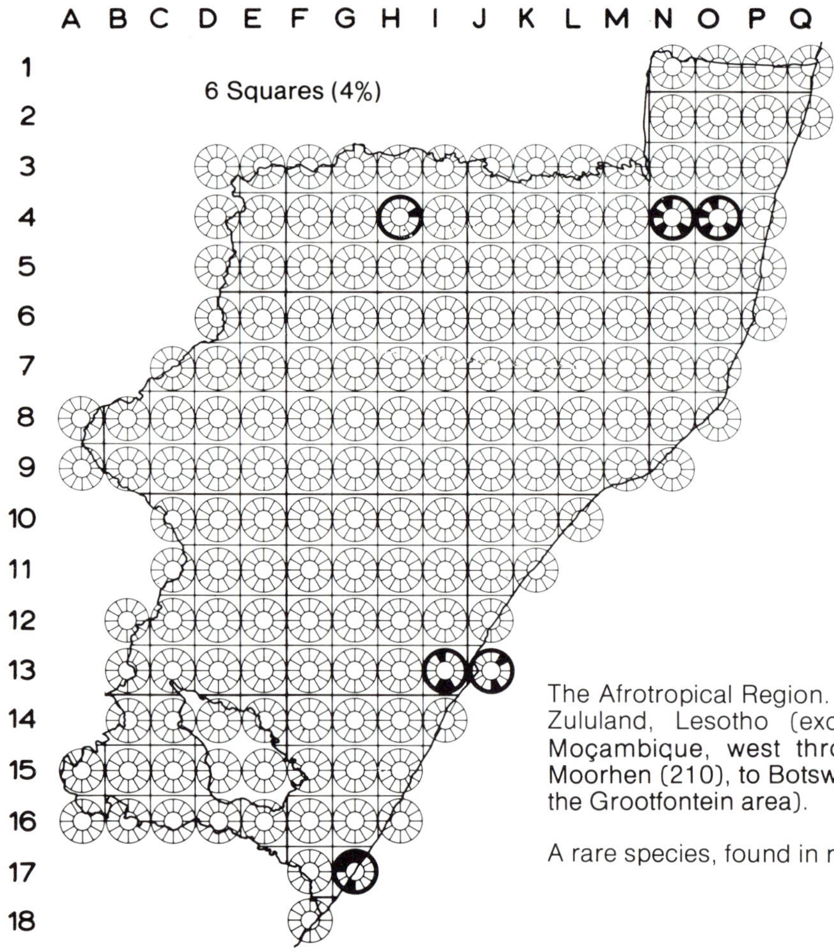

211 Lesser Moorhen

Gallinula angulata

The Afrotropical Region. In Southern Africa, the eastern Cape, Natal and Zululand, Lesotho (exceptional), O.F.S., Transvaal, Swaziland and Moçambique, west through Rhodesia (where more common than Moorhen (210), to Botswana (Ngamiland) and S.W.A. (Ovamboland and the Grootfontein area).

A rare species, found in rank growth and reedbeds surrounding ponds.

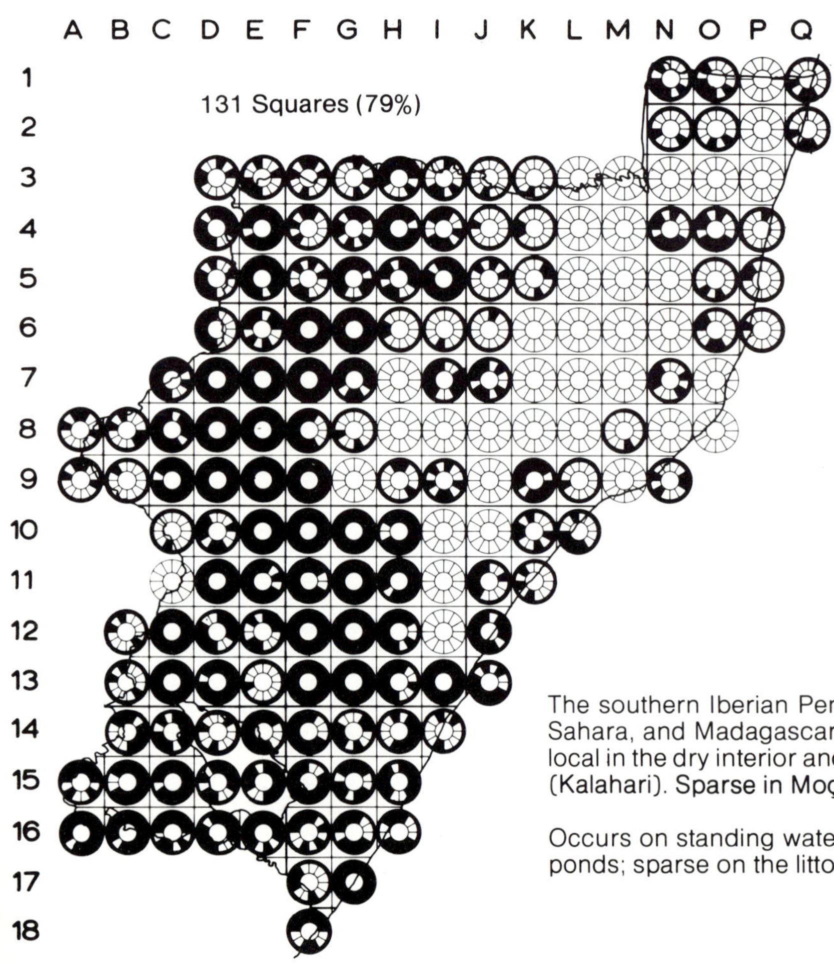

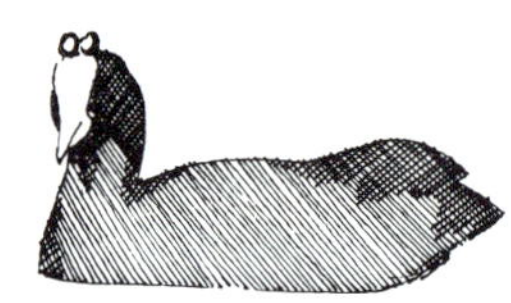

212 Red-knobbed Coot

Fulica cristata

The southern Iberian Peninsula, Africa in the northwest and south of the Sahara, and Madagascar. In Southern Africa, distributed throughout, but local in the dry interior and northern Cape, southern S.W.A. and Botswana (Kalahari). Sparse in Moçambique, where mainly a non-breeding visitor.

Occurs on standing water of any size, from large open dams to secluded ponds; sparse on the littoral plain. Breeds throughout the year.

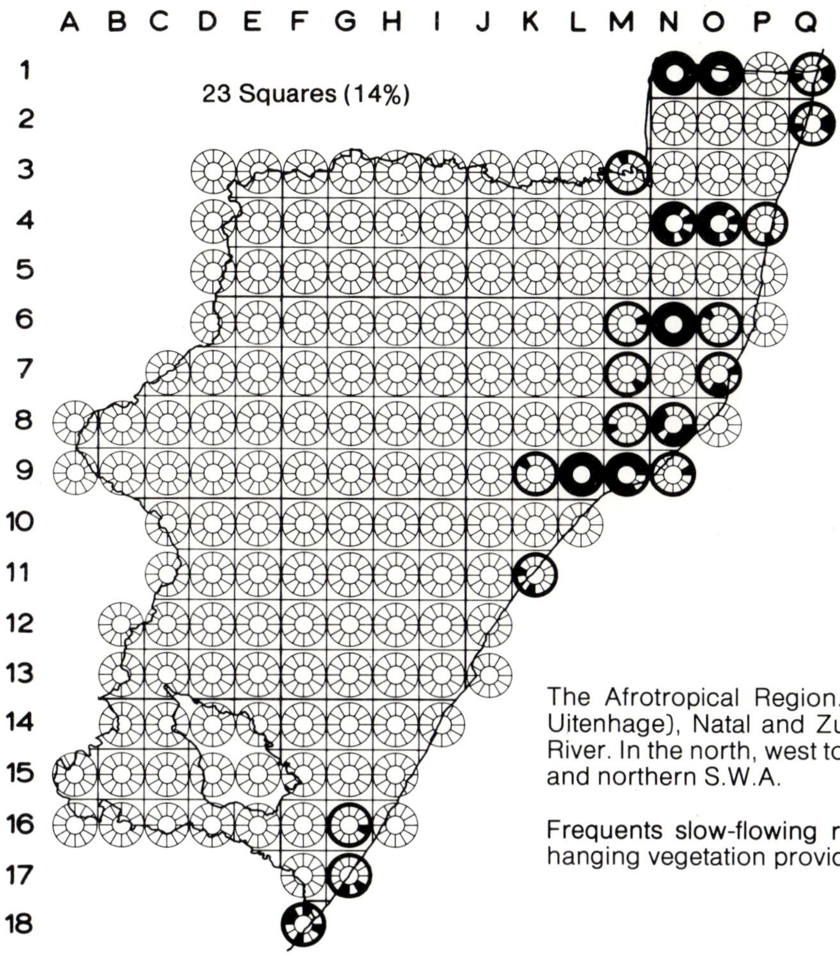

213 African Finfoot

Podica senegalensis

The Afrotropical Region. In Southern Africa, the eastern Cape (from Uitenhage), Natal and Zululand, northwards in the east to the Zambesi River. In the north, west to Rhodesia, northern Botswana, the Caprivi Strip and northern S.W.A.

Frequents slow-flowing rivers, lagoons and ponds, where dense overhanging vegetation provides cover. Breeds during the summer.

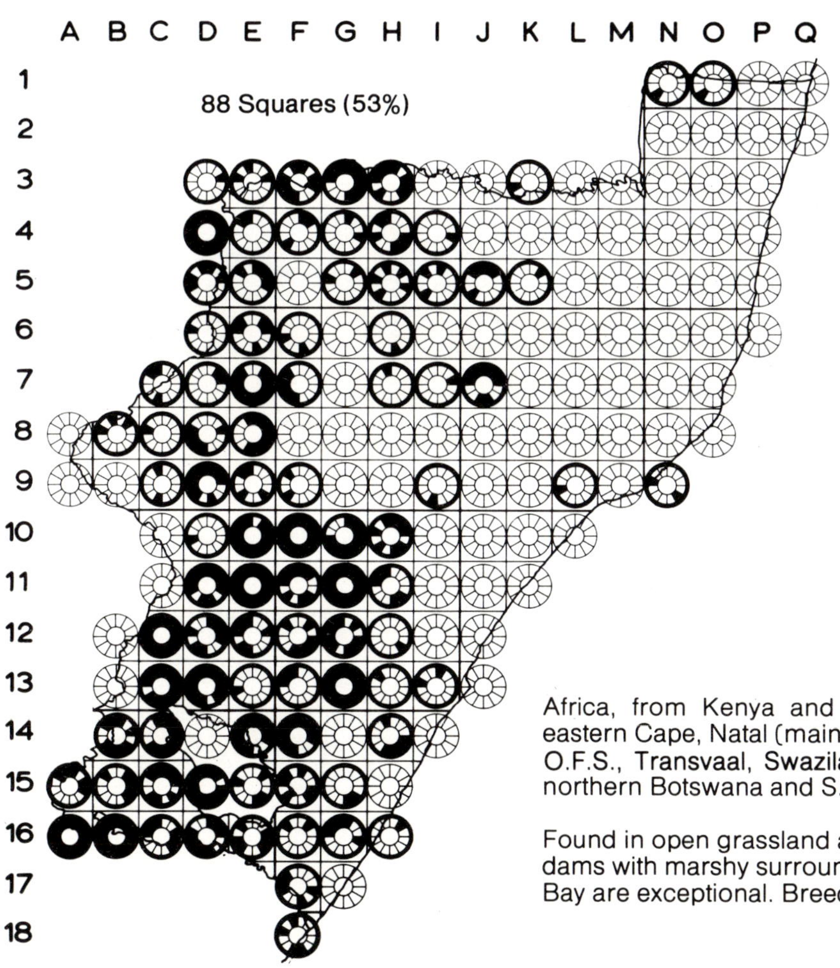

214 Crowned Crane

Balearica regulorum

Africa, from Kenya and Uganda southward. In Southern Africa, the eastern Cape, Natal (mainly interior) and western Zululand, northwards to O.F.S., Transvaal, Swaziland, Moçambique and Rhodesia, and west to northern Botswana and S.W.A.

Found in open grassland and cultivated fields in the vicinity of vleis, and at dams with marshy surroundings. Records from Ndumu G.R. and Richards Bay are exceptional. Breeding recorded from December to April.

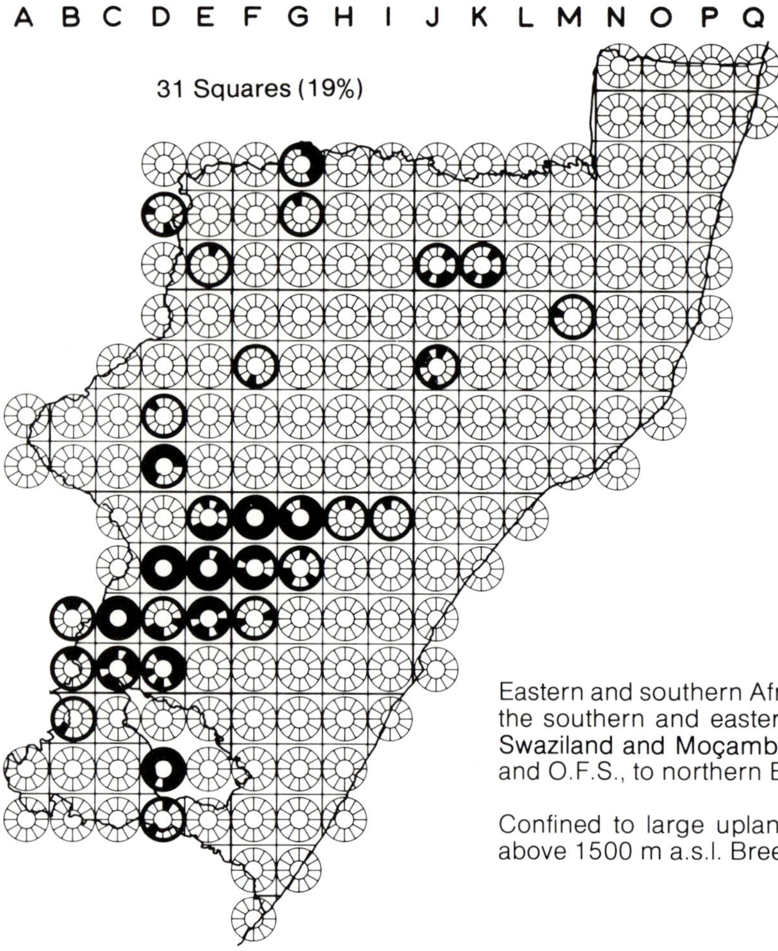

215 Wattled Crane

Grus carunculatus

Eastern and southern Africa from Somalia to the Cape. In Southern Africa, the southern and eastern Cape (rare), Natal interior, western Zululand, Swaziland and Moçambique, west (locally) through Rhodesia, Transvaal and O.F.S., to northern Botswana and northern S.W.A.

Confined to large upland vleis, surrounded by short grassveld, usually above 1500 m a.s.l. Breeds throughout the year.

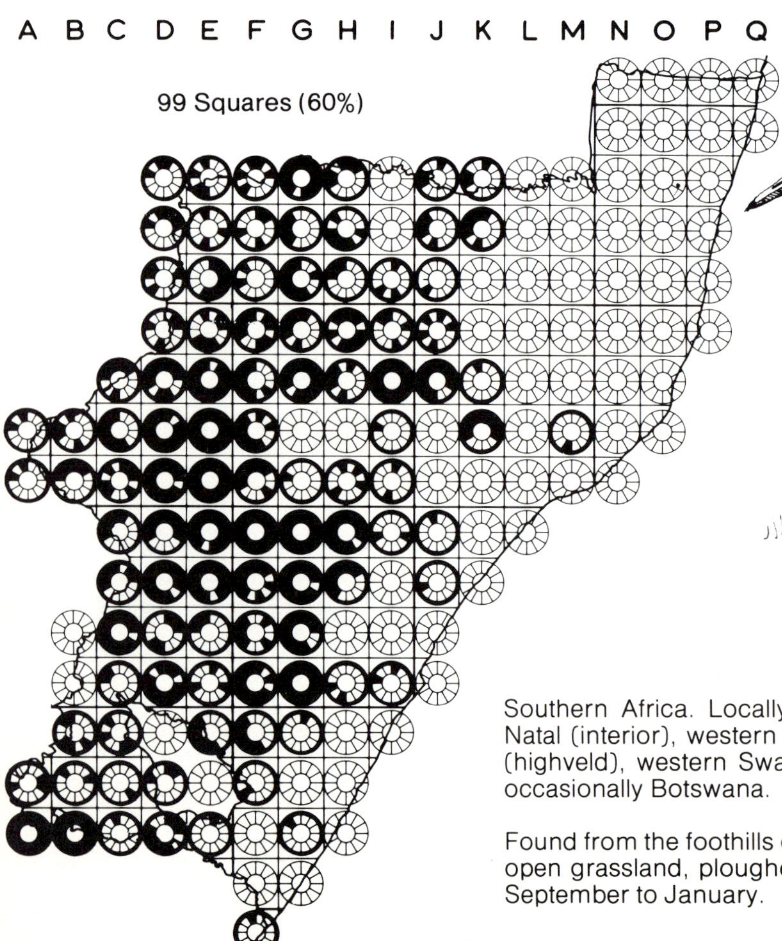

216 Blue Crane

Anthropoides paradisea

Southern Africa. Locally from the south-western and eastern Cape to Natal (interior), western Zululand, Lesotho (lowlands), O.F.S., Transvaal (highveld), western Swaziland, and S.W.A. (south and east of Etosha); occasionally Botswana.

Found from the foothills of the Drakensberg down to about 600 m a.s.l., in open grassland, ploughed fields and old lands. Breeding recorded from September to January.

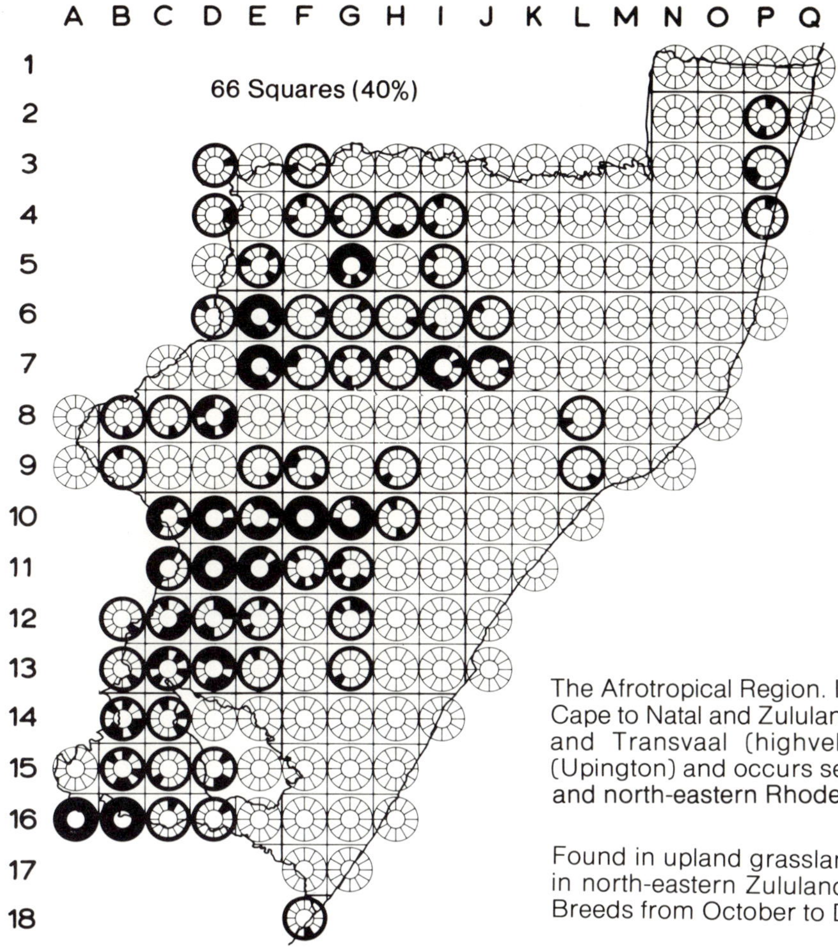

219 Stanley Bustard

Neotis denhami

The Afrotropical Region. In Southern Africa, the south-western and interior Cape to Natal and Zululand, O.F.S., Lesotho (lowlands), western Swaziland and Transvaal (highveld); also recorded from the northern Cape (Upington) and occurs seasonally in northern S.W.A., northern Botswana and north-eastern Rhodesia. Recorded from southern Moçambique.

Found in upland grassland with few trees, and lowland coastal grassland in north-eastern Zululand; in winter often takes to recently burnt slopes. Breeds from October to December.

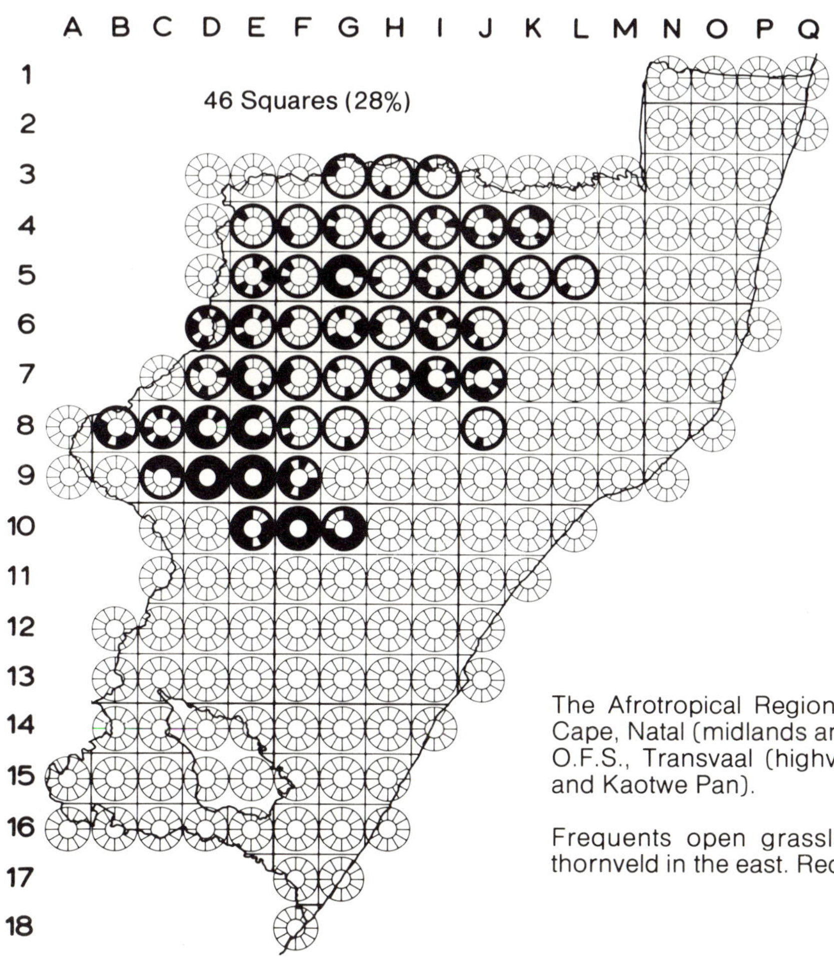

222 White-bellied Korhaan

Eupodotis cafra

The Afrotropical Region. In Southern Africa, the eastern and northern Cape, Natal (midlands and upper districts), Zululand, Lesotho (lowlands), O.F.S., Transvaal (highveld), western Swaziland and Botswana (Kanye and Kaotwe Pan).

Frequents open grassland and cultivated areas, ranging into open thornveld in the east. Recorded breeding from October to January.

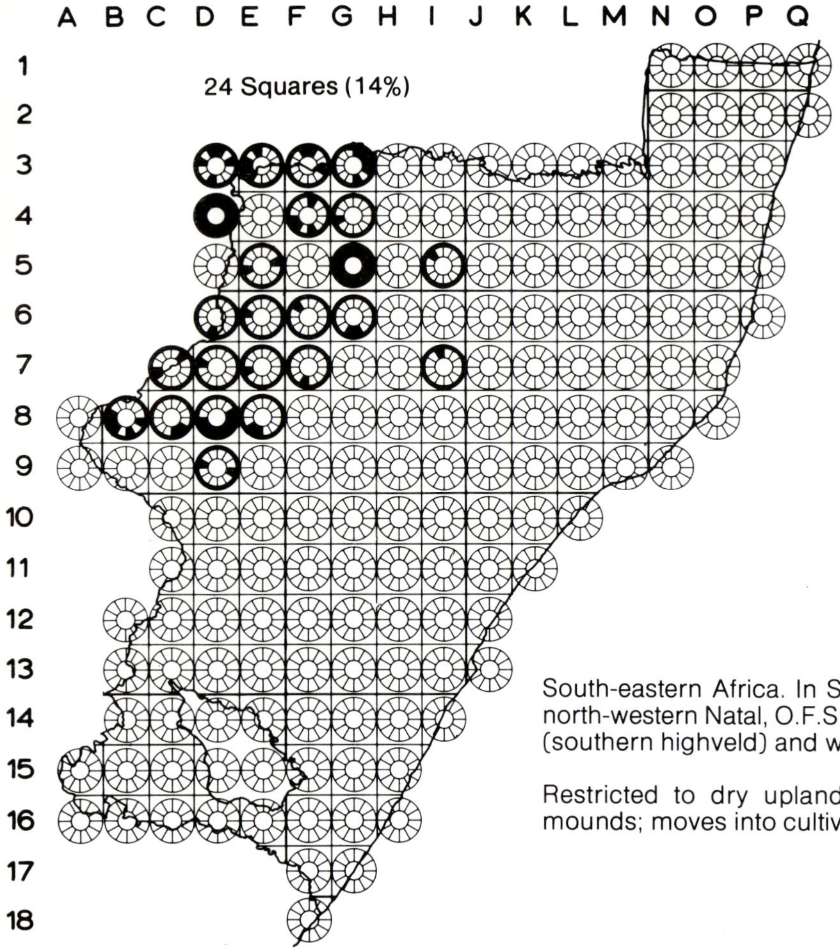

223 Blue Korhaan

Eupodotis caerulescens

South-eastern Africa. In Southern Africa, local from the eastern Cape to north-western Natal, O.F.S., Griqualand West, the northern Cape, Transvaal (southern highveld) and western Swaziland.

Restricted to dry upland grassland, typically with numerous termite mounds; moves into cultivated fields. Breeds mainly during the summer.

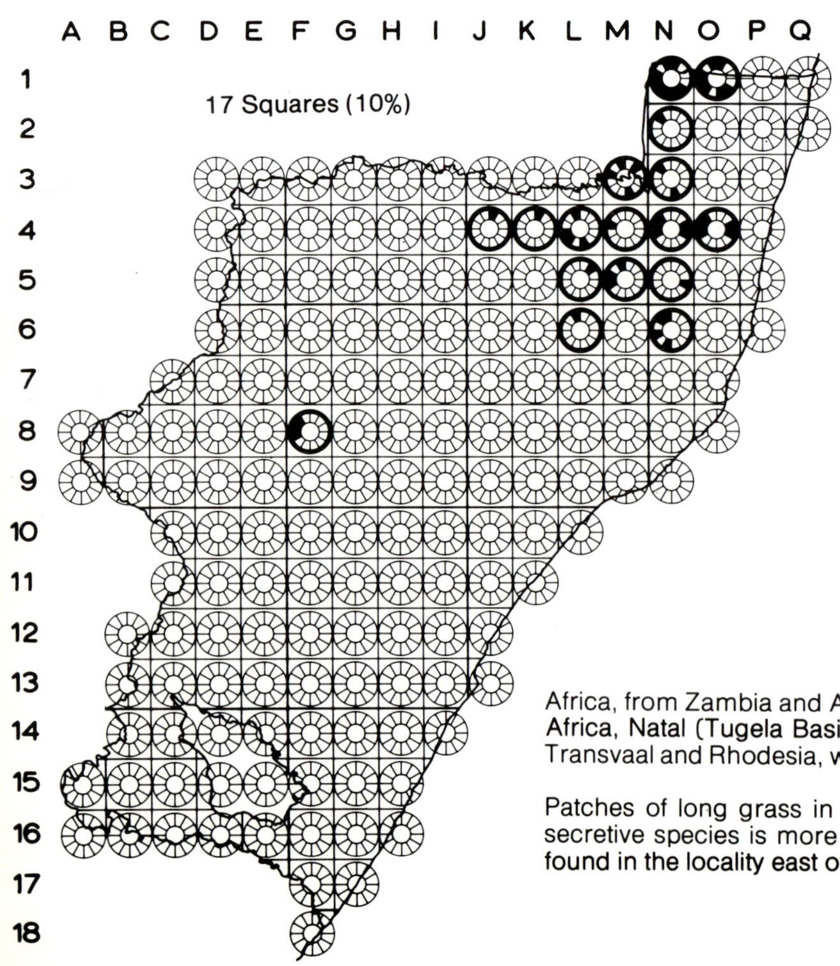

224 Red-crested Korhaan

Eupodotis ruficrista

Africa, from Zambia and Angola south to Natal and the Cape. In Southern Africa, Natal (Tugela Basin), Zululand, eastern Swaziland, Moçambique, Transvaal and Rhodesia, west to the northern Cape, Botswana and S.W.A.

Patches of long grass in open thornveld is the habitat from where this secretive species is more often heard than seen. A few birds have been found in the locality east of Ladysmith. Breeds during the summer.

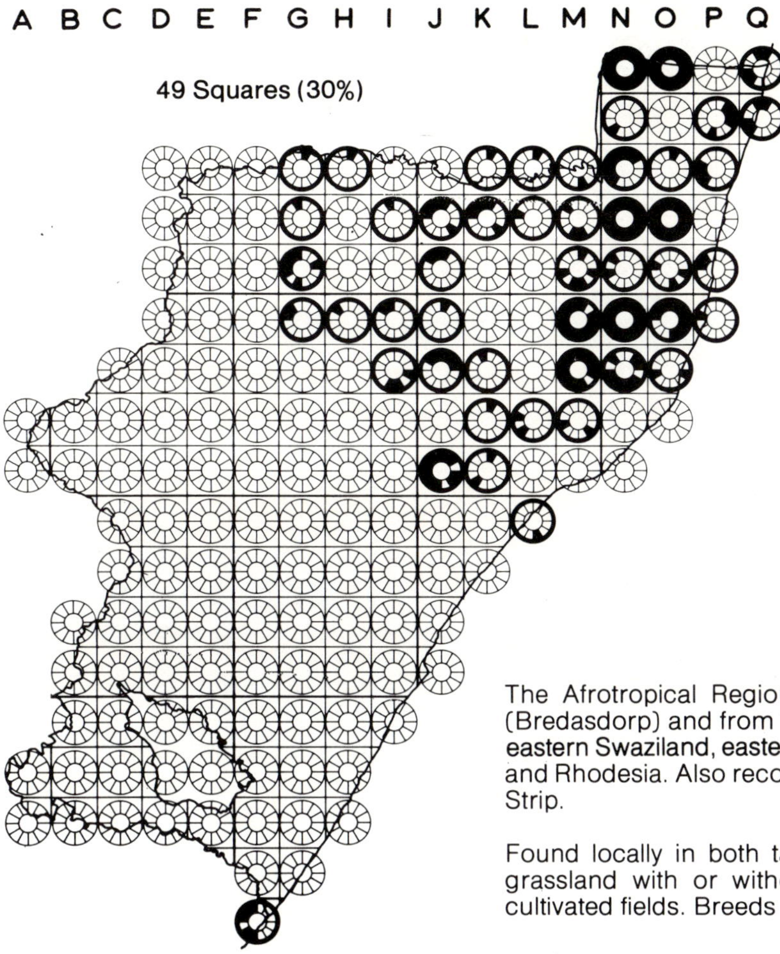

227 Black-bellied Korhaan

Eupodotis melanogaster

The Afrotropical Region. In Southern Africa, the south-western Cape (Bredasdorp) and from Transkei to coastal and northern Natal, Zululand, eastern Swaziland, eastern and northern Transvaal, southern Moçambique, and Rhodesia. Also recorded from Botswana (Ngamiland) and the Caprivi Strip.

Found locally in both tall dry and short moist grass in open savanna, grassland with or without scrub vegetation, vleis, pastures and even cultivated fields. Breeds during the summer.

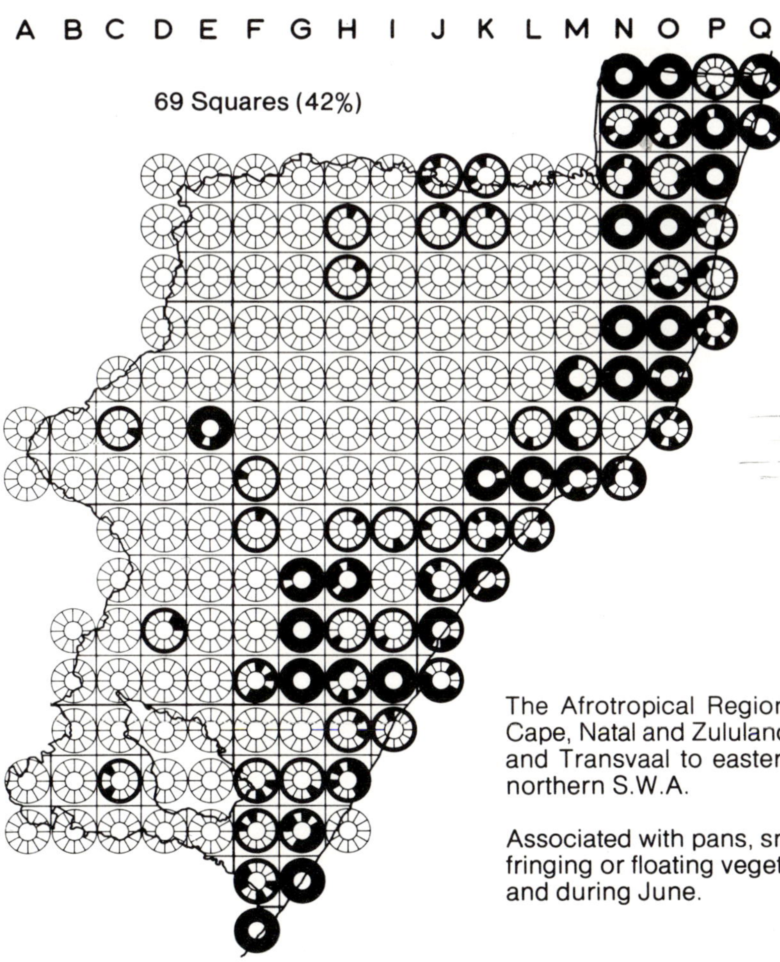

228 African Jacana

Actophilornis africanus

The Afrotropical Region. In Southern Africa, the southern and eastern Cape, Natal and Zululand, northwards. In the north, west through Rhodesia and Transvaal to eastern and northern Botswana, the Caprivi Strip and northern S.W.A.

Associated with pans, small dams, marshes and sewage farm ponds, with fringing or floating vegetation. Recorded breeding from October to March and during June.

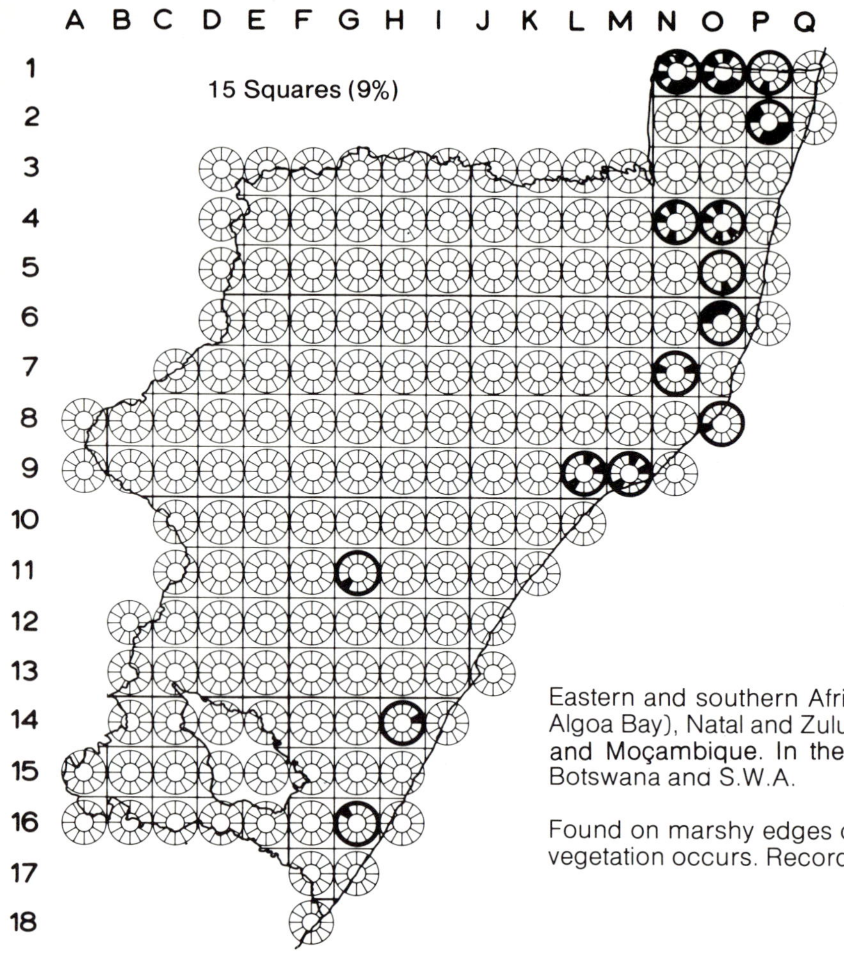

229 Lesser Jacana

Microparra capensis

Eastern and southern Africa. In Southern Africa, the eastern Cape (from Algoa Bay), Natal and Zululand, north in the east through O.F.S., Transvaal and Moçambique. In the north, westwards from Rhodesia to northern Botswana and S.W.A.

Found on marshy edges of coastal lakes, pans and dams, where floating vegetation occurs. Recorded breeding during March.

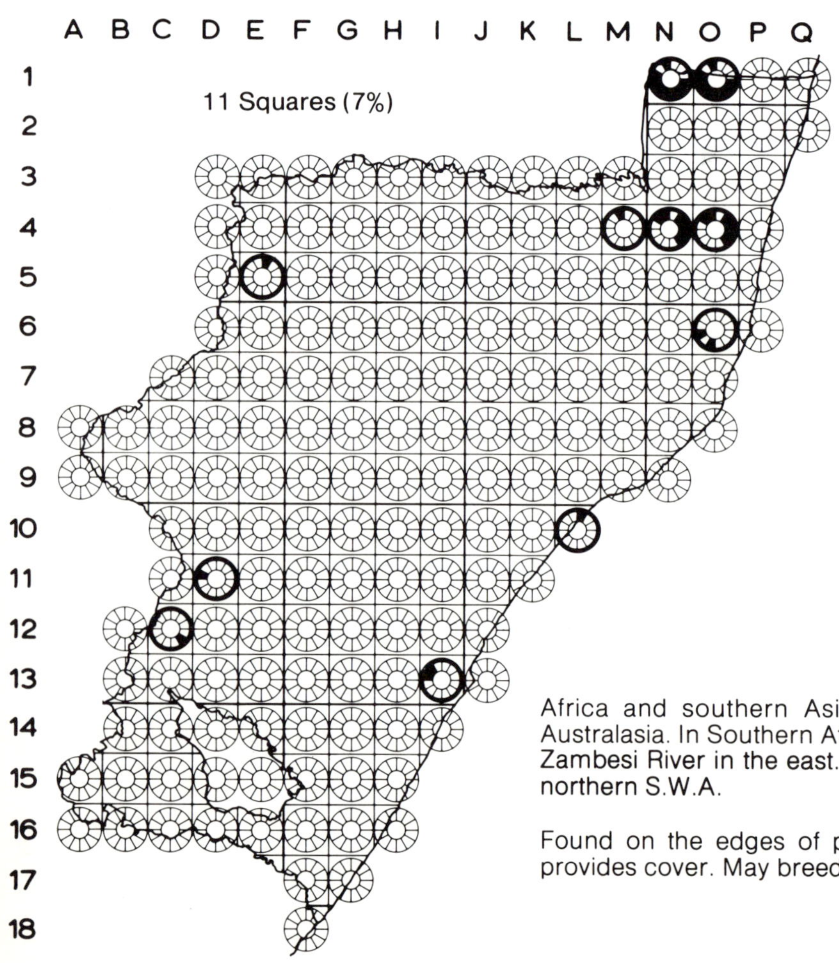

230 Painted Snipe

Rostratula benghalensis

Africa and southern Asia, east to China and Japan, and south to Australasia. In Southern Africa, the Cape, Natal and Zululand, north to the Zambesi River in the east. In the north, west to Rhodesia, Botswana and northern S.W.A.

Found on the edges of ponds and streams where fringing vegetation provides cover. May breed.

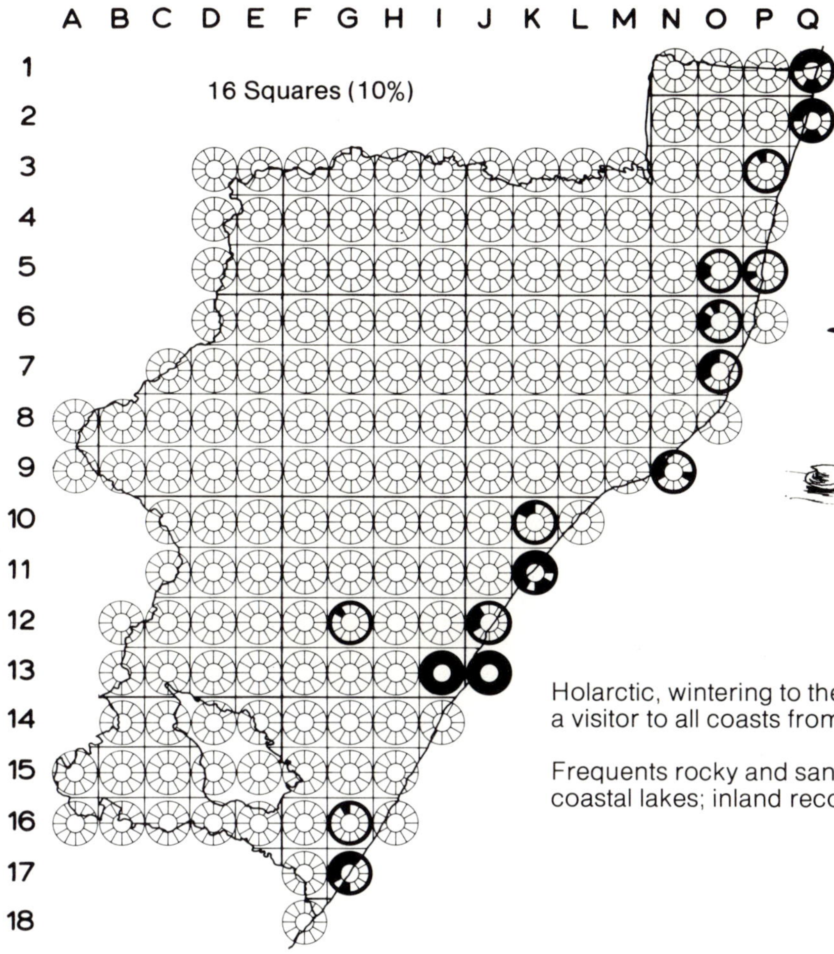

232 Turnstone

Arenaria interpres

Holarctic, wintering to the south of the breeding range. In Southern Africa, a visitor to all coasts from September to April, with some overwintering.

Frequents rocky and sandy shores, tidal mudflats and exposed margins of coastal lakes; inland record exceptional.

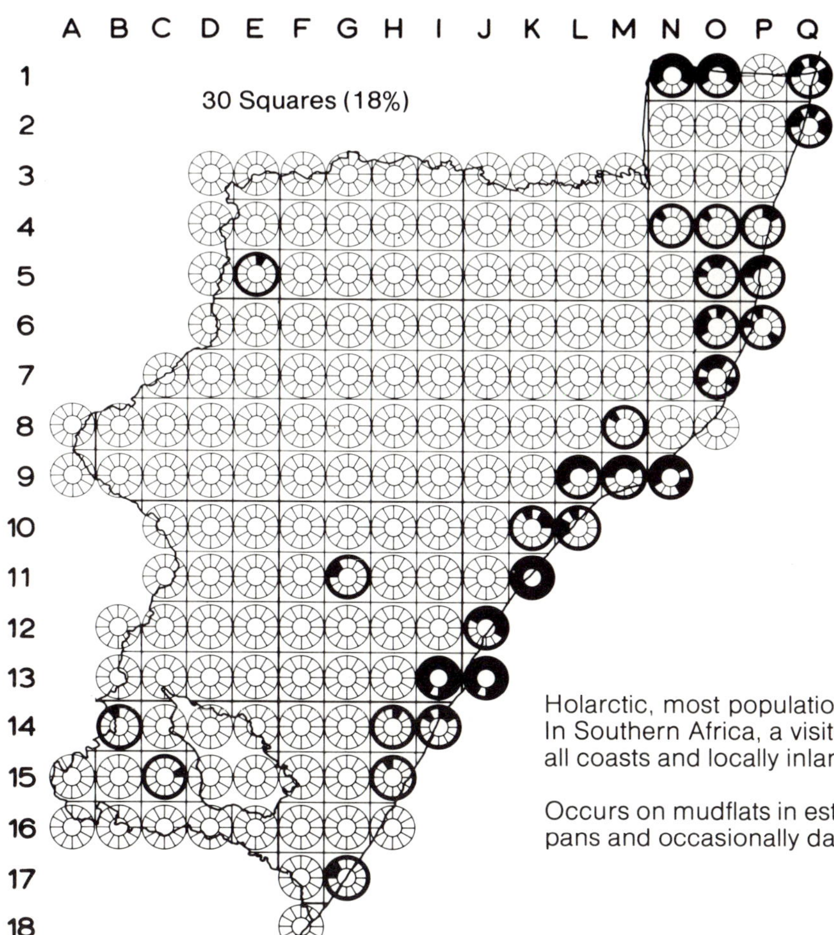

233 Ringed Plover

Charadrius hiaticula

Holarctic, most populations wintering to the south of the breeding range. In Southern Africa, a visitor from late September/October to April/May to all coasts and locally inland; some overwinter.

Occurs on mudflats in estuaries and coastal lakes; also muddy margins of pans and occasionally dams.

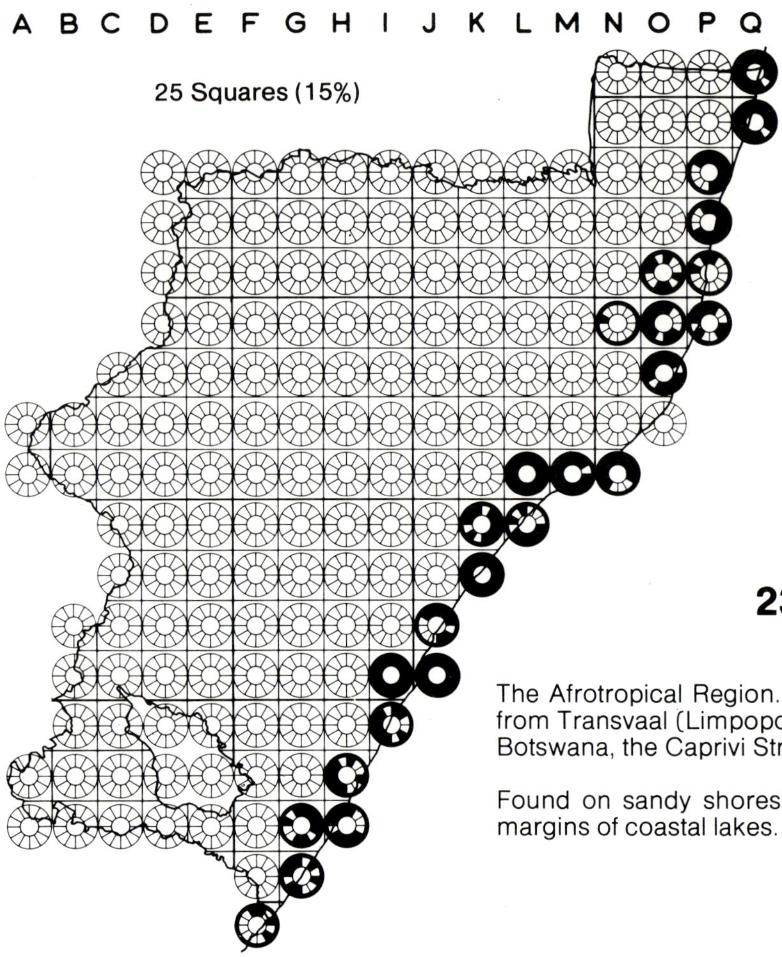

235 White-fronted Sandplover

Charadrius marginatus

The Afrotropical Region. In Southern Africa, along all coasts and inland from Transvaal (Limpopo River) to Rhodesia, northern and north-eastern Botswana, the Caprivi Strip and northern S.W.A. (Etosha Pan).

Found on sandy shores along the coast, in estuaries and on exposed margins of coastal lakes. Breeds throughout the year.

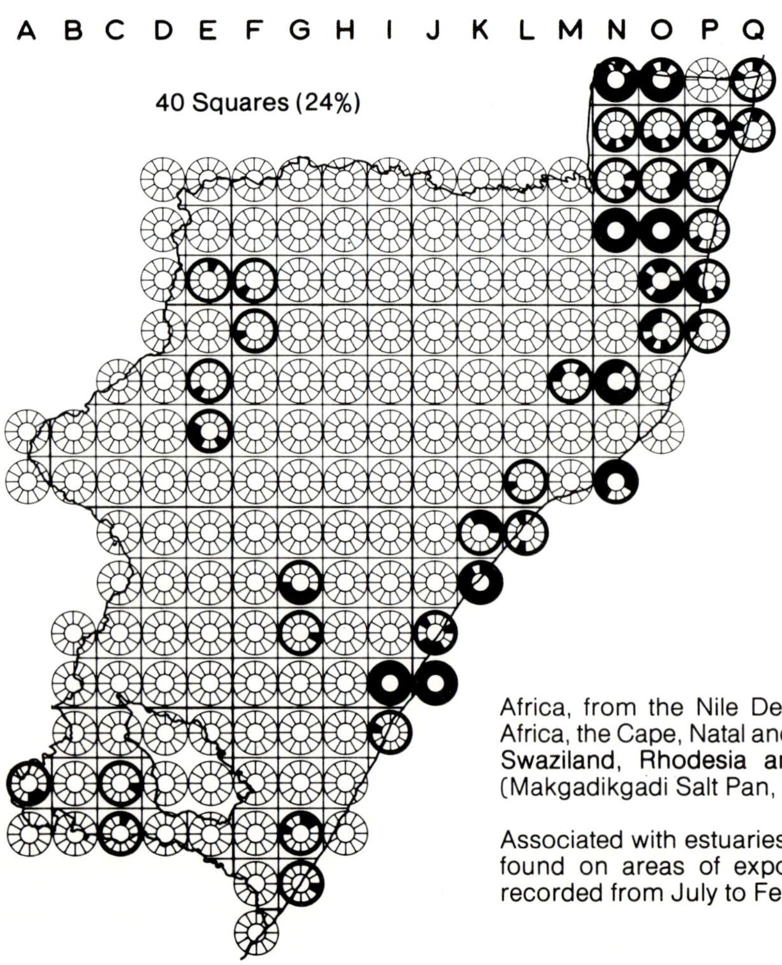

237 Kittlitz's Sandplover

Charadrius pecuarius

Africa, from the Nile Delta to the Cape; also Madagascar. In Southern Africa, the Cape, Natal and Zululand, Lesotho (lowlands), O.F.S., Transvaal, Swaziland, Rhodesia and Moçambique, west to northern Botswana (Makgadikgadi Salt Pan, Lake Dow and Nata River) and northern S.W.A.

Associated with estuaries, coastal lakes, pans and dams, where it may be found on areas of exposed mud or adjacent bare ground. Breeding recorded from July to February.

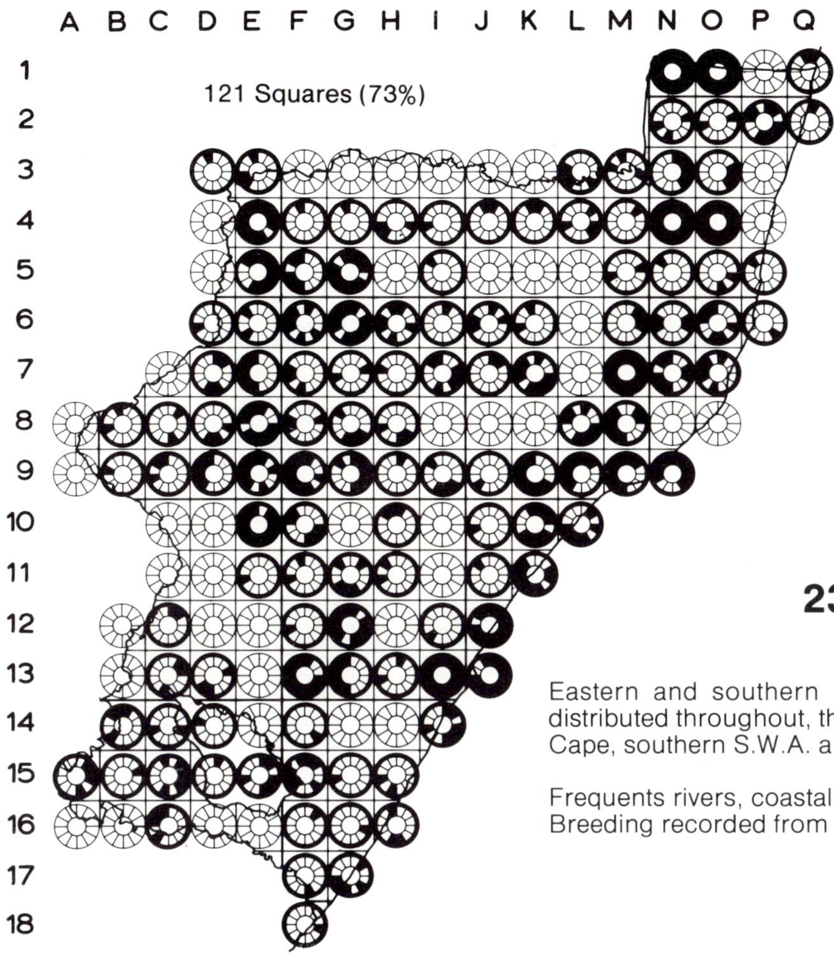

238 Three-banded Sandplover

Charadrius tricollaris

Eastern and southern Africa, and Madagascar. In Southern Africa, distributed throughout, though sparse in parts of the interior and northern Cape, southern S.W.A. and Botswana (Kalahari).

Frequents rivers, coastal lakes, pans, sewage farms, marshes and dams. Breeding recorded from July to December.

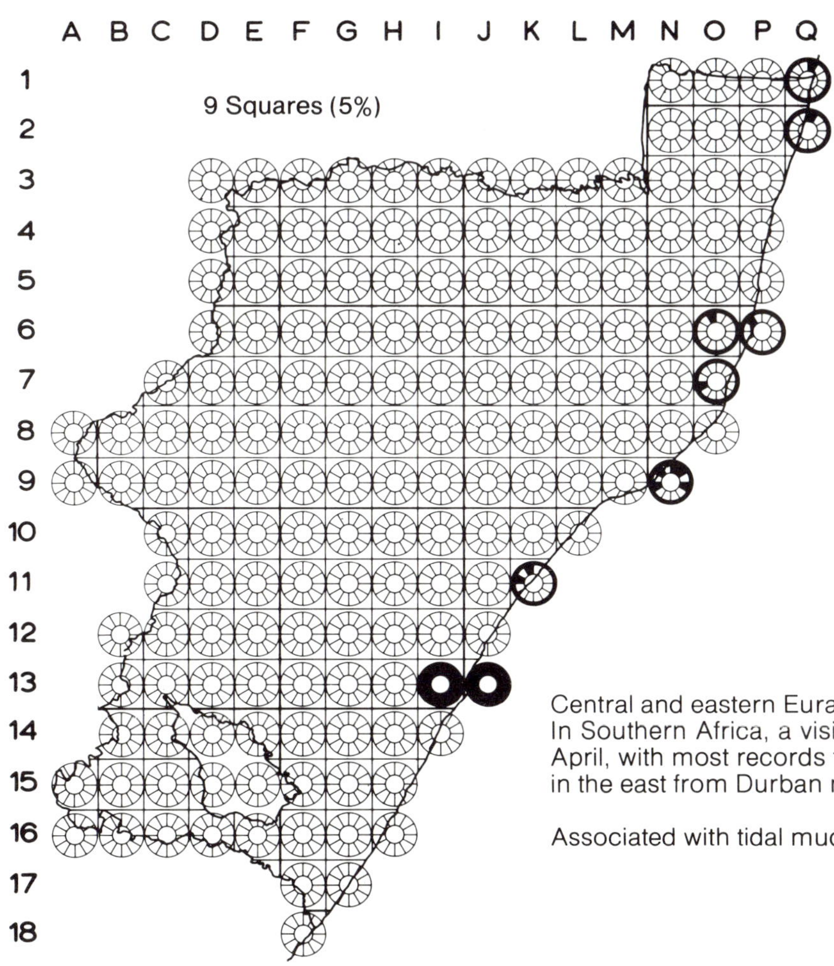

239 Great Sandplover

Charadrius leschenaultii

Central and eastern Eurasia, wintering to the south of the breeding range. In Southern Africa, a visitor to all coasts from late October/November to April, with most records from the west coast of the Cape and S.W.A., and in the east from Durban northwards.

Associated with tidal mudflats.

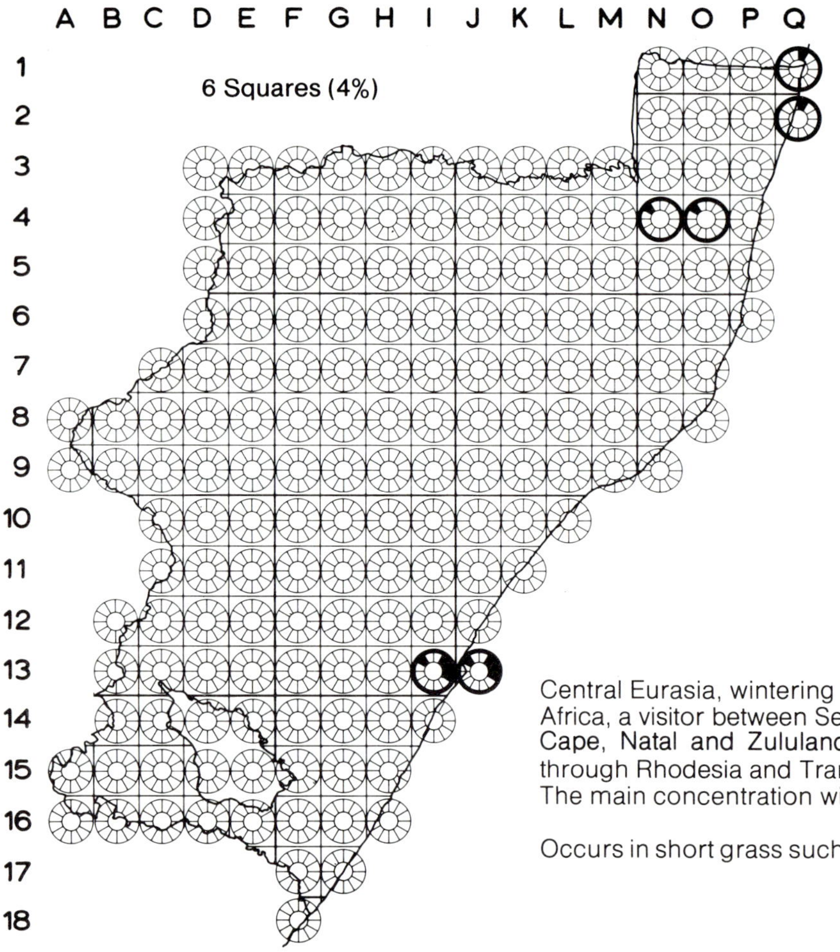

240 Caspian Plover

Charadrius asiaticus

Central Eurasia, wintering to the south of the breeding range. In Southern Africa, a visitor between September and March, recorded from the eastern Cape, Natal and Zululand, and Moçambique; in the north, westwards through Rhodesia and Transvaal to Botswana and S.W.A. (mainly Etosha). The main concentration winters in the dry interior and west.

Occurs in short grass such as airfields; also marshy ground and estuaries.

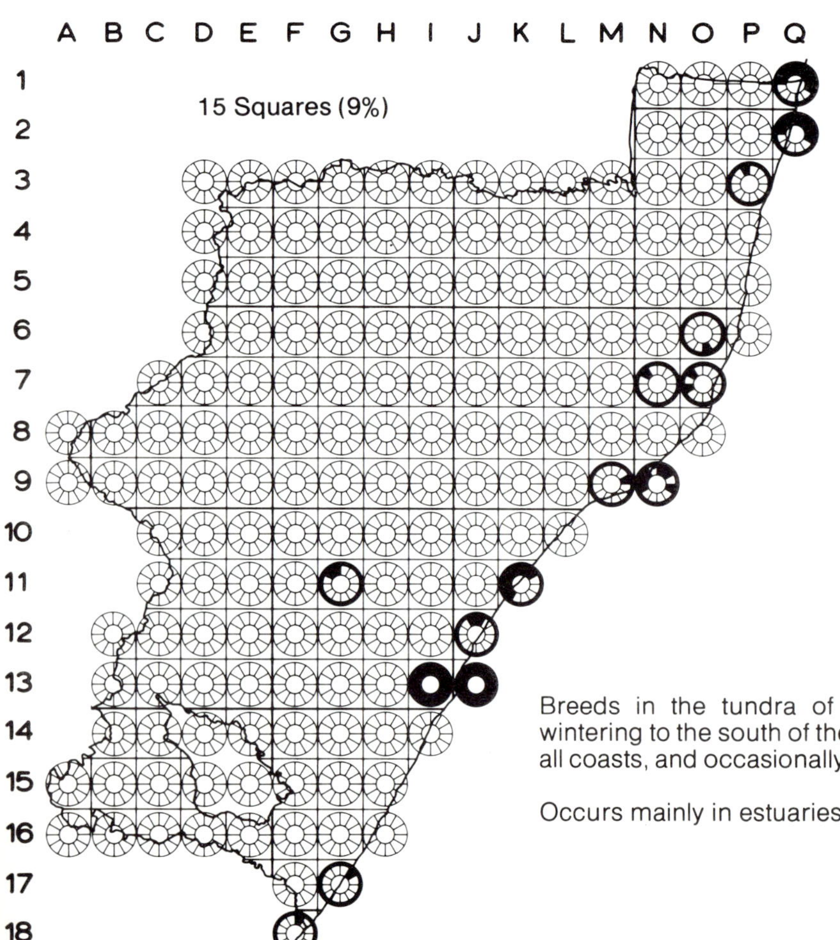

241 Grey Plover

Pluvialis squatarola

Breeds in the tundra of northern Europe, Asia and North America, wintering to the south of the breeding range. In Southern Africa, a visitor to all coasts, and occasionally inland, from August to April; some overwinter.

Occurs mainly in estuaries, and also along the sea shore.

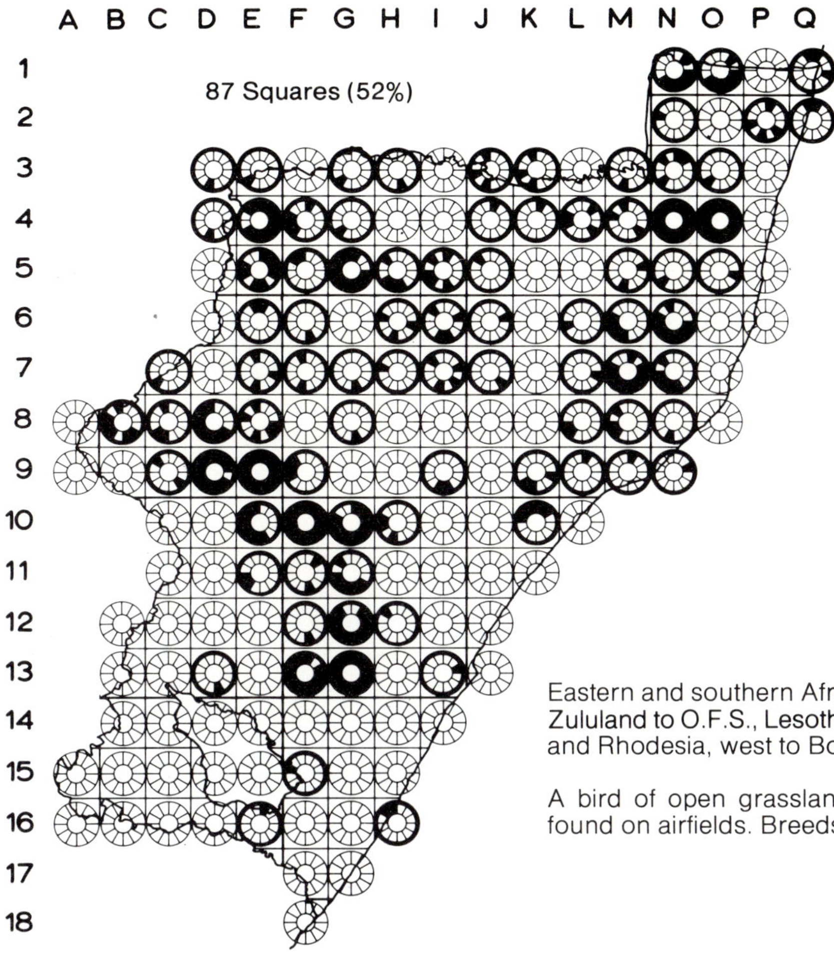

242 Crowned Plover

Vanellus coronatus

Eastern and southern Africa. In Southern Africa, from the Cape, Natal and Zululand to O.F.S., Lesotho (lowlands), Transvaal, Swaziland, Moçambique and Rhodesia, west to Botswana (Ngamiland and Makarikari) and S.W.A.

A bird of open grassland, pastures and lightly-wooded savanna; often found on airfields. Breeds from June to November.

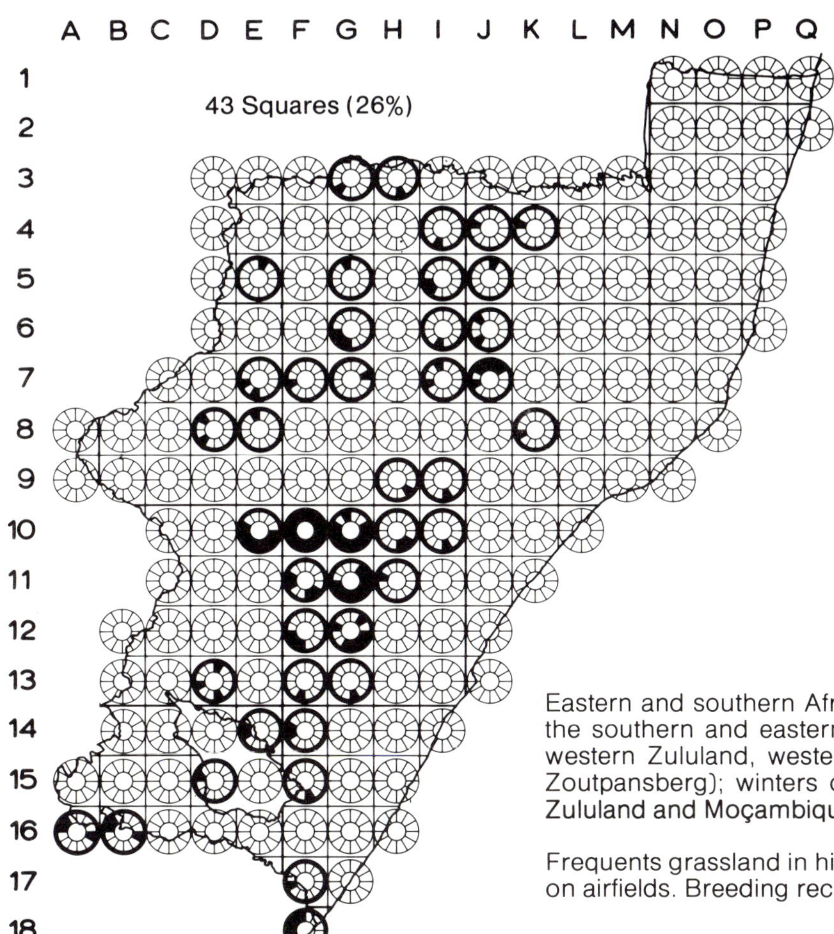

243 Black-winged Plover

Vanellus melanopterus

Eastern and southern Africa. In Southern Africa, breeds in the interior of the southern and eastern Cape, O.F.S. (Bloemfontein), Natal (interior), western Zululand, western Swaziland, and eastern Transvaal (north to Zoutpansberg); winters on the coastal flats of the eastern Cape, Natal, Zululand and Moçambique.

Frequents grassland in hilly country, fallow fields and pastures; often seen on airfields. Breeding recorded from July to October.

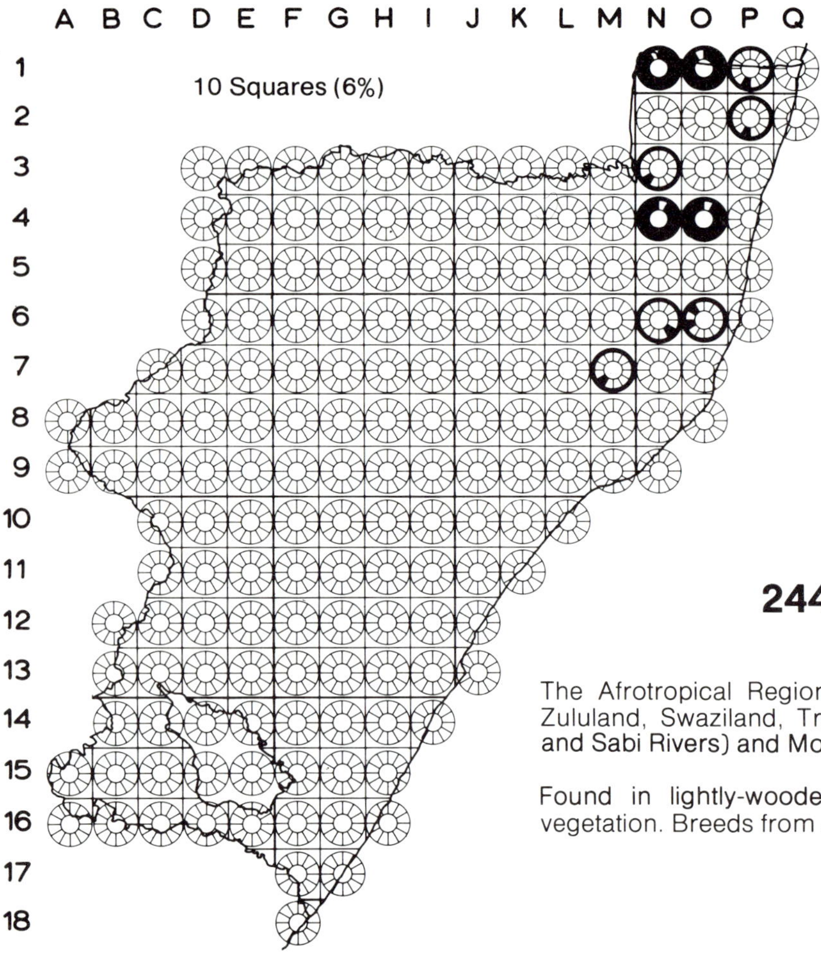

244 Lesser Black-winged Plover

Vanellus lugubris

The Afrotropical Region. In Southern Africa, Natal (seldom), eastern Zululand, Swaziland, Transvaal, Rhodesia (Wankie G.R., mid-Zambesi and Sabi Rivers) and Moçambique.

Found in lightly-wooded savanna, and open grassland with scrub vegetation. Breeds from June to October.

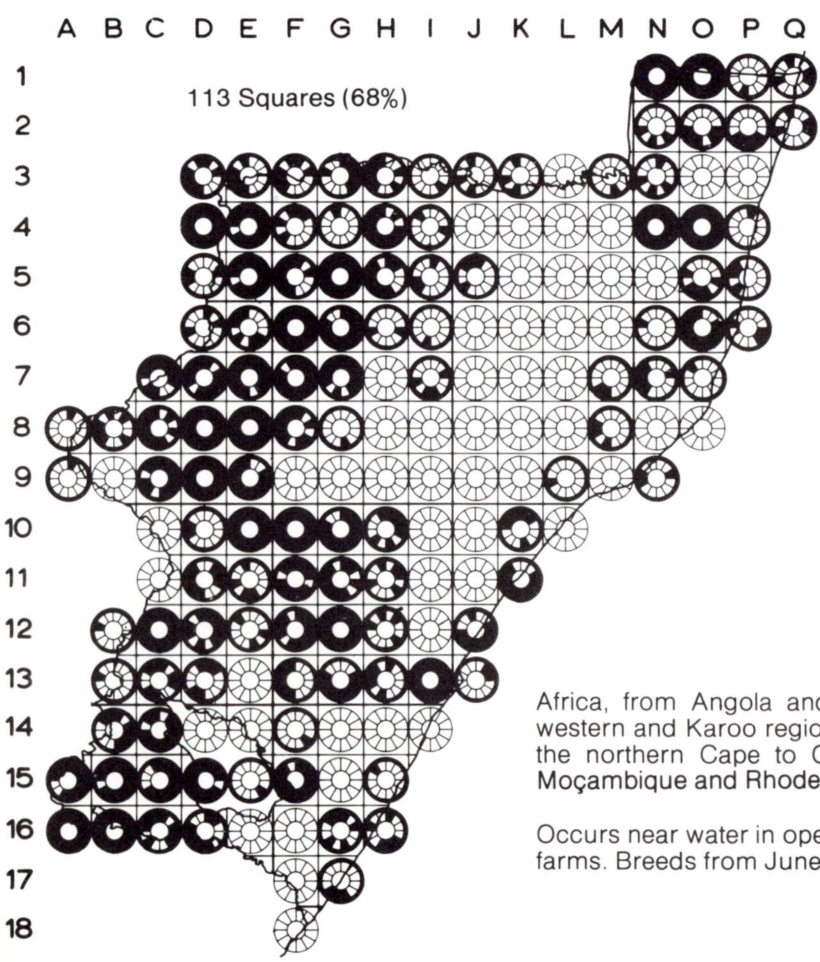

245 Blacksmith Plover

Vanellus armatus

Africa, from Angola and Kenya to the Cape. In Southern Africa, the western and Karoo regions of the Cape and from S.W.A., Botswana and the northern Cape to O.F.S., Lesotho, Natal and Zululand, southern Moçambique and Rhodesia.

Occurs near water in open grassland and cultivated fields; also at sewage farms. Breeds from June to November.

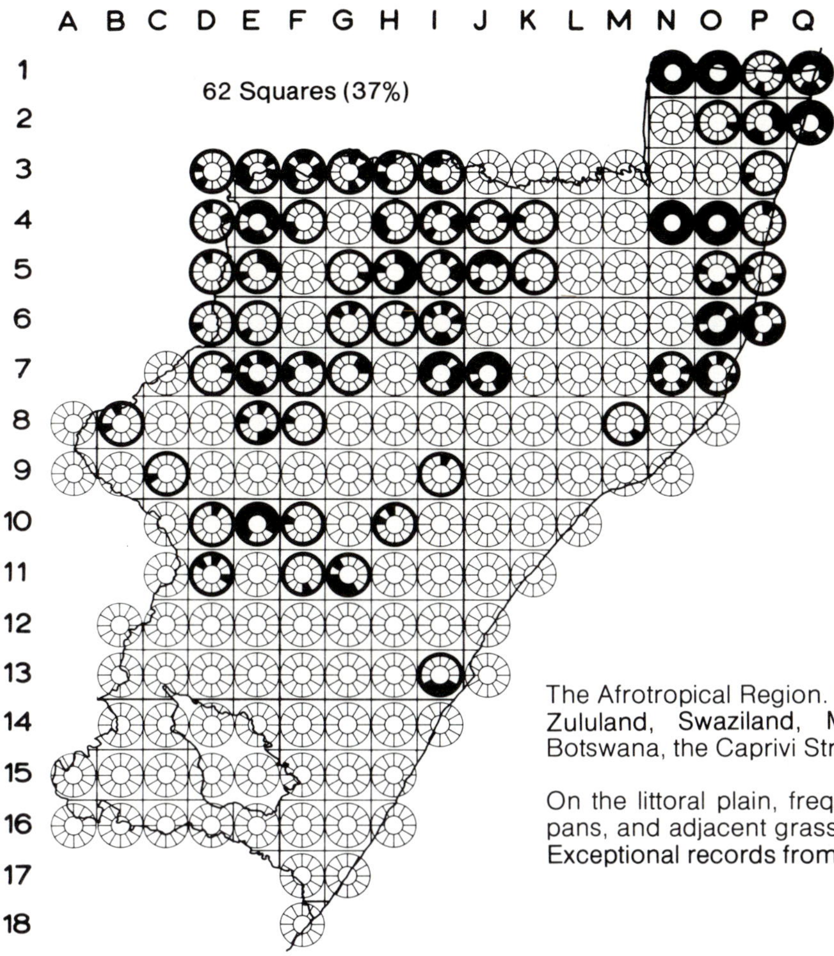

247 Wattled Plover

Vanellus senegallus

The Afrotropical Region. In Southern Africa, midland and northern Natal, Zululand, Swaziland, Moçambique, Transvaal, Rhodesia, northern Botswana, the Caprivi Strip and northern S.W.A.

On the littoral plain, frequents exposed areas around coastal lakes and pans, and adjacent grassland; in the interior, found near vleis and dams. Exceptional records from Durban. Breeds from July to December.

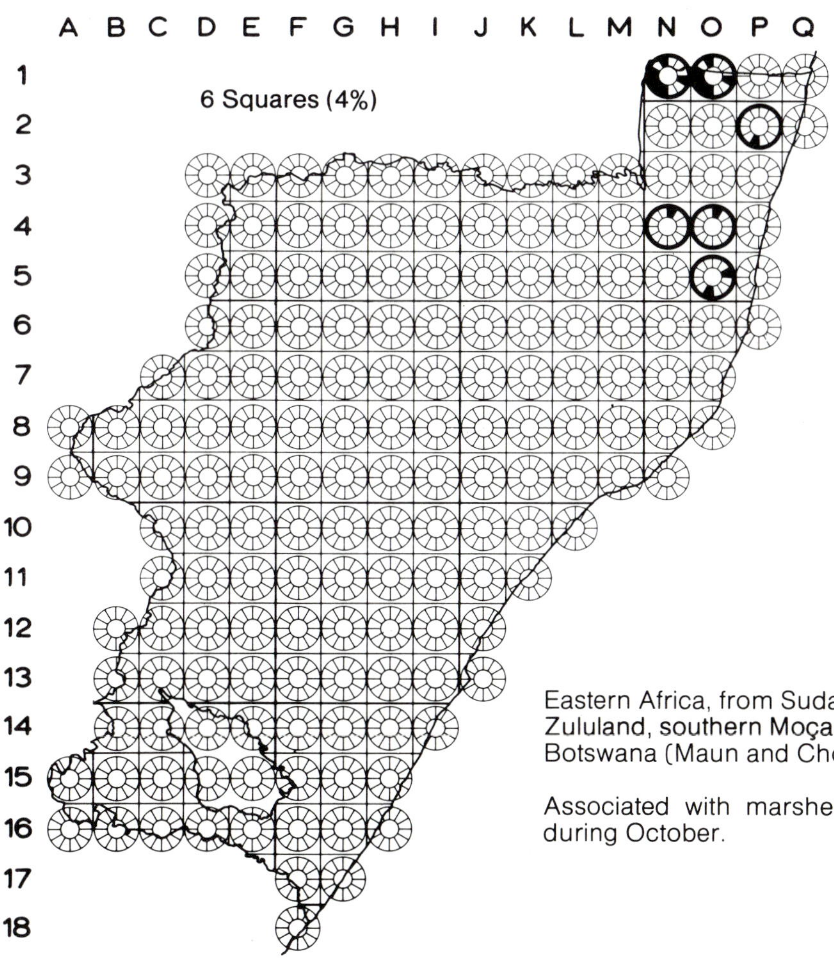

248 Long-toed Plover

Vanellus crassirostris

Eastern Africa, from Sudan to Zululand. In Southern Africa, north-eastern Zululand, southern Moçambique, Rhodesia (Lake Kariba and Sabi River), Botswana (Maun and Chobe) and the eastern Caprivi Strip.

Associated with marshes and flooded grassland. Recorded breeding during October.

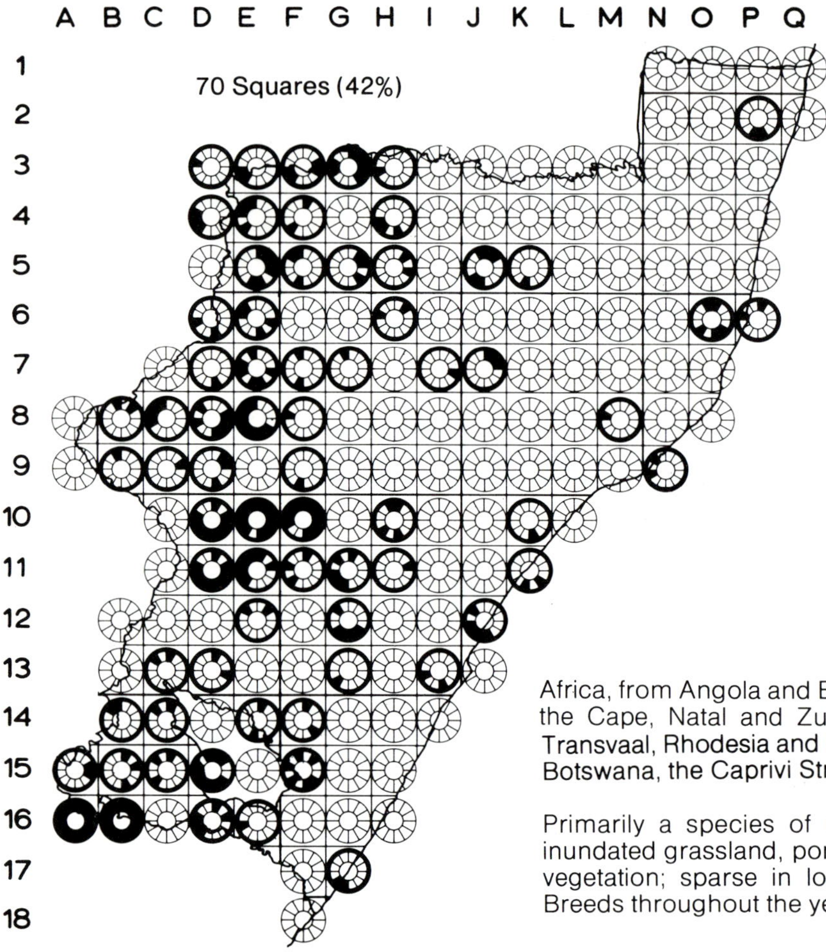

250 Ethiopian Snipe

Gallinago nigripennis

Africa, from Angola and Ethiopia to the Cape. In Southern Africa, locally in the Cape, Natal and Zululand, Swaziland, O.F.S., Lesotho (lowlands), Transvaal, Rhodesia and Moçambique. In the north, westwards to northern Botswana, the Caprivi Strip and northern S.W.A.

Primarily a species of upland grassland, where it inhabits marshes, inundated grassland, ponds and dams with muddy edges and clumps of vegetation; sparse in lowland areas, where associated with marshes. Breeds throughout the year.

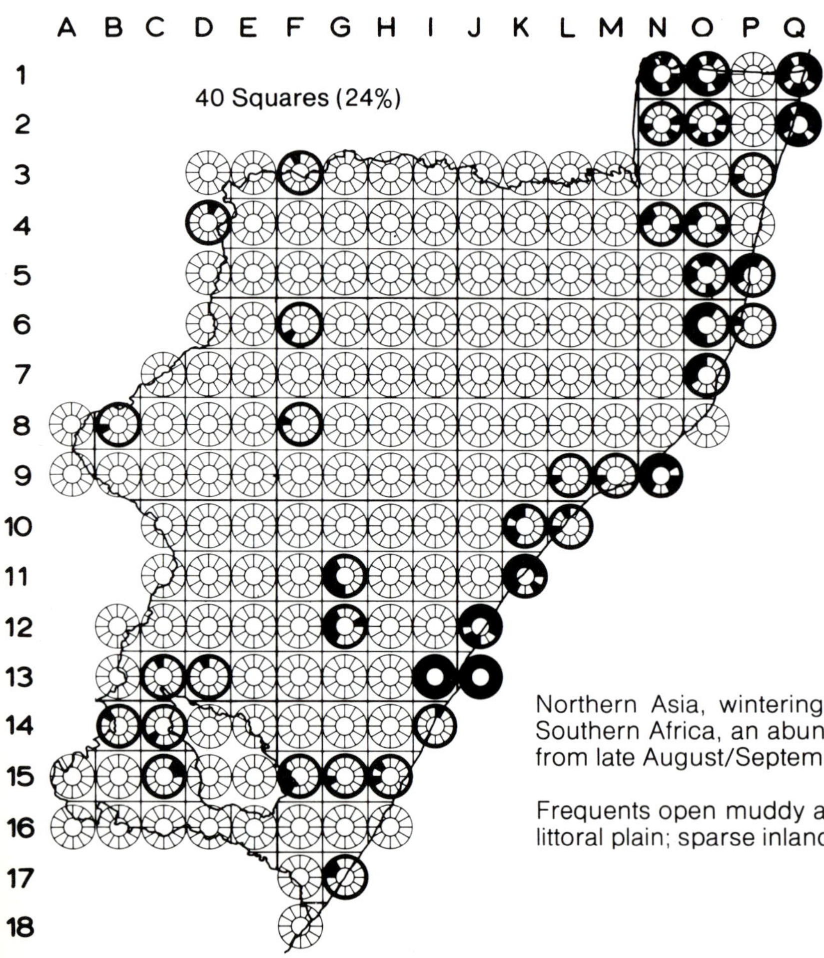

251 Curlew Sandpiper

Calidris ferruginea

Northern Asia, wintering far to the south of the breeding grounds. In Southern Africa, an abundant visitor to coastal and inland waters, mainly from late August/September to April; many overwinter.

Frequents open muddy areas of estuaries, coastal lakes and pans on the littoral plain; sparse inland at dams, vleis and sewage farms.

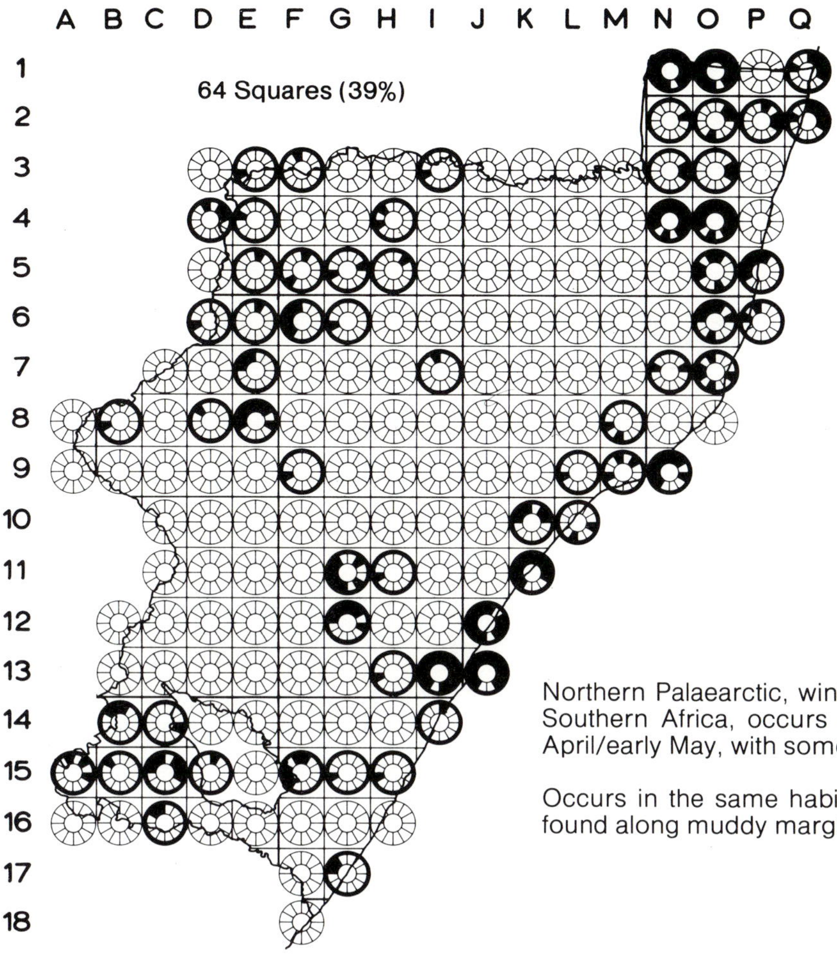

253 Little Stint

Calidris minuta

Northern Palaearctic, wintering to the south of the breeding grounds. In Southern Africa, occurs on inland and coastal waters from August to April/early May, with some overwintering.

Occurs in the same habitats as those of Curlew Sandpiper (251); also found along muddy margins of standing water.

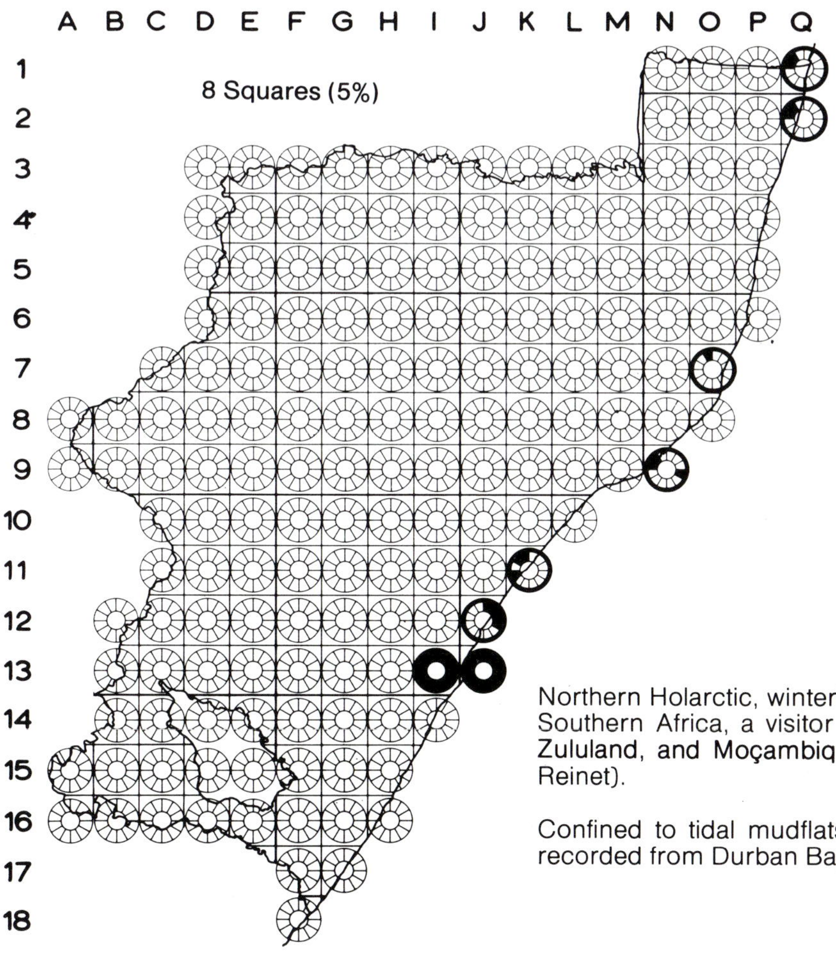

254 Knot

Calidris canutus

Northern Holarctic, wintering far to the south of the breeding grounds. In Southern Africa, a visitor to the coasts of S.W.A., the Cape, Natal and Zululand, and Moçambique. Recorded from the Cape interior (Graaff-Reinet).

Confined to tidal mudflats in estuaries. Flocks of over 100 have been recorded from Durban Bay, where individuals have overwintered.

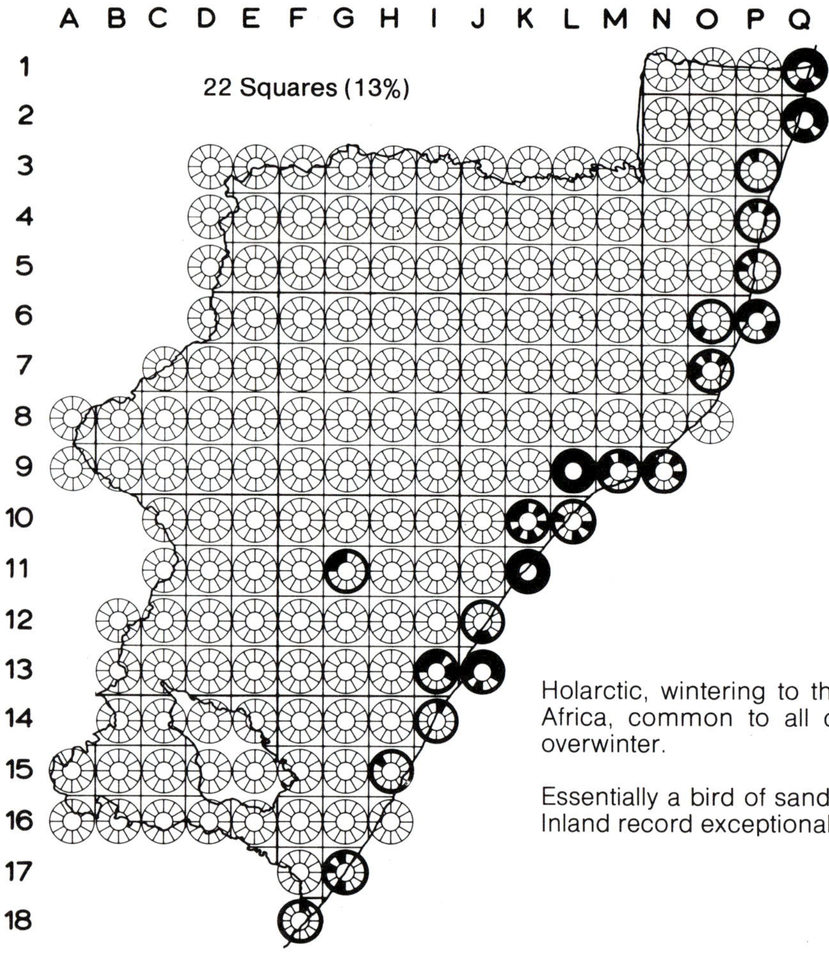

255 Sanderling

Calidris alba

Holarctic, wintering to the south of the breeding grounds. In Southern Africa, common to all coasts from August to April/early May; some overwinter.

Essentially a bird of sandy stretches of the seashore; visits coastal lakes. Inland record exceptional.

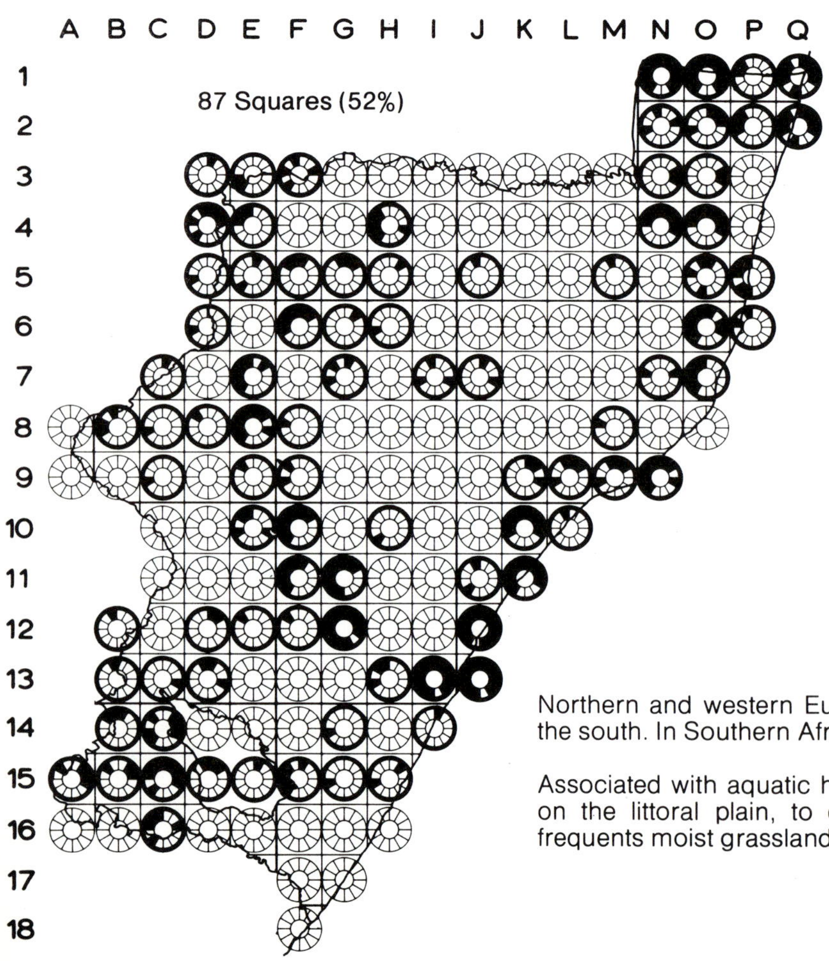

256 Ruff

Philomachus pugnax

Northern and western Europe across northern Eurasia, wintering far to the south. In Southern Africa, a common visitor from August to April.

Associated with aquatic habitats, from estuaries, coastal lakes and pans on the littoral plain, to dams, marshes and sewage farms, where it frequents moist grassland as well as muddy situations.

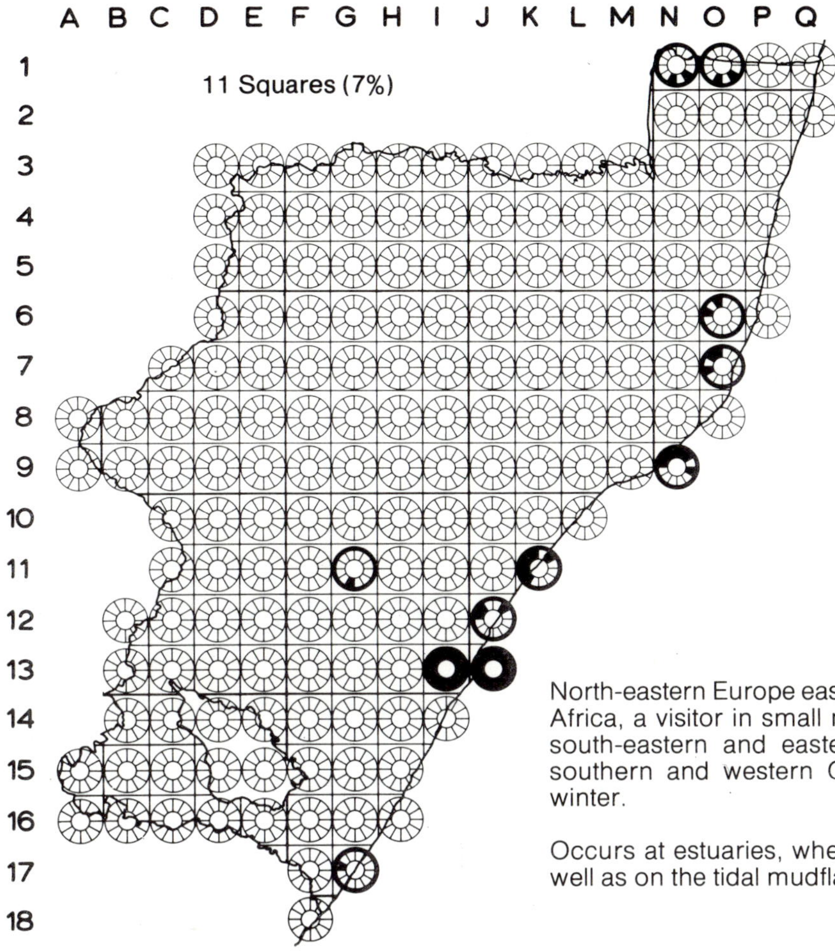

257 Terek Sandpiper

Xenus cinereus

North-eastern Europe east to Siberia, wintering far to the south. In Southern Africa, a visitor in small numbers from October to April/early May to the south-eastern and eastern coasts, and more rarely to coasts of the southern and western Cape, and S.W.A.; seldom inland. Some overwinter.

Occurs at estuaries, where it may be found on muddy banks up-river as well as on the tidal mudflats.

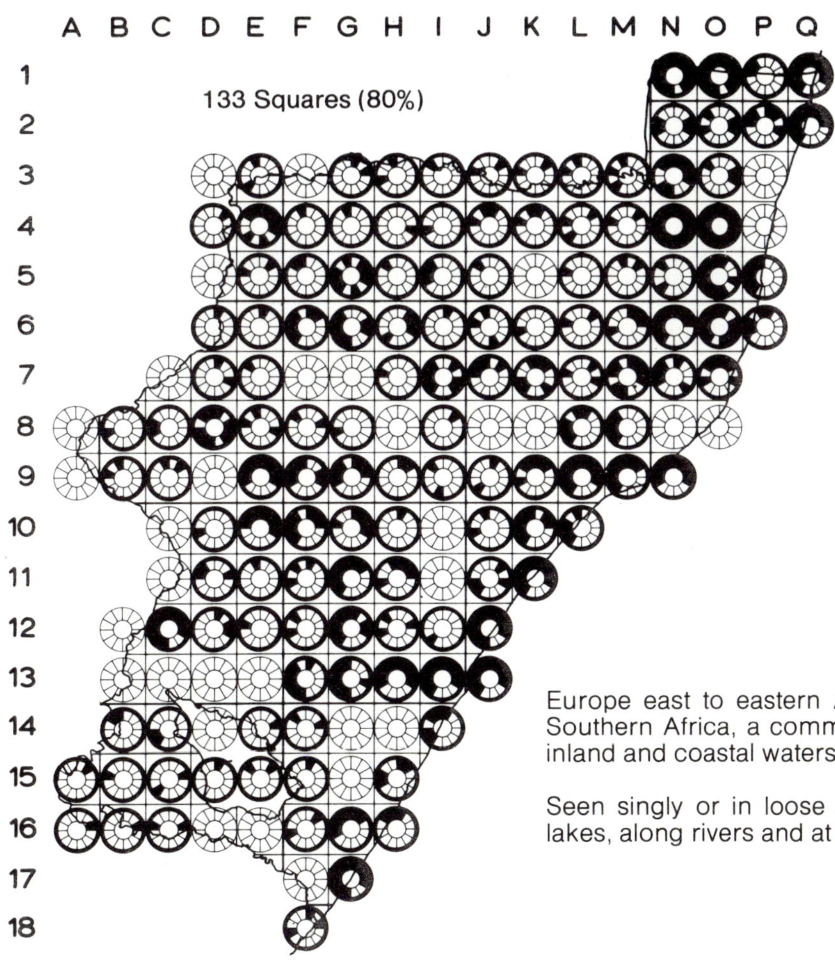

258 Common Sandpiper

Tringa hypoleucos

Europe east to eastern Asia and Japan, wintering far to the south. In Southern Africa, a common visitor from August to March/April, to both inland and coastal waters; some overwinter.

Seen singly or in loose association at estuaries, rocky shores, coastal lakes, along rivers and at pans, dams and sewage farms.

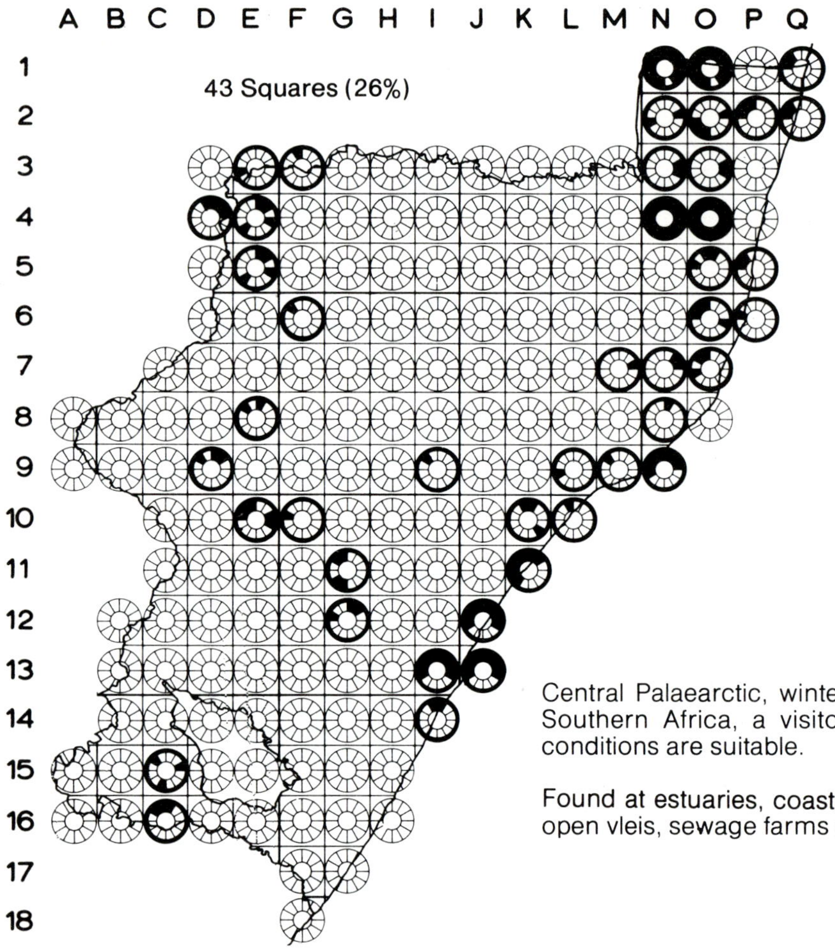

262 Marsh Sandpiper

Tringa stagnatilis

Central Palaearctic, wintering to the south of the breeding grounds. In Southern Africa, a visitor from August to April, occurring wherever conditions are suitable.

Found at estuaries, coastal lakes and pans on the littoral plain; inland at open vleis, sewage farms and dams.

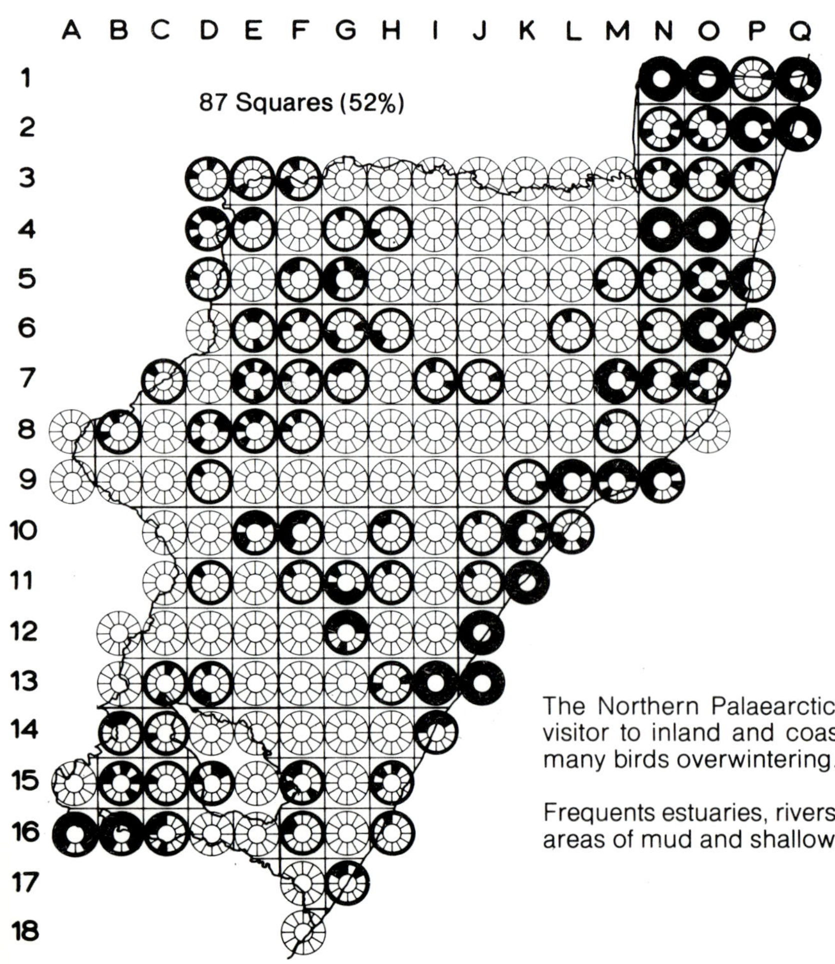

263 Greenshank

Tringa nebularia

The Northern Palaearctic, wintering to the south. In Southern Africa, a visitor to inland and coastal waters from late August to April/May, with many birds overwintering.

Frequents estuaries, rivers, lakes, dams and sewage farms, where exposed areas of mud and shallow water are to be found.

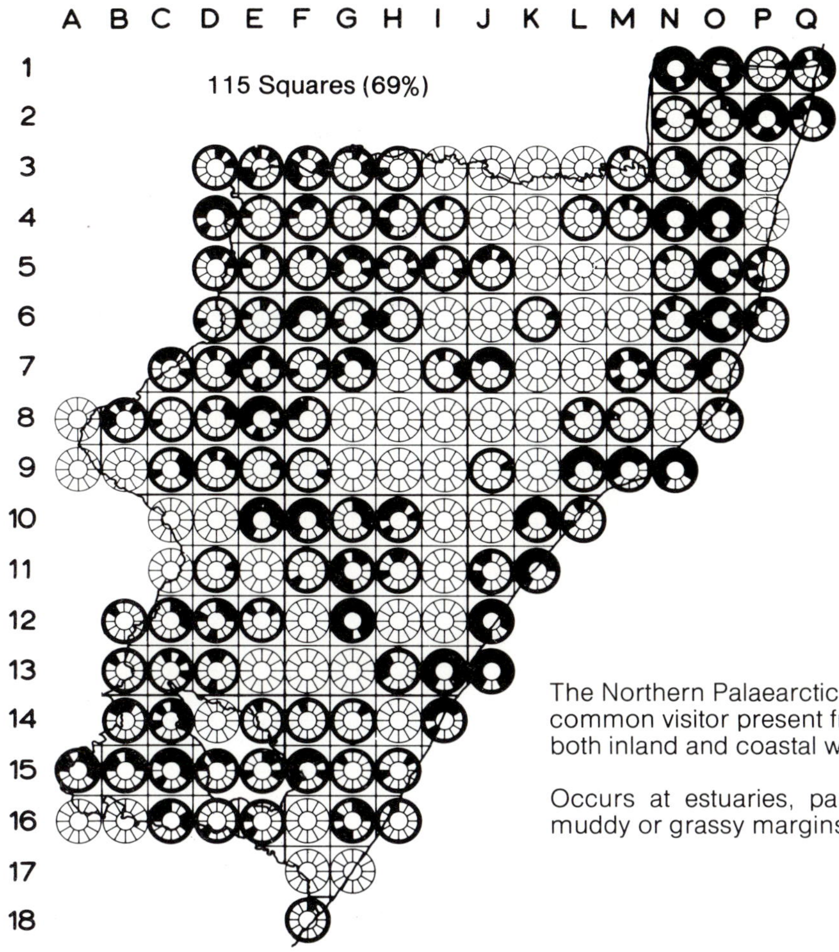

264 Wood Sandpiper

Tringa glareola

The Northern Palaearctic, wintering far to the south. In Southern Africa, a common visitor present from late August to April, generally distributed on both inland and coastal waters. Some overwinter.

Occurs at estuaries, pans, dams, vleis and sewage farms, favouring muddy or grassy margins.

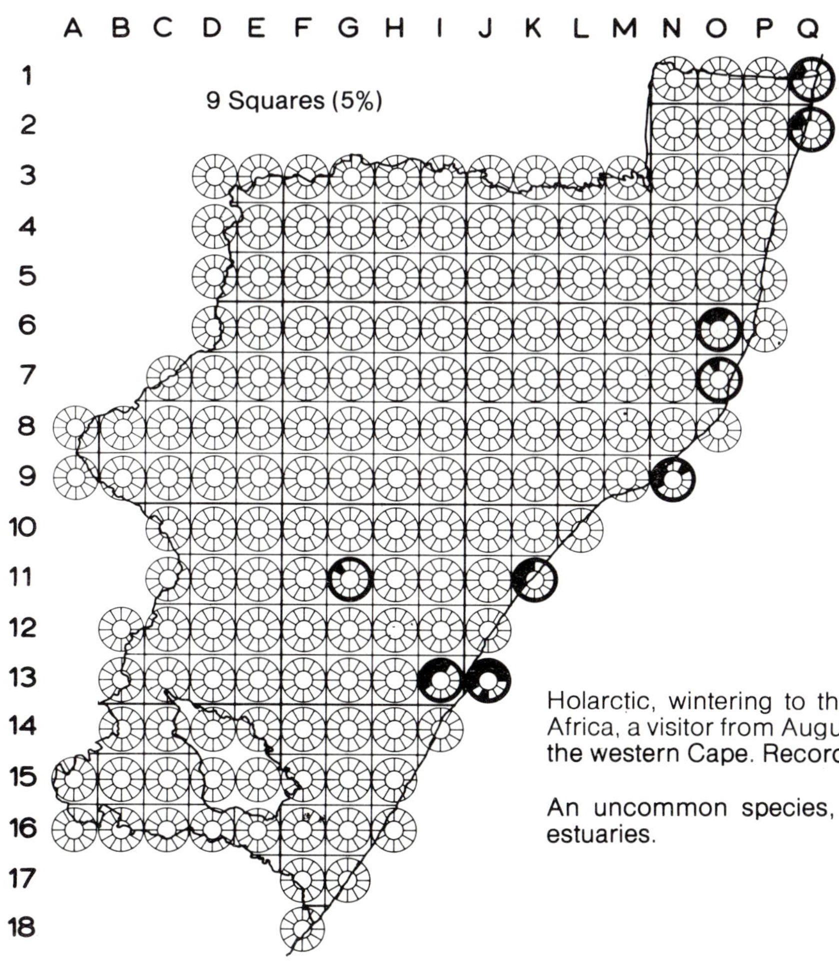

266 Bar-tailed Godwit

Limosa lapponica

Holarctic, wintering to the south of the breeding grounds. In Southern Africa, a visitor from August to April to all coasts, particularly in S.W.A. and the western Cape. Recorded inland in Transvaal and Botswana.

An uncommon species, usually found on tidal mudflats of the larger estuaries.

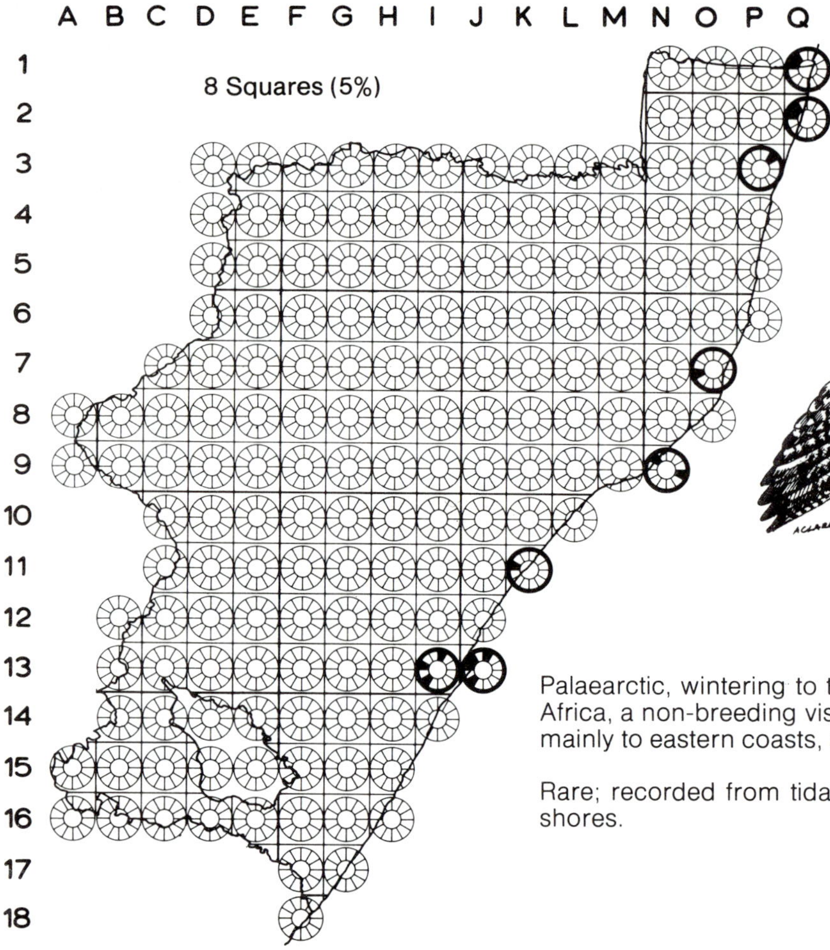

267 Curlew

Numenius arquata

Palaearctic, wintering to the south of the breeding grounds. In Southern Africa, a non-breeding visitor, in small numbers, from September to April mainly to eastern coasts, and locally inland.

Rare; recorded from tidal mudflats in estuaries, and occasionally sandy shores.

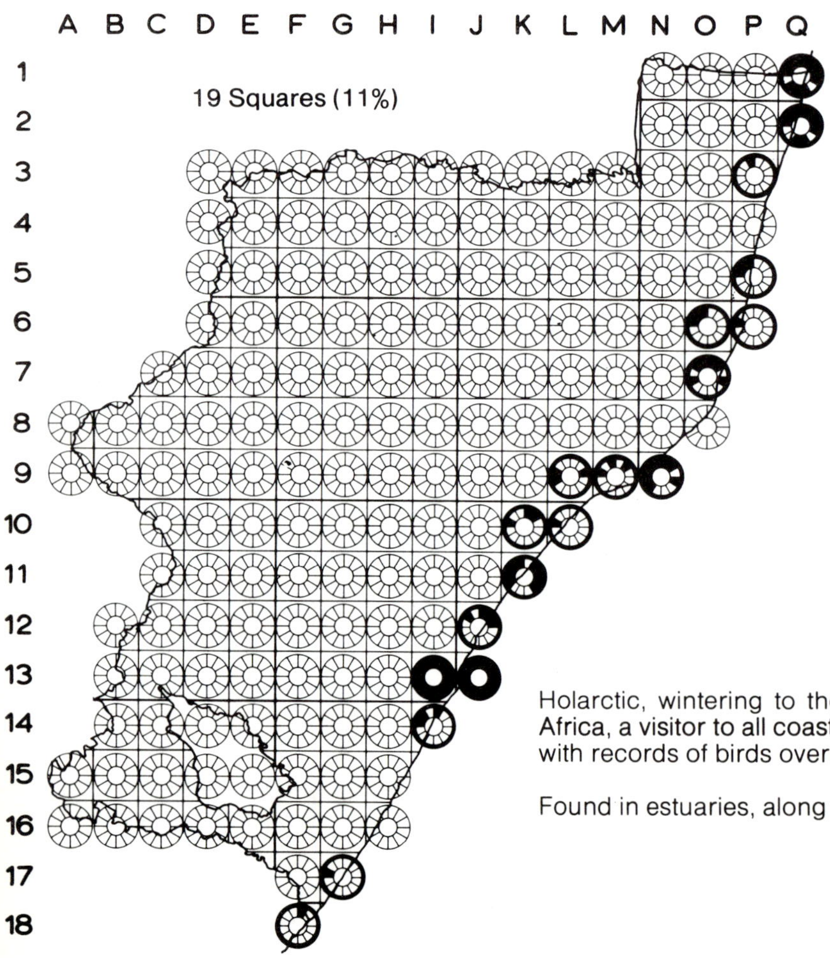

268 Whimbrel

Numenius phaeopus

Holarctic, wintering to the south of the breeding grounds. In Southern Africa, a visitor to all coasts, occasionally inland, from September to April, with records of birds overwintering.

Found in estuaries, along the sea shore and occasionally at coastal lakes.

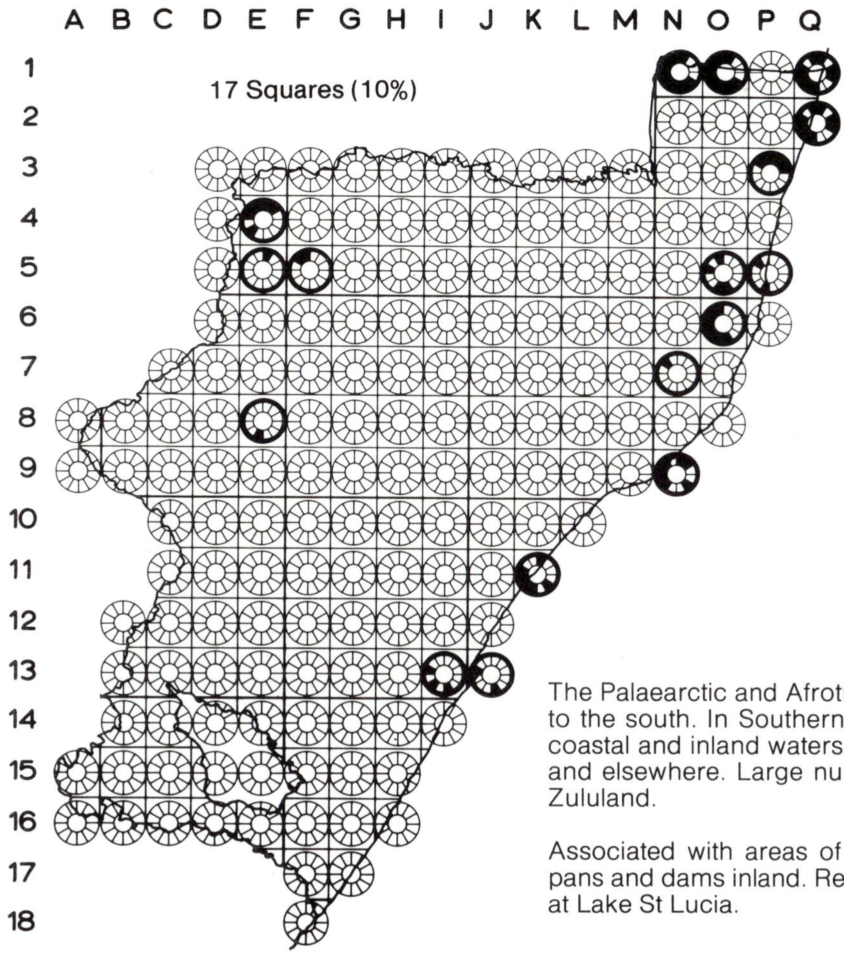

269 Avocet

Recurvirostra avosetta

The Palaearctic and Afrotropical Regions; northern populations wintering to the south. In Southern Africa, mainly a visitor from the Palaearctic to coastal and inland waters, but some breed locally in the Cape, Botswana and elsewhere. Large numbers occasionally recorded at Lake St Lucia, Zululand.

Associated with areas of shallow water in estuaries, coastal lakes, and pans and dams inland. Recorded breeding during October and November at Lake St Lucia.

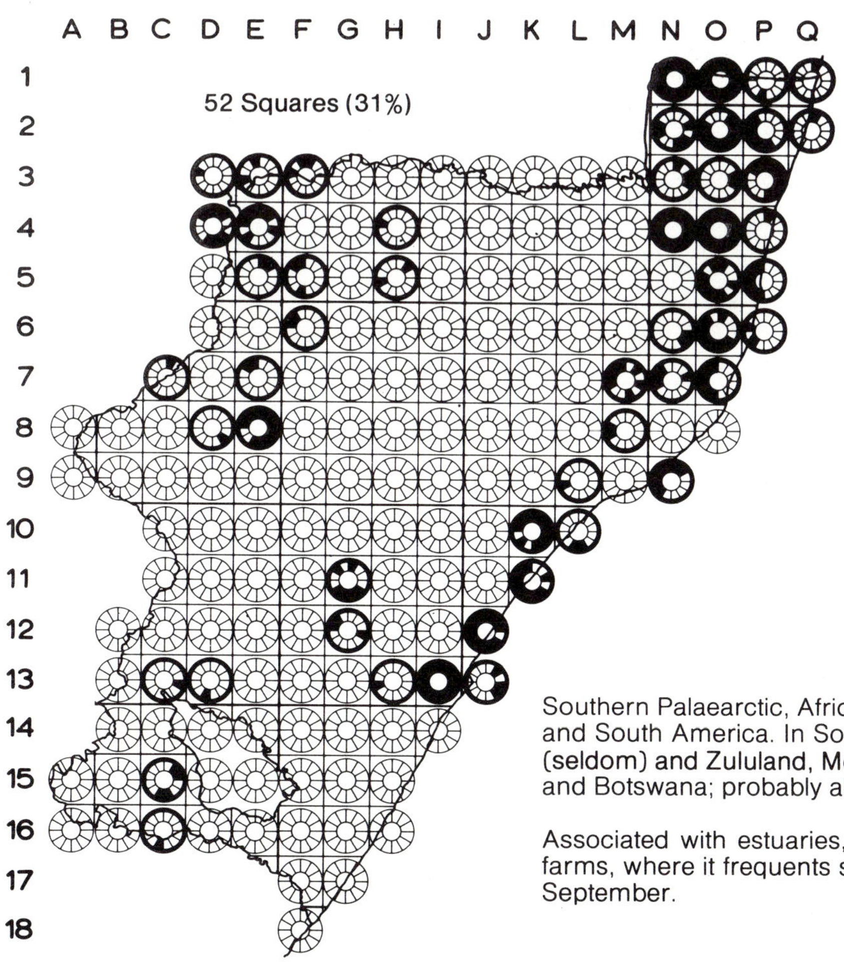

270 Black-winged Stilt

Himantopus himantopus

Southern Palaearctic, Africa and Asia southeast to Australasia; also North and South America. In Southern Africa, breeds locally in the Cape, Natal (seldom) and Zululand, Moçambique, Rhodesia, S.W.A., the Caprivi Strip and Botswana; probably also elsewhere.

Associated with estuaries, coastal lakes, pans, vleis, dams and sewage farms, where it frequents shallow water. Breeding recorded from April to September.

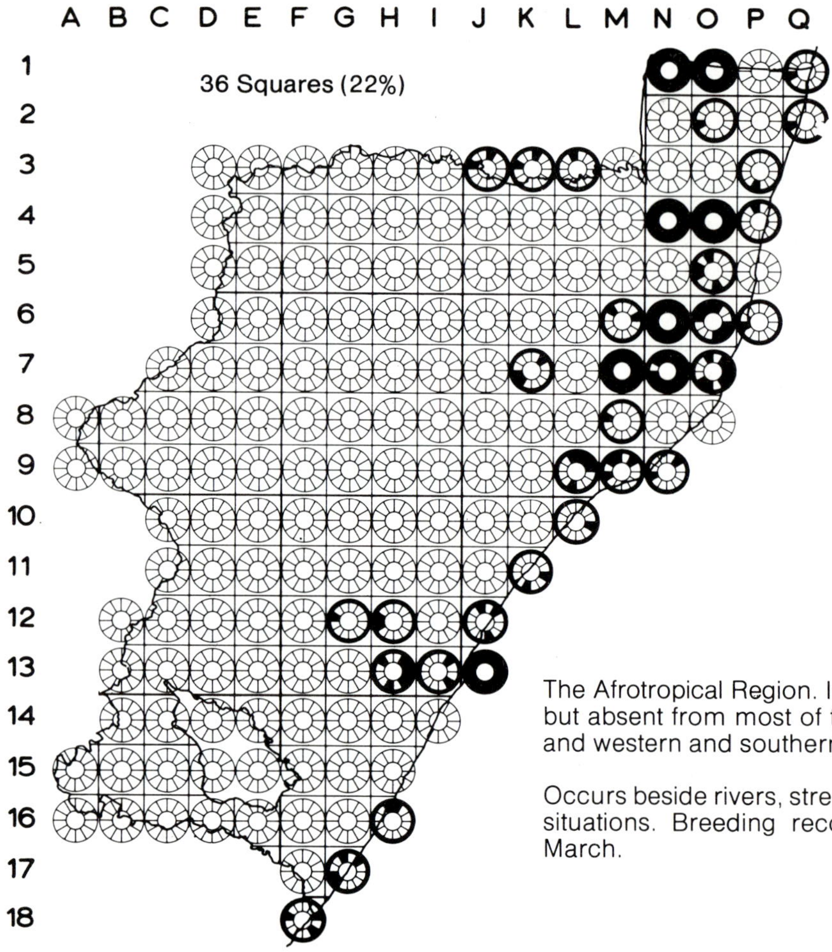

274 Water Dikkop

Burhinus vermiculatus

The Afrotropical Region. In Southern Africa, locally distributed throughout, but absent from most of the interior and northern Cape, much of S.W.A. and western and southern Botswana.

Occurs beside rivers, streams, coastal lakes and pans, usually in sheltered situations. Breeding recorded from August to December and during March.

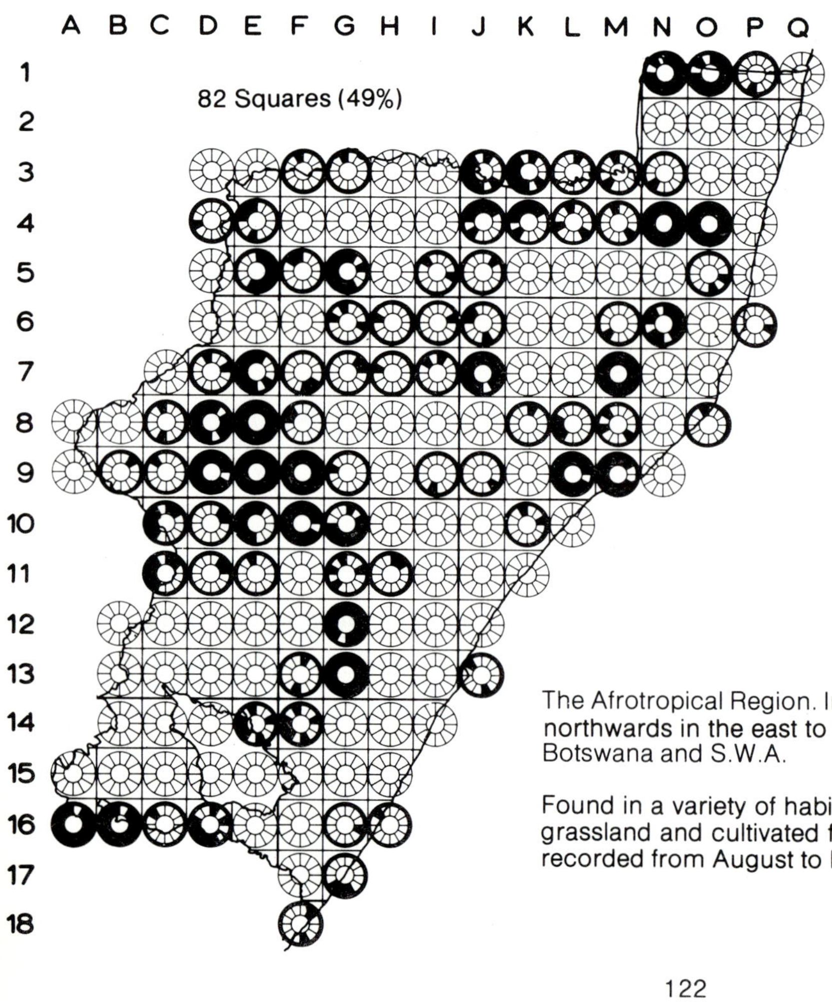

275 Cape Dikkop

Burhinus capensis

The Afrotropical Region. In Southern Africa, the Cape, Natal and Zululand, northwards in the east to Rhodesia and Moçambique; in the northwest to Botswana and S.W.A.

Found in a variety of habitats, from thornveld and open savanna to open grassland and cultivated fields; roosts during the day in shade. Breeding recorded from August to February.

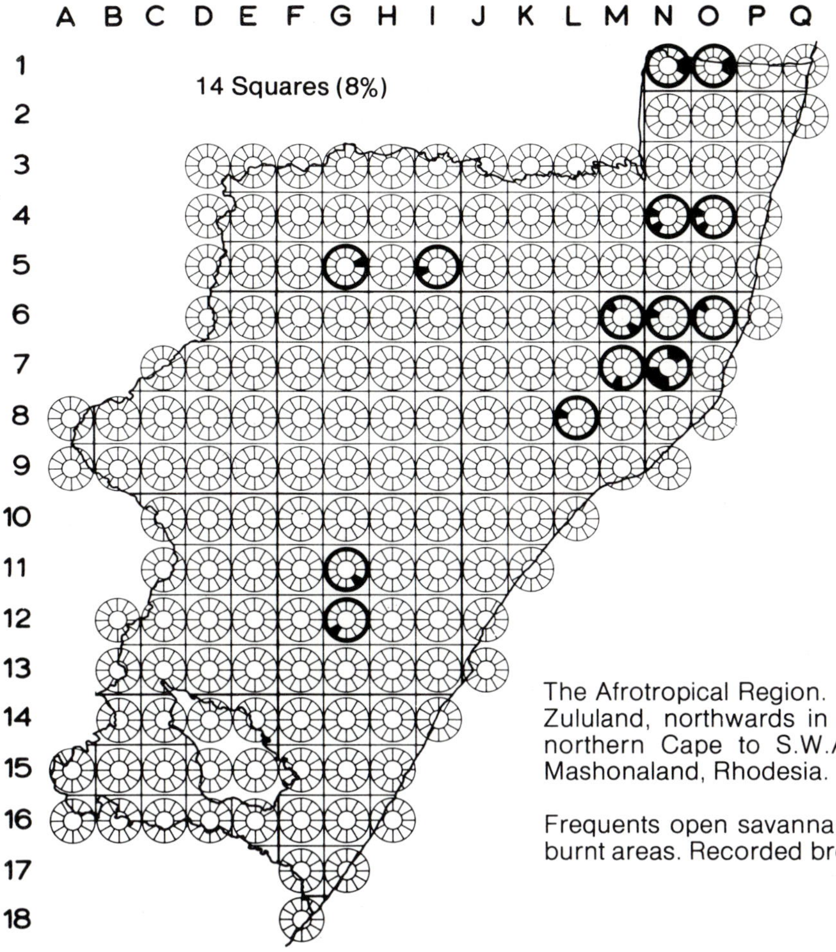

277 Temminck's Courser

Cursorius temminckii

The Afrotropical Region. In Southern Africa, the eastern Cape, Natal and Zululand, northwards in the east, extending across Botswana and the northern Cape to S.W.A. Recorded as a breeding winter visitor to Mashonaland, Rhodesia.

Frequents open savanna, and open grassland such as airfields; favours burnt areas. Recorded breeding during October and November.

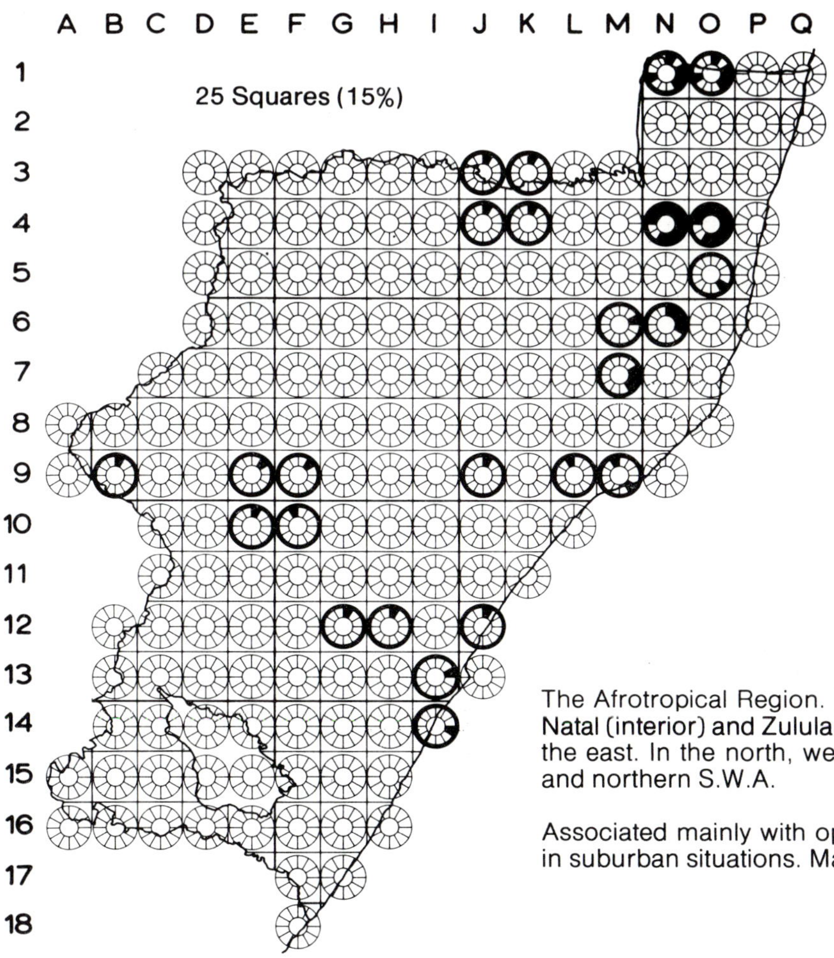

280 Bronze-wing Courser

Rhinoptilus chalcopterus

The Afrotropical Region. In Southern Africa, the eastern Cape (seldom), Natal (interior) and Zululand, northwards to Moçambique and Rhodesia in the east. In the north, west to the northern Cape, Botswana and central and northern S.W.A.

Associated mainly with open savanna and thornveld; has been recorded in suburban situations. May breed.

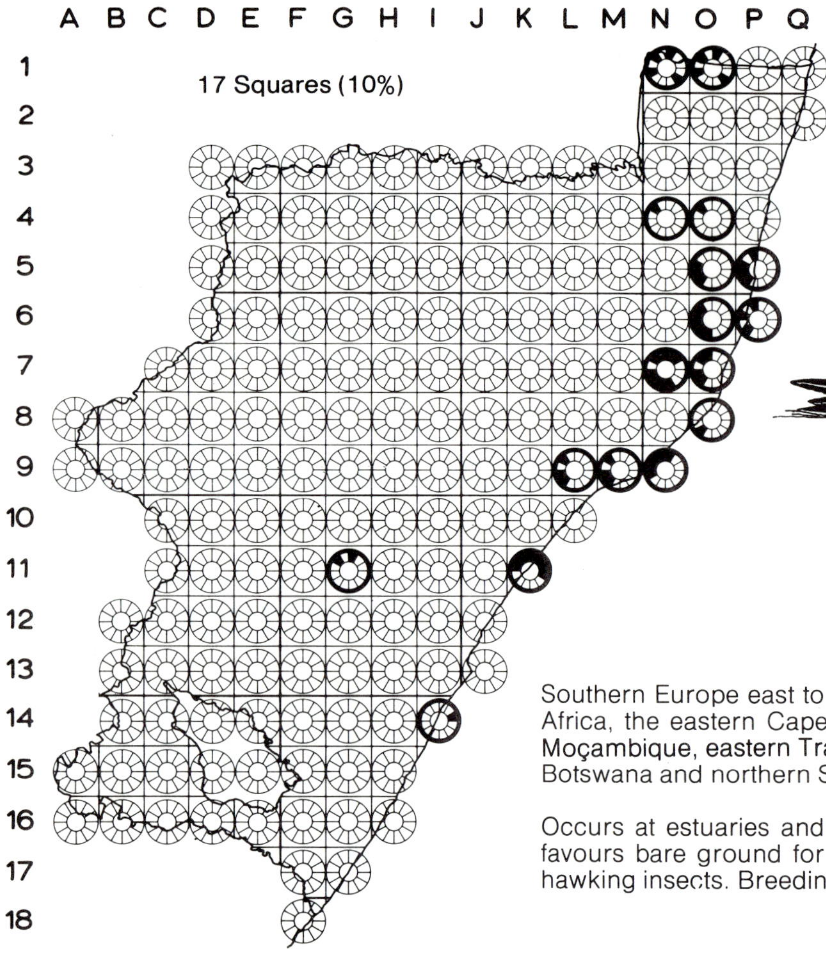

281 Red-winged Pratincole

Glareola pratincola

Southern Europe east to India and south to southern Africa. In Southern Africa, the eastern Cape (seldom), coastal Natal and eastern Zululand, Moçambique, eastern Transvaal, Rhodesia (below 900 m a.s.l.), northern Botswana and northern S.W.A.

Occurs at estuaries and coastal lakes, and occasionally pans, where it favours bare ground for nesting; may be seen over adjacent grassland hawking insects. Breeding recorded from September to December.

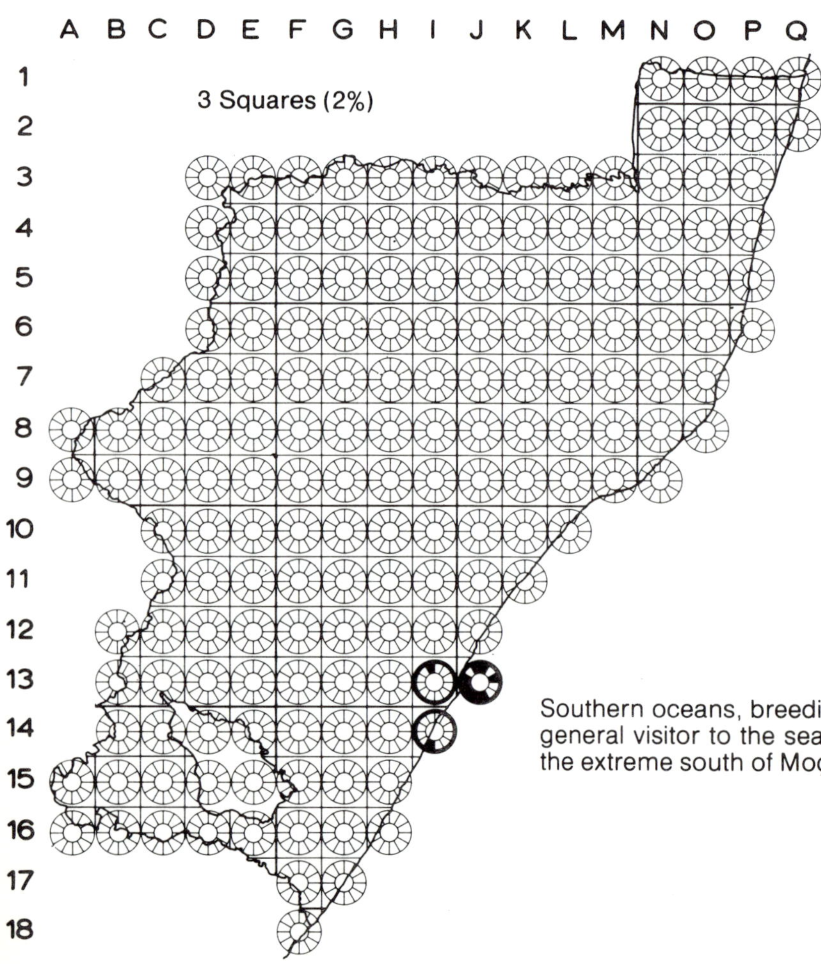

286 Antarctic Skua

Catharacta antarctica

Southern oceans, breeding on subantarctic islands. In Southern Africa, a general visitor to the seas off S.W.A., the Cape, Natal and Zululand, and the extreme south of Moçambique. Most common during the winter.

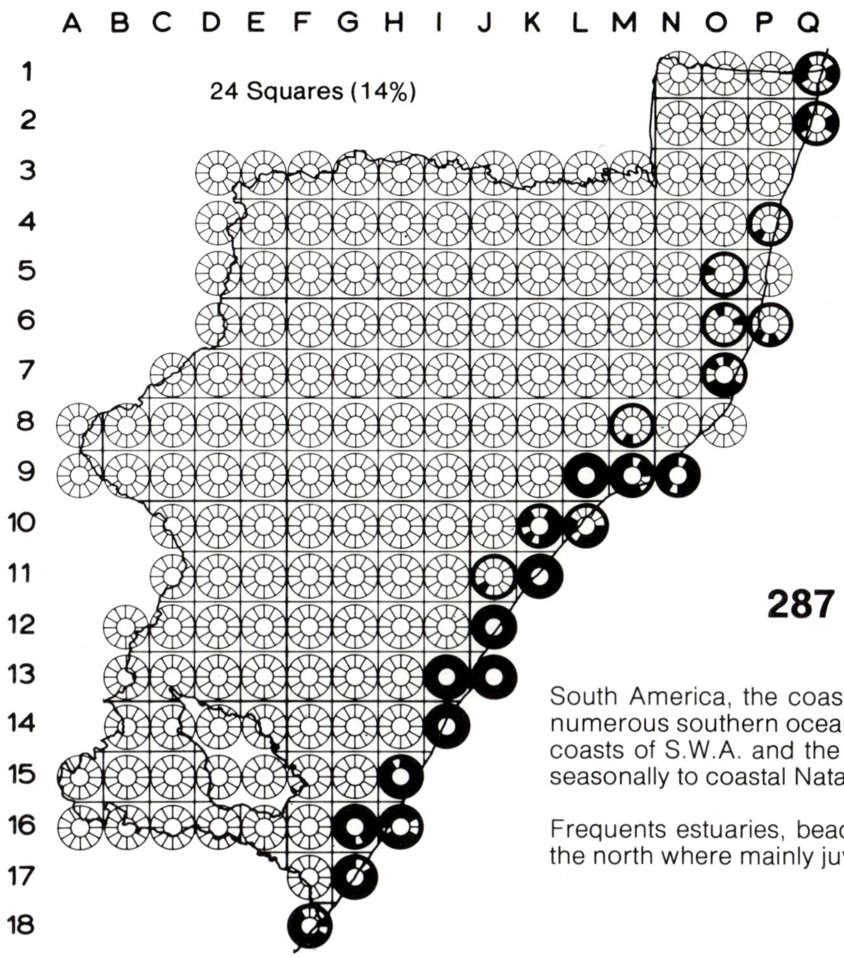

287 Southern Black-backed Gull

Larus dominicanus

South America, the coasts of South Africa, Australia, New Zealand and numerous southern oceanic islands. In Southern Africa, breeds along the coasts of S.W.A. and the Cape (east to Bird Island, Algoa Bay), ranging seasonally to coastal Natal, Zululand and southern Mocambique.

Frequents estuaries, beaches and off-shore waters, becoming sparse in the north where mainly juveniles recorded.

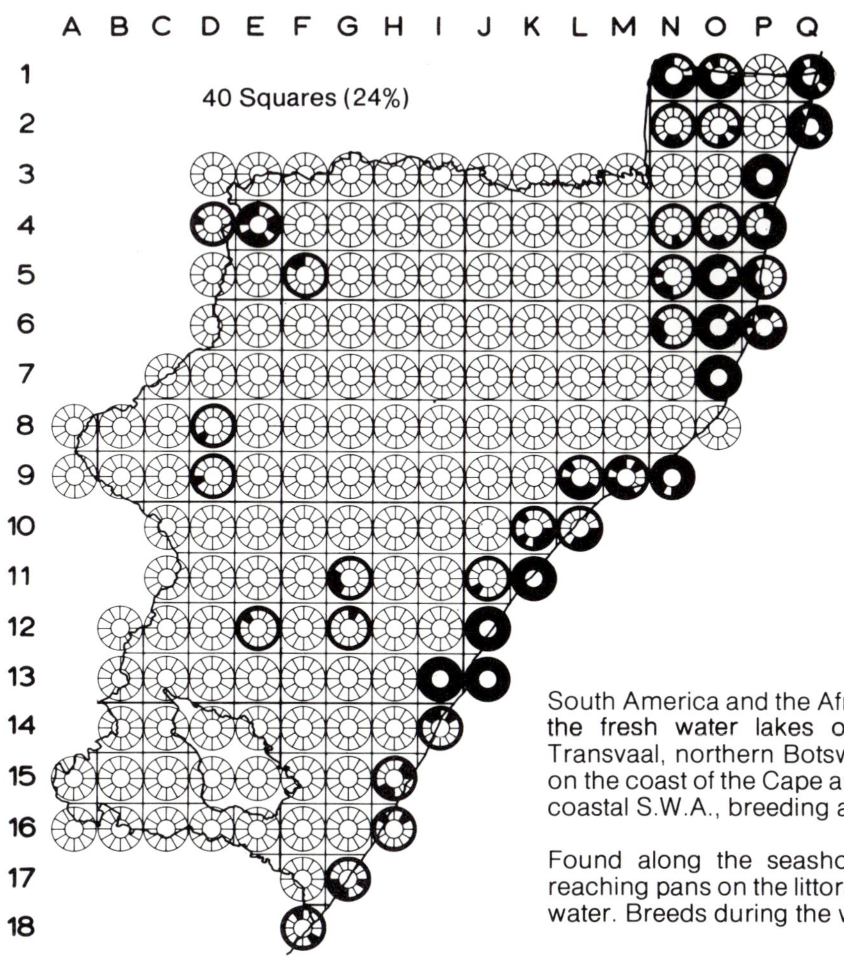

288 Grey-headed Gull

Larus cirrocephalus

South America and the Afrotropical Region. In Southern Africa, breeds on the fresh water lakes of Zululand, Moçambique, the Cape, O.F.S., Transvaal, northern Botswana, and locally elsewhere. Occurs seasonally on the coast of the Cape and Natal, northwards in the east. Recorded from coastal S.W.A., breeding at Walvis Bay.

Found along the seashore, estuaries and coastal lakes, occasionally reaching pans on the littoral plain. Recorded inland from open stretches of water. Breeds during the winter.

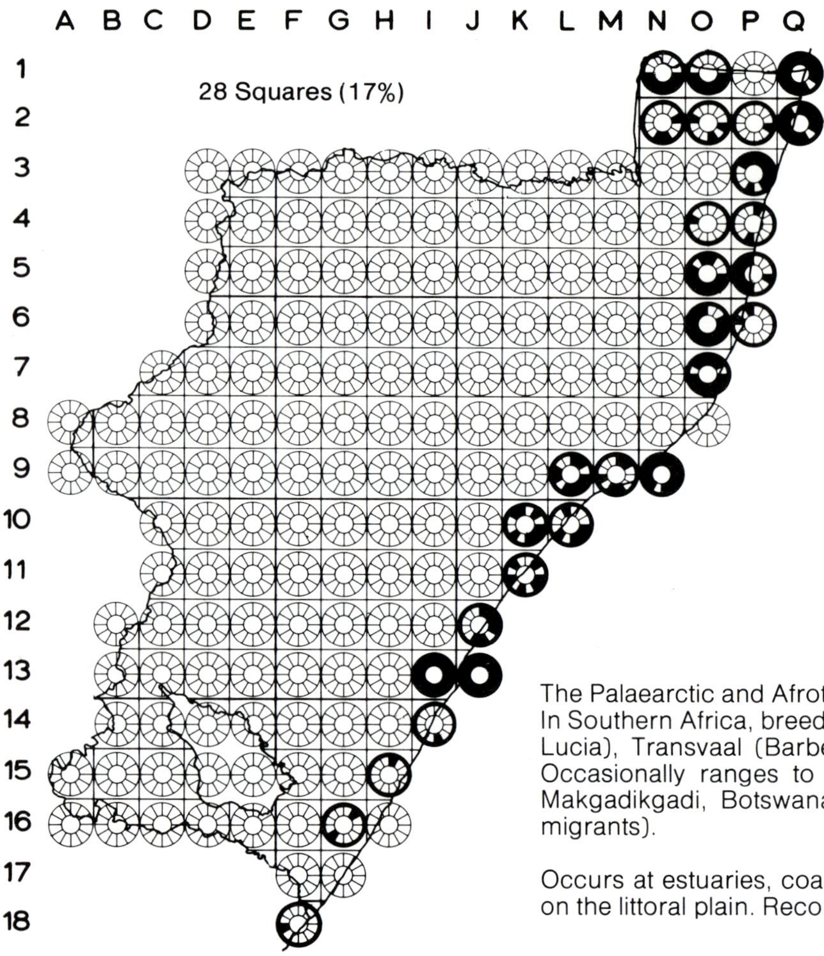

290 Caspian Tern

Hydroprogne caspia

The Palaearctic and Afrotropical Regions, North America and Australasia. In Southern Africa, breeds in S.W.A., the western Cape, Zululand (Lake St Lucia), Transvaal (Barberspan), and the mouth of the Zambesi River. Occasionally ranges to other coastal areas; recorded as common at Makgadikgadi, Botswana, during January and September (Palaearctic migrants).

Occurs at estuaries, coastal lakes and off-shore waters, ranging to pans on the littoral plain. Recorded breeding during June and July.

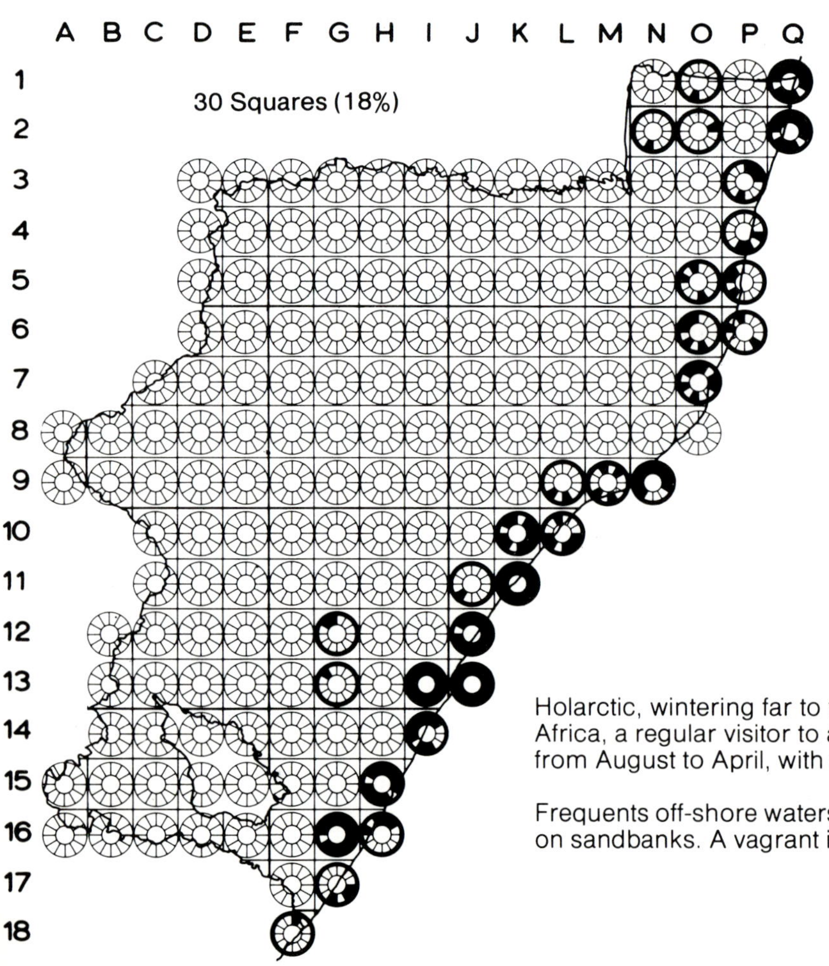

291 Common Tern

Sterna hirundo

Holarctic, wintering far to the south of the breeding grounds. In Sqthern Africa, a regular visitor to all coasts, but mainly the western and southern, from August to April, with some present throughout the year.

Frequents off-shore waters and coastal lakes, coming to estuaries to roost on sandbanks. A vagrant inland.

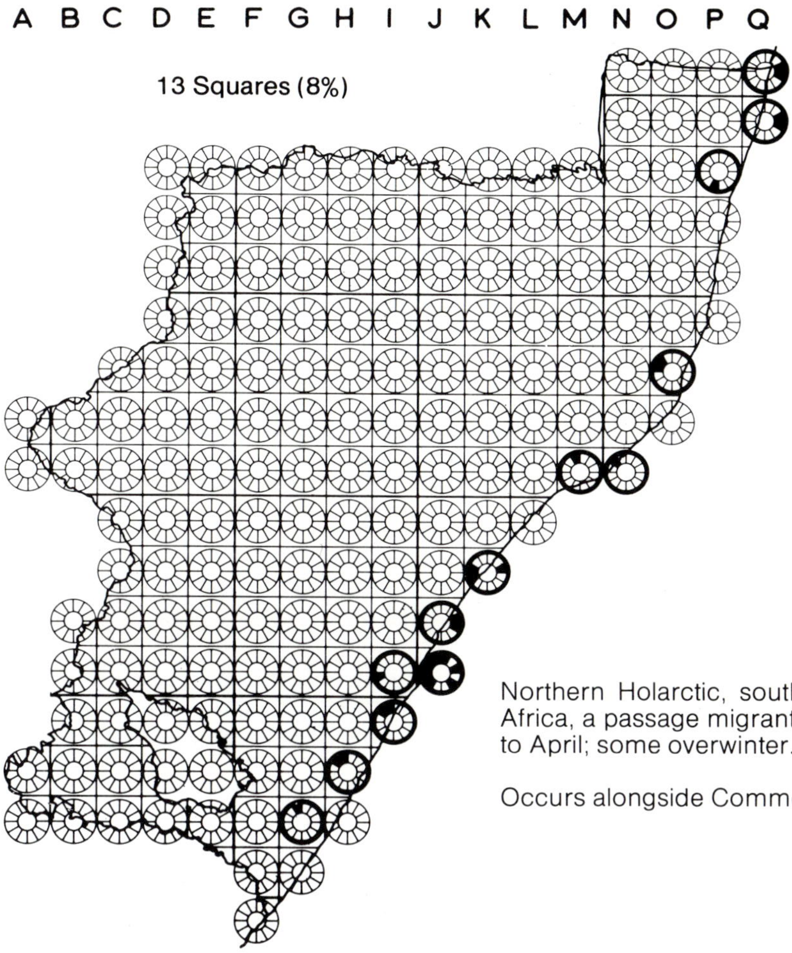

294 Arctic Tern

Sterna paradisaea

Northern Holarctic, south to the Antarctic seas in winter. In Southern Africa, a passage migrant to all coasts (occasionally inland), from August to April; some overwinter.

Occurs alongside Common Tern (291), in much smaller numbers.

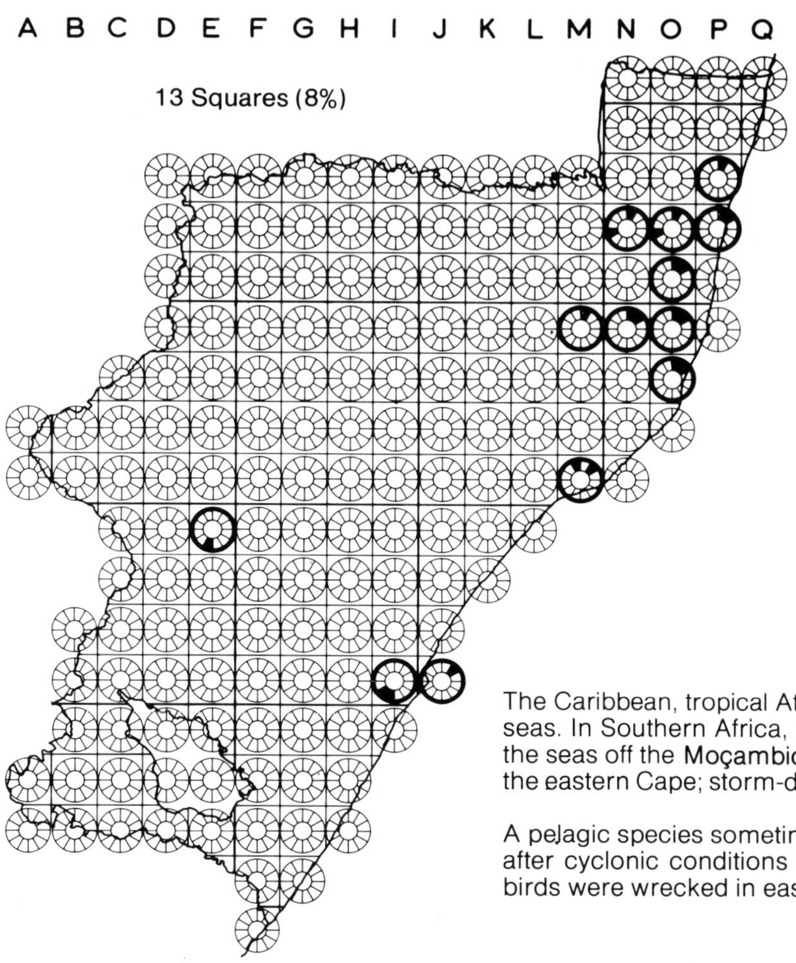

295 Sooty Tern

Sterna fuscata

The Caribbean, tropical Atlantic, Indian and Pacific Oceans, and Australian seas. In Southern Africa, a regular visitor mainly from October to April to the seas off the Moçambique, Zululand and Natal coasts, and less often to the eastern Cape; storm-driven individuals reach Rhodesia and Transvaal.

A pelagic species sometimes found exhausted along the coast and inland after cyclonic conditions in the Moçambique Channel. Several hundred birds were wrecked in eastern Zululand in January, 1976.

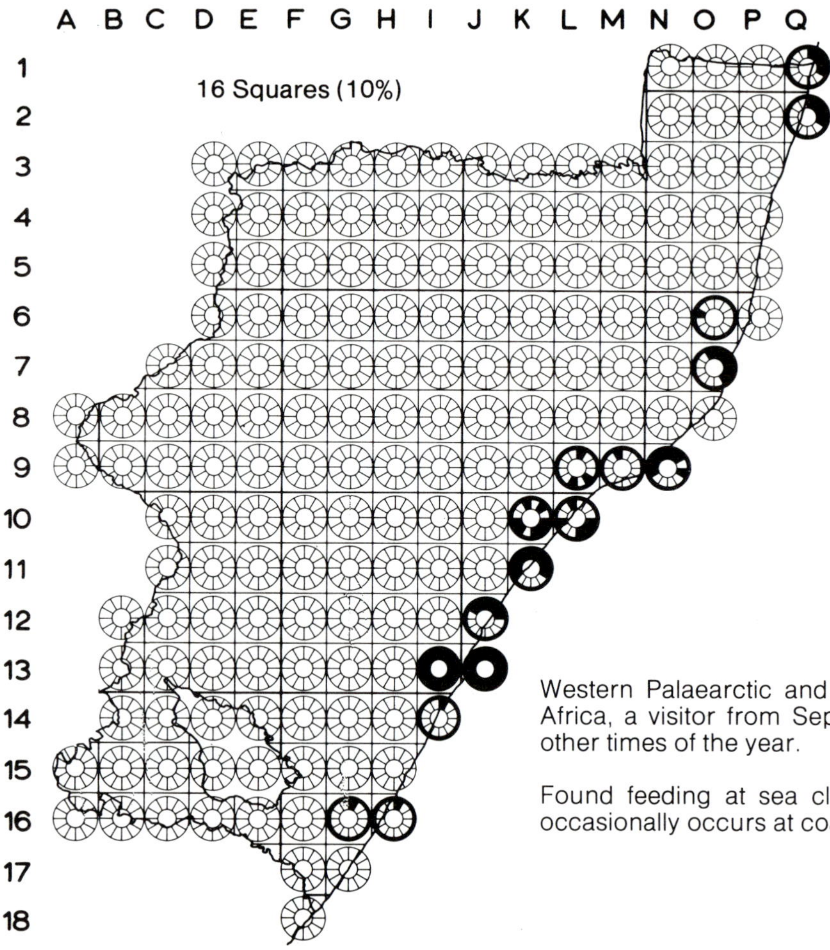

296 Sandwich Tern

Sterna sandvicensis

Western Palaearctic and Nearctic, wintering to the south. In Southern Africa, a visitor from September to April to all coasts; some present at other times of the year.

Found feeding at sea close to the shore, entering estuaries to roost; occasionally occurs at coastal lakes.

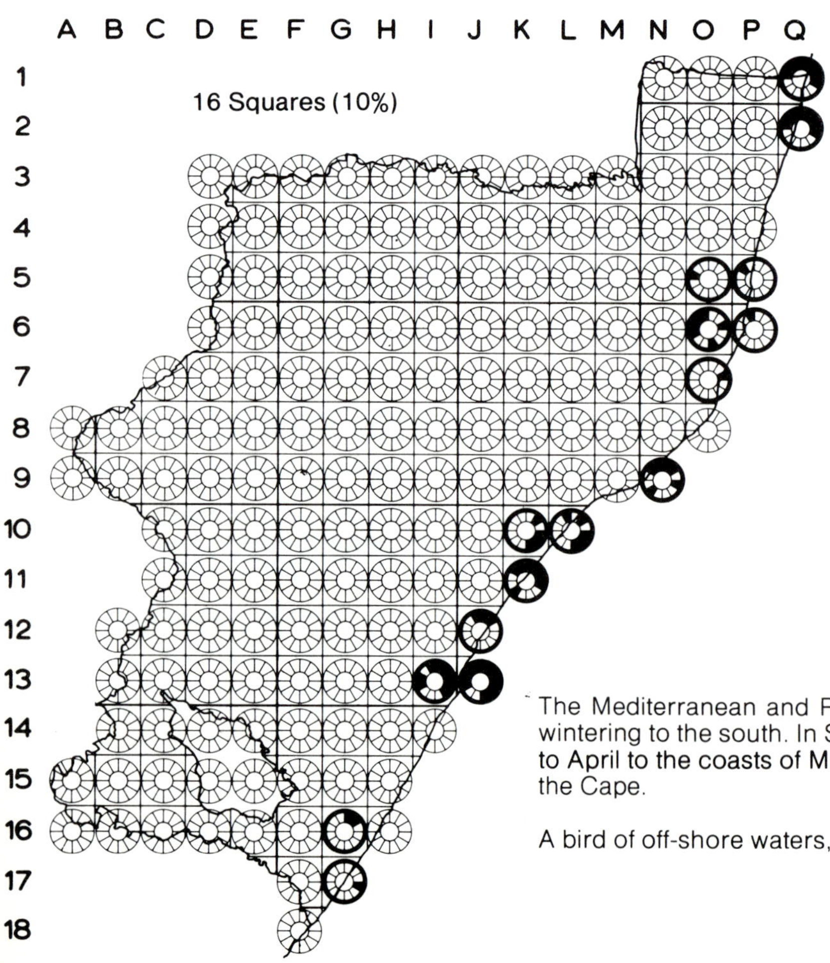

297 Lesser Crested Tern

Sterna bengalensis

The Mediterranean and Red Sea coasts to southern Asia and Australia, wintering to the south. In Southern Africa, a visitor mainly from December to April to the coasts of Moçambique, Natal and Zululand; occasionally to the Cape.

A bird of off-shore waters, visiting coastal lakes and estuaries.

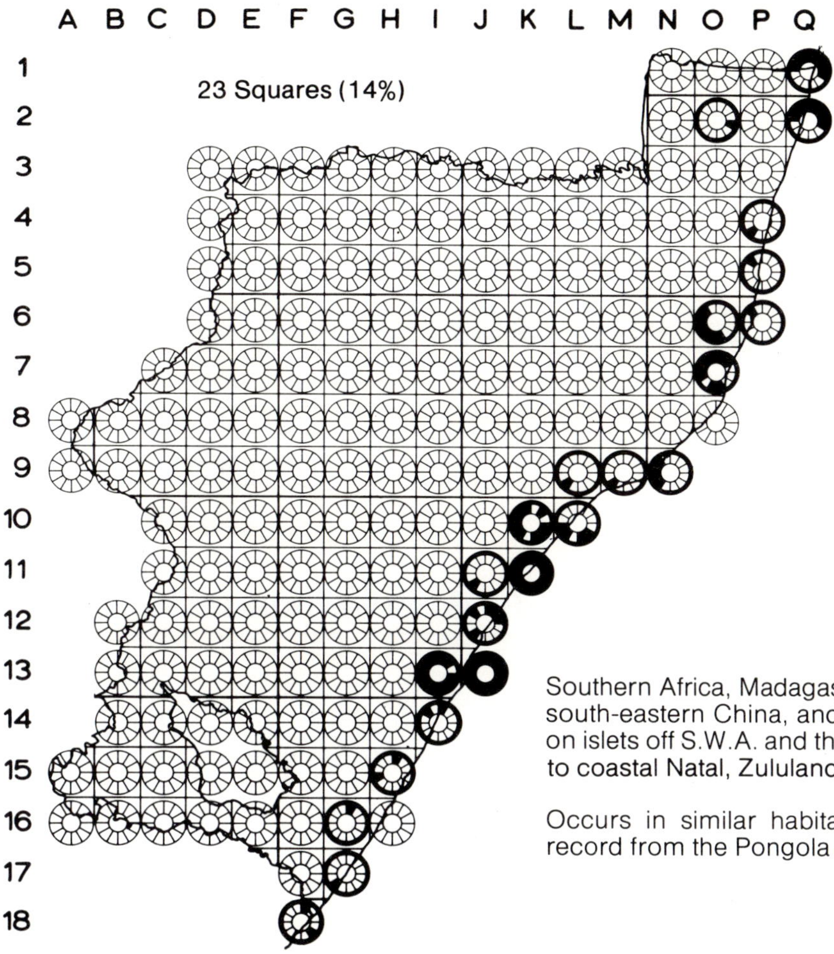

298 Swift Tern

Sterna bergii

Southern Africa, Madagascar, and islands in the Indian Ocean to Australia, south-eastern China, and the western Pacific. In Southern Africa, breeds on islets off S.W.A. and the Cape, east to Algoa Bay, wandering seasonally to coastal Natal, Zululand and southern Moçambique.

Occurs in similar habitats to Lesser Crested Tern (297); exceptional record from the Pongola flood plain.

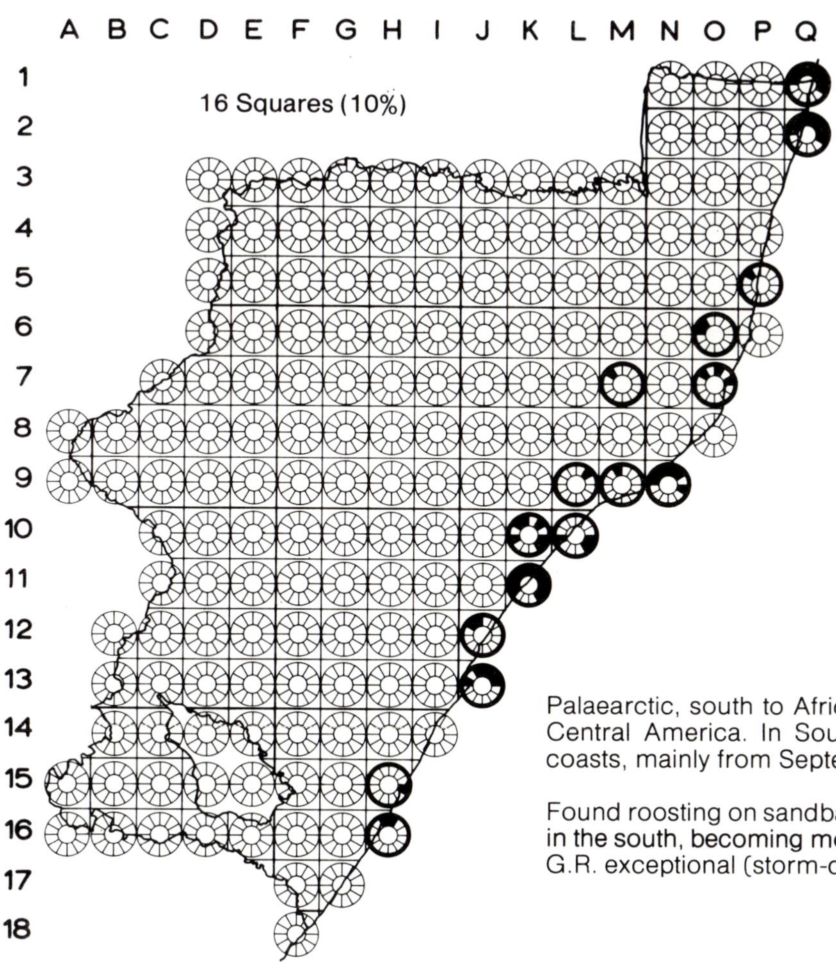

299 Little Tern

Sterna albifrons

Palaearctic, south to Africa, southern Asia and Australia; also North and Central America. In Southern Africa, a visitor in small numbers to all coasts, mainly from September to April.

Found roosting on sandbanks at estuaries, flying out to sea to feed. Sparse in the south, becoming more common farther north. Record from Umfolozi G.R. exceptional (storm-driven).

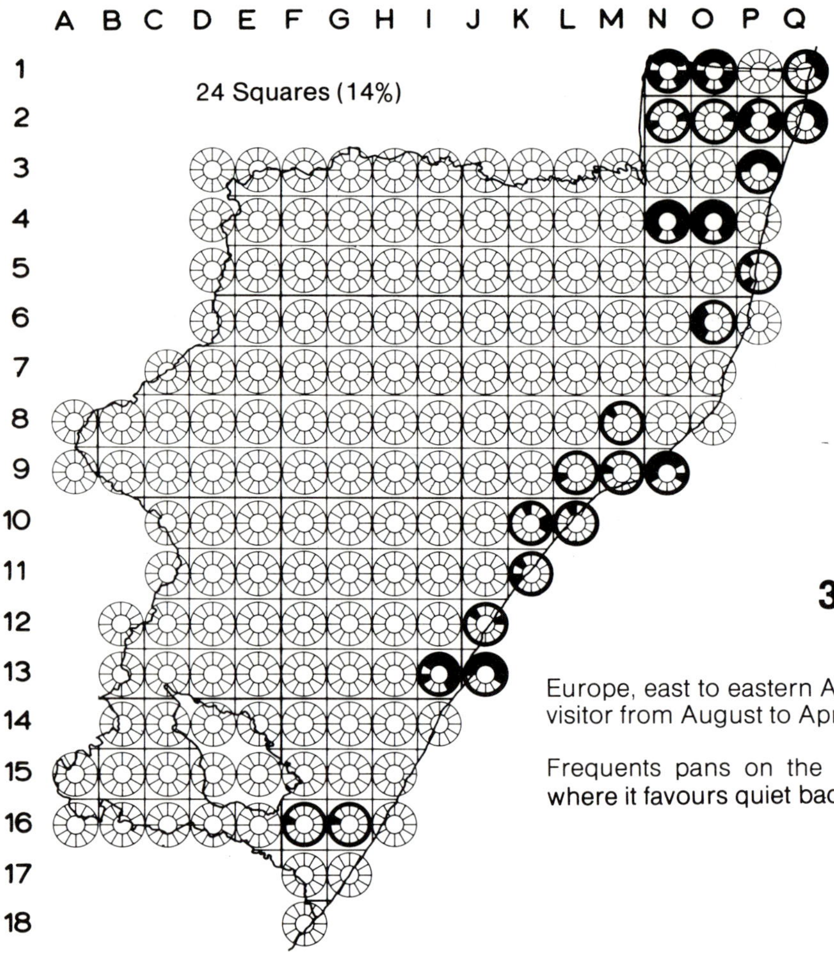

304 White-winged Black Tern

Chlidonias leucopterus

Europe, east to eastern Asia, wintering to the south. In Southern Africa, a visitor from August to April, recorded throughout (numbers fluctuate).

Frequents pans on the littoral plain and coastal lakes; also estuaries, where it favours quiet backwaters.

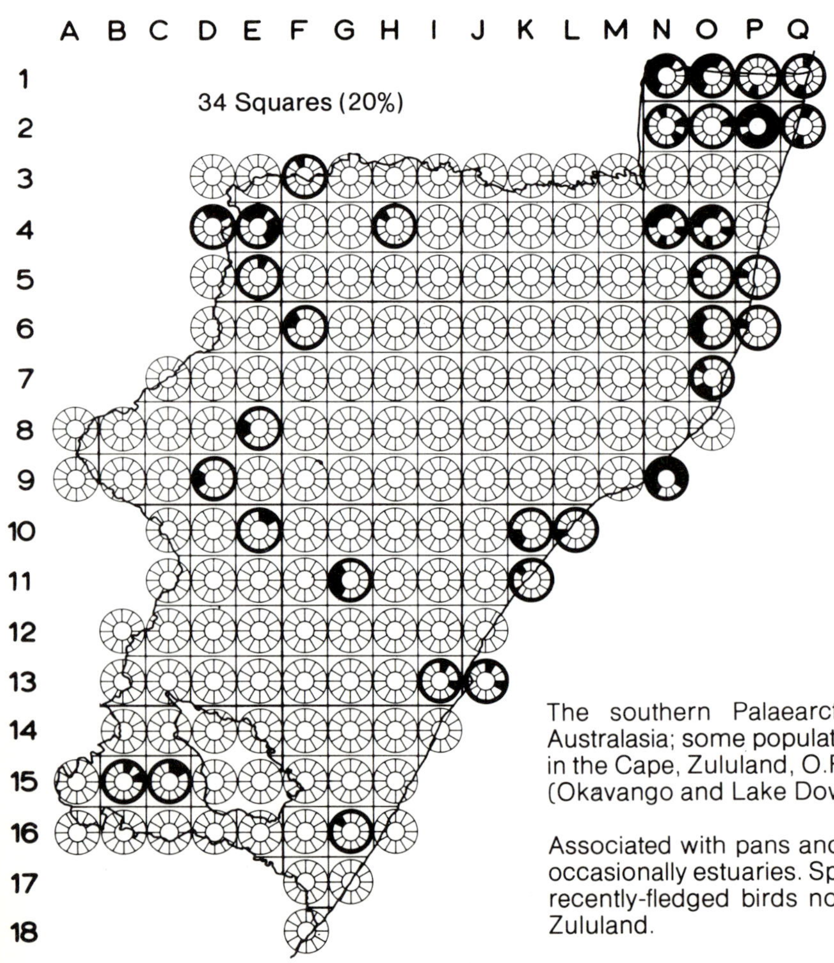

305 Whiskered Tern

Chlidonias hybridus

The southern Palaearctic, Africa and southern Asia southeast to Australasia; some populations migratory. In Southern Africa, breeds locally in the Cape, Zululand, O.F.S., Transvaal, western Rhodesia and Botswana (Okavango and Lake Dow); probably elsewhere.

Associated with pans and marshes on the littoral plain, coastal lakes and occasionally estuaries. Sparse inland at open stretches of water. Breeding: recently-fledged birds noted in March 1979 at Pelindaba, north-eastern Zululand.

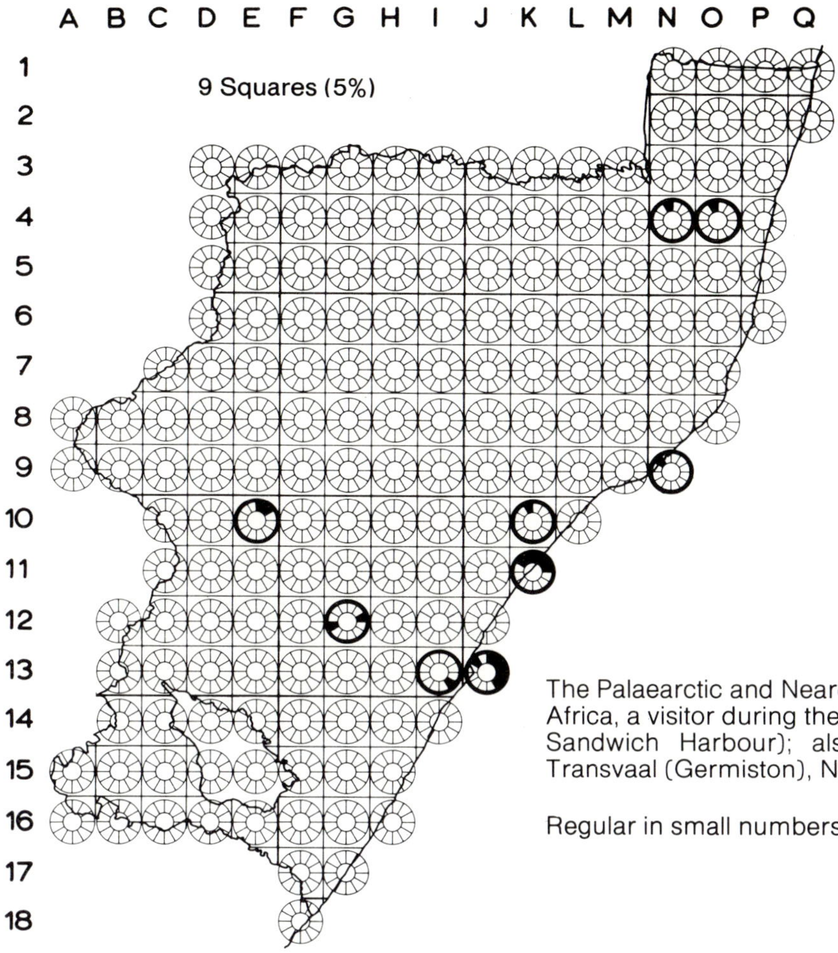

305X Black Tern

Chlidonias niger

The Palaearctic and Nearctic Regions, wintering to the south. In Southern Africa, a visitor during the northern winter to the coast of S.W.A. (south to Sandwich Harbour); also recorded from Botswana (Lake Ngami), Transvaal (Germiston), Natal and the eastern Cape (Port Elizabeth).

Regular in small numbers to estuaries; rare at inland waters.

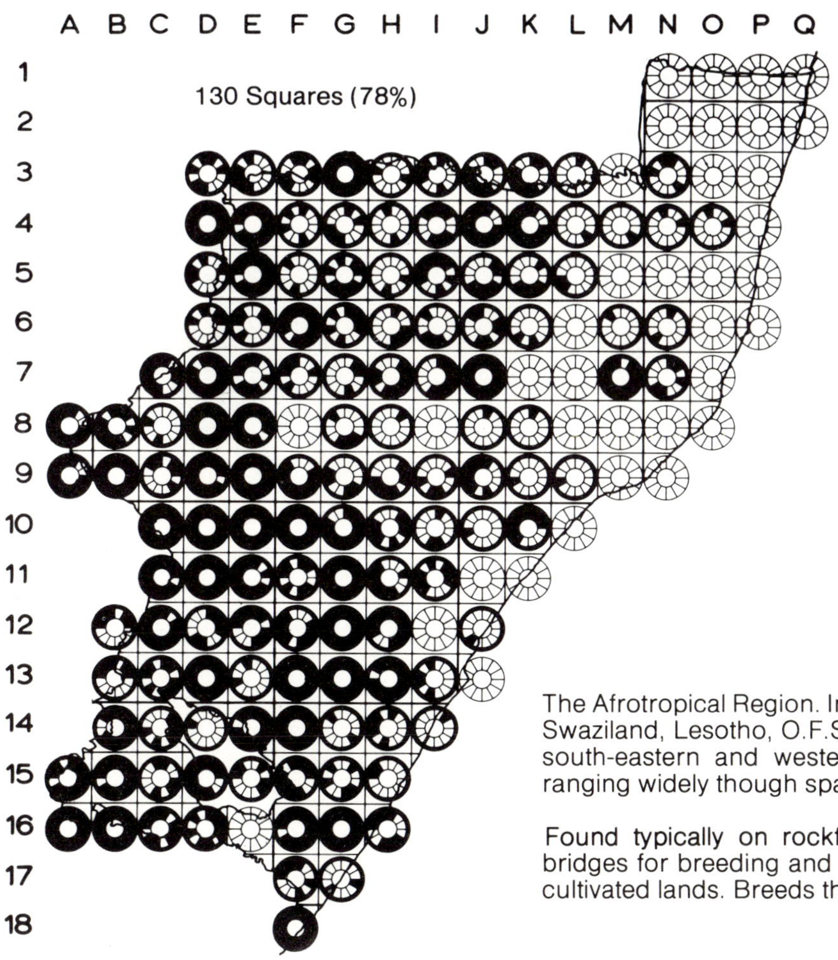

311 Rock Pigeon

Columba guinea

The Afrotropical Region. In Southern Africa, the Cape, Natal and Zululand, Swaziland, Lesotho, O.F.S., Transvaal and Rhodesia, westwards through south-eastern and western Botswana to S.W.A. A common species, ranging widely though sparse on the coast.

Found typically on rockfaces, but has also adapted to buildings and bridges for breeding and roosting. Often seen feeding in large flocks on cultivated lands. Breeds throughout the year.

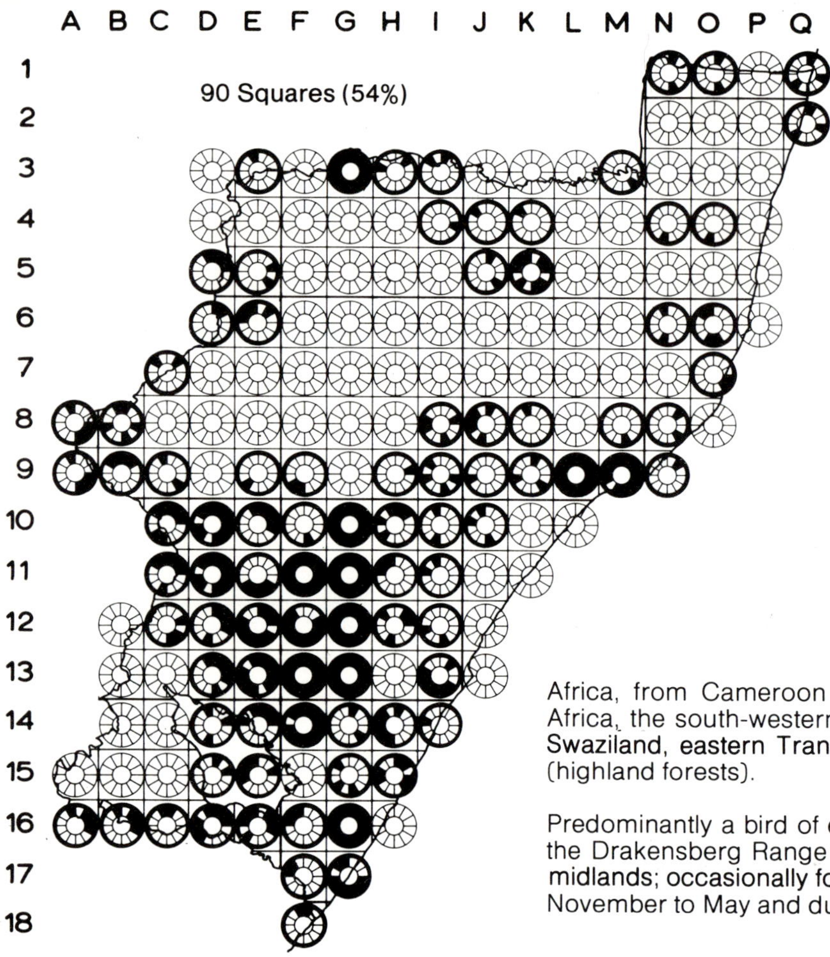

312 Rameron Pigeon

Columba arquatrix

Africa, from Cameroon and Ethiopia south to the Cape. In Southern Africa, the south-western Cape to Natal and Zululand, O.F.S. (seldom), Swaziland, eastern Transvaal and eastern Rhodesia; also Moçambique (highland forests).

Predominantly a bird of evergreen forests, ranging from high altitudes in the Drakensberg Range down to the coast, being most common in the midlands; occasionally found in riparian growth. Recorded breeding from November to May and during July.

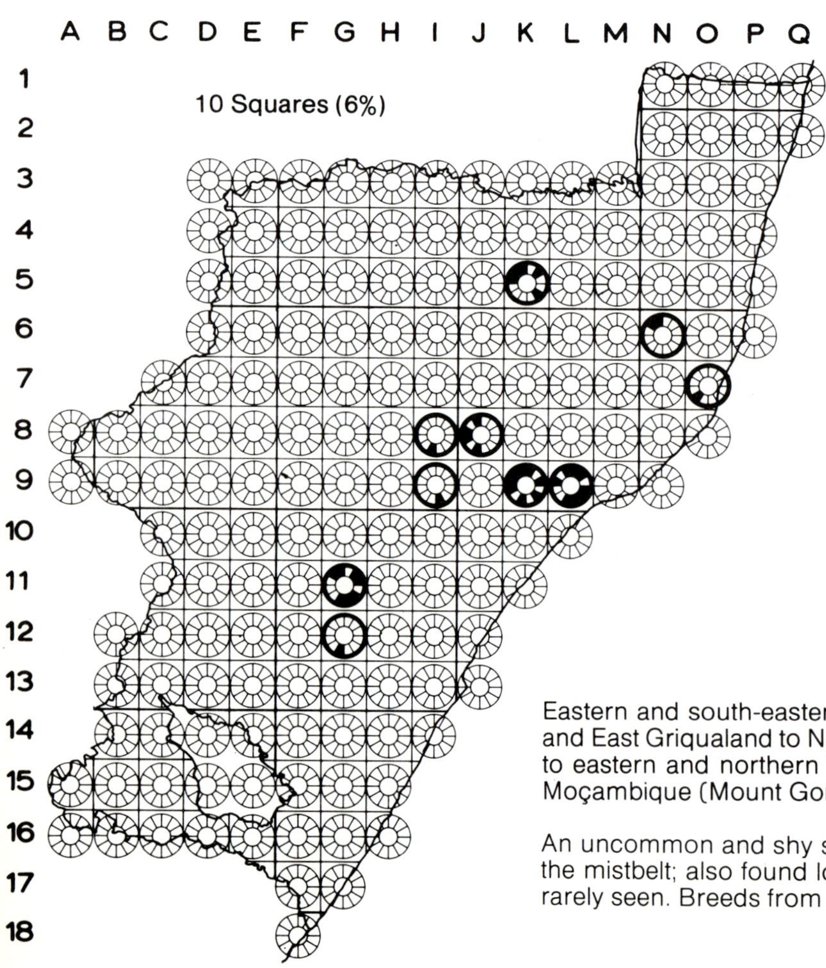

313 Bronze-naped Pigeon

Columba delegorguei

Eastern and south-eastern Africa. In Southern Africa, forests of Transkei and East Griqualand to Natal (mainly interior) and western Zululand, north to eastern and northern Transvaal, eastern Rhodesia (Lusitu River) and Moçambique (Mount Gorongoza).

An uncommon and shy species, restricted to climax evergreen forests in the mistbelt; also found locally on the coast. Recorded mainly on call, as rarely seen. Breeds from November to April.

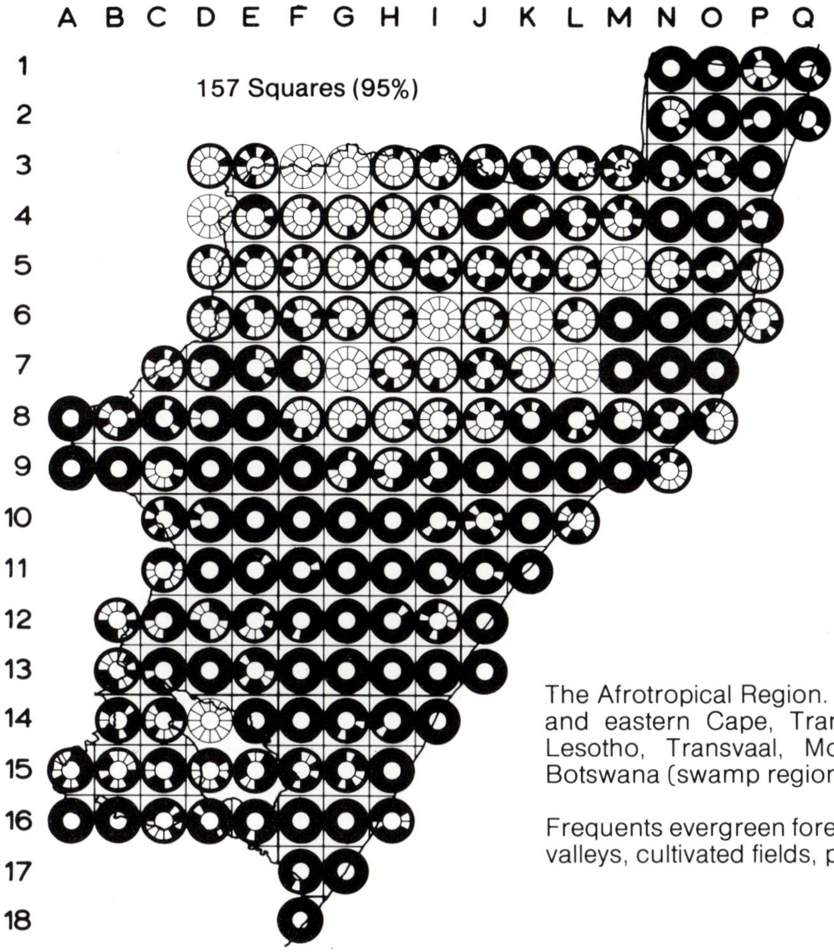

314 Red-eyed Turtle Dove

Streptopelia semitorquata

The Afrotropical Region. In Southern Africa, the south-western, southern and eastern Cape, Transkei, Natal and Zululand, Swaziland, O.F.S., Lesotho, Transvaal, Moçambique and Rhodesia, west to northern Botswana (swamp region), the Caprivi Strip and north-eastern S.W.A.

Frequents evergreen forest, riverine and mixed woodland, thickly-wooded valleys, cultivated fields, parks and gardens. Breeds throughout the year.

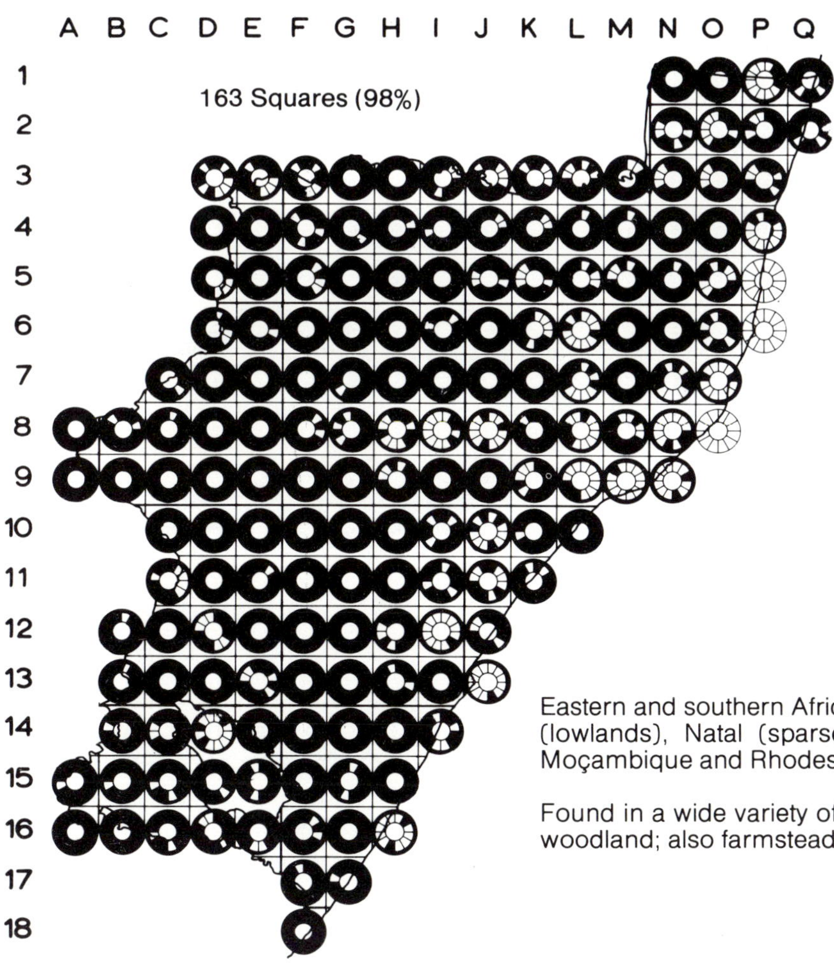

316 Cape Turtle Dove

Streptopelia capicola

Eastern and southern Africa. In Southern Africa, the Cape, O.F.S., Lesotho (lowlands), Natal (sparse on coast), Zululand, Swaziland, Transvaal, Moçambique and Rhodesia, west to Botswana and S.W.A.

Found in a wide variety of wooded habitats, favouring thornveld and light woodland; also farmsteads and gardens. Breeds throughout the year.

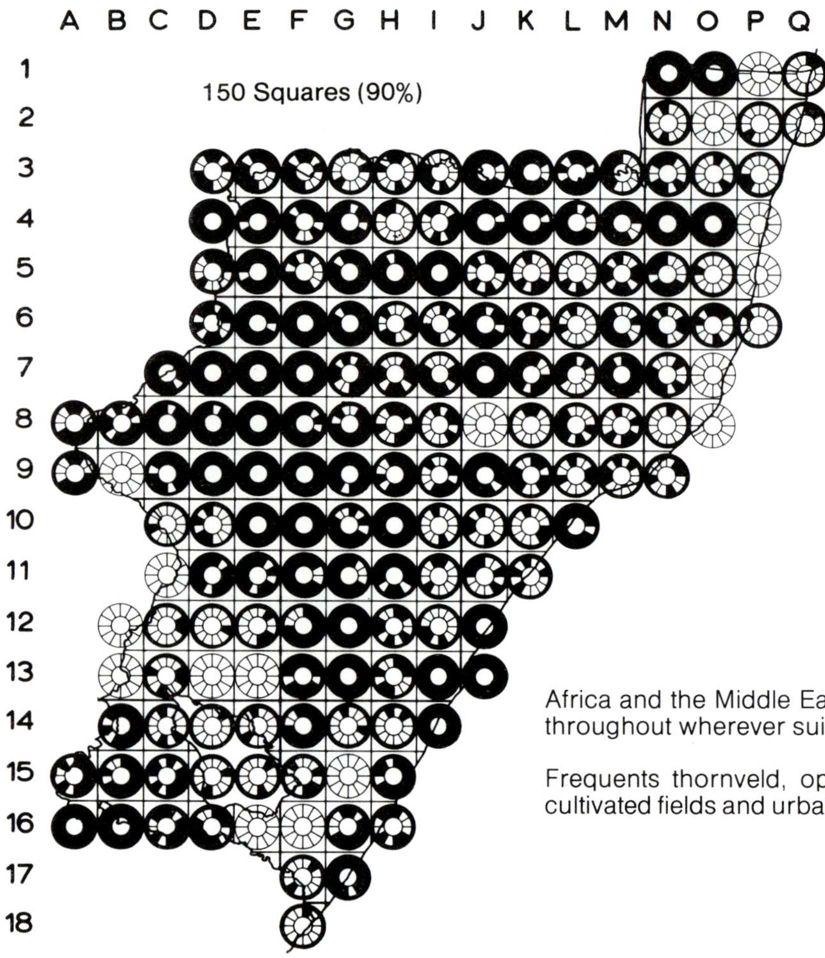

317 Laughing Dove

Streptopelia senegalensis

Africa and the Middle East east to central Asia. In Southern Africa, found throughout wherever suitable habitat occurs.

Frequents thornveld, open grassland with scattered trees, open bush, cultivated fields and urban areas. Recorded breeding throughout the year.

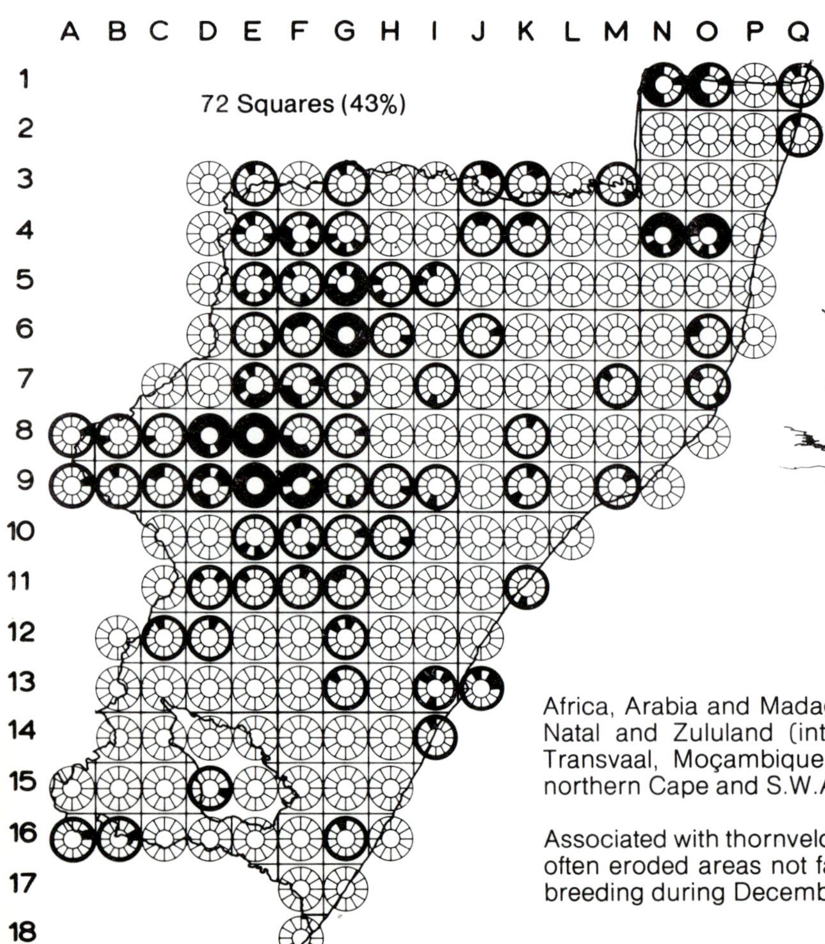

318 Namaqua Dove

Oena capensis

Africa, Arabia and Madagascar. In Southern Africa, the western Cape to Natal and Zululand (interior), O.F.S., Lesotho (lowlands), Swaziland, Transvaal, Moçambique (local) and Rhodesia, west to Botswana, the northern Cape and S.W.A.

Associated with thornveld and open grassland, where it is found in dry and often eroded areas not far from water; sporadic on the coast. Recorded breeding during December.

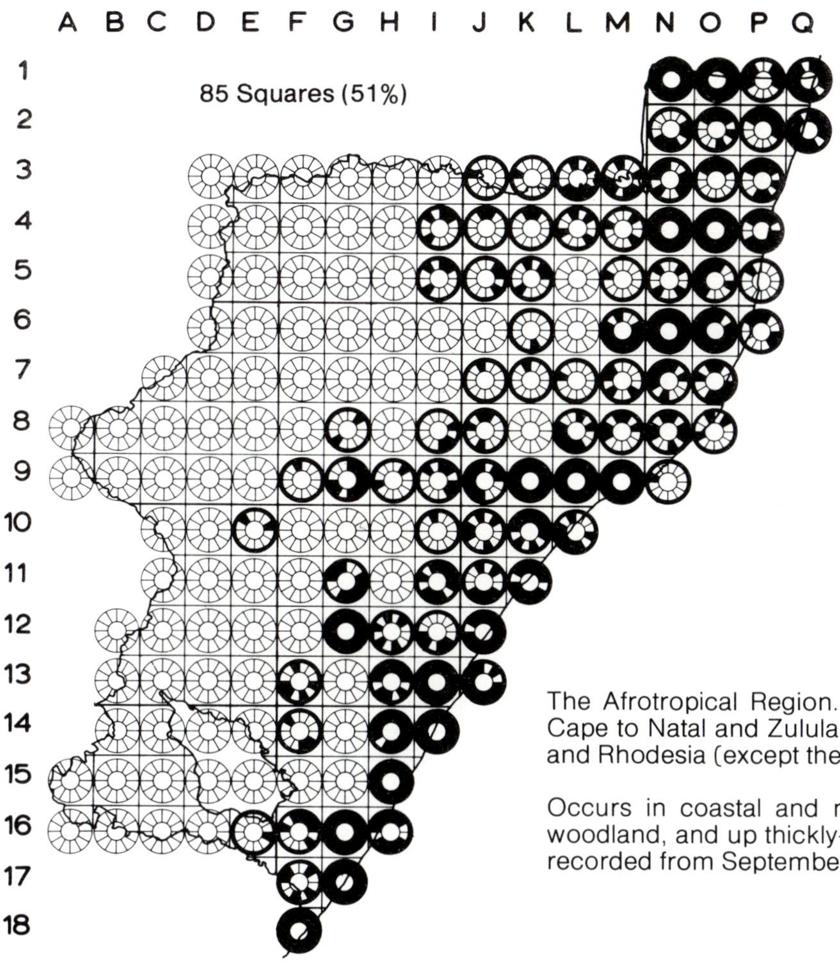

319 Tambourine Dove

Turtur tympanistria

The Afrotropical Region. In Southern Africa, the southern and eastern Cape to Natal and Zululand, Swaziland, eastern Transvaal, Moçambique and Rhodesia (except the dry west).

Occurs in coastal and mistbelt evergreen forests, riverine and mixed woodland, and up thickly-wooded valleys, to about 900 m a.s.l. Breeding recorded from September to March and during June.

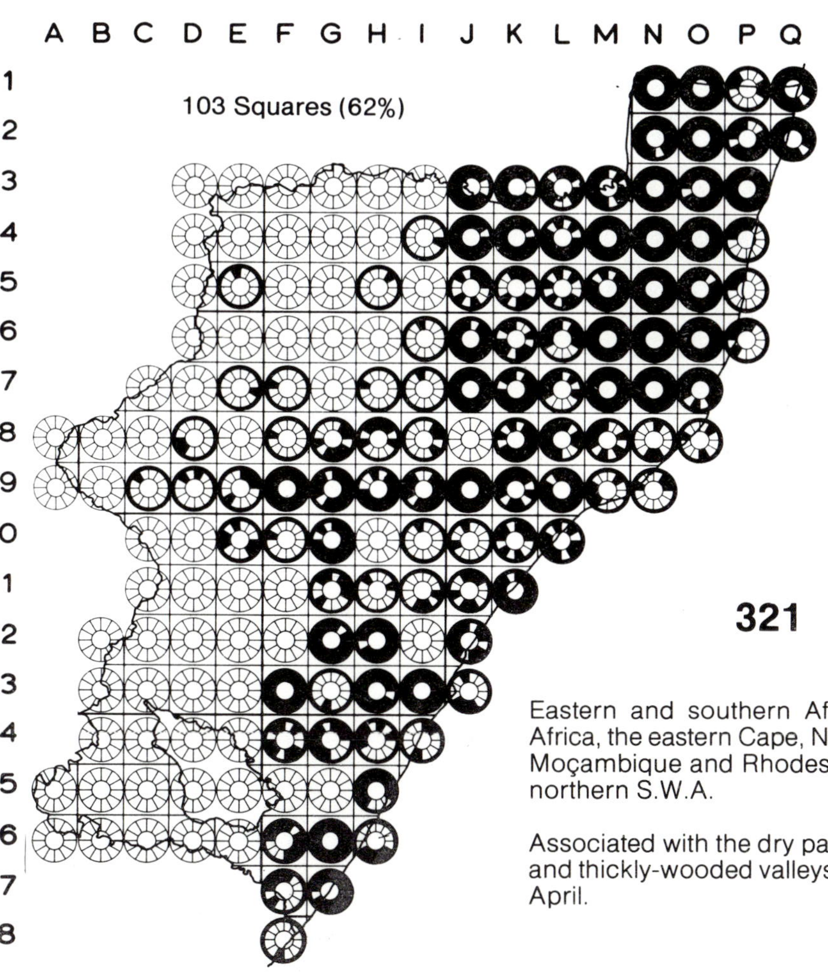

321 Emerald-spotted Wood Dove

Turtur chalcospilos

Eastern and southern Africa, from Ethiopia to the Cape. In Southern Africa, the eastern Cape, Natal and Zululand, western Swaziland, Transvaal, Moçambique and Rhodesia, west to northern and eastern Botswana and northern S.W.A.

Associated with the dry parts of coastal scrub, mixed woodland, thornveld and thickly-wooded valleys, to about 1200 m a.s.l. Breeds from October to April.

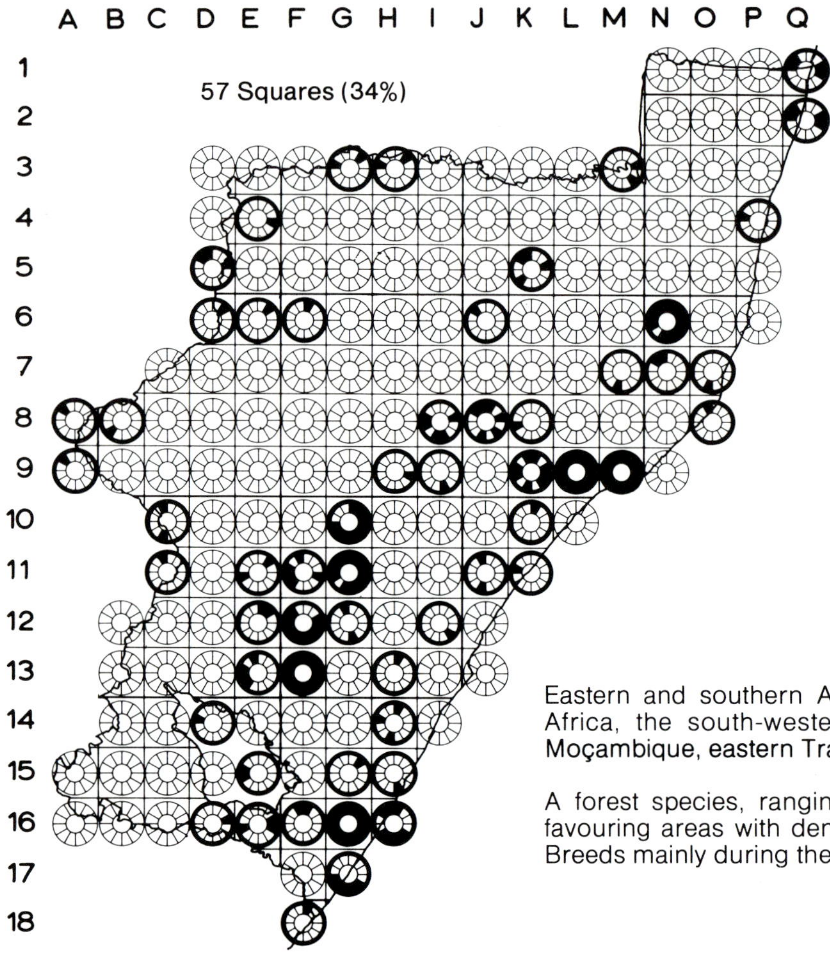

322 Cinnamon Dove

Aplopelia larvata

Eastern and southern Africa, from Ethiopia to the Cape. In Southern Africa, the south-western Cape to Natal and Zululand, Swaziland, Moçambique, eastern Transvaal and eastern Rhodesia.

A forest species, ranging from the coast to the Drakensberg Range, favouring areas with dense undergrowth where it feeds on the ground. Breeds mainly during the summer.

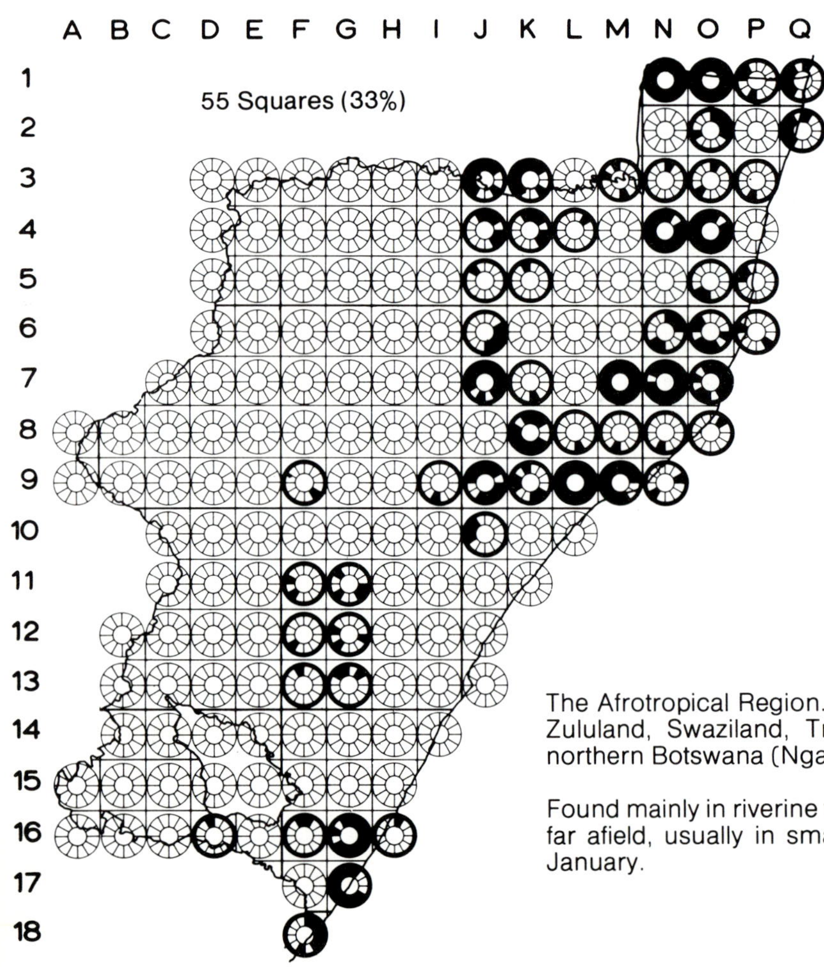

323 Green Pigeon

Treron calva

The Afrotropical Region. In Southern Africa, the eastern Cape, Natal and Zululand, Swaziland, Transvaal, Moçambique and Rhodesia, west to northern Botswana (Ngamiland), the Caprivi Strip and northern S.W.A.

Found mainly in riverine vegetation where wild figs abound; often wanders far afield, usually in small parties. Recorded breeding from October to January.

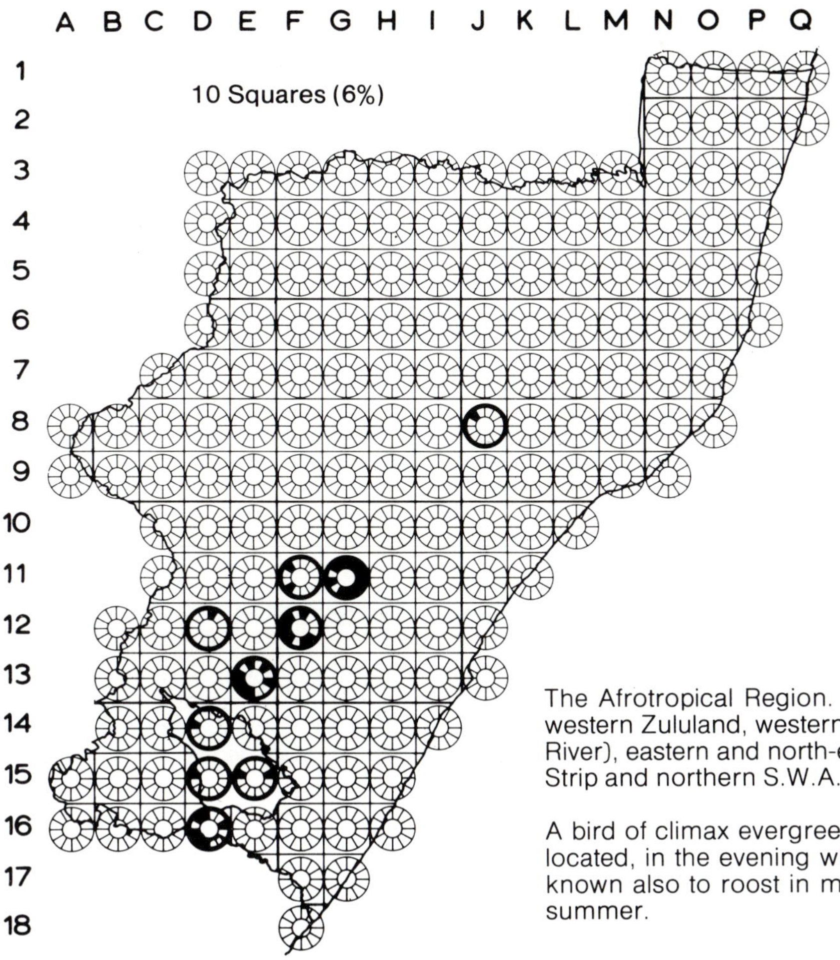

326 Brown-necked Parrot

Poicephalus robustus

The Afrotropical Region. In Southern Africa, the eastern Cape to Natal, western Zululand, western Swaziland, Moçambique (north of the Limpopo River), eastern and north-eastern Transvaal and Rhodesia; also the Caprivi Strip and northern S.W.A.

A bird of climax evergreen forests in the mistbelt. Most active, and easily located, in the evening when gathering to roost, often in large numbers; known also to roost in mature stands of exotic trees. Breeds during the summer.

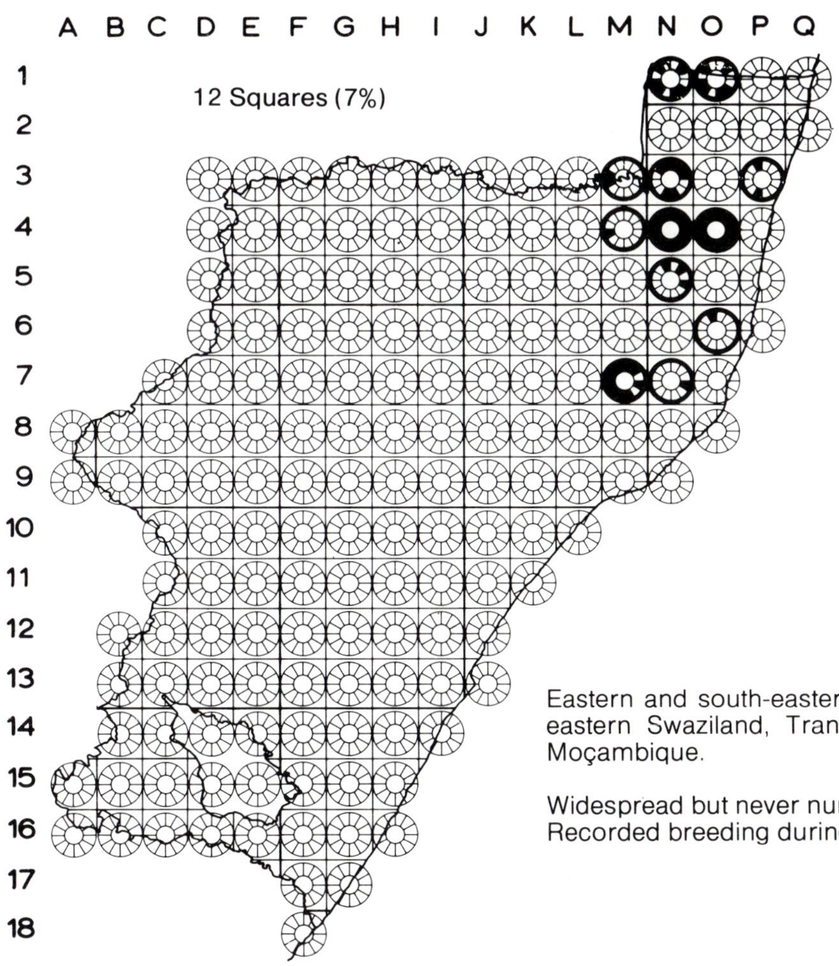

328 Brown-headed Parrot

Poicephalus cryptoxanthus

Eastern and south-eastern Africa. In Southern Africa, eastern Zululand, eastern Swaziland, Transvaal (lowveld), south-eastern Rhodesia, and Moçambique.

Widespread but never numerous, inhabiting thornveld and dry woodland. Recorded breeding during June and July.

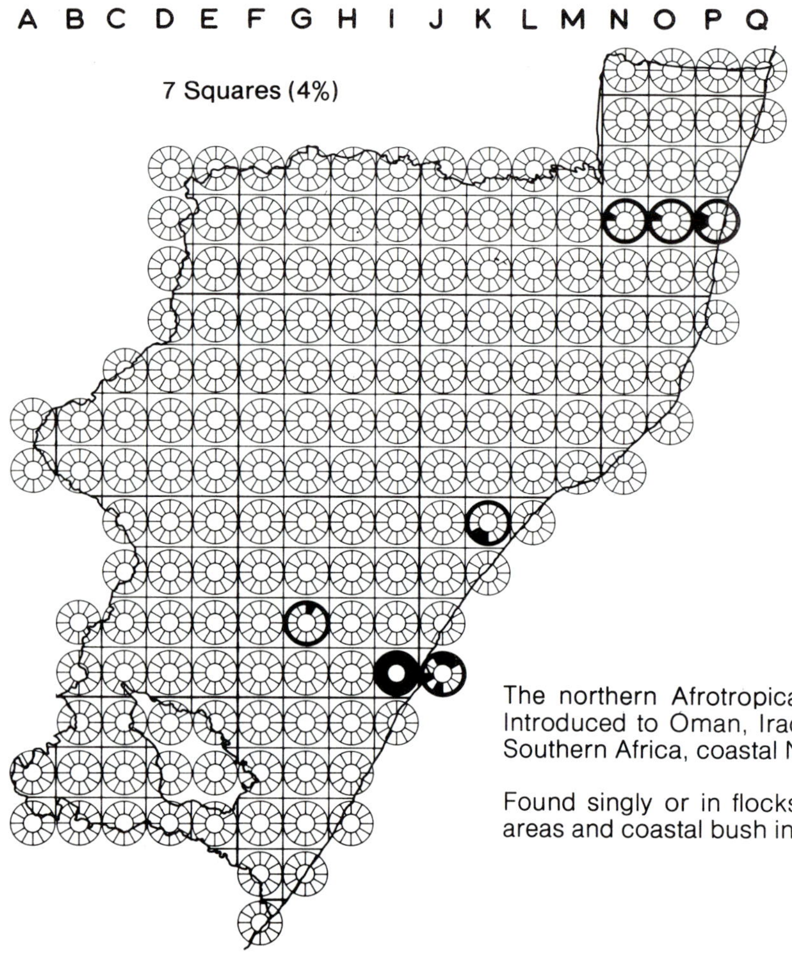

329X Ring-necked Parakeet

Psittacula krameri

The northern Afrotropical Region east to India, Sri Lanka and China. Introduced to Oman, Iraq and Egypt; naturalised on Zanzibar Island. In Southern Africa, coastal Natal and Zululand (introduced).

Found singly or in flocks in parks and gardens; also in lightly-wooded areas and coastal bush in Zululand. Has bred.

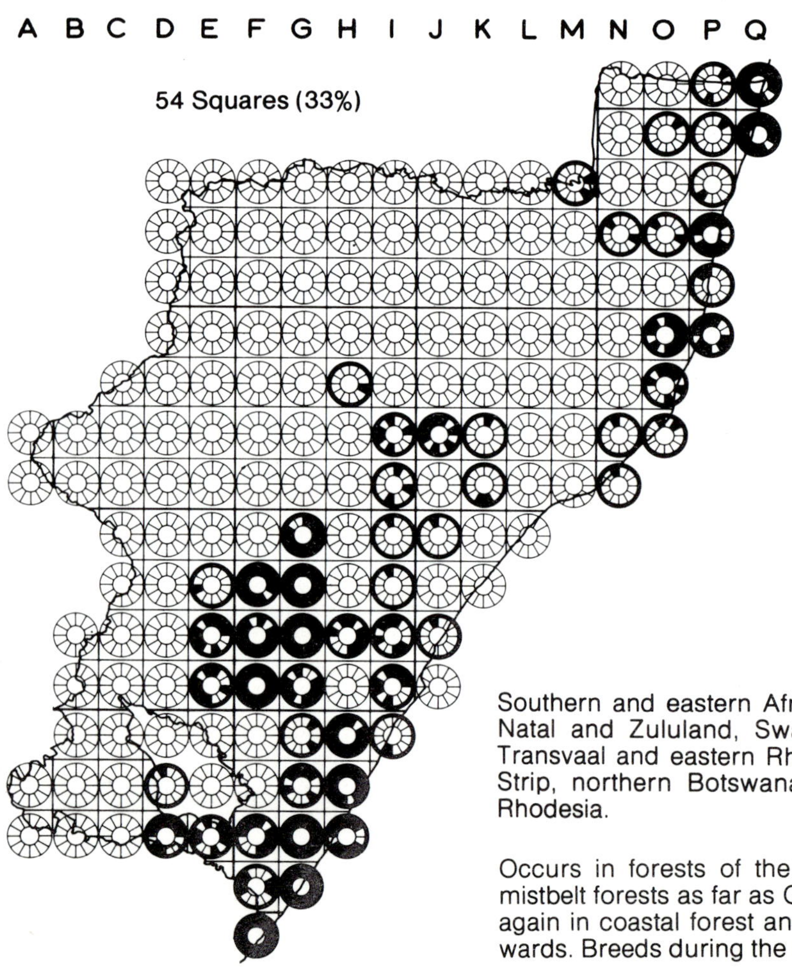

336 Knysna Loerie

Tauraco corythaix

Southern and eastern Africa. In Southern Africa, the southern Cape to Natal and Zululand, Swaziland, Moçambique, eastern and northern Transvaal and eastern Rhodesia; also north-eastern S.W.A., the Caprivi Strip, northern Botswana (Chobe River) and extreme north-western Rhodesia.

Occurs in forests of the south coast, extending through the midland mistbelt forests as far as Qudeni and Nkandla in south-western Zululand; again in coastal forest and riparian woodland from Richards Bay northwards. Breeds during the summer.

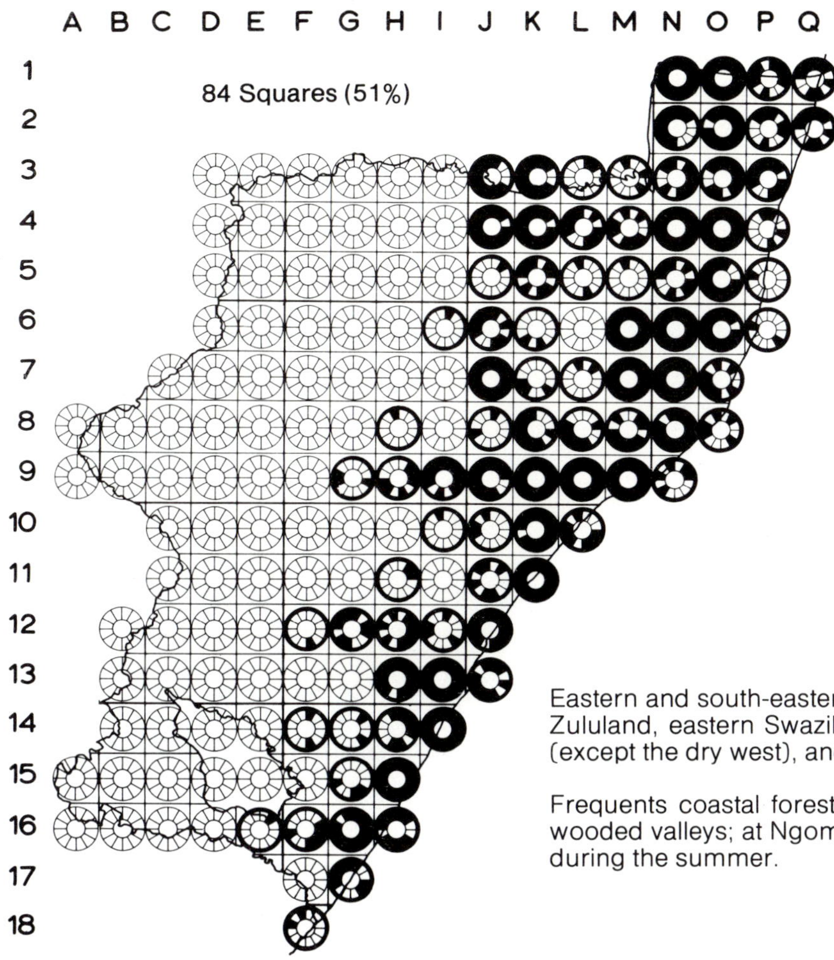

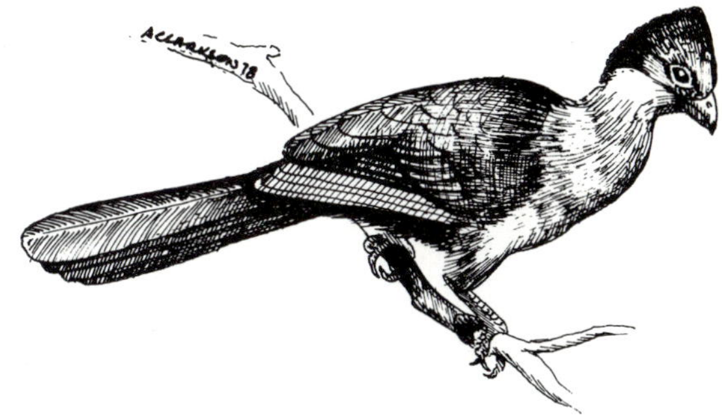

337 Purple-crested Loerie

Tauraco porphyreolophus

Eastern and south-eastern Africa. In Southern Africa, Transkei, Natal and Zululand, eastern Swaziland, eastern and northern Transvaal, Rhodesia (except the dry west), and southern Moçambique.

Frequents coastal forests and riverine vegetation, extending up thickly-wooded valleys; at Ngome, common in mistbelt evergreen forest. Breeds during the summer.

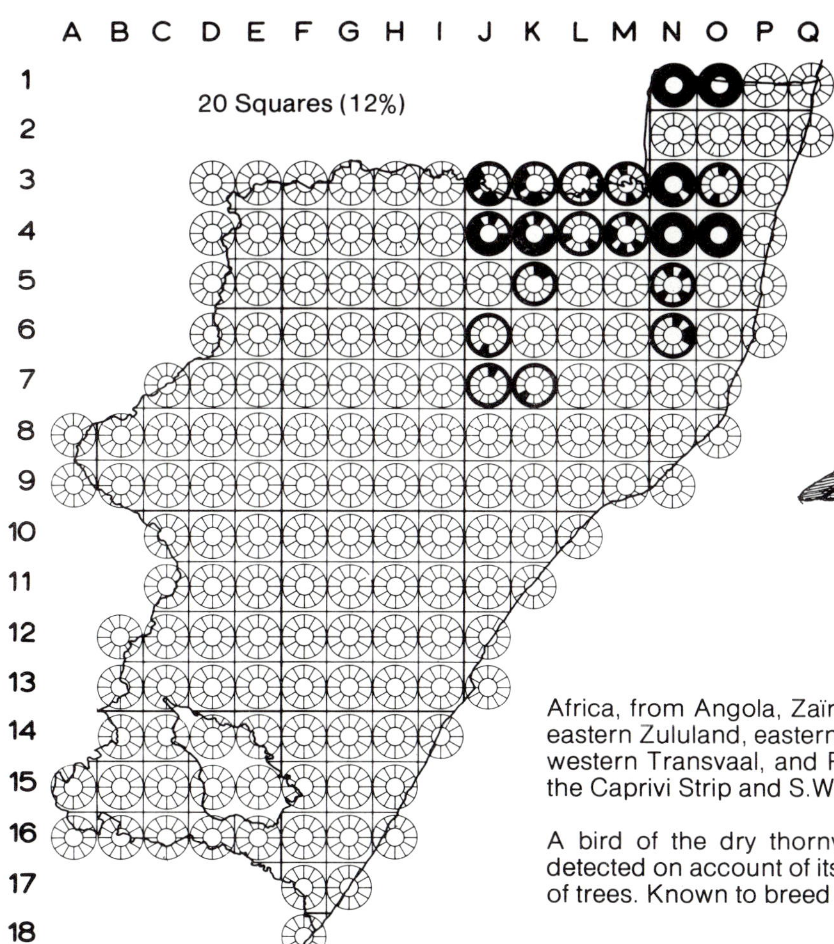

339 Grey Loerie

Corythaixoides concolor

Africa, from Angola, Zaïre and Tanzania to Zululand. In Southern Africa, eastern Zululand, eastern Swaziland, Moçambique, eastern, northern and western Transvaal, and Rhodesia west to the northern half of Botswana, the Caprivi Strip and S.W.A. (north from Seeheim, Great Namaqualand).

A bird of the dry thornveld regions of the northeast, which is readily detected on account of its characteristic call and habit of perching on tops of trees. Known to breed during the summer.

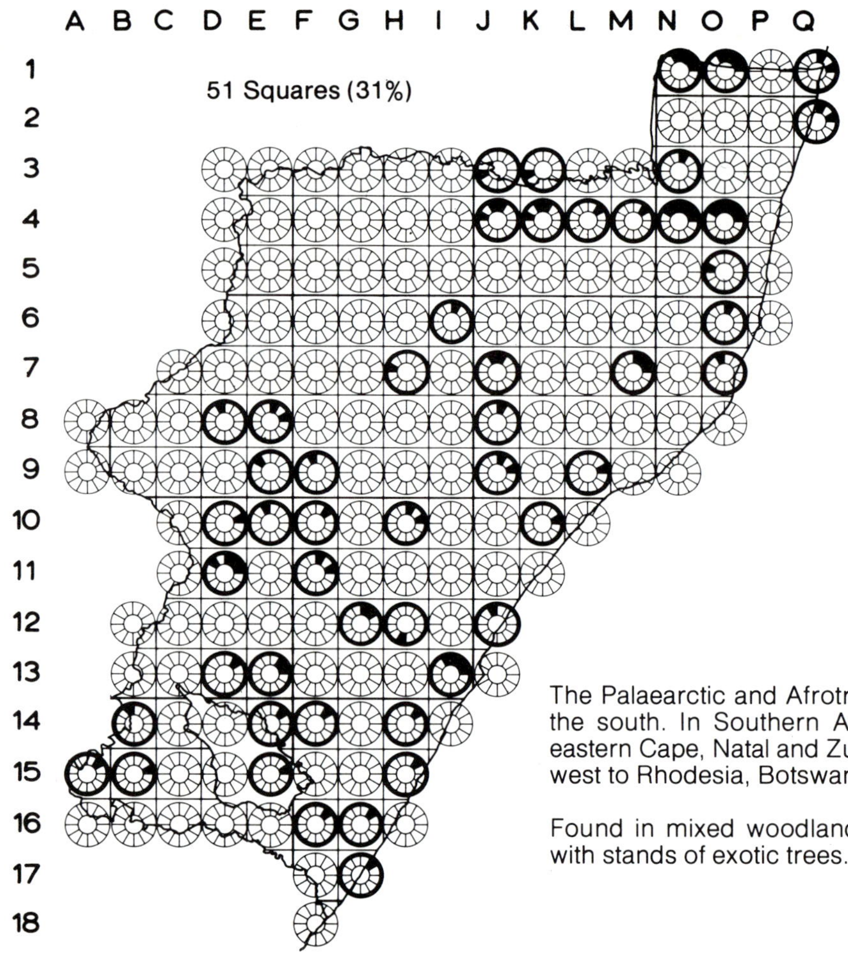

340 Cuckoo

Cuculus canorus

The Palaearctic and Afrotropical Regions; northern populations winter to the south. In Southern Africa, a visitor from October to March to the eastern Cape, Natal and Zululand, Swaziland, Transvaal and Moçambique, west to Rhodesia, Botswana and S.W.A.

Found in mixed woodland, thornveld, open bush, and grassland areas with stands of exotic trees. May breed.

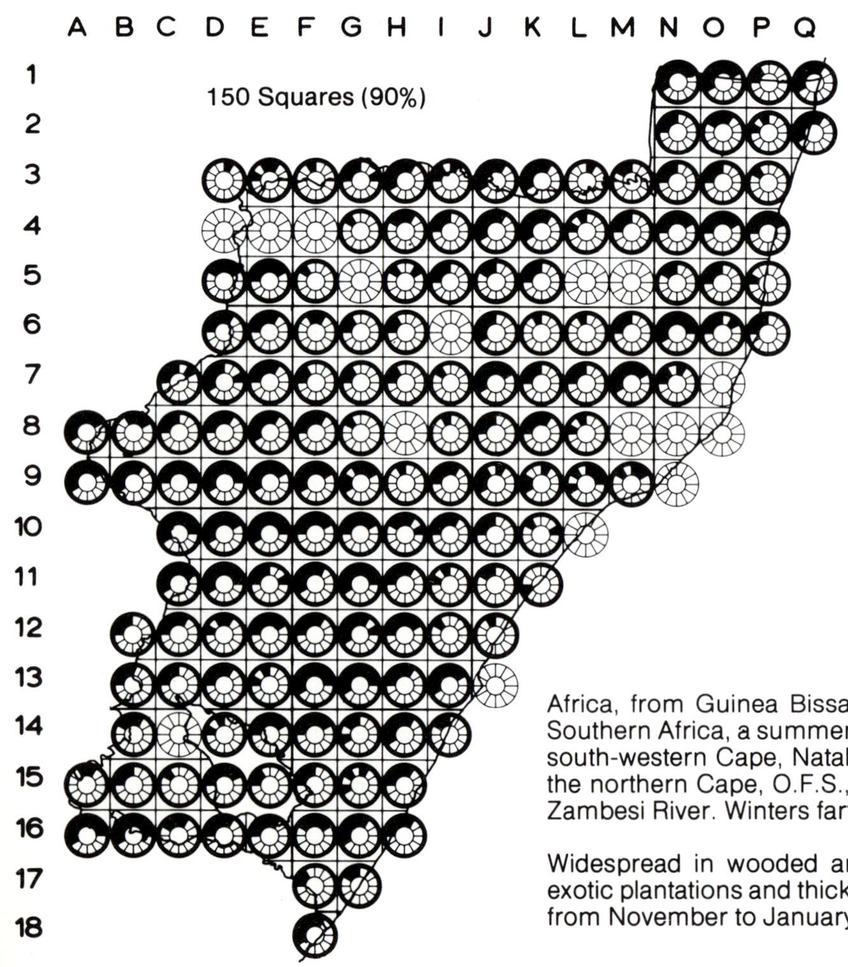

343 Red-chested Cuckoo

Cuculus solitarius

Africa, from Guinea Bissau, Sudan and Ethiopia south to the Cape. In Southern Africa, a summer resident from September to March/April in the south-western Cape, Natal and Zululand, northwards in the east through the northern Cape, O.F.S., Transvaal, Rhodesia and Moçambique, to the Zambesi River. Winters farther north.

Widespread in wooded areas with a preference for evergreen forests, exotic plantations and thick bush; sparse on the coast. Recorded breeding from November to January.

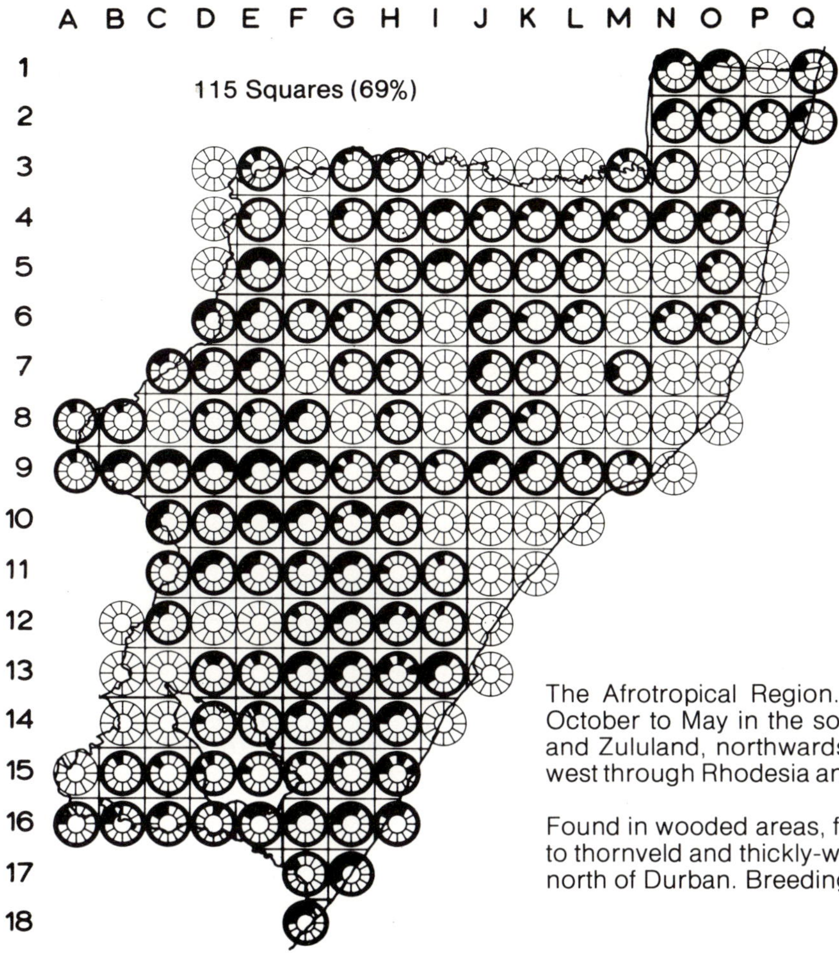

344 Black Cuckoo

Cuculus clamosus

The Afrotropical Region. In Southern Africa, a summer resident from October to May in the south-western, southern and eastern Cape, Natal and Zululand, northwards to the Zambesi River in the east. In the north, west through Rhodesia and Botswana to S.W.A. Winters north to Ethiopia.

Found in wooded areas, from edges of evergreen forests and plantations, to thornveld and thickly-wooded valleys; largely absent from coastal forest north of Durban. Breeding recorded from October to December.

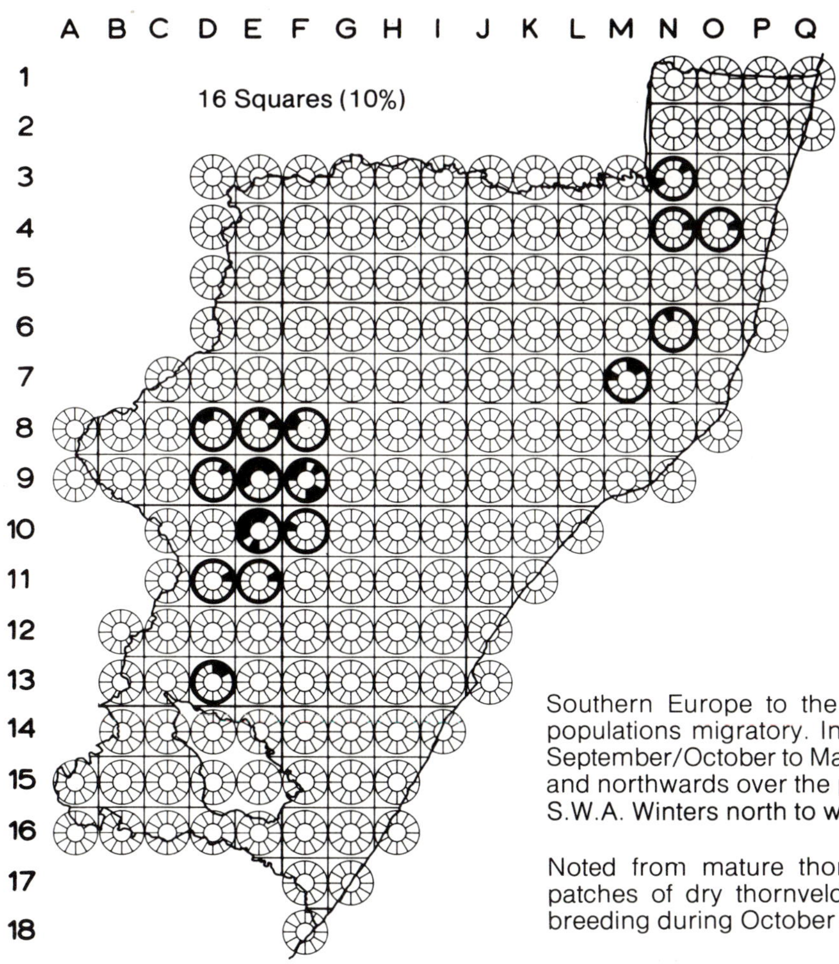

346 Great Spotted Cuckoo

Clamator glandarius

Southern Europe to the Middle East and south through Africa; most populations migratory. In Southern Africa, a summer resident from late September/October to March in the eastern Cape, Natal (interior), Zululand and northwards over the plateau in the east; in the north, west to northern S.W.A. Winters north to western Sudan.

Noted from mature thornveld in Zululand and fringes of the denser patches of dry thornveld in Natal; occasionally overwinters. Recorded breeding during October and November.

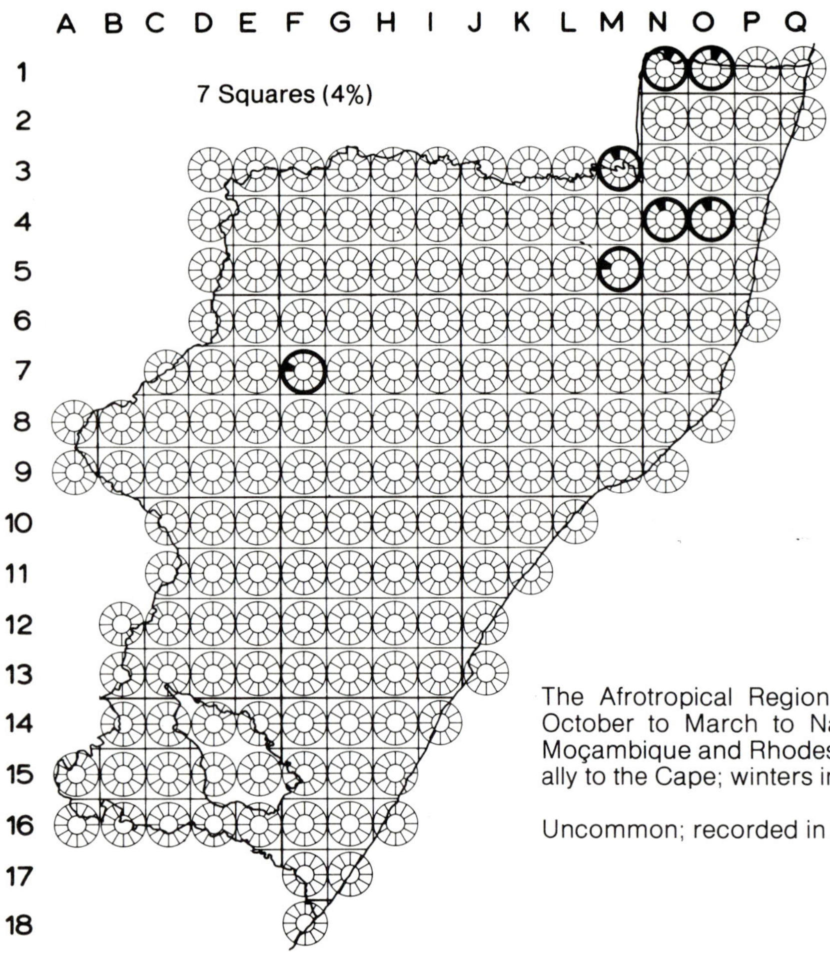

347 Striped Cuckoo

Clamator levaillantii

The Afrotropical Region. In Southern Africa, a breeding visitor from October to March to Natal (interior), Zululand, Swaziland, Transvaal, Moçambique and Rhodesia, west to Botswana and S.W.A., and occasionally to the Cape; winters in tropical Africa.

Uncommon; recorded in scrub as well as mature thornveld.

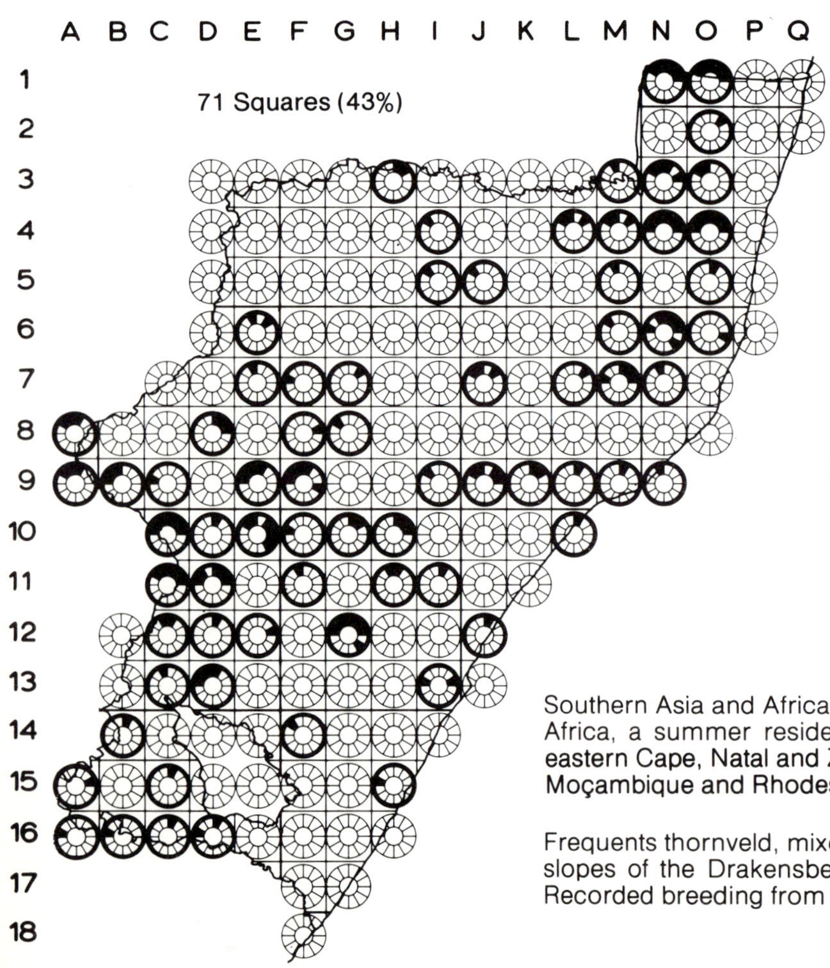

348 Jacobin Cuckoo

Clamator jacobinus

Southern Asia and Africa, most populations being migratory. In Southern Africa, a summer resident from October to April in the southern and eastern Cape, Natal and Zululand, northwards through O.F.S., Transvaal, Moçambique and Rhodesia, west to Botswana and northern S.W.A.

Frequents thornveld, mixed woodland, and scrub vegetation on the lower slopes of the Drakensberg Range; occasionally in parks and gardens. Recorded breeding from October to January.

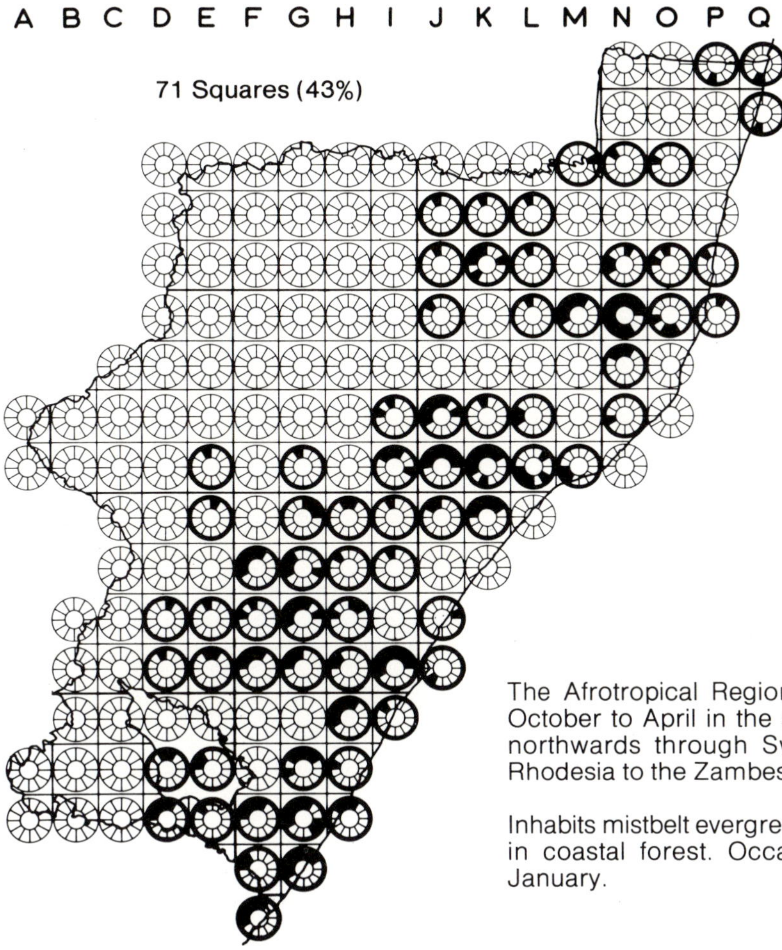

350 Emerald Cuckoo

Chrysococcyx cupreus

The Afrotropical Region. In Southern Africa, a summer resident from October to April in the southern and eastern Cape, Natal and Zululand, northwards through Swaziland, Transvaal, Moçambique and eastern Rhodesia to the Zambesi River.

Inhabits mistbelt evergreen forest and thickly-wooded valleys; less common in coastal forest. Occasionally overwinters. Breeds from October to January.

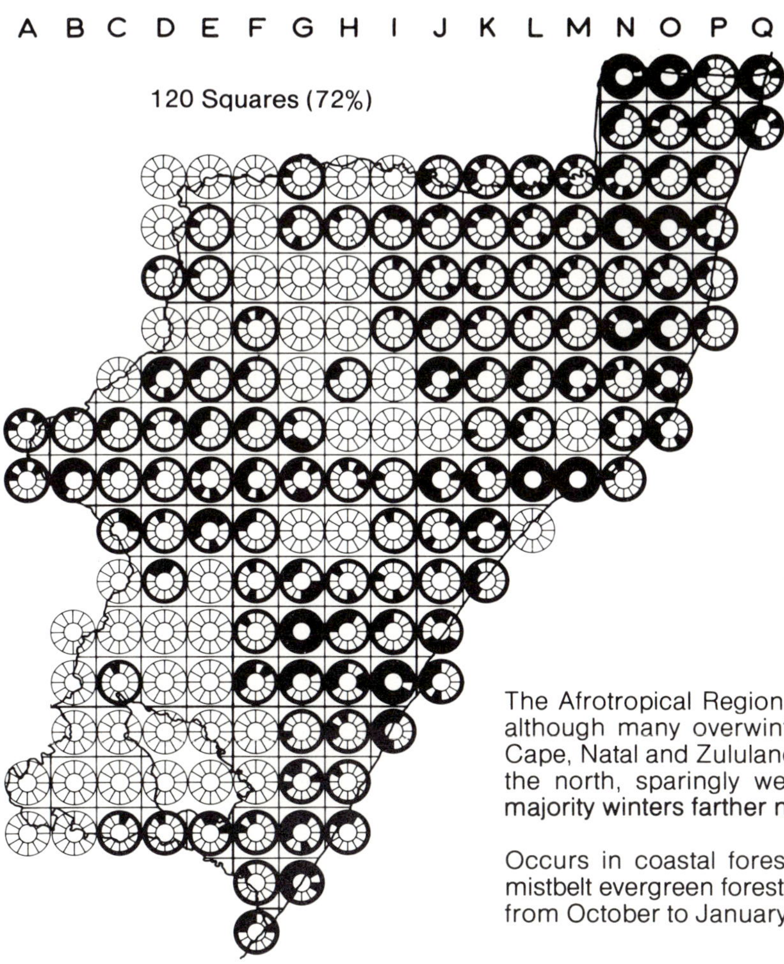

351 Klaas's Cuckoo

Chrysococcyx klaas

The Afrotropical Region. In Southern Africa, mainly a summer resident, although many overwinter, in the south-western, southern and eastern Cape, Natal and Zululand, northwards to the Zambesi River in the east. In the north, sparingly west through Botswana (Sepopa) to S.W.A. The majority winters farther north in Africa.

Occurs in coastal forest, mixed and riverine woodland, thornveld and mistbelt evergreen forest; occasionally in exotic trees and gardens. Breeds from October to January.

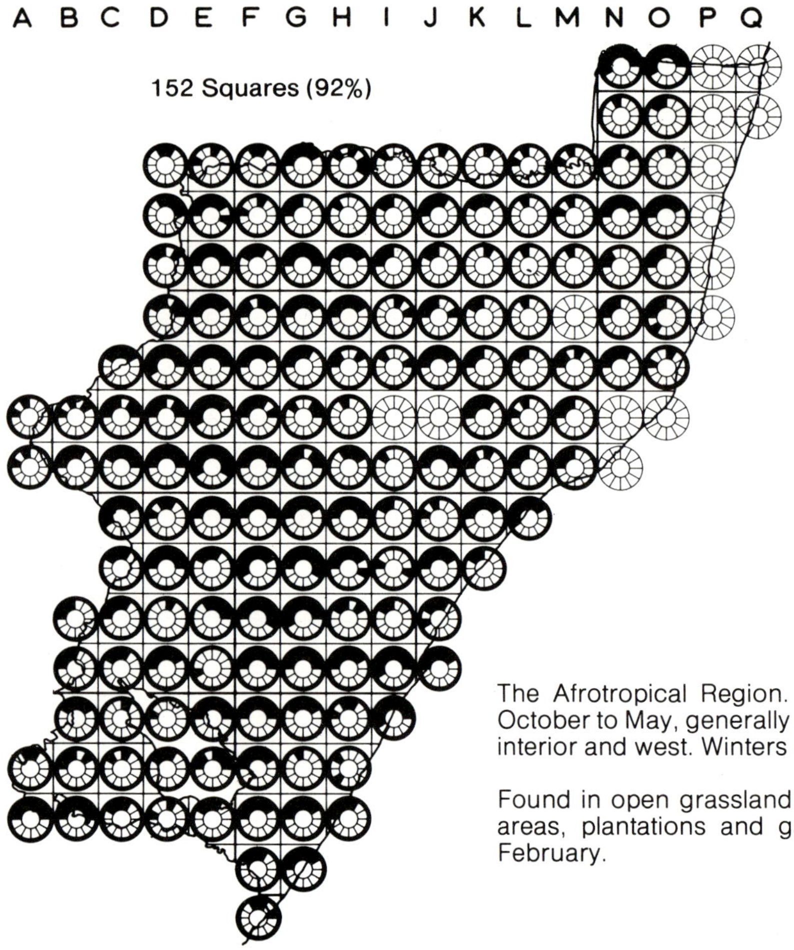

352 Diederik Cuckoo

Chrysococcyx caprius

The Afrotropical Region. In Southern Africa, a summer resident from October to May, generally distributed though sparse over much of the arid interior and west. Winters in the African tropics.

Found in open grassland with scattered trees, reedbeds, lightly-wooded areas, plantations and gardens. Recorded breeding from October to February.

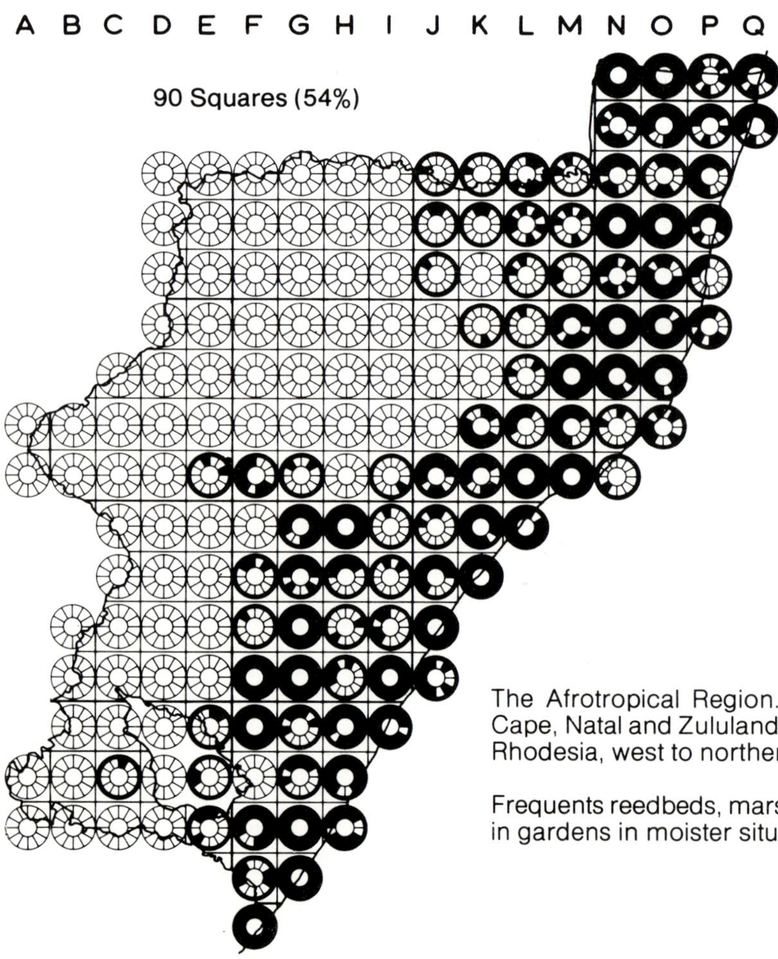

356 White-browed Coucal

Centropus superciliosus

The Afrotropical Region. In Southern Africa, the southern and eastern Cape, Natal and Zululand, Swaziland, O.F.S., Transvaal, Moçambique and Rhodesia, west to northern Botswana and the Caprivi Strip.

Frequents reedbeds, marshes, riverine thickets and rank vegetation; occurs in gardens in moister situations. Breeds from September to February.

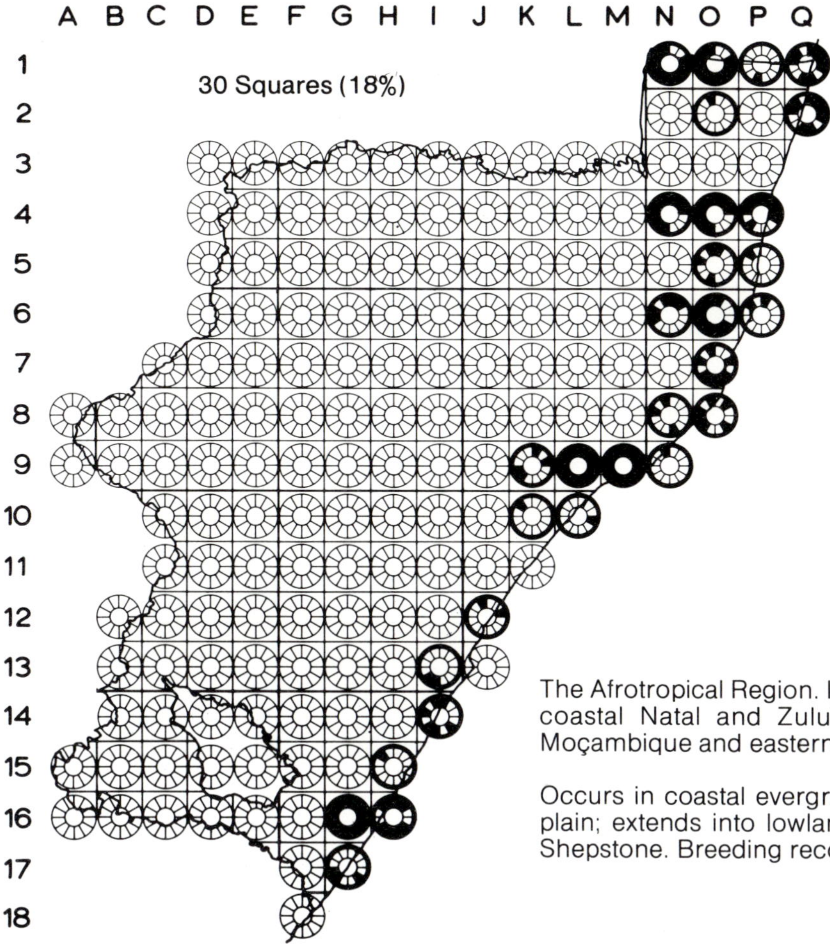

358 Green Coucal

Ceuthmochares aereus

The Afrotropical Region. In Southern Africa, the Cape (from Uitenhage) to coastal Natal and Zululand, eastern Swaziland (Lebombo Range), Moçambique and eastern Rhodesia (at low levels).

Occurs in coastal evergreen forest and riverine woodland on the littoral plain; extends into lowland evergreen forests around Eshowe and Port Shepstone. Breeding recorded from October to December.

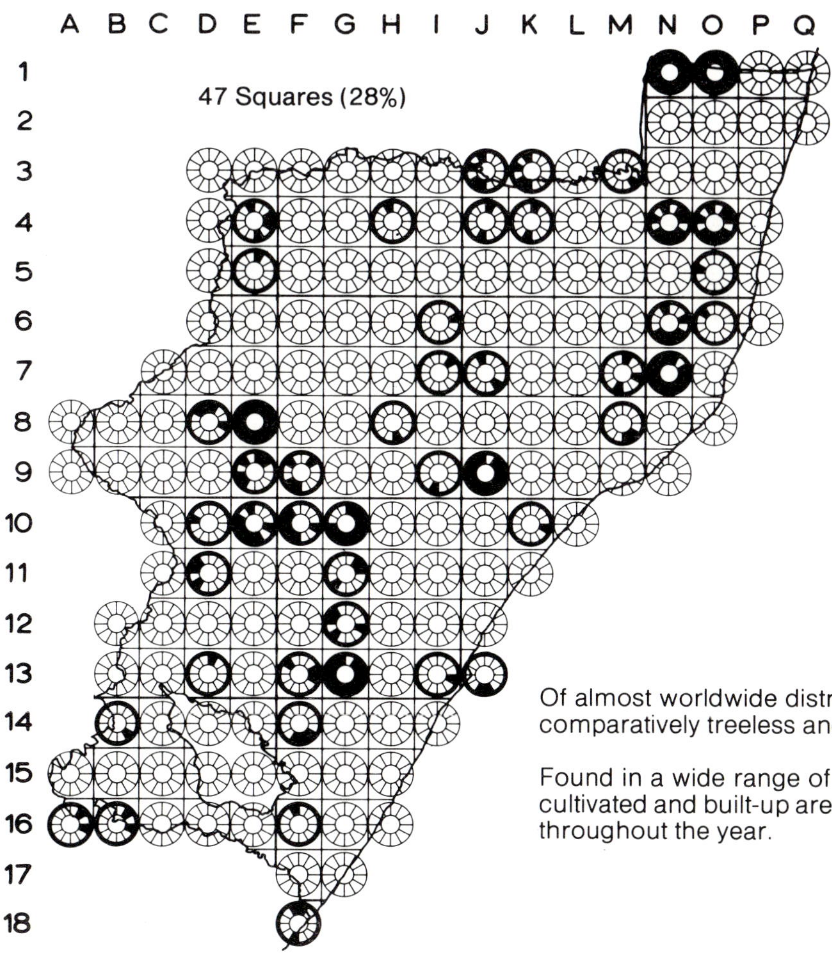

359 Barn Owl

Tyto alba

Of almost worldwide distribution. In Southern Africa, throughout, even in comparatively treeless and desert regions.

Found in a wide range of habitats, from thornveld and open grassland to cultivated and built-up areas, wherever suitable roosting sites exist. Breeds throughout the year.

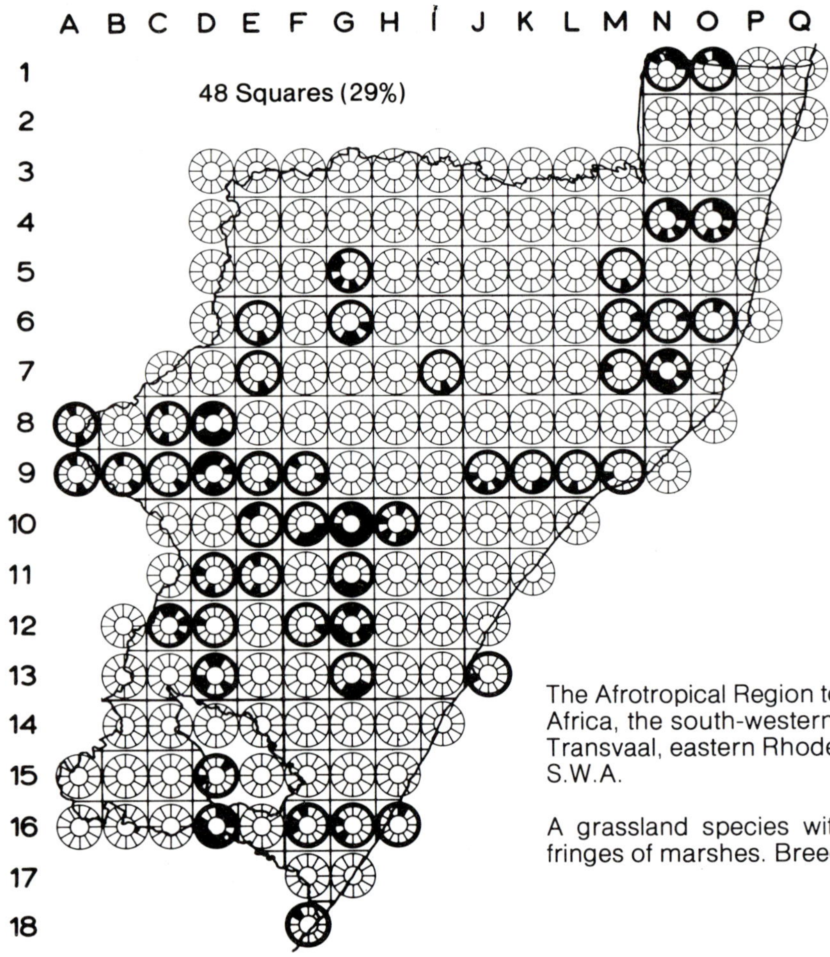

360 Grass Owl

Tyto capensis

The Afrotropical Region to eastern Asia, southeast to Australia. In Southern Africa, the south-western Cape to Natal and Zululand, Swaziland, O.F.S., Transvaal, eastern Rhodesia, and Moçambique; also central and northern S.W.A.

A grassland species with a preference for moister situations such as fringes of marshes. Breeds throughout the year.

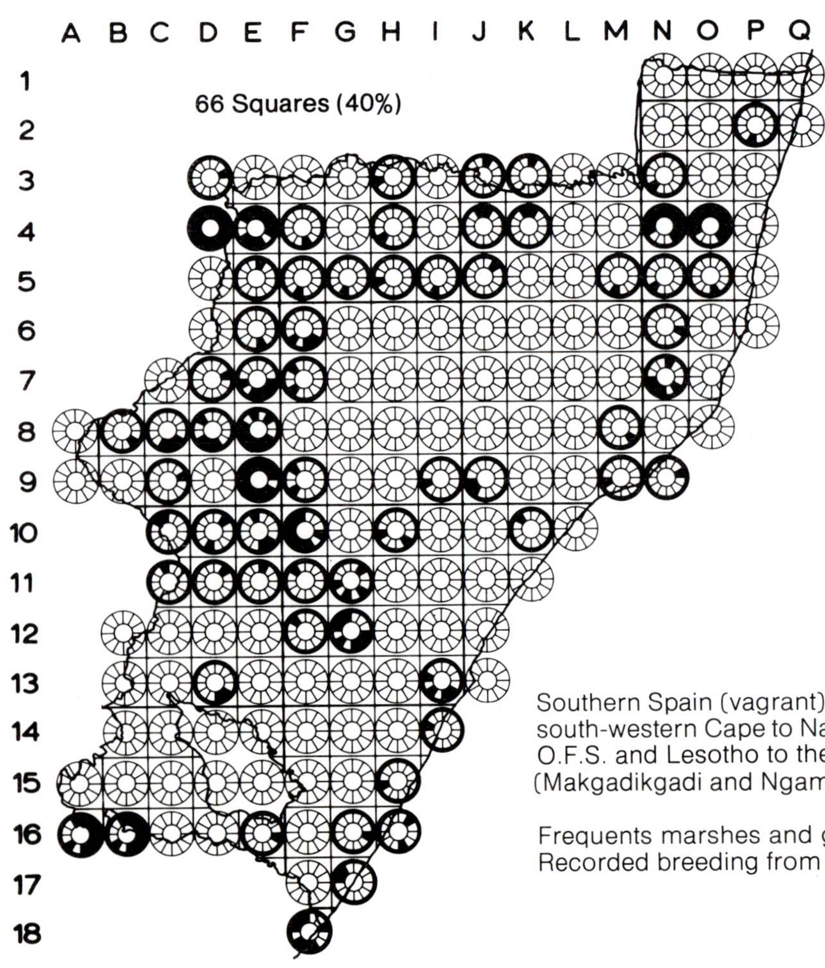

361 Marsh Owl

Asio capensis

Southern Spain (vagrant), Africa and Madagascar. In Southern Africa, the south-western Cape to Natal and Zululand, northwards in the east through O.F.S. and Lesotho to the Zambesi River. In the north, West to Botswana (Makgadikgadi and Ngamiland) and northern S.W.A.

Frequents marshes and grassland adjacent to standing water and rivers. Recorded breeding from February to June.

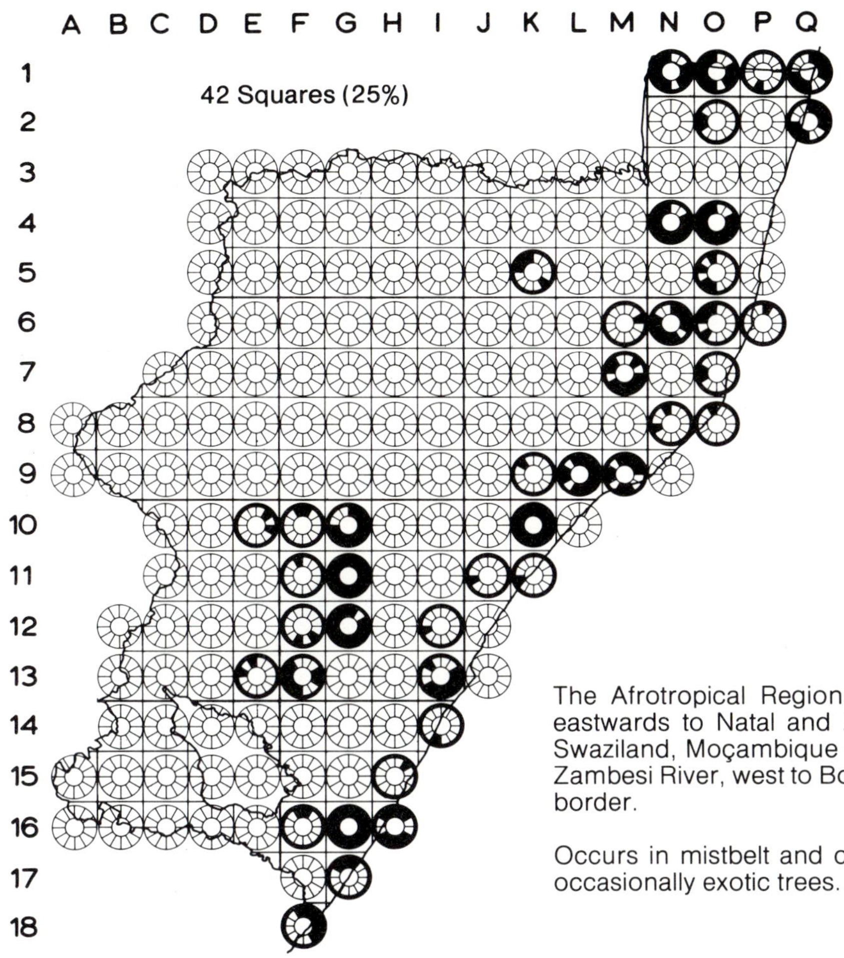

362 Wood Owl

Strix woodfordii

The Afrotropical Region. In Southern Africa, the south-western Cape eastwards to Natal and Zululand, north in the east through Transvaal, Swaziland, Moçambique and Rhodesia (absent from the dry west) to the Zambesi River, west to Botswana, the Caprivi Strip and the Angola/S.W.A. border.

Occurs in mistbelt and coastal evergreen forest, riverine woodland and occasionally exotic trees. Breeds from August to October.

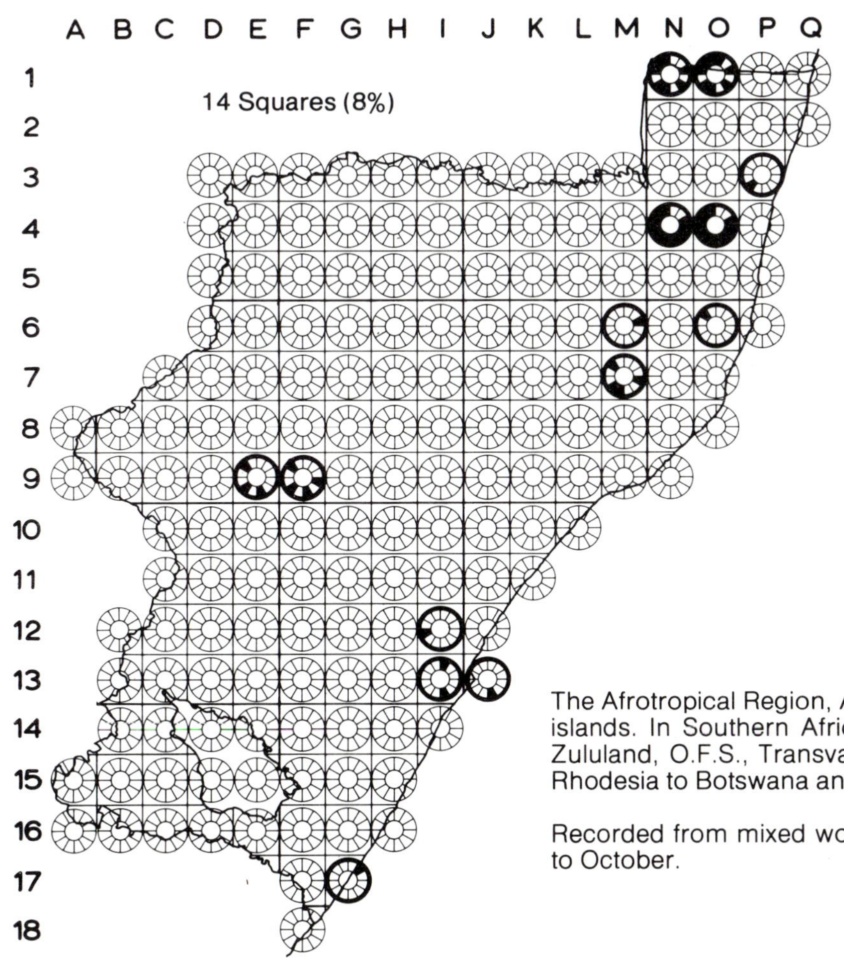

363 African Scops Owl

Otus senegalensis

The Afrotropical Region, Arabia, Socotra and other western Indian Ocean islands. In Southern Africa, the southern and eastern Cape, Natal and Zululand, O.F.S., Transvaal, south-eastern Swaziland, Moçambique and Rhodesia to Botswana and northern S.W.A.

Recorded from mixed woodland and dry thornveld. Breeds from August to October.

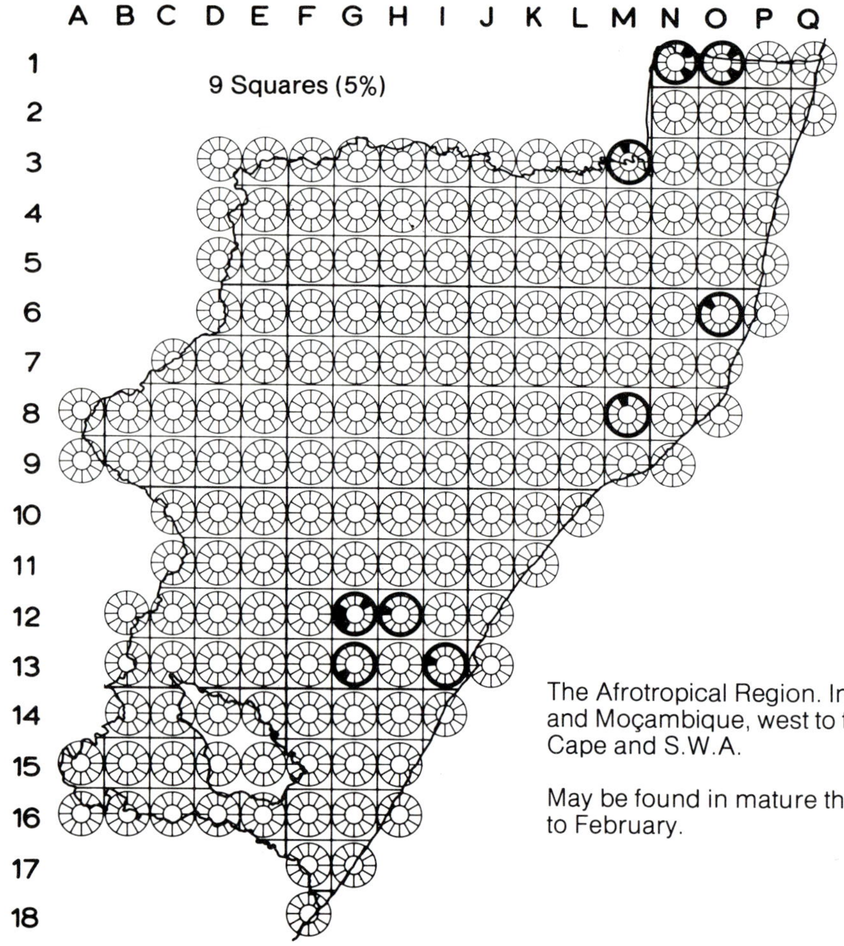

364 White-faced Owl

Otus leucotis

The Afrotropical Region. In Southern Africa, Natal and Zululand, Swaziland and Moçambique, west to the Transvaal, Rhodesia, Botswana, the northern Cape and S.W.A.

May be found in mature thornveld and mixed woodland. Breeds from July to February.

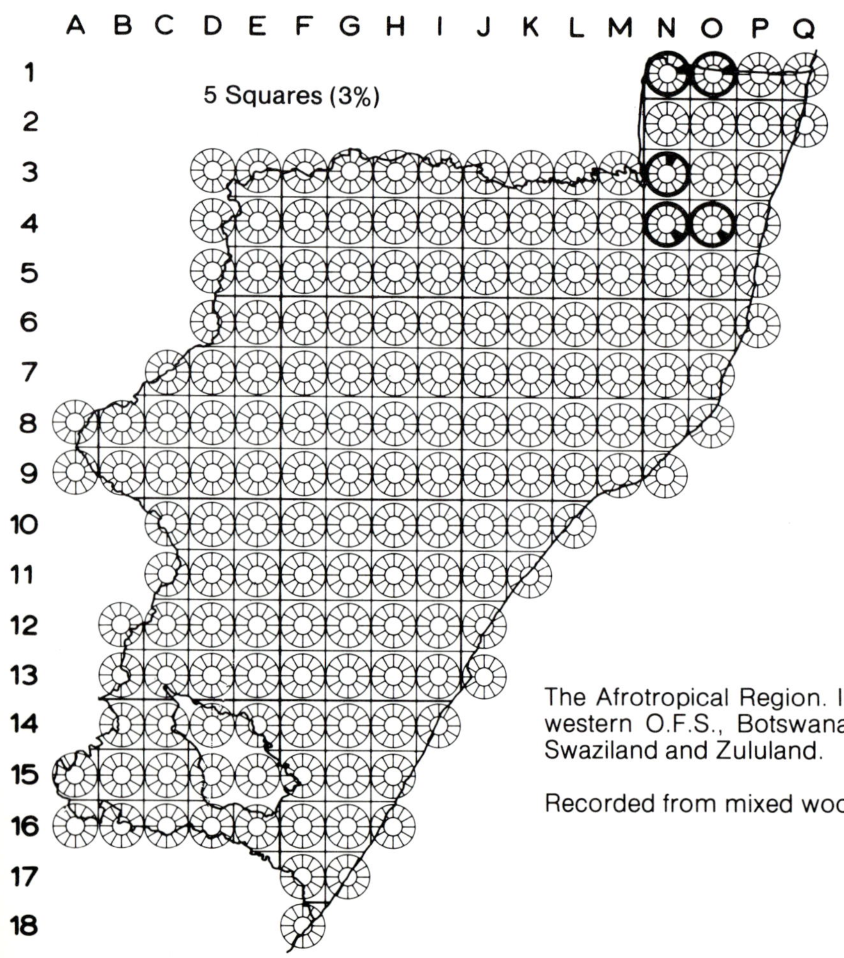

365 Pearl-spotted Owl

Glaucidium perlatum

The Afrotropical Region. In Southern Africa, S.W.A., the northern Cape, western O.F.S., Botswana, Rhodesia, Moçambique, Transvaal, eastern Swaziland and Zululand.

Recorded from mixed woodland on the littoral plain.

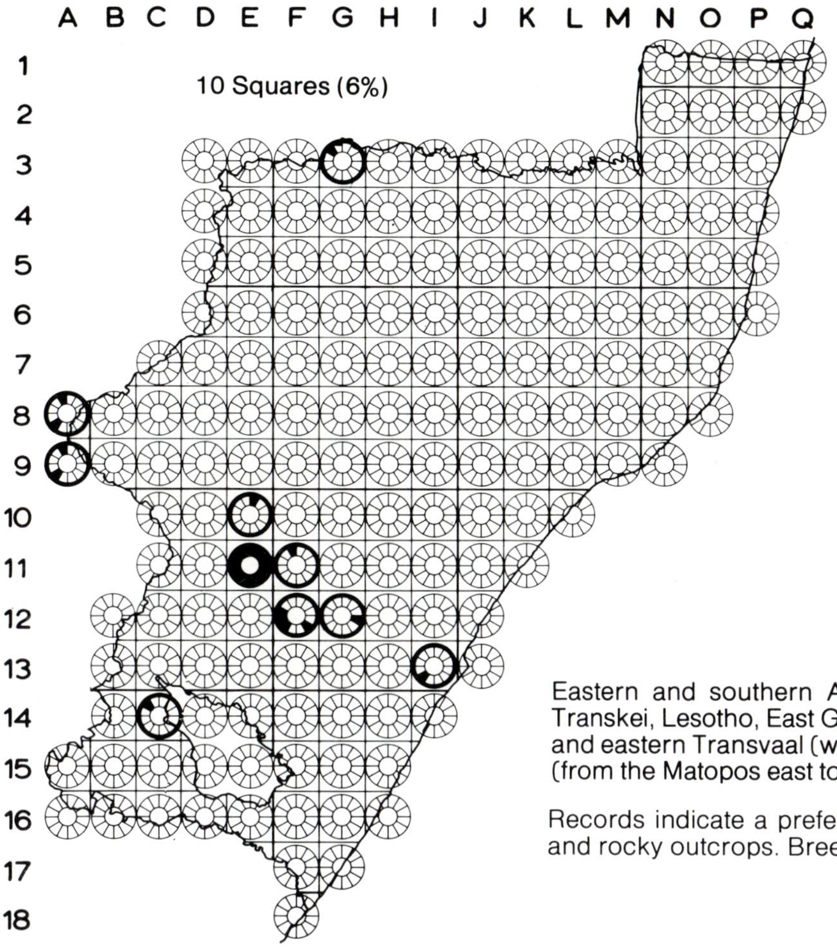

367 Cape Eagle Owl

Bubo capensis

Eastern and southern Africa. In Southern Africa, the Cape to O.F.S., Transkei, Lesotho, East Griqualand, Natal and Zululand, western Swaziland and eastern Transvaal (west to Krugersdorp); also the plateau of Rhodesia (from the Matopos east to the Eastern Districts) and adjacent Moçambique.

Records indicate a preference for upland grassland with scattered trees and rocky outcrops. Breeds throughout the year.

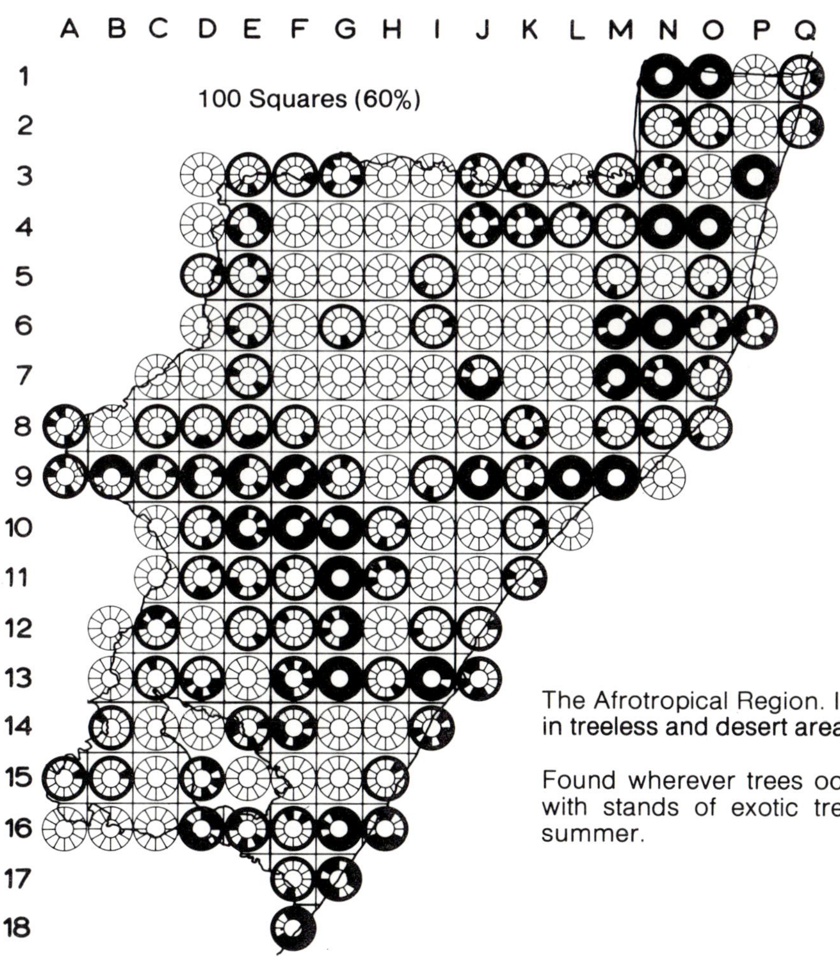

368 Spotted Eagle Owl

Bubo africanus

The Afrotropical Region. In Southern Africa, distributed throughout, even in treeless and desert areas of the interior and west.

Found wherever trees occur, from mixed woodland to open grassland with stands of exotic trees. Breeds during the late winter and early summer.

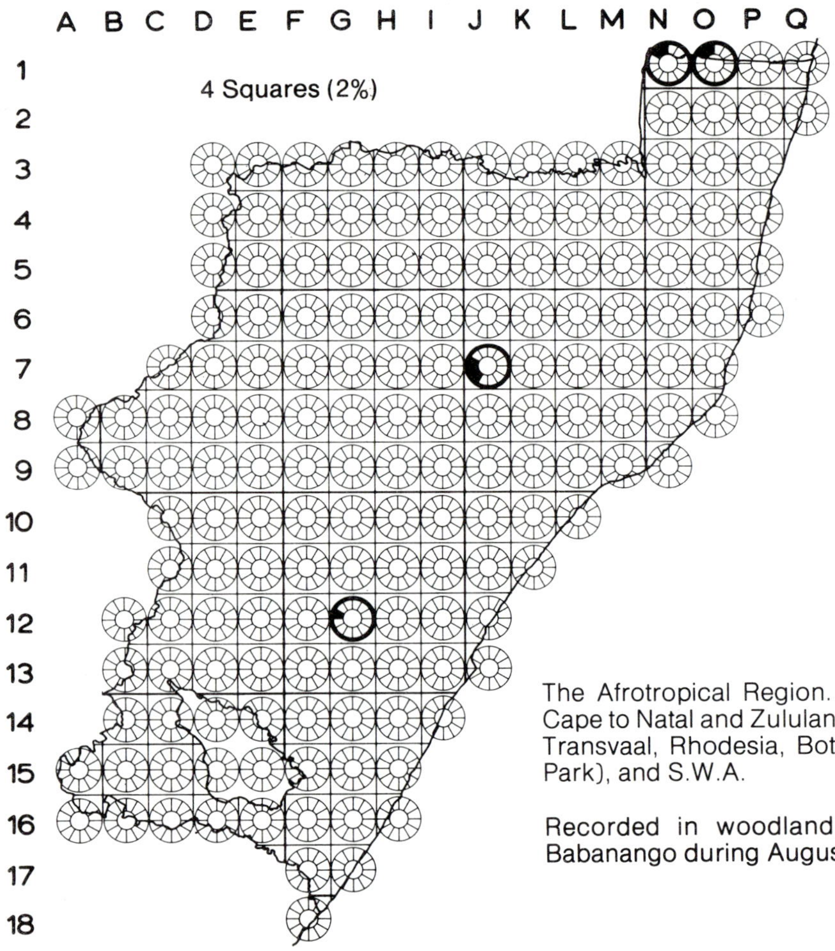

369 Giant Eagle Owl

Bubo lacteus

The Afrotropical Region. In Southern Africa, the southern and eastern Cape to Natal and Zululand (sparse), Swaziland and Moçambique, west to Transvaal, Rhodesia, Botswąna, the northern Cape (Gemsbok National Park), and S.W.A.

Recorded in woodland and open savanna. Found breeding near Babanango during August 1977.

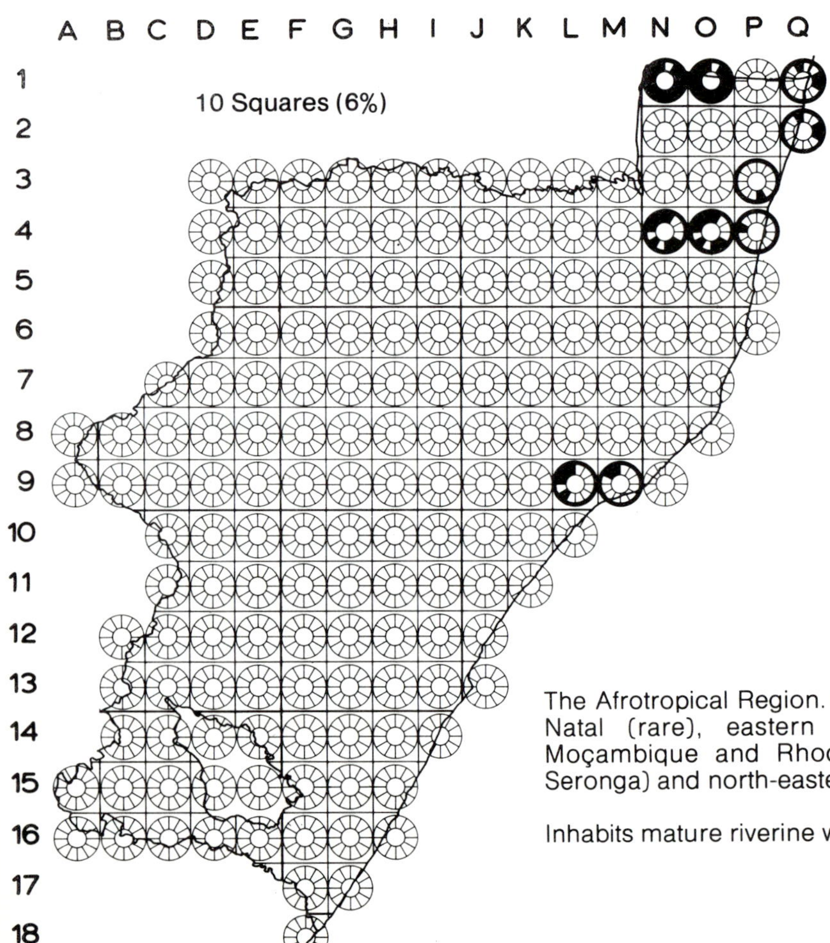

370 Fishing Owl

Scotopelia peli

The Afrotropical Region. In Southern Africa, the eastern Cape, Transkei, Natal (rare), eastern Zululand, Swaziland, Transvaal (lowveld), Moçambique and Rhodesia, west to northern Botswana (Sepopa, Seronga) and north-eastern S.W.A. (Okavango River).

Inhabits mature riverine woodland. Breeds during the late summer.

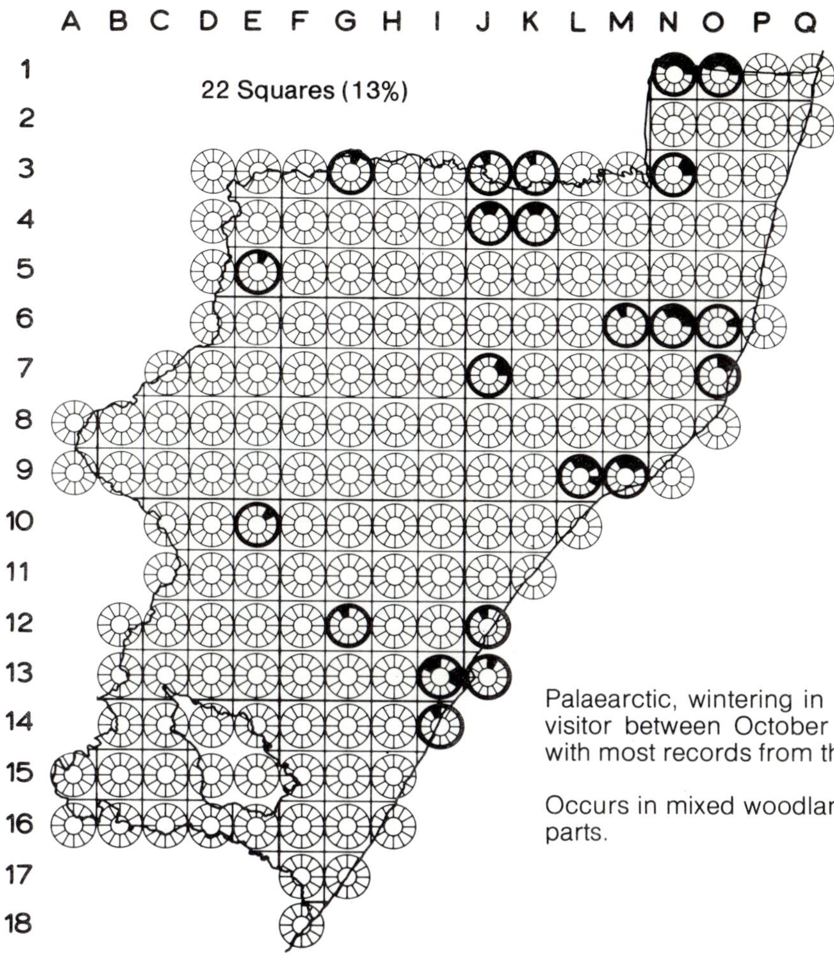

371 European Nightjar

Caprimulgus europaeus

Palaearctic, wintering in Africa and southern Asia. In Southern Africa, a visitor between October and March/early April, of general distribution, with most records from the south and east.

Occurs in mixed woodland and coastal bush, generally in the more open parts.

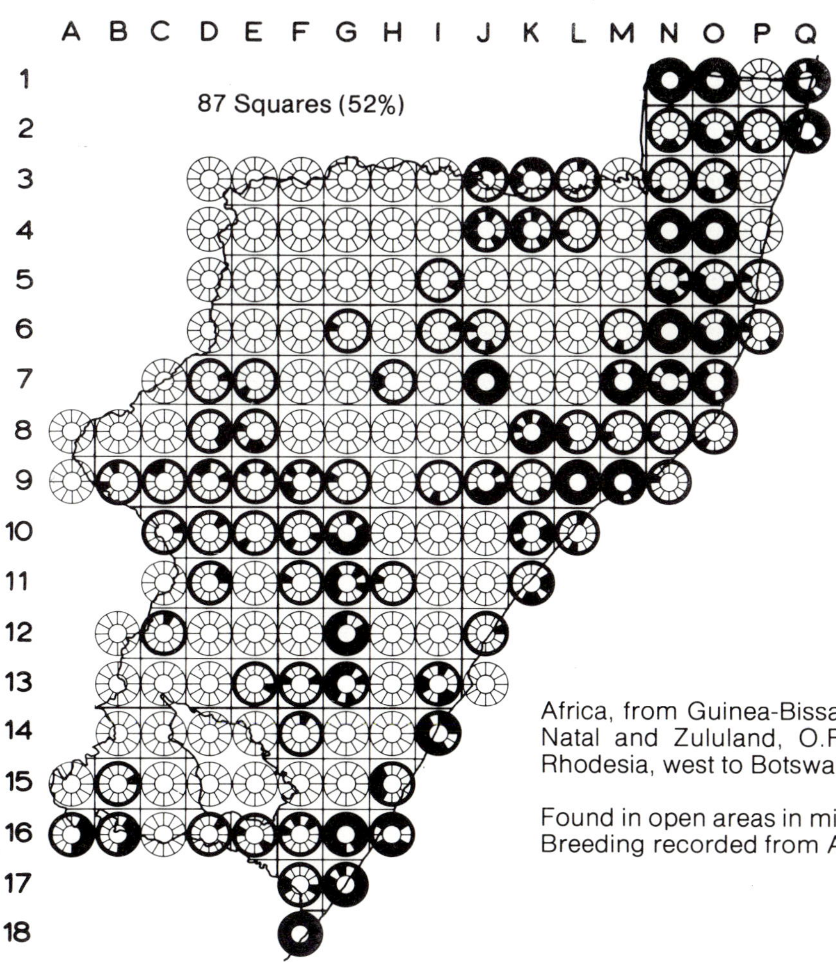

373 Fiery-necked Nightjar

Caprimulgus pectoralis

Africa, from Guinea-Bissau and Nigeria to the Cape. In Southern Africa, Natal and Zululand, O.F.S., Swaziland, Moçambique, Transvaal and Rhodesia, west to Botswana and S.W.A.

Found in open areas in mixed woodland, thornveld and exotic plantations. Breeding recorded from August to November.

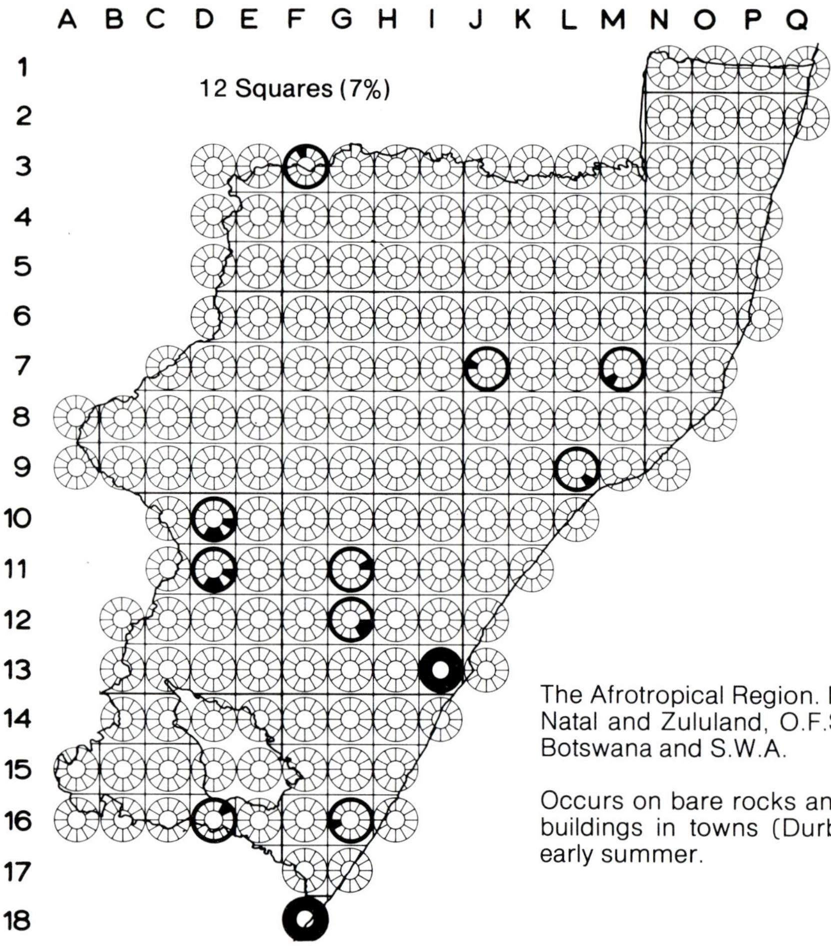

374 Freckled Nightjar

Caprimulgus tristigma

The Afrotropical Region. In Southern Africa, the Cape, Transkei, Lesotho, Natal and Zululand, O.F.S., Transvaal, western Moçambique, Rhodesia, Botswana and S.W.A.

Occurs on bare rocks and boulder-strewn hillsides; occasionally around buildings in towns (Durban and Pietermaritzburg). Breeds during the early summer.

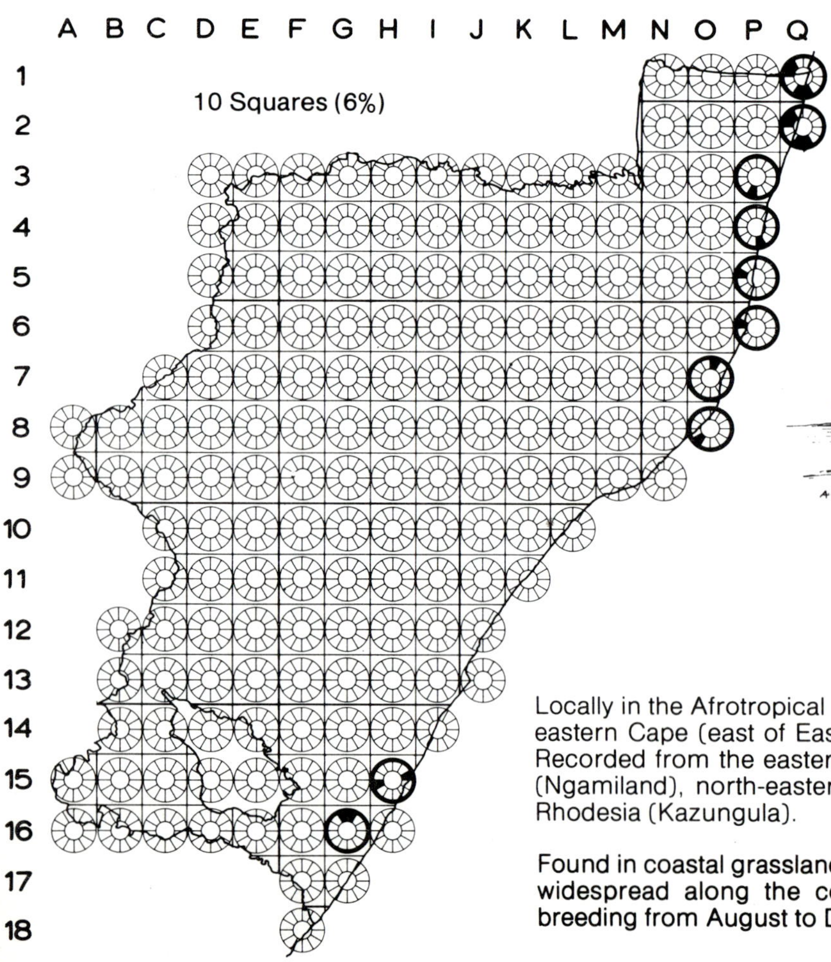

375 Natal Nightjar

Caprimulgus natalensis

Locally in the Afrotropical Region. In Southern Africa, coastal areas of the eastern Cape (east of East London), Natal, Zululand, and Moçambique. Recorded from the eastern Transvaal (Kruger National Park), Botswana (Ngamiland), north-eastern S.W.A., the Caprivi Strip and north-western Rhodesia (Kazungula).

Found in coastal grassland with scattered clumps of bush. Probably more widespread along the coast than records would indicate. Recorded breeding from August to December.

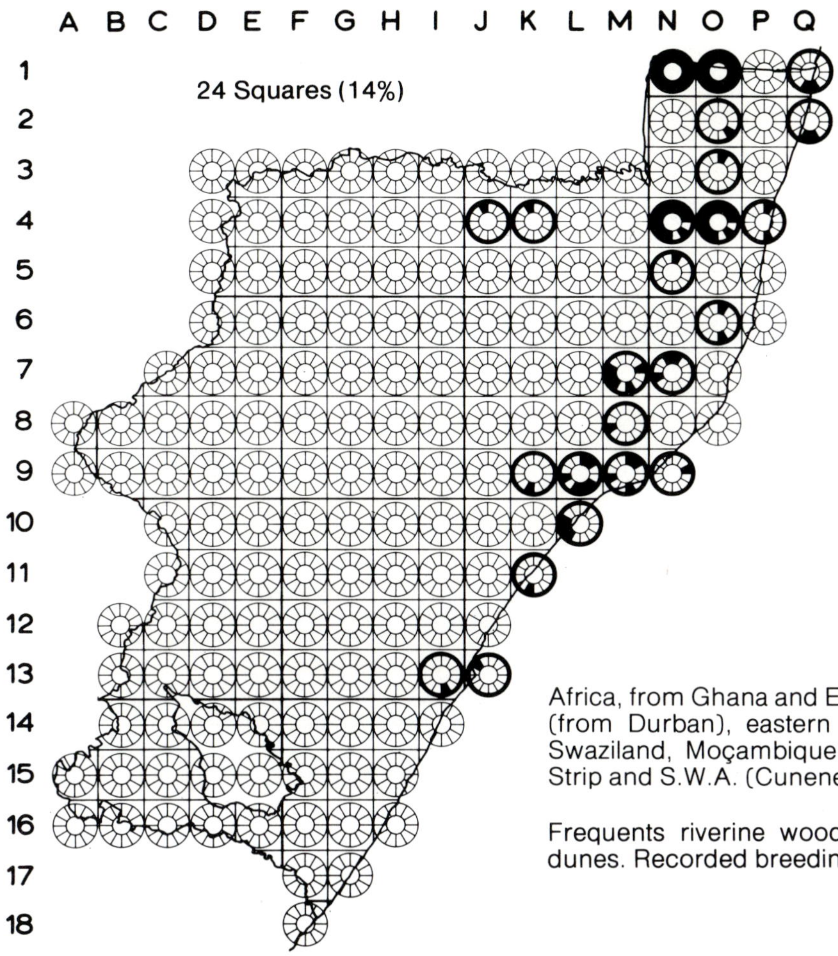

376 Moçambique Nightjar

Caprimulgus fossii

Africa, from Ghana and Ethiopia to Natal. In Southern Africa, coastal Natal (from Durban), eastern Zululand, northern O.F.S., Transvaal, eastern Swaziland, Moçambique and Rhodesia, west to Botswana, the Caprivi Strip and S.W.A. (Cunene and Okavango Rivers).

Frequents riverine woodland, coastal forest and scrub vegetation on dunes. Recorded breeding from August to November.

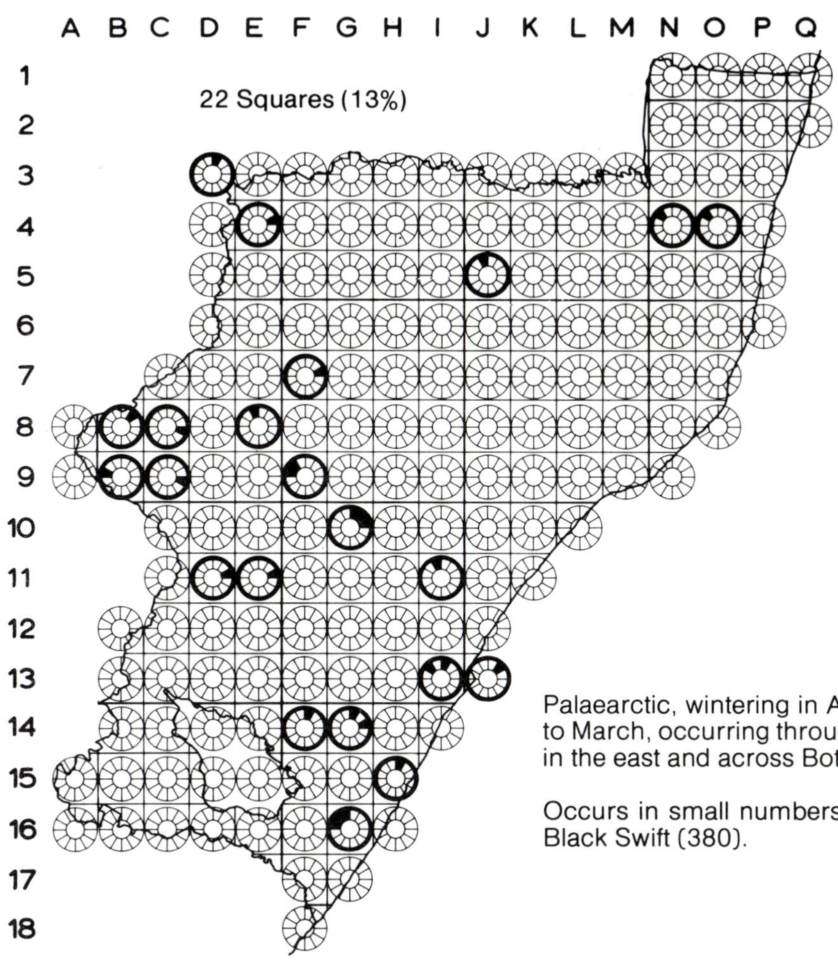

378 European Swift

Apus apus

Palaearctic, wintering in Africa. In Southern Africa, a visitor from October to March, occurring throughout, with most records from Natal northwards in the east and across Botswana to S.W.A.

Occurs in small numbers throughout the area, often in association with Black Swift (380).

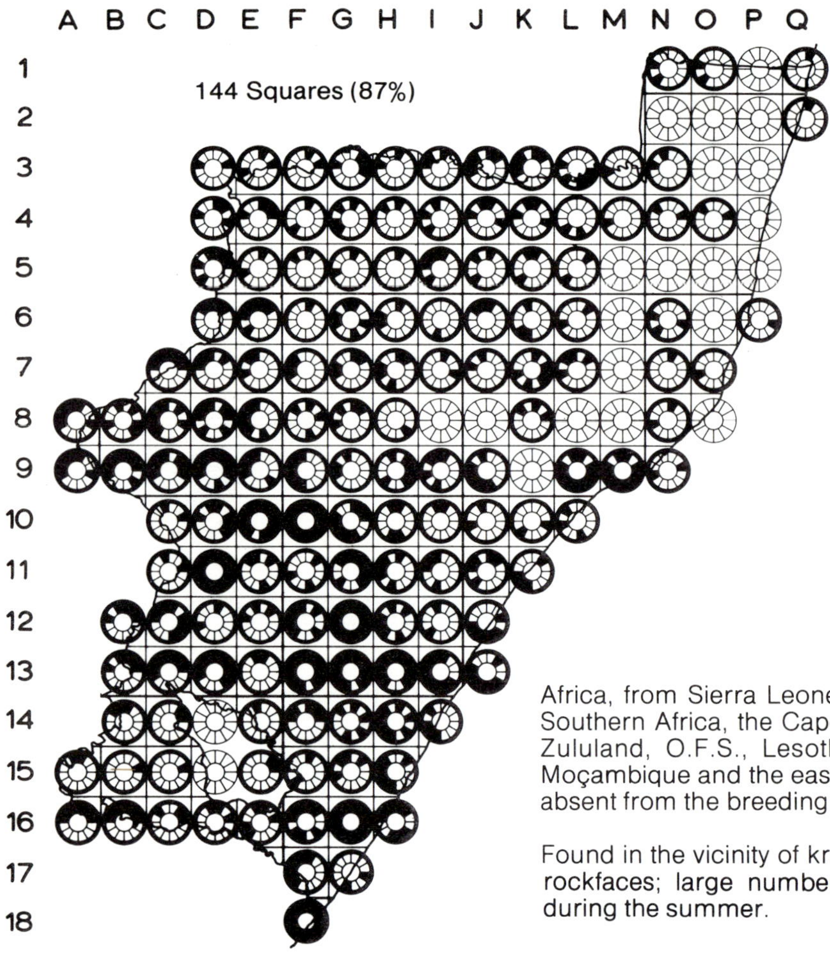

380 Black Swift

Apus barbatus

Africa, from Sierra Leone and Kenya to the Cape; also Madagascar. In Southern Africa, the Cape (south of the Orange River) east to Natal and Zululand, O.F.S., Lesotho, Swaziland and Transvaal; also southern Moçambique and the eastern plateau and highlands of Rhodesia. Largely absent from the breeding grounds during the winter.

Found in the vicinity of kranses and gorges, where it nests in crevices on rockfaces; large numbers move to coastal regions in winter. Breeds during the summer.

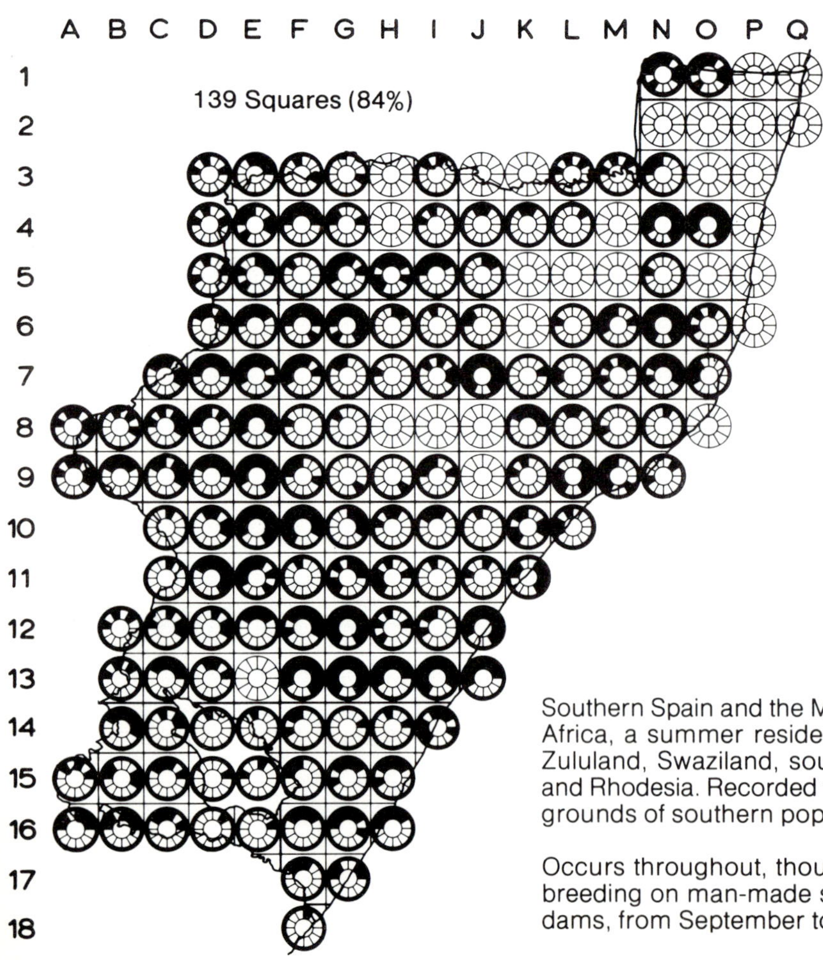

383 White-rumped Swift

Apus caffer

Southern Spain and the Maghreb, and the Afrotropical Region. In Southern Africa, a summer resident from August to April, in the Cape, Natal and Zululand, Swaziland, southern Moçambique, Lesotho, O.F.S., Transvaal and Rhodesia. Recorded from Botswana (Kanye) and S.W.A. The wintering grounds of southern populations are unknown.

Occurs throughout, though sparse in thickly-wooded regions. Recorded breeding on man-made structures such as bridges and concrete walls of dams, from September to March.

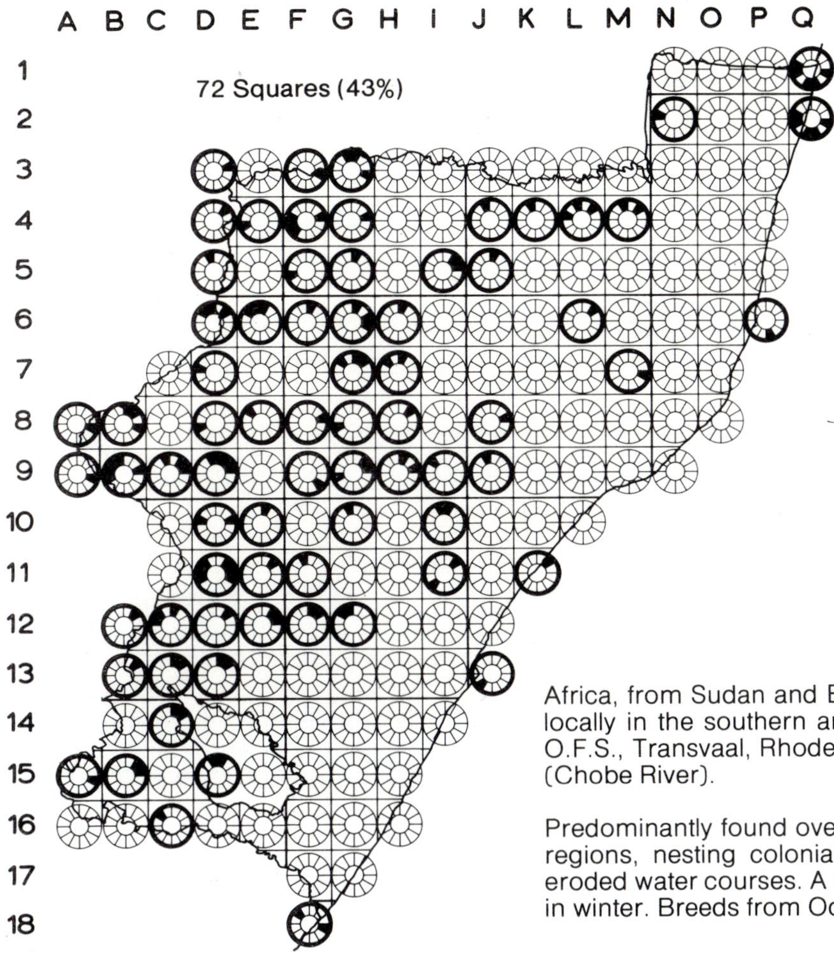

384 Horus Swift

Apus horus

Africa, from Sudan and Ethiopia to the Cape. In Southern Africa, breeds locally in the southern and eastern Cape, Natal and Zululand, Lesotho, O.F.S., Transvaal, Rhodesia and Moçambique; recorded from Botswana (Chobe River).

Predominantly found over grassland areas, but also recorded in wooded regions, nesting colonially in sand banks, road cuttings and sides of eroded water courses. A summer visitor to the interior; noted on the coast in winter. Breeds from October to April.

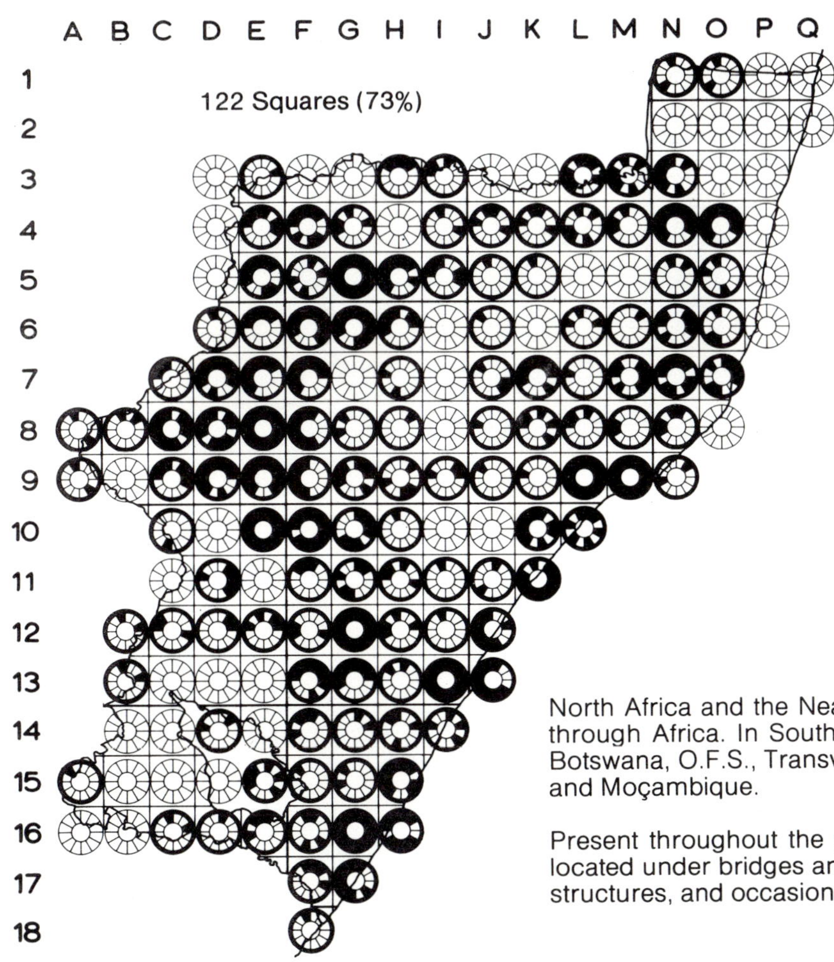

385 Little Swift

Apus affinis

North Africa and the Near East to China, Malaysia and Indonesia; south through Africa. In Southern Africa, locally in the Cape, S.W.A., eastern Botswana, O.F.S., Transvaal and Rhodesia; also Natal, eastern Swaziland and Moçambique.

Present throughout the year in the vicinity of breeding sites, which are located under bridges and eaves of buildings, on concrete silos and dam structures, and occasionally cliff faces. Breeds during the summer.

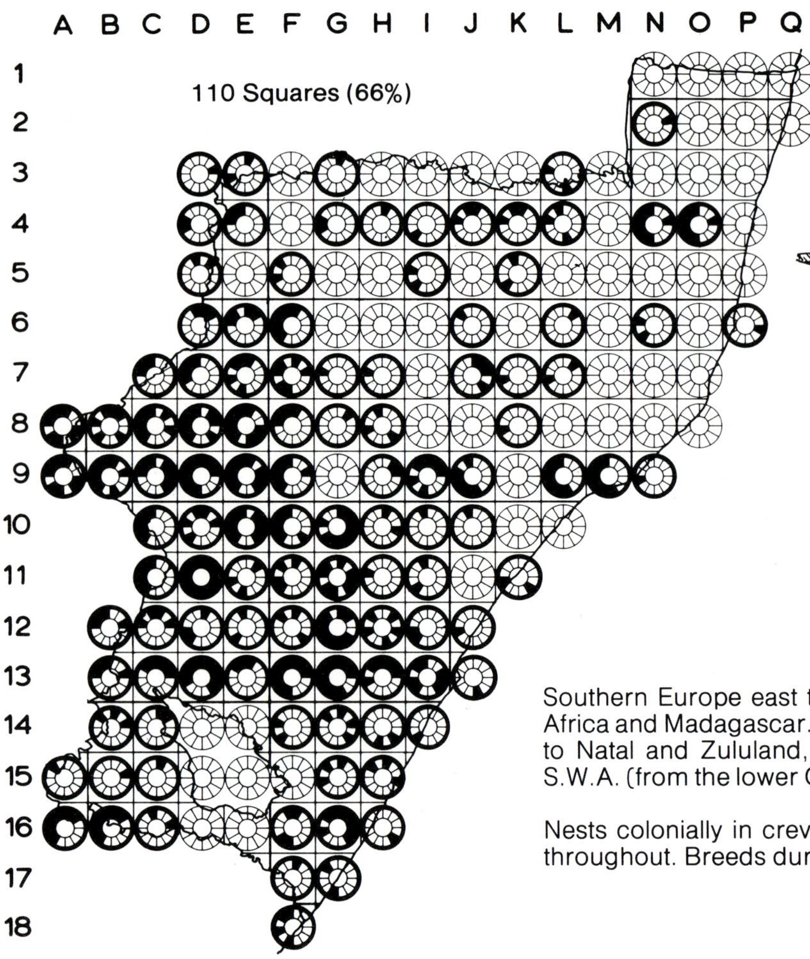

386 Alpine Swift

Apus melba

Southern Europe east to Turkestan, India and Sri Lanka; south through Africa and Madagascar. In Southern Africa, breeds from the western Cape to Natal and Zululand, Swaziland, Lesotho, O.F.S. and Transvaal; also S.W.A. (from the lower Orange River north to the Damaraland highlands).

Nests colonially in crevices on cliff faces and gorges, but may be found throughout. Breeds during the summer.

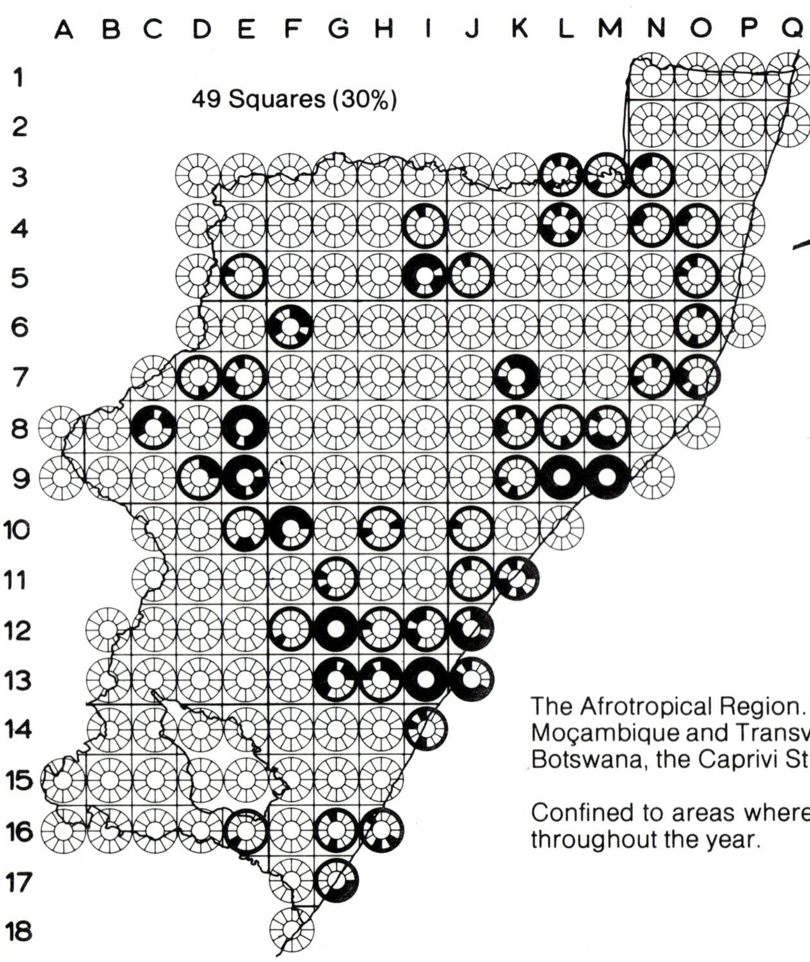

387 Palm Swift

Cypsiurus parvus

The Afrotropical Region. In Southern Africa, Natal and Zululand, southern Moçambique and Transvaal (plateau), north to Rhodesia, west to northern Botswana, the Caprivi Strip and northern S.W.A.

Confined to areas where tall palm trees provide nesting sites. May breed throughout the year.

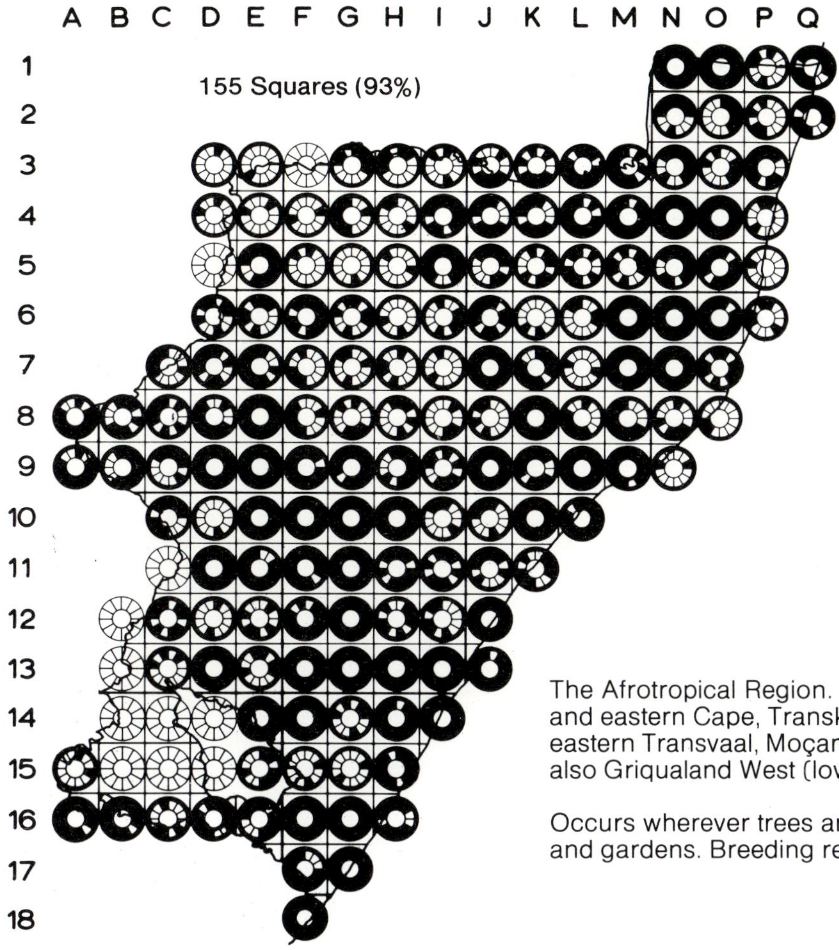

390 Speckled Mousebird

Colius striatus

The Afrotropical Region. In Southern Africa, the south-western, southern and eastern Cape, Transkei, Natal and Zululand, Swaziland, northern and eastern Transvaal, Moçambique and south-eastern and eastern Rhodesia; also Griqualand West (lower Vaal River) and western O.F.S.

Occurs wherever trees and scrub vegetation are found; often in orchards and gardens. Breeding recorded from August to May.

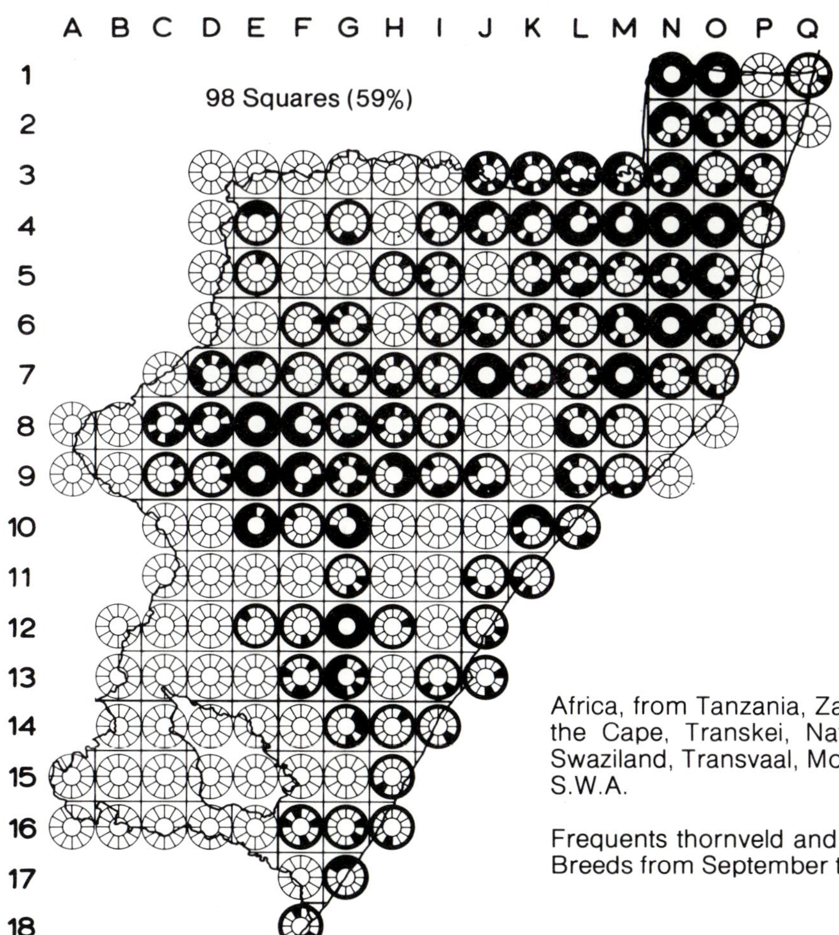

392 Red-faced Mousebird

Colius indicus

Africa, from Tanzania, Zaïre and Angola to the Cape. In Southern Africa, the Cape, Transkei, Natal and Zululand, O.F.S., Lesotho (lowlands), Swaziland, Transvaal, Moçambique and Rhodesia, west to Botswana and S.W.A.

Frequents thornveld and wooded valleys, ranging to the coast in winter. Breeds from September to March.

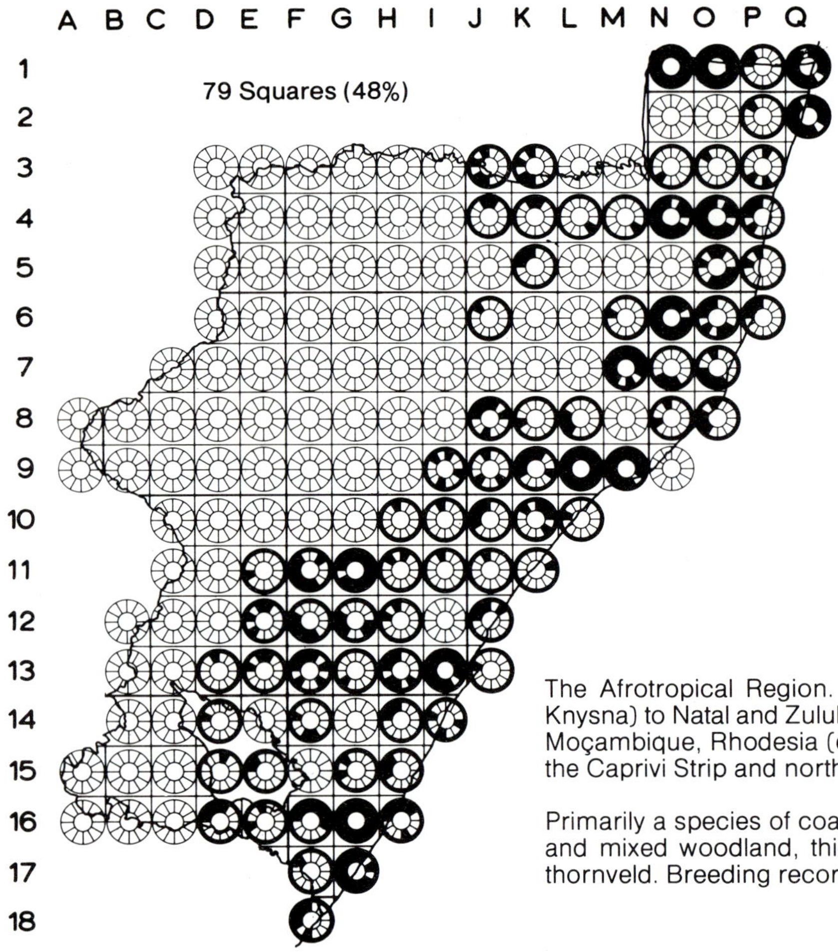

393 Narina Trogon

Apaloderma narina

The Afrotropical Region. In Southern Africa, the southern Cape (from Knysna) to Natal and Zululand, Swaziland, eastern and northern Transvaal, Moçambique, Rhodesia (eastern highlands), west to northern Botswana, the Caprivi Strip and northern S.W.A.

Primarily a species of coastal and evergreen forest, also found in riverine and mixed woodland, thickly-wooded valleys and occasionally in moist thornveld. Breeding recorded from October to January.

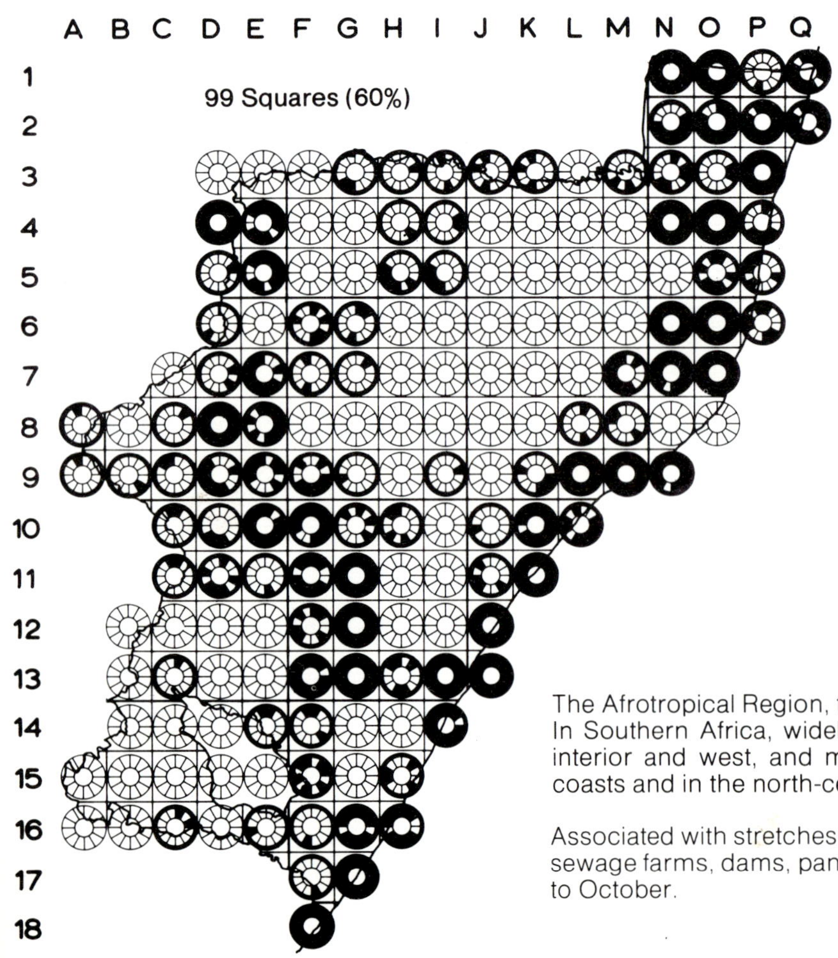

394 Pied Kingfisher

Ceryle rudis

The Afrotropical Region, the Nile valley and the Middle East, east to China. In Southern Africa, widely distributed, but absent from much of the dry interior and west, and most abundant along the southern and eastern coasts and in the north-central swamp region.

Associated with stretches of open water and wide, slow-flowing rivers; also sewage farms, dams, pans and estuaries. Breeding recorded from August to October.

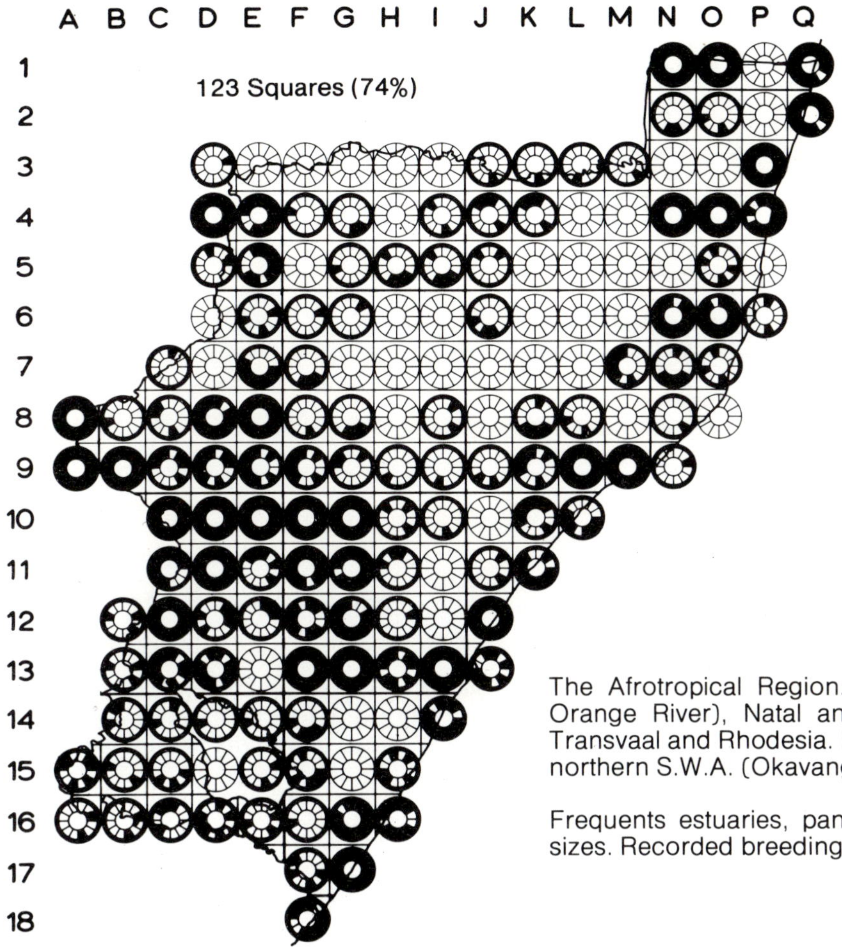

395 Giant Kingfisher

Ceryle maxima

The Afrotropical Region. In Southern Africa, the Cape (north of the Orange River), Natal and Zululand, Swaziland, Moçambique, O.F.S., Transvaal and Rhodesia. In the north, west to Botswana (Ngamiland) and northern S.W.A. (Okavango and Cunene Rivers).

Frequents estuaries, pans, dams, sewage farms and rivers of varying sizes. Recorded breeding from August to October.

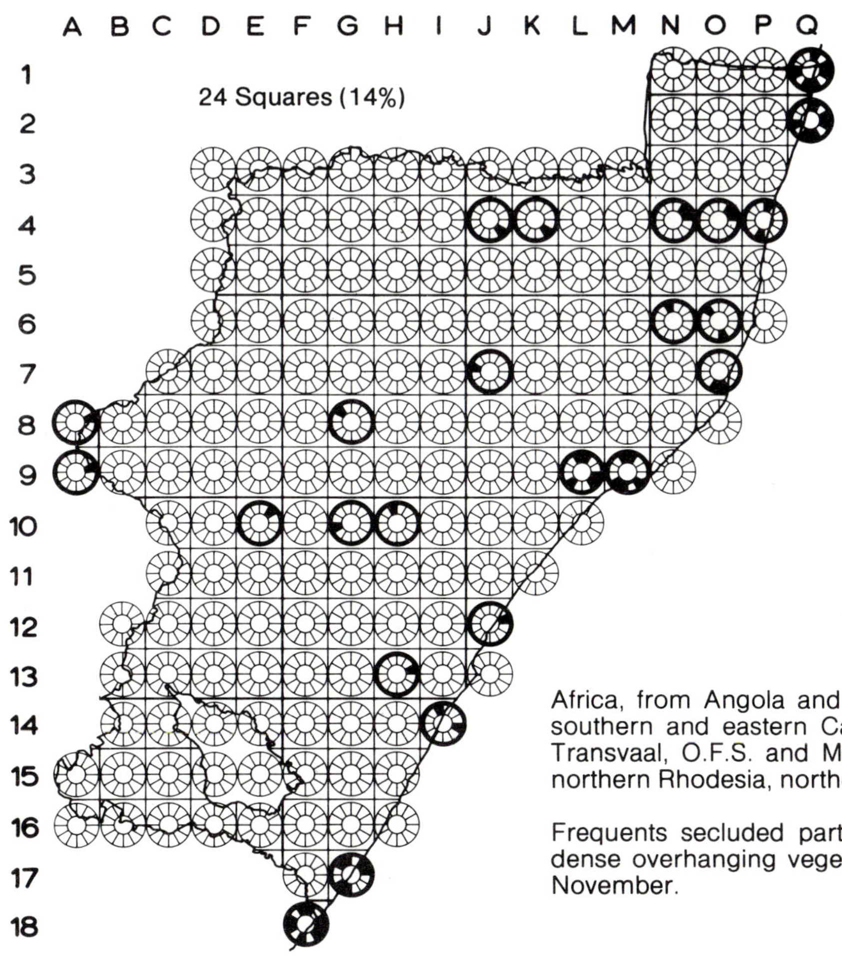

396 Half-collared Kingfisher

Alcedo semitorquata

Africa, from Angola and Ethiopia to the Cape. In Southern Africa, the southern and eastern Cape, Transkei, Natal and Zululand, Swaziland, Transvaal, O.F.S. and Moçambique; in the north, west to eastern and northern Rhodesia, northern Botswana and the Caprivi Strip.

Frequents secluded parts of estuaries, rivers and small streams, with dense overhanging vegetation. Breeding recorded during October and November.

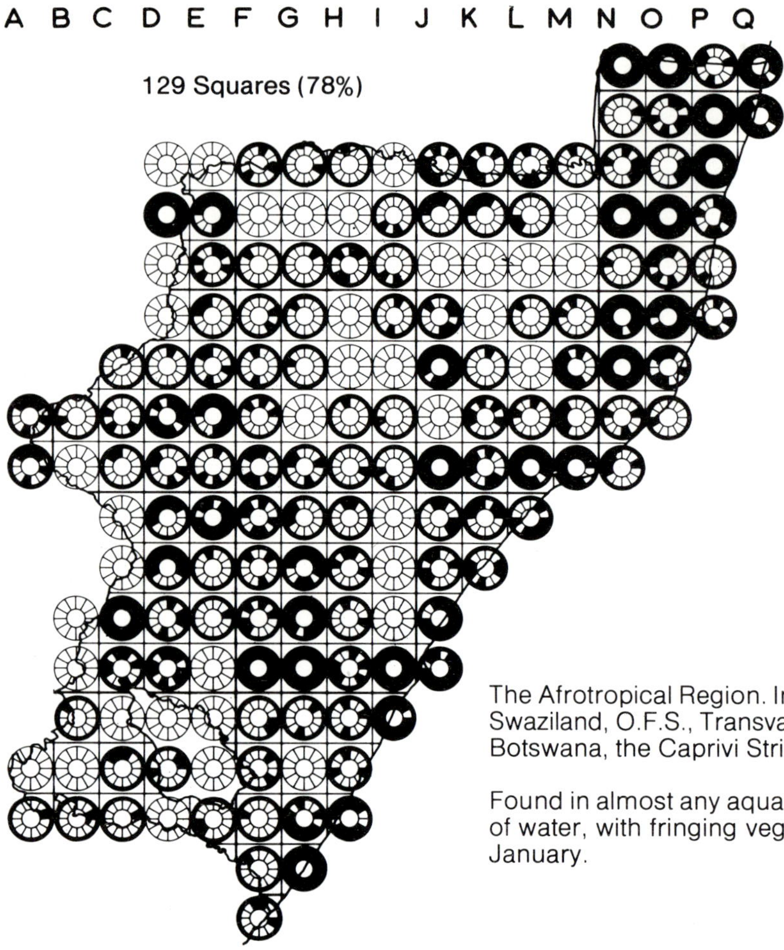

397 Malachite Kingfisher

Alcedo cristata

The Afrotropical Region. In Southern Africa, the Cape, Natal and Zululand, Swaziland, O.F.S., Transvaal, Moçambique and Rhodesia, west to northern Botswana, the Caprivi Strip and S.W.A. (absent from dry central parts).

Found in almost any aquatic habitat, from small streams to open stretches of water, with fringing vegetation. Recorded breeding from September to January.

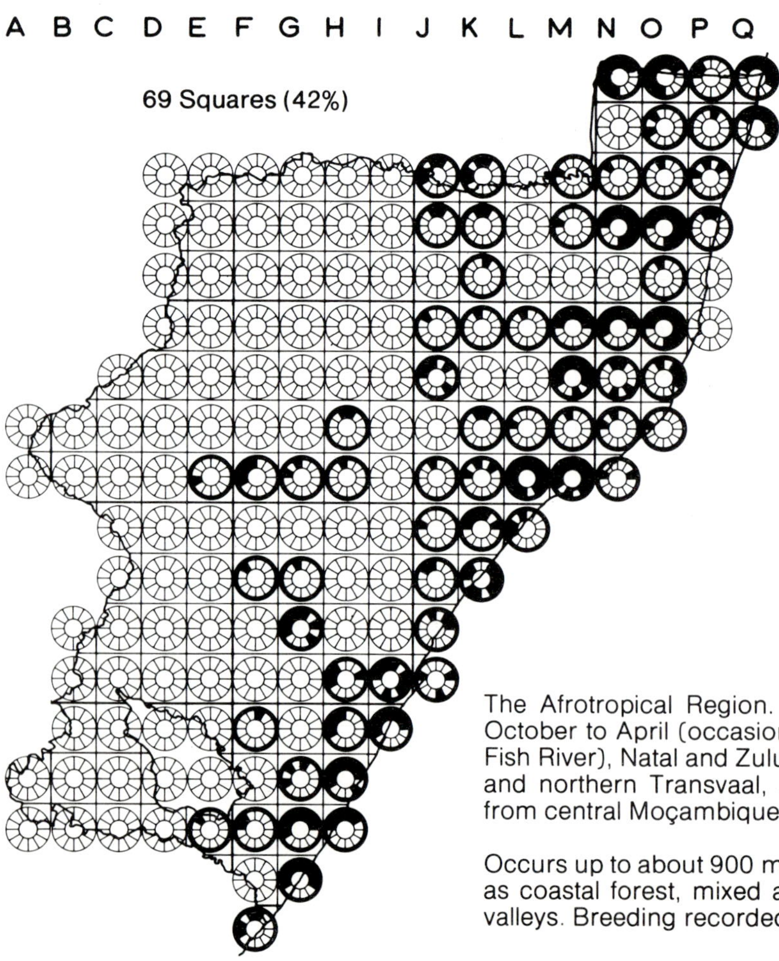

398 Pygmy Kingfisher

Ispidina picta

The Afrotropical Region. In Southern Africa, a summer resident from October to April (occasionally May) in the eastern Cape (from the Great Fish River), Natal and Zululand, eastern Swaziland, Moçambique, eastern and northern Transvaal, and Rhodesia. Winters in the eastern tropics, from central Moçambique northwards.

Occurs up to about 900 m a.s.l. in a wide variety of wooded habitats, such as coastal forest, mixed and riverine woodland, and up thickly-wooded valleys. Breeding recorded from September to February.

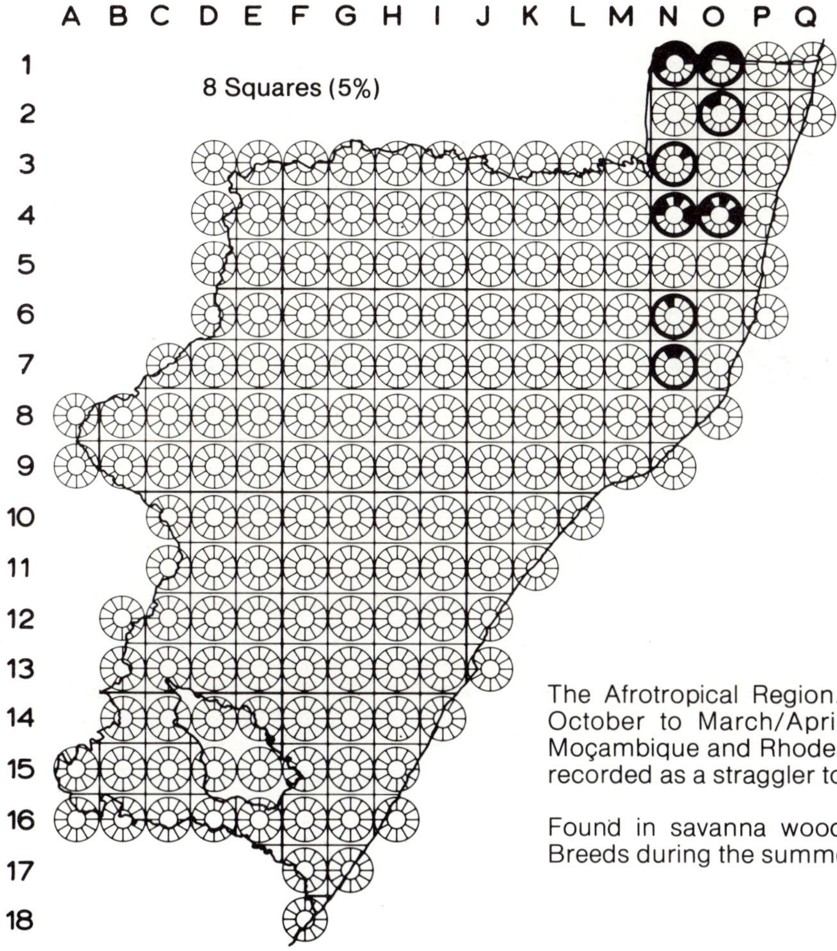

399 Woodland Kingfisher

Halcyon senegalensis

The Afrotropical Region. In Southern Africa, a summer resident from October to March/April in eastern Zululand, Swaziland, Transvaal, Moçambique and Rhodesia, west to Botswana and northern S.W.A. Also recorded as a straggler to Transkei.

Found in savanna woodland and on the edge of riverine vegetation. Breeds during the summer.

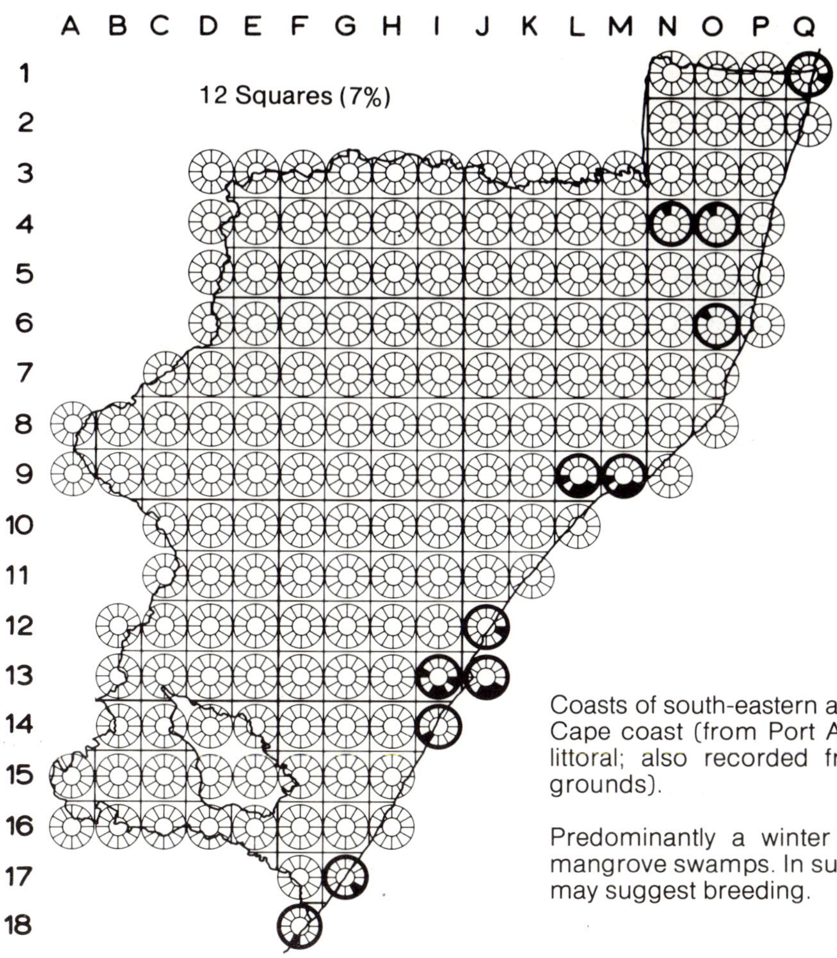

400 Mangrove Kingfisher

Halcyon senegaloides

Coasts of south-eastern and eastern Africa. In Southern Africa, the eastern Cape coast (from Port Alfred) to Natal, Zululand, and the Moçambique littoral; also recorded from eastern Transvaal (possibly on breeding grounds).

Predominantly a winter visitor to the coast, where it is restricted to mangrove swamps. In summer, records in thornveld away from the coast may suggest breeding.

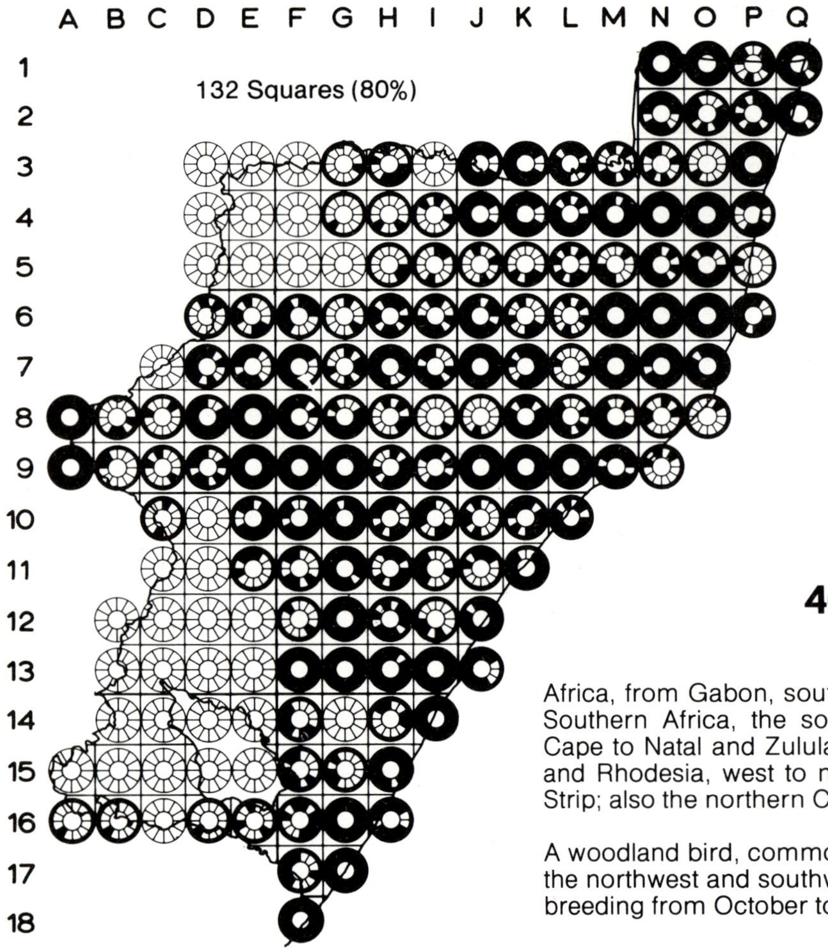

402 Brown-hooded Kingfisher

Halcyon albiventris

Africa, from Gabon, southern Zaïre, Kenya and Somalia to the Cape. In Southern Africa, the south-western (from Swellendam) and southern Cape to Natal and Zululand, Swaziland, O.F.S., Transvaal, Moçambique and Rhodesia, west to northern and eastern Botswana and the Caprivi Strip; also the northern Cape.

A woodland bird, common and widespread except in the upland areas of the northwest and southwest; freely enters parks and gardens. Recorded breeding from October to December.

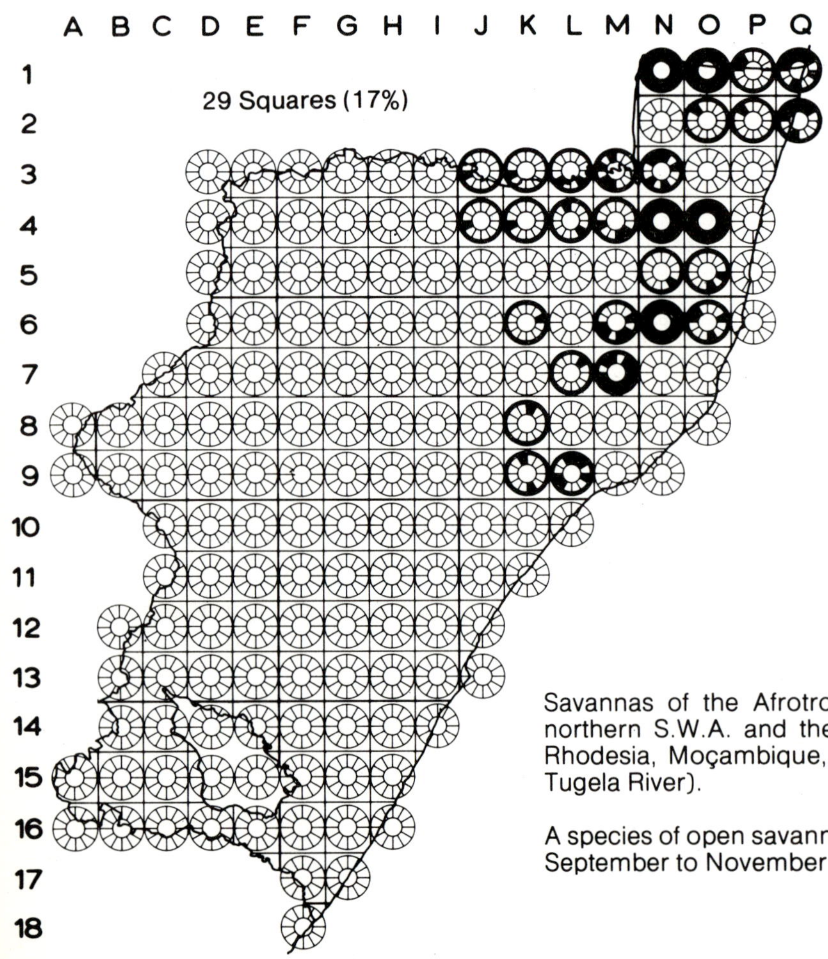

403 Striped Kingfisher

Halcyon chelicuti

Savannas of the Afrotropical Region. In Southern Africa, central and northern S.W.A. and the northern Cape, east to Botswana, Transvaal, Rhodesia, Moçambique, Swaziland and eastern Zululand (south to the Tugela River).

A species of open savanna and mixed woodland. Breeding recorded from September to November.

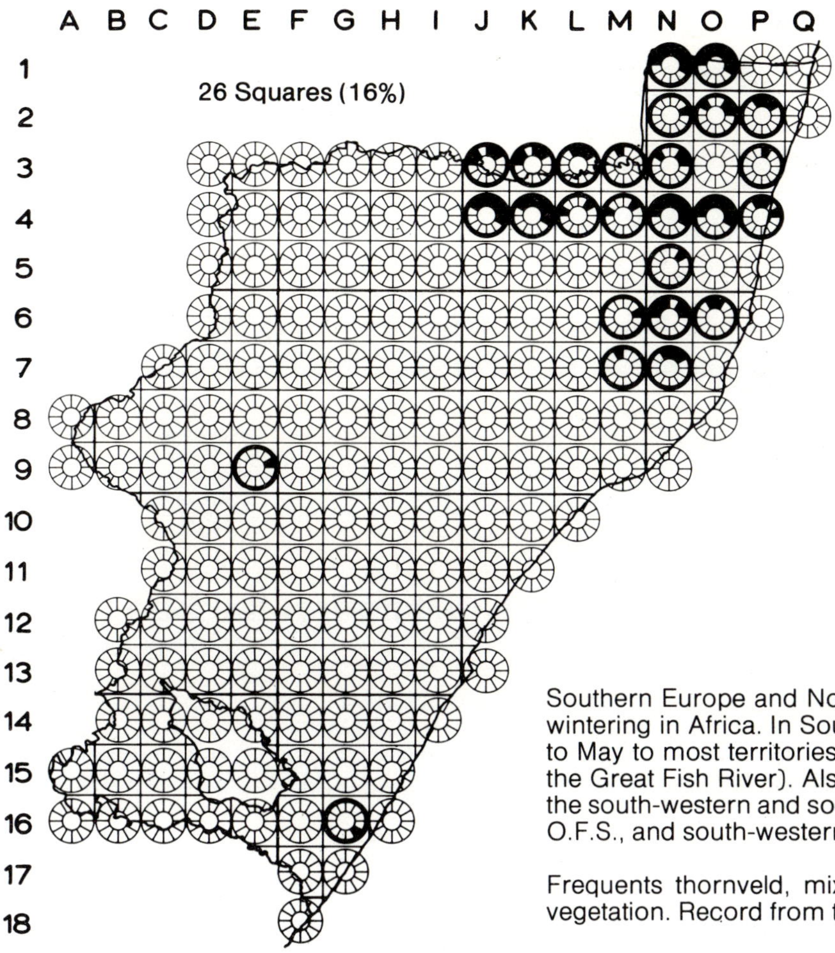

404 European Bee-eater

Merops apiaster

Southern Europe and North Africa, east to the north-western Himalayas, wintering in Africa. In Southern Africa, a visitor from September/October to May to most territories, but uncommon in Natal and the Cape (east of the Great Fish River). Also breeds locally from September to February in the south-western and southern Cape, north to Griqualand West, adjacent O.F.S., and south-western Transvaal.

Frequents thornveld, mixed woodland and open grassland with scrub vegetation. Record from the south coast exceptional.

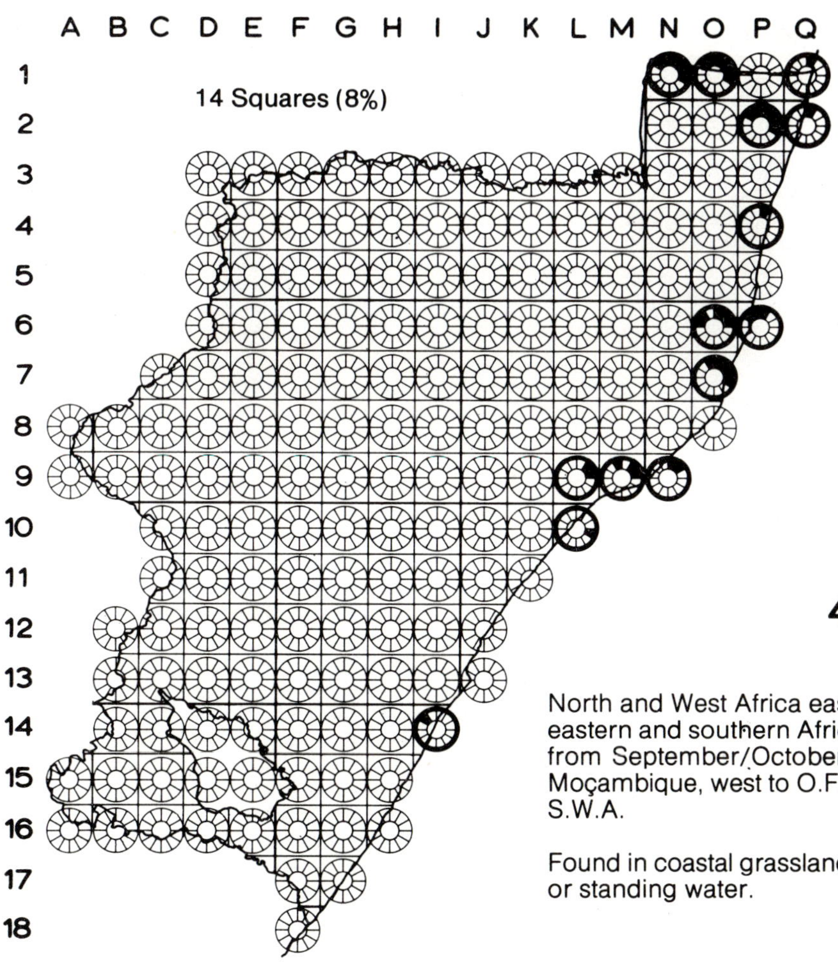

406 Blue-cheeked Bee-eater

Merops superciliosus

North and West Africa east to the Middle East, India and central Asia; also eastern and southern Africa, and Madagascar. In Southern Africa, a visitor from September/October to May to Natal and Zululand, Swaziland and Moçambique, west to O.F.S., Transvaal, Rhodesia, Botswana and northern S.W.A.

Found in coastal grassland with scattered bushes, in the vicinity of marshes or standing water.

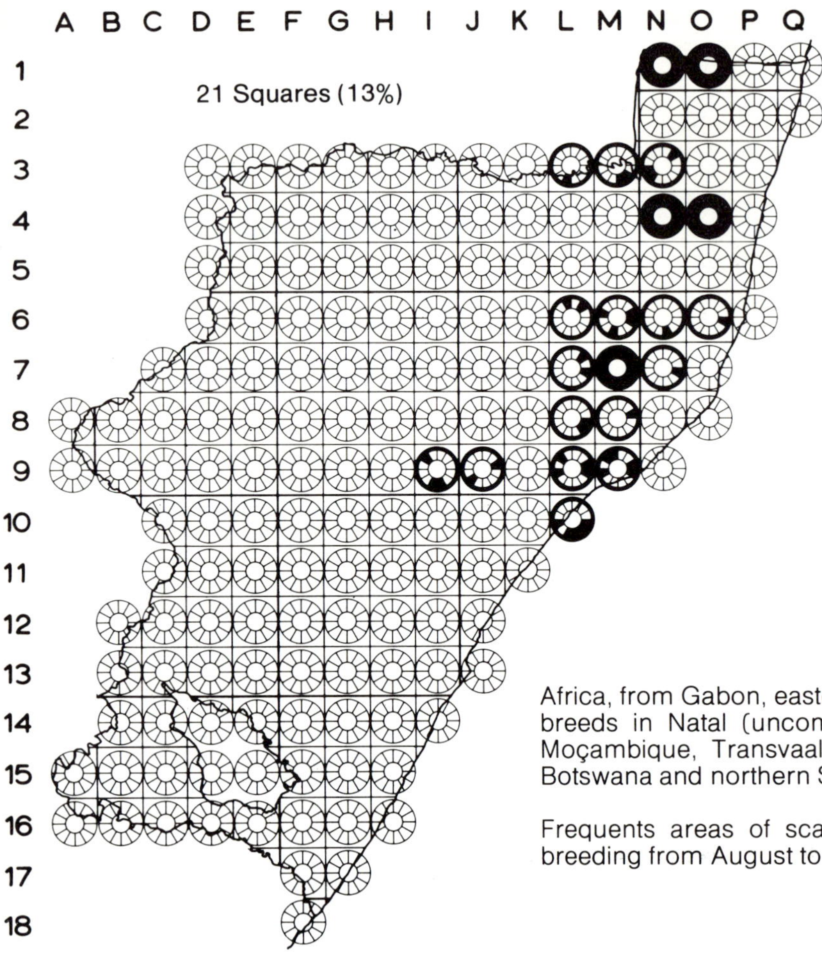

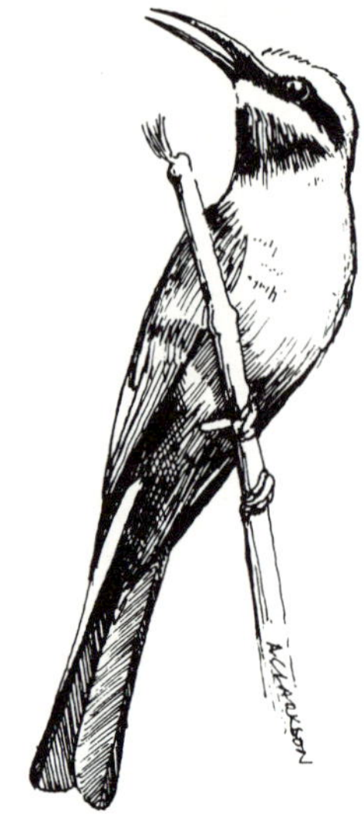

409 White-fronted Bee-eater

Merops bullockoides

Africa, from Gabon, eastern Zaïre and Kenya to Natal. In Southern Africa, breeds in Natal (uncommon), eastern Zululand, Swaziland, southern Moçambique, Transvaal and Rhodesia, west to eastern and northern Botswana and northern S.W.A.

Frequents areas of scattered trees usually close to water. Recorded breeding from August to October.

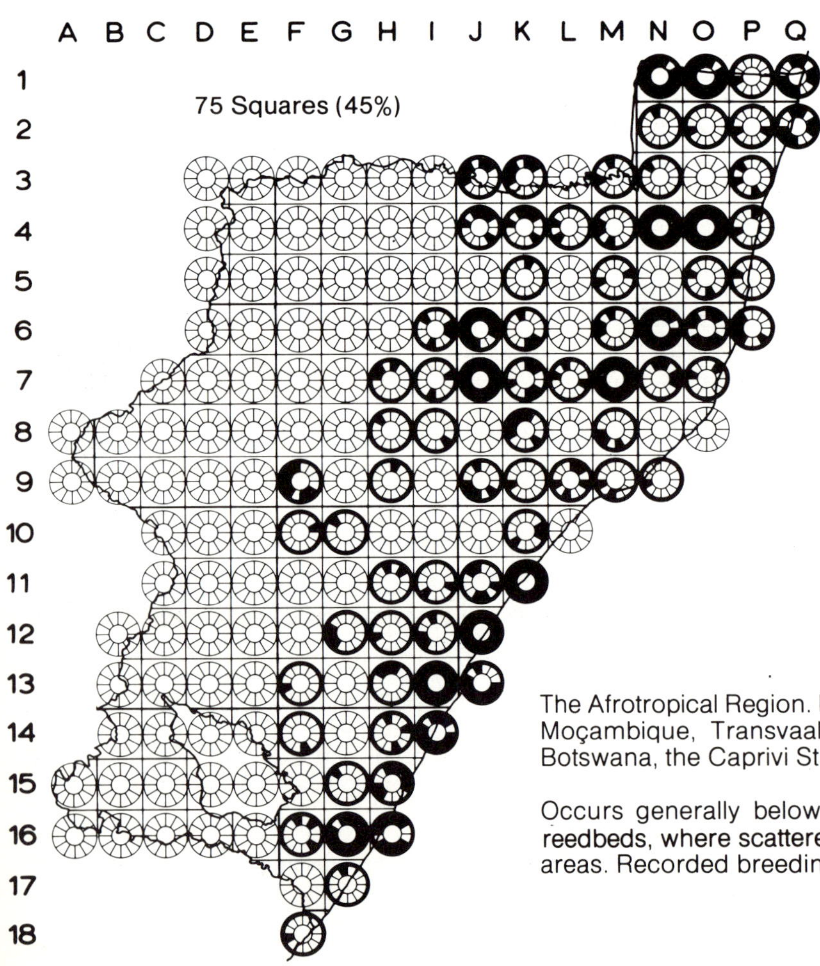

410 Little Bee-eater

Merops pusillus

The Afrotropical Region. In Southern Africa, Natal and Zululand, Swaziland, Moçambique, Transvaal and Rhodesia, west to northern and eastern Botswana, the Caprivi Strip and northern S.W.A.

Occurs generally below 900 m a.s.l. on the edges of vleis, rivers and reedbeds, where scattered bushes provide perches; also in lightly-wooded areas. Recorded breeding from October to January.

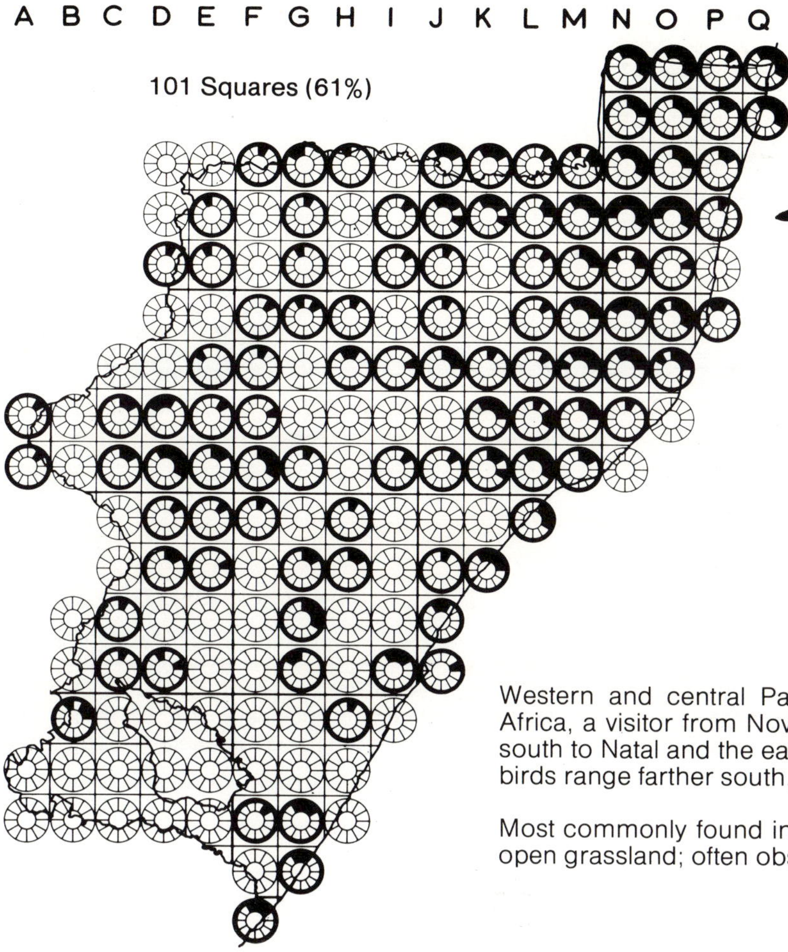

412 European Roller

Coracias garrulus

Western and central Palaearctic, wintering to the south. In Southern Africa, a visitor from November to March to northern and eastern parts, south to Natal and the eastern Cape. Numbers fluctuate, and some years birds range farther south, even reaching Cape Town.

Most commonly found in thornveld and mixed woodland, extending into open grassland; often observed perched on telegraph poles.

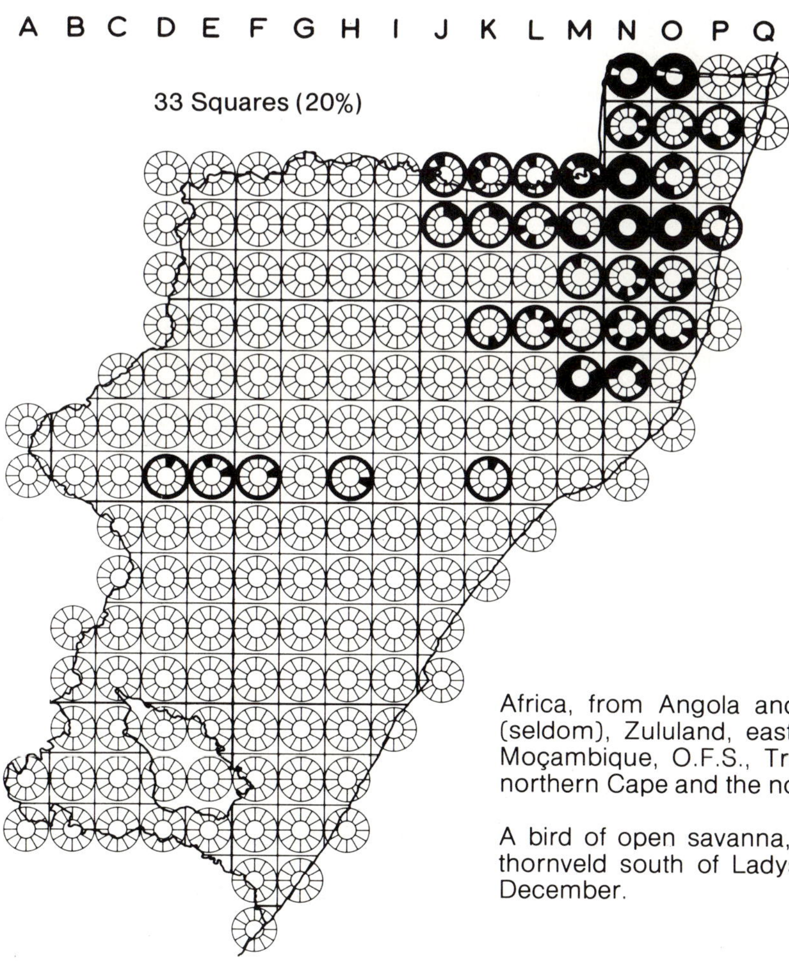

413 Lilac-breasted Roller

Coracias caudata

Africa, from Angola and Ethiopia to Natal. In Southern Africa, Natal (seldom), Zululand, eastern Swaziland, extreme southern and coastal Moçambique, O.F.S., Transvaal and Rhodesia, west to Botswana, the northern Cape and the northern half of S.W.A.

A bird of open savanna, usually seen singly or in pairs. Records from thornveld south of Ladysmith are exceptional. Breeds from August to December.

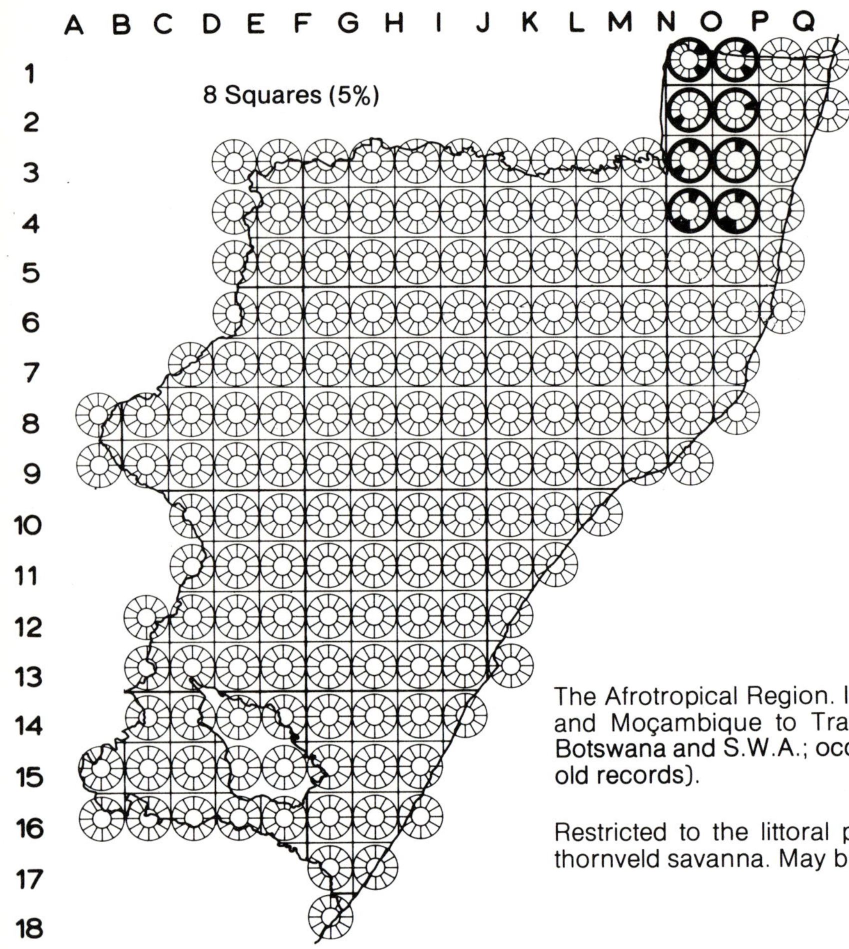

415 Purple Roller

Coracias naevia

The Afrotropical Region. In Southern Africa, eastern Zululand, Swaziland and Moçambique to Transvaal, O.F.S., Rhodesia, the northern Cape, Botswana and S.W.A.; occasionally to Natal and the eastern Cape (mainly old records).

Restricted to the littoral plain, where it occurs in open, well-developed thornveld savanna. May breed.

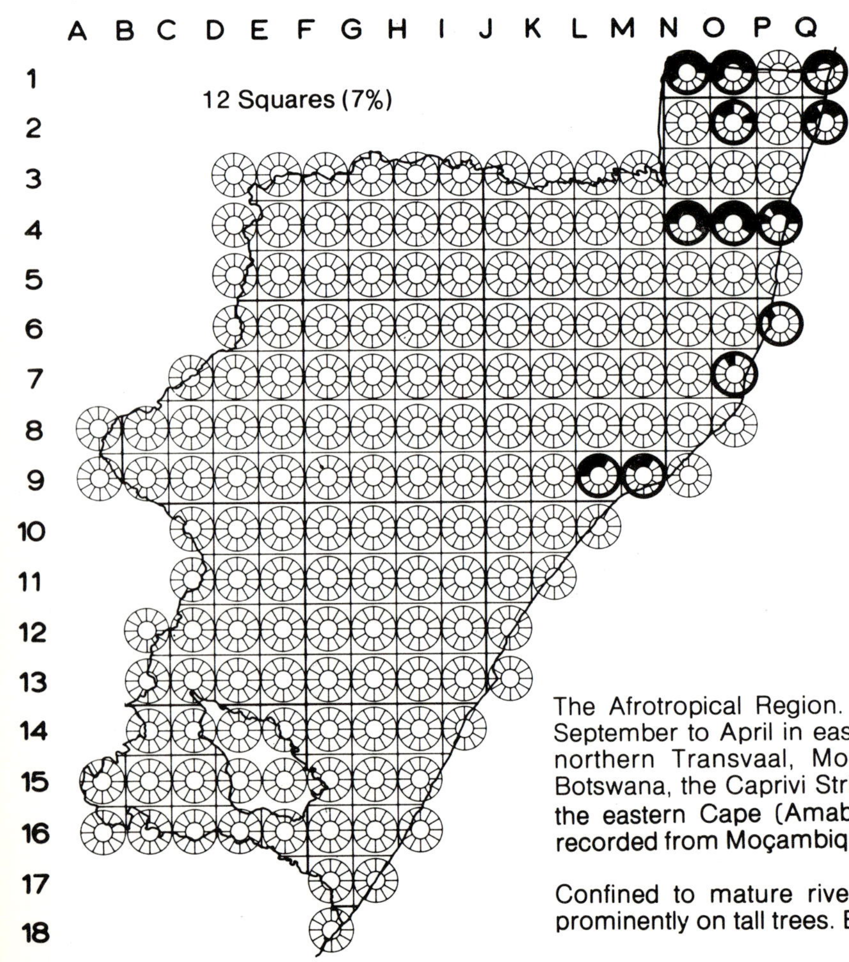

416 Broad-billed Roller

Eurystomus glaucurus

The Afrotropical Region. In Southern Africa, a summer resident from September to April in eastern Zululand, eastern Swaziland, eastern and northern Transvaal, Moçambique and Rhodesia, west to northern Botswana, the Caprivi Strip and northern S.W.A. (Okavango River); also the eastern Cape (Amabele). Non-breeding visitors from Madagascar recorded from Moçambique (Beira) and Rhodesia (March to September).

Confined to mature riverine woodland, where usually seen perched prominently on tall trees. Breeds from October to December.

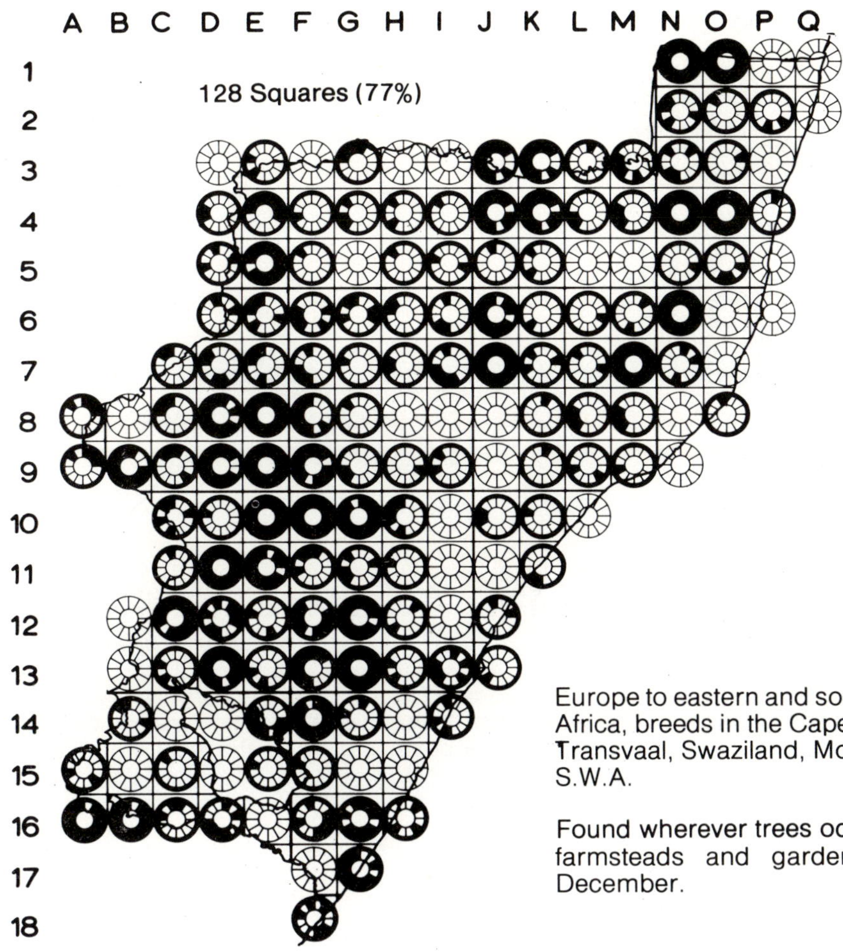

418 Hoopoe

Upupa epops

Europe to eastern and southern Asia, Africa and Madagascar. In Southern Africa, breeds in the Cape, Natal and Zululand, O.F.S., Lesotho (lowlands), Transvaal, Swaziland, Moçambique and Rhodesia, west to Botswana and S.W.A.

Found wherever trees occur, from thornveld and lightly-wooded areas to farmsteads and gardens. Breeding recorded from September to December.

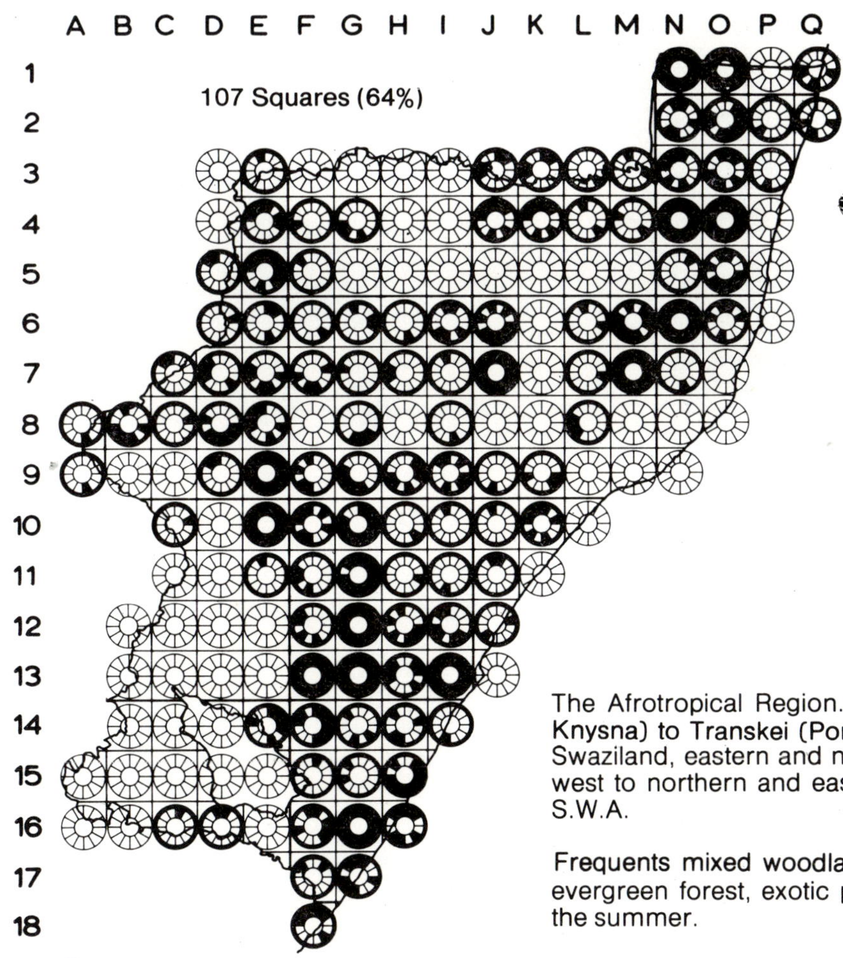

419 Red-billed Hoopoe

Phoeniculus purpureus

The Afrotropical Region. In Southern Africa, the southern Cape (from Knysna) to Transkei (Pondoland), East Griqualand, Natal and Zululand, Swaziland, eastern and northern Transvaal, Moçambique and Rhodesia, west to northern and eastern Botswana, the Caprivi Strip and northern S.W.A.

Frequents mixed woodland and thornveld with tall trees; also edges of evergreen forest, exotic plantations and gardens. Breeds mainly during the summer.

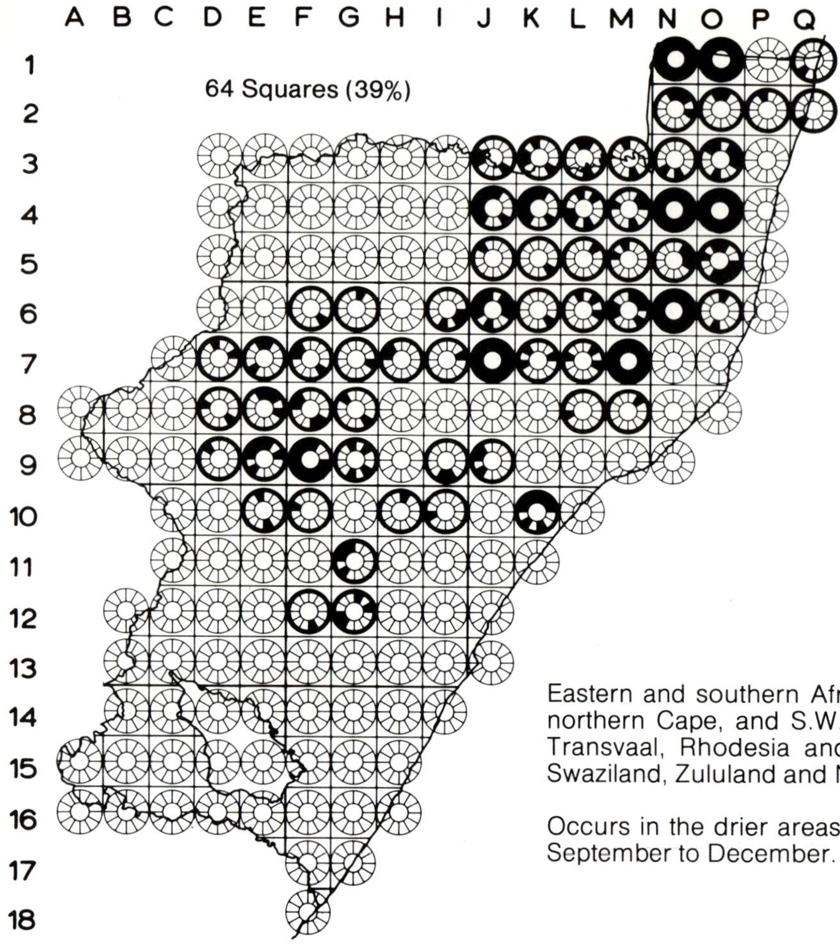

421 Scimitar-billed Hoopoe

Phoeniculus cyanomelas

Eastern and southern Africa. In Southern Africa, the north-western and northern Cape, and S.W.A., east through Botswana to western O.F.S., Transvaal, Rhodesia and Moçambique, south in the east to eastern Swaziland, Zululand and Natal (interior).

Occurs in the drier areas of thornveld and light woodland. Breeds from September to December.

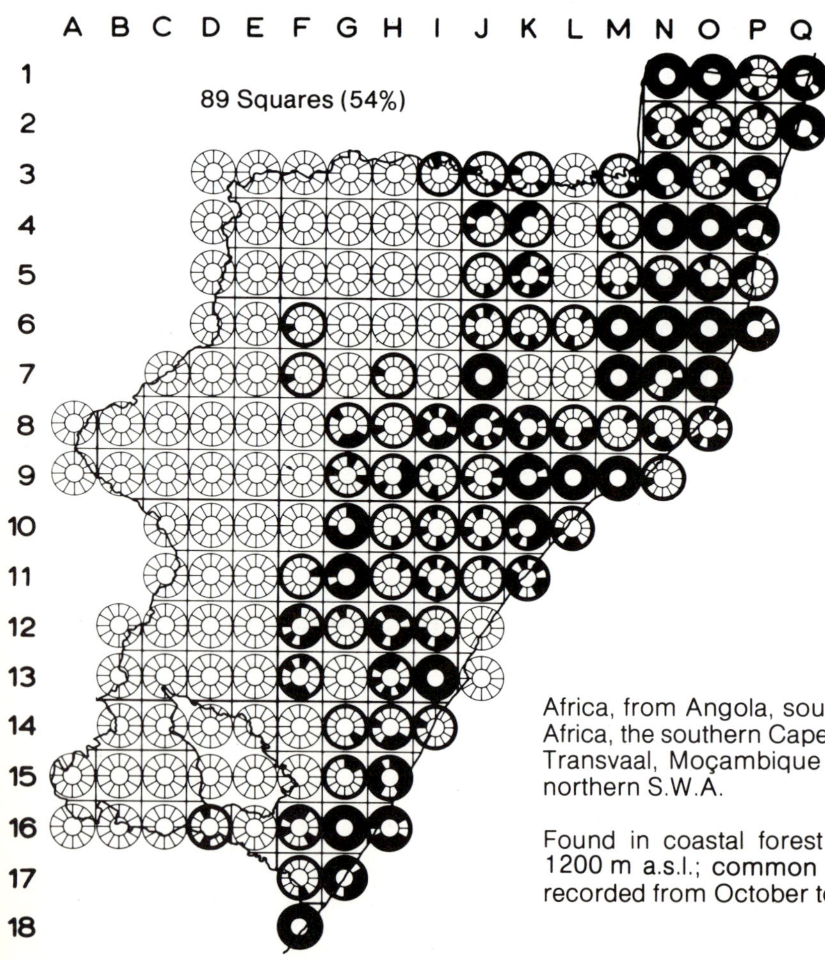

422 Trumpeter Hornbill

Bycanistes bucinator

Africa, from Angola, southern Zaïre and Kenya to the Cape. In Southern Africa, the southern Cape (from Knysna) to Natal and Zululand, Swaziland, Transvaal, Moçambique and Rhodesia, west to northern Botswana and northern S.W.A.

Found in coastal forest and up thickly-wooded river valleys to about 1200 m a.s.l.; common in evergreen forests of the midlands. Breeding recorded from October to January.

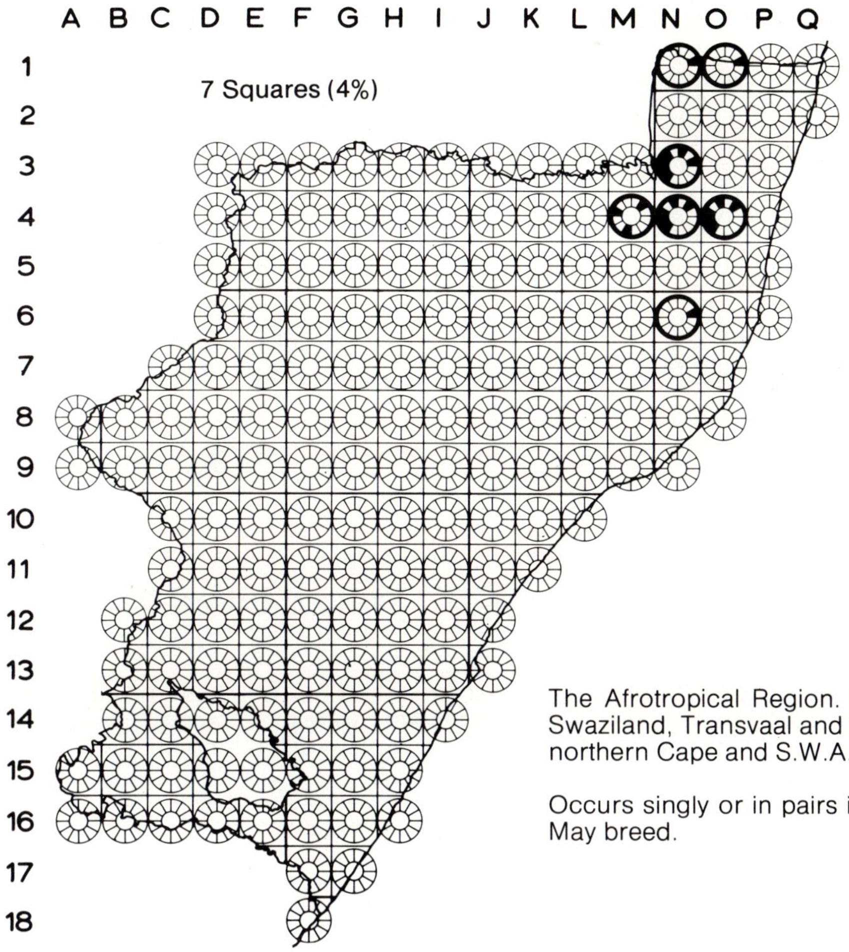

424 Grey Hornbill

Tockus nasutus

The Afrotropical Region. In Southern Africa, eastern Zululand, eastern Swaziland, Transvaal and Moçambique, west to Rhodesia, Botswana, the northern Cape and S.W.A.

Occurs singly or in pairs in the drier parts of mature thornveld savanna. May breed.

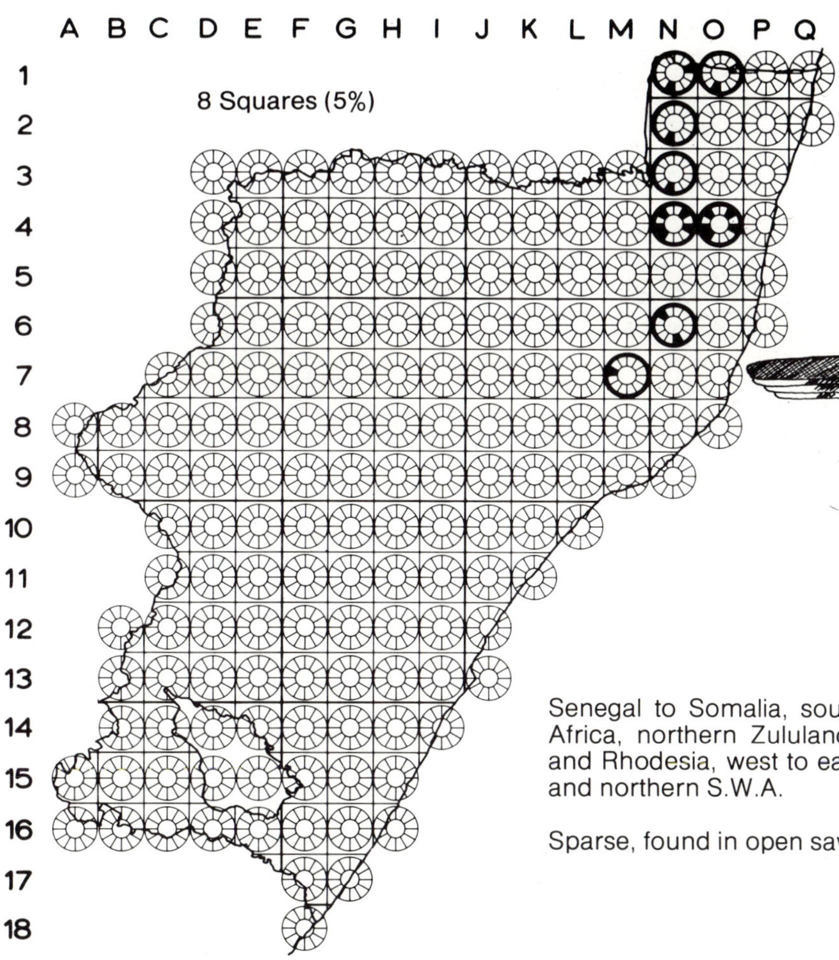

425 Red-billed Hornbill

Tockus erythrorhynchus

Senegal to Somalia, south in the east to southern Africa. In Southern Africa, northern Zululand, eastern Swaziland, Transvaal, Moçambique and Rhodesia, west to eastern and northern Botswana, the Caprivi Strip and northern S.W.A.

Sparse, found in open savanna. Breeds during the summer.

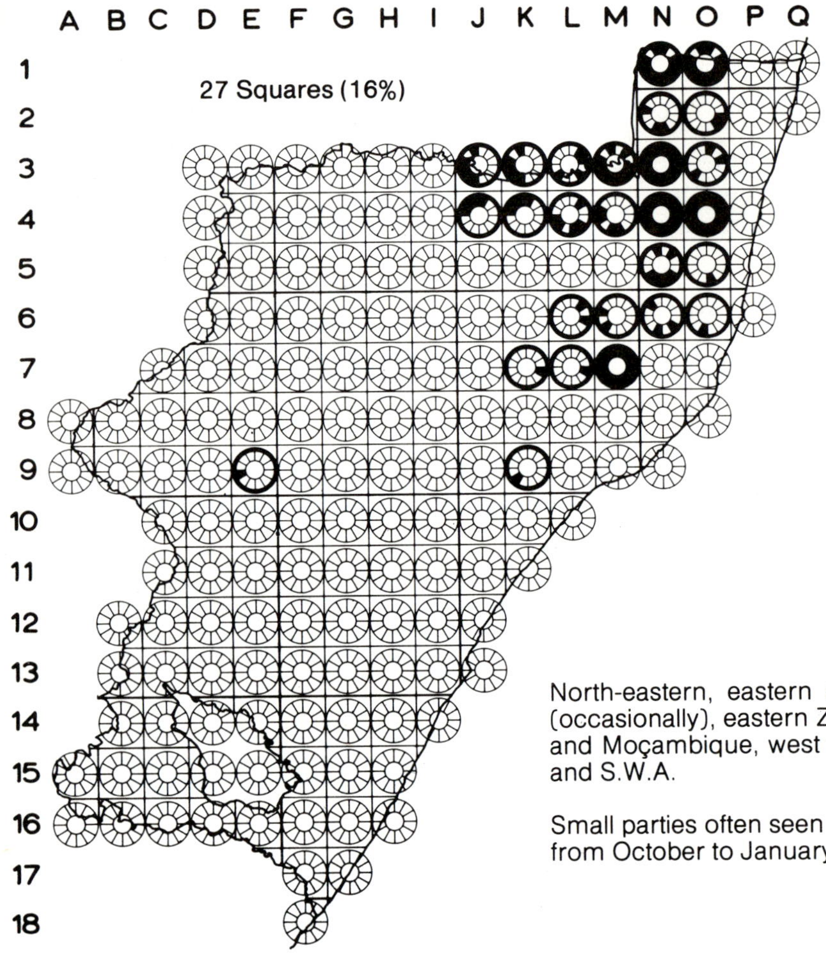

426 Yellow-billed Hornbill

Tockus flavirostris

North-eastern, eastern and southern Africa. In Southern Africa, Natal (occasionally), eastern Zululand, eastern Swaziland, Transvaal, Rhodesia and Moçambique, west to Botswana, western O.F.S., the northern Cape and S.W.A.

Small parties often seen in mature trees in dry thornveld savanna. Breeds from October to January.

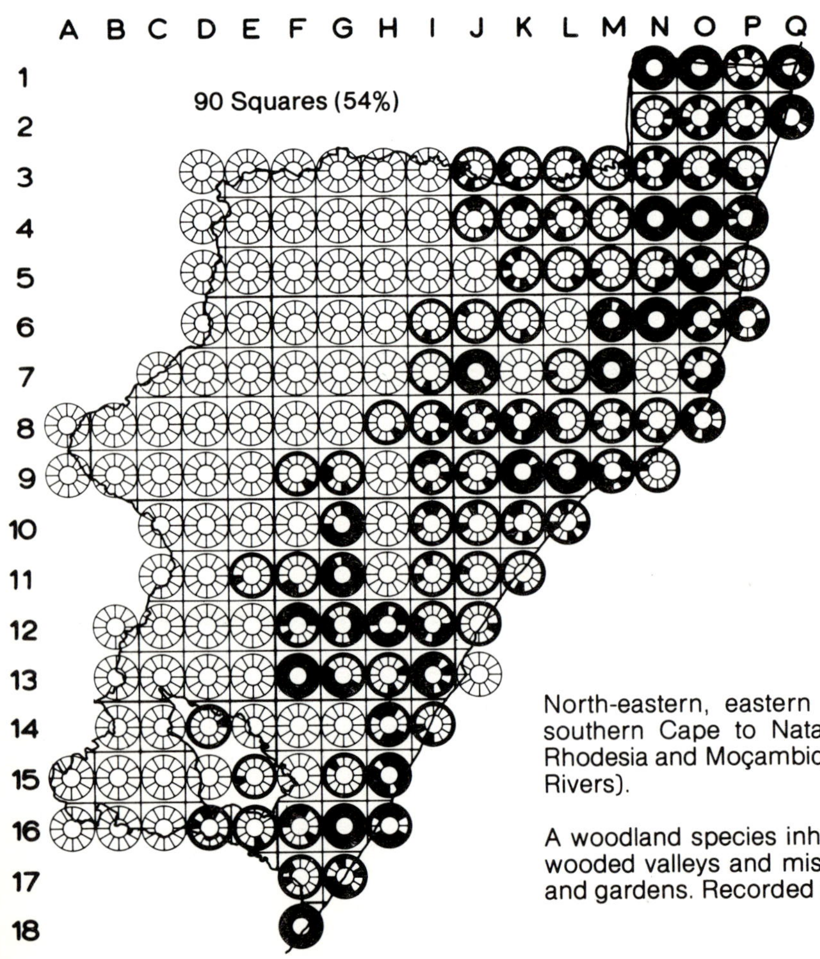

427 Crowned Hornbill

Tockus alboterminatus

North-eastern, eastern and southern Africa. In Southern Africa, the southern Cape to Natal and Zululand, Swaziland, eastern Transvaal, Rhodesia and Moçambique; also northern S.W.A. (Cunene and Okavango Rivers).

A woodland species inhabiting coastal forest, riverine woodland, thickly-wooded valleys and mistbelt evergreen forest; occasionally enters parks and gardens. Recorded breeding during November and December.

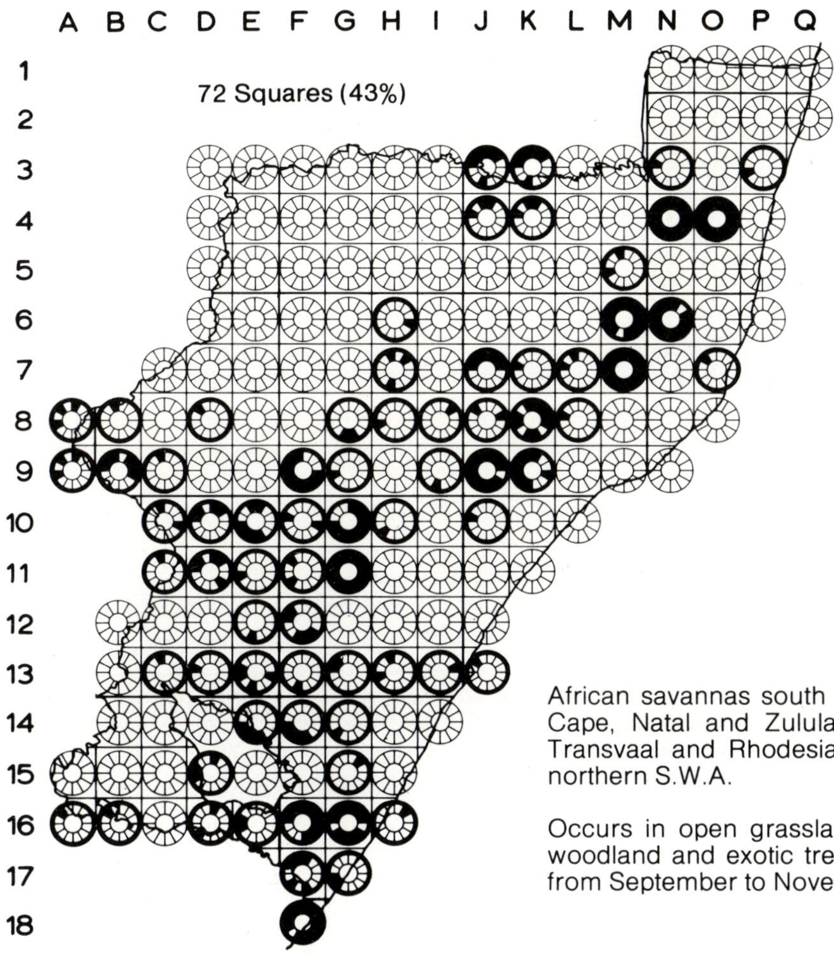

430 Ground Hornbill

Bucorvus leadbeateri

African savannas south of the Equator. In Southern Africa, the eastern Cape, Natal and Zululand, Swaziland, Moçambique, O.F.S. (Kestell), Transvaal and Rhodesia, west to eastern and northern Botswana and northern S.W.A.

Occurs in open grassland in the vicinity of evergreen forest, savanna woodland and exotic trees; usually in small parties. Recorded breeding from September to November.

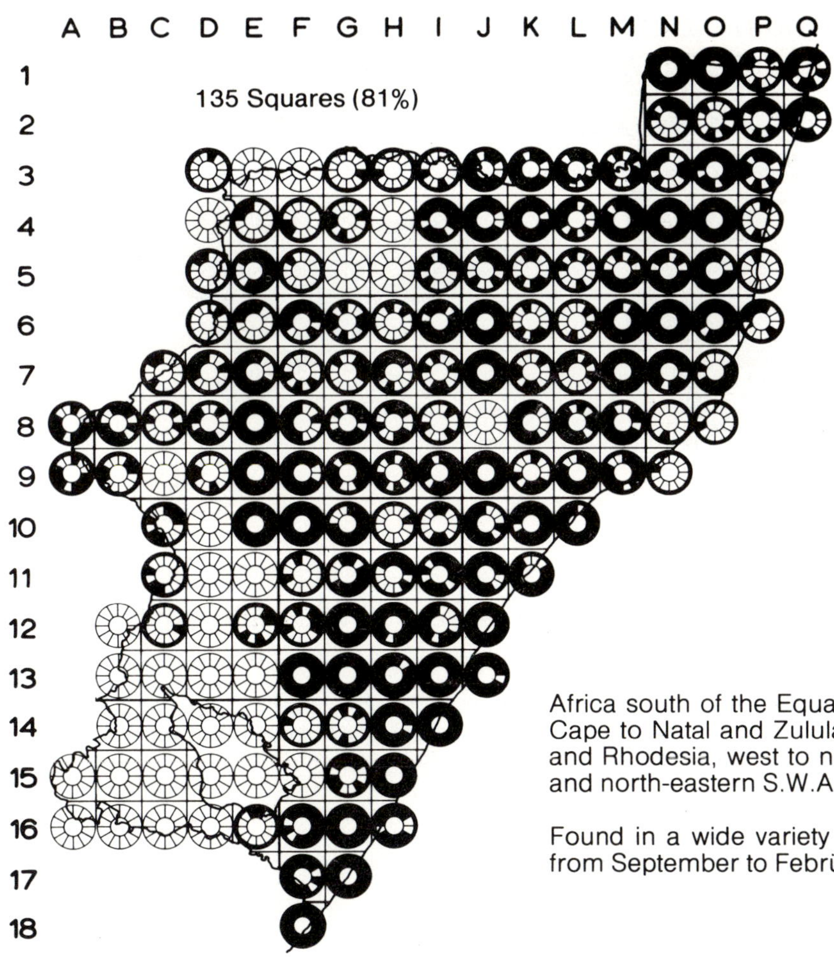

431 Black-collared Barbet

Lybius torquatus

Africa south of the Equator to the Cape. In Southern Africa, the eastern Cape to Natal and Zululand, Swaziland, O.F.S., Transvaal, Moçambique and Rhodesia, west to northern and eastern Botswana, the Caprivi Strip and north-eastern S.W.A.

Found in a wide variety of habitats wherever trees are present. Breeds from September to February.

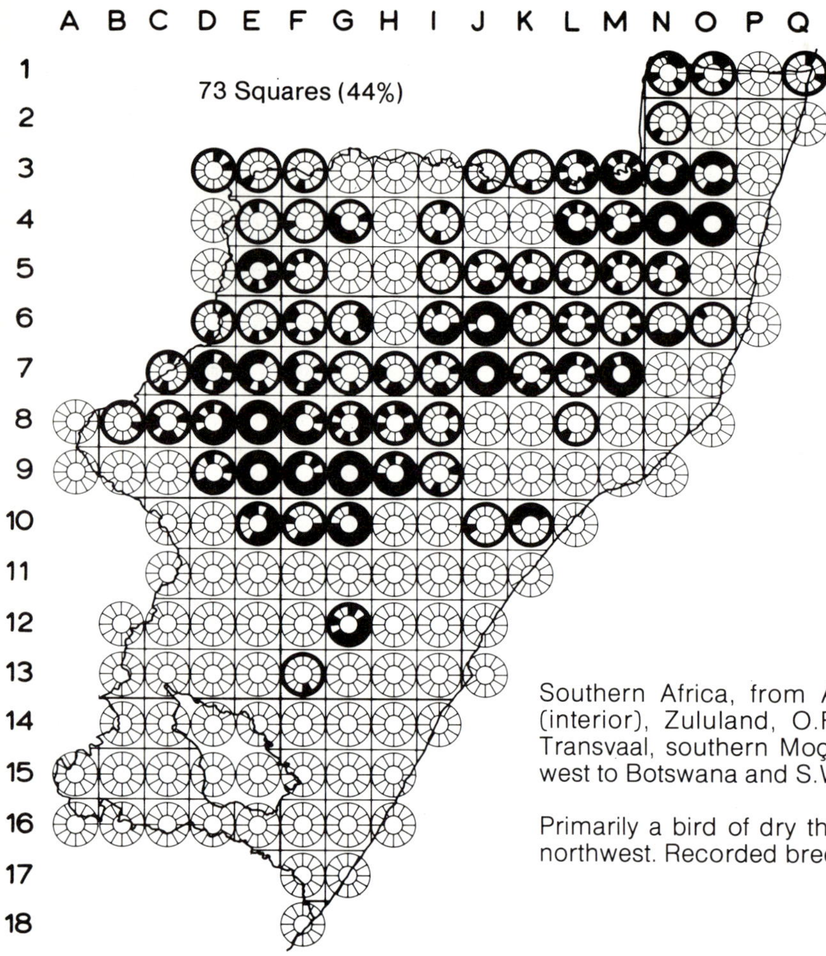

432 Pied Barbet

Lybius leucomelas

Southern Africa, from Angola southwards. From the Cape to Natal (interior), Zululand, O.F.S., Lesotho (lowlands), eastern Swaziland, Transvaal, southern Moçambique and southern and western Rhodesia, west to Botswana and S.W.A.

Primarily a bird of dry thornveld, extending into exotic vegetation in the northwest. Recorded breeding from October to December.

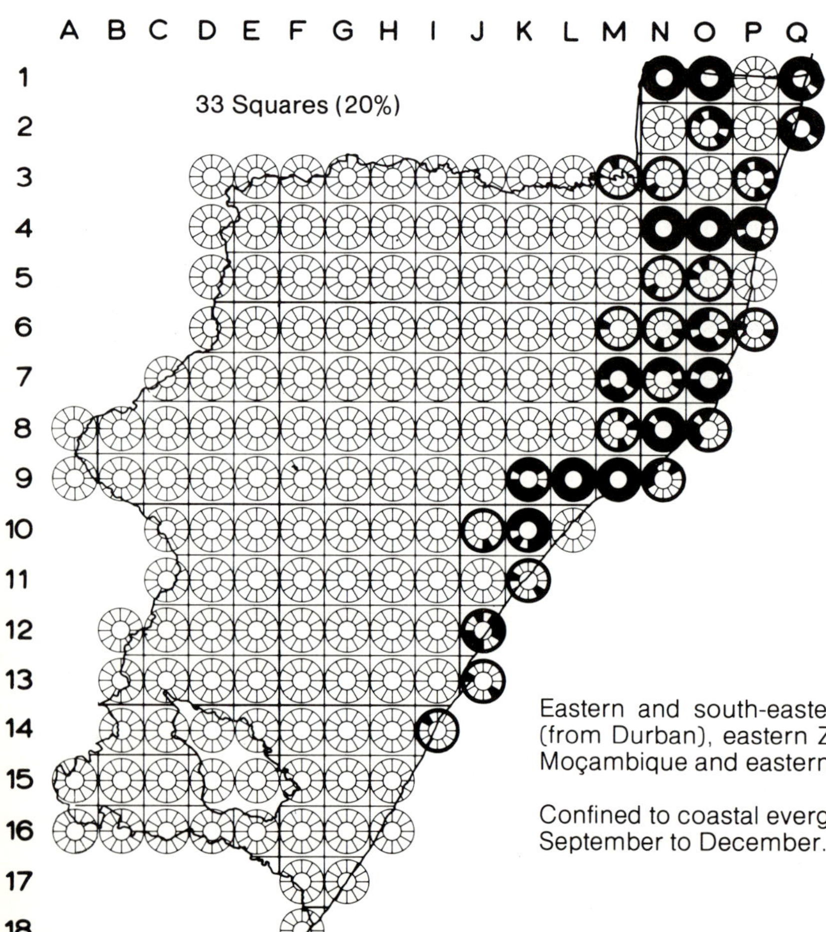

433 White-eared Barbet

Stactolaema leucotis

Eastern and south-eastern Africa. In Southern Africa, the Natal coast (from Durban), eastern Zululand, eastern Swaziland (Lebombo Range), Moçambique and eastern Rhodesia.

Confined to coastal evergreen forest and riverine woodland. Breeds from September to December.

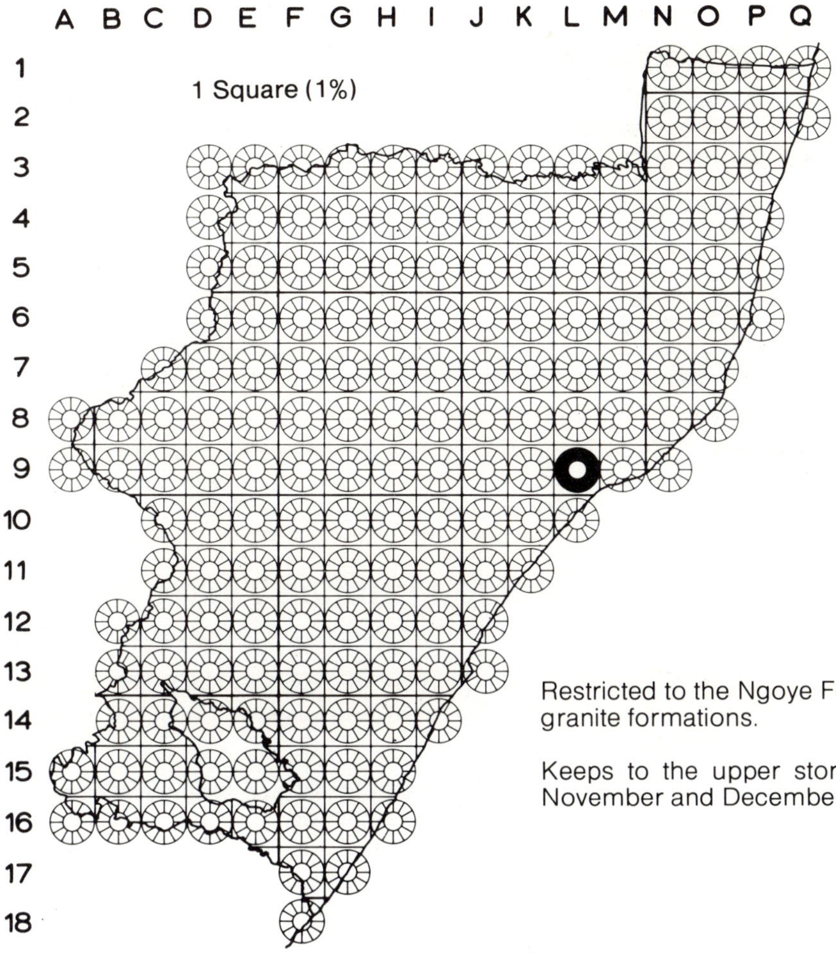

435 Green Barbet

Stactolaema woodwardi

Restricted to the Ngoye Forest, Zululand – a unique forest developed on granite formations.

Keeps to the upper storey of the canopy. Recorded breeding during November and December.

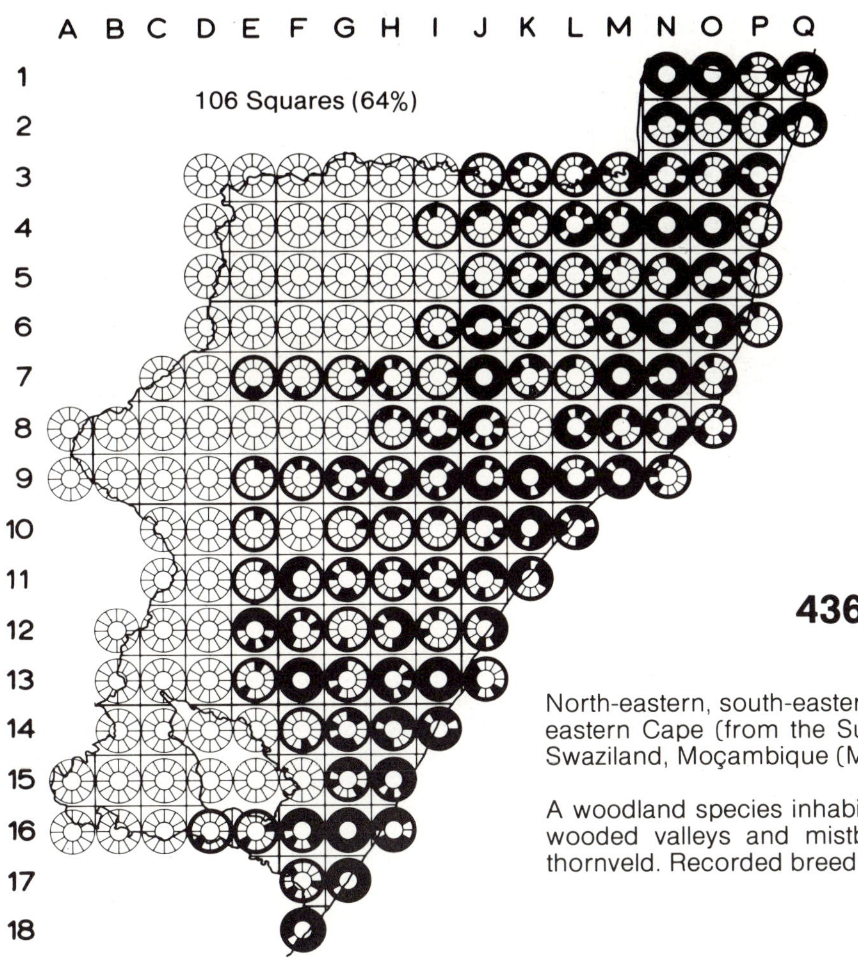

436 Red-fronted Tinker Barbet

Pogoniulus pusillus

North-eastern, south-eastern and southern Africa. In Southern Africa, the eastern Cape (from the Sundays River) to Natal and Zululand, eastern Swaziland, Moçambique (Maputo district) and south-eastern Transvaal.

A woodland species inhabiting coastal forest, riverine woodland, thickly-wooded valleys and mistbelt evergreen forest; occasionally in dense thornveld. Recorded breeding from October to December.

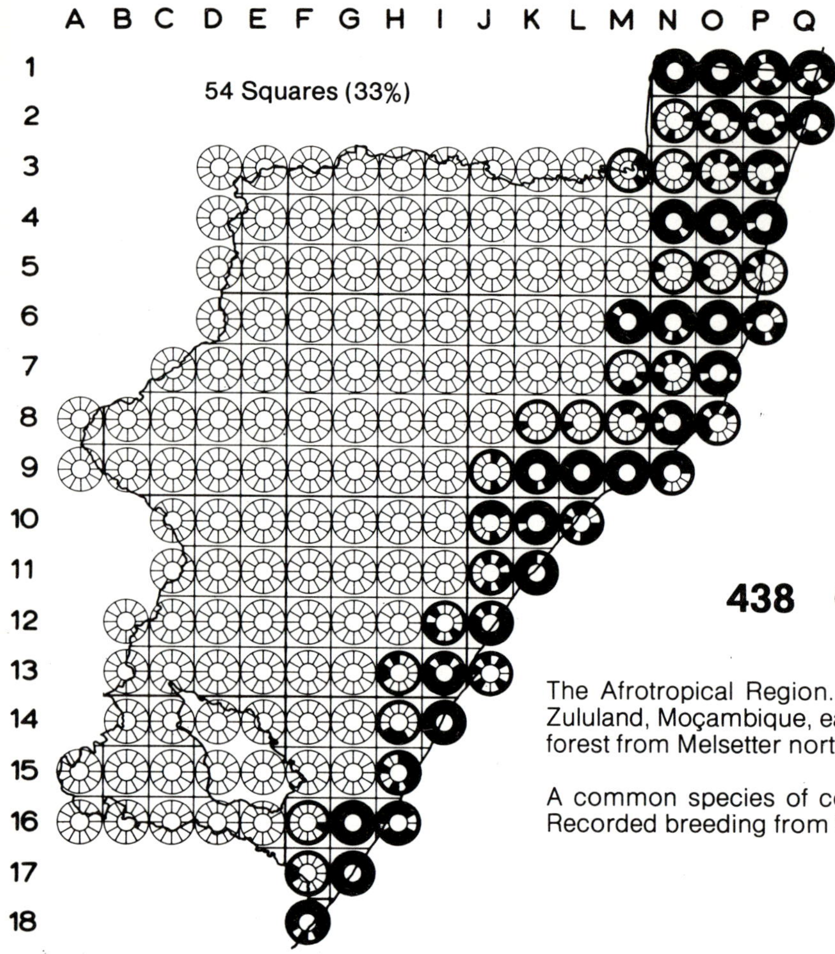

438 Golden-rumped Tinker Barbet

Pogoniulus bilineatus

The Afrotropical Region. In Southern Africa, coastal Natal and eastern Zululand, Moçambique, eastern Transvaal and eastern Rhodesia (highland forest from Melsetter north to Inyanga).

A common species of coastal evergreen forest and riverine woodland. Recorded breeding from September to December.

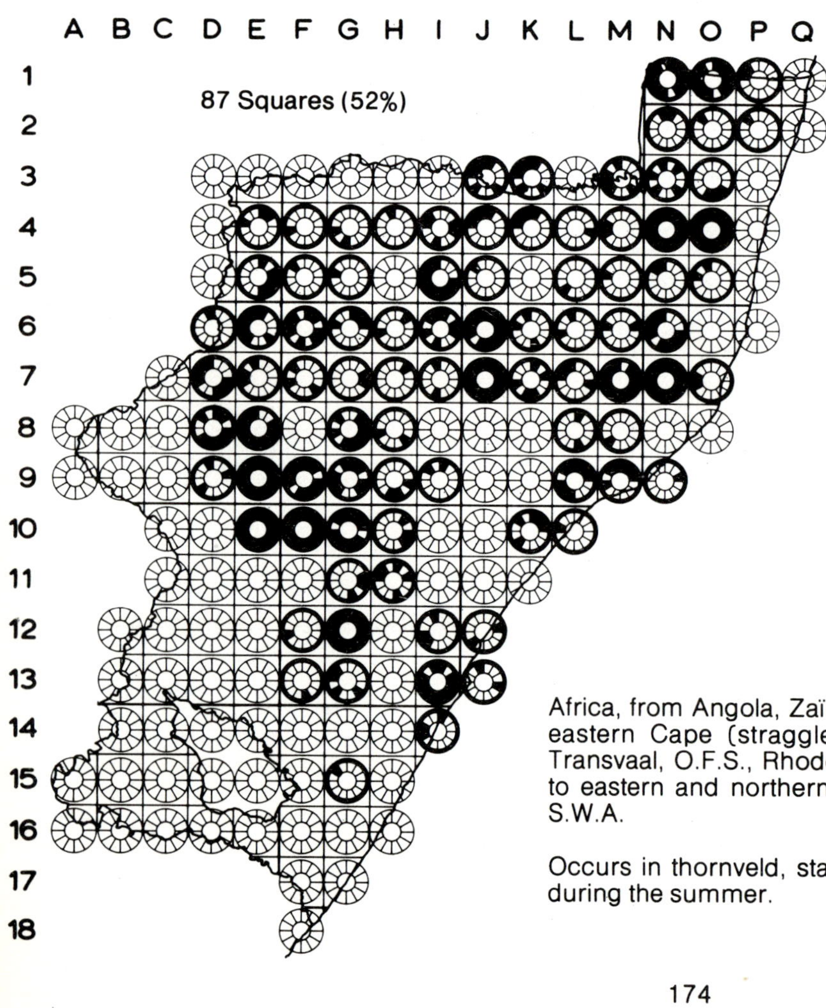

439 Crested Barbet

Trachyphonus vaillantii

Africa, from Angola, Zaïre and Tanzania to Natal. In Southern Africa, the eastern Cape (straggler), Natal (local) and Zululand to Swaziland, Transvaal, O.F.S., Rhodesia and Moçambique (Tete district), westwards to eastern and northern Botswana, the Caprivi Strip and north-eastern S.W.A.

Occurs in thornveld, stands of exotic trees, parks and gardens. Breeds during the summer.

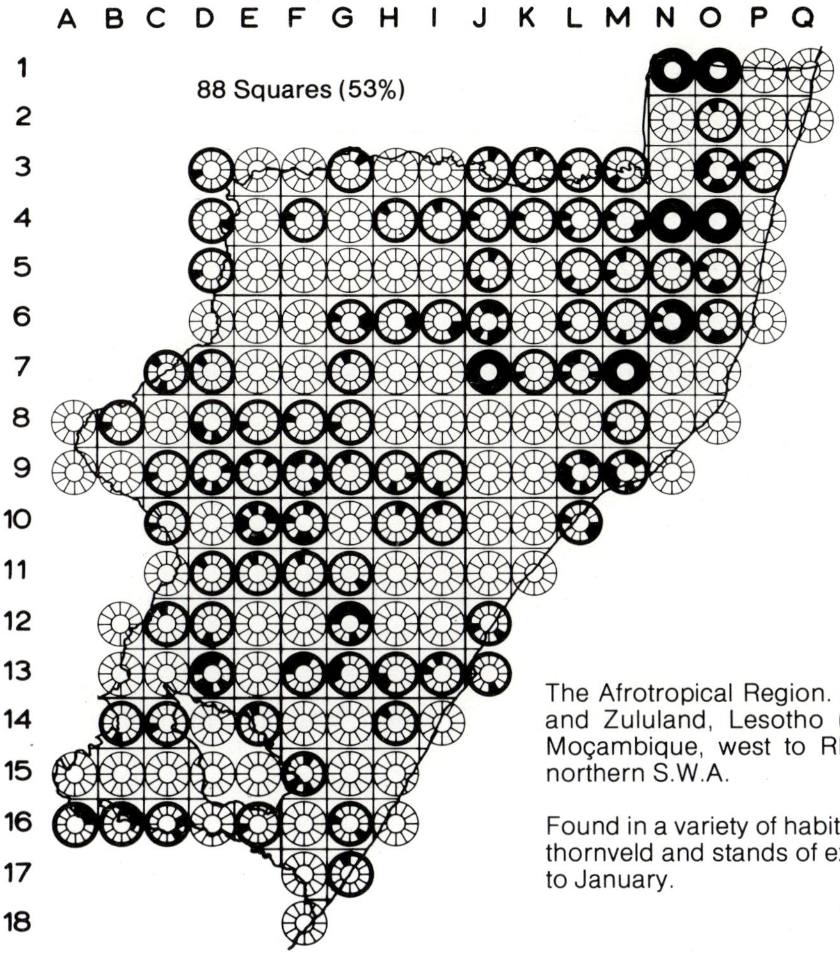

440 Greater Honeyguide

Indicator indicator

The Afrotropical Region. In Southern Africa, the southern Cape to Natal and Zululand, Lesotho (lowlands), O.F.S., Transvaal, Swaziland and Moçambique, west to Rhodesia, eastern and northern Botswana and northern S.W.A.

Found in a variety of habitats, from coastal forest and riverine woodland to thornveld and stands of exotic trees. Recorded breeding from September to January.

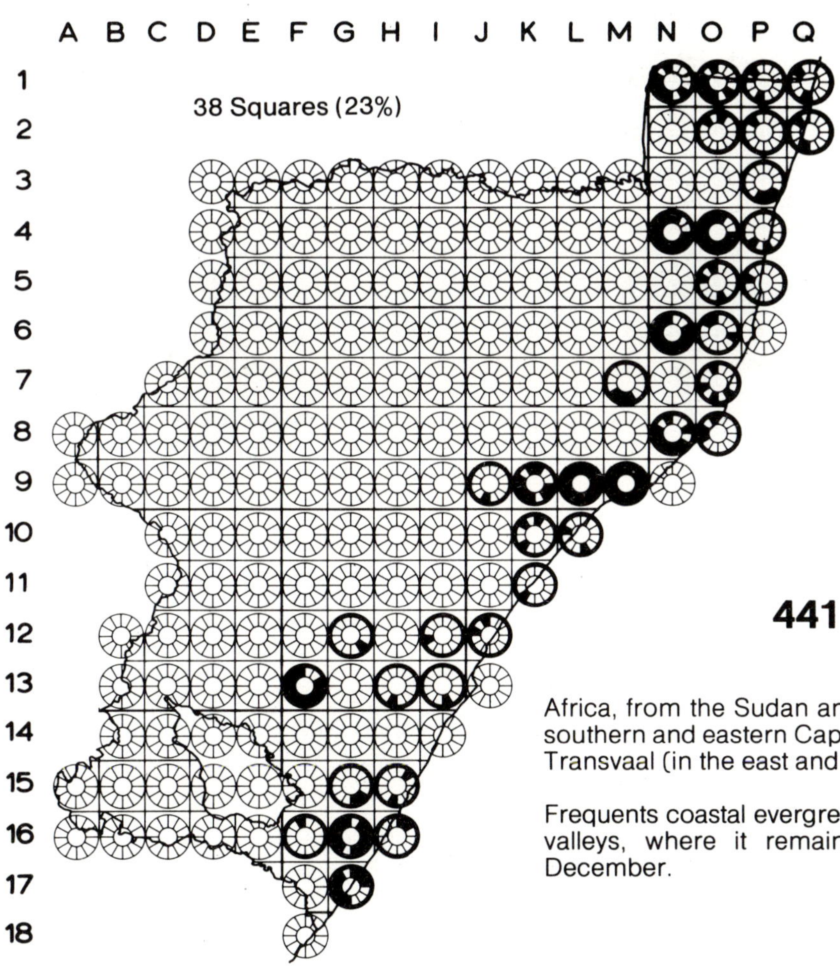

441 Scaly-throated Honeyguide

Indicator variegatus

Africa, from the Sudan and Ethiopia to the Cape. In Southern Africa, the southern and eastern Cape, Natal and Zululand, Swaziland, Moçambique, Transvaal (in the east and north) and eastern Rhodesia.

Frequents coastal evergreen forest, riverine woodland and thickly-wooded valleys, where it remains in the canopy. Breeds from October to December.

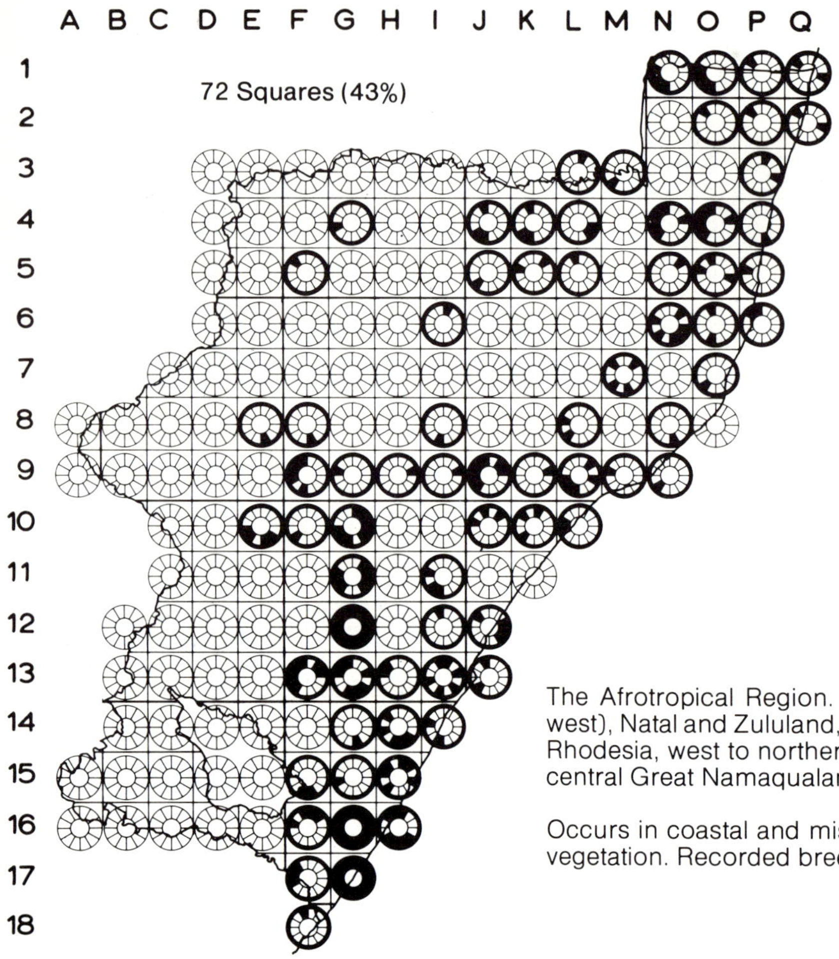

442 Lesser Honeyguide

Indicator minor

The Afrotropical Region. In Southern Africa, the Cape (except extreme west), Natal and Zululand, O.F.S., Swaziland, Moçambique, Transvaal and Rhodesia, west to northern Botswana, the Caprivi Strip and S.W.A. (from central Great Namaqualand northwards).

Occurs in coastal and mistbelt forest, exotic trees, thornveld and riverine vegetation. Recorded breeding from September to February.

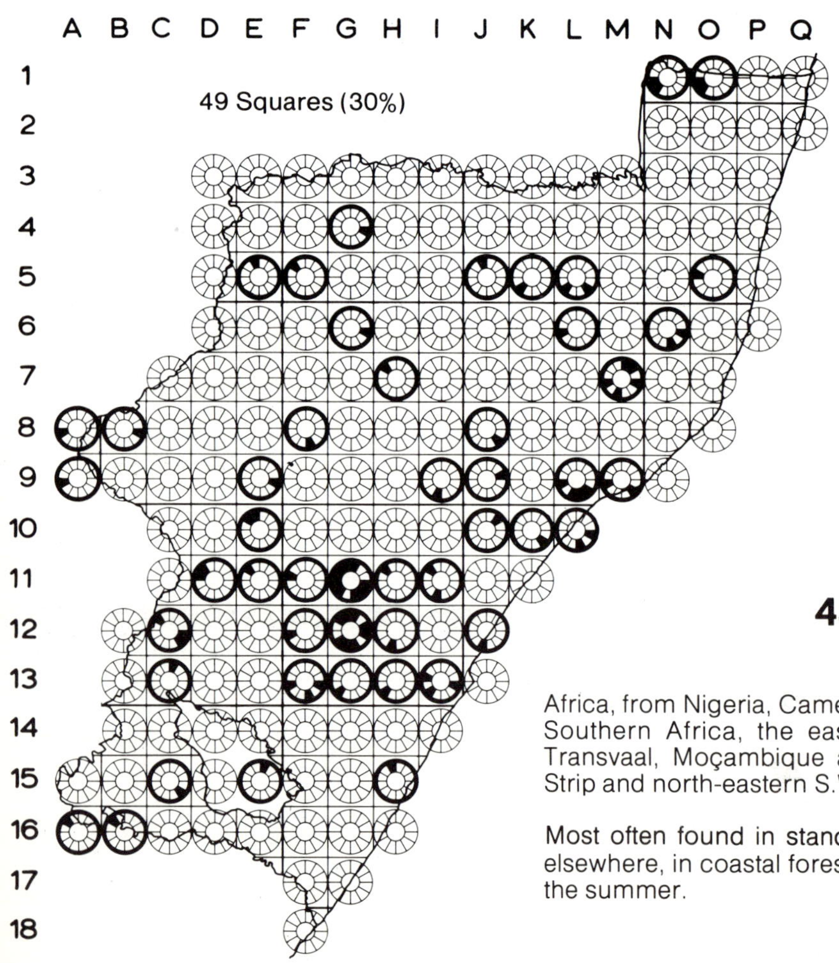

433 Sharp-billed Honeyguide

Prodotiscus regulus

Africa, from Nigeria, Cameroon, Sudan and Ethiopia south to the Cape. In Southern Africa, the eastern Cape, Natal and Zululand, Swaziland, Transvaal, Moçambique and Rhodesia, west to Botswana, the Caprivi Strip and north-eastern S.W.A.

Most often found in stands of exotic trees in midland and upper Natal; elsewhere, in coastal forest and thornveld in the northeast. Breeds during the summer.

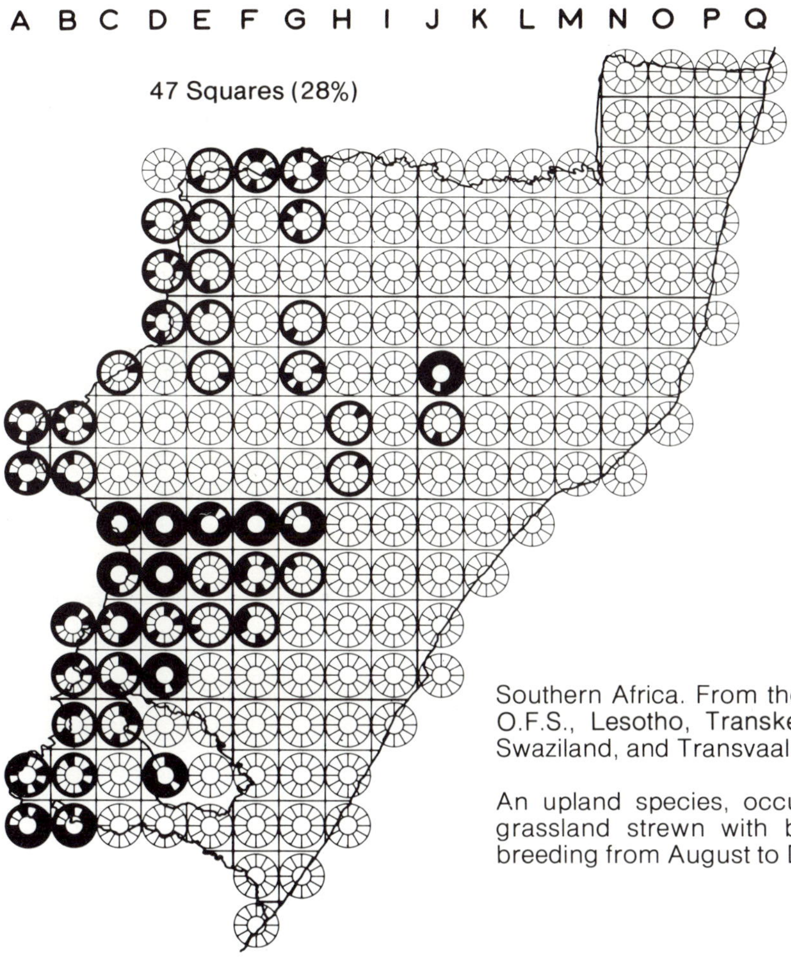

445 Ground Woodpecker

Geocolaptes olivaceus

Southern Africa. From the north-western, southern and eastern Cape to O.F.S., Lesotho, Transkei, Natal (interior), western Zululand, western Swaziland, and Transvaal (highveld north to the Zoutpansberg).

An upland species, occurring down to about 1200 m a.s.l.; found in grassland strewn with boulders, and on rocky outcrops. Recorded breeding from August to December.

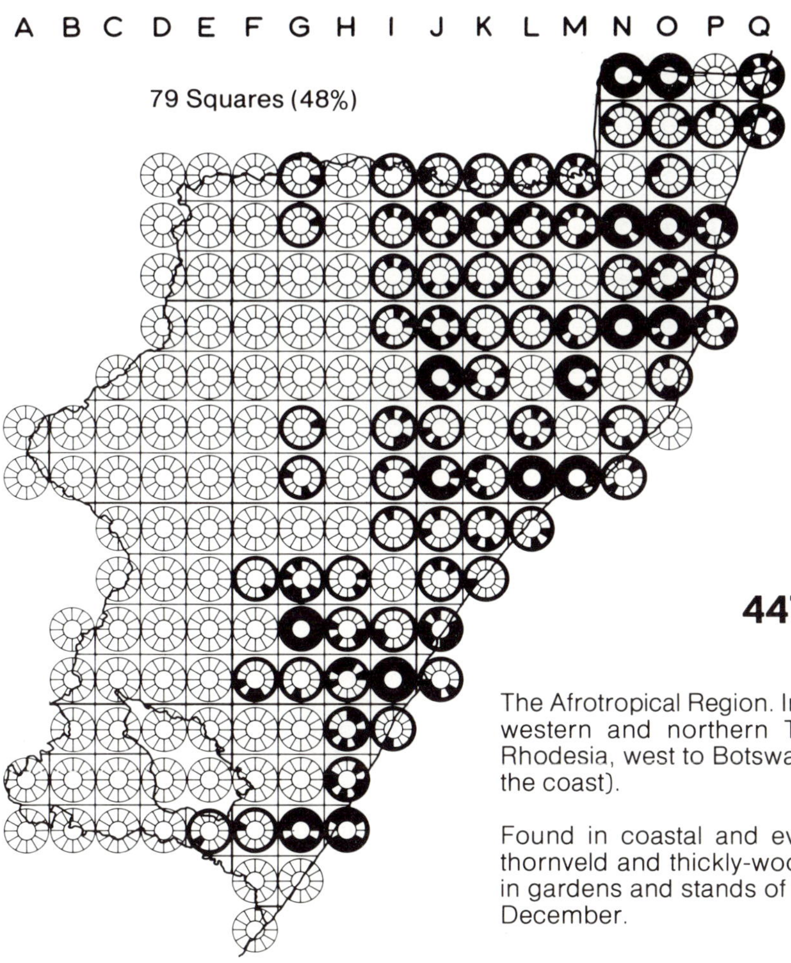

447 Golden-tailed Woodpecker

Campethera abingoni

The Afrotropical Region. In Southern Africa, Natal and Zululand, Swaziland, western and northern Transvaal, adjacent O.F.S., Moçambique and Rhodesia, west to Botswana, the northern Cape and S.W.A. (absent from the coast).

Found in coastal and evergreen forest, mixed and riverine woodland, thornveld and thickly-wooded valleys, to about 900 m a.s.l.; occasionally in gardens and stands of exotic trees. Recorded breeding from August to December.

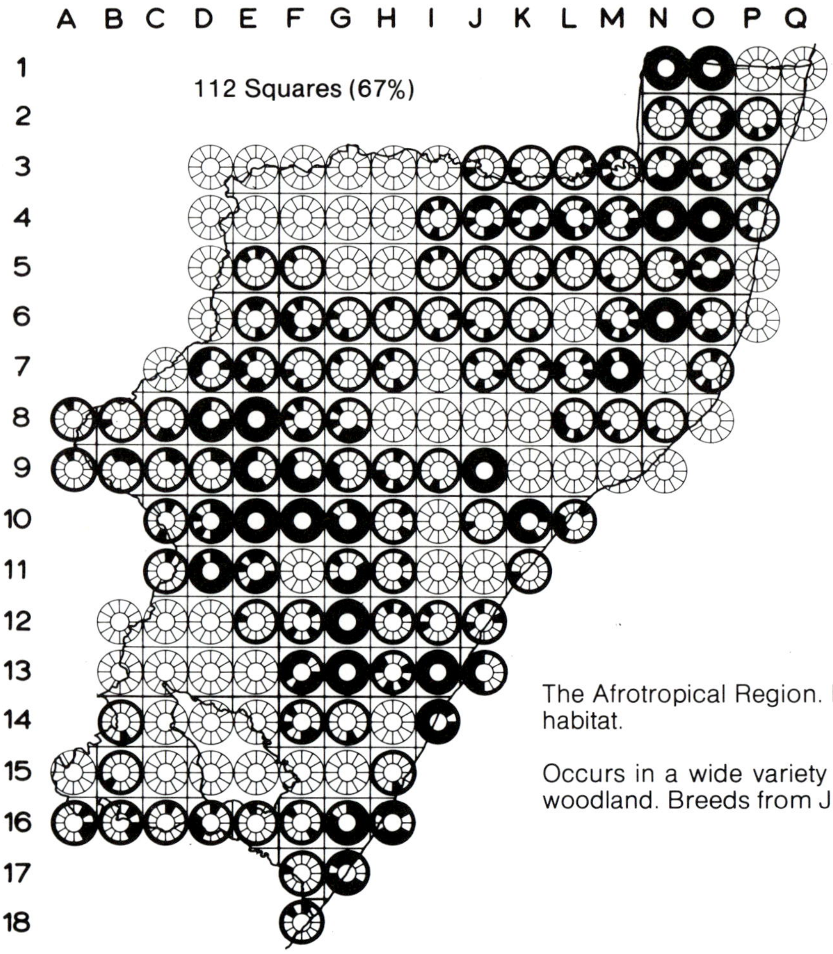

450 Cardinal Woodpecker

Dendropicos fuscenscens

The Afrotropical Region. In Southern Africa, found throughout in suitable habitat.

Occurs in a wide variety of wooded habitats, mainly thornveld and light woodland. Breeds from July to November.

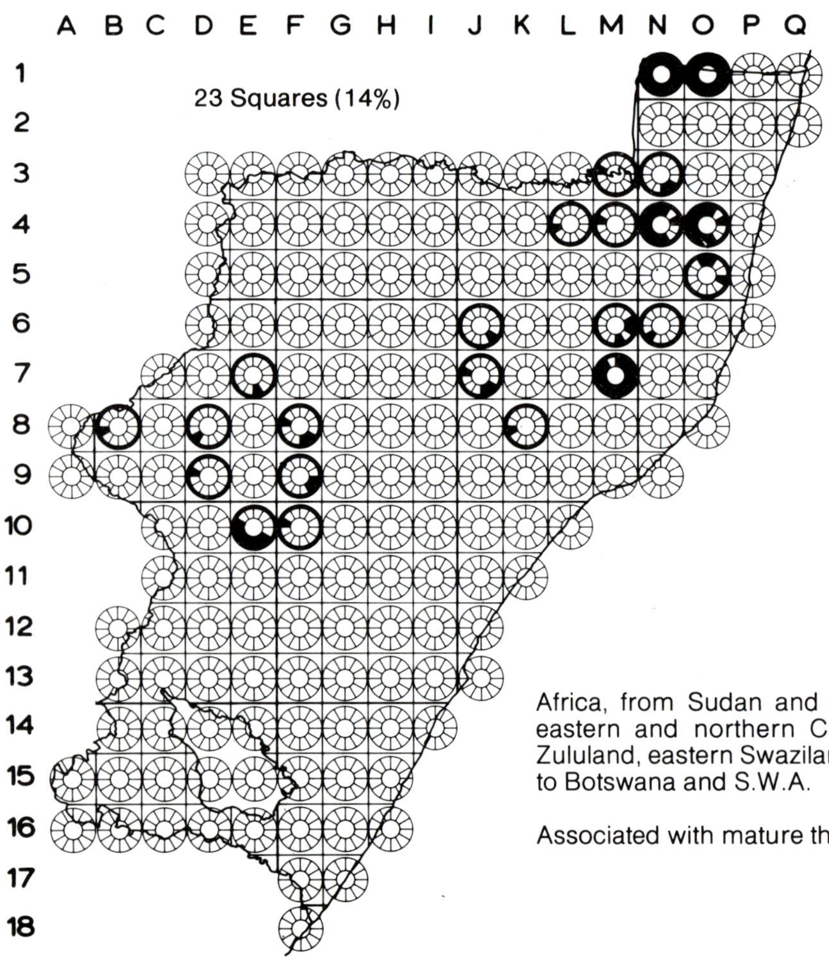

451 Bearded Woodpecker

Thripias namaquus

Africa, from Sudan and Ethiopia to the Cape. In Southern Africa, the eastern and northern Cape, Transkei (coast) to Natal (interior) and Zululand, eastern Swaziland, Transvaal, Moçambique and Rhodesia, west to Botswana and S.W.A.

Associated with mature thornveld. Breeding recorded during November.

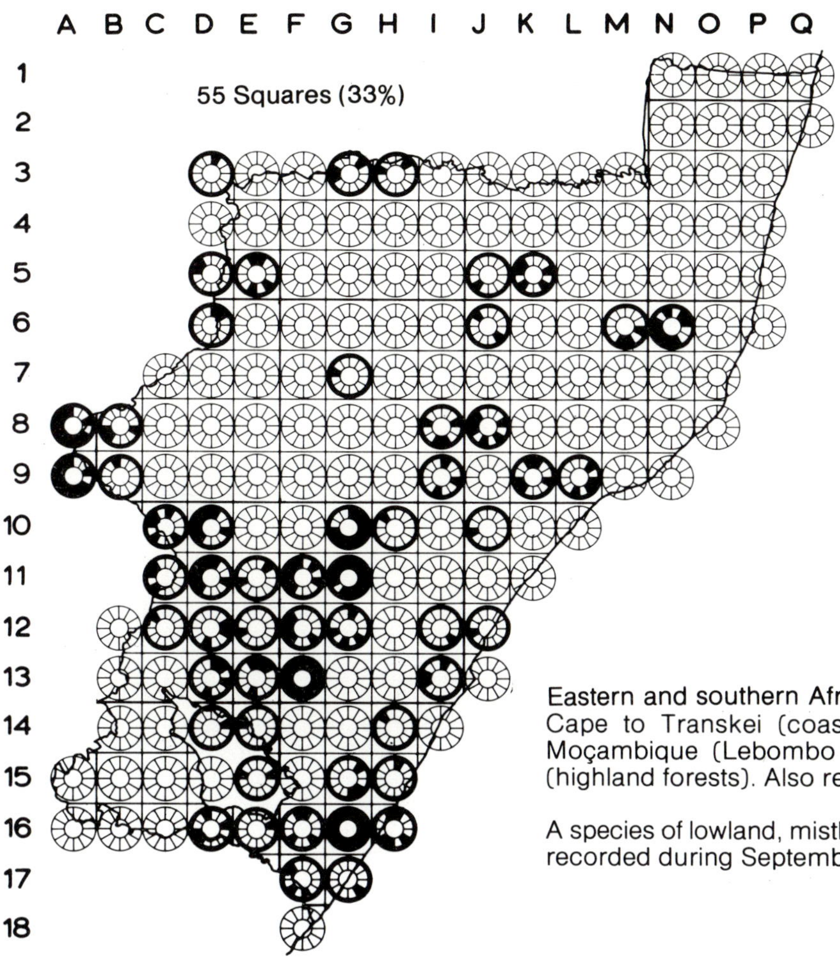

452 Olive Woodpecker

Mesopicos griseocephalus

Eastern and southern Africa. In Southern Africa, from the south-western Cape to Transkei (coast), Natal and Zululand, Swaziland, southern Moçambique (Lebombo Range) and eastern and northern Transvaal (highland forests). Also recorded from the Caprivi Strip (Katima Mulilo).

A species of lowland, mistbelt and Drakensberg evergreen forest. Breeding recorded during September and October.

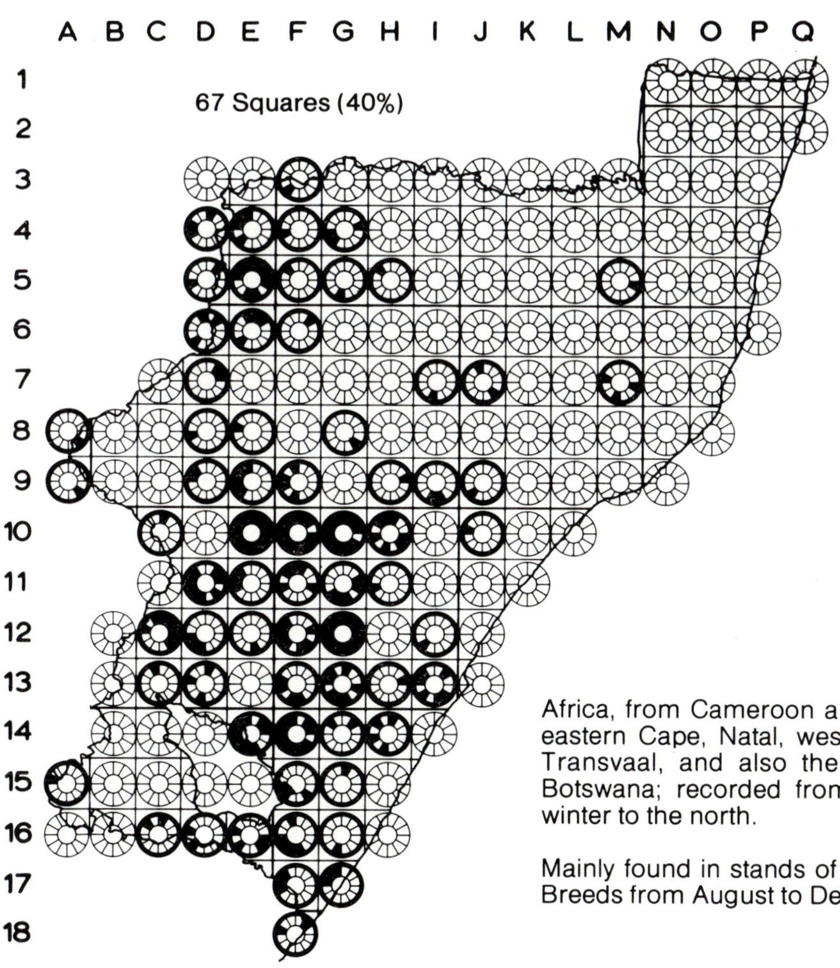

453 Red-throated Wryneck

Jynx ruficollis

Africa, from Cameroon and Sudan to the Cape. In Southern Africa, the eastern Cape, Natal, western Zululand, western Swaziland, O.F.S. and Transvaal, and also the northern Cape and possibly south-eastern Botswana; recorded from southern Moçambique (Save River). Some winter to the north.

Mainly found in stands of exotic trees; also in thornveld and open bush. Breeds from August to December.

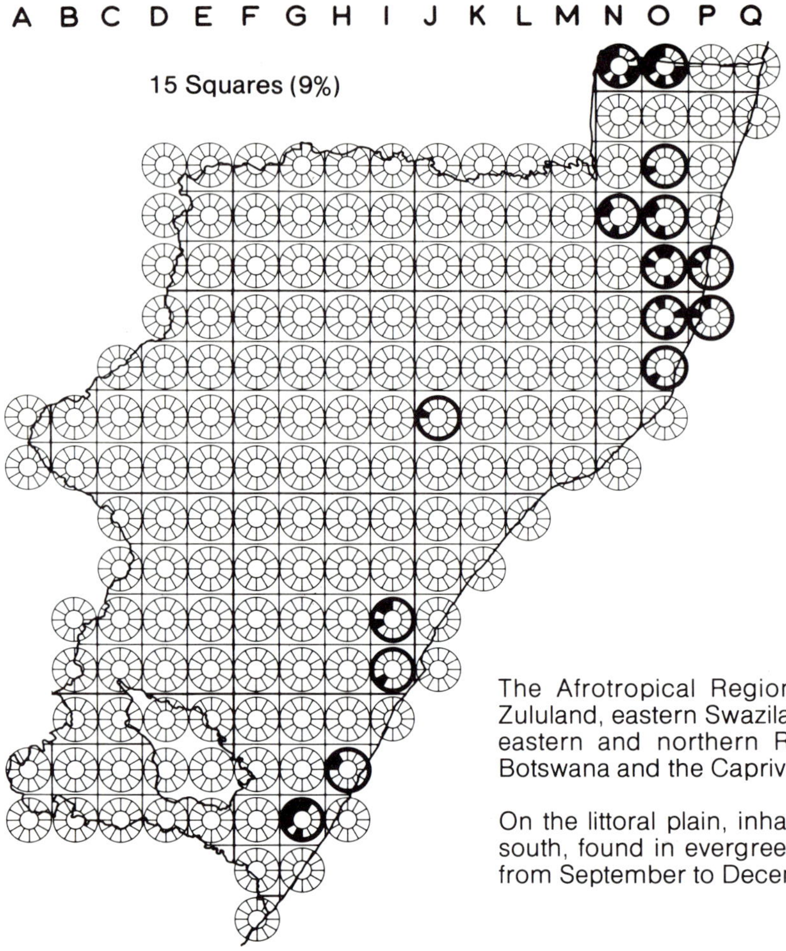

454 Broadbill

Smithornis capensis

The Afrotropical Region. In Southern Africa, coastal Natal, eastern Zululand, eastern Swaziland (Lebombo Range), north-eastern Transvaal, eastern and northern Rhodesia and Moçambique, west to northern Botswana and the Caprivi Strip.

On the littoral plain, inhabits dense parts of dry mixed woodland; to the south, found in evergreen forest and coastal scrub. Recorded breeding from September to December.

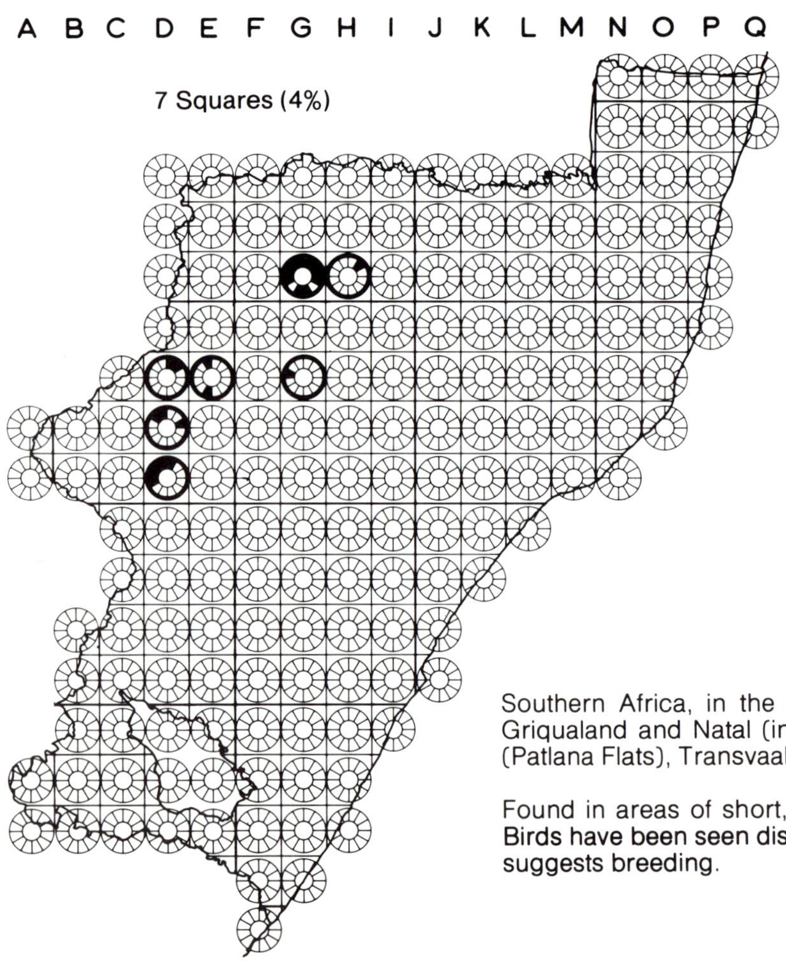

456 Singing Bush Lark

Mirafra cheniana

Southern Africa, in the eastern interior. From the eastern Cape, East Griqualand and Natal (interior) to O.F.S., the northern Cape, Botswana (Patlana Flats), Transvaal and Rhodesia (Matabeleland).

Found in areas of short, dry grass, occasionally with scattered bushes. Birds have been seen displaying during November near Winterton, which suggests breeding.

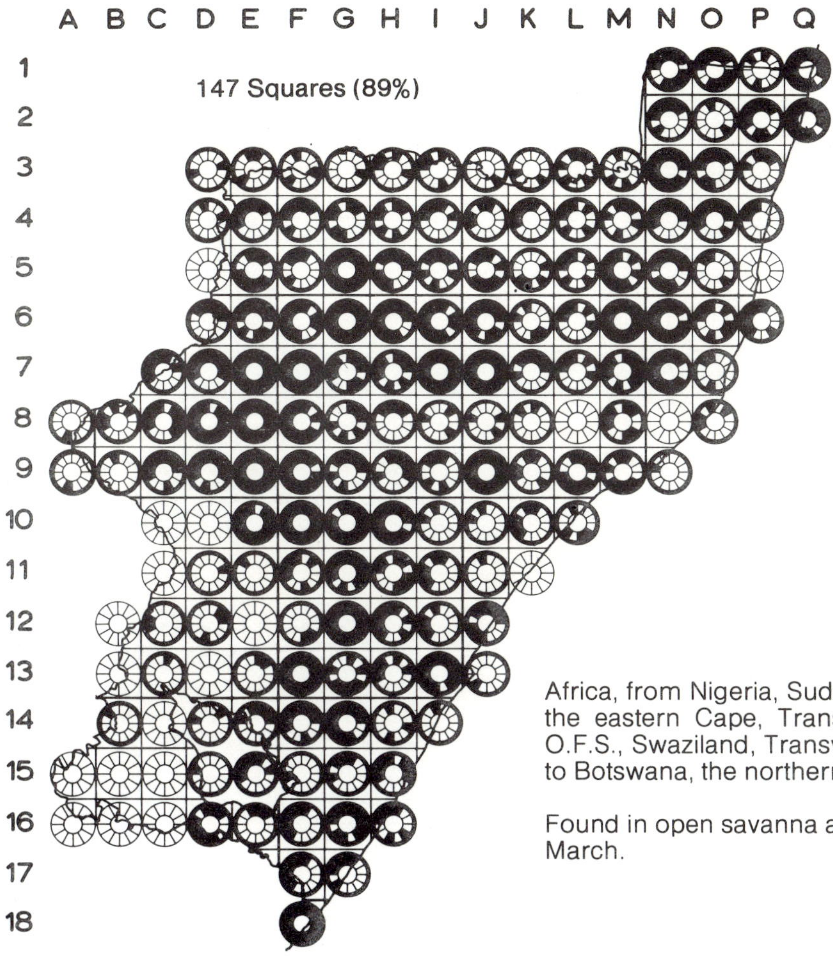

458 Rufous-naped Lark

Mirafra africana

Africa, from Nigeria, Sudan and Somalia to the Cape. In Southern Africa, the eastern Cape, Transkei, Natal and Zululand, Lesotho (lowlands), O.F.S., Swaziland, Transvaal, southern Moçambique and Rhodesia, west to Botswana, the northern Cape and S.W.A.

Found in open savanna and bush, and grassland. Breeds from August to March.

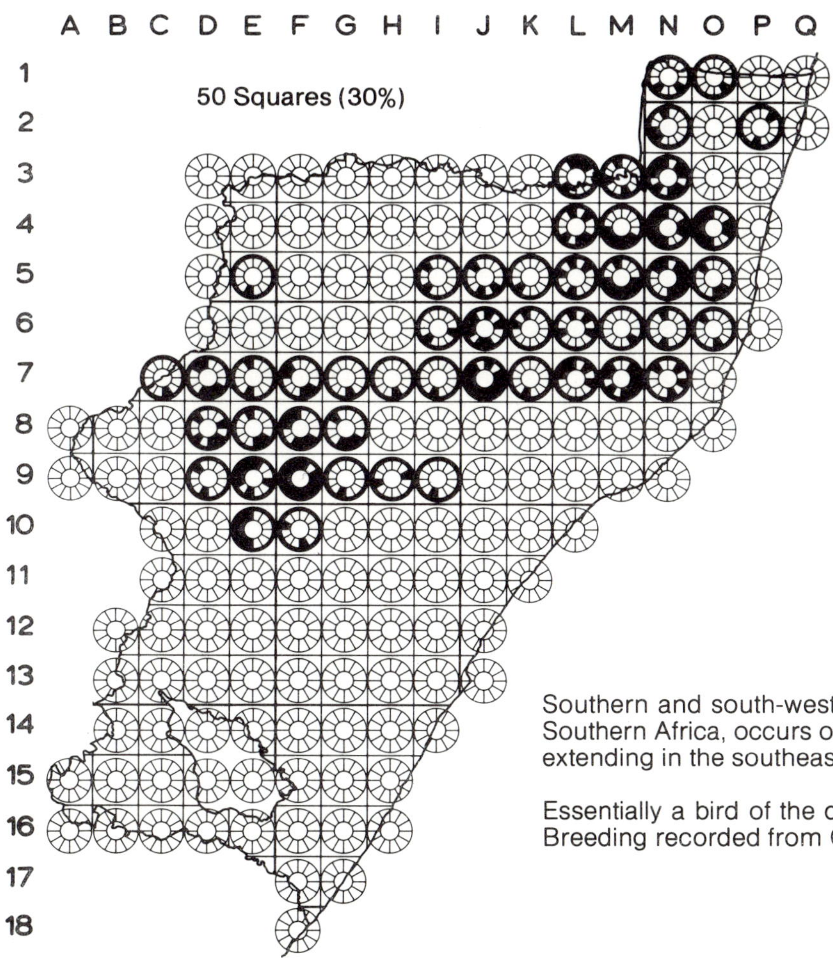

460 Sabota Lark

Mirafra sabota

Southern and south-western Africa, from Cabinda south to the Cape. In Southern Africa, occurs over much of the dry interior and western regions, extending in the southeast to Natal.

Essentially a bird of the dry thornveld, from low scrub to open savanna. Breeding recorded from October to February.

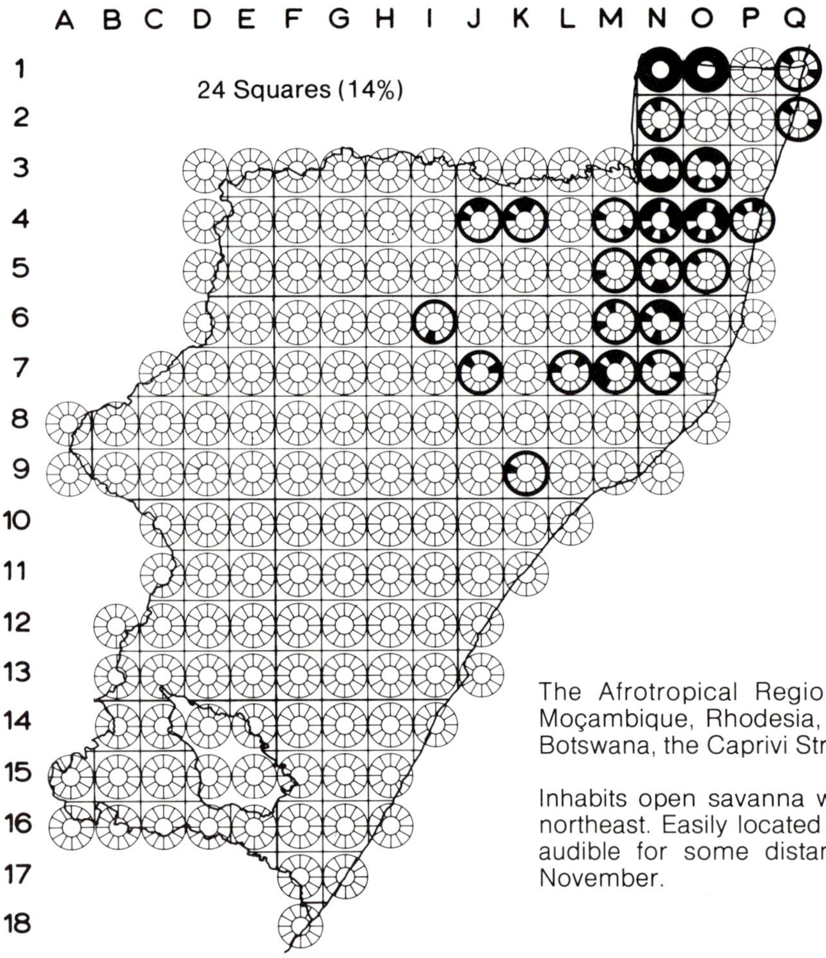

468 Flappet Lark

Mirafra rufocinnamomea

The Afrotropical Region. In Southern Africa, Zululand, Swaziland, Moçambique, Rhodesia, eastern and north-western Transvaal, northern Botswana, the Caprivi Strip and north-eastern S.W.A.

Inhabits open savanna woodland extending to coastal grassland in the northeast. Easily located during aerial display, when its wing-clapping is audible for some distance. Recorded breeding during October and November.

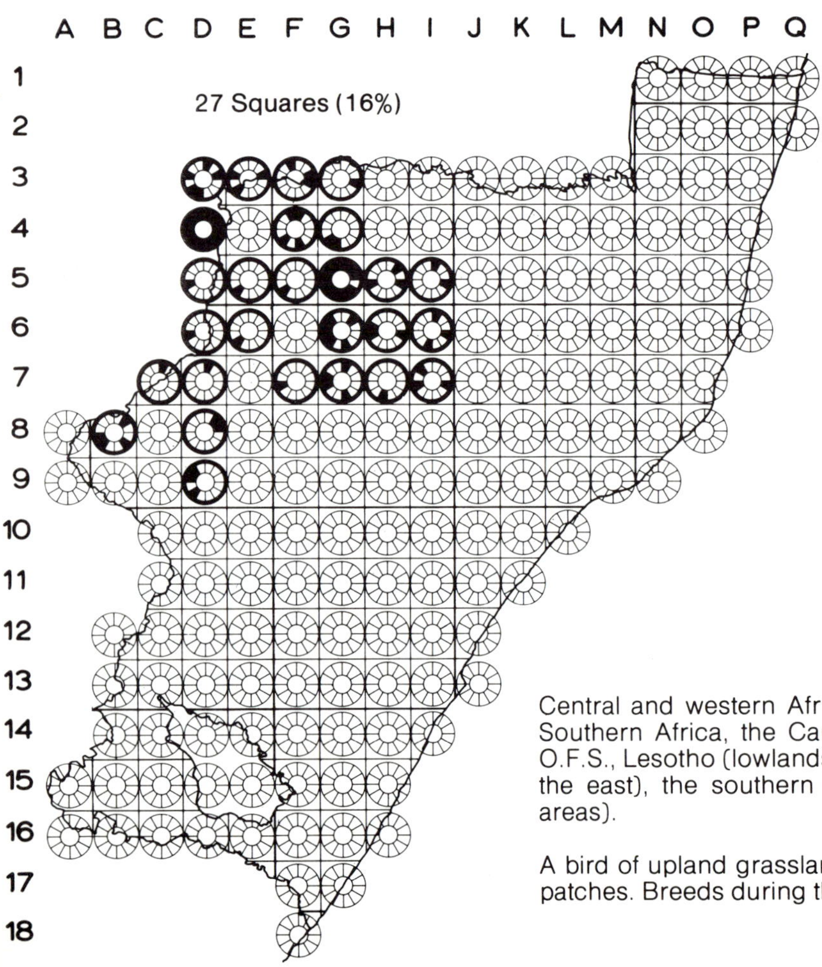

474 Spike-heeled Lark

Chersomanes albofaciata

Central and western Africa, from Angola and Tanzania to the Cape. In Southern Africa, the Cape, Transkei (east to the upper Bashee River), O.F.S., Lesotho (lowlands), western and upper Natal, Transvaal (except in the east), the southern half of Botswana, and S.W.A. (except coastal areas).

A bird of upland grassland, inhabiting areas of short dry grass with bare patches. Breeds during the summer.

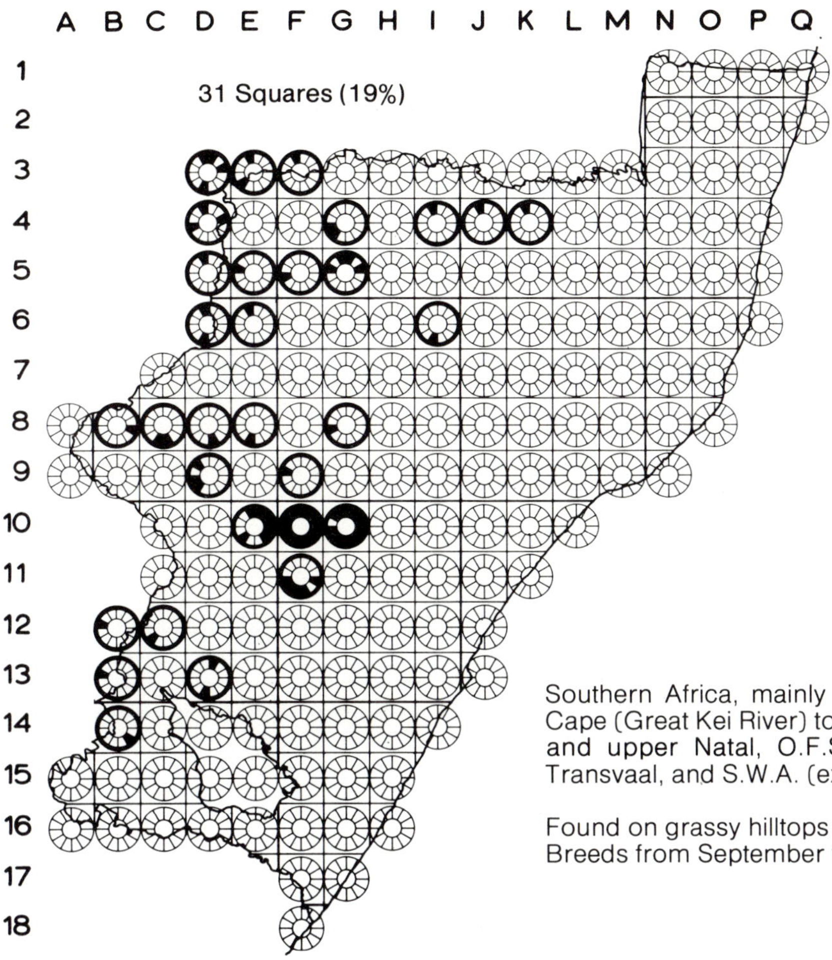

475 Long-billed Lark

Certhilauda curvirostris

Southern Africa, mainly in the mountainous and drier parts. From the Cape (Great Kei River) to interior and western Transkei, Lesotho, western and upper Natal, O.F.S., western Swaziland, central and southern Transvaal, and S.W.A. (except the northeast).

Found on grassy hilltops strewn with rocks, generally above 1200 m a.s.l. Breeds from September to December.

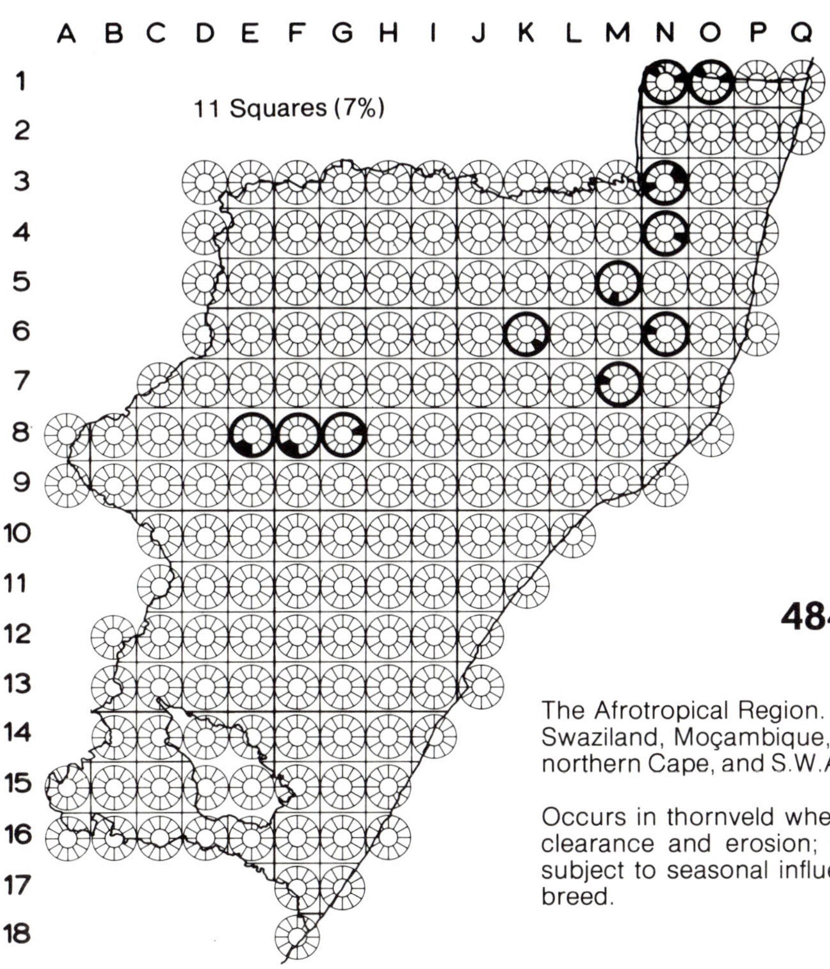

484 Chestnut-backed Finchlark

Eremopterix leucotis

The Afrotropical Region. In Southern Africa, Natal and Zululand, eastern Swaziland, Moçambique, Transvaal and Rhodesia, west to Botswana, the northern Cape, and S.W.A. (northwards from the Damaraland highlands).

Occurs in thornveld where large bare patches have resulted from bush clearance and erosion; also on exposed roadside verges. Apparently subject to seasonal influences associated with drought conditions. May breed.

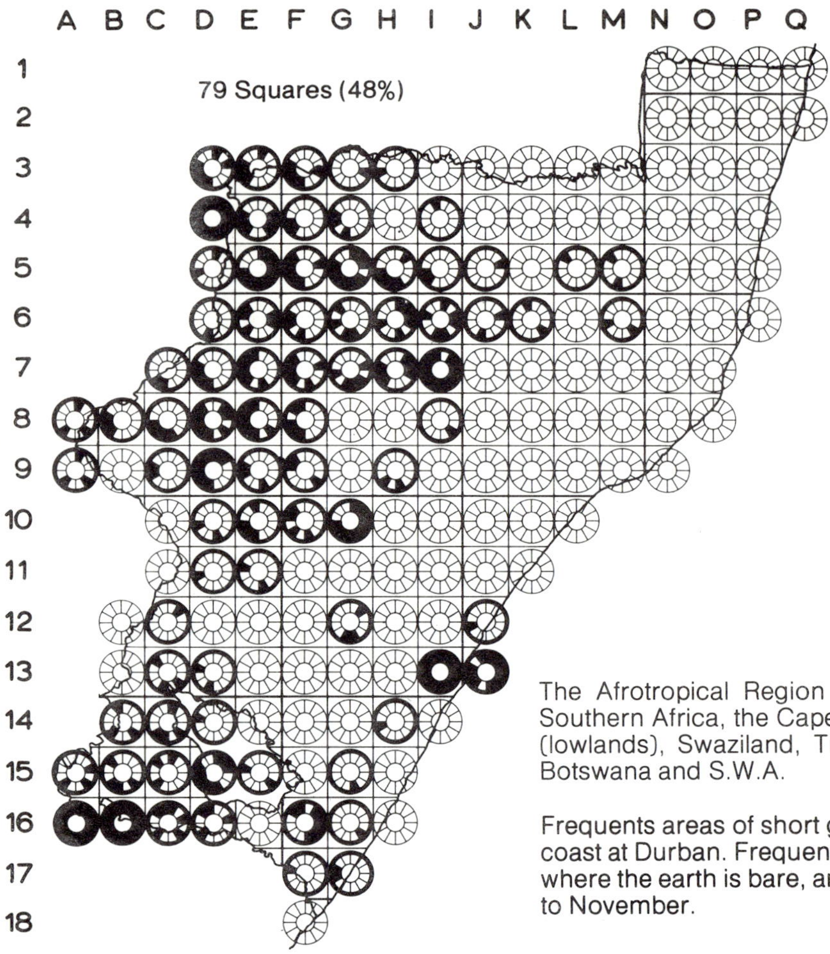

488 Red-capped Lark

Calandrella cinerea

The Afrotropical Region and southern Europe, east to Manchuria. In Southern Africa, the Cape, Transkei, Natal and Zululand, O.F.S., Lesotho (lowlands), Swaziland, Transvaal, Moçambique and Rhodesia, west to Botswana and S.W.A.

Frequents areas of short grass, mainly in upland regions but reaching the coast at Durban. Frequently observed along dirt roads, near cattle troughs where the earth is bare, and on air strips. Recorded breeding from August to November.

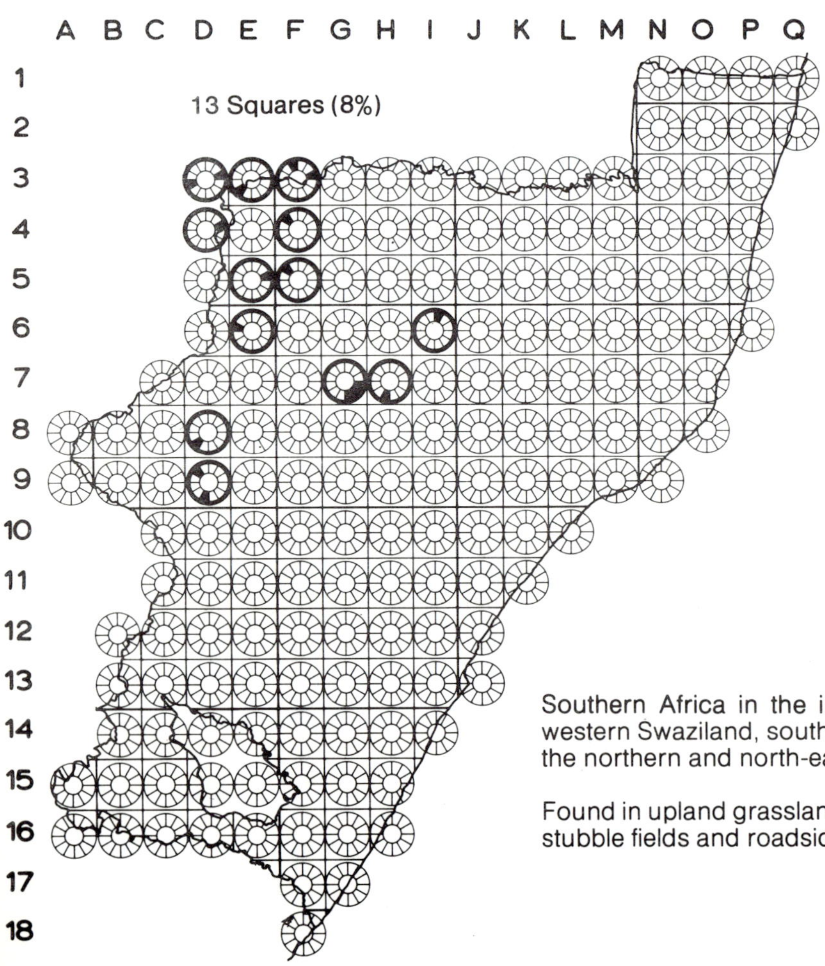

490 Pink-billed Lark

Calandrella conirostris

Southern Africa in the interior and west. From upper Natal to O.F.S., western Swaziland, southern and western Transvaal, southern Botswana, the northern and north-eastern Cape, and S.W.A.

Found in upland grassland, frequenting short dry grass with bare patches, stubble fields and roadsides. Breeds from October to May.

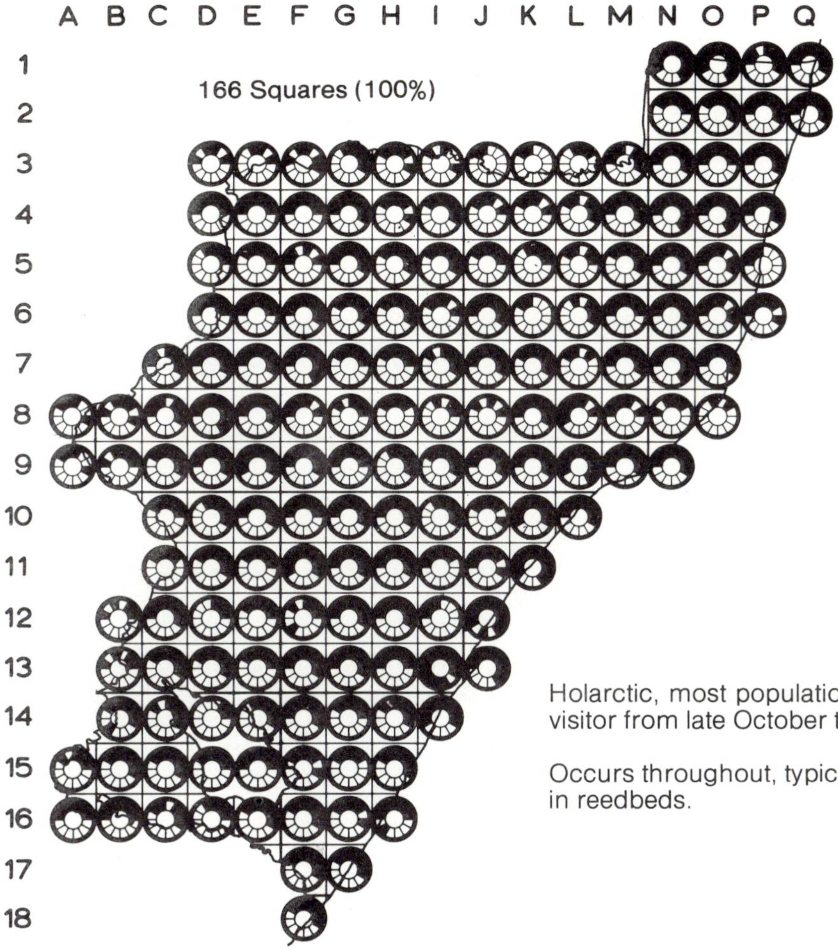

493 European Swallow

Hirundo rustica

Holarctic, most populations wintering to the south. In Southern Africa, a visitor from late October to April over the whole region.

Occurs throughout, typically perching along telegraph wires and roosting in reedbeds.

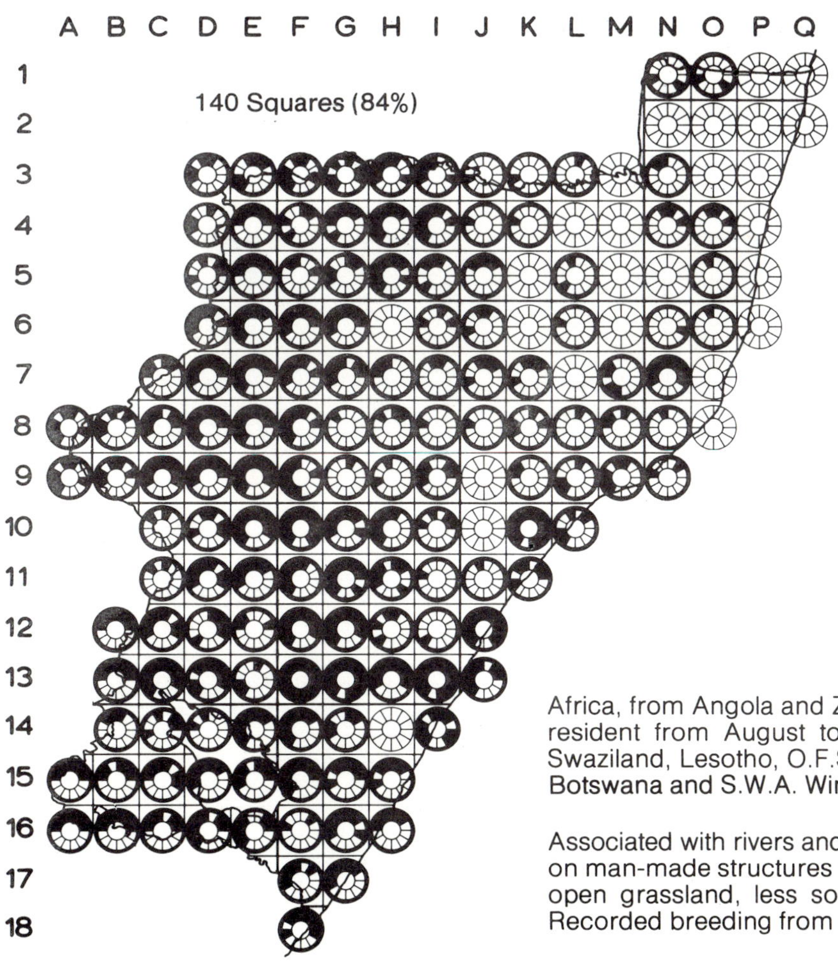

495 White-throated Swallow

Hirundo albigularis

Africa, from Angola and Zaïre to the Cape. In Southern Africa, a summer resident from August to April/May in the Cape, Natal and Zululand, Swaziland, Lesotho, O.F.S., Transvaal and Rhodesia; also recorded from Botswana and S.W.A. Winters north to Angola, Zambia and Zaïre.

Associated with rivers and standing water, where it may be found breeding on man-made structures such as bridges and farm buildings. Common in open grassland, less so in thornveld and sparse on the littoral plain. Recorded breeding from September to March.

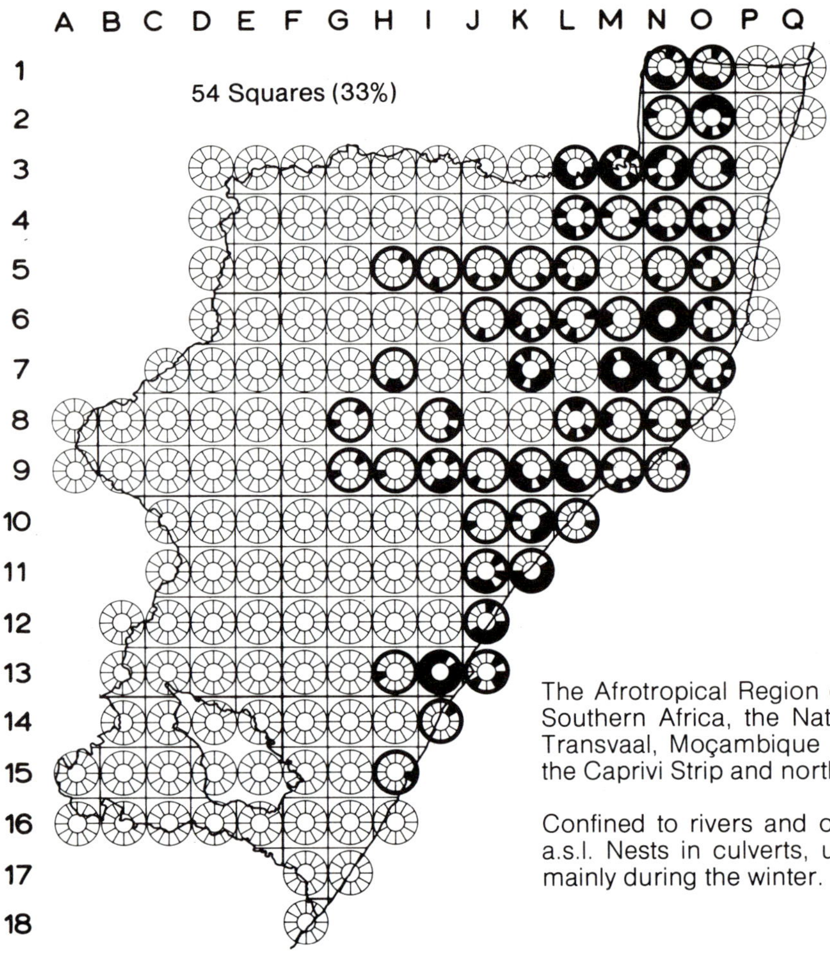

496 Wire-tailed Swallow

Hirundo smithii

The Afrotropical Region east to central Asia, Burma and Indo-China. In Southern Africa, the Natal coast, Zululand, eastern Swaziland, eastern Transvaal, Moçambique and Rhodesia, west to Botswana (Ngamiland), the Caprivi Strip and northern S.W.A. (Okavango River).

Confined to rivers and open stretches of water generally below 900 m a.s.l. Nests in culverts, under bridges and occasionally on rock faces, mainly during the winter.

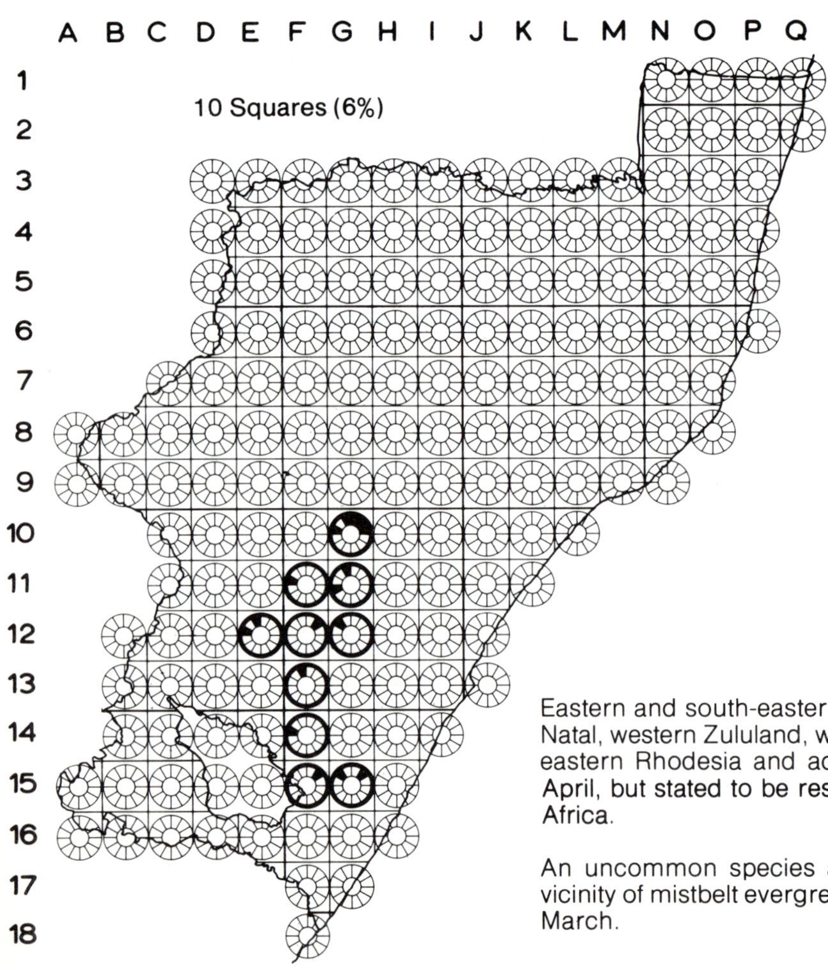

497 Blue Swallow

Hirundo atrocaerulea

Eastern and south-eastern Africa. In Southern Africa, local and sparse in Natal, western Zululand, western Swaziland, eastern Transvaal (highlands), eastern Rhodesia and adjacent Moçambique. Present from October to April, but stated to be resident in Rhodesia. Winters north to east-central Africa.

An uncommon species associated with moist grassland usually in the vicinity of mistbelt evergreen forest. Breeding recorded from November to March.

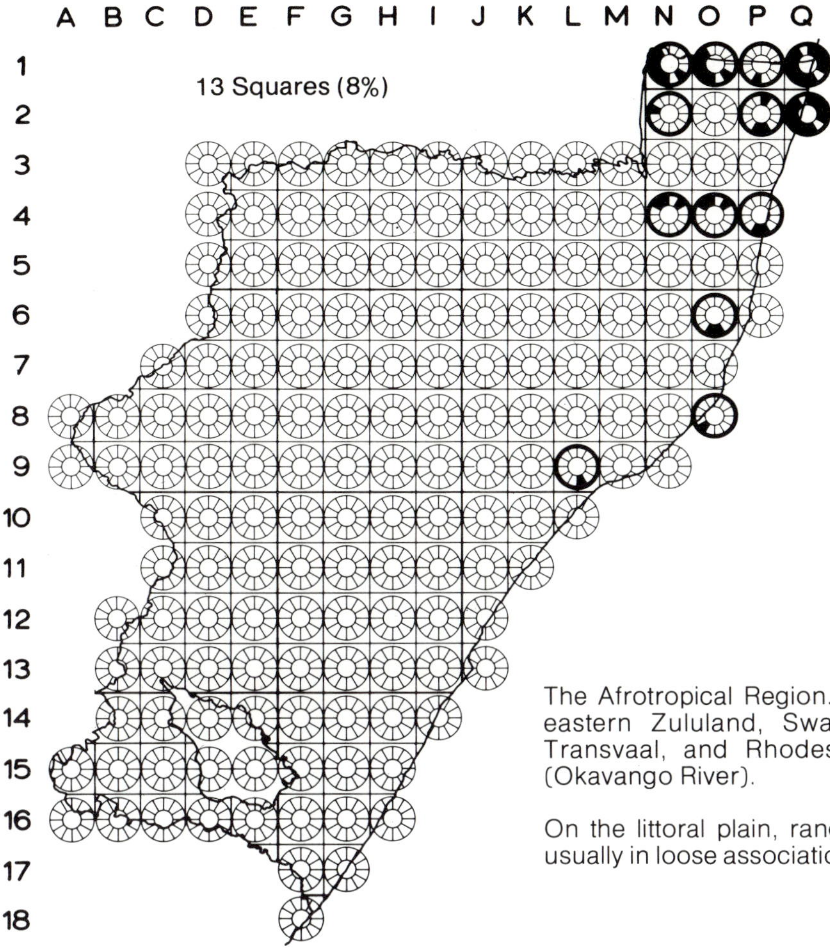

499 Grey-rumped Swallow

Hirundo griseopyga

The Afrotropical Region. In Southern Africa, the Natal coast (formerly), eastern Zululand, Swaziland, Moçambique, eastern and northern Transvaal, and Rhodesia, west to northern Botswana and S.W.A. (Okavango River).

On the littoral plain, ranges over open grassland with marshy patches, usually in loose associations. Breeds during the winter and early summer.

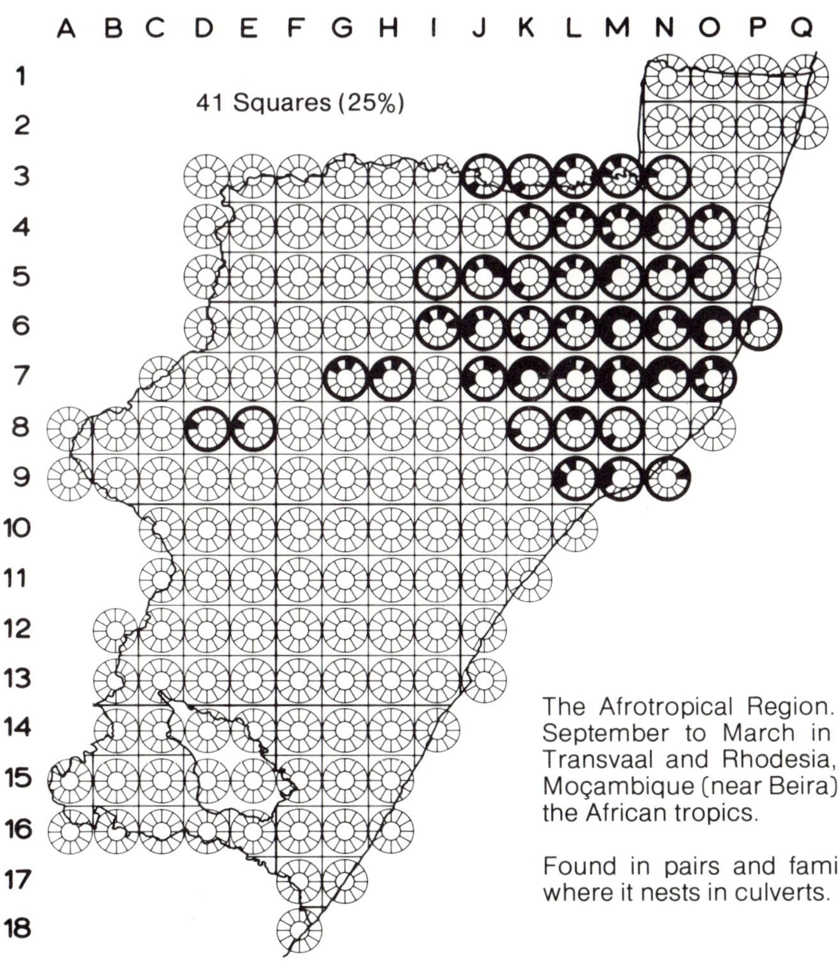

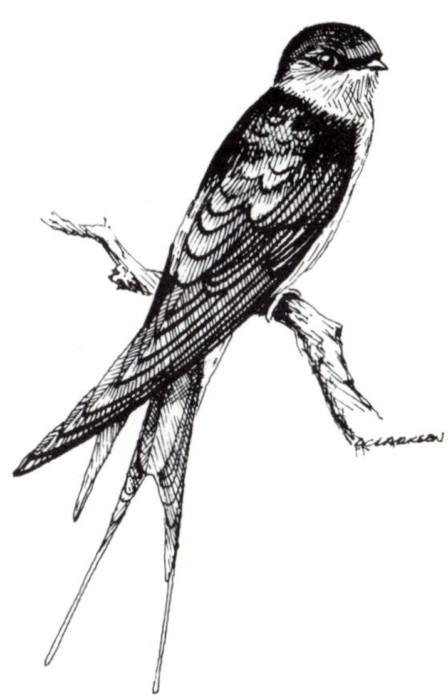

501 Red-breasted Swallow

Hirundo semirufa

The Afrotropical Region. In Southern Africa, a summer resident from September to March in Natal (interior), Zululand, Swaziland, O.F.S., Transvaal and Rhodesia, west to the northern Cape, and S.W.A.; also Moçambique (near Beira) and recorded from the eastern Cape. Winters in the African tropics.

Found in pairs and family parties in thornveld, often by the roadside, where it nests in culverts. Breeds from September to March.

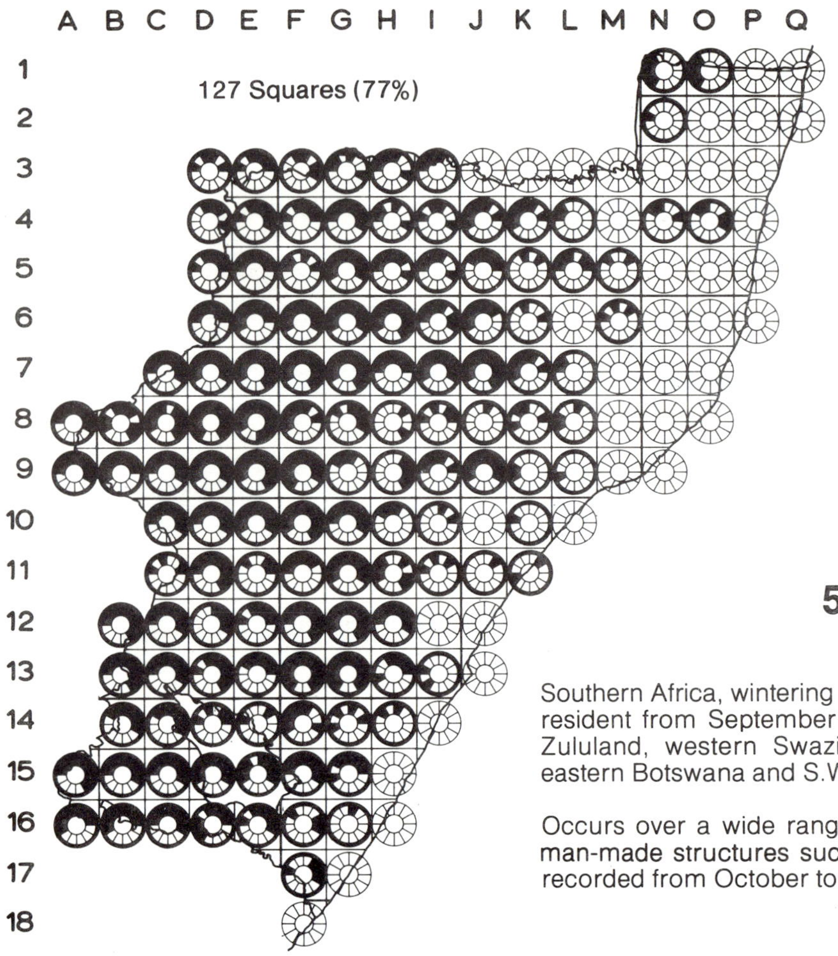

502 Greater Striped Swallow

Hirundo cucullata

Southern Africa, wintering north to Angola, Zaïre and Tanzania. A summer resident from September to April in the Cape, Natal (interior), western Zululand, western Swaziland, Lesotho, O.F.S., Transvaal, Rhodesia, eastern Botswana and S.W.A.

Occurs over a wide range of habitats away from the coast, nesting on man-made structures such as farm buildings and in culverts. Breeding recorded from October to March.

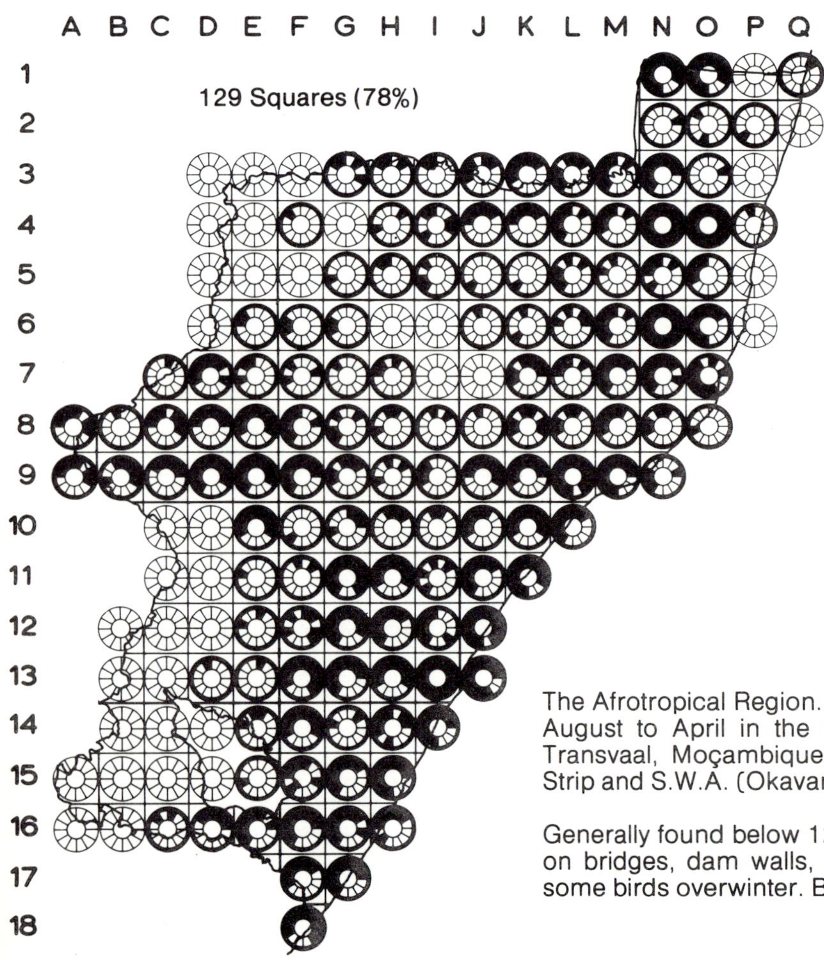

503 Lesser Striped Swallow

Hirundo abyssinica

The Afrotropical Region. In Southern Africa, a summer resident from late August to April in the eastern Cape, Natal and Zululand, Swaziland, Transvaal, Moçambique and Rhodesia, west to Botswana, the Caprivi Strip and S.W.A. (Okavango River). Winters in the African tropics.

Generally found below 1200 m a.s.l. in a variety of habitats, where it nests on bridges, dam walls, under eaves of farm buildings and in culverts; some birds overwinter. Breeding recorded from September to February.

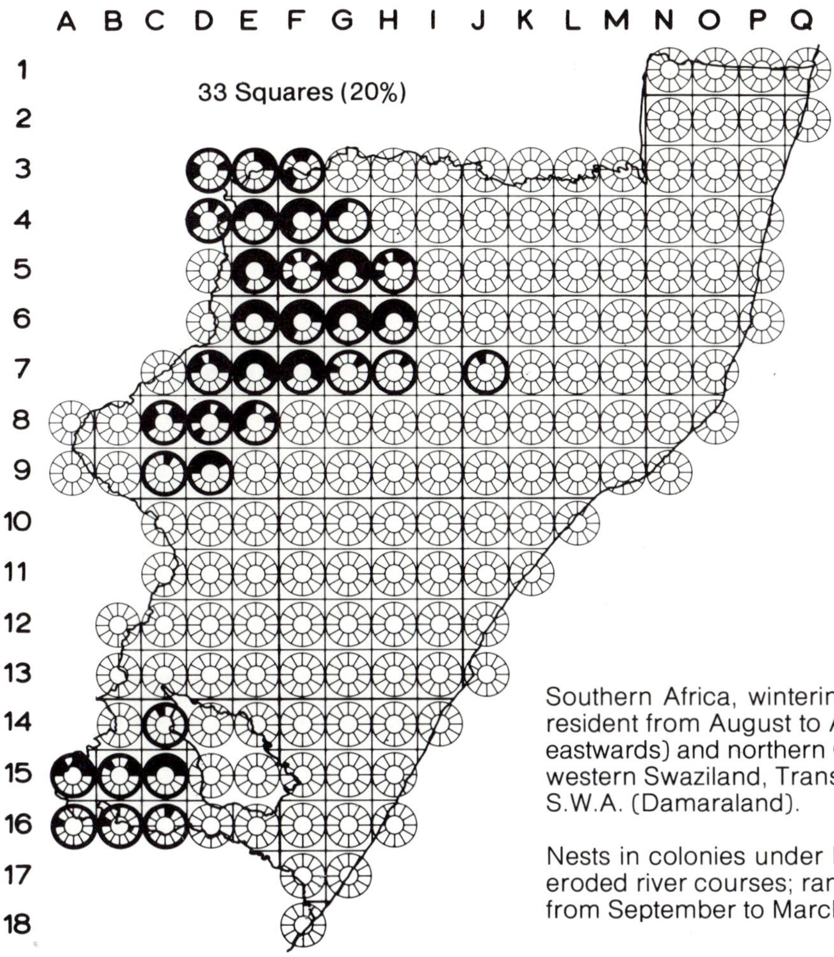

504 Cliff Swallow

Hirundo spilodera

Southern Africa, wintering north to the middle Congo River. A summer resident from August to April in the Karoo districts (from about Brandvlei eastwards) and northern Cape, O.F.S., Lesotho (lowlands), Natal (interior), western Swaziland, Transvaal and south-western Rhodesia; also locally in S.W.A. (Damaraland).

Nests in colonies under bridges of any size and occasionally on sides of eroded river courses; ranges over surrounding grassland to feed. Breeds from September to March.

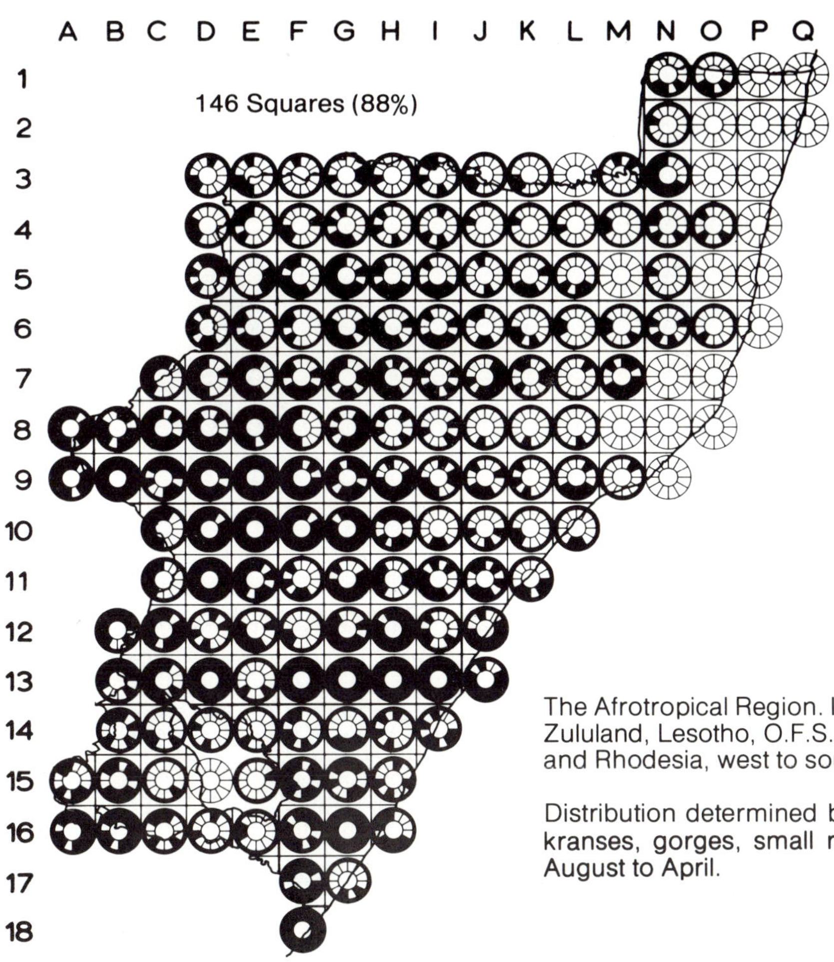

506 Rock Martin

Hirundo fuligula

The Afrotropical Region. In Southern Africa, the Cape, Transkei, Natal and Zululand, Lesotho, O.F.S., Swaziland, Transvaal, Moçambique (highlands) and Rhodesia, west to south-eastern Botswana and S.W.A.

Distribution determined by the availability of nesting sites, which include kranses, gorges, small rockfaces, bridges and buildings. Breeds from August to April.

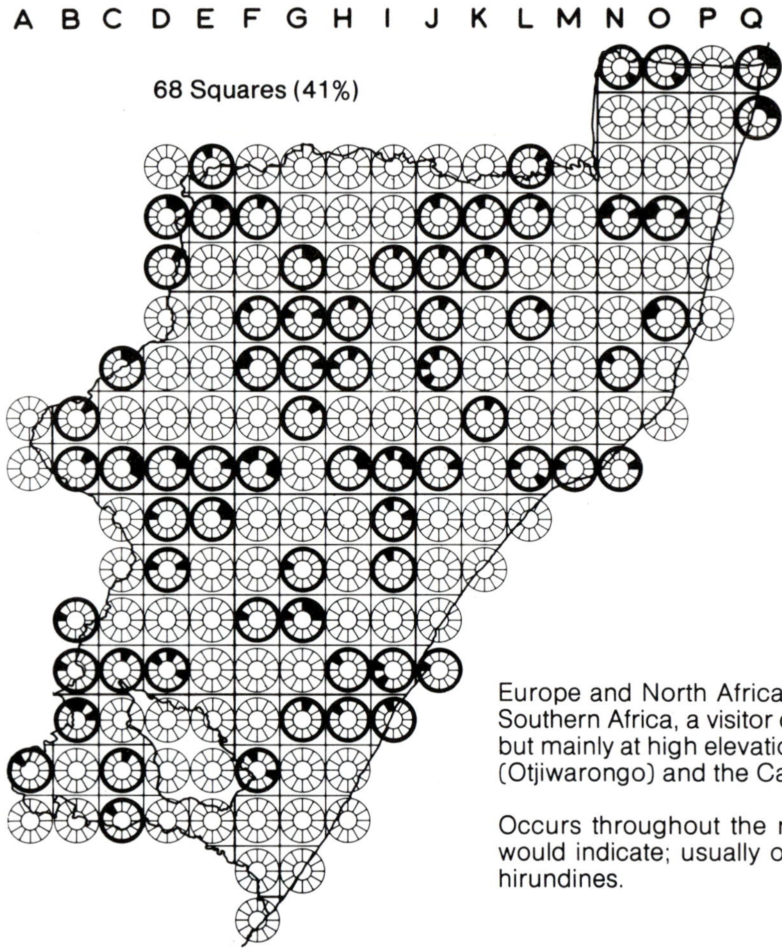

507 House Martin

Delichon urbica

Europe and North Africa, east to eastern Asia, wintering to the south. In Southern Africa, a visitor during the northern winter, occurring throughout, but mainly at high elevations in the east and southeast. Has bred in S.W.A. (Otjiwarongo) and the Cape.

Occurs throughout the region, probably more commonly than records would indicate; usually observed in flight, often in association with other hirundines.

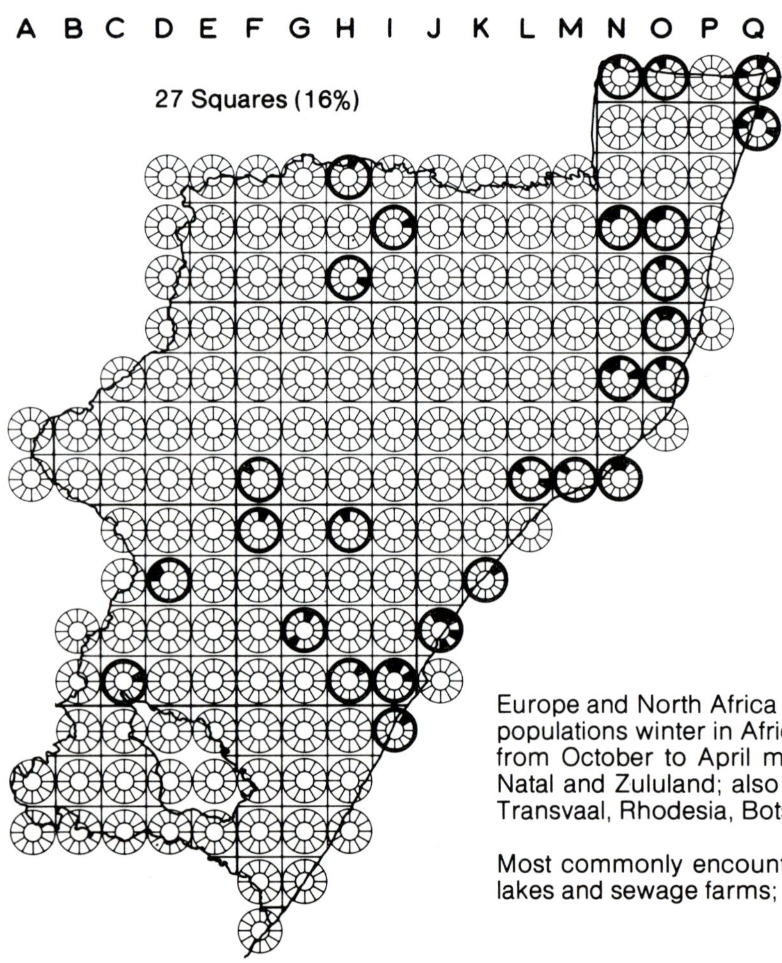

508 European Sand Martin

Riparia riparia

Europe and North Africa to eastern Asia; also North America. Palaearctic populations winter in Africa and Madagascar. In Southern Africa, a visitor from October to April mainly along the littoral in eastern Moçambique, Natal and Zululand; also recorded from the eastern and northern Cape, Transvaal, Rhodesia, Botswana (Lake Dow) and S.W.A. (Okahanja).

Most commonly encountered in the coastal region, at estuaries, coastal lakes and sewage farms; also found inland near standing water.

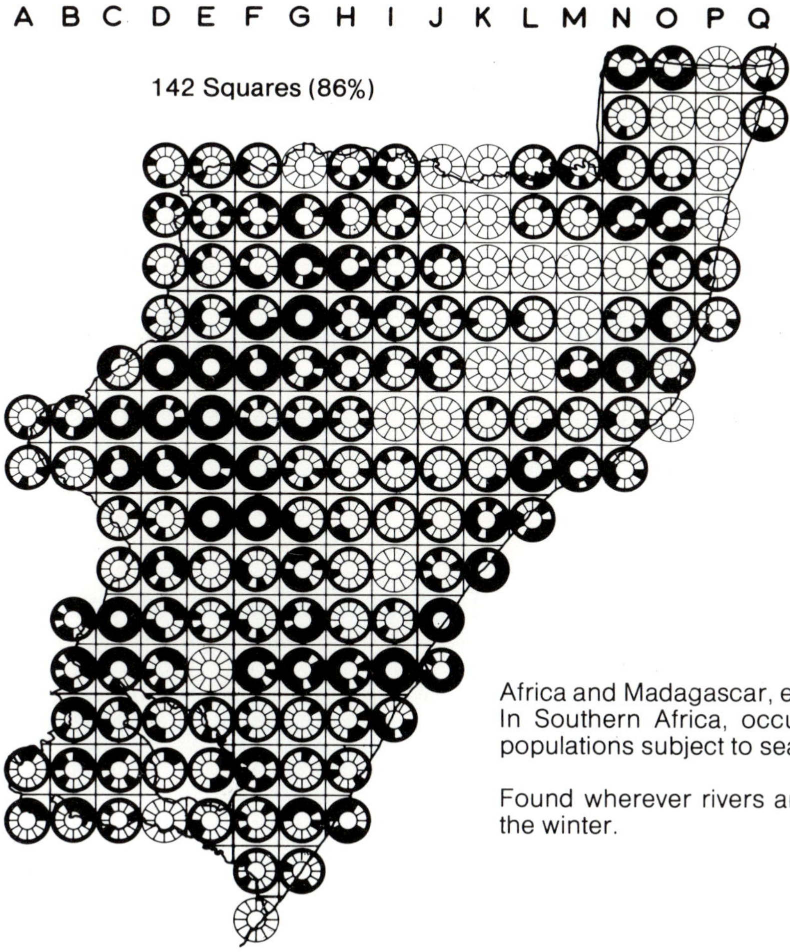

509 African Sand Martin

Riparia paludicola

Africa and Madagascar, east to southern Asia, Taiwan and the Philippines. In Southern Africa, occurs locally throughout. Resident, though Cape populations subject to seasonal movement.

Found wherever rivers and standing water occur. Breeds mainly during the winter.

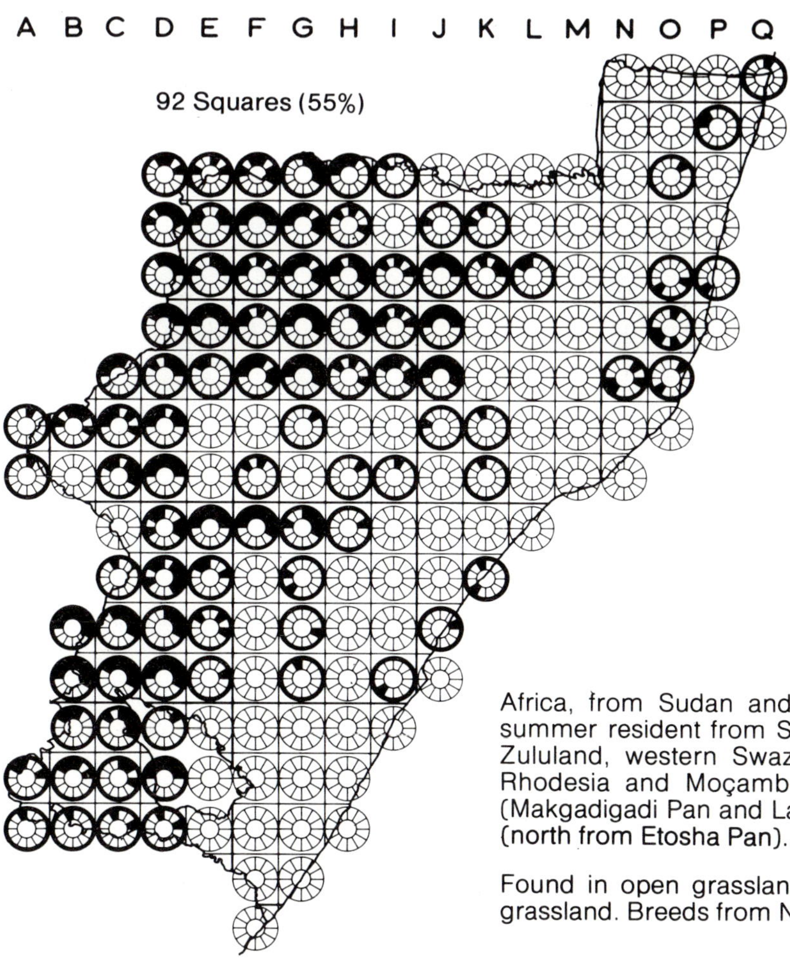

510 Banded Sand Martin

Riparia cincta

Africa, from Sudan and Ethiopia to the Cape. In Southern Africa, a summer resident from September to April in the Cape, Natal (interior), Zululand, western Swaziland, O.F.S., Transvaal, Lesotho (lowlands), Rhodesia and Moçambique (highlands); also in northern Botswana (Makgadigadi Pan and Lake Dow), the Caprivi Strip and northern S.W.A. (north from Etosha Pan). Winters to the north in Angola and Zaïre.

Found in open grassland above 900 m a.s.l. and, sparsely, in coastal grassland. Breeds from November to February.

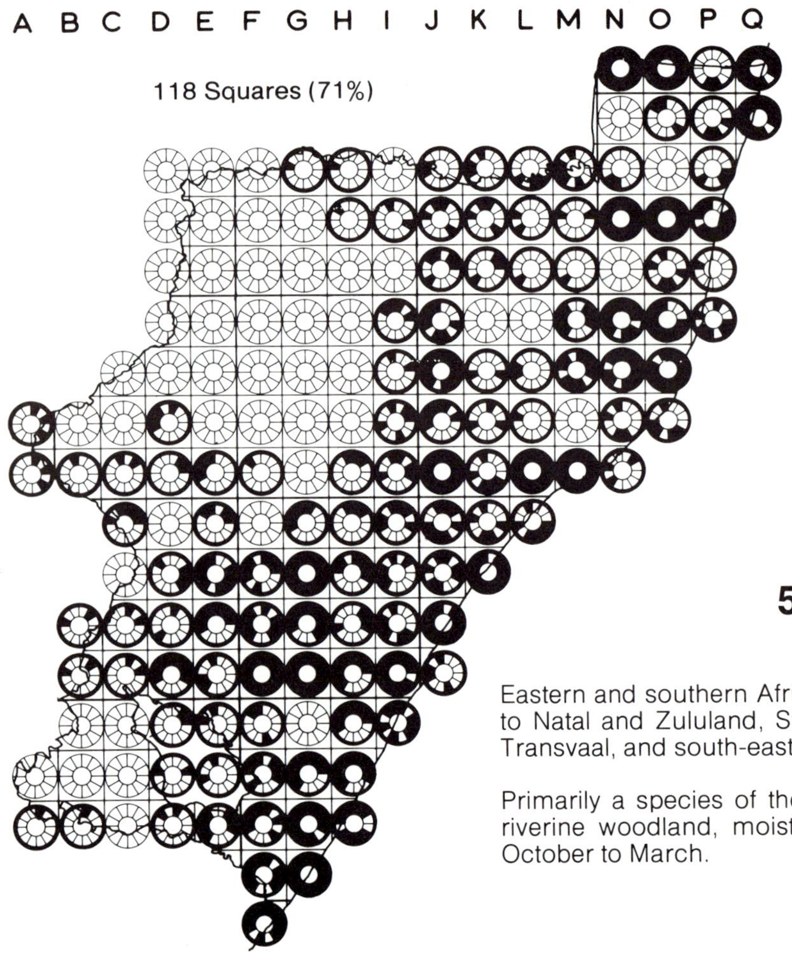

511 Black Saw-wing Swallow

Psalidoprocne holomelaena

Eastern and southern Africa. In Southern Africa, the south-western Cape to Natal and Zululand, Swaziland, Moçambique (Sul do Save), eastern Transvaal, and south-eastern Rhodesia.

Primarily a species of the edge of evergreen forest; also in mixed and riverine woodland, moist thornveld and wooded valleys. Breeds from October to March.

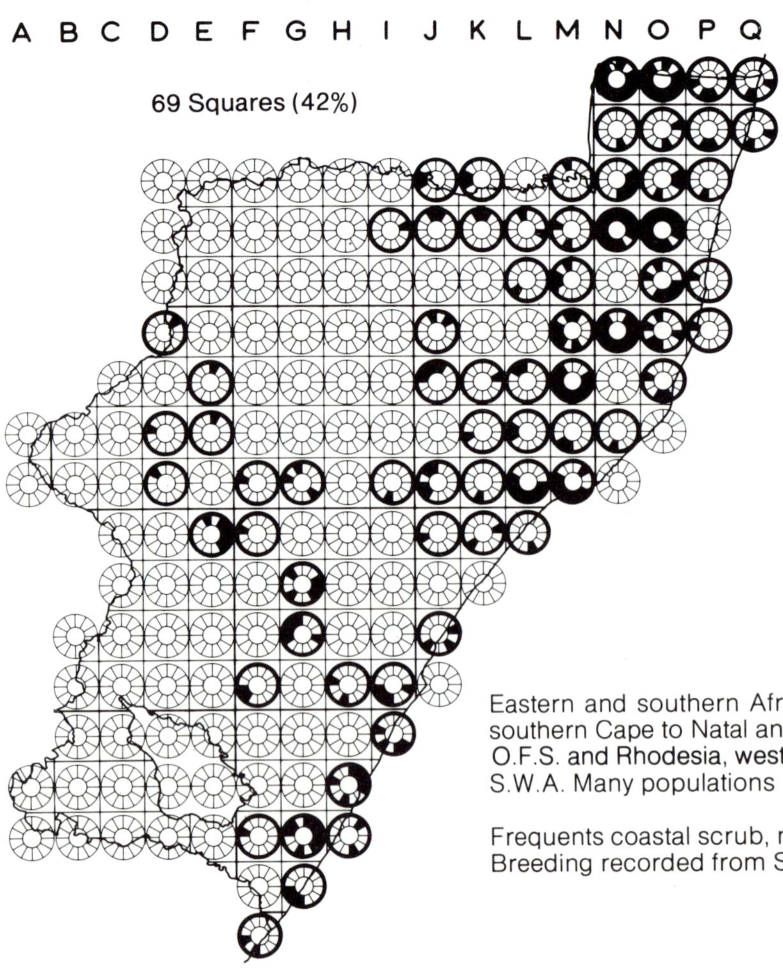

513 Black Cuckooshrike

Campephaga flava

Eastern and southern Africa. In Southern Africa, the south-western and southern Cape to Natal and Zululand, Swaziland, Moçambique, Transvaal, O.F.S. and Rhodesia, west to eastern and northern Botswana and northern S.W.A. Many populations migratory.

Frequents coastal scrub, mixed woodland, thornveld and wooded valleys. Breeding recorded from September to December.

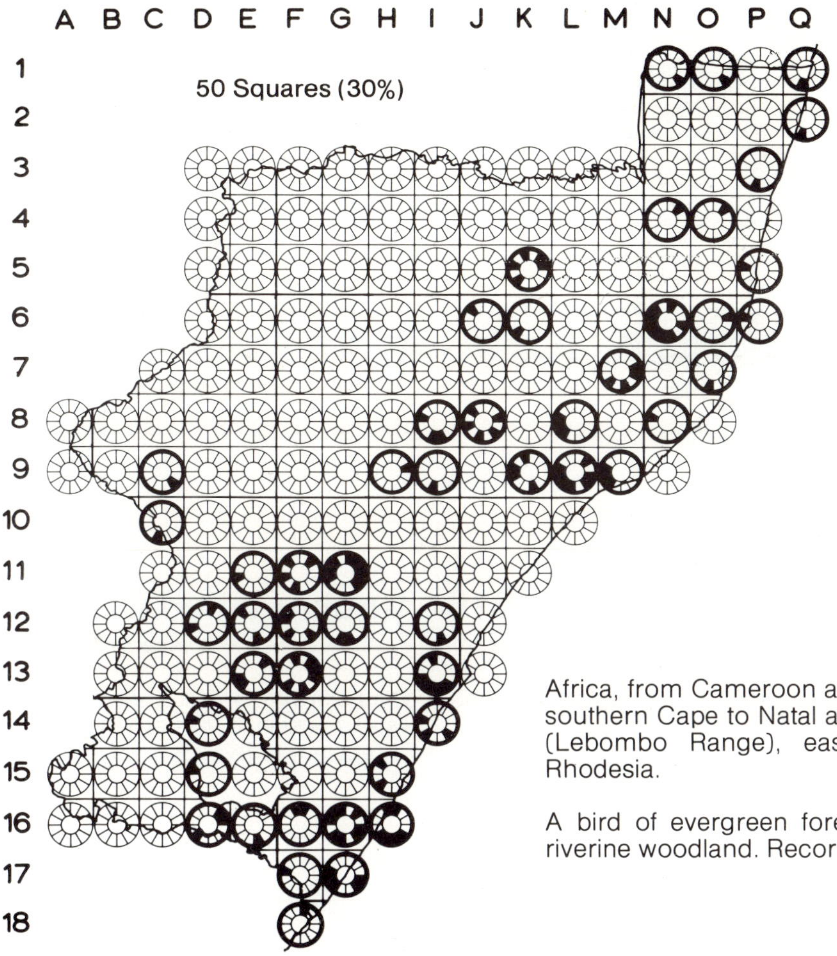

516 Grey Cuckooshrike

Coracina caesia

Africa, from Cameroon and Ethiopia to the Cape. In Southern Africa, the southern Cape to Natal and Zululand, Swaziland, southern Moçambique (Lebombo Range), eastern and northern Transvaal, and eastern Rhodesia.

A bird of evergreen forest, extending into thickly-wooded valleys and riverine woodland. Recorded breeding during December.

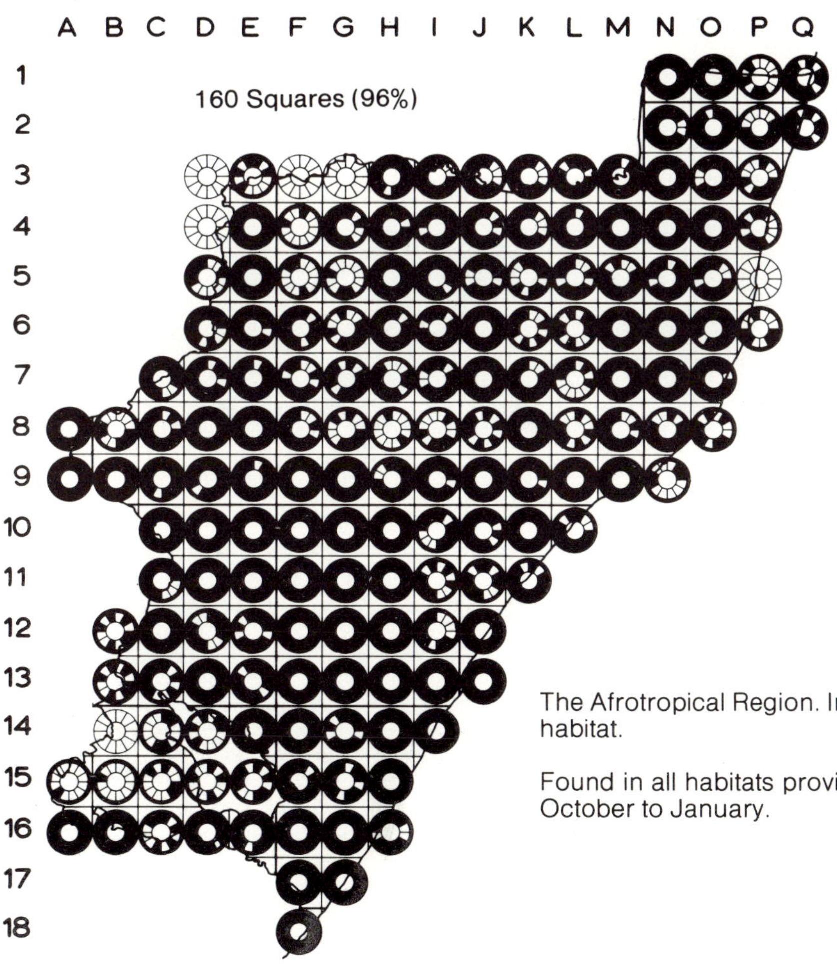

517 Fork-tailed Drongo

Dicrurus adsimilis

The Afrotropical Region. In Southern Africa, occurs throughout in suitable habitat.

Found in all habitats provided trees are present. Breeding recorded from October to January.

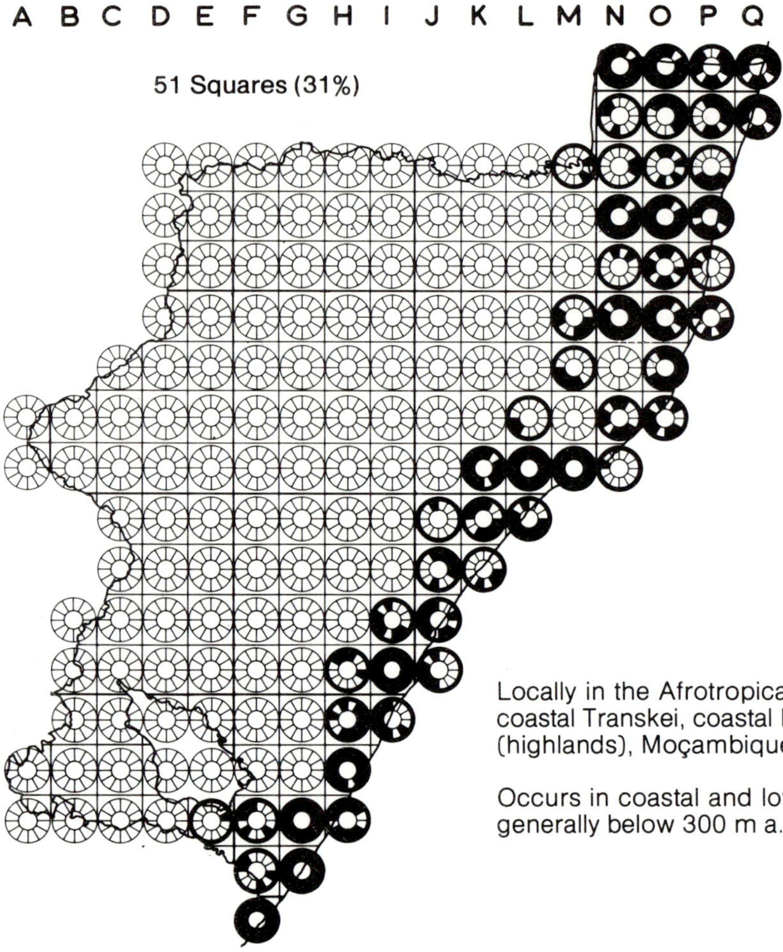

518 Square-tailed Drongo

Dicrurus ludwigii

Locally in the Afrotropical Region. In Southern Africa, the eastern Cape, coastal Transkei, coastal Natal and Zululand, Swaziland, eastern Transvaal (highlands), Moçambique and Rhodesia (eastern highlands).

Occurs in coastal and lowland forest, and riverine and mixed woodland, generally below 300 m a.s.l. Breeds from October to April.

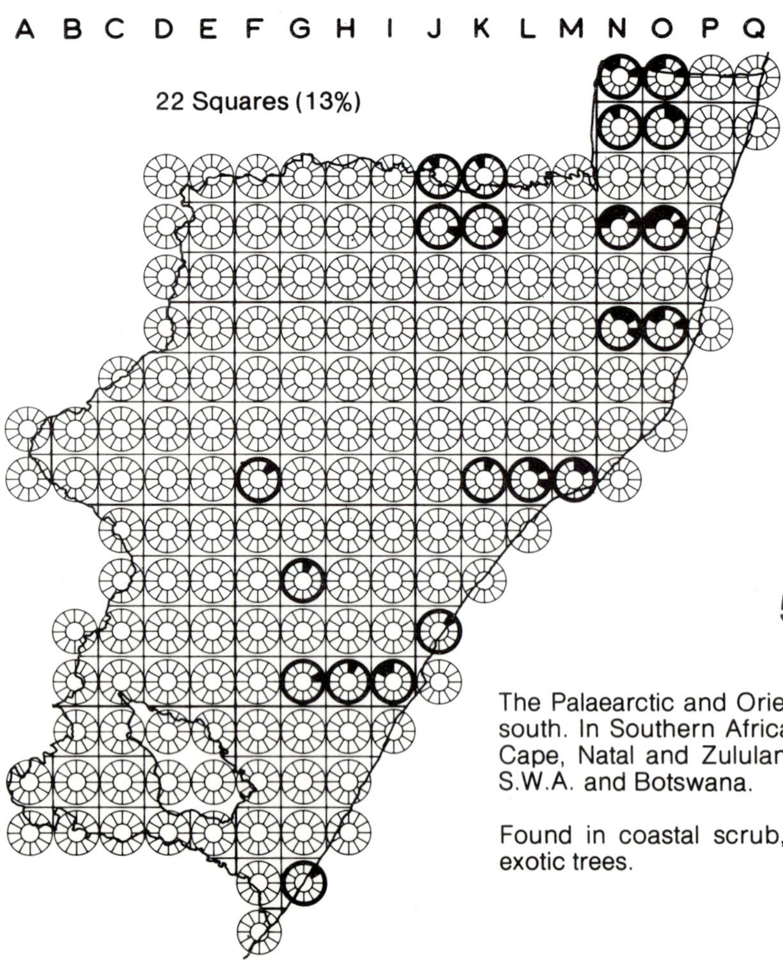

519 European Golden Oriole

Oriolus oriolus

The Palaearctic and Oriental Regions, most populations wintering to the south. In Southern Africa, a visitor from October to April to the eastern Cape, Natal and Zululand, northwards in the east; also recorded from S.W.A. and Botswana.

Found in coastal scrub, mixed and riverine woodland, thornveld and exotic trees.

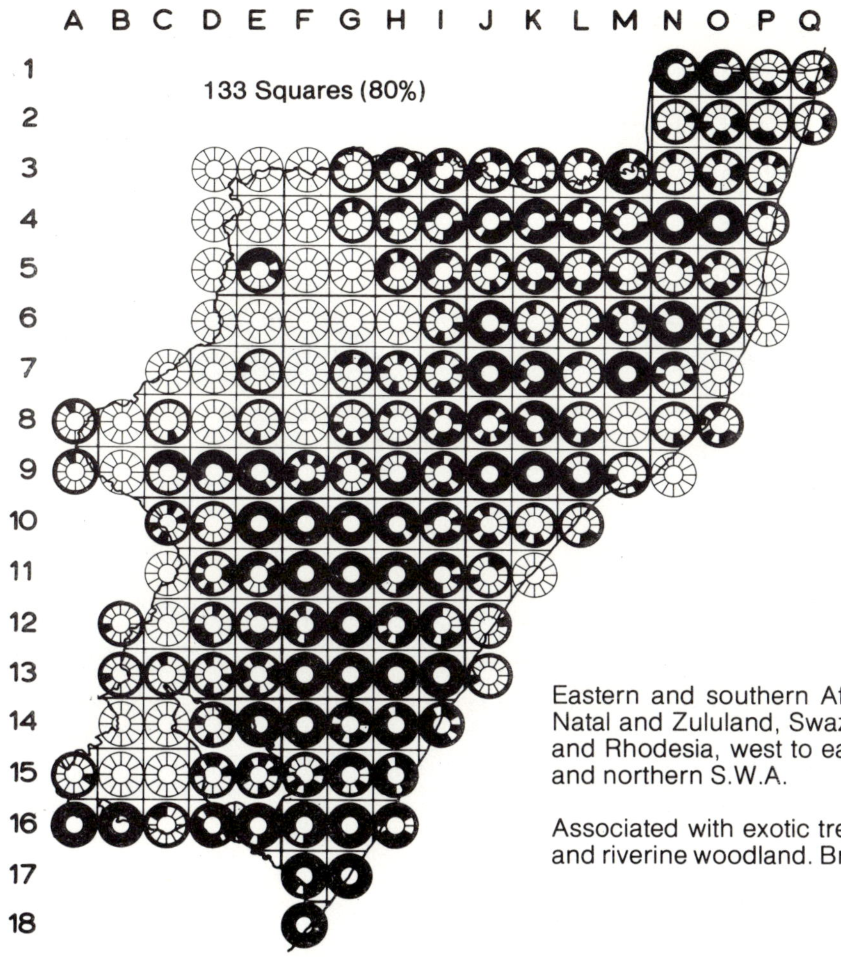

521 Black-headed Oriole

Oriolus larvatus

Eastern and southern Africa. In Southern Africa, the southern Cape to Natal and Zululand, Swaziland, northern O.F.S., Transvaal, Moçambique and Rhodesia, west to eastern and northern Botswana, the Caprivi Strip and northern S.W.A.

Associated with exotic trees, evergreen forest, open savanna, and mixed and riverine woodland. Breeding recorded from September to December.

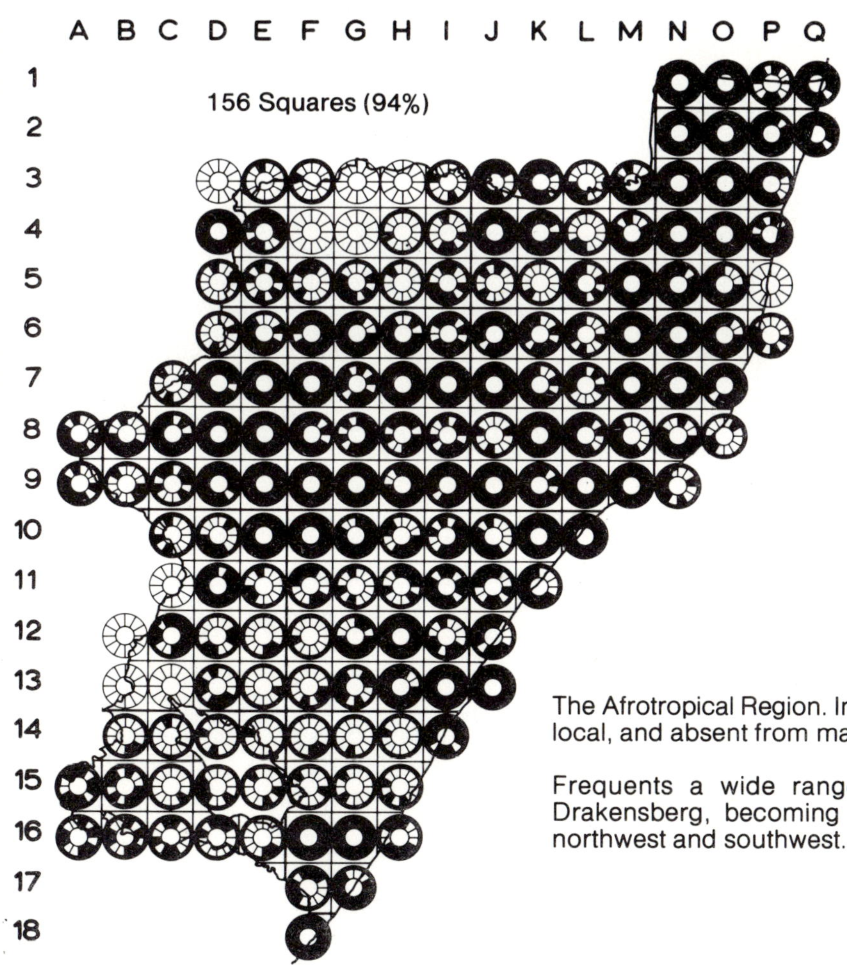

522 Pied Crow

Corvus albus

The Afrotropical Region. In Southern Africa, distributed throughout, though local, and absent from many districts.

Frequents a wide range of habitats, from the coast to the lower Drakensberg, becoming scarce in the upland grassland areas of the northwest and southwest. Recorded breeding from August to November.

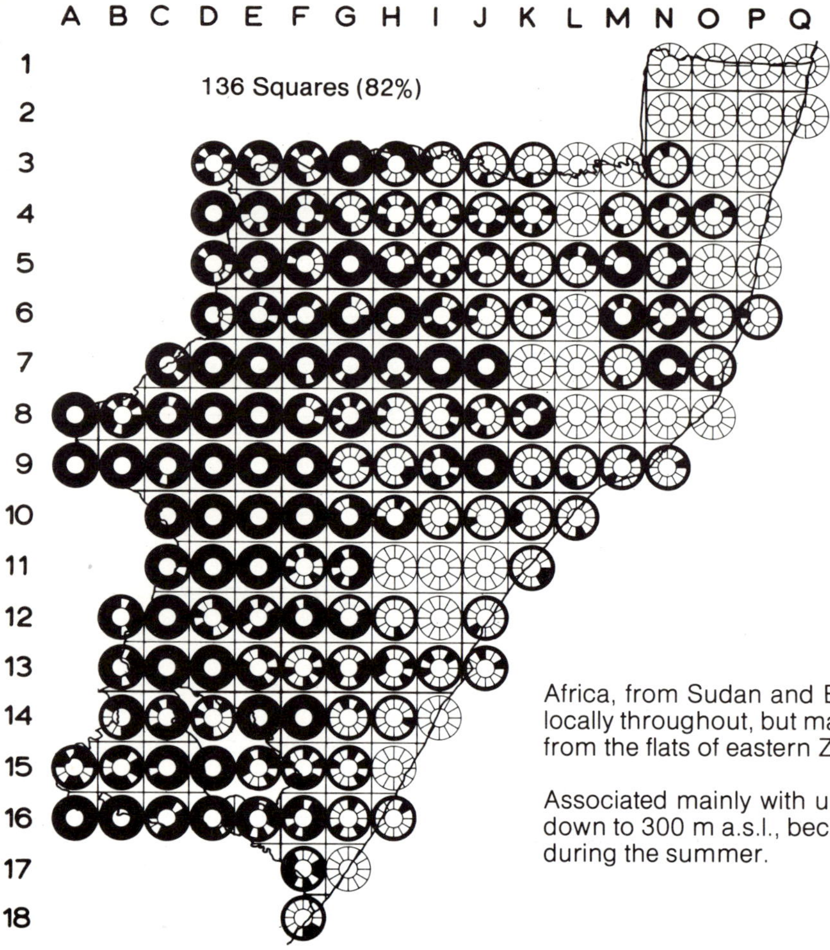

523 Black Crow

Corvus capensis

Africa, from Sudan and Ethiopia to the Cape. In Southern Africa, occurs locally throughout, but mainly in the higher interior grasslands, and absent from the flats of eastern Zululand and Moçambique.

Associated mainly with upland grassland and cultivated areas, extending down to 300 m a.s.l., becoming scarcer towards the coast. Breeds mainly during the summer.

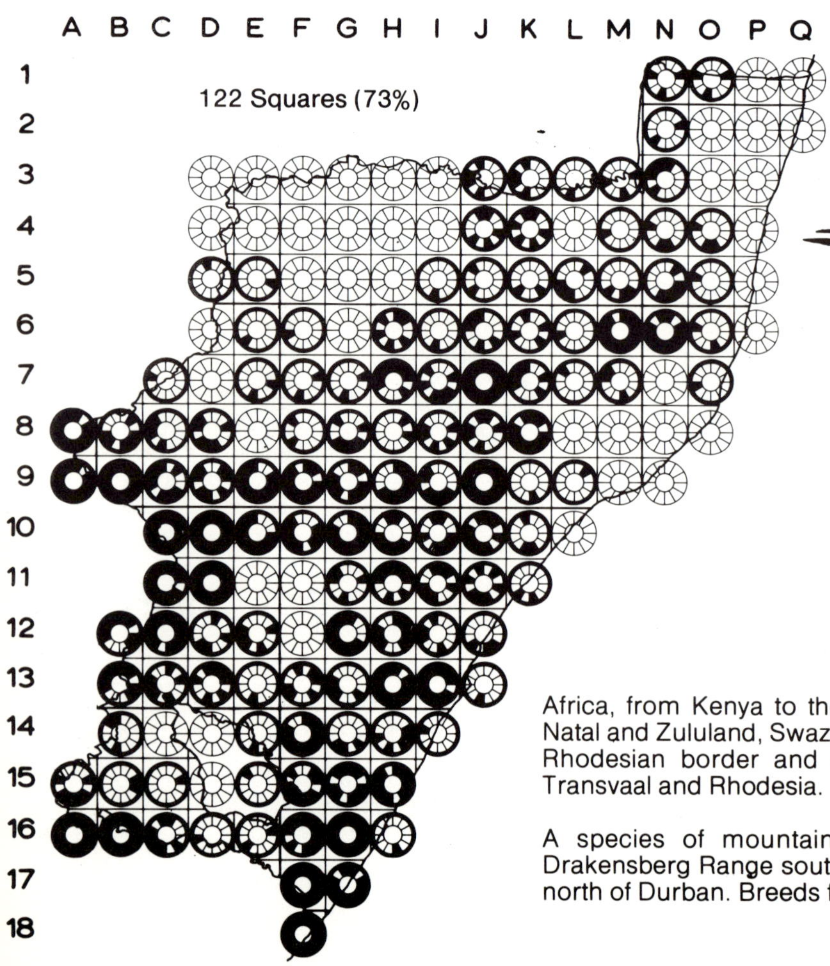

524 White-necked Raven

Corvus albicollis

Africa, from Kenya to the Cape. In Southern Africa, S.W.A., the Cape, Natal and Zululand, Swaziland, Moçambique (Lebombo Range, along the Rhodesian border and on the Manica Plateau), to Lesotho, O.F.S., Transvaal and Rhodesia.

A species of mountainous and hilly country, extending from the Drakensberg Range south of Van Reenen to the coast, where it is sparse north of Durban. Breeds from August to October.

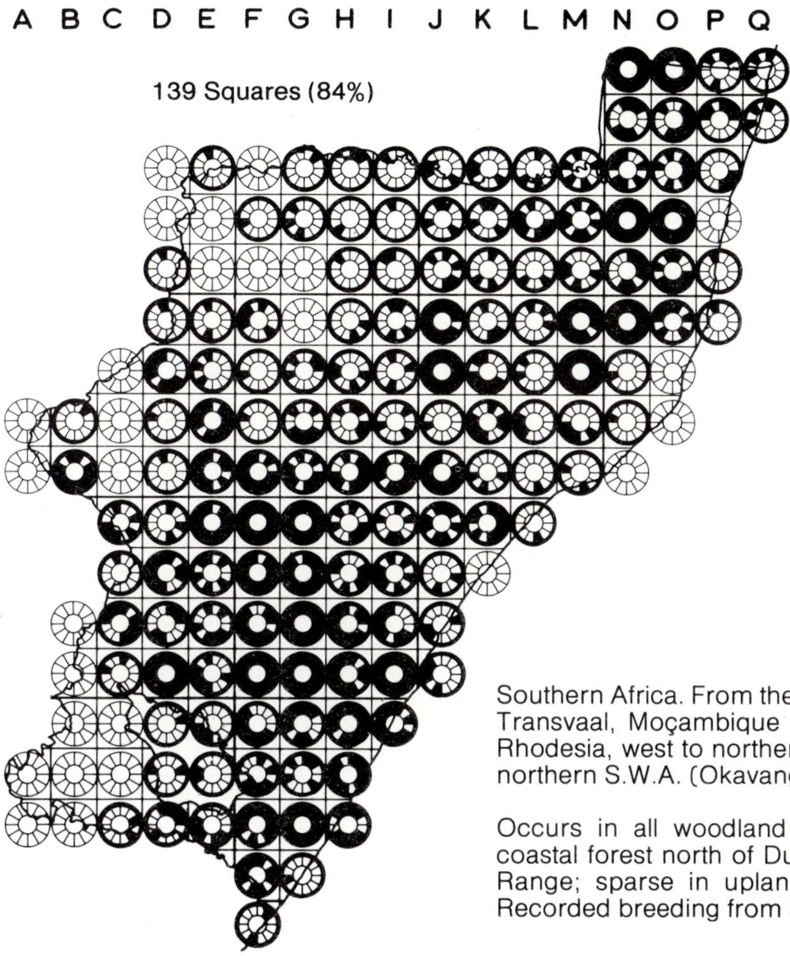

527 Southern Black Tit

Parus niger

Southern Africa. From the eastern Cape to Natal and Zululand, Swaziland, Transvaal, Moçambique (lower Limpopo River to the Lurio River) and Rhodesia, west to northern and eastern Botswana, the Caprivi Strip, and northern S.W.A. (Okavango River).

Occurs in all woodland habitats, including exotic plantations, except coastal forest north of Durban and evergreen forest of the Drakensberg Range; sparse in upland grassland where exotic trees are present. Recorded breeding from September to December.

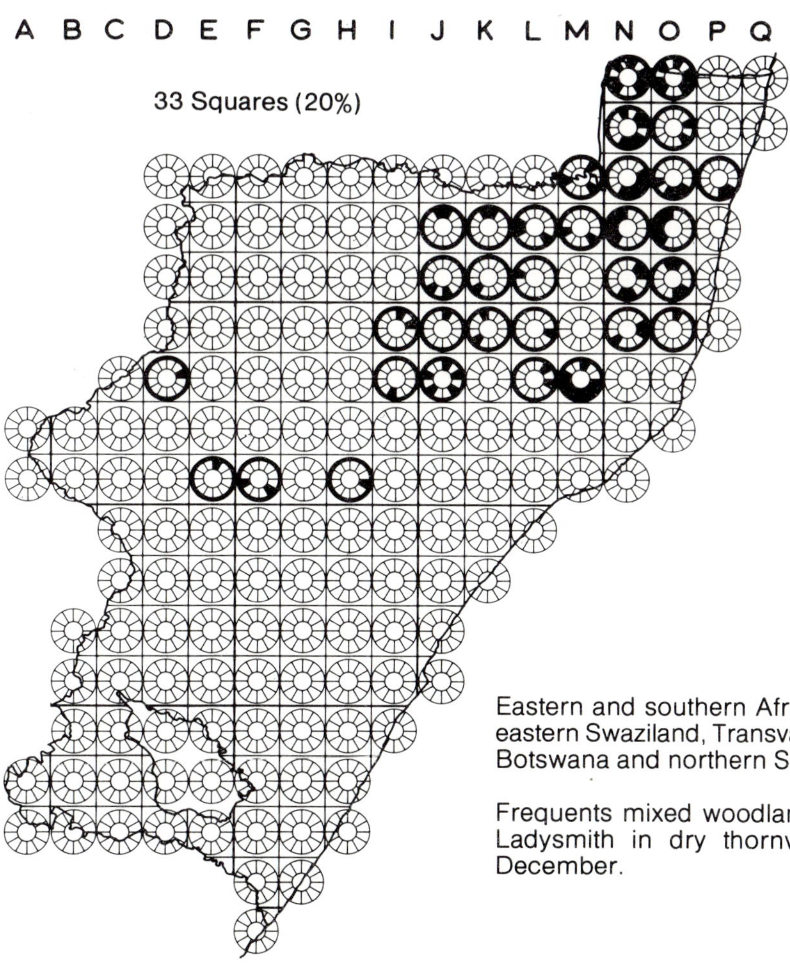

530 Grey Penduline Tit

Anthoscopus caroli

Eastern and southern Africa. In Southern Africa, Natal interior, Zululand, eastern Swaziland, Transvaal, Moçambique and Rhodesia, west to northern Botswana and northern S.W.A.

Frequents mixed woodland on the littoral plain, extending westwards to Ladysmith in dry thornveld. Breeding recorded from September to December.

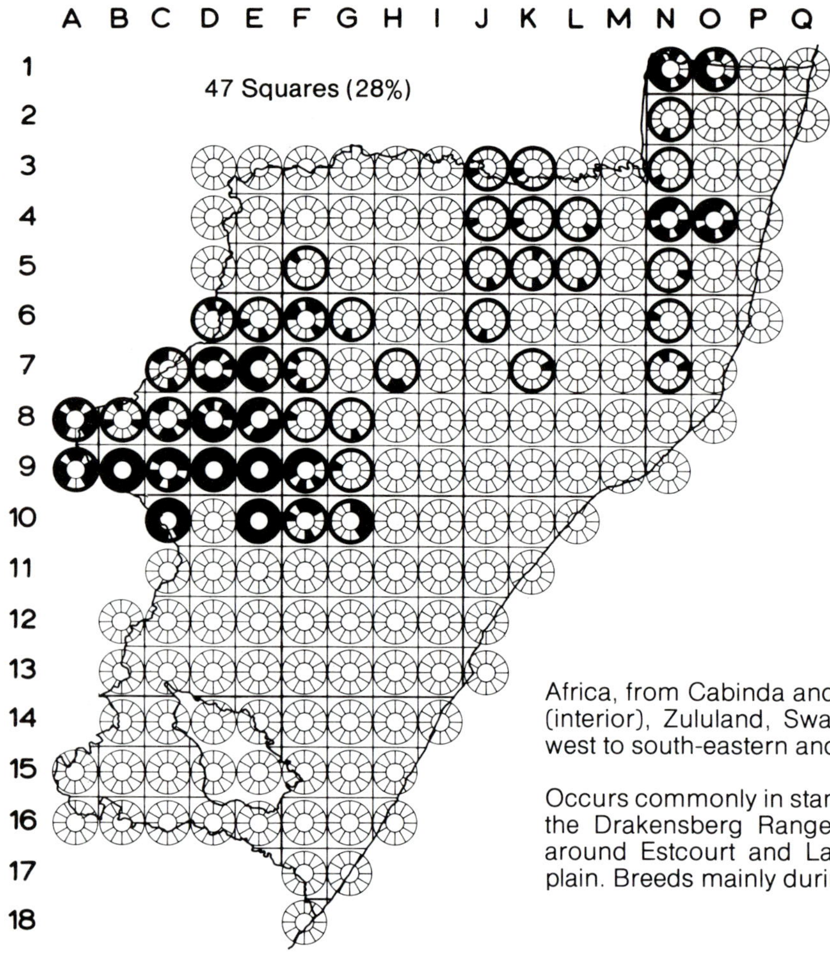

533 Arrow-marked Babbler

Turdoides jardineii

Africa, from Cabinda and Kenya south to Natal. In Southern Africa, Natal (interior), Zululand, Swaziland, Transvaal, Moçambique and Rhodesia, west to south-eastern and northern Botswana and northern S.W.A.

Occurs commonly in stands of exotic trees and low scrub in the foothills of the Drakensberg Range, extending into dry thornveld on hilly slopes around Estcourt and Ladysmith and sparsely eastwards to the littoral plain. Breeds mainly during the summer.

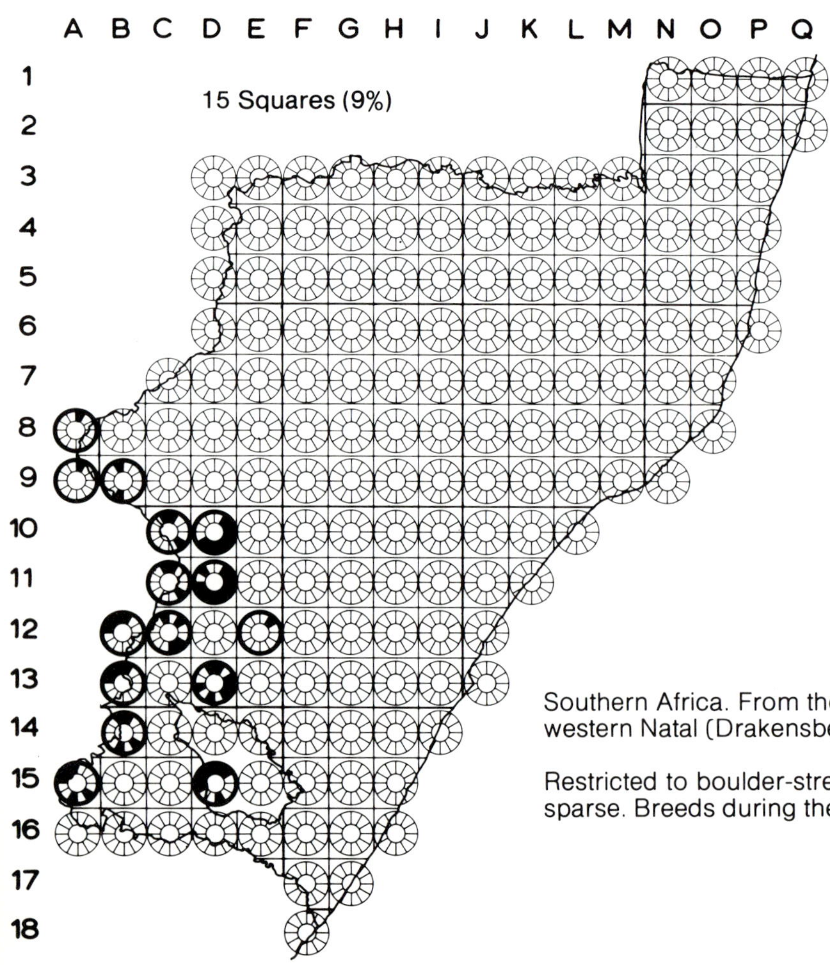

540B Rockjumper

Chaetops aurantius

Southern Africa. From the eastern Cape to East Griqualand, Lesotho and western Natal (Drakensberg Range and outliers).

Restricted to boulder-strewn slopes at high altitudes, where vegetation is sparse. Breeds during the summer.

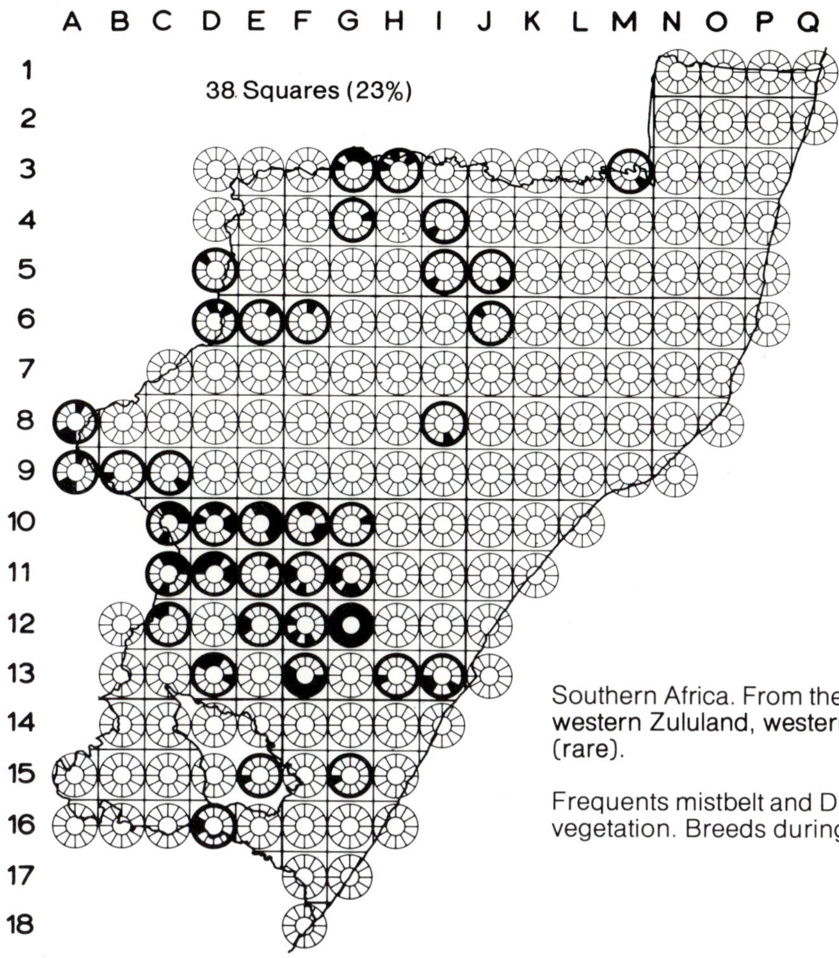

542 Bush Blackcap

Lioptilus nigricapillus

Southern Africa. From the eastern Cape to East Griqualand, Natal (interior), western Zululand, western Swaziland, and eastern and northern Transvaal (rare).

Frequents mistbelt and Drakensberg evergreen forest, and adjacent scrub vegetation. Breeds during the summer.

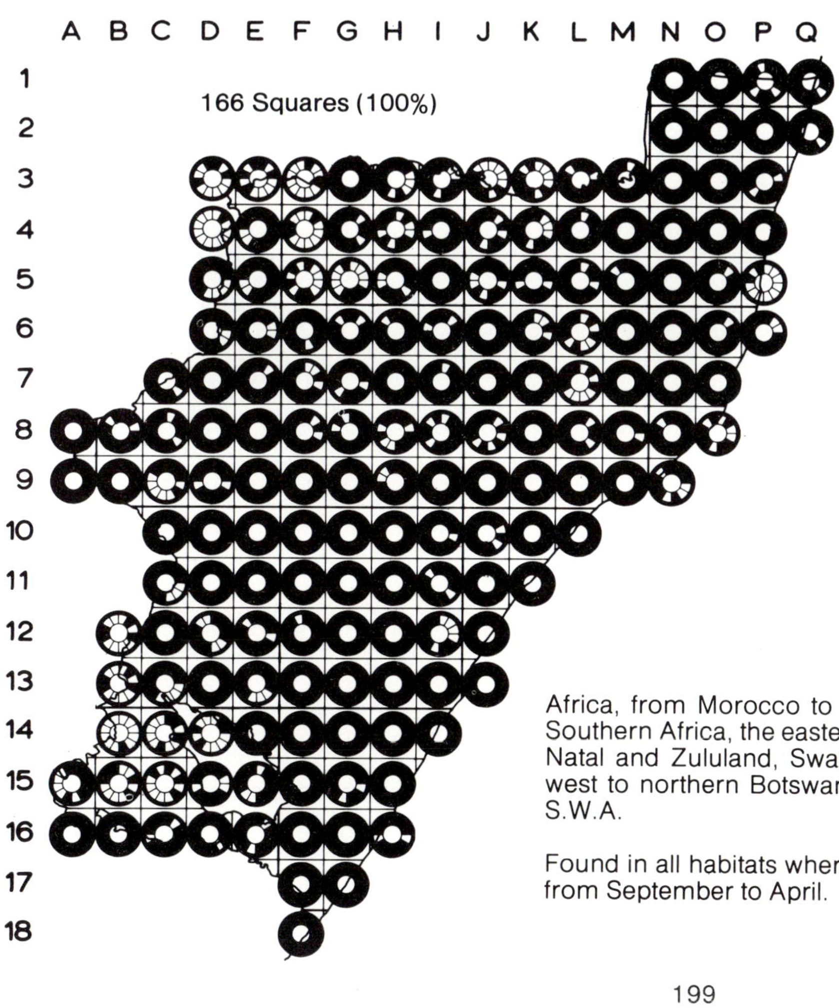

545 Black-eyed Bulbul

Pycnonotus barbatus

Africa, from Morocco to the Cape; also the Middle East and Arabia. In Southern Africa, the eastern Cape, Transkei, northern and eastern O.F.S., Natal and Zululand, Swaziland, Transvaal, Moçambique and Rhodesia, west to northern Botswana (Ngamiland), the Caprivi Strip and northern S.W.A.

Found in all habitats wherever trees or bushes occur. Breeding recorded from September to April.

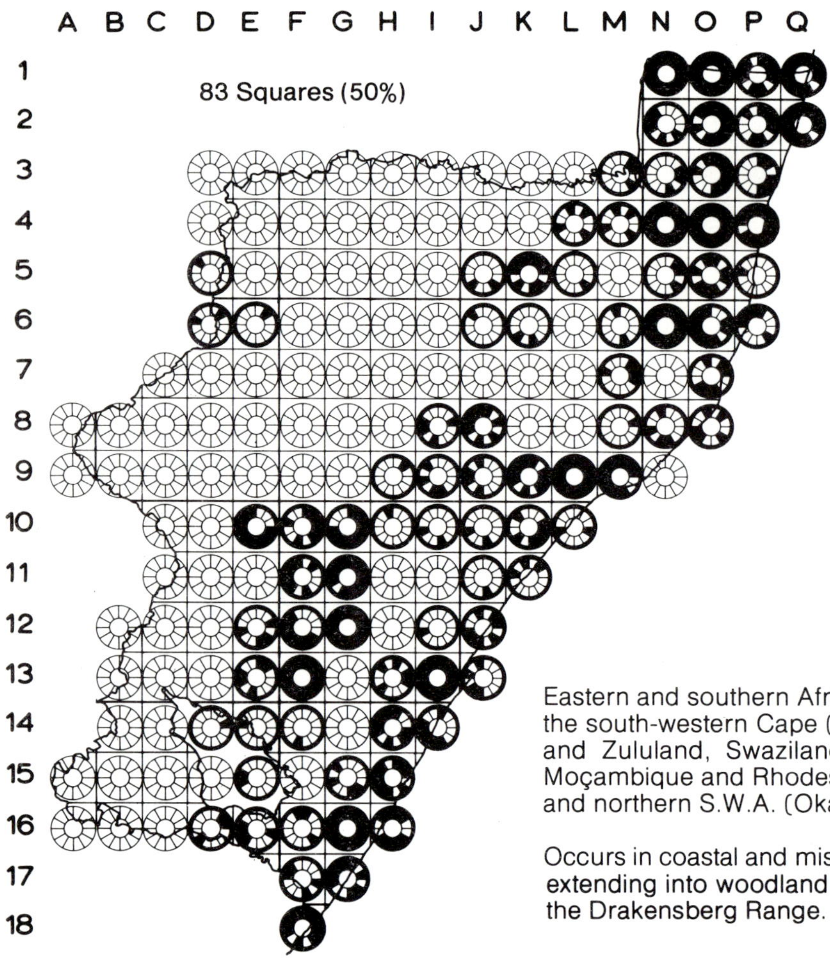

546 Terrestrial Bulbul

Phyllastrephus terrestris

Eastern and southern Africa, from Kenya to the Cape. In Southern Africa, the south-western Cape (from Swellendam) eastwards to Transkei, Natal and Zululand, Swaziland, eastern and northern Transvaal, southern Moçambique and Rhodesia, west to northern Botswana, the Caprivi Strip and northern S.W.A. (Okavango River).

Occurs in coastal and mistbelt evergreen forest and dense riparian growth, extending into woodland thickets in the northeast; also in some forests in the Drakensberg Range. Recorded breeding from October to January.

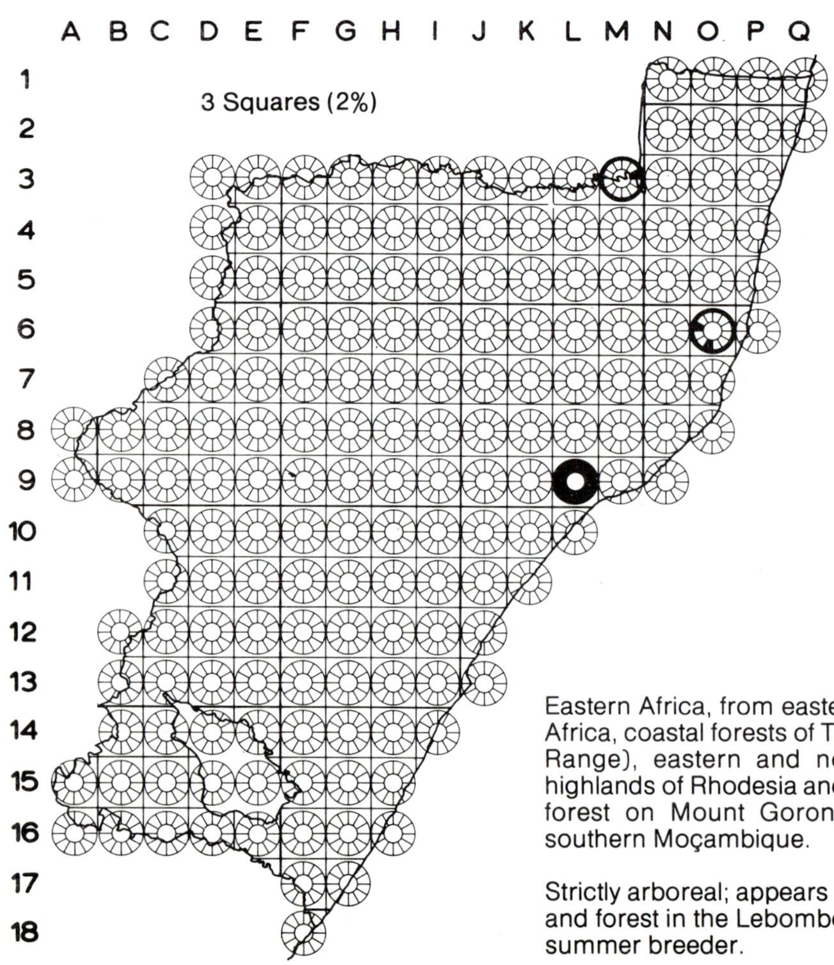

547 Yellow-streaked Bulbul

Phyllastrephus flavostriatus

Eastern Africa, from eastern Zaïre and Tanzania to Transkei. In Southern Africa, coastal forests of Transkei, Zululand, eastern Swaziland (Lebombo Range), eastern and northern Transvaal (highland forest), eastern highlands of Rhodesia and adjacent highland Moçambique; also montane forest on Mount Gorongosa, and lowland forest near Inhamitanga, southern Moçambique.

Strictly arboreal; appears to be confined to the Ngoye Forest near Eshowe and forest in the Lebombo Range. Isolated records from Lake St Lucia. A summer breeder.

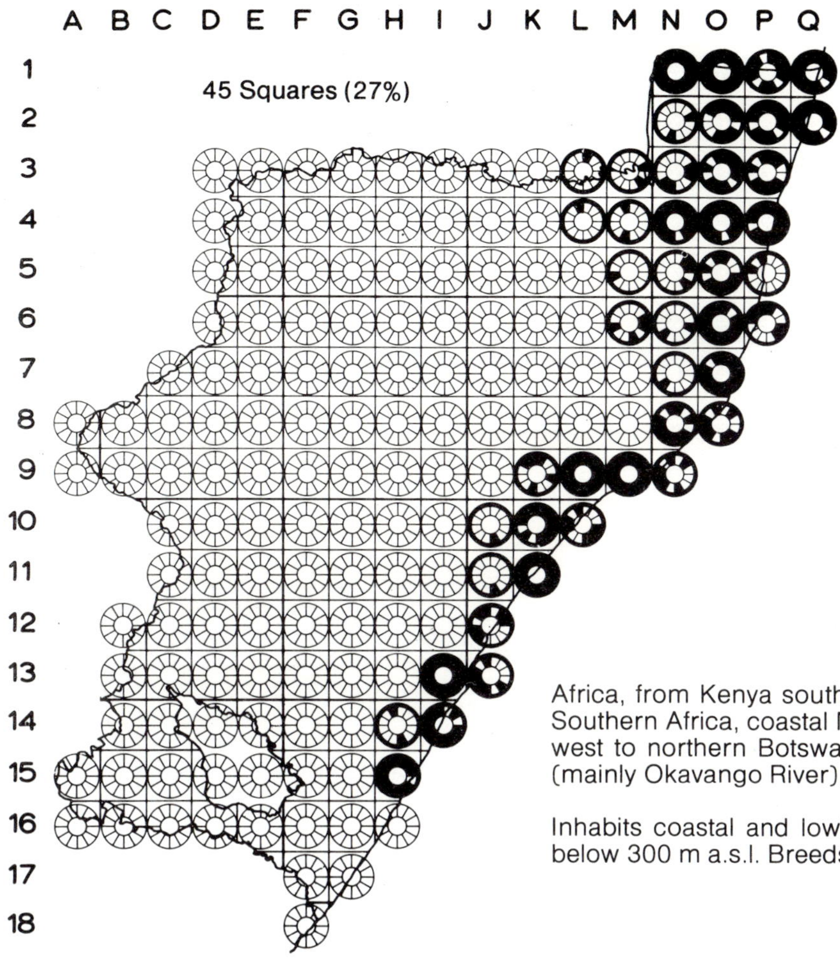

550 Yellow-bellied Bulbul

Chlorocichla flaviventris

Africa, from Kenya south to Natal in the east and S.W.A. in the west. In Southern Africa, coastal Natal and Zululand, Moçambique and Rhodesia, west to northern Botswana, the Caprivi Strip and north-western S.W.A. (mainly Okavango River).

Inhabits coastal and lowland forest, and riverine and mixed woodland, below 300 m a.s.l. Breeds during the summer.

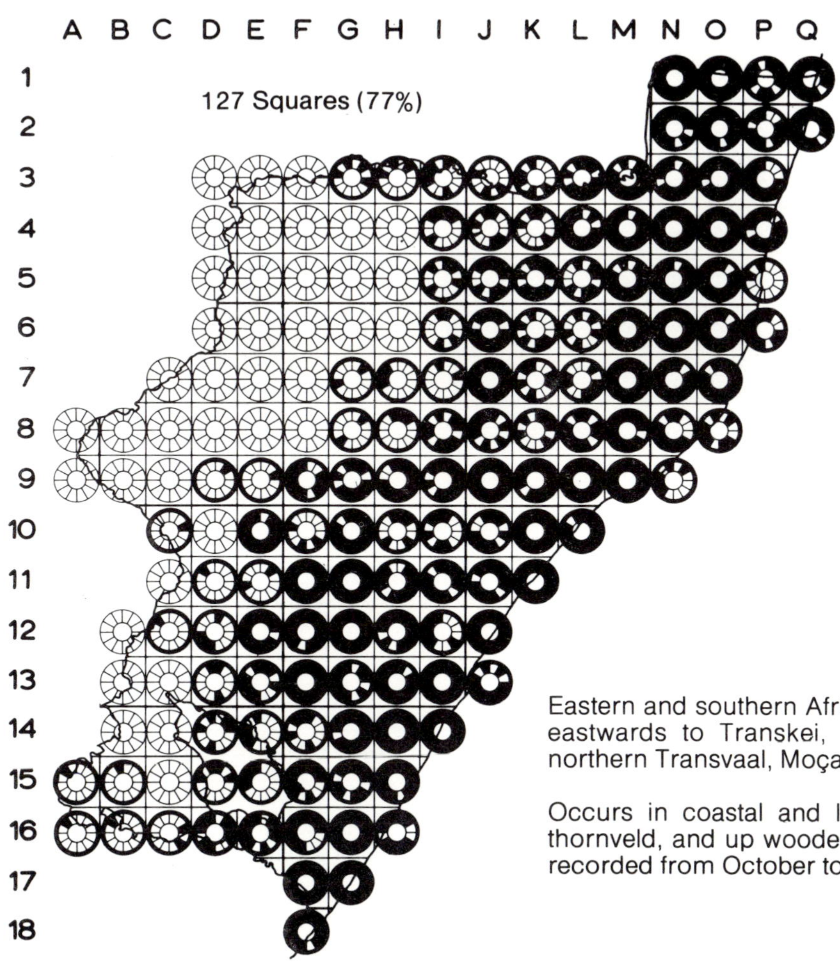

551 Sombre Bulbul

Andropadus importunus

Eastern and southern Africa. In Southern Africa, the south-western Cape eastwards to Transkei, Natal and Zululand, Swaziland, eastern and northern Transvaal, Moçambique and eastern Rhodesia.

Occurs in coastal and lowland forest, mixed and riverine woodland, thornveld, and up wooded valleys to mistbelt evergreen forest. Breeding recorded from October to April.

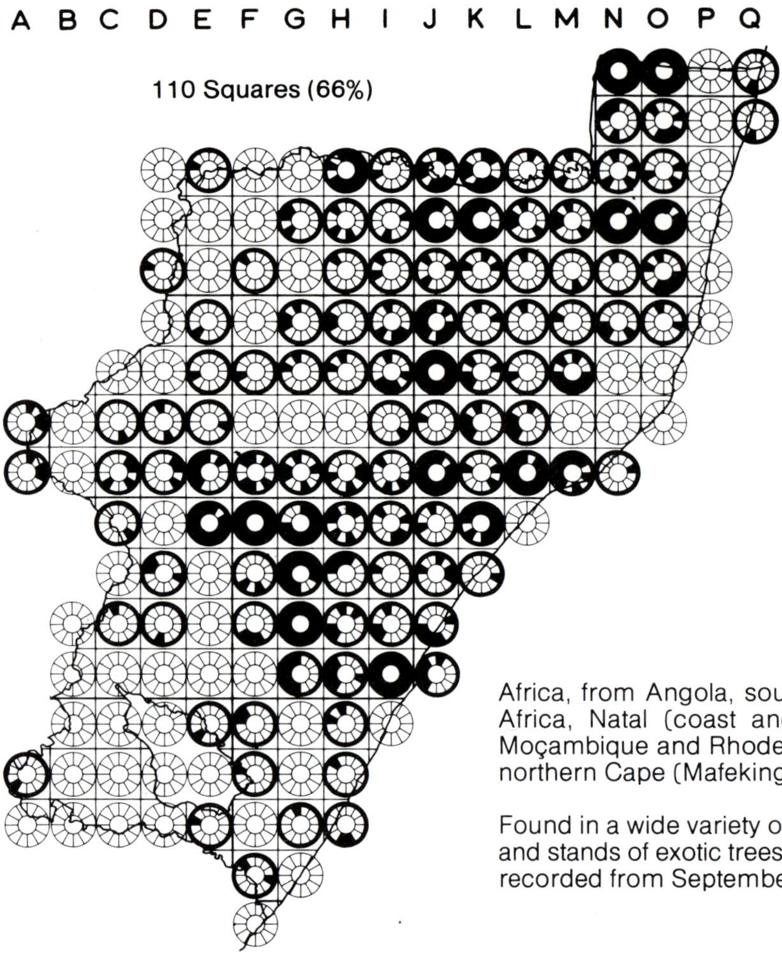

552 Kurrichane Thrush

Turdus libonyana

Africa, from Angola, southern Zaïre and Tanzania to Natal. In Southern Africa, Natal (coast and midlands), Zululand, Swaziland, Transvaal, Moçambique and Rhodesia, west to northern and eastern Botswana, the northern Cape (Mafeking), the Caprivi Strip and northern S.W.A.

Found in a wide variety of wooded habitats, such as thornveld, open bush and stands of exotic trees; prefers the taller trees in dry situations. Breeding recorded from September to January.

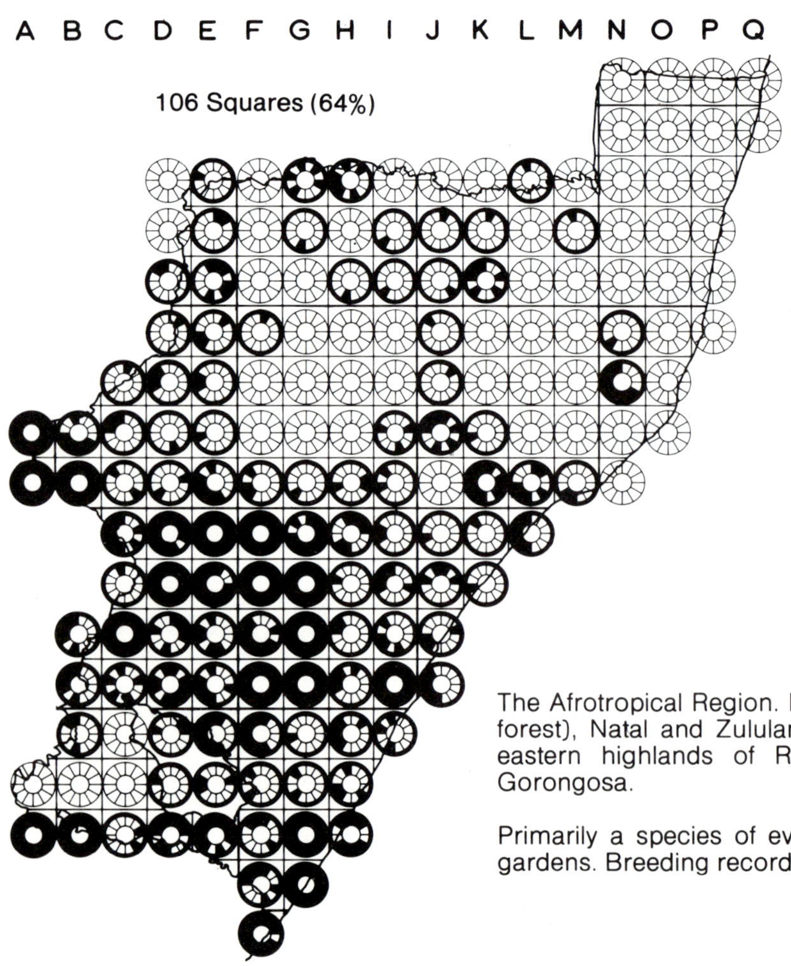

553 Olive Thrush

Turdus olivaceus

The Afrotropical Region. In Southern Africa, the Cape, Transkei (coastal forest), Natal and Zululand, O.F.S., Swaziland and Transvaal; also the eastern highlands of Rhodesia, adjacent Moçambique and Mount Gorongosa.

Primarily a species of evergreen forests; enters exotic plantations and gardens. Breeding recorded from September to January.

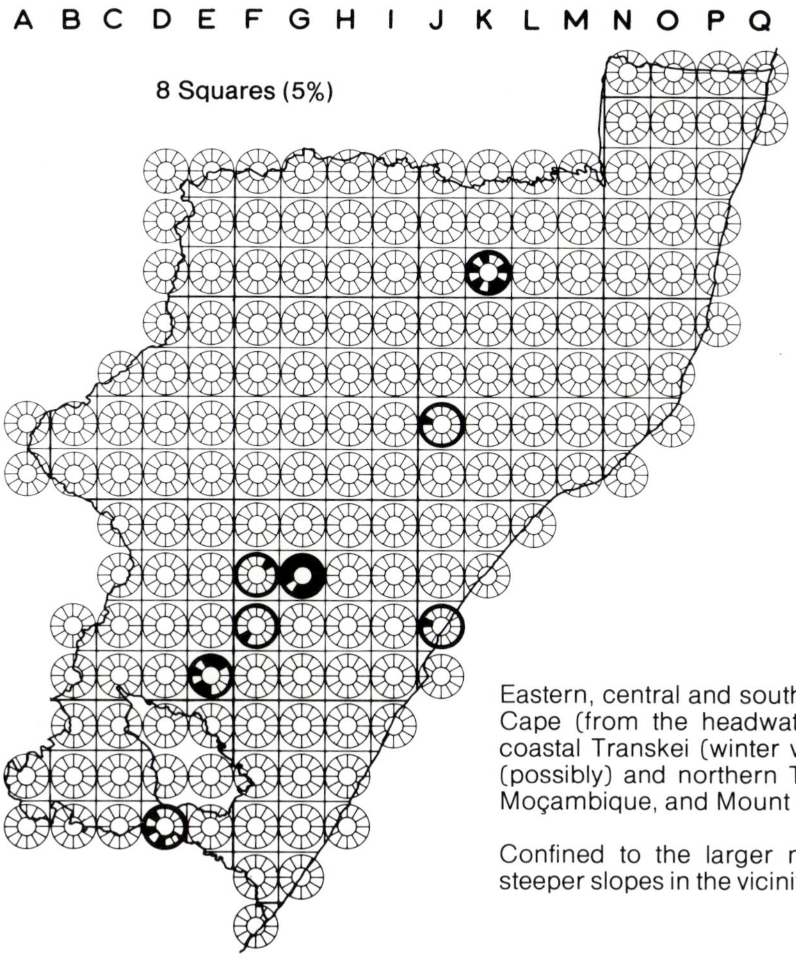

556 Orange Thrush

Turdus gurneyi

Eastern, central and south-eastern Africa. In Southern Africa, the eastern Cape (from the headwaters of the Buffalo River) to East Griqualand, coastal Transkei (winter visitor), Natal and western Zululand, Swaziland (possibly) and northern Transvaal; also eastern Rhodesia and adjacent Moçambique, and Mount Gorongosa.

Confined to the larger mistbelt evergreen forests, where it frequents steeper slopes in the vicinity of streams. Breeds from October to January.

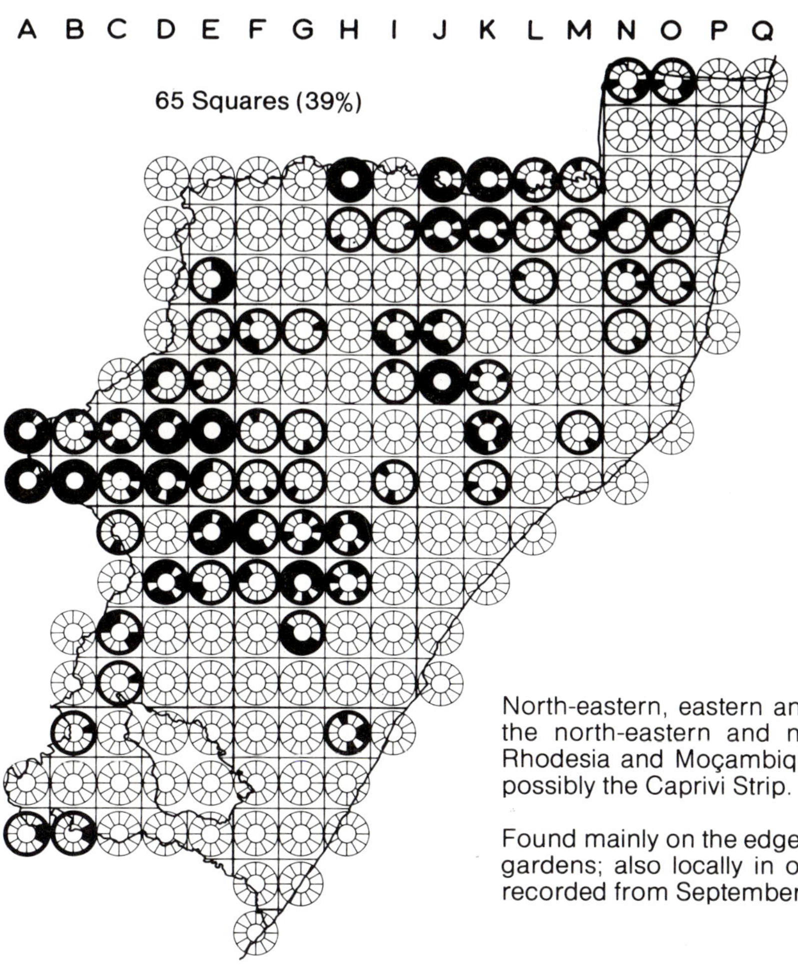

557 Groundscraper Thrush

Turdus litsitsirupa

North-eastern, eastern and southern Africa. In Southern Africa, S.W.A., the north-eastern and northern Cape, Botswana, O.F.S., Transvaal, Rhodesia and Moçambique to Swaziland, Natal (interior) and Zululand; possibly the Caprivi Strip.

Found mainly on the edge of exotic plantations, around farmsteads and in gardens; also locally in open savanna woodland in Zululand. Breeding recorded from September to November.

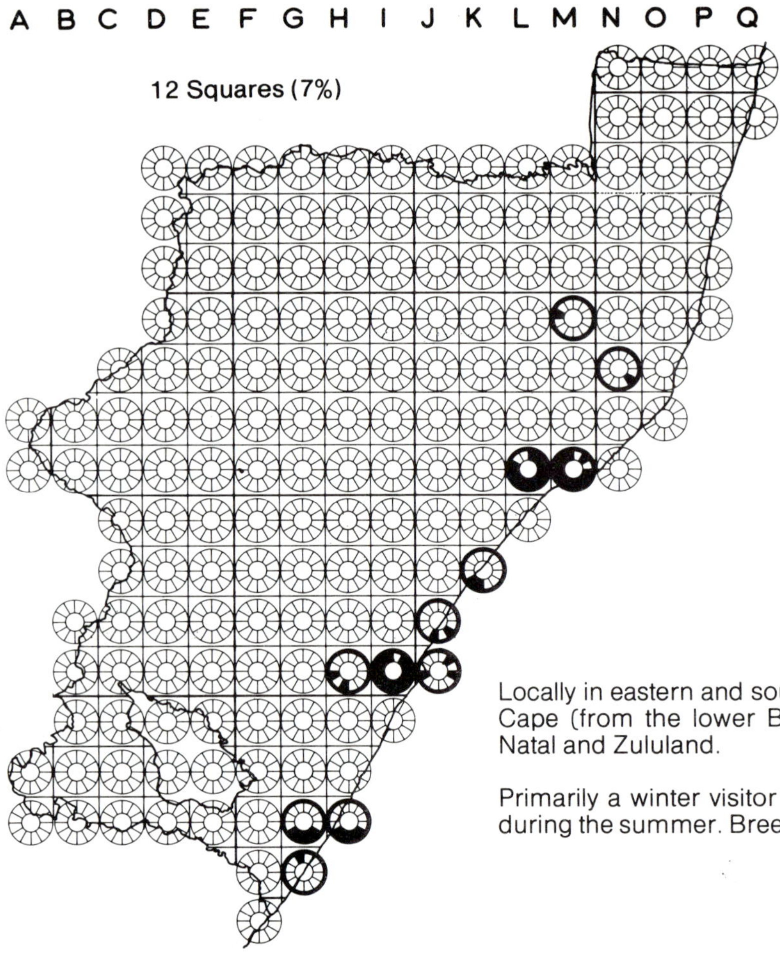

558 Natal Thrush

Turdus fischeri

Locally in eastern and south-eastern Africa. In Southern Africa, the eastern Cape (from the lower Buffalo River) to coastal forests of the Transkei, Natal and Zululand.

Primarily a winter visitor to coastal evergreen forest; some birds present during the summer. Breeding recorded during November and December.

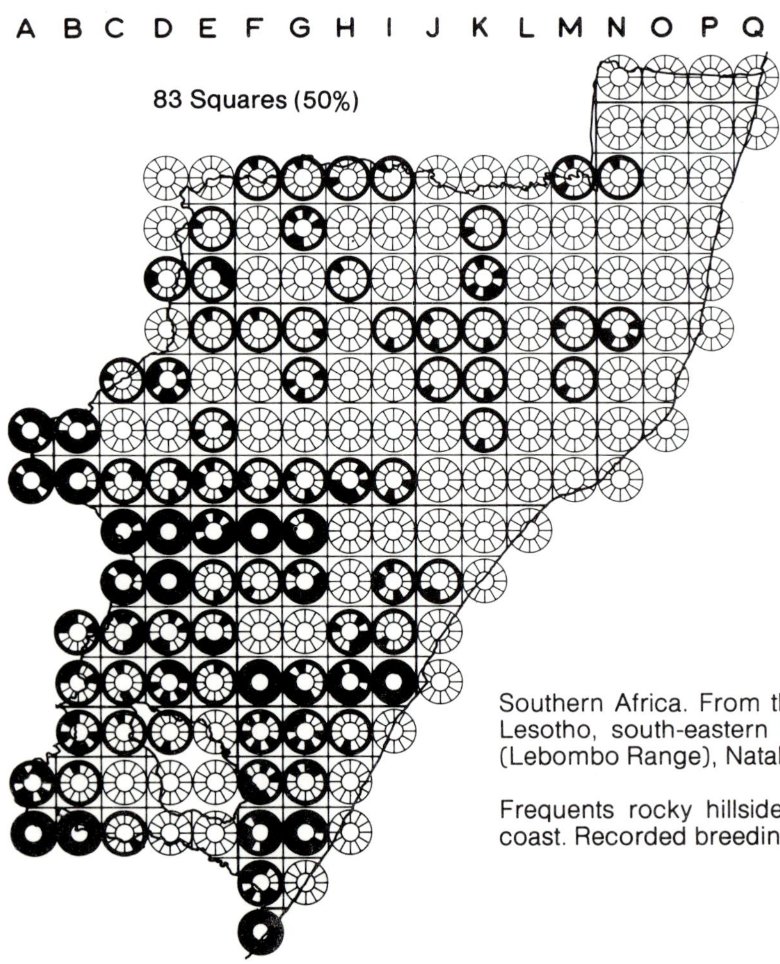

559 Cape Rock Thrush

Monticola rupestris

Southern Africa. From the Cape (south of the Orange River) to O.F.S., Lesotho, south-eastern Botswana, Transvaal, Swaziland, Moçambique (Lebombo Range), Natal and Zululand.

Frequents rocky hillsides at higher altitudes, and gorges towards the coast. Recorded breeding from October to January.

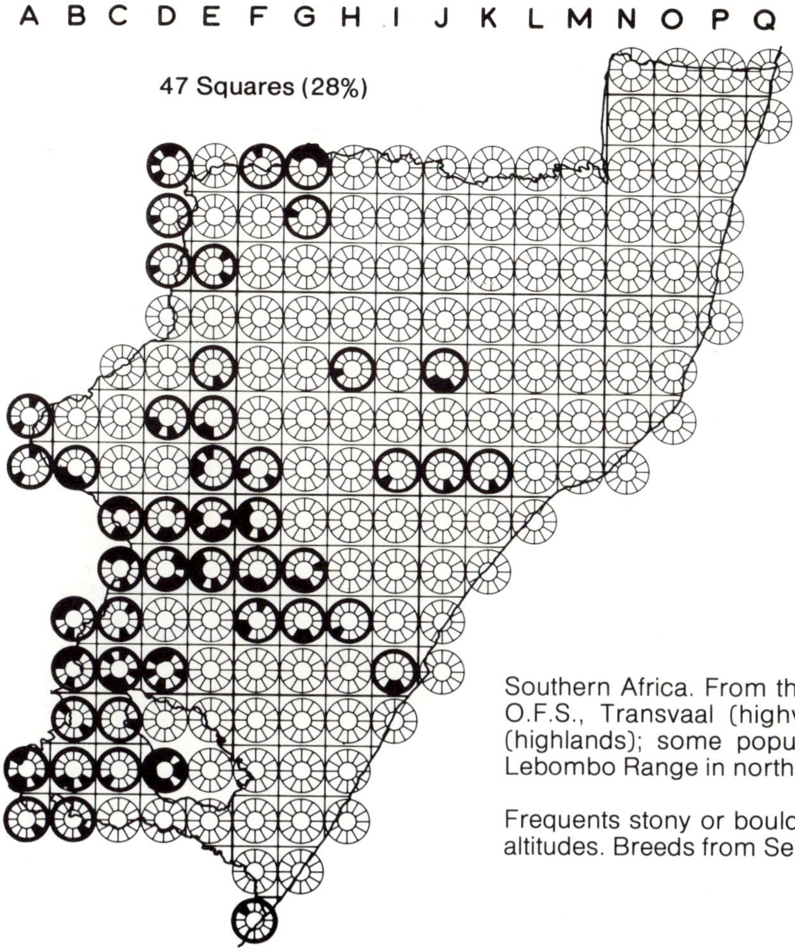

560 Sentinel Rock Thrush

Monticola explorator

Southern Africa. From the mountains of the south-western Cape east to O.F.S., Transvaal (highveld), western Swaziland, Natal and Lesotho (highlands); some populations winter on the Natal coast, and in the Lebombo Range in north-eastern Zululand and Moçambique.

Frequents stony or boulder-strewn hillsides, and open grassland at high altitudes. Breeds from September to December.

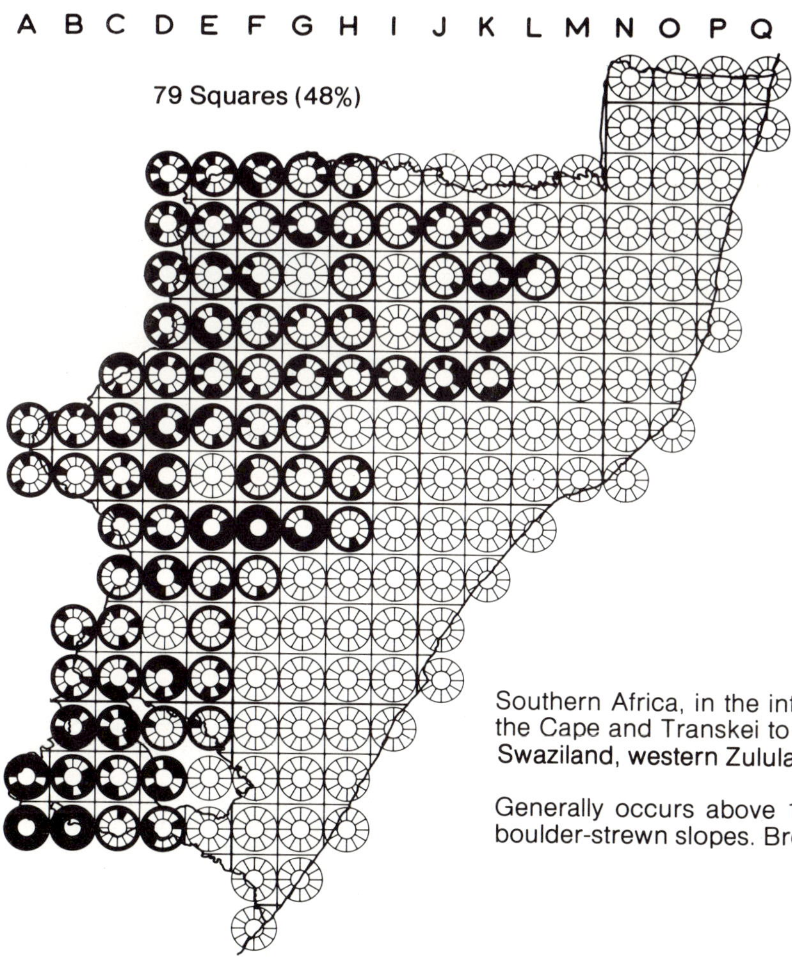

564 Mountain Chat

Oenanthe monticola

Southern Africa, in the interior and west, north to Angola. From S.W.A., the Cape and Transkei to O.F.S., Lesotho, Transvaal (highveld), western Swaziland, western Zululand, and midland and upper Natal.

Generally occurs above 1200 m a.s.l., in areas of open grassland with boulder-strewn slopes. Breeds from August to January.

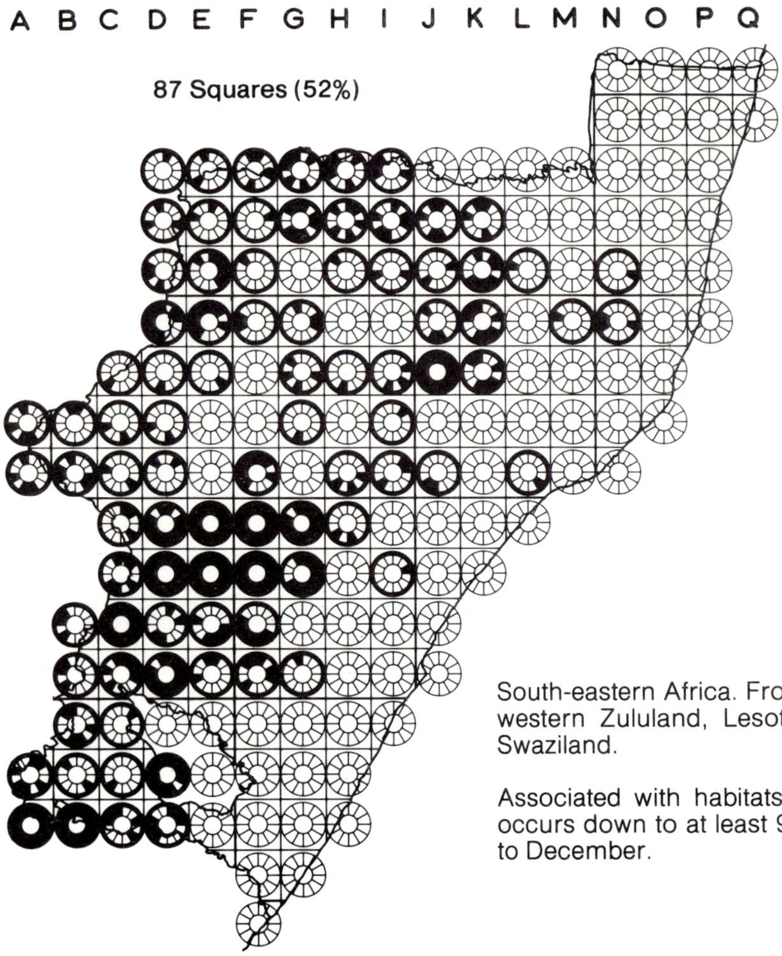

569 Buff-streaked Chat

Oenanthe bifasciata

South-eastern Africa. From the eastern Cape to Transkei, Natal (interior), western Zululand, Lesotho, O.F.S., Transvaal (highveld) and western Swaziland.

Associated with habitats similar to those of Mountain Chat (564), but occurs down to at least 900 m a.s.l. Breeding recorded from September to December.

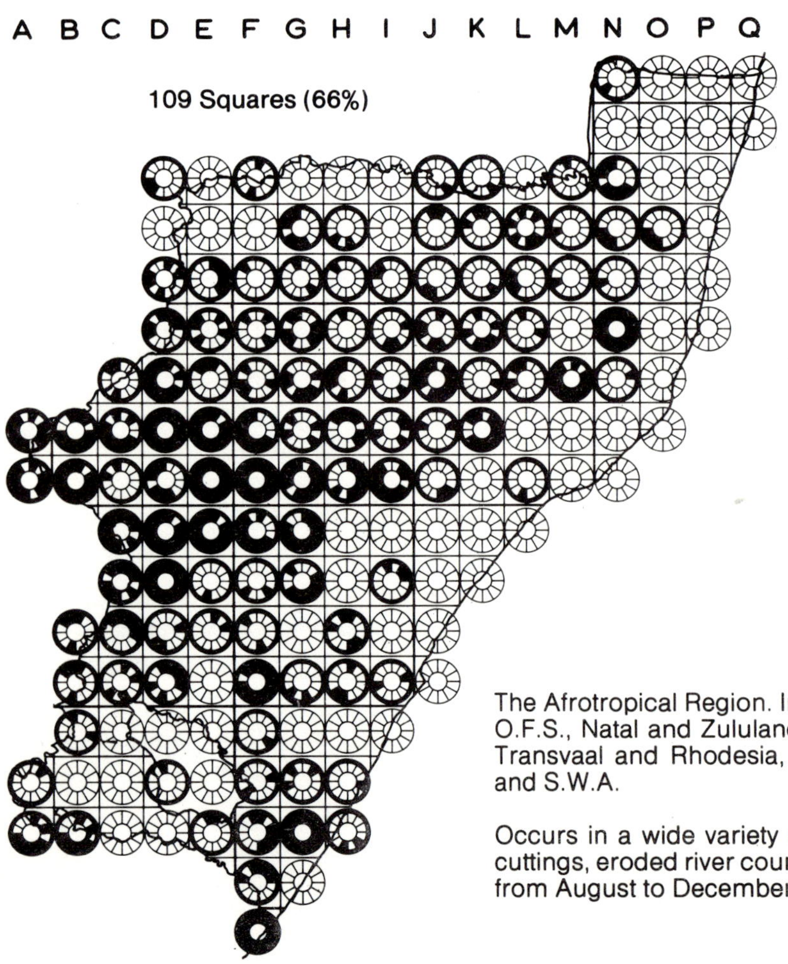

570 Familiar Chat

Cercomela familiaris

The Afrotropical Region. In Southern Africa, the Cape, Transkei, Lesotho, O.F.S., Natal and Zululand, Moçambique (Lebombo Range), Swaziland, Transvaal and Rhodesia, west to south-eastern and western Botswana and S.W.A.

Occurs in a wide variety of open habitats, such as rocky hillsides, road cuttings, eroded river courses and around farmsteads. Breeding recorded from August to December.

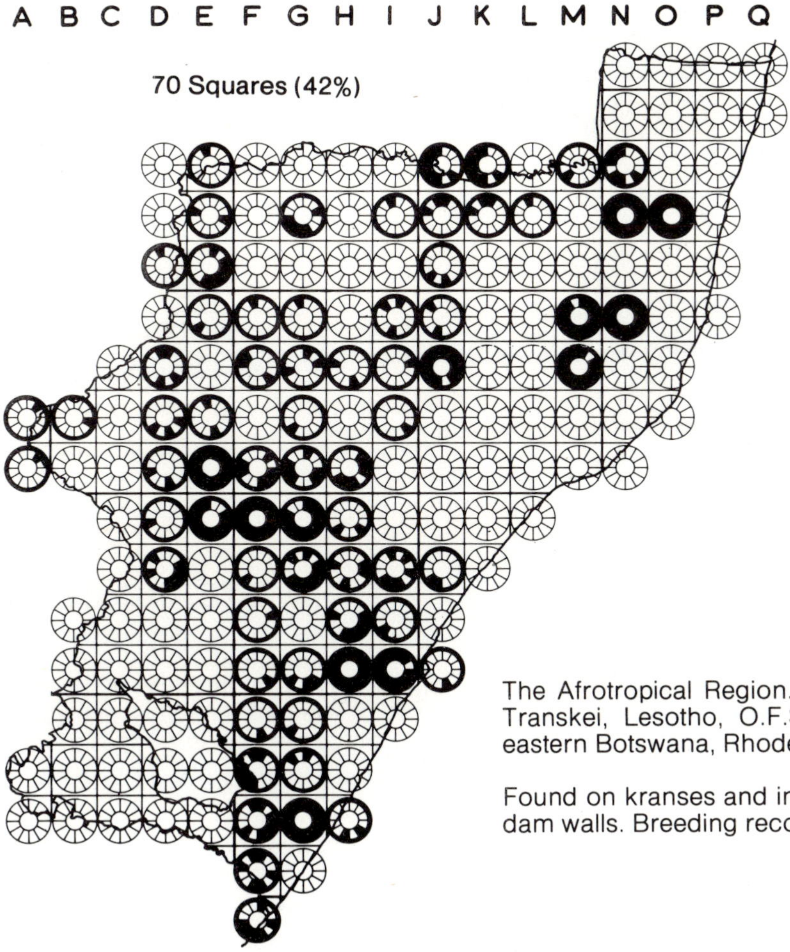

573 Mocking Chat

Thamnolaea cinnamomeiventris

The Afrotropical Region. In Southern Africa, the eastern Cape, coastal Transkei, Lesotho, O.F.S., Natal and Zululand, Swaziland, Transvaal, eastern Botswana, Rhodesia and Moçambique (highland).

Found on kranses and in gorges; also in deep dongas and around large dam walls. Breeding recorded from September to December.

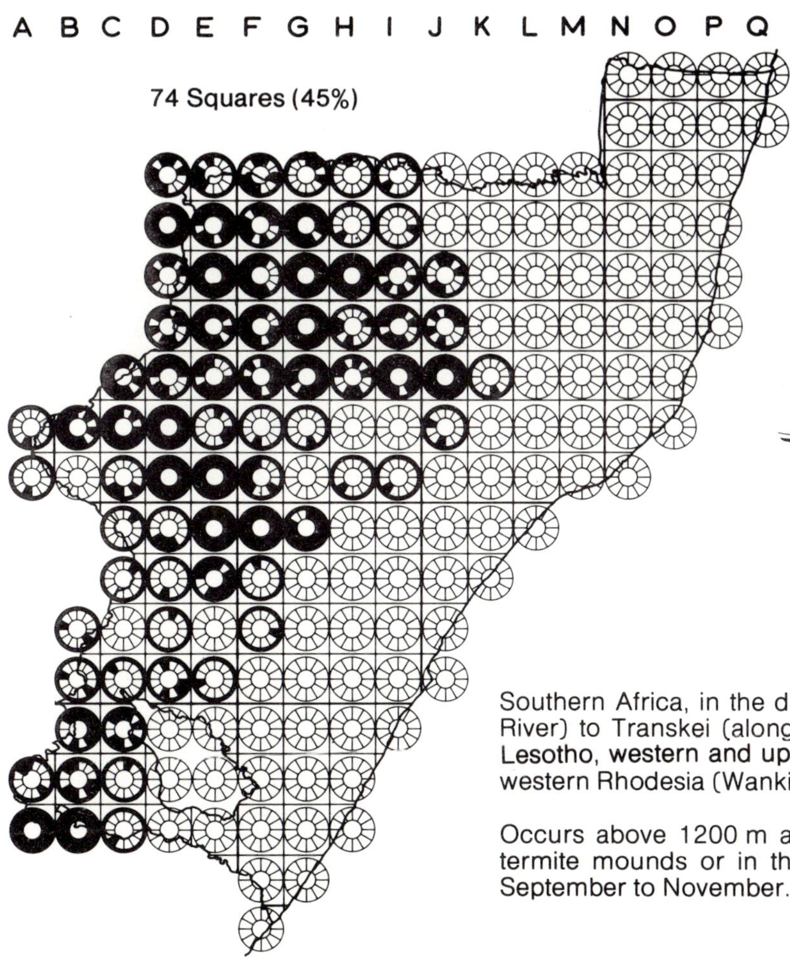

575 Ant-eating Chat

Myrmecocichla formicivora

Southern Africa, in the drier parts. From the Cape (from the Great Fish River) to Transkei (along the base of the Drakensberg Range), O.F.S., Lesotho, western and upper Natal, western Swaziland, Transvaal, north-western Rhodesia (Wankie), Botswana and S.W.A.

Occurs above 1200 m a.s.l. in open areas of short grass usually with termite mounds or in the vicinity of dongas. Breeding recorded from September to November.

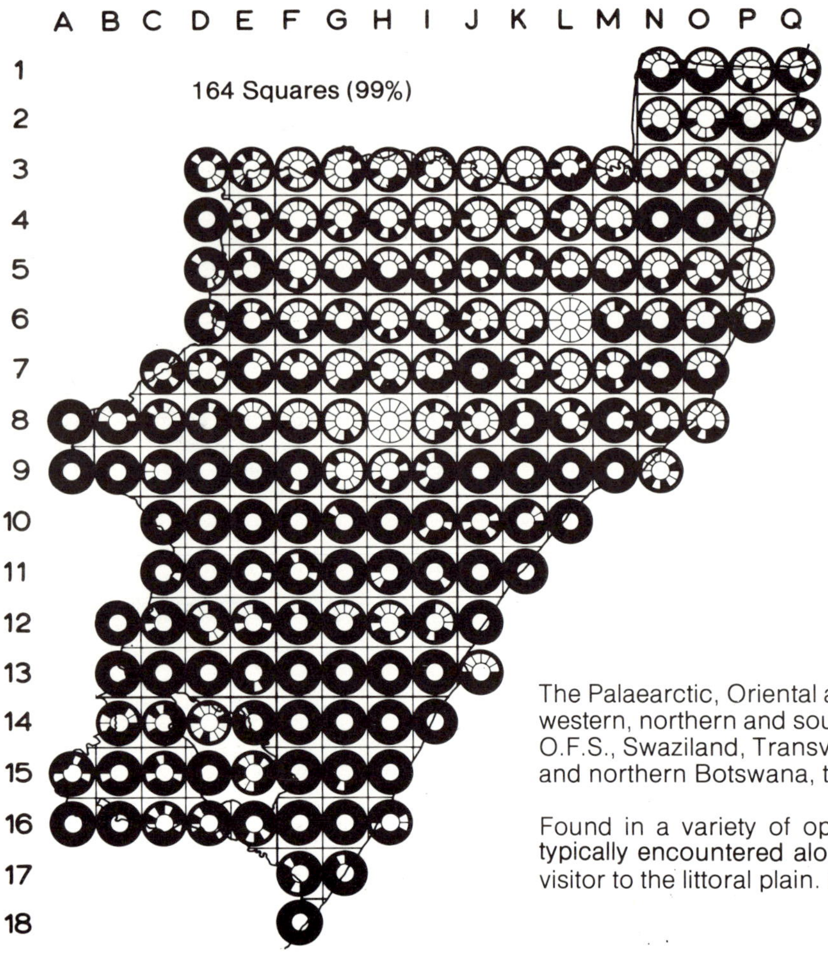

576 Stonechat

Saxicola torquata

The Palaearctic, Oriental and Afrotropical Regions. In Southern Africa, the western, northern and south-western Cape to Natal and Zululand, Lesotho, O.F.S., Swaziland, Transvaal and Moçambique, west to Rhodesia, eastern and northern Botswana, the Caprivi Strip and north-eastern S.W.A.

Found in a variety of open habitats, from grassland to open savanna; typically encountered along roadsides and edges of canefields. A winter visitor to the littoral plain. Recorded breeding from August to December.

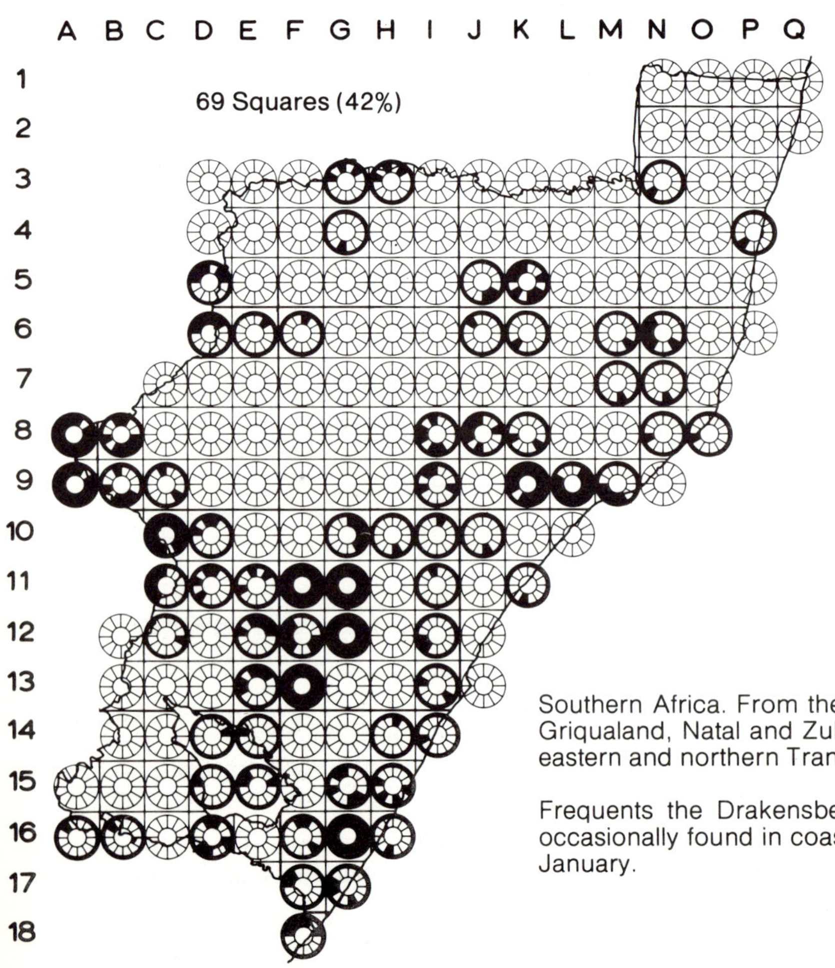

578 Chorister Robin

Cossypha dichroa

Southern Africa. From the southern and eastern Cape to Transkei, East Griqualand, Natal and Zululand, eastern O.F.S., western Swaziland, and eastern and northern Transvaal.

Frequents the Drakensberg and mistbelt evergreen forests; in winter, occasionally found in coastal forest. Recorded breeding from October to January.

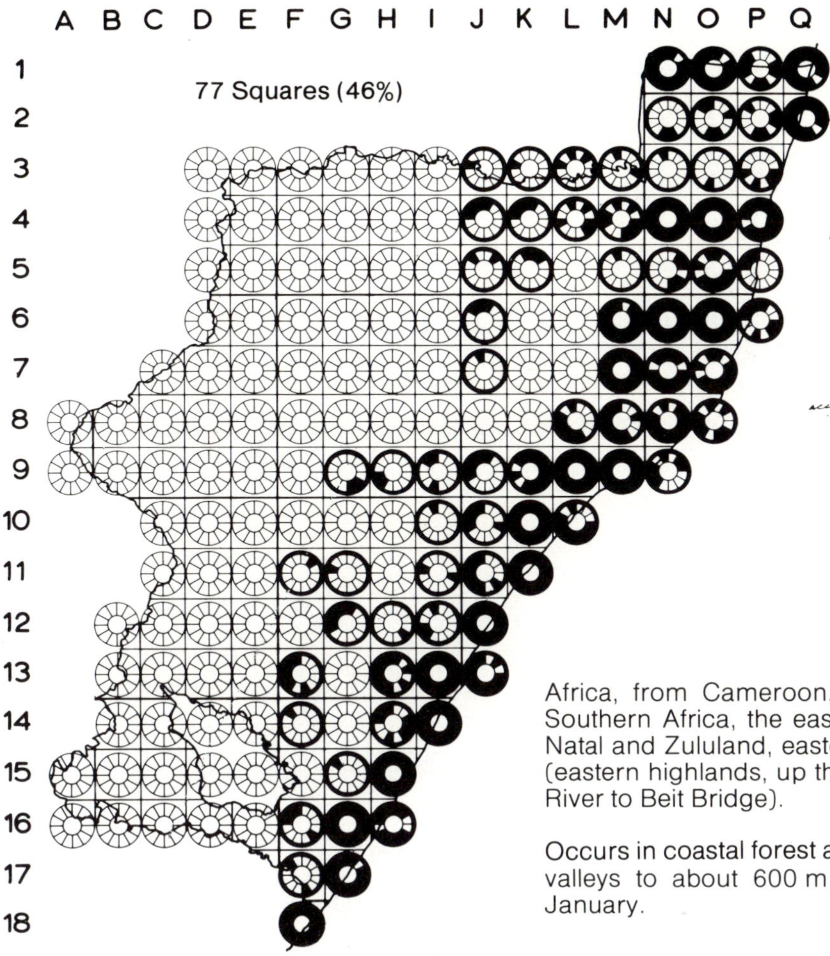

579 Natal Robin

Cossypha natalensis

Africa, from Cameroon, southern Sudan and Kenya to the Cape. In Southern Africa, the eastern Cape (from the lower Great Fish River) to Natal and Zululand, eastern Swaziland, eastern Transvaal, and Rhodesia (eastern highlands, up the Zambesi River to Chirundu, and the Limpopo River to Beit Bridge).

Occurs in coastal forest and scrub, riverine woodland and thickly-wooded valleys to about 600 m a.s.l. Recorded breeding from September to January.

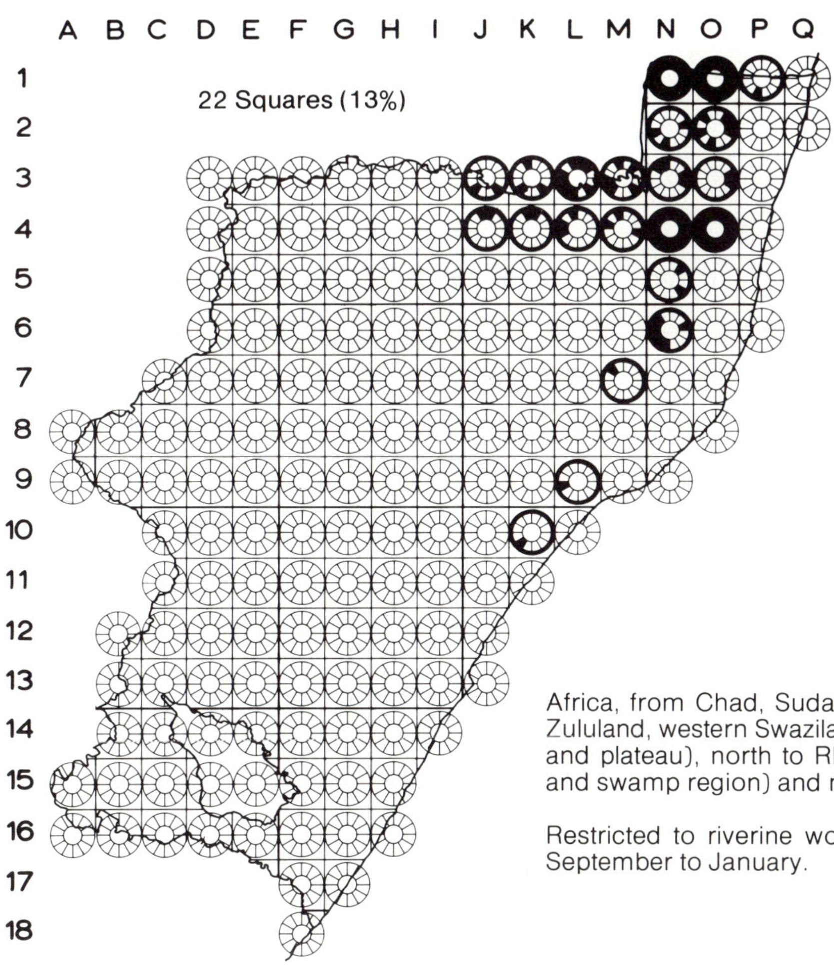

580 Heuglin's Robin

Cossypha heuglini

Africa, from Chad, Sudan and Ethiopia to Zululand. In Southern Africa, Zululand, western Swaziland, Moçambique and Transvaal (eastern lowveld and plateau), north to Rhodesia and west to Botswana (Limpopo River and swamp region) and north-eastern S.W.A.

Restricted to riverine woodland and thickets. Recorded breeding from September to January.

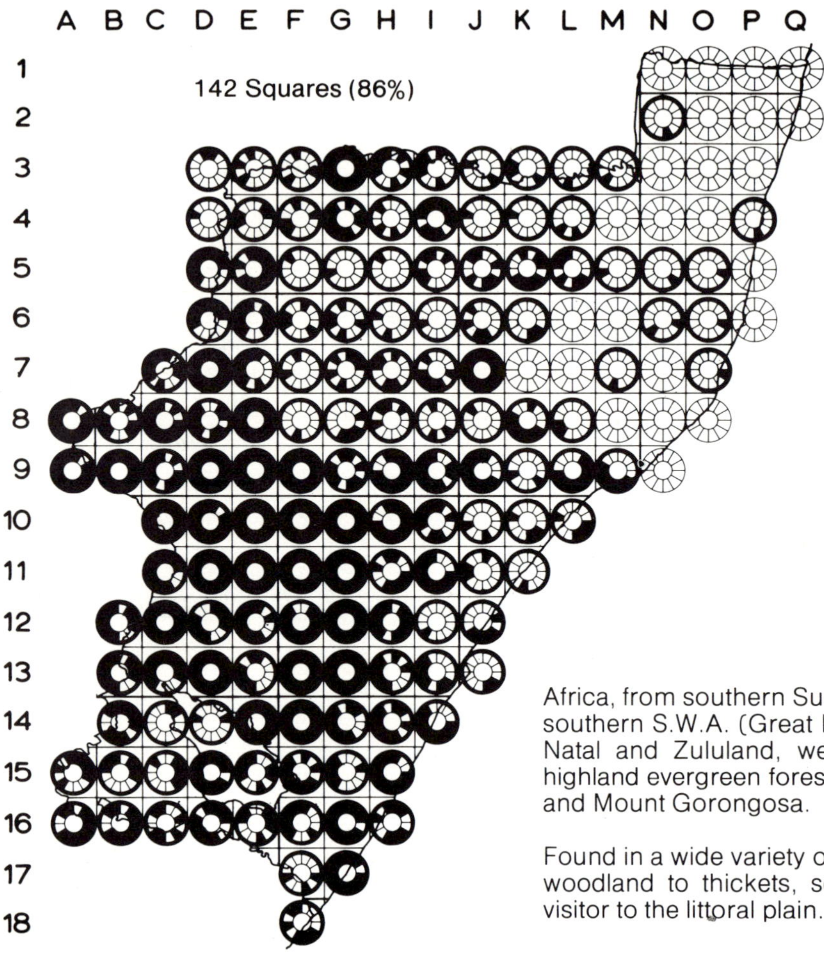

581 Cape Robin

Cossypha caffra

Africa, from southern Sudan and Uganda to the Cape. In Southern Africa, southern S.W.A. (Great Fish River), the Cape, Transkei, Lesotho, O.F.S., Natal and Zululand, western Swaziland, and Transvaal; again in the highland evergreen forest of eastern Rhodesia and adjacent Moçambique, and Mount Gorongosa.

Found in a wide variety of habitats, from evergreen forest, plantations and woodland to thickets, scrub vegetation and gardens. A scarce winter visitor to the littoral plain. Breeding recorded from August to December.

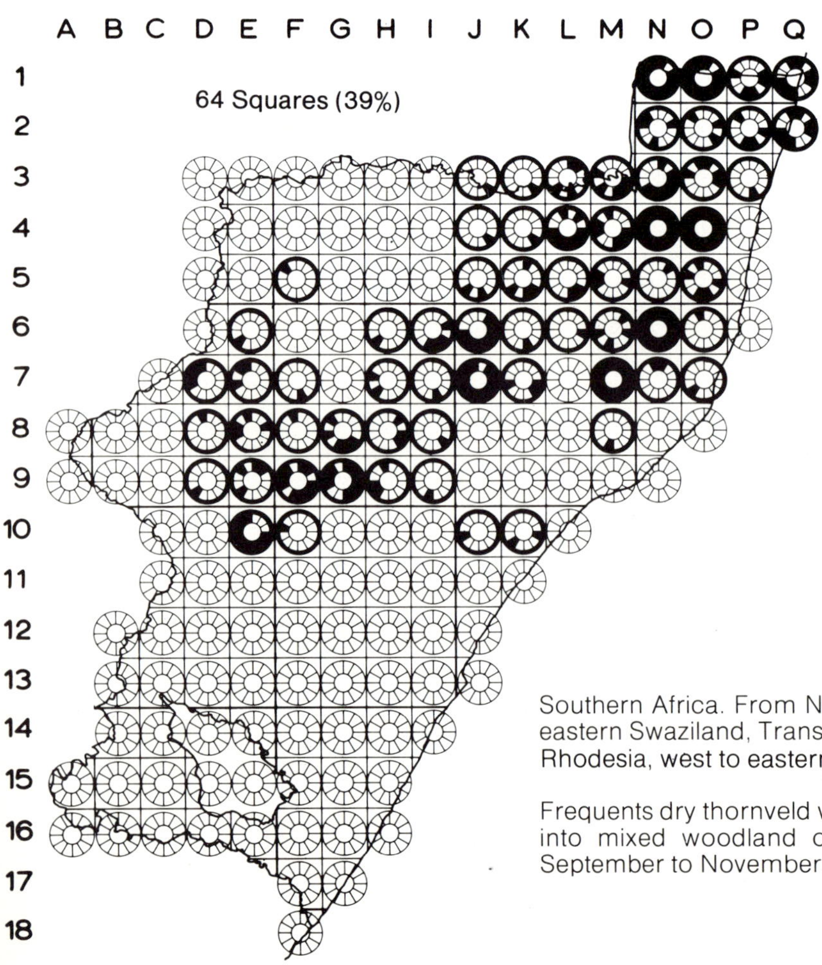

582 White-throated Robin

Cossypha humeralis

Southern Africa. From Natal and Zululand (south to the Tugela Basin) to eastern Swaziland, Transvaal, Moçambique (littoral to the Save River) and Rhodesia, west to eastern Botswana and adjacent northern Cape.

Frequents dry thornveld with a preference for the denser patches; extends into mixed woodland on the littoral plain. Recorded breeding from September to November.

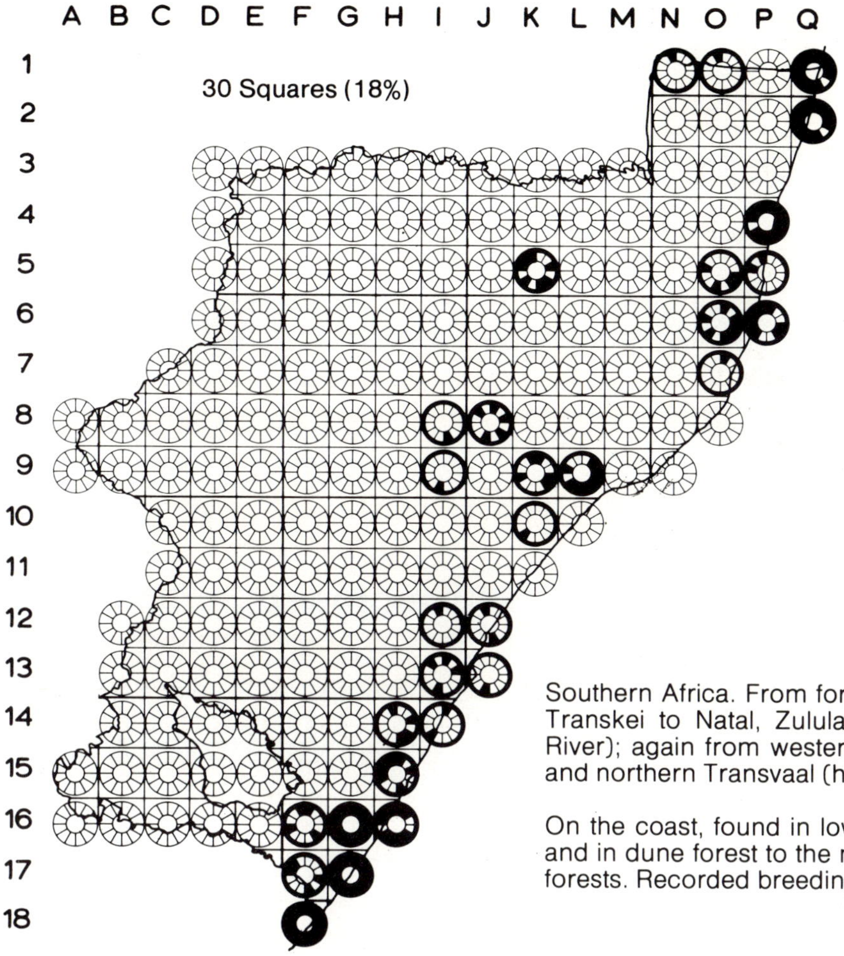

584 Brown Robin

Erythropygia signata

Southern Africa. From forests of the eastern Cape and coastal forests of Transkei to Natal, Zululand and Moçambique (north to the Limpopo River); again from western Zululand to western Swaziland, and eastern and northern Transvaal (highland forest).

On the coast, found in lowland evergreen forest south of Richards Bay, and in dune forest to the north; inland in the northern mistbelt evergreen forests. Recorded breeding from September to November.

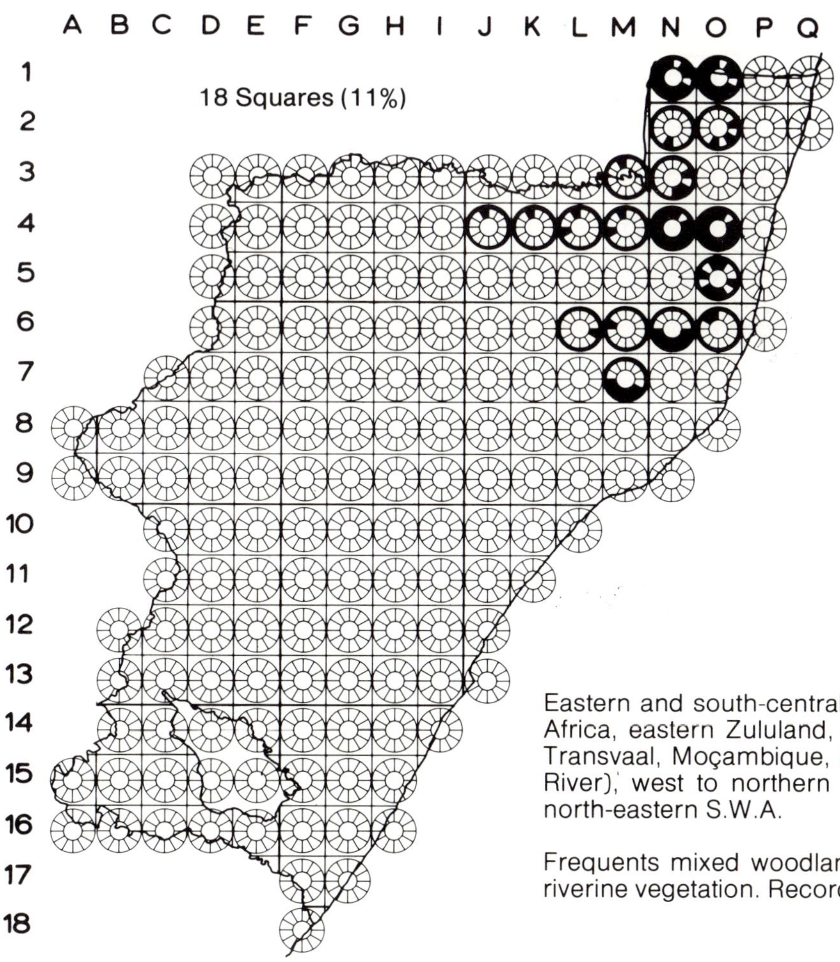

585 Bearded Robin

Erythropygia quadrivirgata

Eastern and south-central Africa, from Somalia to Zululand. In Southern Africa, eastern Zululand, eastern Swaziland, south-eastern and eastern Transvaal, Moçambique, eastern and north-western Rhodesia (Zambesi River), west to northern Botswana (Ngamiland), the Caprivi Strip and north-eastern S.W.A.

Frequents mixed woodland on the littoral plain, extending westwards in riverine vegetation. Recorded breeding from September to November.

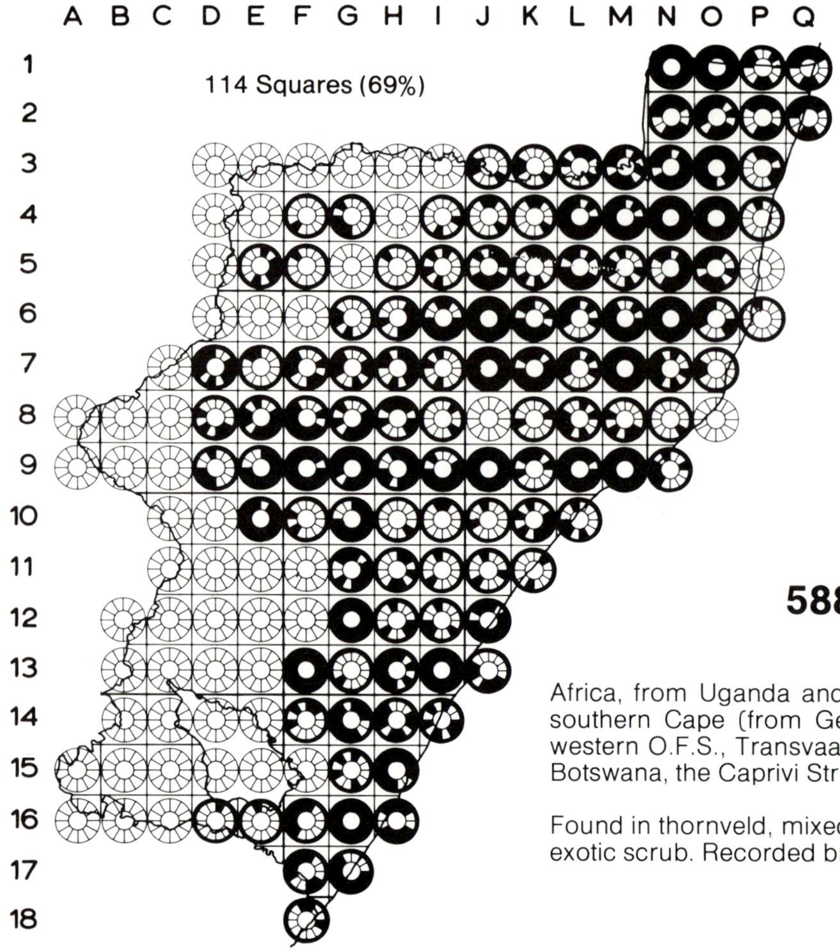

588 White-browed Scrub Robin

Erythropygia leucophrys

Africa, from Uganda and Ethiopia to the Cape. In Southern Africa, the southern Cape (from George) east to Natal and Zululand, Swaziland, western O.F.S., Transvaal and Moçambique, west to Rhodesia, northern Botswana, the Caprivi Strip and central and northern S.W.A.

Found in thornveld, mixed woodland and wooded valleys; occasionally in exotic scrub. Recorded breeding from September to December.

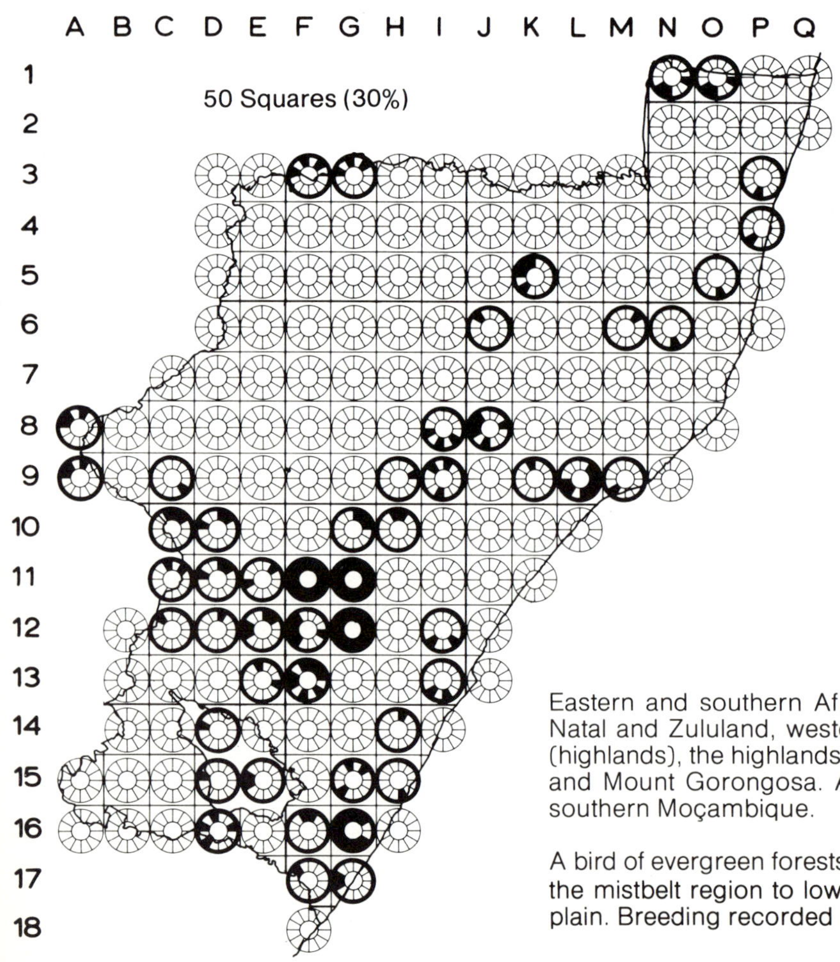

589 Starred Robin

Pogonocichla stellata

Eastern and southern Africa. In Southern Africa, the southern Cape to Natal and Zululand, western Swaziland, eastern and northern Transvaal (highlands), the highlands of eastern Rhodesia and adjacent Moçambique, and Mount Gorongosa. A non-breeding visitor to the coastal forests of southern Moçambique.

A bird of evergreen forests, found from the southern Drakensberg through the mistbelt region to lowland areas; a scarce winter visitor to the littoral plain. Breeding recorded from October to December.

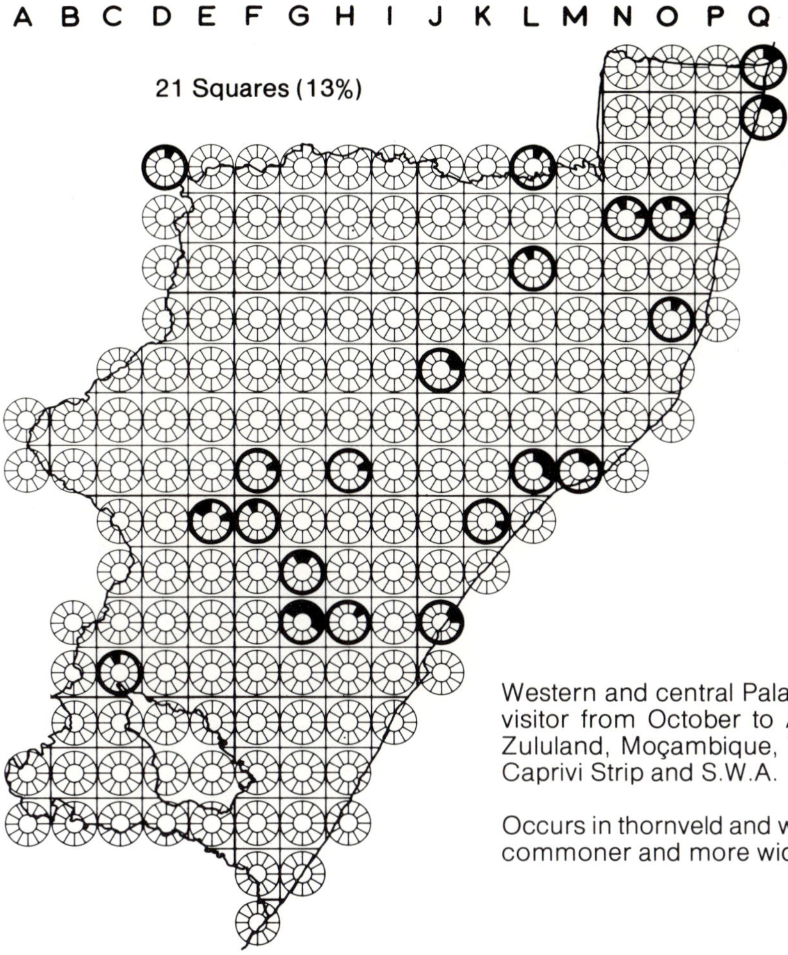

595 Garden Warbler

Sylvia borin

Western and central Palaearctic, wintering in Africa. In Southern Africa, a visitor from October to April, to the eastern Cape, Transkei, Natal and Zululand, Moçambique, Transvaal and Rhodesia, west to Botswana, the Caprivi Strip and S.W.A.

Occurs in thornveld and woodland thickets. This retiring species is probably commoner and more widely distributed than records would indicate.

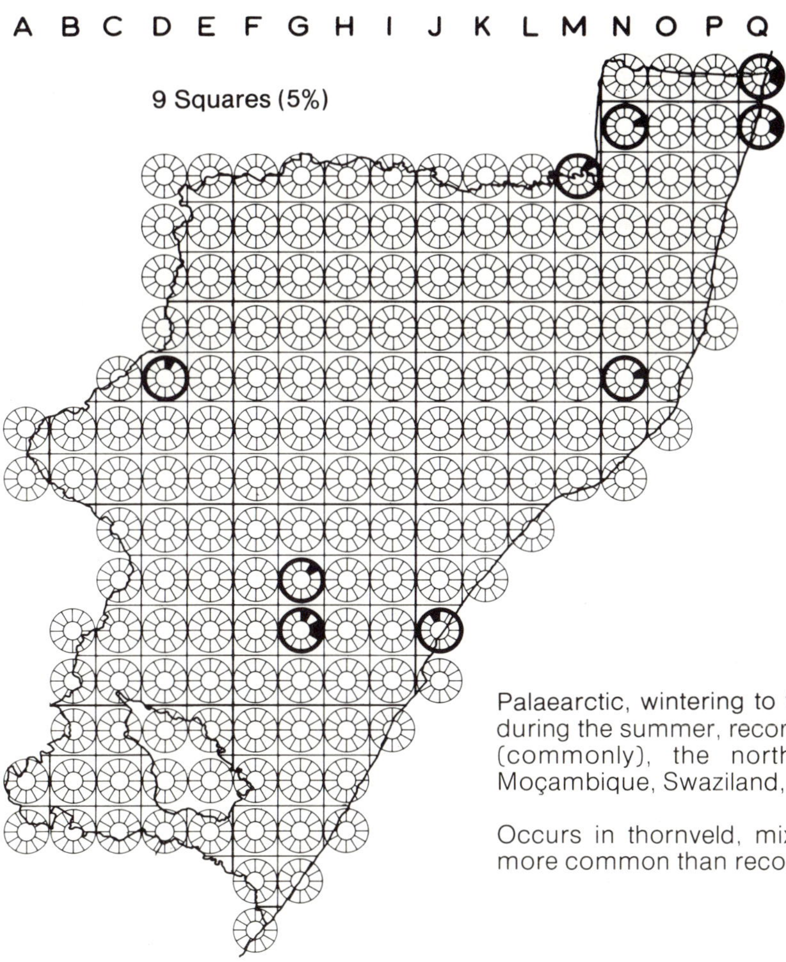

596 Icterine Warbler

Hippolais icterina

Palaearctic, wintering to the south in Africa. In Southern Africa, a visitor during the summer, recorded from central and northern S.W.A., Botswana (commonly), the northern Cape, Rhodesia, Transvaal, southern Moçambique, Swaziland, Natal (interior) and Zululand.

Occurs in thornveld, mixed woodland and exotic vegetation; probably more common than records would indicate.

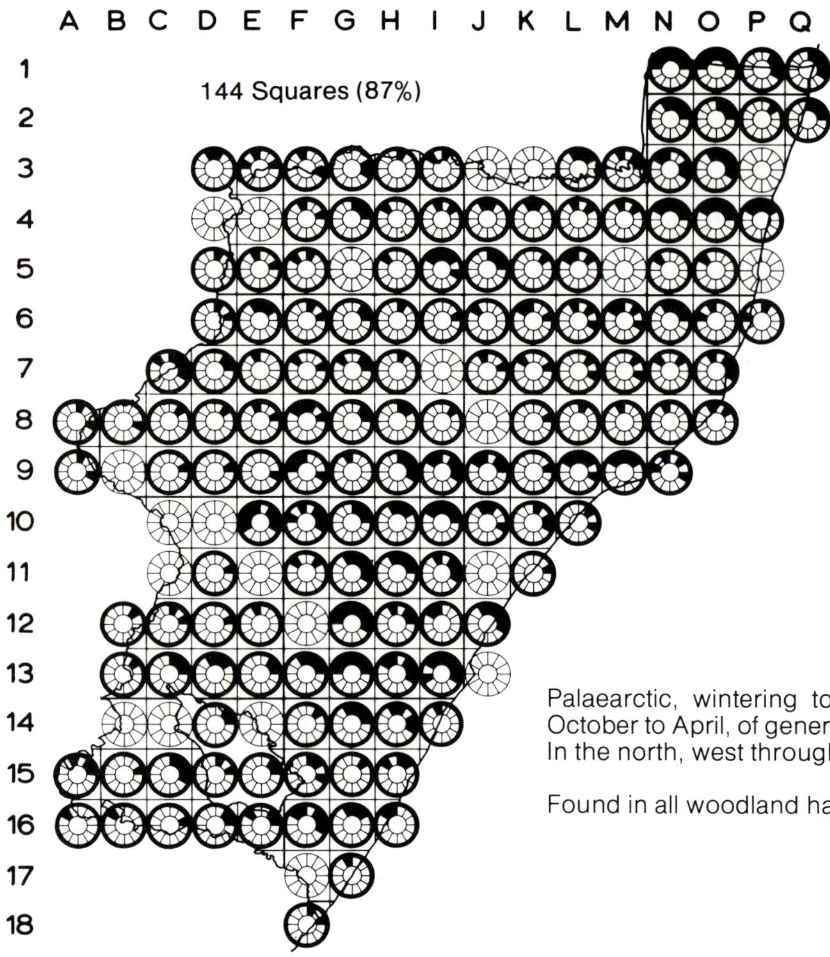

599 Willow Warbler

Phylloscopus trochilus

Palaearctic, wintering to the south. In Southern Africa, a visitor from October to April, of general occurrence to the eastern half of the subregion. In the north, west through Botswana, the Caprivi Strip and S.W.A.

Found in all woodland habitats, including exotic trees and gardens.

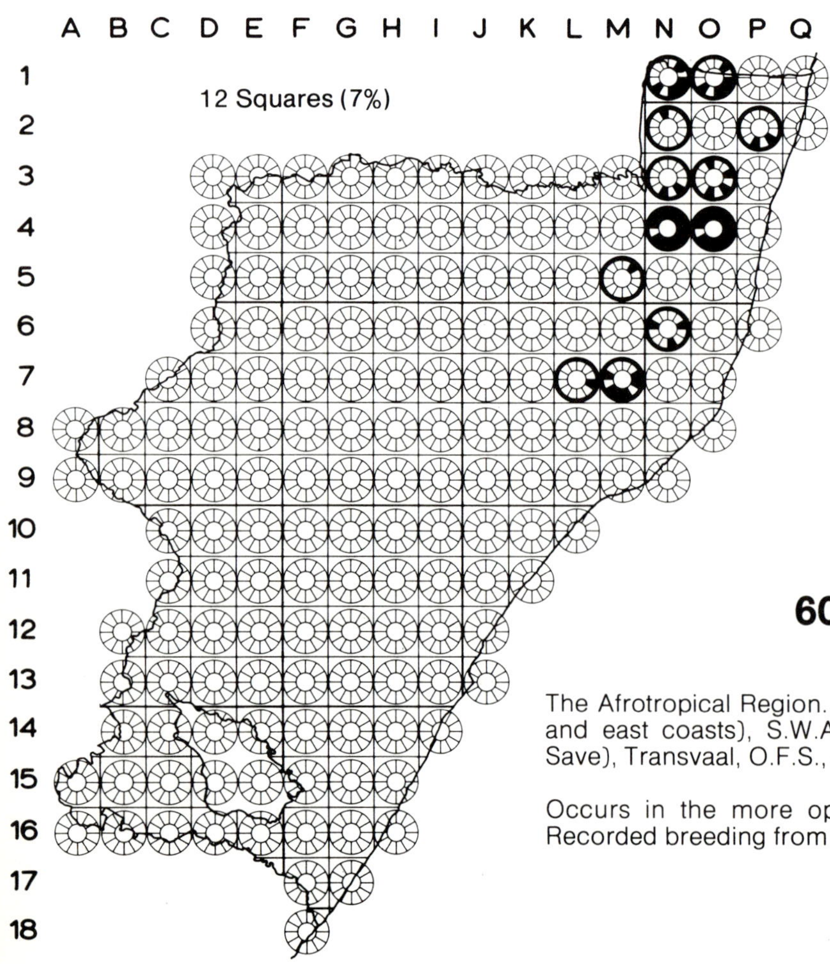

600 Yellow-bellied Eremomela

Eremomela icteropygialis

The Afrotropical Region. In Southern Africa, the Cape (except the south and east coasts), S.W.A., Botswana, Rhodesia, Moçambique (Sul do Save), Transvaal, O.F.S., Swaziland and eastern Zululand.

Occurs in the more open parts of mixed woodland and thornveld. Recorded breeding from October to December.

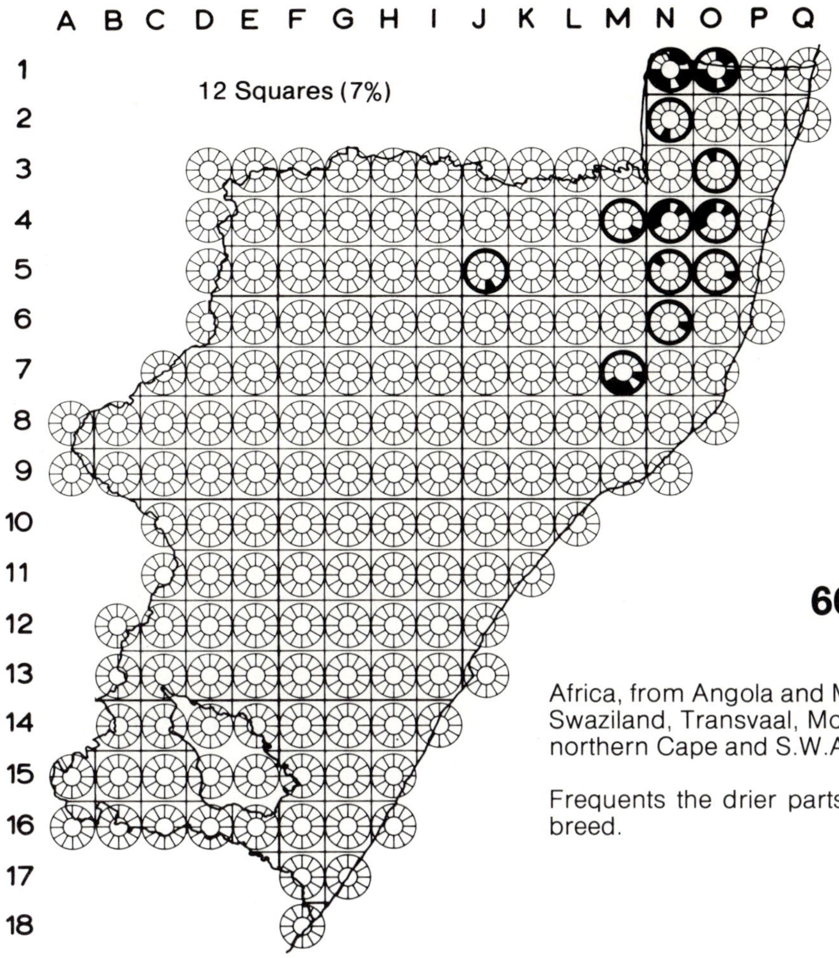

601 Burnt-necked Eremomela

Eremomela usticollis

Africa, from Angola and Malawi to Zululand. In Southern Africa, Zululand, Swaziland, Transvaal, Moçambique and Rhodesia, west to Botswana, the northern Cape and S.W.A.

Frequents the drier parts of open thornveld and mixed woodland. May breed.

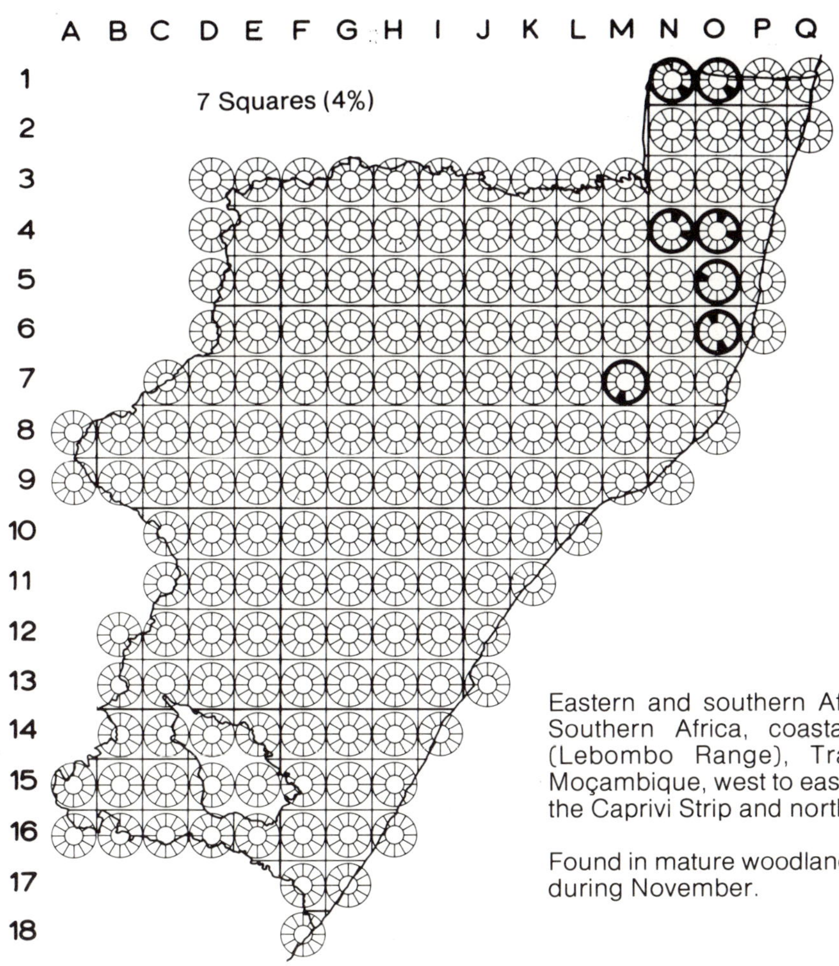

602 Greencap Eremomela

Eremomela scotops

Eastern and southern Africa, from Kenya and Zaïre south to Natal. In Southern Africa, coastal Natal (rare), eastern Zululand, Swaziland (Lebombo Range), Transvaal (lowveld), Rhodesia and southern Moçambique, west to eastern and northern Botswana, western Transvaal, the Caprivi Strip and north-eastern S.W.A.

Found in mature woodland and occasionally thornveld. Recorded breeding during November.

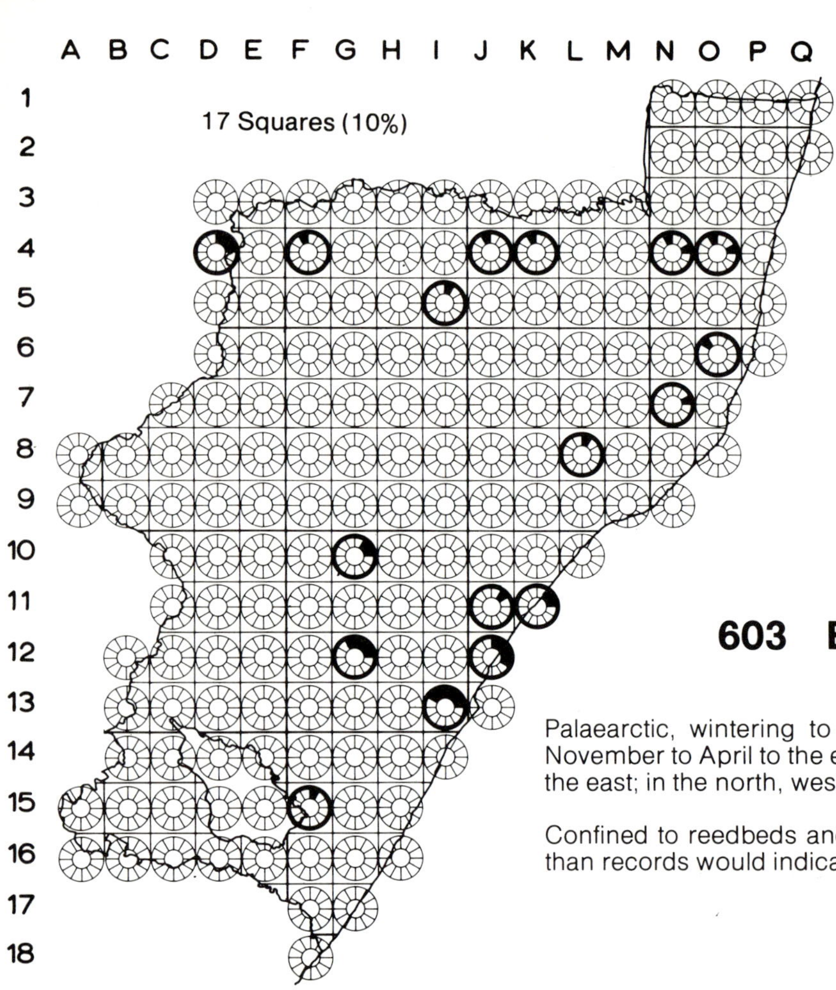

603 European Great Reed Warbler

Acrocephalus arundinaceus

Palaearctic, wintering to the south. In Southern Africa, a visitor from November to April to the eastern Cape, Natal and Zululand, northwards in the east; in the north, west to Botswana, the northern Cape and S.W.A.

Confined to reedbeds and adjacent thickets. Probably more widespread than records would indicate.

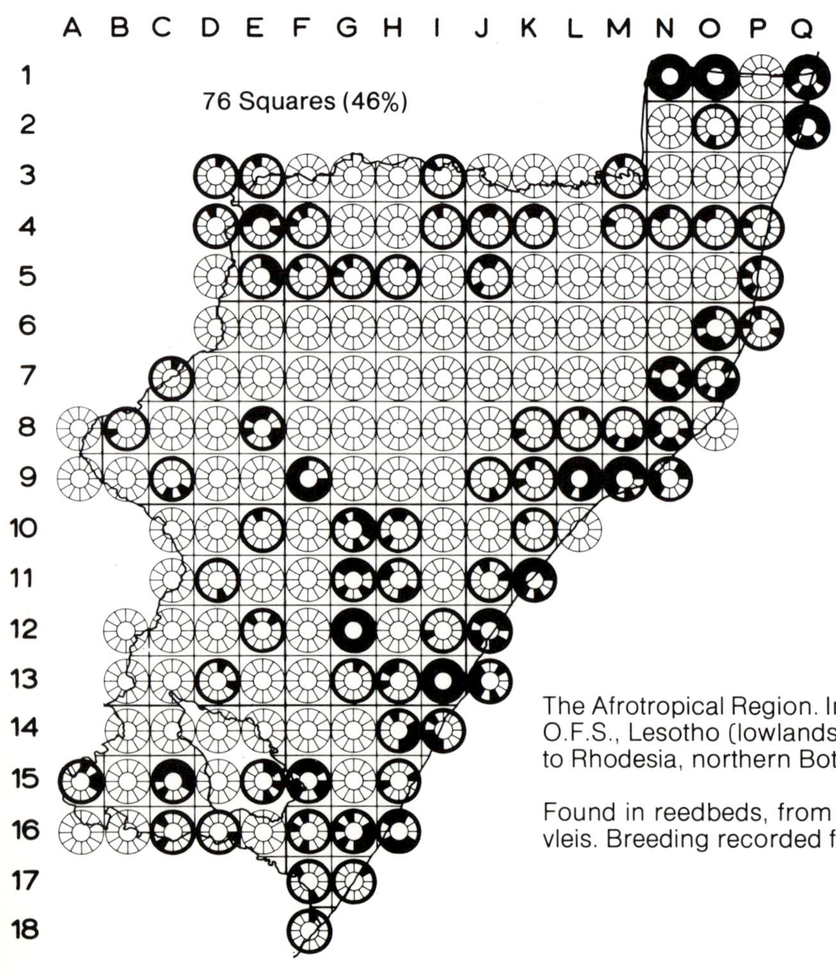

604 Cape Reed Warbler

Acrocephalus gracilirostris

The Afrotropical Region. In Southern Africa, the Cape, Natal and Zululand, O.F.S., Lesotho (lowlands), Swaziland, Transvaal and Moçambique, west to Rhodesia, northern Botswana, the Caprivi Strip and northern S.W.A.

Found in reedbeds, from estuaries and coastal lakes to pans, dams and vleis. Breeding recorded from September to January.

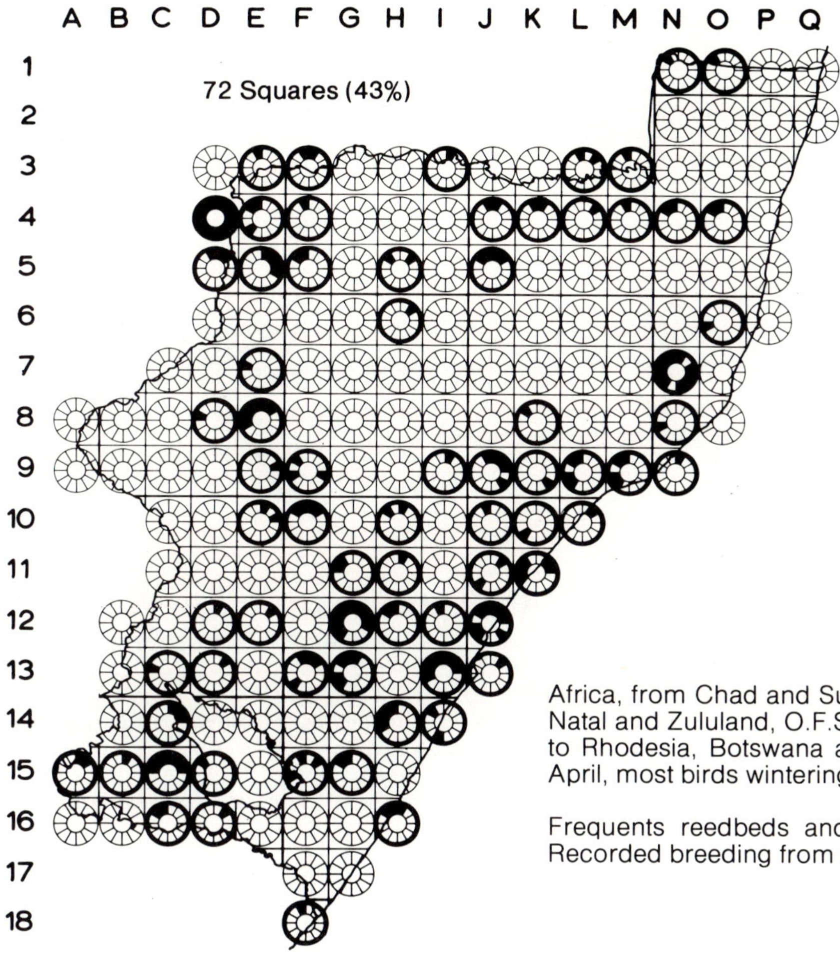

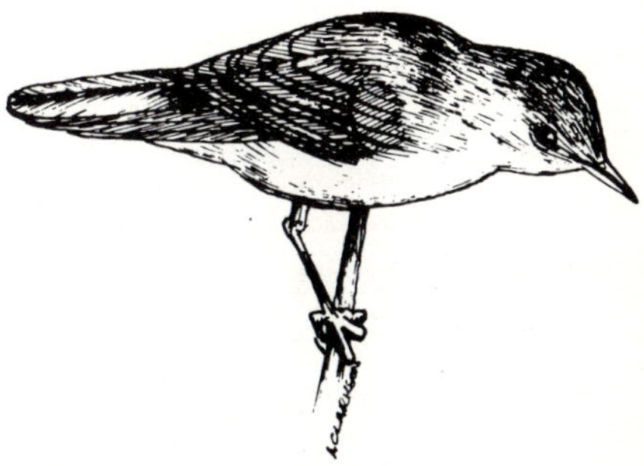

606 African Marsh Warbler

Acrocephalus baeticatus

Africa, from Chad and Sudan to the Cape. In Southern Africa, the Cape, Natal and Zululand, O.F.S., Swaziland, Transvaal and Moçambique, west to Rhodesia, Botswana and S.W.A. Present mainly from September to April, most birds wintering in east-coastal and south-central Africa.

Frequents reedbeds and scrub vegetation in the proximity of water. Recorded breeding from November to January.

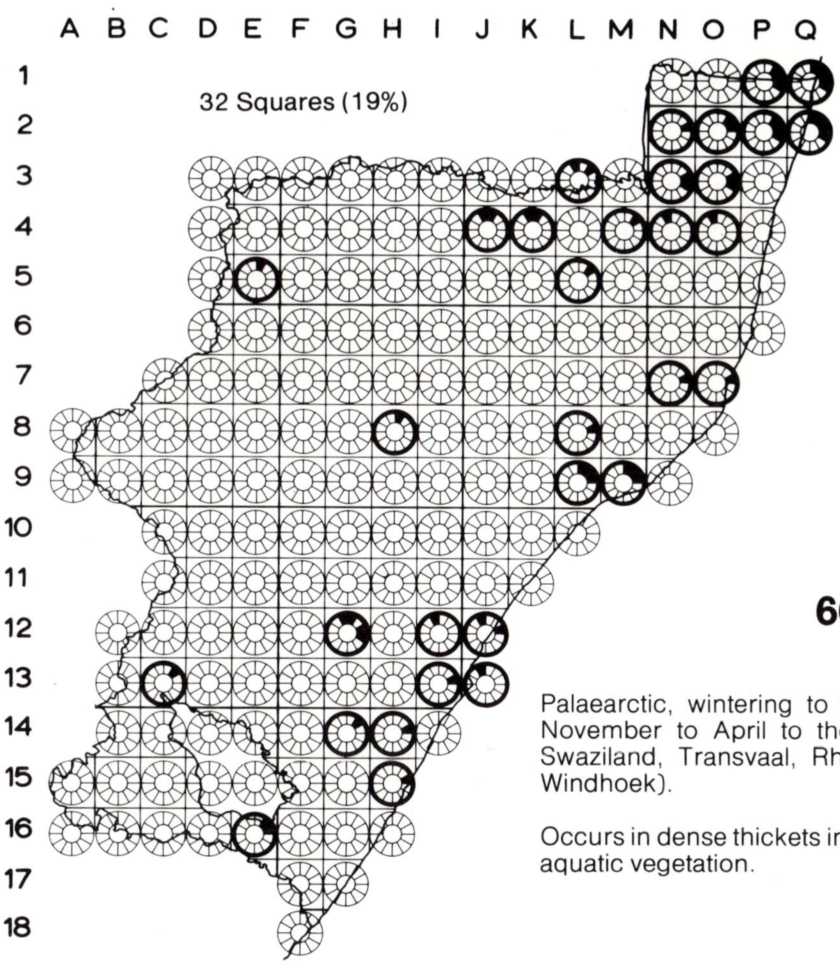

607 European Marsh Warbler

Acrocephalus palustris

Palaearctic, wintering to the south. In Southern Africa, a visitor from November to April to the eastern Cape, Natal and Zululand, O.F.S., Swaziland, Transvaal, Rhodesia and Moçambique; also S.W.A. (near Windhoek).

Occurs in dense thickets in thornveld and mixed woodland; also in tangled aquatic vegetation.

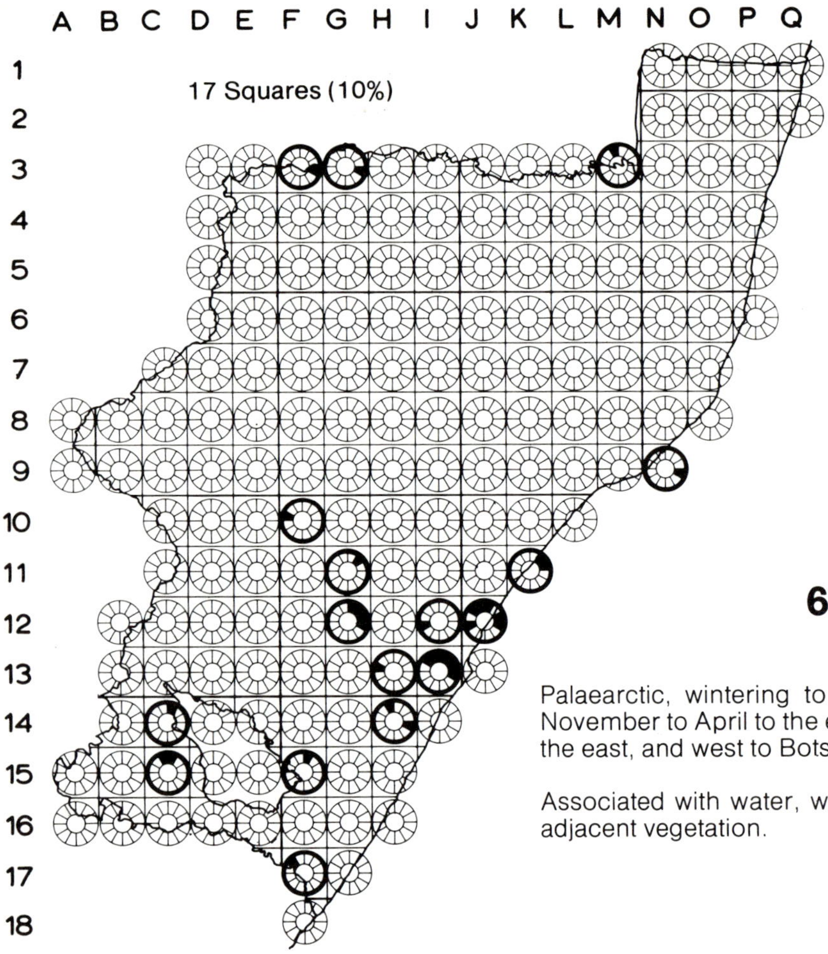

608 European Sedge Warbler

Acrocephalus schoenobaenus

Palaearctic, wintering to the south. In Southern Africa, a visitor from November to April to the eastern Cape, Natal and Zululand, northwards in the east, and west to Botswana and S.W.A.

Associated with water, where it inhabits the fringes of reedbeds and the adjacent vegetation.

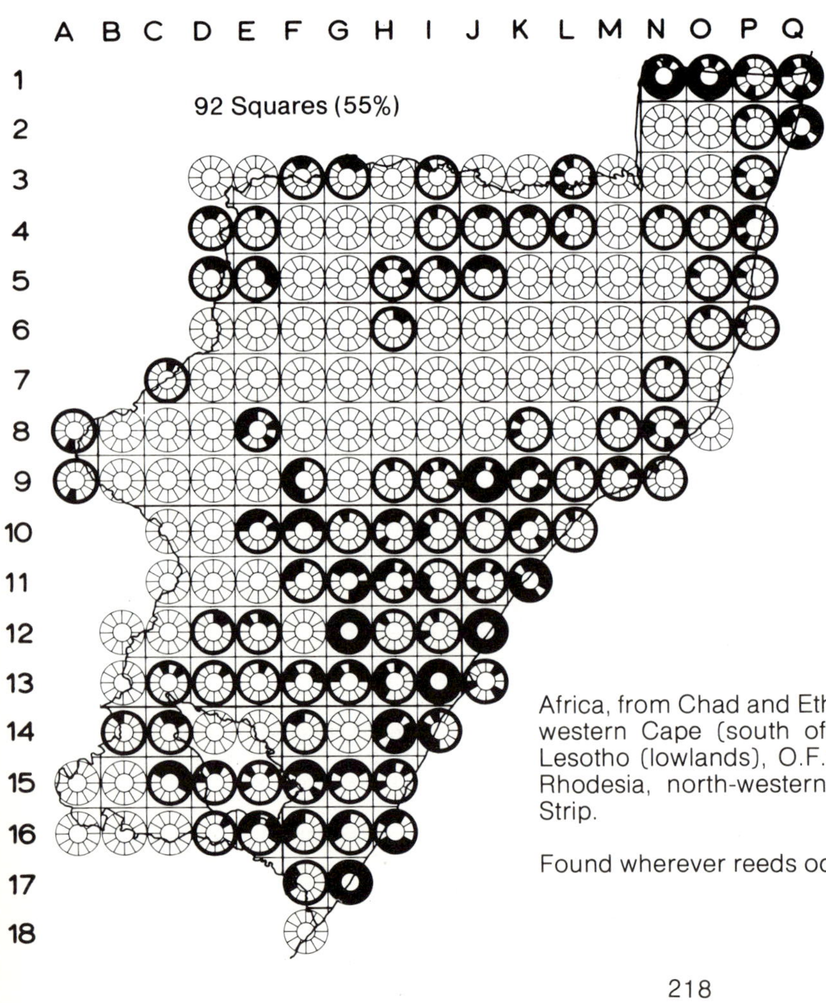

609 African Sedge Warbler

Bradypterus baboecala

Africa, from Chad and Ethiopia to the Cape. In Southern Africa, the south-western Cape (south of the Berg River) east to Natal and Zululand, Lesotho (lowlands), O.F.S., Swaziland, Transvaal, eastern and northern Rhodesia, north-western Botswana (swamp regions) and the Caprivi Strip.

Found wherever reeds occur. Breeds during the summer.

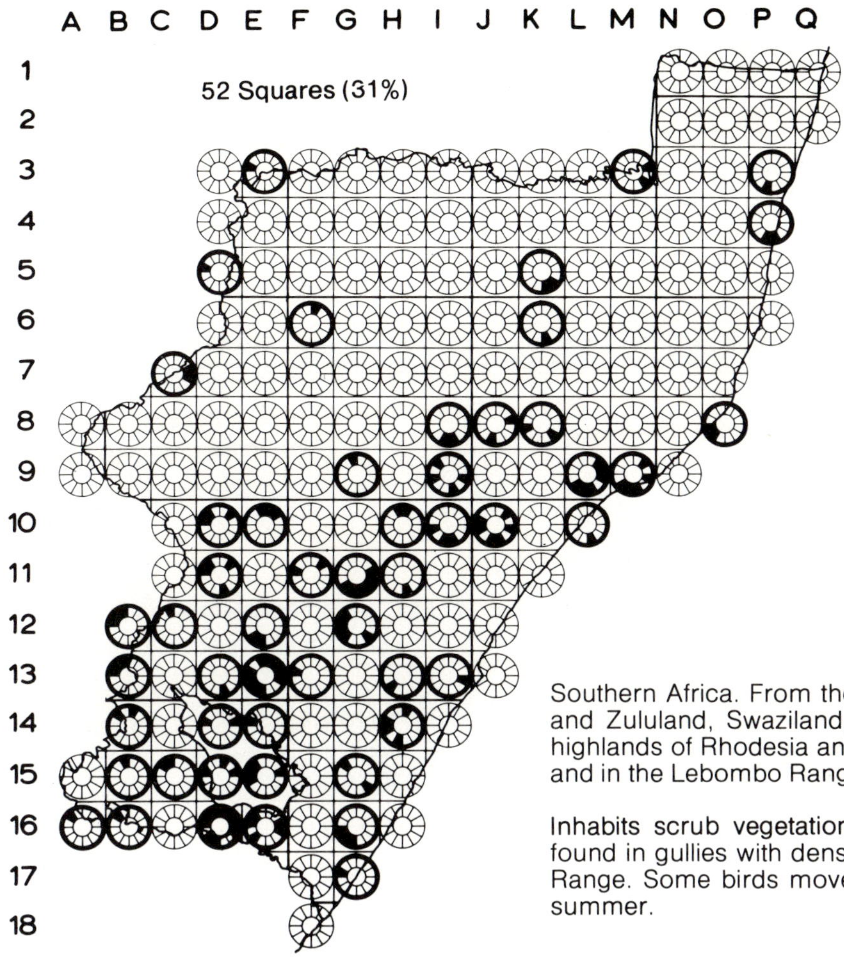

610 Scrub Warbler

Bradypterus barratti

Southern Africa. From the eastern Cape (Great Fish River) east to Natal and Zululand, Swaziland, eastern and northern Transvaal, the eastern highlands of Rhodesia and adjacent Moçambique, on Mount Gorongosa and in the Lebombo Range.

Inhabits scrub vegetation on the periphery of evergreen forests; also found in gullies with dense vegetation in the foothills of the Drakensberg Range. Some birds move to the coast in the winter. Breeds during the summer.

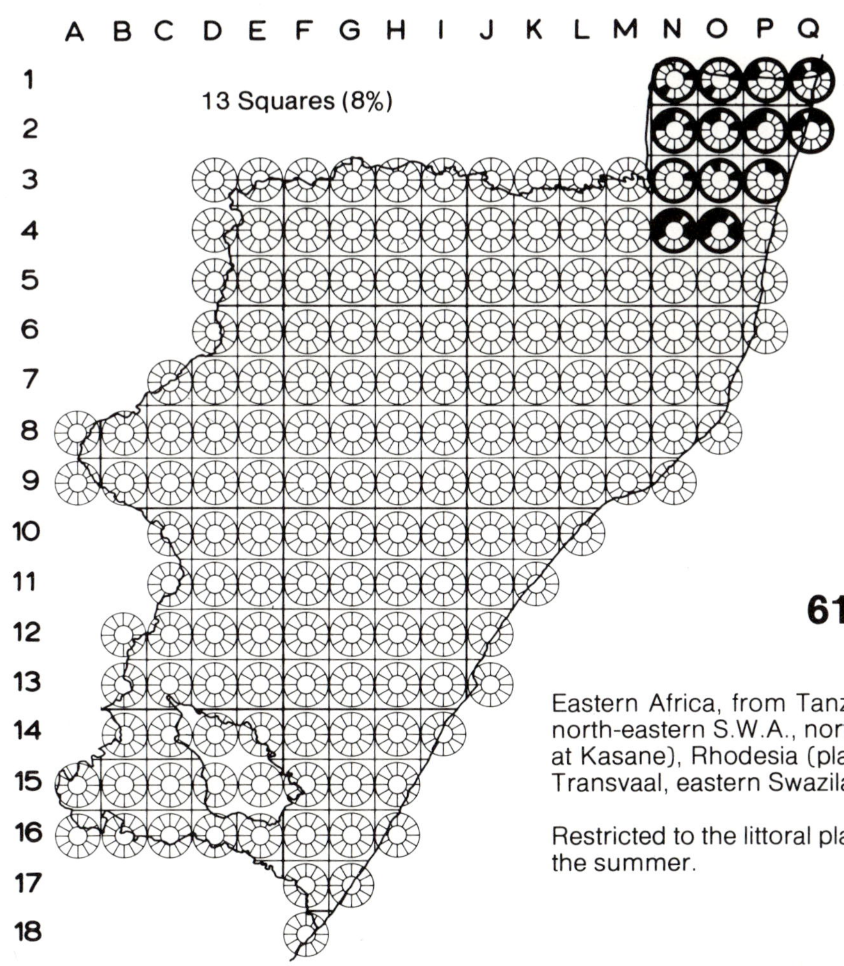

615 Stierling's Barred Warbler

Camaroptera stierlingi

Eastern Africa, from Tanzania to Zululand. In Southern Africa, locally in north-eastern S.W.A., northern Botswana (north-east of Francistown, and at Kasane), Rhodesia (plateau), Moçambique, north-eastern and eastern Transvaal, eastern Swaziland and northern Zululand.

Restricted to the littoral plain, frequenting mixed woodland. Breeds during the summer.

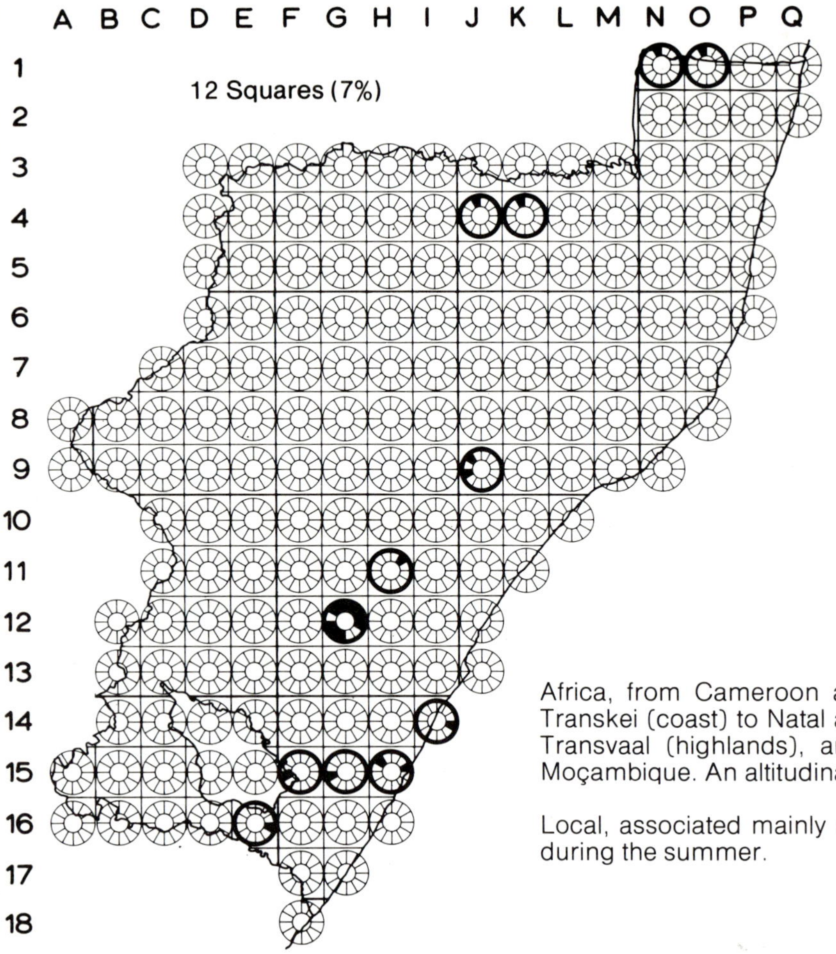

616 Fan-tailed Warbler

Schoenicola brevirostris

Africa, from Cameroon and Ethiopia to Transkei. In Southern Africa, Transkei (coast) to Natal and Zululand, Swaziland, eastern and northern Transvaal (highlands), and again in eastern Rhodesia and adjacent Moçambique. An altitudinal migrant in parts of its range.

Local, associated mainly with rank grass in the vicinity of water. Breeds during the summer.

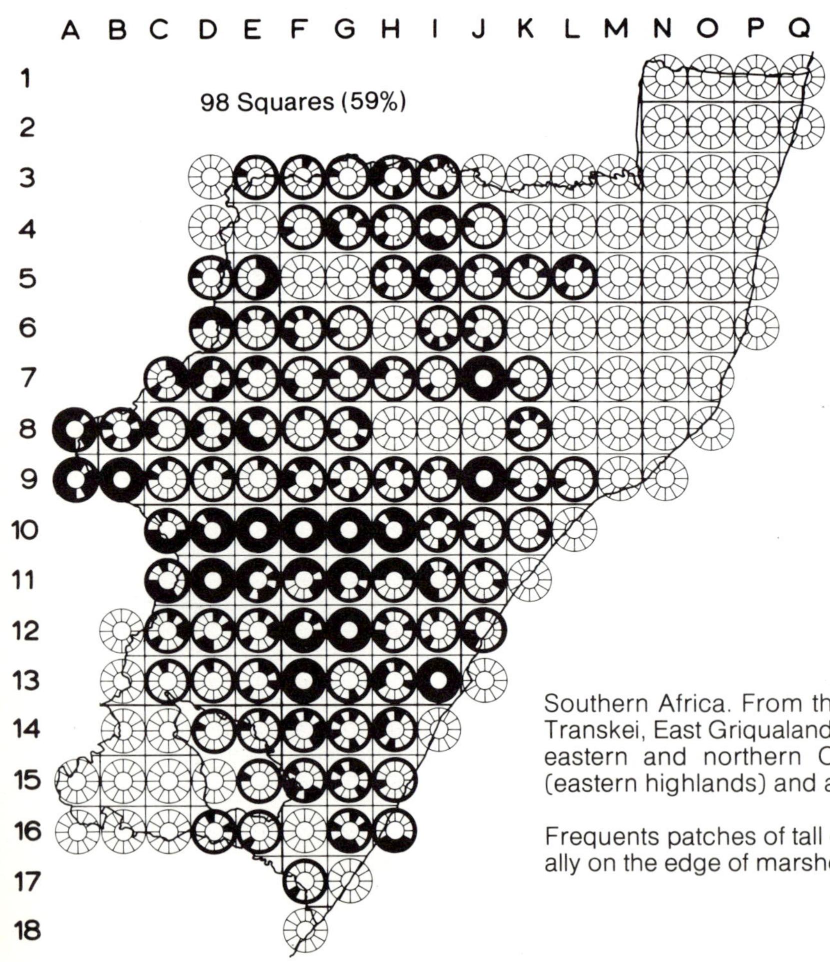

618 Grassbird

Sphenoeacus afer

Southern Africa. From the south-western, southern and eastern Cape to Transkei, East Griqualand, Natal and western Zululand, western Swaziland, eastern and northern O.F.S., northern Lesotho, Transvaal, Rhodesia (eastern highlands) and adjacent Moçambique.

Frequents patches of tall grass on hillsides, beside streams and occasionally on the edge of marshes. Breeding recorded from October to February.

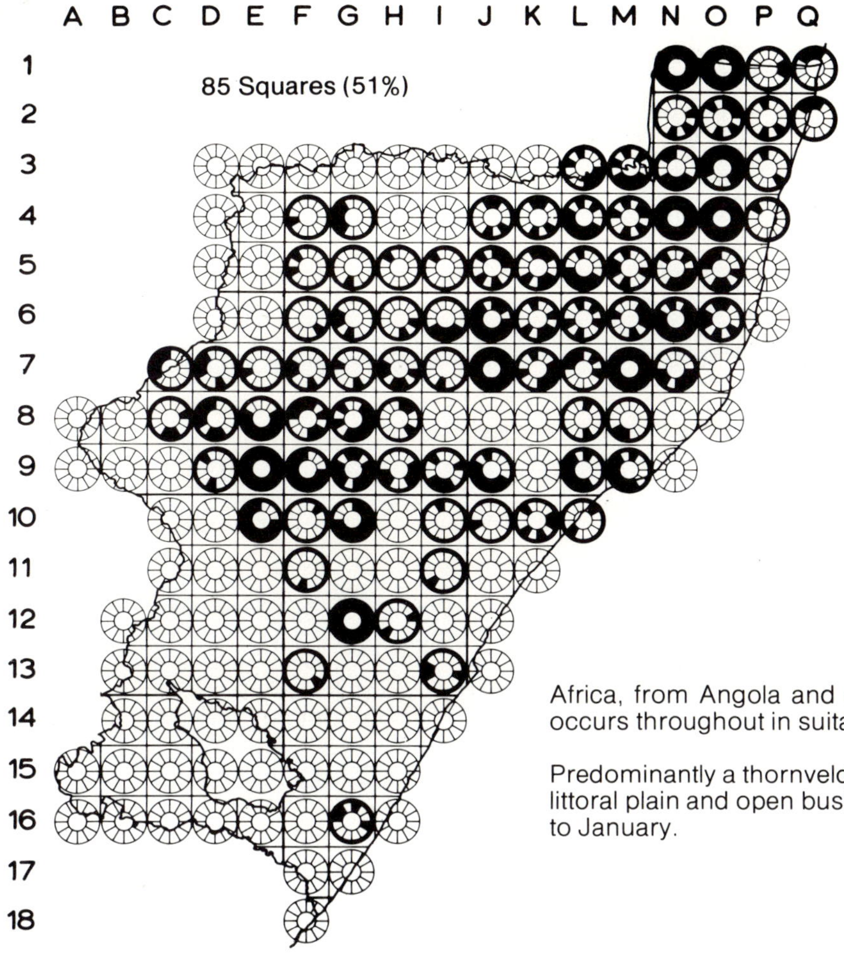

621 Crombec

Sylvietta rufescens

Africa, from Angola and eastern Zaïre to the Cape. In Southern Africa, occurs throughout in suitable habitat.

Predominantly a thornveld species, extending into mixed woodland on the littoral plain and open bush in the south. Breeding recorded from October to January.

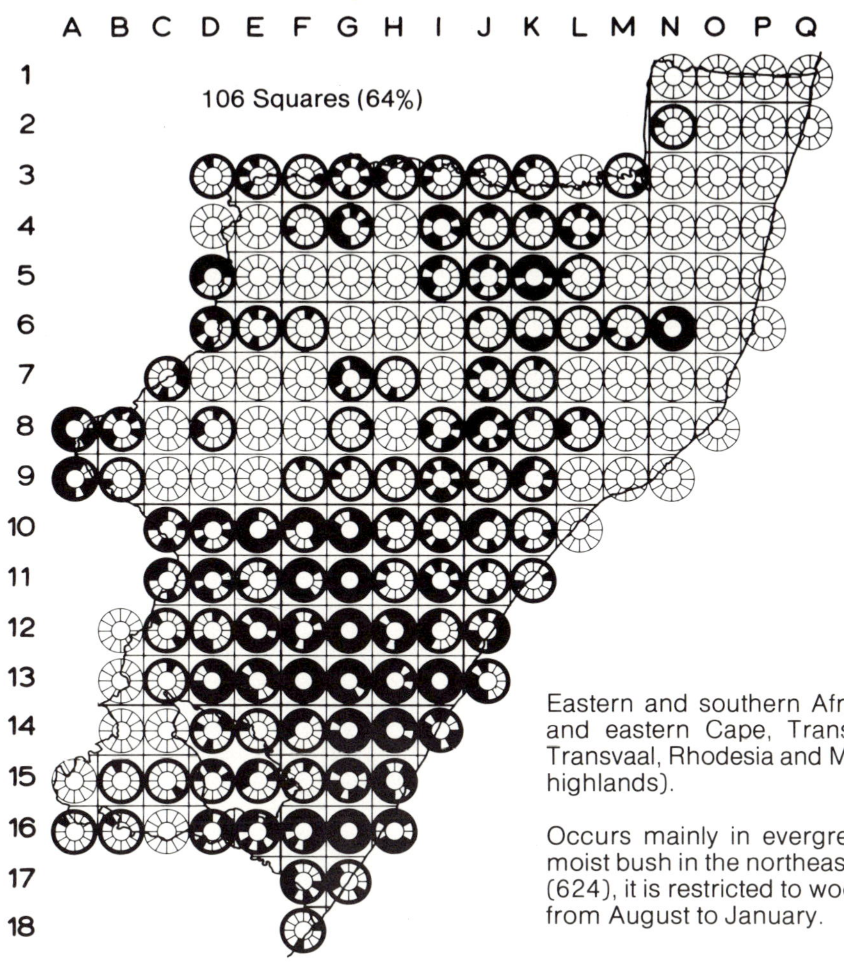

622 Bar-throated Apalis

Apalis thoracica

Eastern and southern Africa. In Southern Africa, the western, southern and eastern Cape, Transkei, Natal and Zululand, eastern Swaziland, Transvaal, Rhodesia and Moçambique (Mount Gorongosa and the western highlands).

Occurs mainly in evergreen forest and wooded gullies, extending into moist bush in the northeast. Where its range overlaps that of Rudd's Apalis (624), it is restricted to wooded areas on high ground. Breeding recorded from August to January.

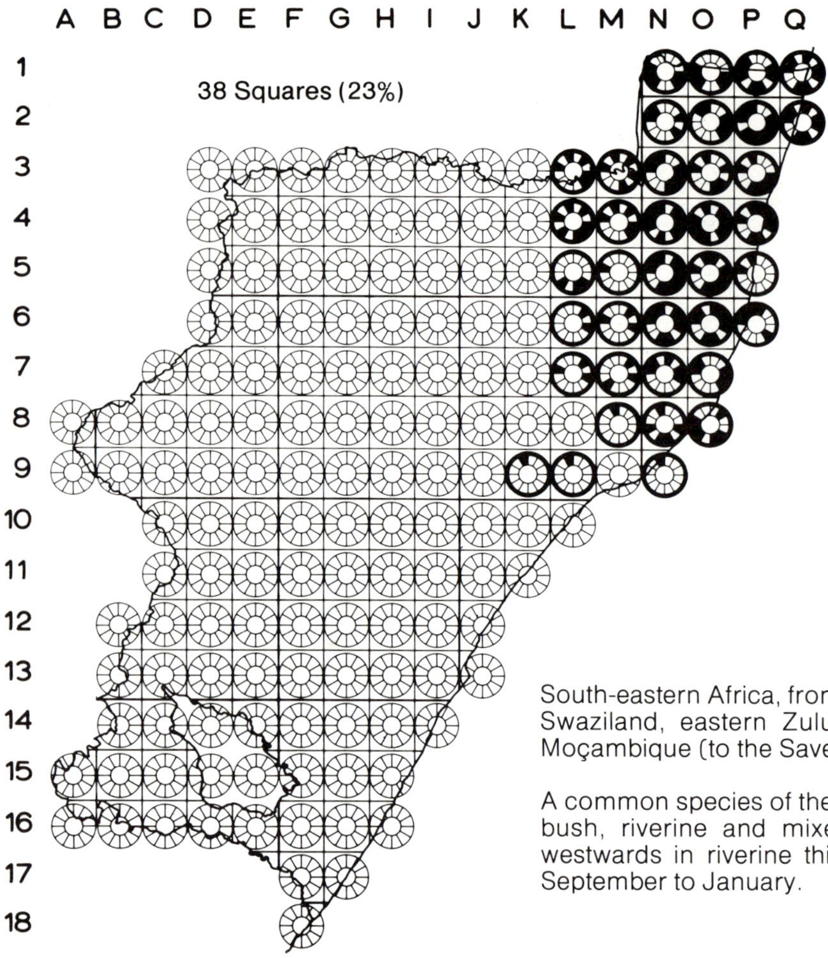

624 Rudd's Apalis

Apalis ruddi

South-eastern Africa, from Malawi to Zululand. In Southern Africa, eastern Swaziland, eastern Zululand (north from Empangeni) and southern Moçambique (to the Save River).

A common species of the littoral plain, where it inhabits coastal forest and bush, riverine and mixed woodland and mature thornveld. Extends westwards in riverine thickets and even dry thorn scrub. Breeds from September to January.

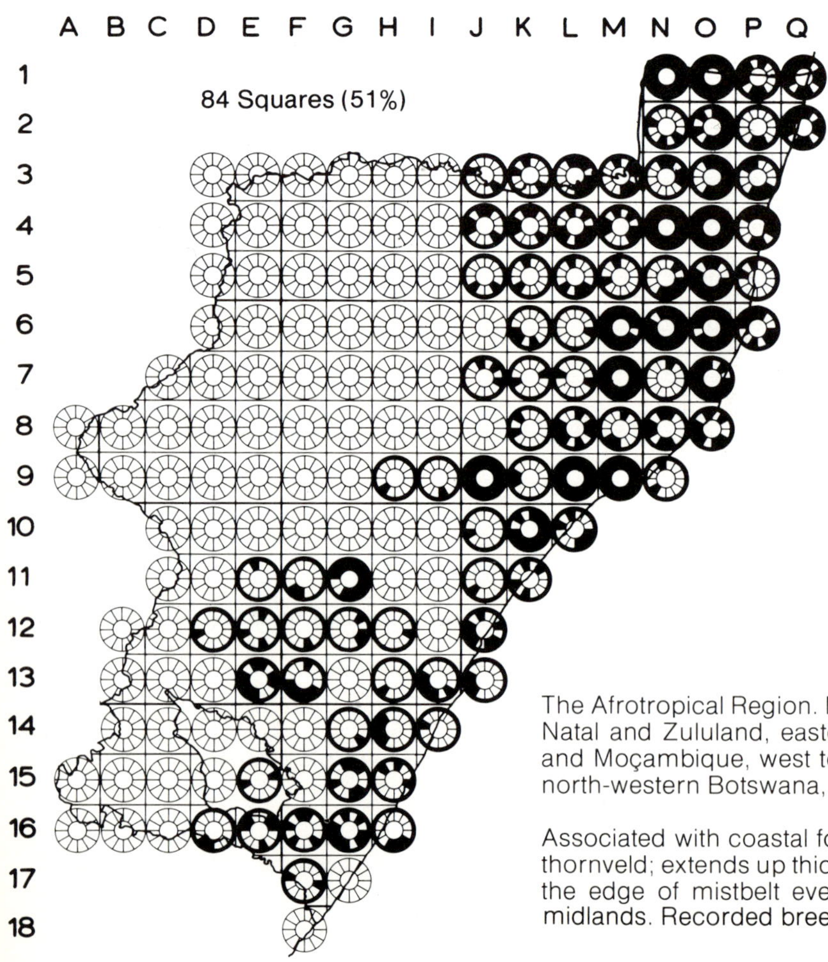

625 Yellow-breasted Apalis

Apalis flavida

The Afrotropical Region. In Southern Africa, the eastern Cape to Transkei, Natal and Zululand, eastern Swaziland, eastern and northern Transvaal and Moçambique, west to eastern and northern Rhodesia, northern and north-western Botswana, the Caprivi Strip and northern S.W.A.

Associated with coastal forest, riverine and mixed woodland, and mature thornveld; extends up thickly-wooded valleys to 900 m a.s.l. Also found on the edge of mistbelt evergreen forest, reaching higher altitudes in the midlands. Recorded breeding from October to January.

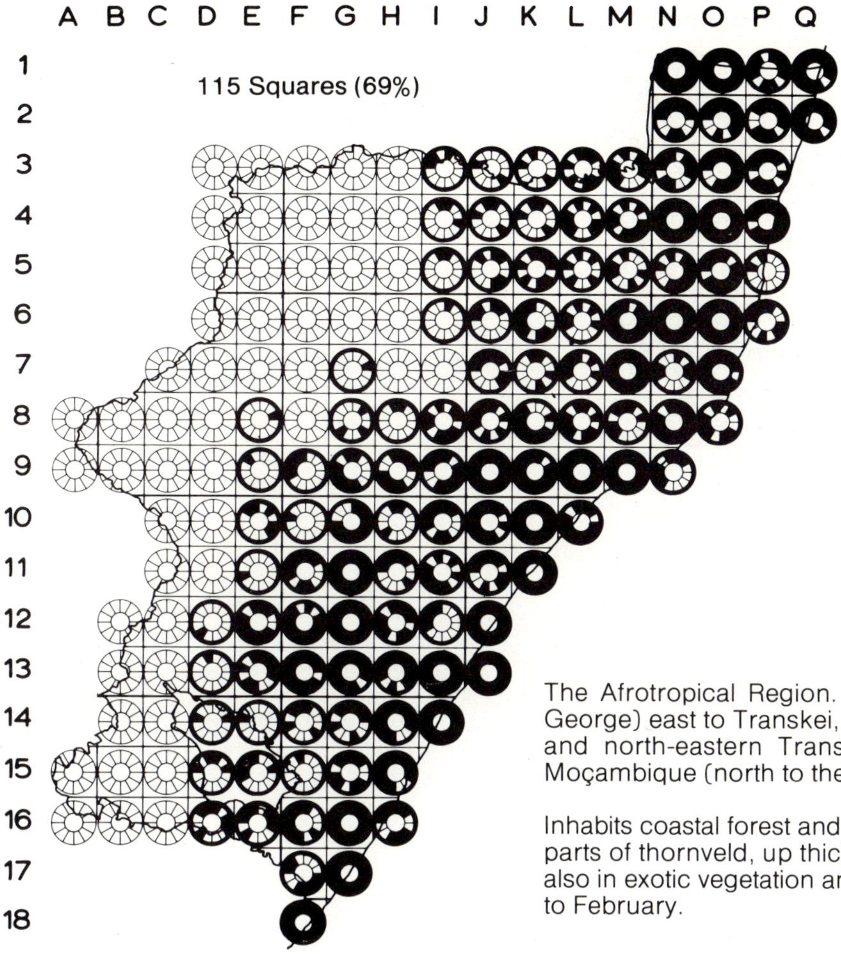

627 Bleating Bush Warbler

Camaroptera brachyura

The Afrotropical Region. In Southern Africa, the southern Cape (from George) east to Transkei, Natal and Zululand, eastern Swaziland, eastern and north-eastern Transvaal, south-eastern Rhodesia, and southern Moçambique (north to the Save River).

Inhabits coastal forest and bush, mixed and riverine woodland, the moister parts of thornveld, up thickly-wooded valleys to mistbelt evergreen forest; also in exotic vegetation and gardens. Breeding recorded from September to February.

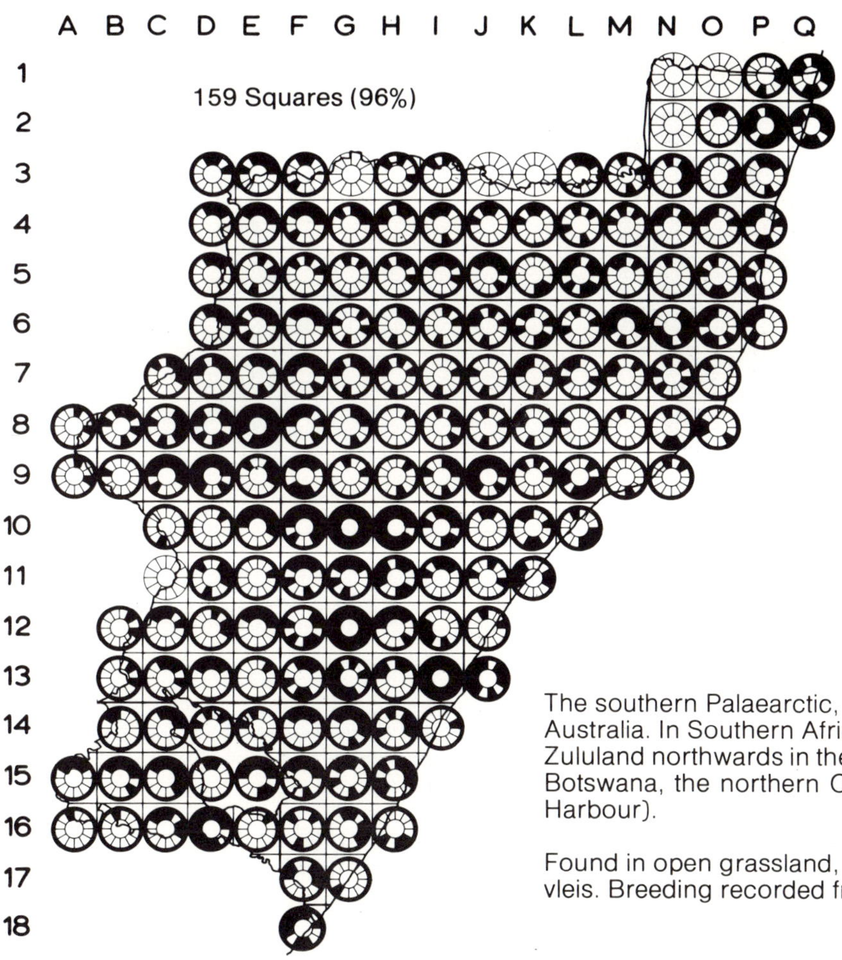

629 Fantail Cisticola

Cisticola juncidis

The southern Palaearctic, Afrotropical and Oriental Regions, southeast to Australia. In Southern Africa, the southern and eastern Cape to Natal and Zululand northwards in the east; in the north, west to eastern and northern Botswana, the northern Cape and north-eastern S.W.A. (also Sandwich Harbour).

Found in open grassland, fallow fields, waste ground and on the edges of vleis. Breeding recorded from September to March.

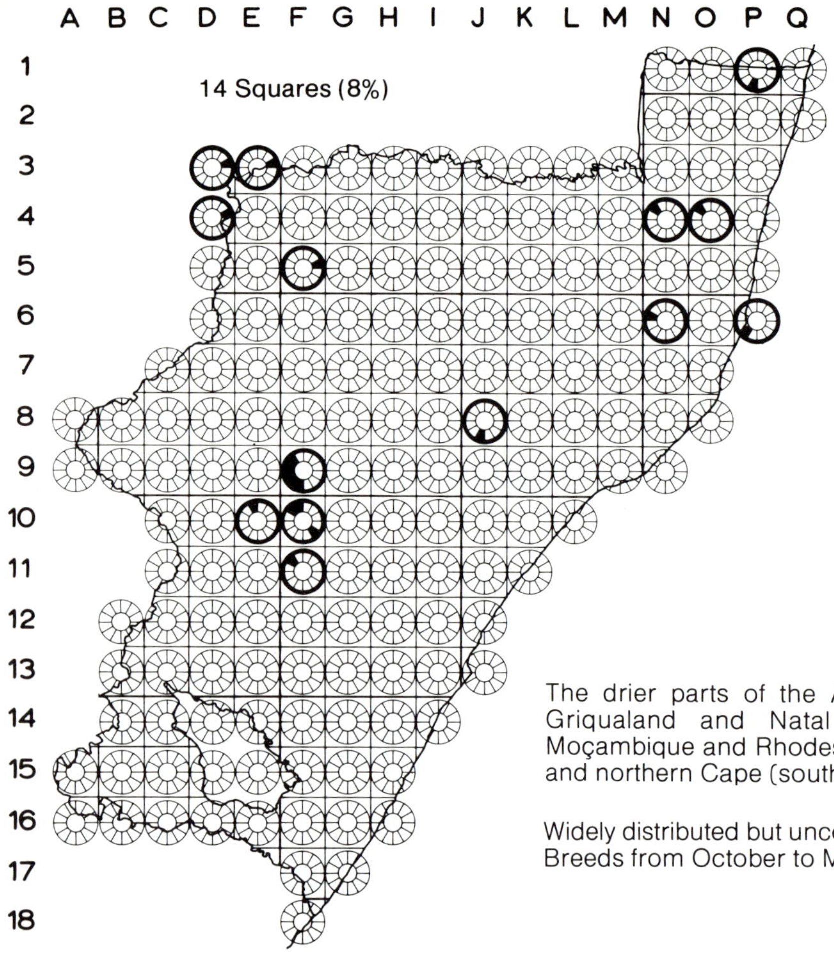

630 Desert Cisticola

Cisticola aridula

The drier parts of the Afrotropical Region. In Southern Africa, East Griqualand and Natal (interior), Zululand, Swaziland, southern Moçambique and Rhodesia, west to extreme western O.F.S., the eastern and northern Cape (south to Cradock), Botswana and S.W.A.

Widely distributed but uncommon, favouring drier areas of open grassland. Breeds from October to March.

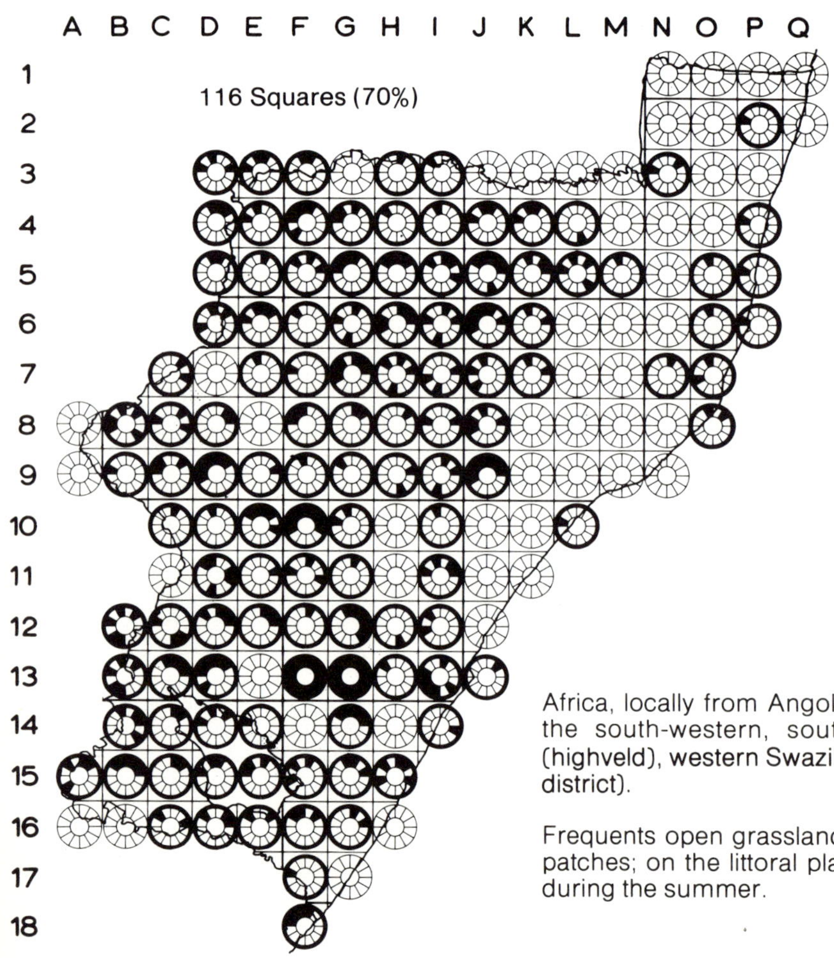

631 Cloud Cisticola

Cisticola textrix

Africa, locally from Angola and Zambia to the Cape. In Southern Africa, the south-western, southern and eastern Cape, O.F.S., Transvaal (highveld), western Swaziland, Natal, Zululand and Moçambique (Maputo district).

Frequents open grassland, preferring short grass interspersed with bare patches; on the littoral plain, found in areas of tall, coarse grass. Breeds during the summer.

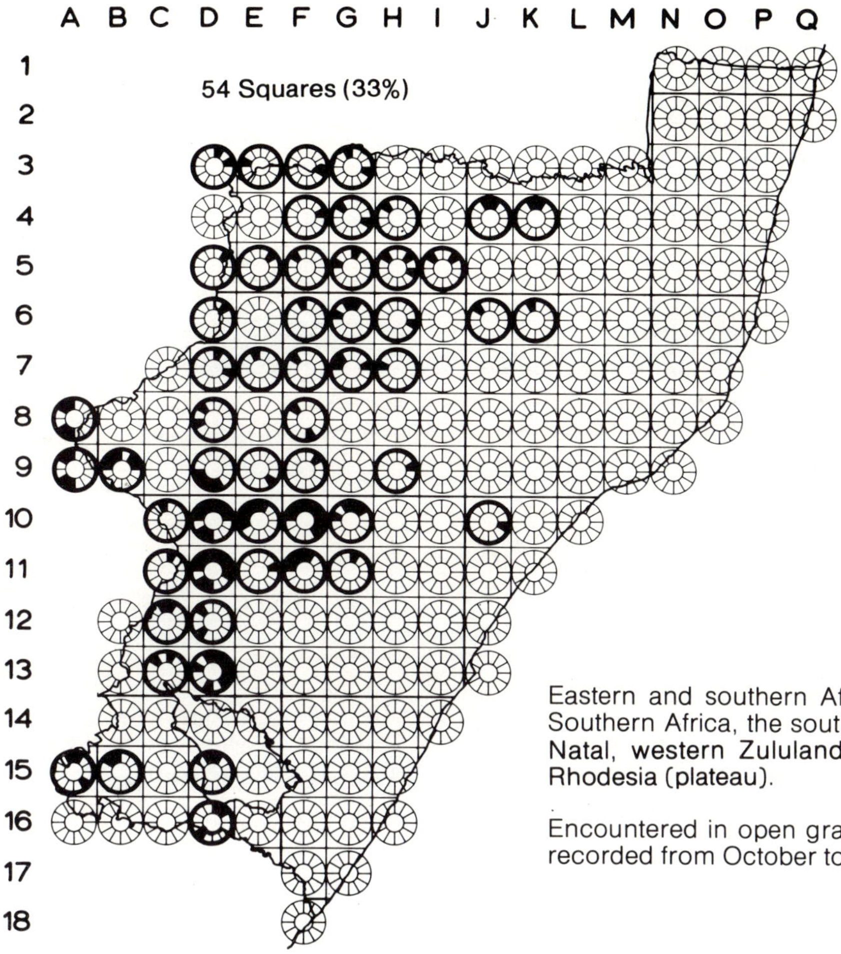

634 Ayres' Cloud Cisticola

Cisticola ayresii

Eastern and southern Africa, from Gabon and Sudan to the Cape. In Southern Africa, the south-eastern and eastern Cape to O.F.S., Lesotho, Natal, western Zululand, western Swaziland, Transvaal and eastern Rhodesia (plateau).

Encountered in open grassland generally above 1200 m a.s.l. Breeding recorded from October to March.

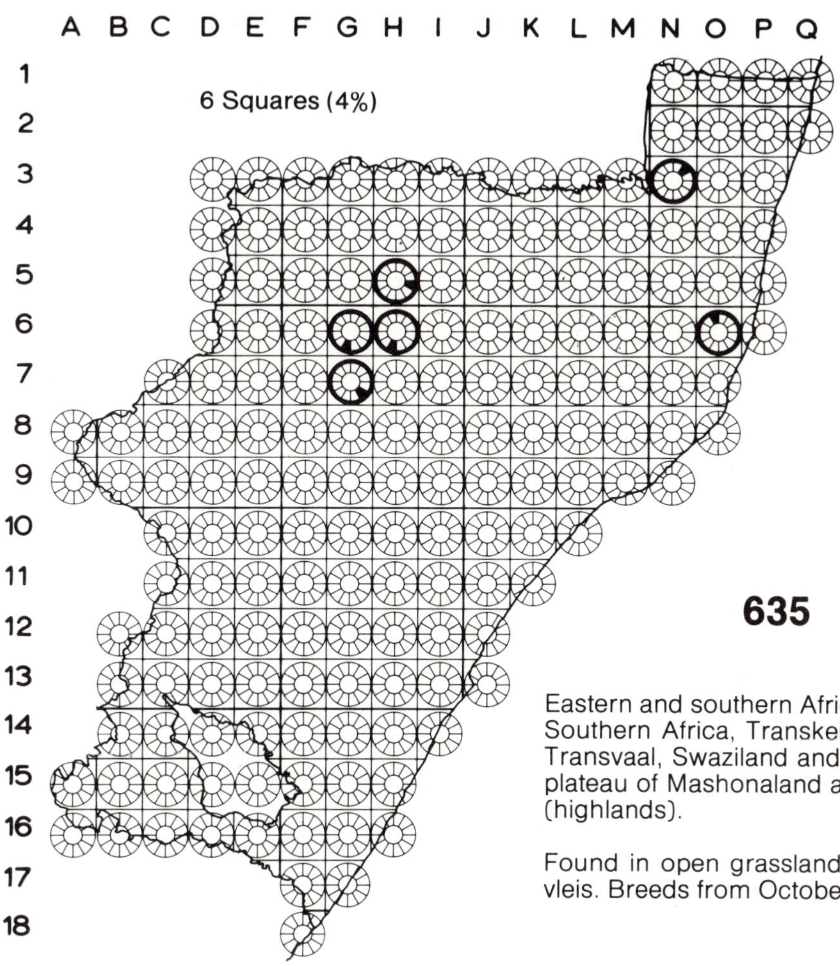

635 Pale-crowned Cloud Cisticola

Cisticola brunnescens

Eastern and southern Africa, from Cameroon and Ethiopia to Transkei. In Southern Africa, Transkei, East Griqualand, Natal and Zululand, O.F.S., Transvaal, Swaziland and southern Moçambique; also Rhodesia (on the plateau of Mashonaland above 1200 m a.s.l.) and adjacent Moçambique (highlands).

Found in open grassland, generally in moist situations such as around vleis. Breeds from October to March.

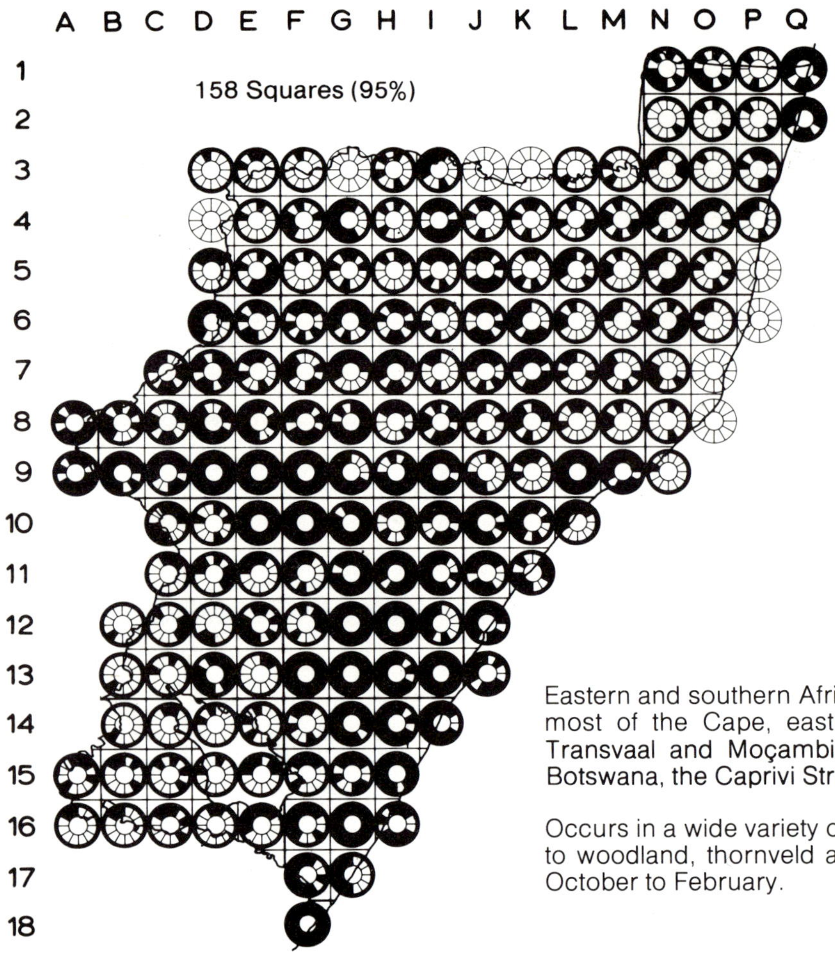

637 Neddicky

Cisticola fulvicapilla

Eastern and southern Africa, from Kenya to the Cape. In Southern Africa, most of the Cape, east to Natal and Zululand, Lesotho, Swaziland, Transvaal and Moçambique, west to Rhodesia, eastern and northern Botswana, the Caprivi Strip and northern S.W.A.; also western O.F.S.

Occurs in a wide variety of habitats, from open scrub and cultivated fields to woodland, thornveld and exotic vegetation. Recorded breeding from October to February.

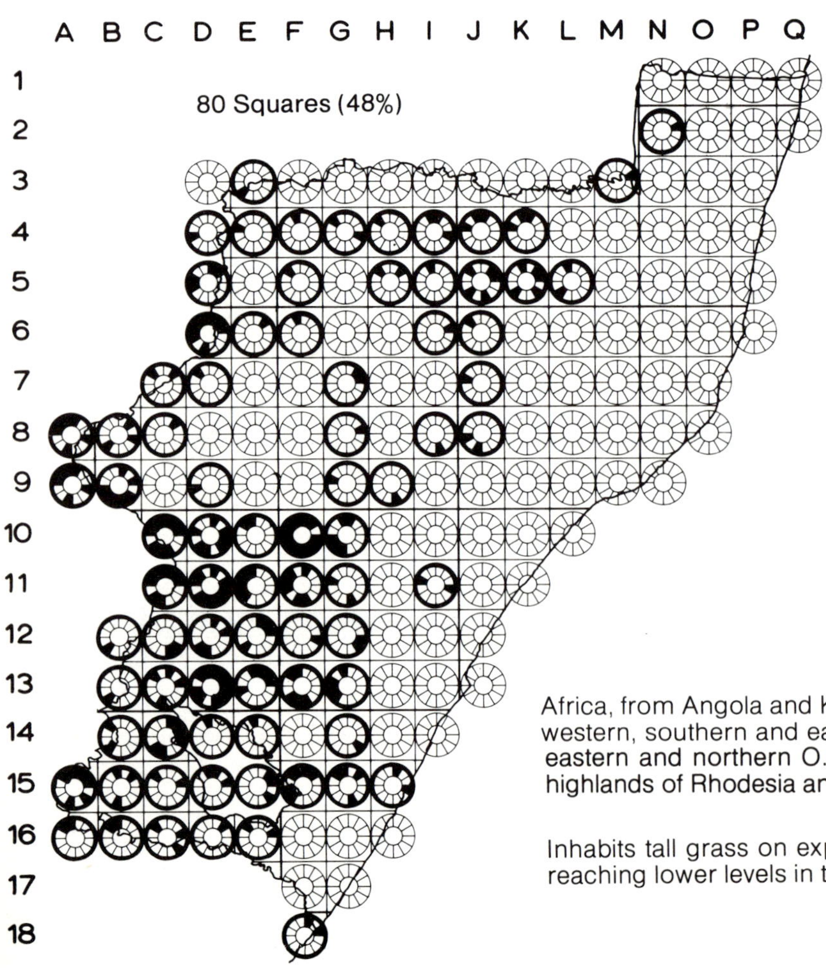

639 Wailing Cisticola

Cisticola lais

Africa, from Angola and Kenya to the Cape. In Southern Africa, the south-western, southern and eastern Cape to Natal and Zululand, Lesotho and eastern and northern O.F.S., north to Transvaal, Swaziland, the eastern highlands of Rhodesia and adjacent Moçambique, and Mount Gorongosa.

Inhabits tall grass on exposed hillsides, generally down to 1200 m a.s.l., reaching lower levels in the south. Breeds from September to January.

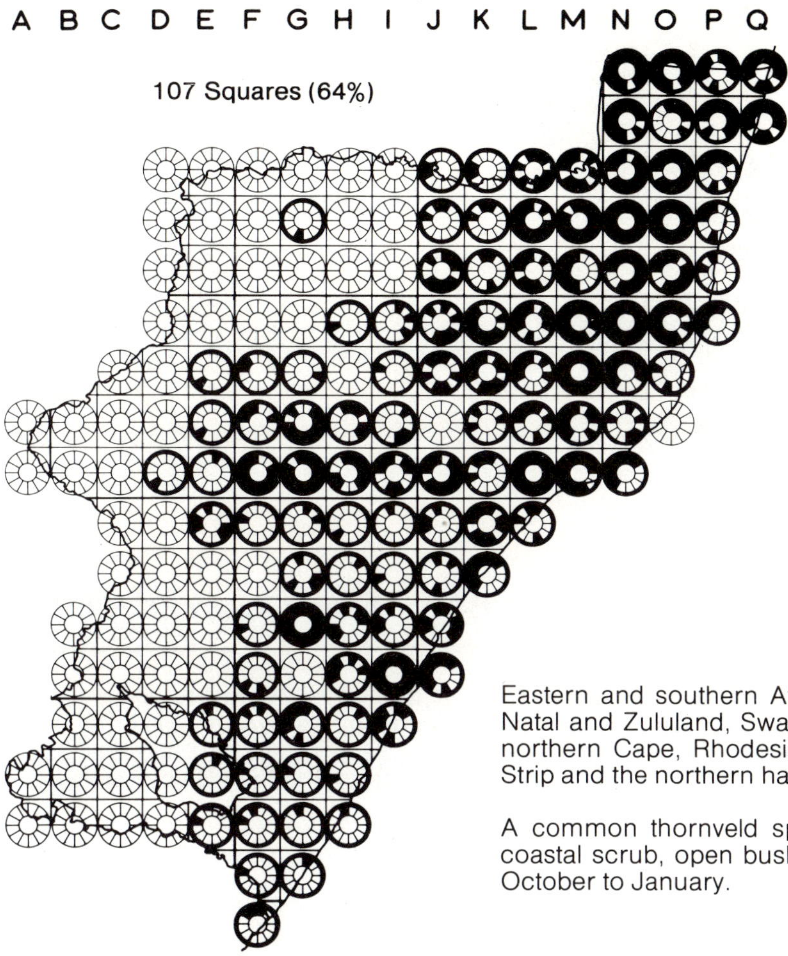

642 Rattling Cisticola

Cisticola chiniana

Eastern and southern Africa, from Eritrea to Natal. In Southern Africa, Natal and Zululand, Swaziland and Moçambique, west to Transvaal, the northern Cape, Rhodesia, eastern and northern Botswana, the Caprivi Strip and the northern half of S.W.A.

A common thornveld species which also occurs in mixed woodland, coastal scrub, open bush and wooded valleys. Breeding recorded from October to January.

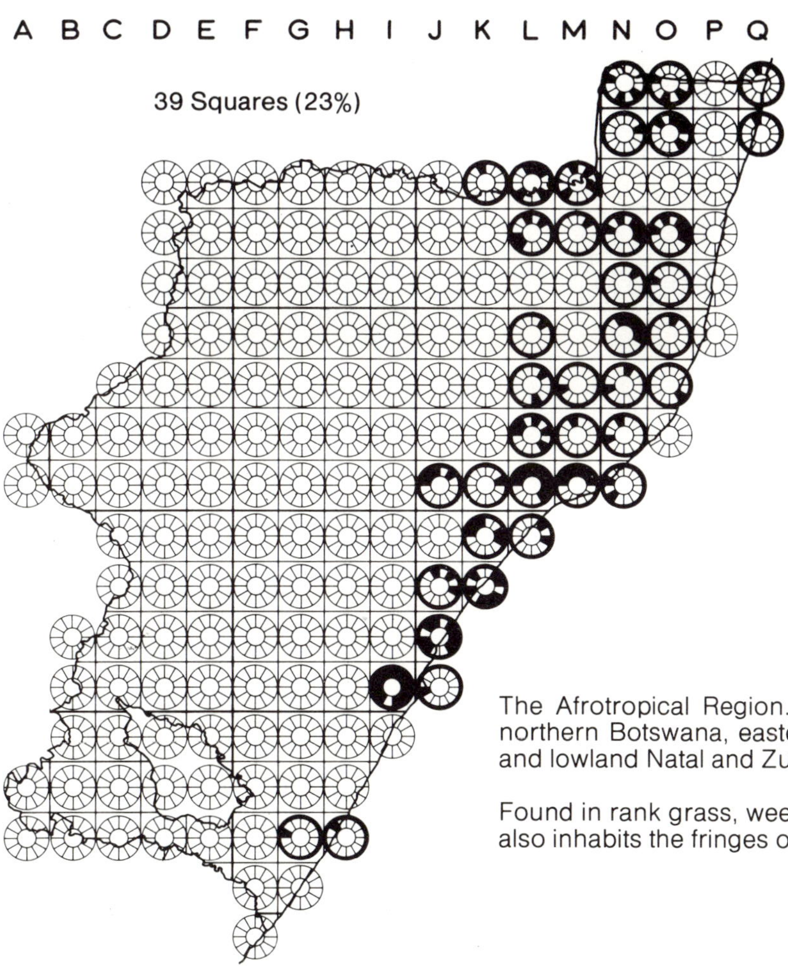

644 Red-faced Cisticola

Cisticola erythrops

The Afrotropical Region. In Southern Africa, the eastern Caprivi Strip, northern Botswana, eastern Transvaal, Moçambique, eastern Swaziland and lowland Natal and Zululand.

Found in rank grass, weeds and stands of reeds along the edge of rivers; also inhabits the fringes of cane fields. Breeds during the summer.

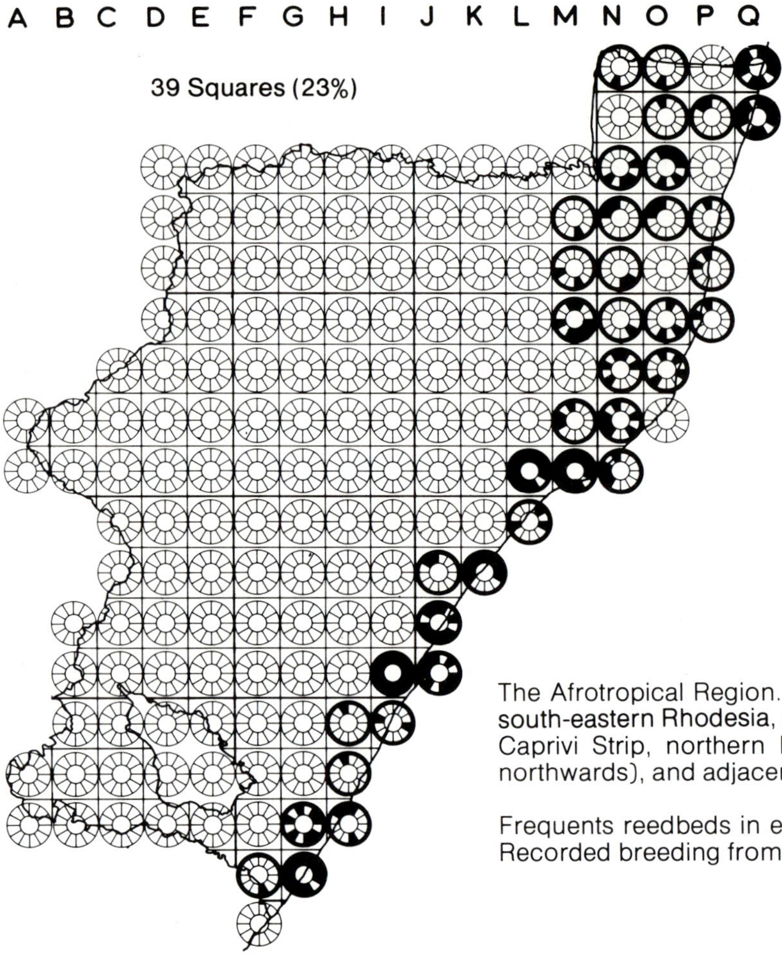

645 Black-backed Cisticola

Cisticola galactotes

The Afrotropical Region. In Southern Africa, coastal Natal and Zululand, south-eastern Rhodesia, and Moçambique; also north-eastern S.W.A., the Caprivi Strip, northern Botswana (Lake Ngami and the Botletle River northwards), and adjacent north-western Rhodesia (Zambesi River).

Frequents reedbeds in estuaries, rivers, lakes and pans; also canefields. Recorded breeding from October to February.

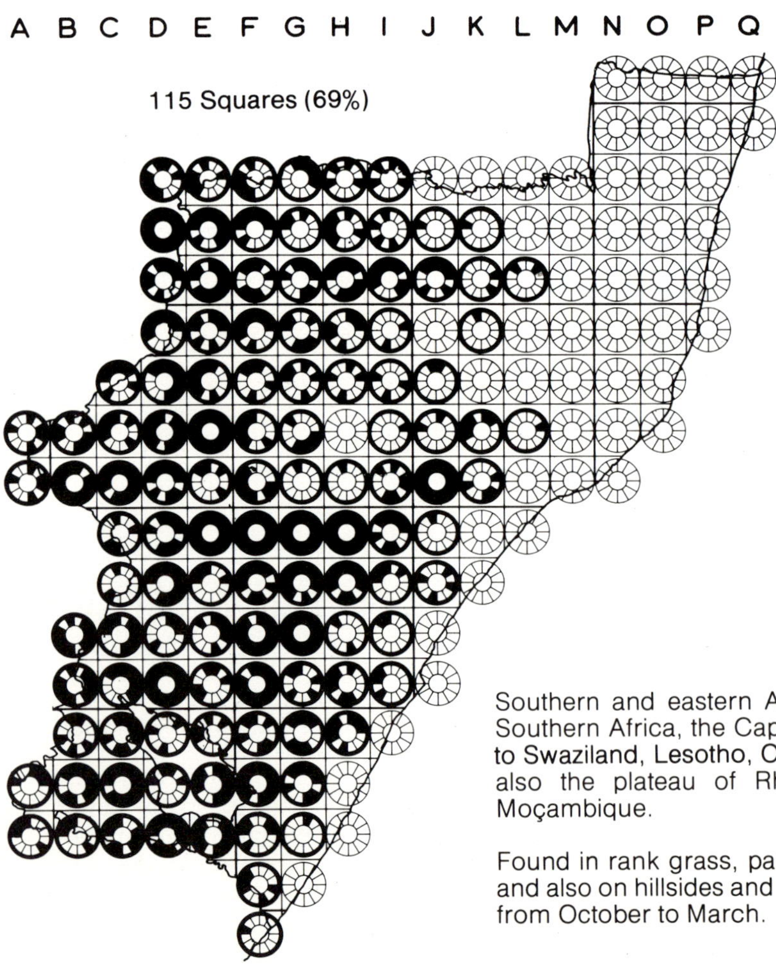

646 Le Vaillant's Cisticola

Cisticola tinniens

Southern and eastern Africa, from Kenya and Angola to the Cape. In Southern Africa, the Cape (except interior and west), Natal and Zululand to Swaziland, Lesotho, O.F.S., Transvaal and possibly eastern Botswana; also the plateau of Rhodesia, the eastern highlands and adjacent Moçambique.

Found in rank grass, particularly around vleis, dams and sewage farms, and also on hillsides and the edges of cultivated fields. Recorded breeding from October to March.

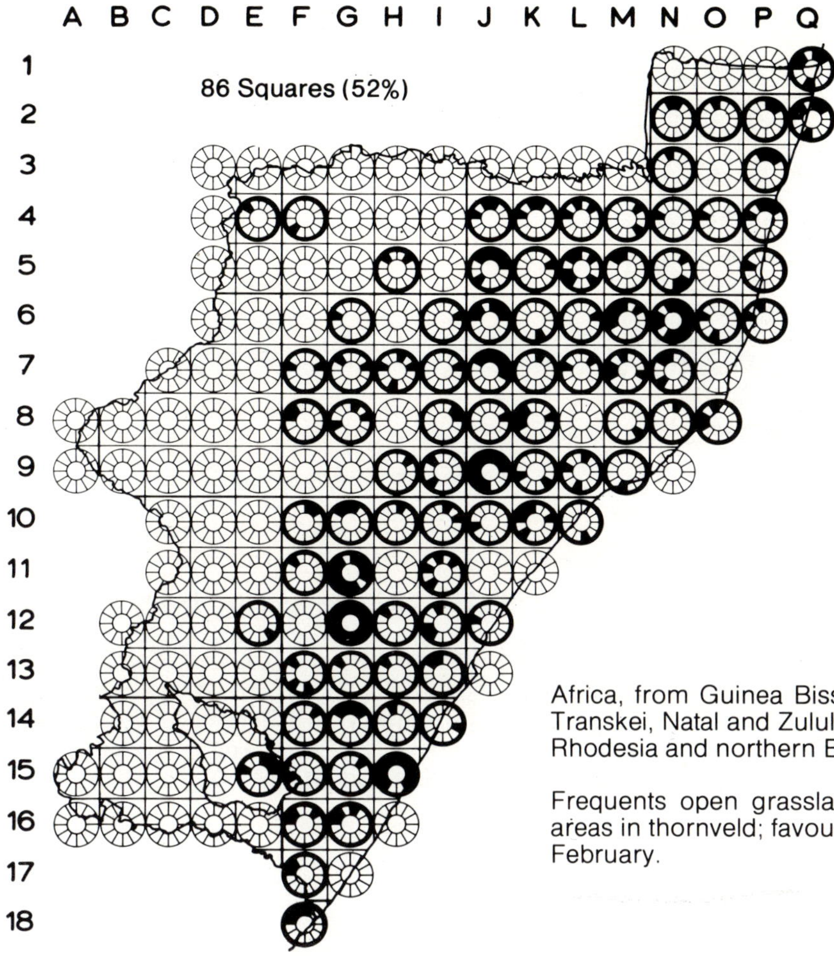

647 Croaking Cisticola

Cisticola natalensis

Africa, from Guinea Bissau and Sudan to Transkei. In Southern Africa, Transkei, Natal and Zululand, Swaziland, eastern Transvaal, Moçambique, Rhodesia and northern Botswana.

Frequents open grassland with scattered bushes and the more open areas in thornveld; favours tall grass. Breeding recorded from October to February.

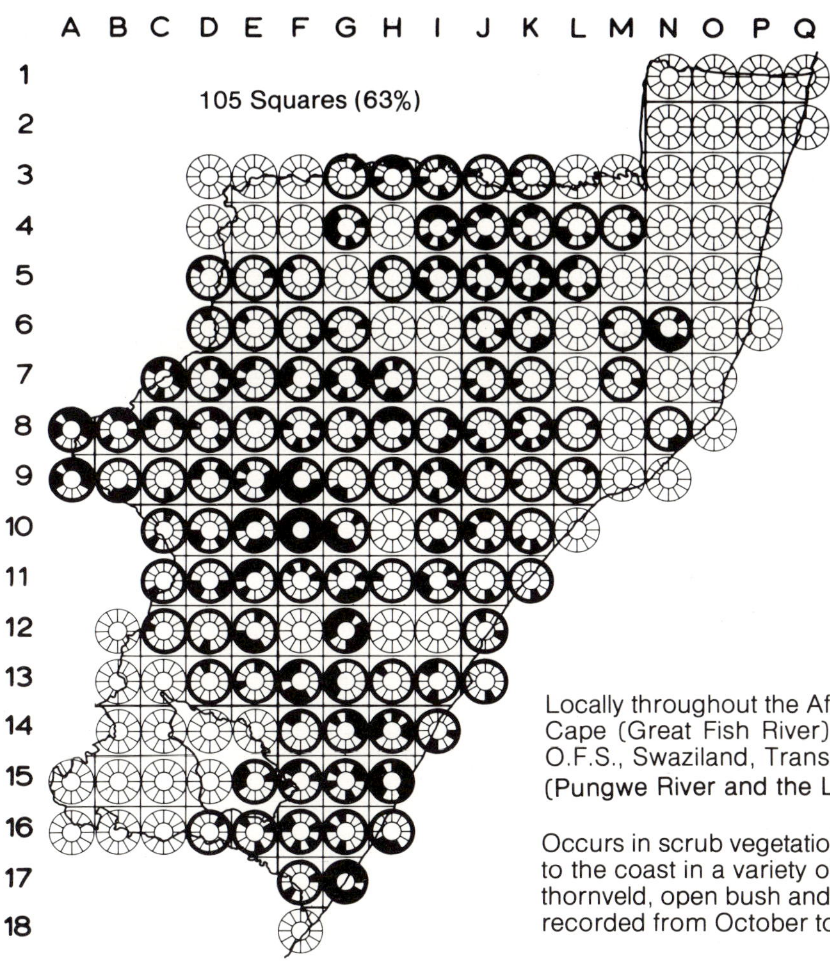

648 Lazy Cisticola

Cisticola aberrans

Locally throughout the Afrotropical Region. In Southern Africa, the eastern Cape (Great Fish River), east to Transkei, Natal and Zululand, eastern O.F.S., Swaziland, Transvaal, Rhodesia (plateau), adjacent Moçambique (Pungwe River and the Lebombo Range); also south-eastern Botswana.

Occurs in scrub vegetation in gullies of the Drakensberg Range, extending to the coast in a variety of habitats, such as the edge of evergreen forest, thornveld, open bush and tall grass; absent from the littoral plain. Breeding recorded from October to February.

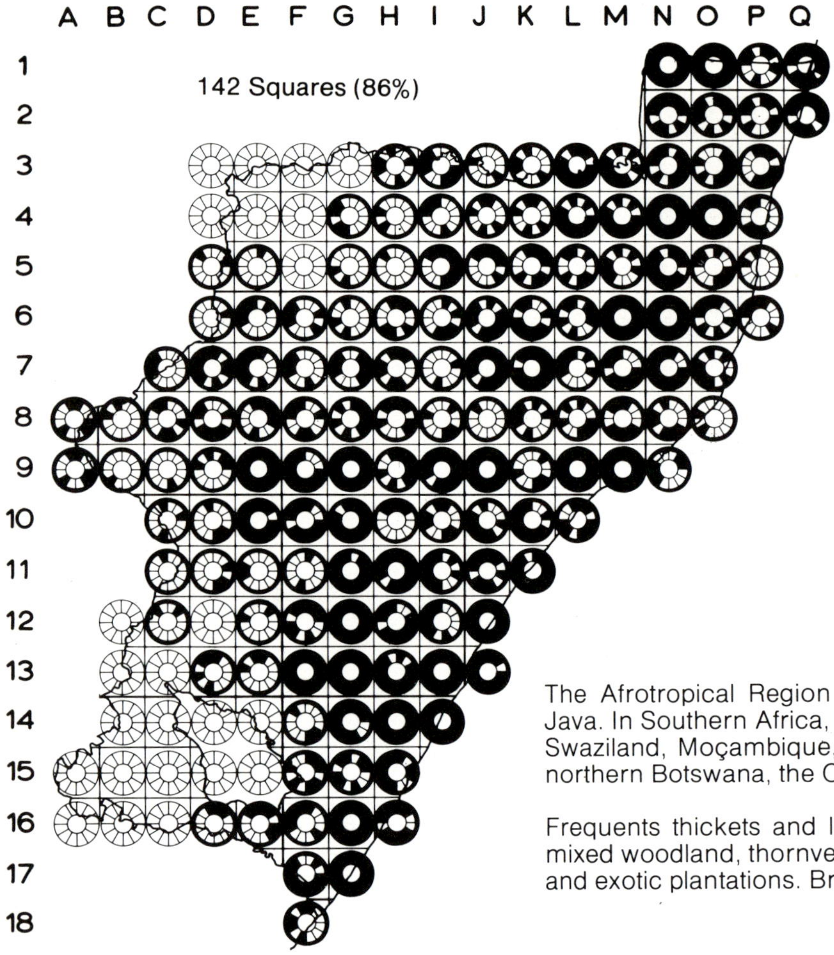

649 Tawny-flanked Prinia

Prinia subflava

The Afrotropical Region and southern and eastern Asia, southeast to Java. In Southern Africa, the eastern Cape, Transkei, Natal and Zululand, Swaziland, Moçambique, Transvaal and Rhodesia, west to eastern and northern Botswana, the Caprivi Strip and northern S.W.A.

Frequents thickets and long grass in the moister parts of riverine and mixed woodland, thornveld and open bush; also gardens, cultivated fields and exotic plantations. Breeding recorded from September to March.

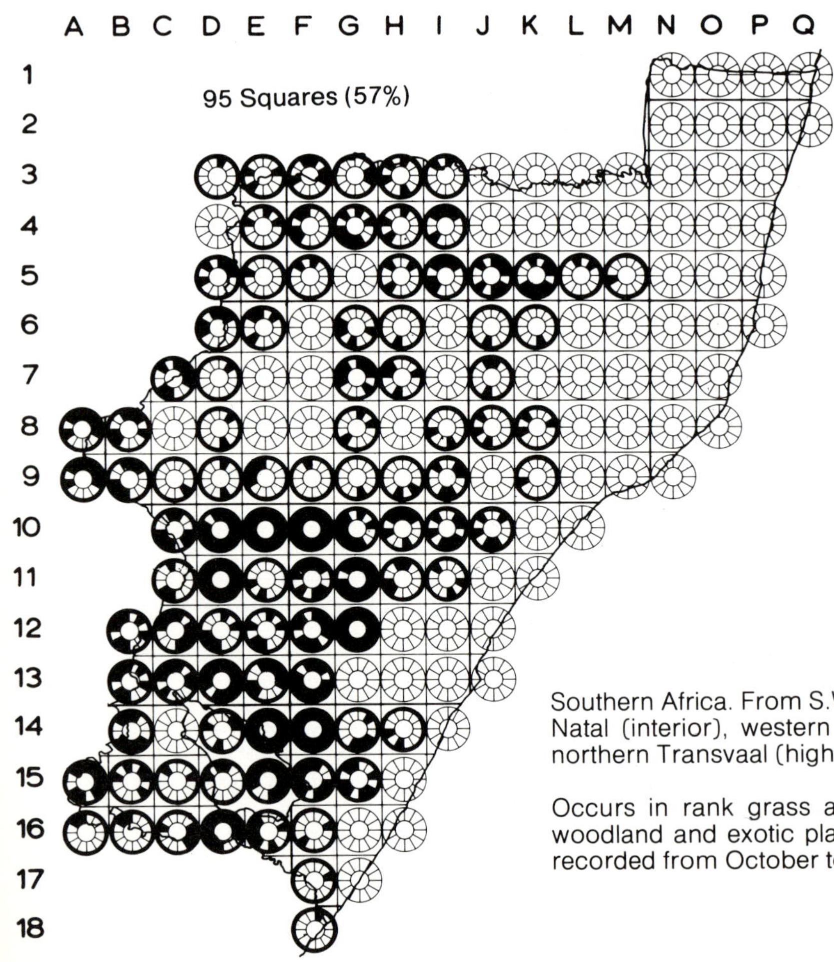

651 Karoo Prinia

Prinia maculosa

Southern Africa. From S.W.A. (south of Aus), to the Cape, O.F.S., Lesotho, Natal (interior), western Zululand, western Swaziland and eastern and northern Transvaal (highlands).

Occurs in rank grass and thickets, often on the edges of forest, dry woodland and exotic plantations, generally above 600 m a.s.l. Breeding recorded from October to February.

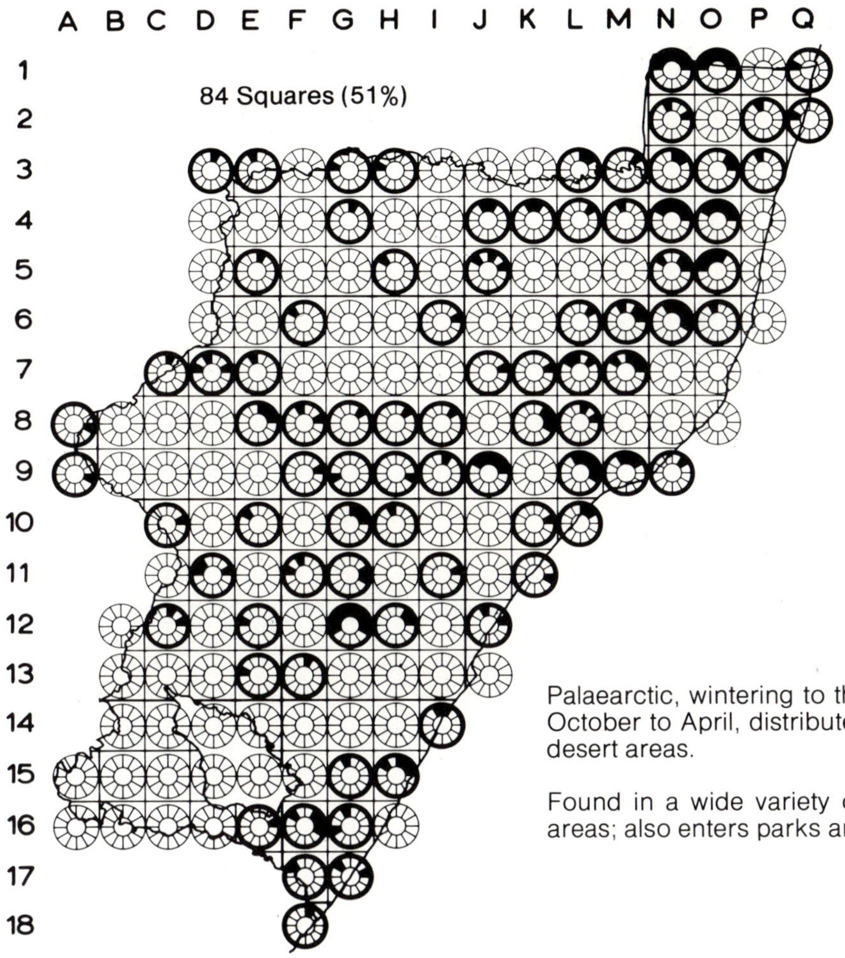

654 Spotted Flycatcher

Muscicapa striata

Palaearctic, wintering to the south. In Southern Africa, a visitor from late October to April, distributed throughout though sparse in the Karoo and desert areas.

Found in a wide variety of wooded habitats, usually in the more open areas; also enters parks and gardens.

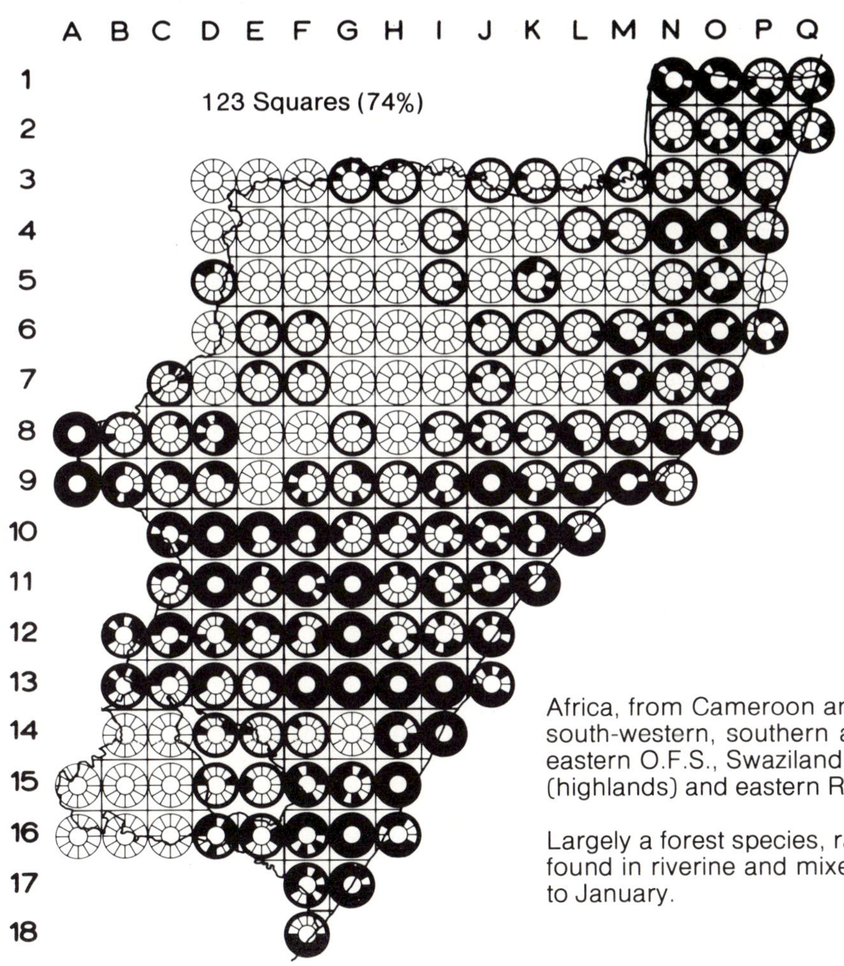

655 Dusky Flycatcher

Muscicapa adusta

Africa, from Cameroon and Ethiopia to the Cape. In Southern Africa, the south-western, southern and eastern Cape, east to Natal and Zululand, eastern O.F.S., Swaziland, Moçambique, eastern and northern Transvaal (highlands) and eastern Rhodesia.

Largely a forest species, ranging from the coast to the Drakensberg; also found in riverine and mixed woodland. Breeding recorded from October to January.

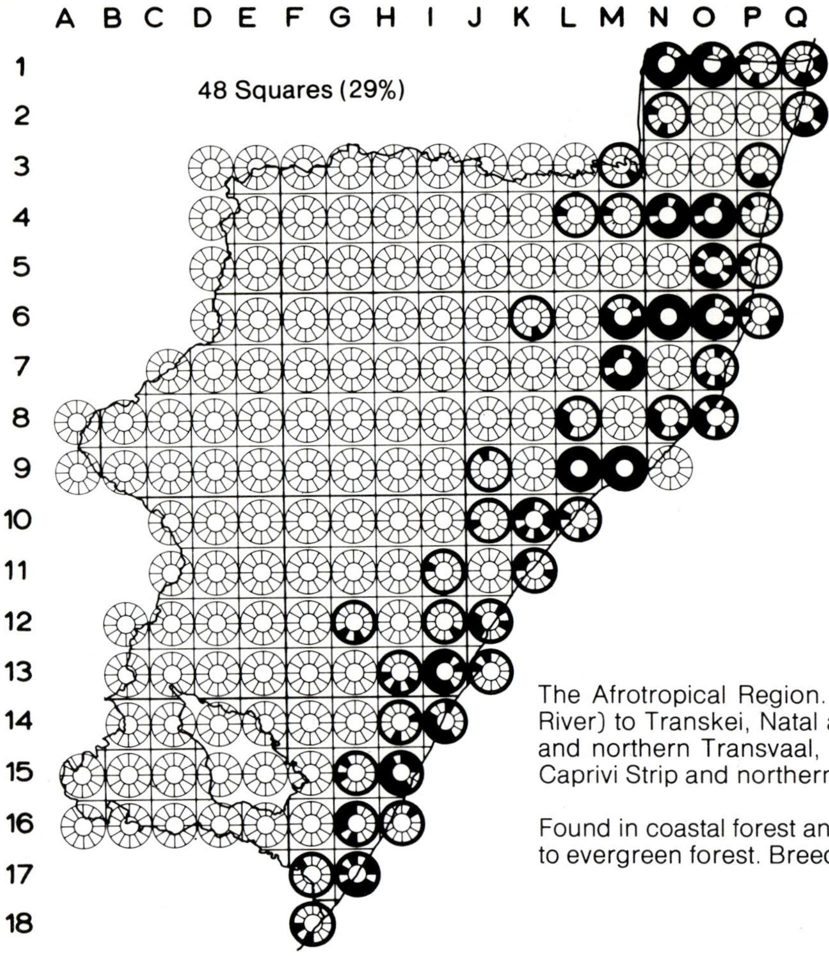

656 Blue-grey Flycatcher

Muscicapa caerulescens

The Afrotropical Region. In Southern Africa, the eastern Cape (Buffalo River) to Transkei, Natal and Zululand, Swaziland, Moçambique, eastern and northern Transvaal, and Rhodesia, west to northern Botswana, the Caprivi Strip and northern S.W.A.

Found in coastal forest and riverine woodland, extending inland marginally to evergreen forest. Breeding recorded from October to December.

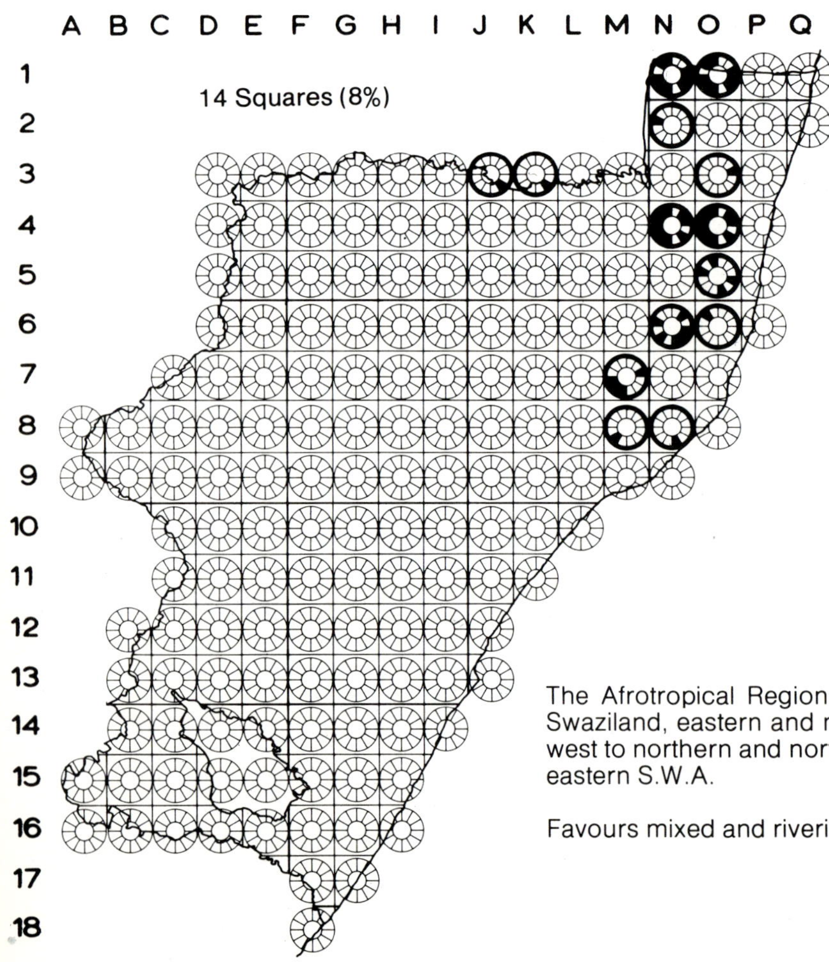

657 Grey Titbabbler

Myioparus plumbeus

The Afrotropical Region. In Southern Africa, eastern Zululand, eastern Swaziland, eastern and northern Transvaal, Moçambique and Rhodesia, west to northern and north-eastern Botswana, the Caprivi Strip and north-eastern S.W.A.

Favours mixed and riverine woodland. Breeds during the summer.

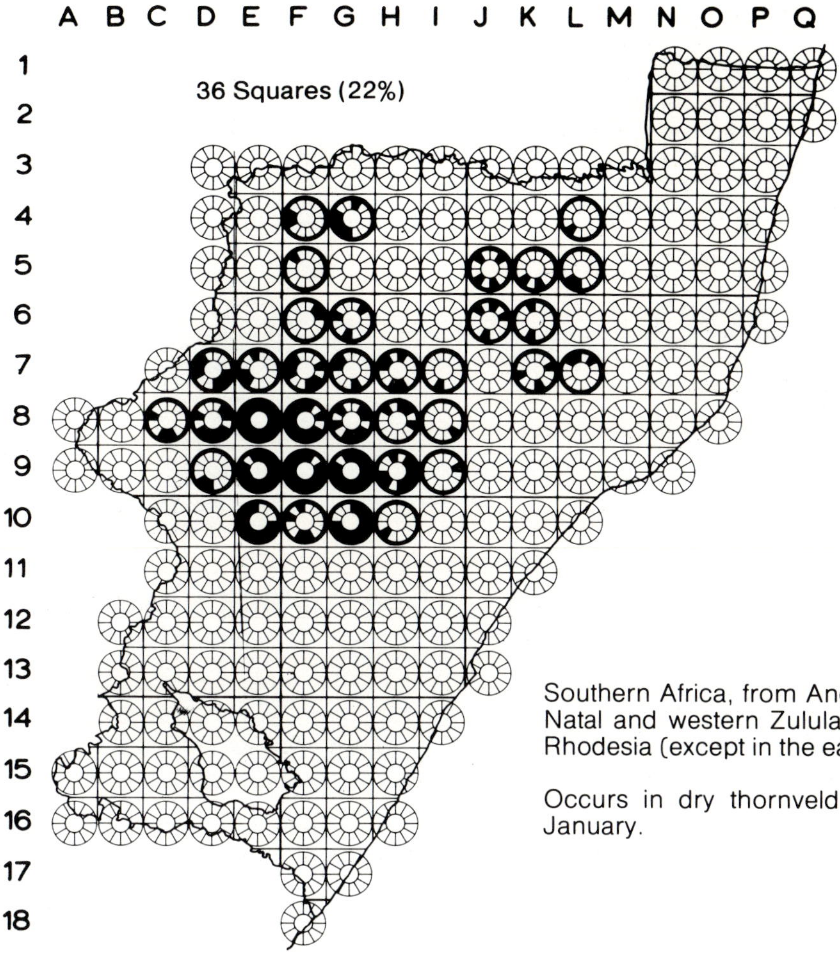

658 Titbabbler

Parisoma subcaeruleum

Southern Africa, from Angola to the Cape. In Southern Africa, the Cape, Natal and western Zululand, Lesotho (lowlands), O.F.S., Transvaal and Rhodesia (except in the east), west to Botswana and S.W.A.

Occurs in dry thornveld scrub. Recorded breeding from October to January.

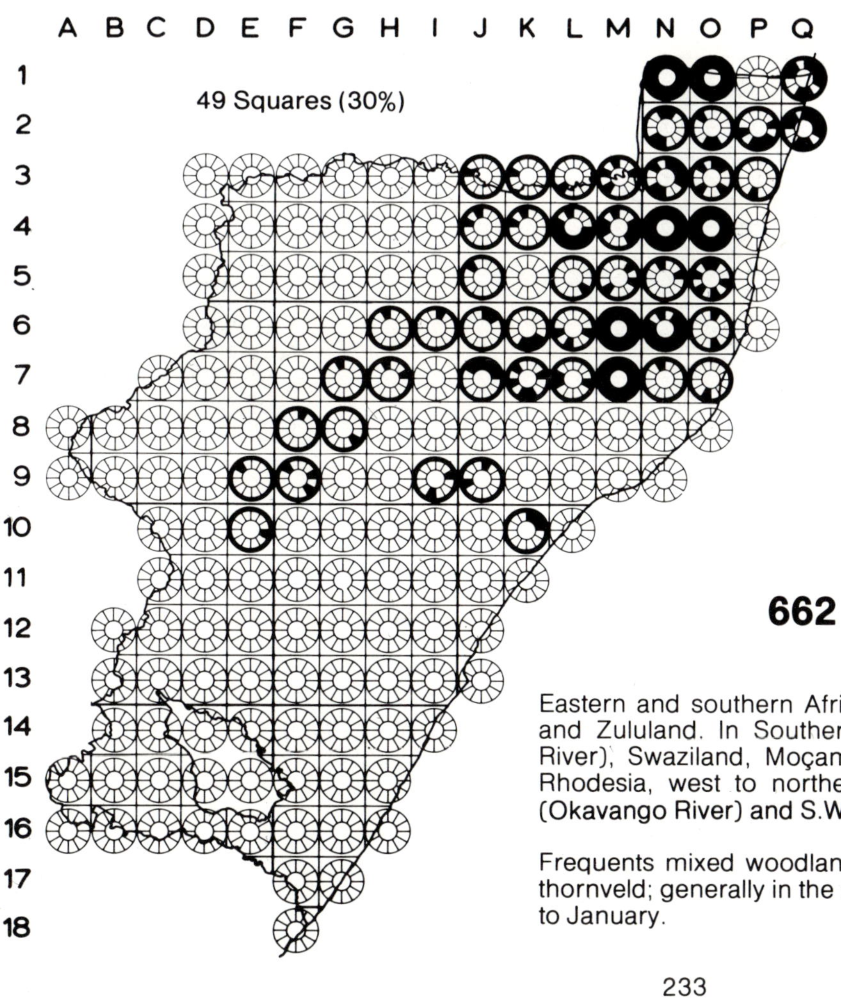

662 Mouse-coloured Flycatcher

Melaenornis pallidus

Eastern and southern Africa, from Sudan and Ethiopia south to S.W.A. and Zululand. In Southern Africa, eastern Zululand (from the Tugela River), Swaziland, Moçambique, eastern and northern Transvaal, and Rhodesia, west to northern Botswana (Ngamiland), the Caprivi Strip (Okavango River) and S.W.A. (Waterberg district).

Frequents mixed woodland on the littoral plain, extending westwards in thornveld; generally in the more mature patches. Breeds from September to January.

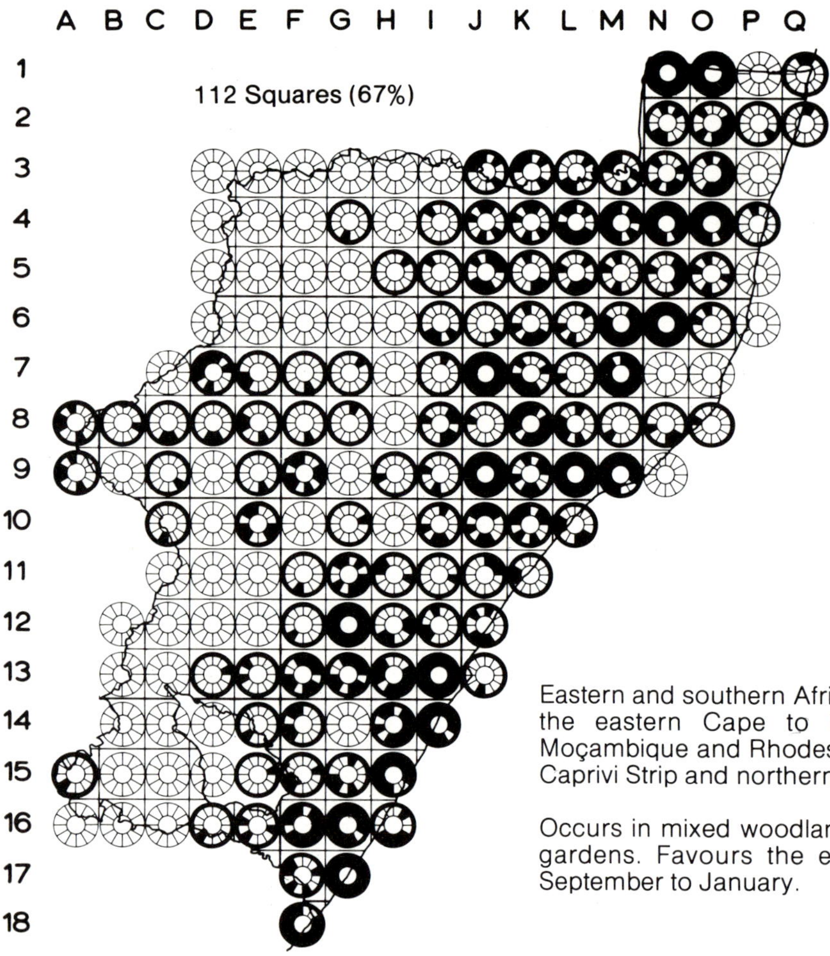

664 Black Flycatcher

Melaenornis pammelaina

Eastern and southern Africa, from Kenya to the Cape. In Southern Africa, the eastern Cape to Natal and Zululand, Swaziland, Transvaal, Moçambique and Rhodesia, west to northern and eastern Botswana, the Caprivi Strip and northern S.W.A.

Occurs in mixed woodland and thornveld; also in exotic plantations and gardens. Favours the edges of clearings. Recorded breeding from September to January.

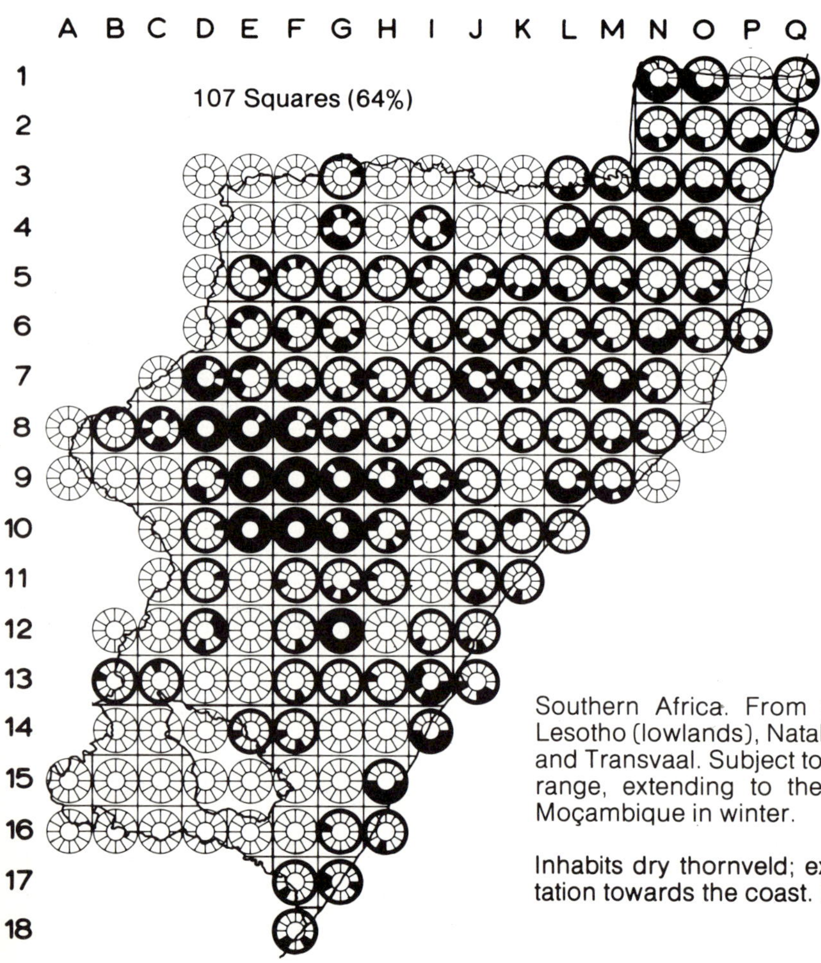

665 Fiscal Flycatcher

Sigelus silens

Southern Africa. From the Cape to south-eastern Botswana, O.F.S., Lesotho (lowlands), Natal (interior), western Zululand, western Swaziland, and Transvaal. Subject to much post-breeding movement in the east of its range, extending to the Natal coast, eastern Zululand and southern Moçambique in winter.

Inhabits dry thornveld; extending into mixed woodland and scrub vegetation towards the coast. Recorded breeding from October to December.

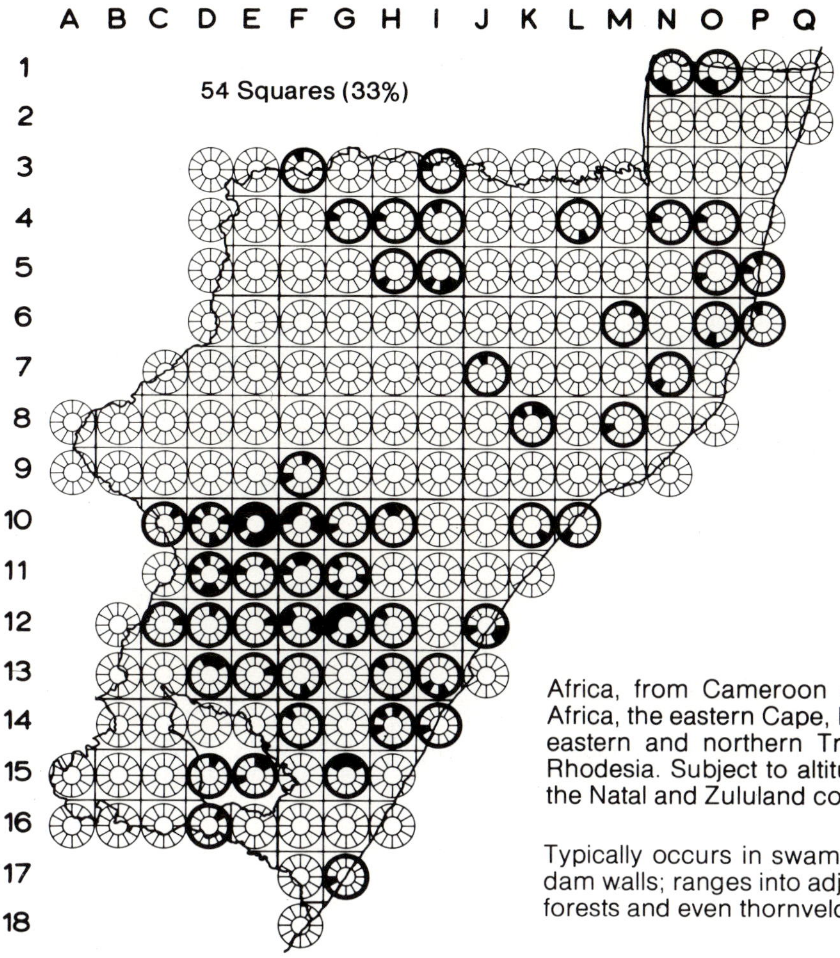

666 Yellow Warbler

Chloropeta natalensis

Africa, from Cameroon and Ethiopia south to the Cape. In Southern Africa, the eastern Cape, East Griqualand, Natal and Zululand, Swaziland, eastern and northern Transvaal, and again in the eastern districts of Rhodesia. Subject to altitudinal movement after breeding, then reaching the Natal and Zululand coasts and the Moçambique littoral.

Typically occurs in swampy areas with taller vegetation, as found below dam walls; ranges into adjacent habitats, such as along streams, edges of forests and even thornveld. Breeds from October to February.

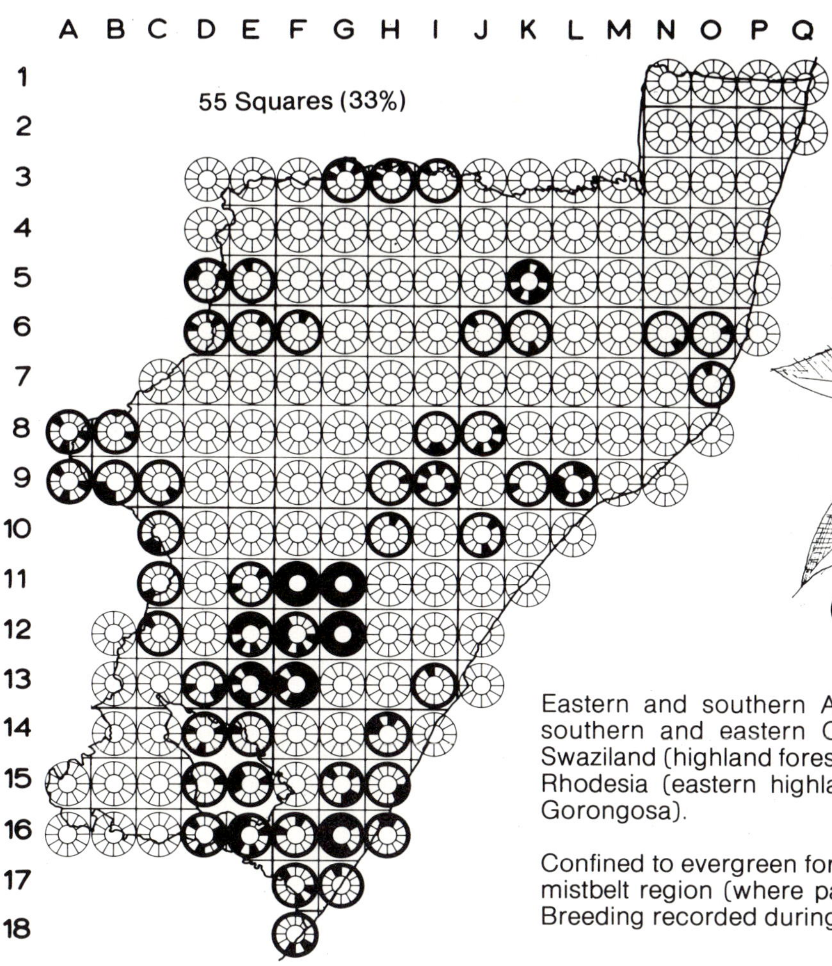

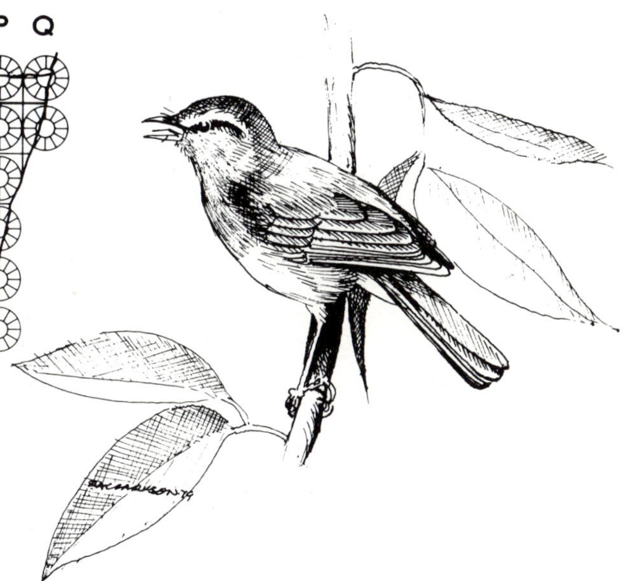

671 Yellow-throated Warbler

Seicercus ruficapillus

Eastern and southern Africa. In Southern Africa, the south-western, southern and eastern Cape, Transkei, Natal and Zululand, western Swaziland (highland forests) and eastern and northern Transvaal; again in Rhodesia (eastern highlands) and adjacent Moçambique (also Mount Gorongosa).

Confined to evergreen forests, from the Drakensberg Range, through the mistbelt region (where particularly common) and sparsely to the coast. Breeding recorded during October and November.

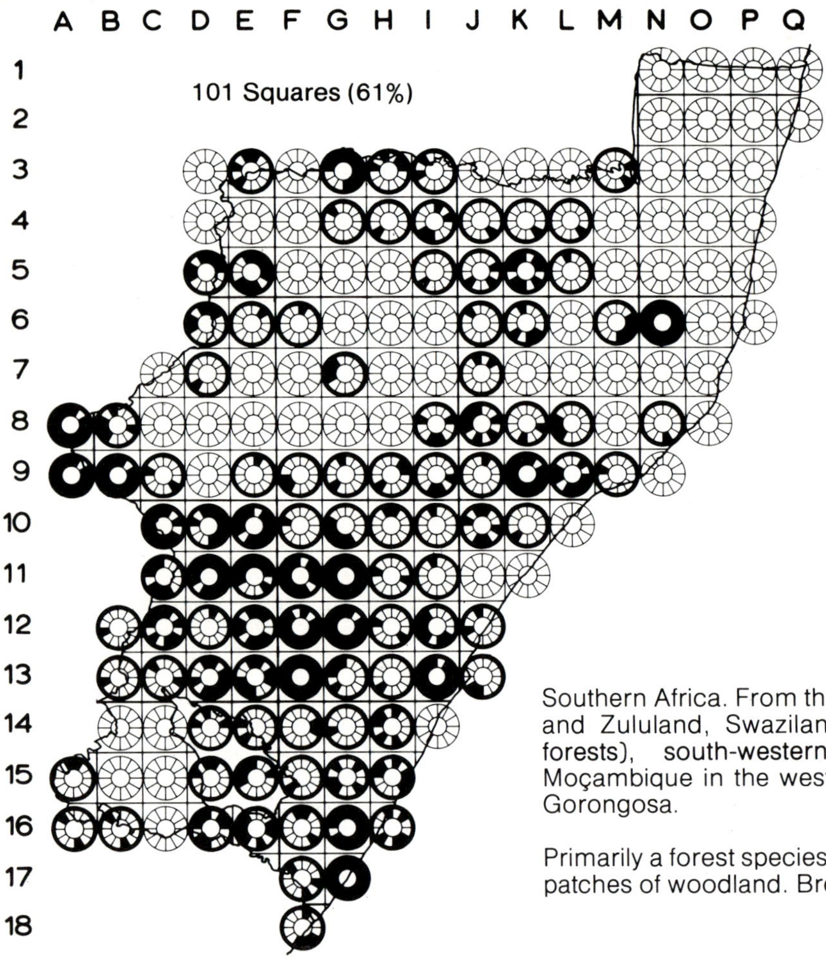

672 Cape Batis

Batis capensis

Southern Africa. From the south-western and southern Cape east to Natal and Zululand, Swaziland, eastern and northern Transvaal (highland forests), south-western and eastern Rhodesia (highland); also Moçambique in the western highlands, Lebombo Range and on Mount Gorongosa.

Primarily a forest species, which extends into exotic plantations and dense patches of woodland. Breeding recorded from September to December.

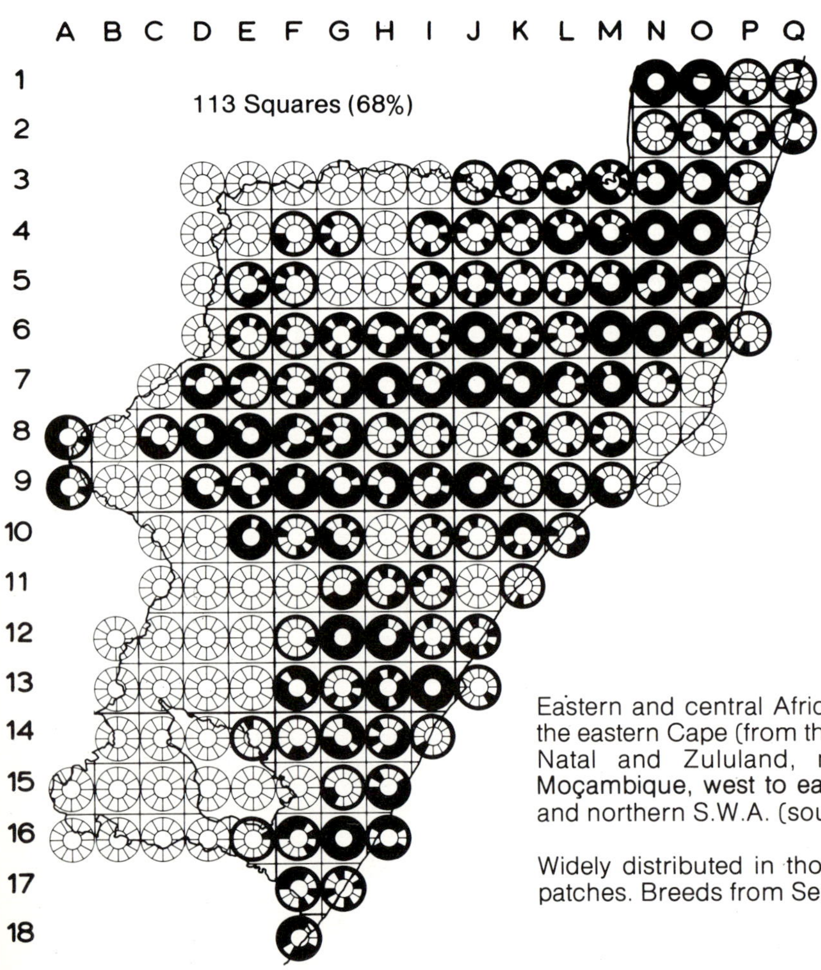

673 Chin Spot Batis

Batis molitor

Eastern and central Africa, from Kenya to the Cape. In Southern Africa, the eastern Cape (from the middle and lower Great Fish River) to Transkei, Natal and Zululand, northern O.F.S., Swaziland, Transvaal and Moçambique, west to eastern and northern Botswana, the Caprivi Strip and northern S.W.A. (south to Etosha).

Widely distributed in thornveld and dry woodland, even in the smallest patches. Breeds from September to January.

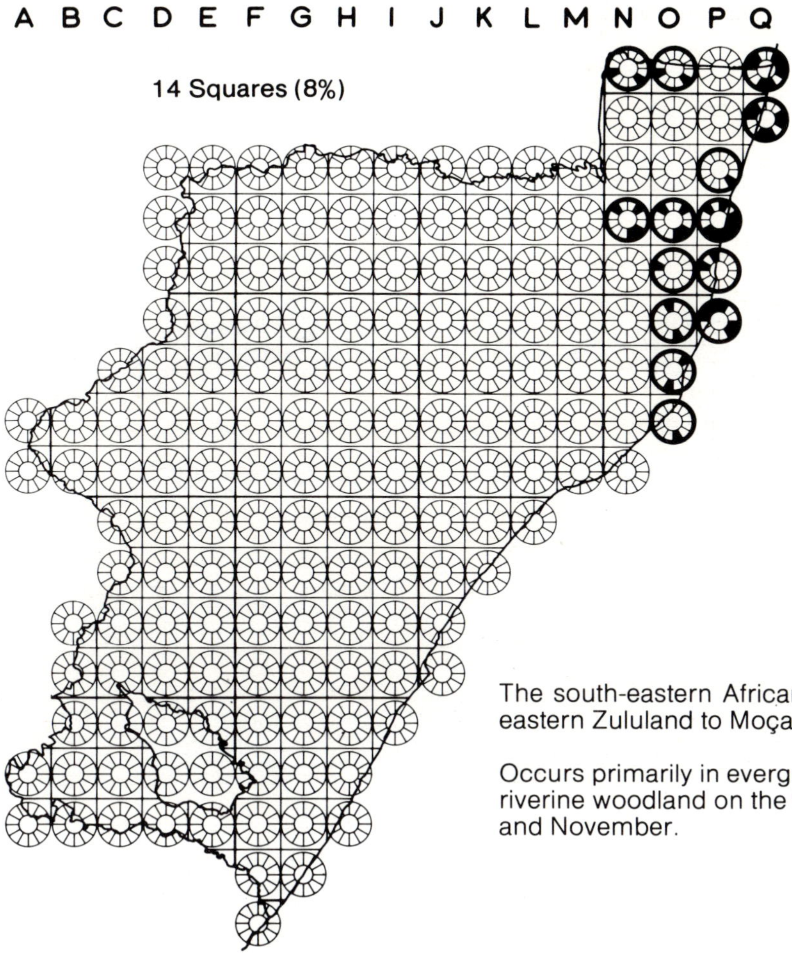

676 Woodwards' Batis

Batis fratrum

The south-eastern African littoral south to Zululand. In Southern Africa, eastern Zululand to Moçambique.

Occurs primarily in evergreen forest on the coastal dunes; occasionally in riverine woodland on the littoral plain. Breeding recorded during October and November.

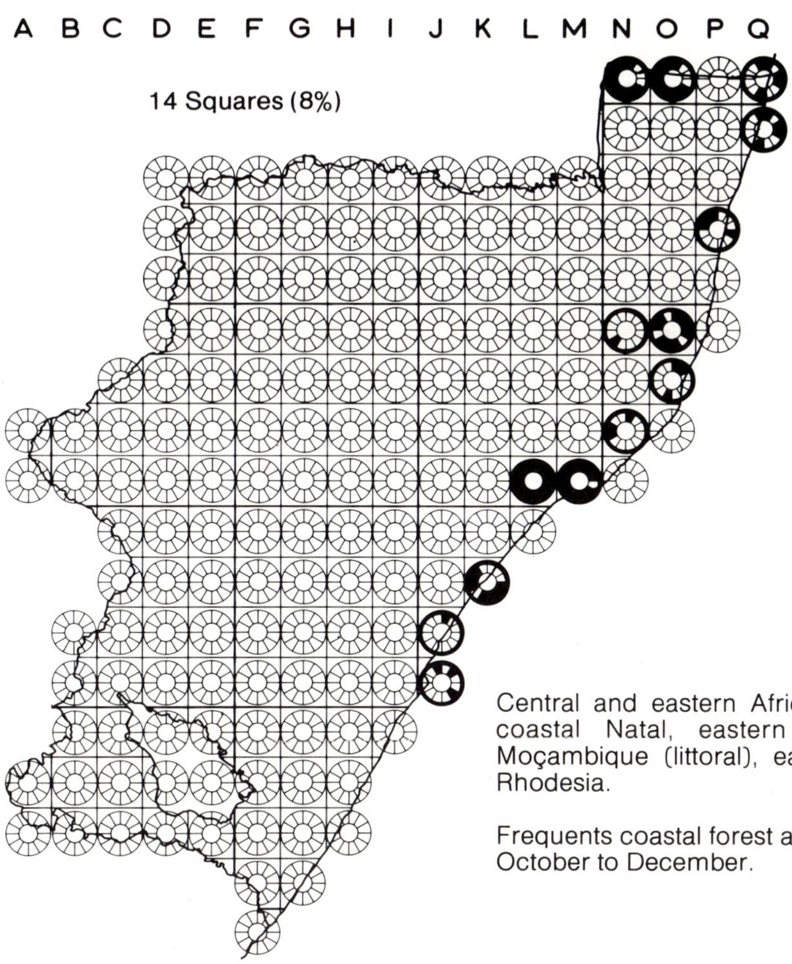

677 Wattle-eyed Flycatcher

Platysteira peltata

Central and eastern Africa, from Kenya to Natal. In Southern Africa, coastal Natal, eastern Zululand, Swaziland (Lebombo Range), Moçambique (littoral), eastern Transvaal (lowveld) and south-eastern Rhodesia.

Frequents coastal forest and riverine woodland. Recorded breeding from October to December.

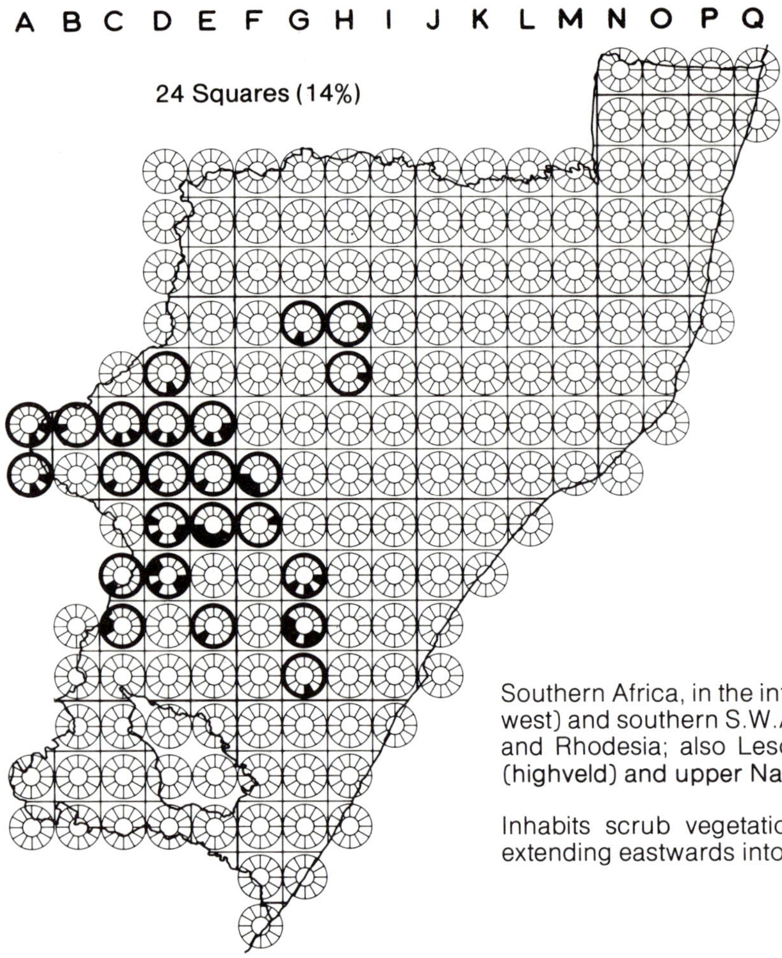

678 Fairy Flycatcher

Stenostira scita

Southern Africa, in the interior and west. The Cape (to the coast only in the west) and southern S.W.A., wintering north to northern S.W.A., Botswana, and Rhodesia; also Lesotho (highlands), wintering in O.F.S., Transvaal (highveld) and upper Natal.

Inhabits scrub vegetation in the foothills of the Drakensberg Range, extending eastwards into thornveld and exotic vegetation.

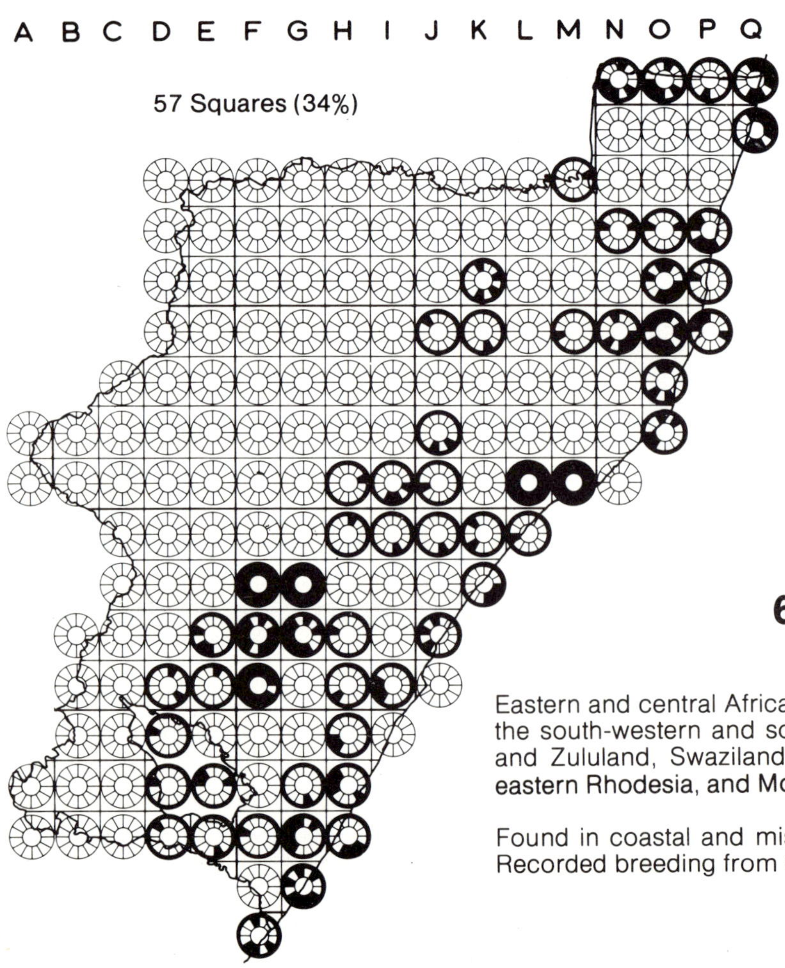

680 Blue-mantled Flycatcher

Trochocercus cyanomelas

Eastern and central Africa, from Somalia to the Cape. In Southern Africa, the south-western and southern Cape (from Swellendam) east to Natal and Zululand, Swaziland, eastern and northern Transvaal (highlands), eastern Rhodesia, and Moçambique.

Found in coastal and mistbelt evergreen forest, and riverine woodland. Recorded breeding from October to December.

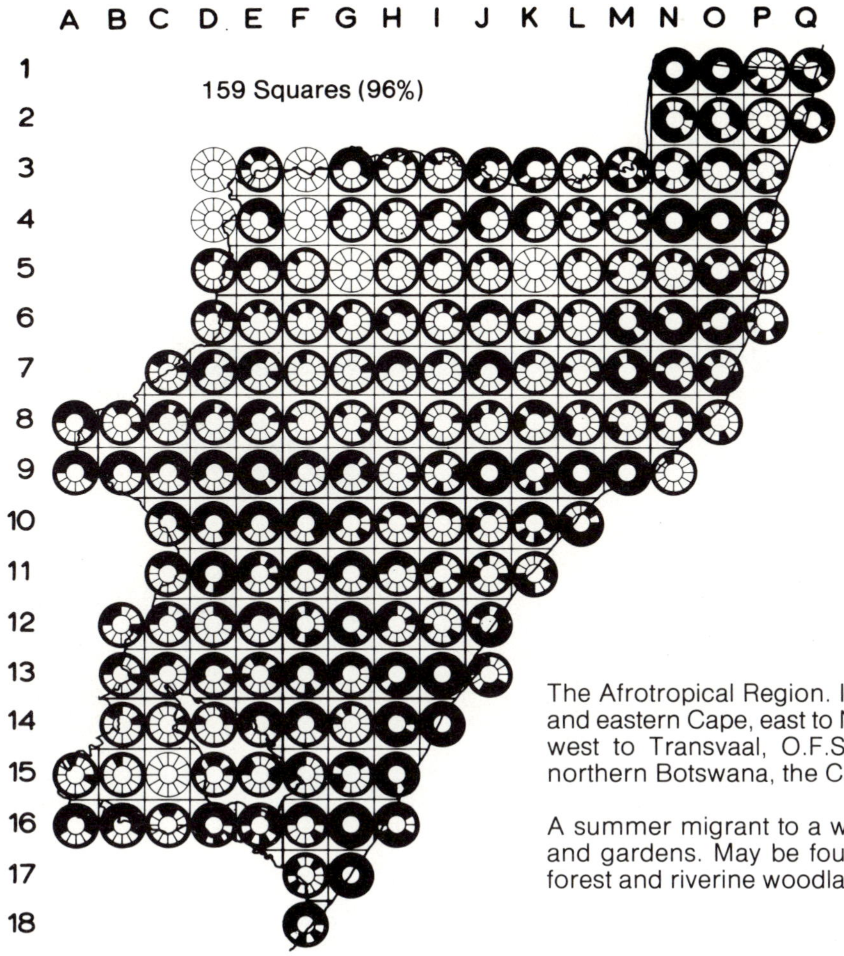

682 Paradise Flycatcher

Terpsiphone viridis

The Afrotropical Region. In Southern Africa, the south-western, southern and eastern Cape, east to Natal and Zululand, Swaziland and Moçambique, west to Transvaal, O.F.S., the northern Cape, Rhodesia, eastern and northern Botswana, the Caprivi Strip and northern S.W.A.

A summer migrant to a wide variety of wooded habitats, including parks and gardens. May be found wintering on the coastal plain in evergreen forest and riverine woodland. Breeds from September to March.

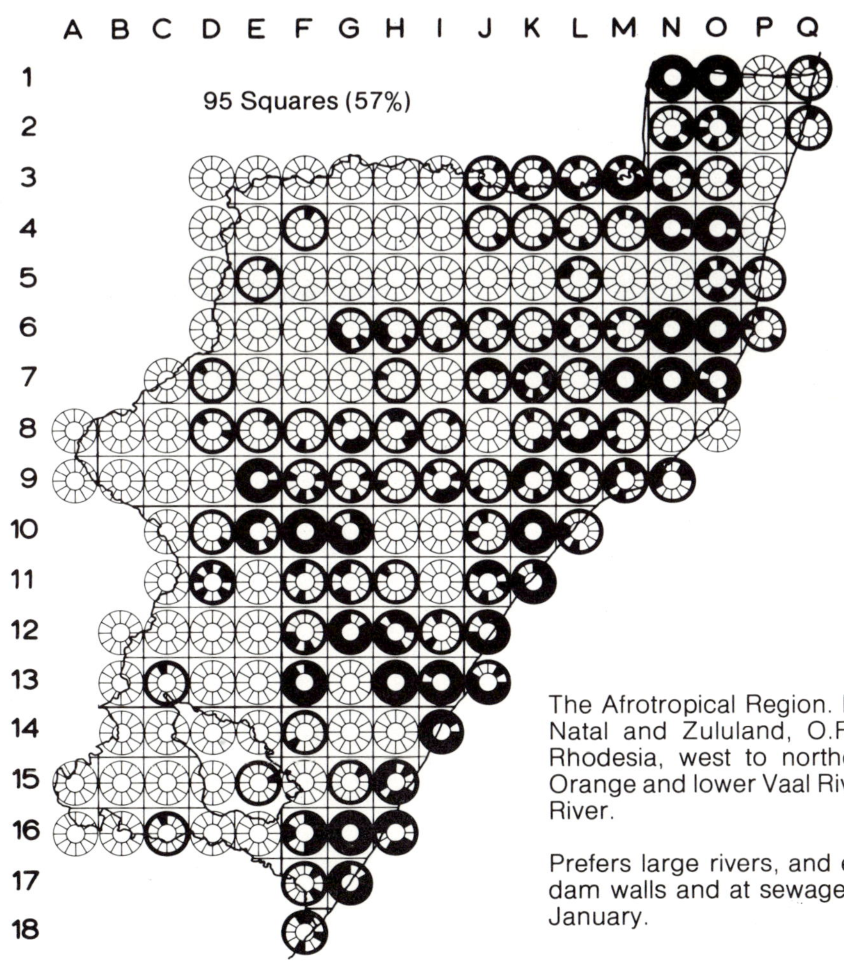

685 African Pied Wagtail

Motacilla aguimp

The Afrotropical Region. In Southern Africa, the eastern Cape, Transkei, Natal and Zululand, O.F.S., Swaziland, Moçambique, Transvaal and Rhodesia, west to northern Botswana and northern S.W.A.; also the Orange and lower Vaal Rivers in the Cape, south in the west to the Olifants River.

Prefers large rivers, and estuaries with exposed margins; also found on dam walls and at sewage farms. Breeding recorded from September to January.

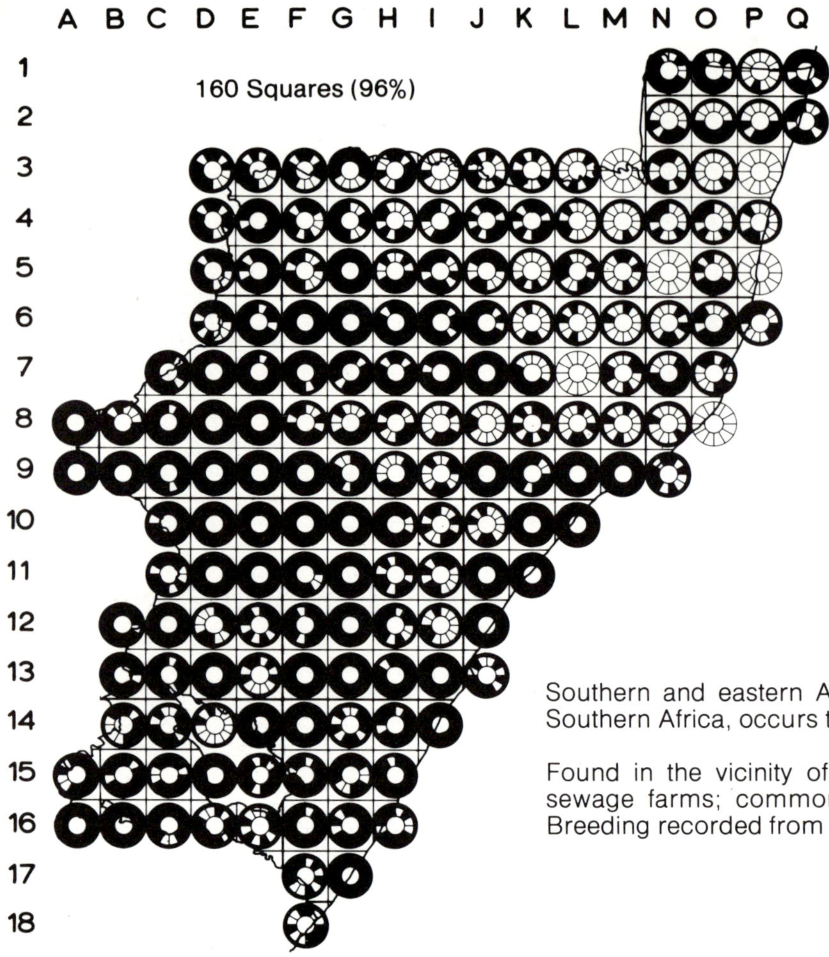

686 Cape Wagtail

Motacilla capensis

Southern and eastern Africa, from Angola and Kenya to the Cape. In Southern Africa, occurs throughout in suitable habitat.

Found in the vicinity of water, with a preference for dams, vleis and sewage farms; commonly encountered at farmsteads and in towns. Breeding recorded from July to January.

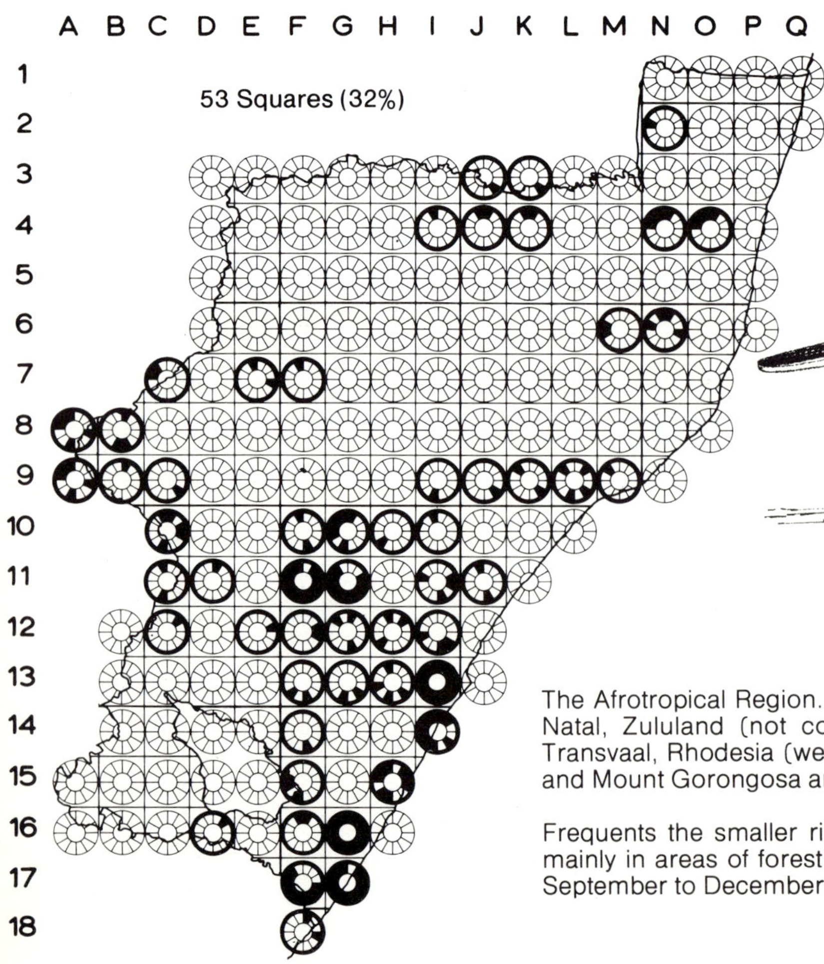

688 Long-tailed Wagtail

Motacilla clara

The Afrotropical Region. In Southern Africa, the eastern Cape, Transkei, Natal, Zululand (not coastal flats), Swaziland, eastern and northern Transvaal, Rhodesia (west to Sengwa River) and adjacent Moçambique, and Mount Gorongosa and the Lebombo Range.

Frequents the smaller rivers and streams with overhanging vegetation, mainly in areas of forest or riverine woodland. Recorded breeding from September to December.

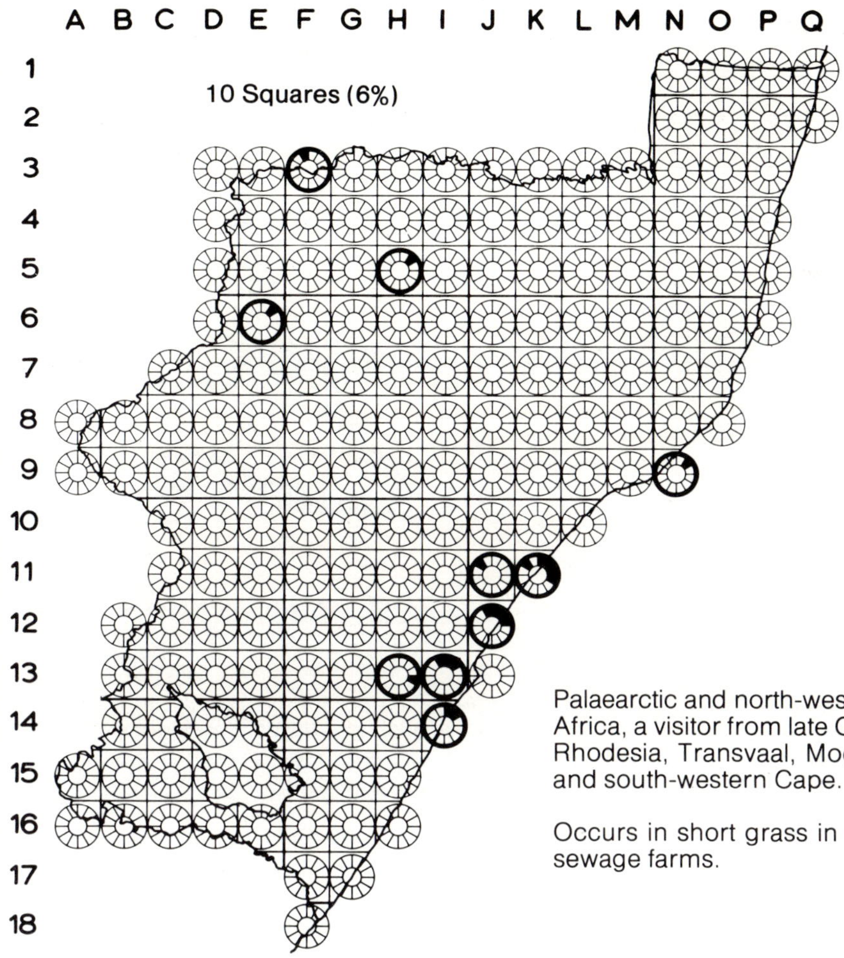

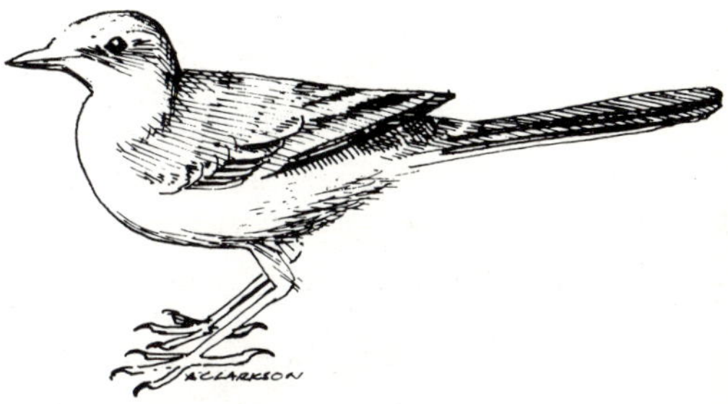

689 Yellow Wagtail

Motacilla flava

Palaearctic and north-western Nearctic, wintering to the south. In Southern Africa, a visitor from late October/November to April to S.W.A., Botswana, Rhodesia, Transvaal, Moçambique, Natal and Zululand, and the eastern and south-western Cape.

Occurs in short grass in moist situations, such as at estuaries, vleis and sewage farms.

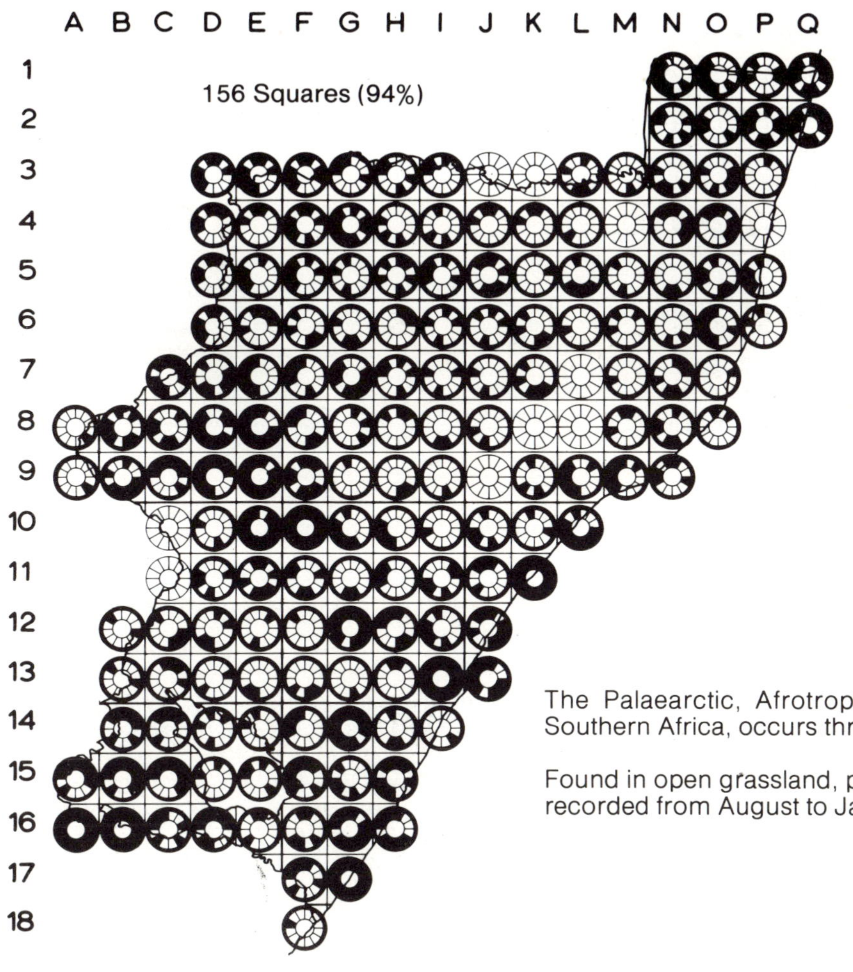

692 Richard's Pipit

Anthus novaeseelandiae

The Palaearctic, Afrotropical, Oriental and Australasian Regions. In Southern Africa, occurs throughout in suitable habitat.

Found in open grassland, preferring short grass on flat ground. Breeding recorded from August to January.

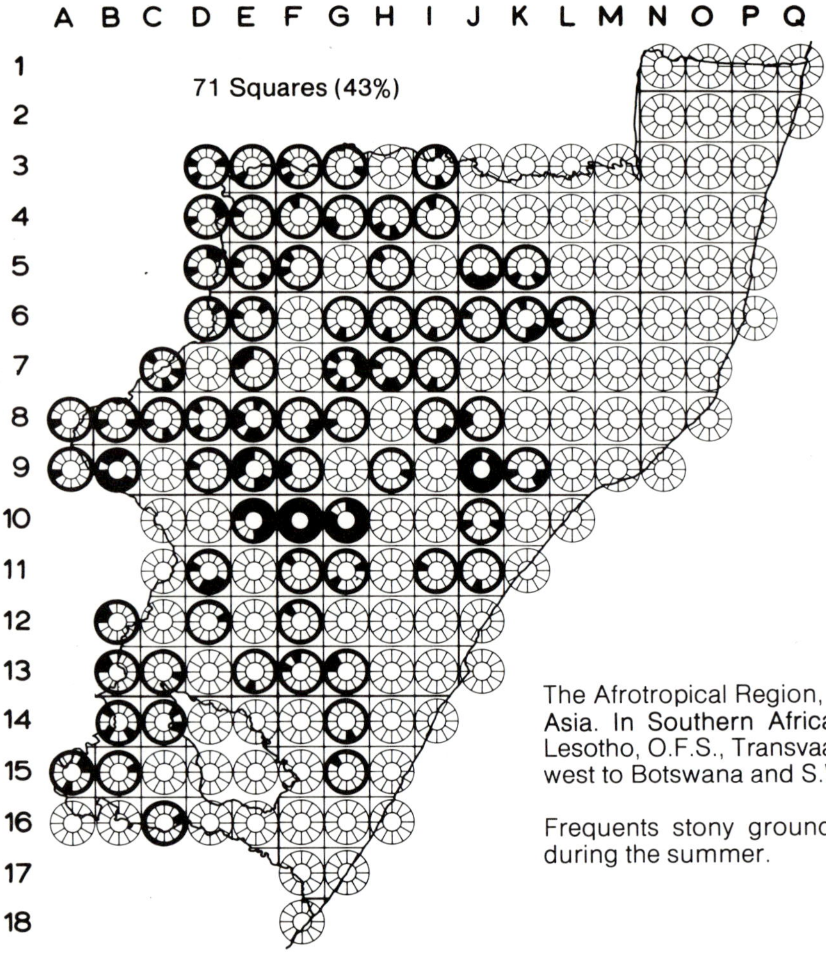

693 Nicholson's Pipit

Anthus similis

The Afrotropical Region, the Middle East, and south-central and southern Asia. In Southern Africa, the Cape east to western and upper Natal, Lesotho, O.F.S., Transvaal, Rhodesia (plateau) and adjacent Moçambique, west to Botswana and S.W.A.

Frequents stony ground in open grassland, often on hilltops. Breeds during the summer.

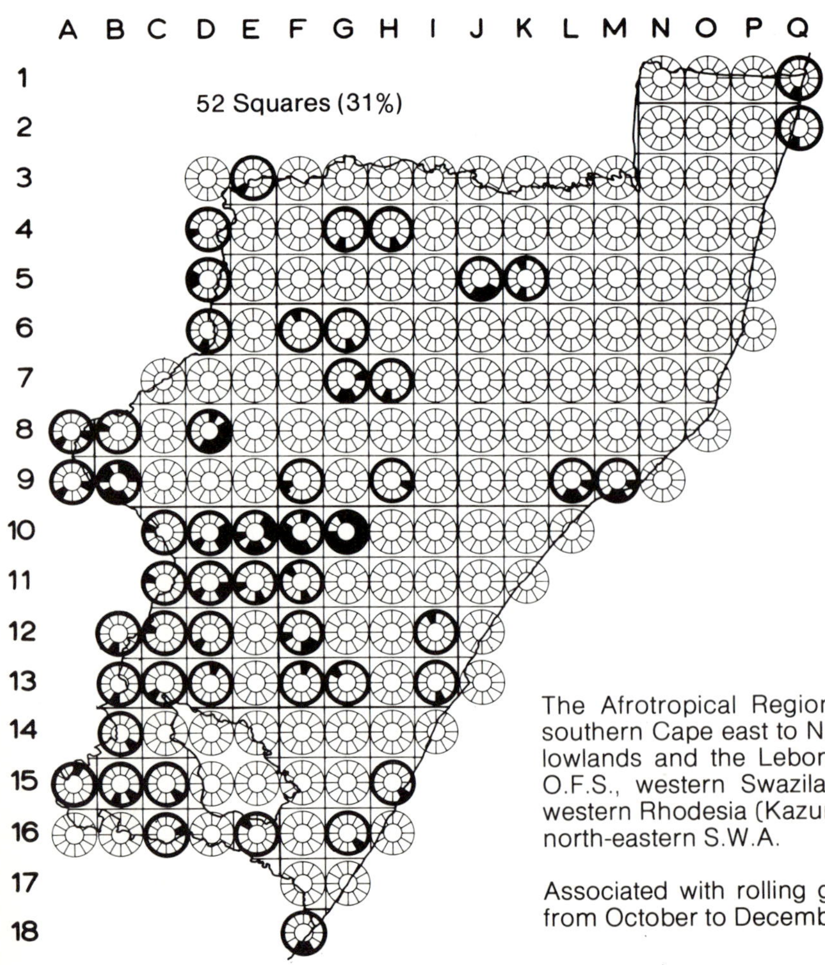

694 Plain-backed Pipit

Anthus leucophrys

The Afrotropical Region. In Southern Africa, the south-western and southern Cape east to Natal and western Zululand (reaching the eastern lowlands and the Lebombo Range in the winter), Lesotho (lowlands), O.F.S., western Swaziland and Transvaal (highveld). Again in north-western Rhodesia (Kazungula), northern Botswana, the Caprivi Strip and north-eastern S.W.A.

Associated with rolling grassland generally above 1200 m a.s.l. Breeds from October to December.

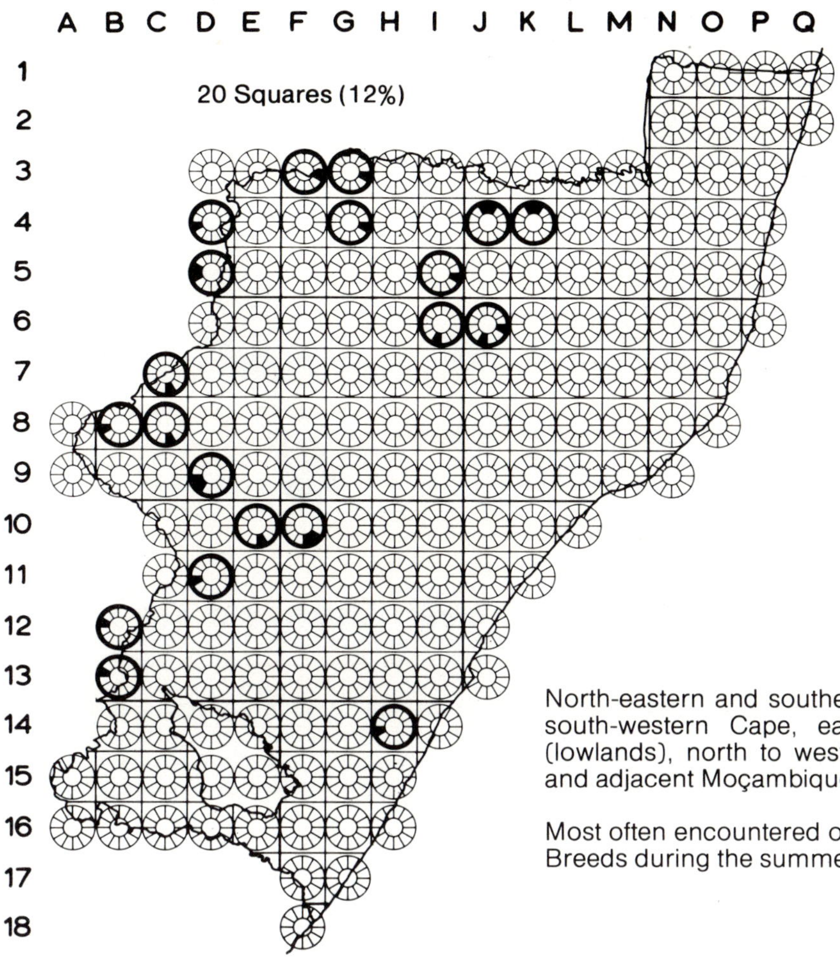

695 Buffy Pipit

Anthus vaalensis

North-eastern and southern Africa. In Southern Africa, the western and south-western Cape, east to O.F.S., Transkei, Natal and Lesotho (lowlands), north to western Swaziland, Transvaal, Rhodesia (plateau) and adjacent Moçambique, west to Botswana and S.W.A.

Most often encountered on hillsides with short grass and rocky outcrops. Breeds during the summer.

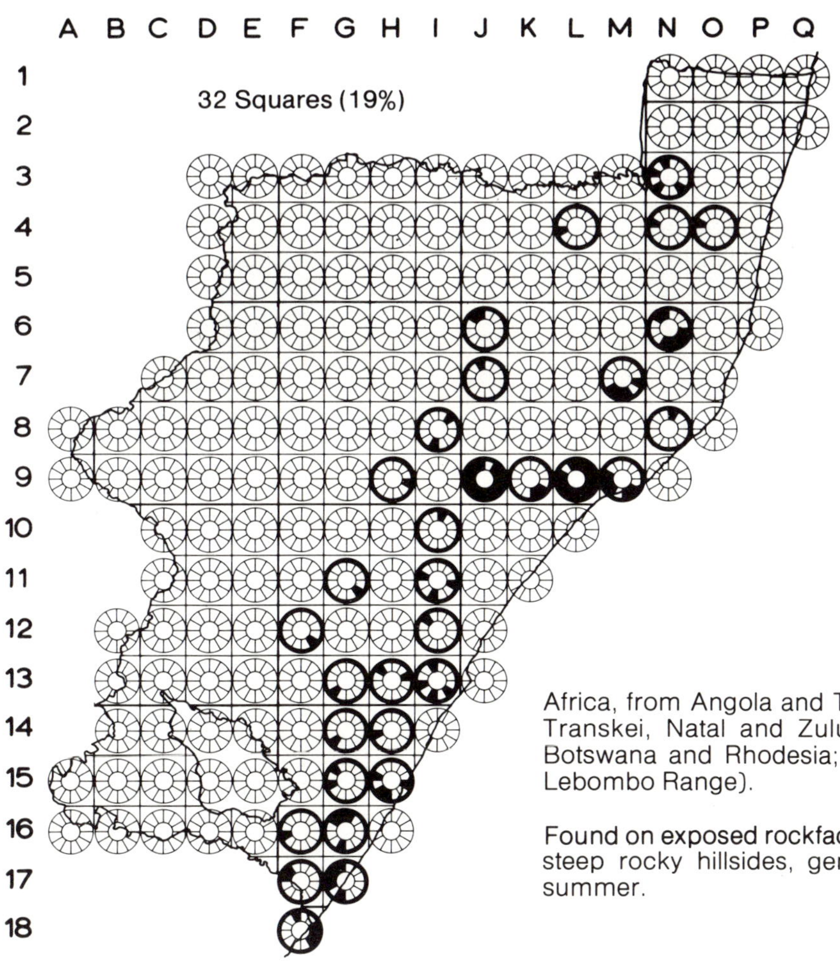

696 Striped Pipit

Anthus lineiventris

Africa, from Angola and Tanzania south to Transkei. In Southern Africa, Transkei, Natal and Zululand, eastern Swaziland, Transvaal, eastern Botswana and Rhodesia; also in Moçambique (western highlands and Lebombo Range).

Found on exposed rockfaces, such as in gorges and road cuttings, and on steep rocky hillsides, generally below 900 m a.s.l. Breeds during the summer.

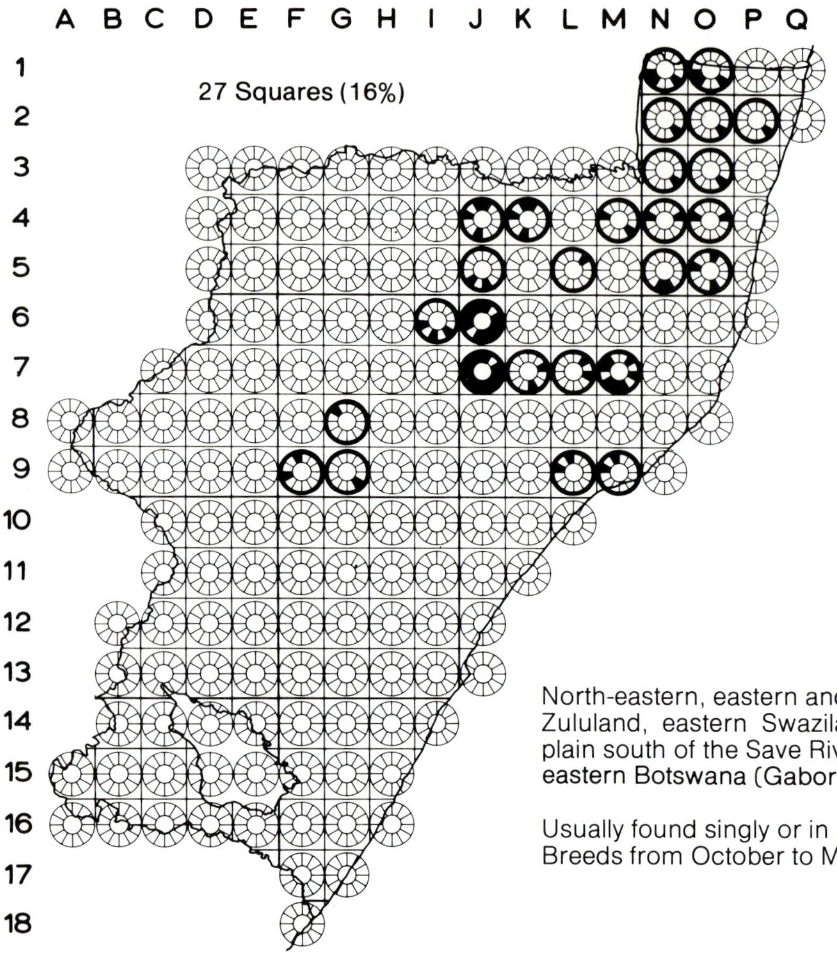

699 Bushveld Tree Pipit

Anthus caffer

North-eastern, eastern and southern Africa. In Southern Africa, Natal and Zululand, eastern Swaziland, Transvaal (bushveld), Moçambique (the plain south of the Save River), Rhodesia (plateau) and eastern and north-eastern Botswana (Gaborone north to Nata).

Usually found singly or in pairs in open savanna thornveld and woodland. Breeds from October to March.

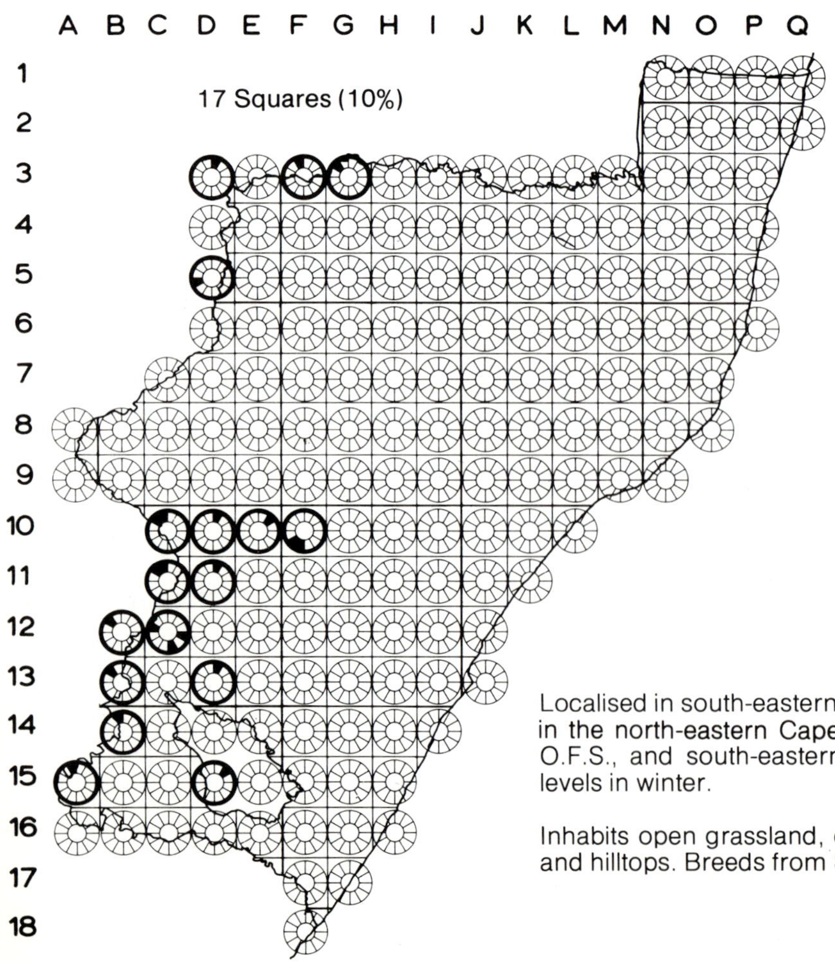

701 Yellow-breasted Pipit

Anthus chloris

Localised in south-eastern Africa. In Southern Africa, breeds at high levels in the north-eastern Cape, western and upper Natal, Lesotho, adjacent O.F.S., and south-eastern Transvaal (Wakkerstroom); ranges to lower levels in winter.

Inhabits open grassland, often with rocky outcrops, on mountain slopes and hilltops. Breeds from September to January.

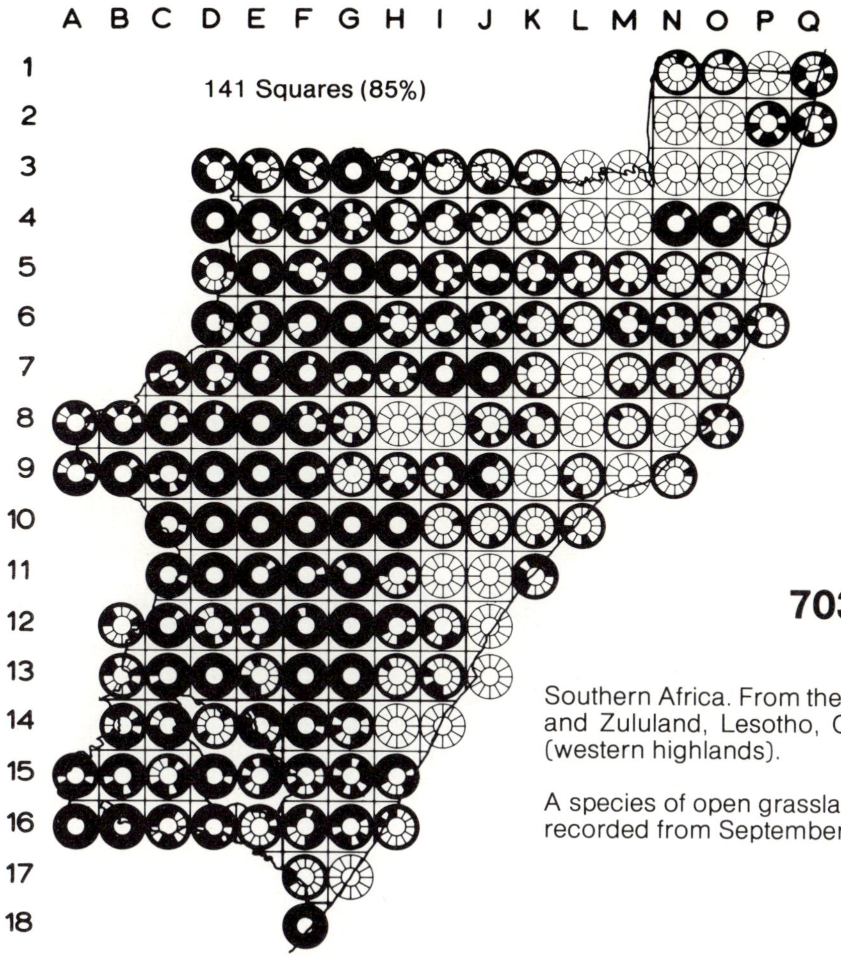

703 Orange-throated Longclaw

Macronyx capensis

Southern Africa. From the south-western and southern Cape, east to Natal and Zululand, Lesotho, O.F.S., Swaziland, Transvaal and Moçambique (western highlands).

A species of open grassland, most common above 600 m a.s.l. Breeding recorded from September to March.

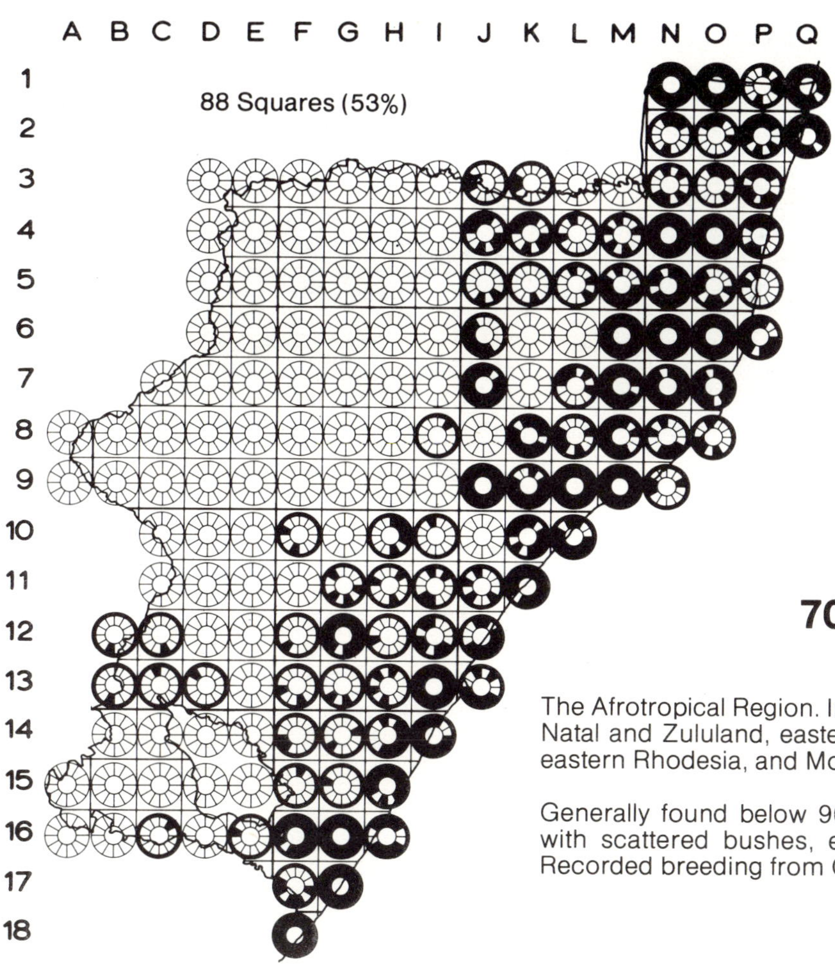

704 Yellow-throated Longclaw

Macronyx croceus

The Afrotropical Region. In Southern Africa, coastal districts of Transkei to Natal and Zululand, eastern Swaziland, eastern and northern Transvaal, eastern Rhodesia, and Moçambique.

Generally found below 900 m a.s.l., where it frequents open grassland with scattered bushes, extending into light woodland and thornveld. Recorded breeding from October to February.

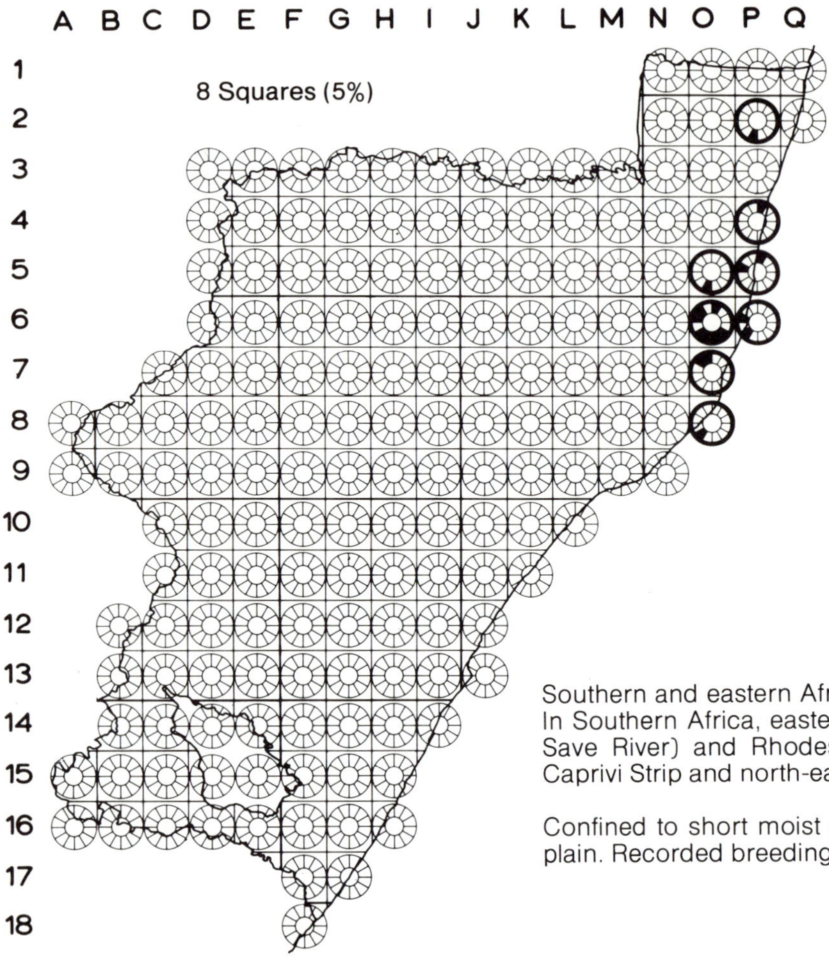

705 Pink-throated Longclaw

Macronyx ameliae

Southern and eastern Africa, from Angola and Kenya southeast to Natal. In Southern Africa, eastern Zululand, coastal Moçambique (south of the Save River) and Rhodesia (plateau), west to northern Botswana, the Caprivi Strip and north-eastern S.W.A.

Confined to short moist grassland surrounding marshes on the coastal plain. Recorded breeding from October to December.

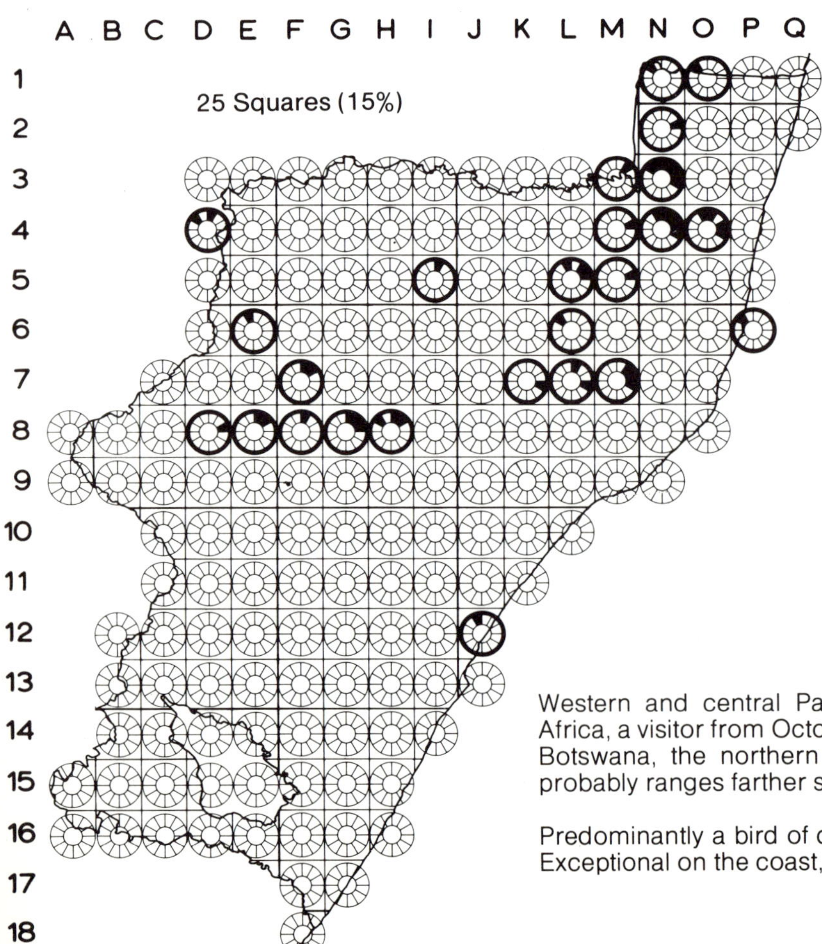

706 Lesser Grey Shrike

Lanius minor

Western and central Palaearctic, wintering to the south. In Southern Africa, a visitor from October to April to the northern half of S.W.A., east to Botswana, the northern Cape, O.F.S., Transvaal, Natal and Zululand; probably ranges farther south.

Predominantly a bird of dry thornveld, often on the edges of open areas. Exceptional on the coast, in open situations.

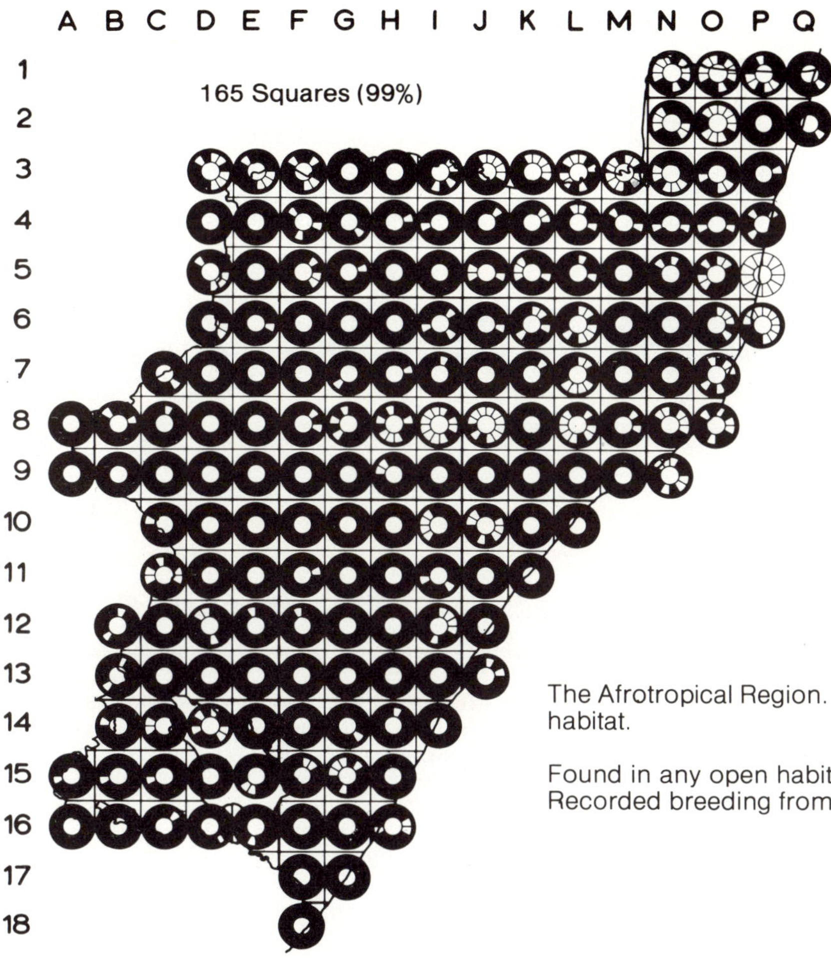

707 Fiscal Shrike

Lanius collaris

The Afrotropical Region. In Southern Africa, occurs throughout in suitable habitat.

Found in any open habitat provided some trees are available for nesting. Recorded breeding from June to January.

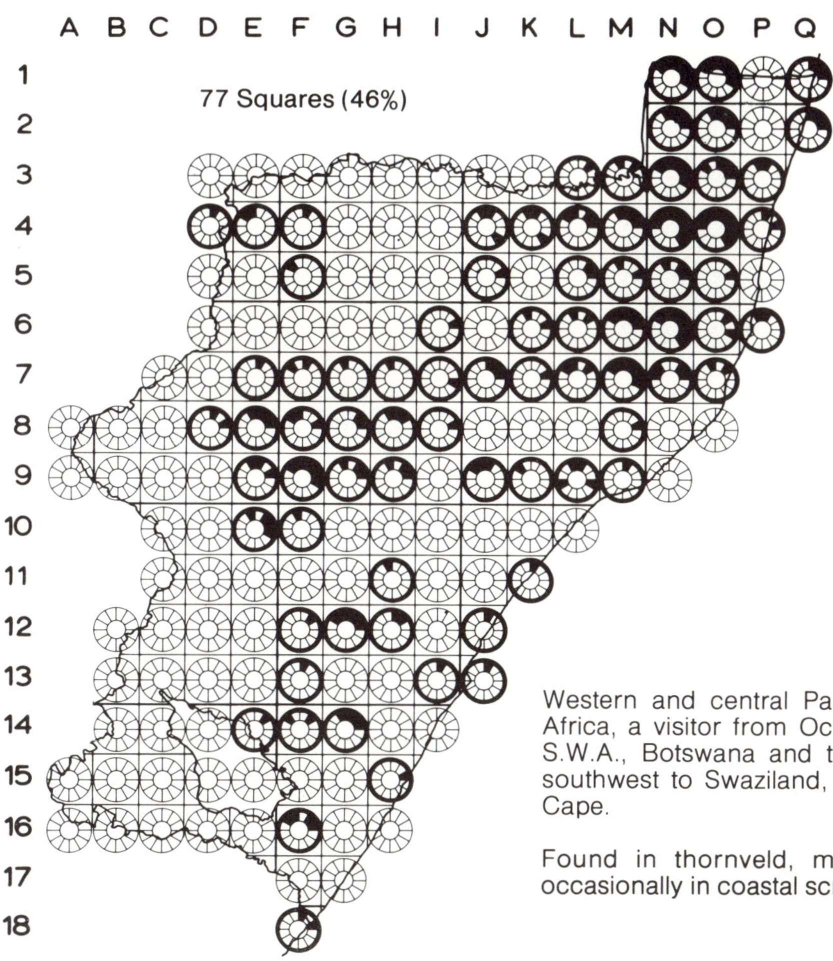

708 Red-backed Shrike

Lanius collurio

Western and central Palaearctic, wintering to the south. In Southern Africa, a visitor from October/November to April, occasionally May, to S.W.A., Botswana and the northern Cape, east to Moçambique, and southwest to Swaziland, Natal and Zululand, Transkei and the eastern Cape.

Found in thornveld, mixed woodland and thickly-wooded valleys; occasionally in coastal scrub.

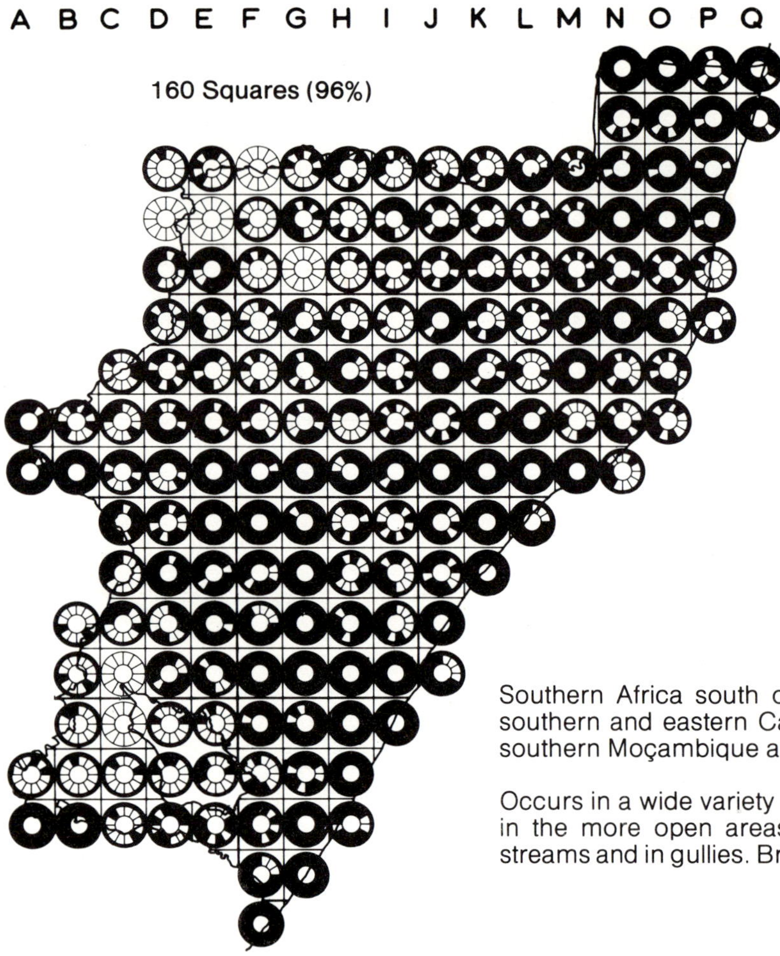

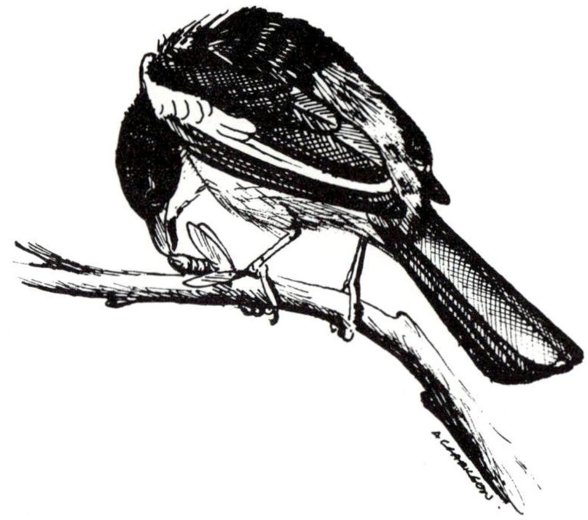

709 Boubou Shrike

Laniarius ferrugineus

Southern Africa south of the Limpopo River. From the south-western, southern and eastern Cape to Transkei, Natal and Zululand, Swaziland, southern Moçambique and Transvaal.

Occurs in a wide variety of wooded habitats, preferring dense vegetation; in the more open areas, confined to thickets and dense scrub along streams and in gullies. Breeding recorded from September to December.

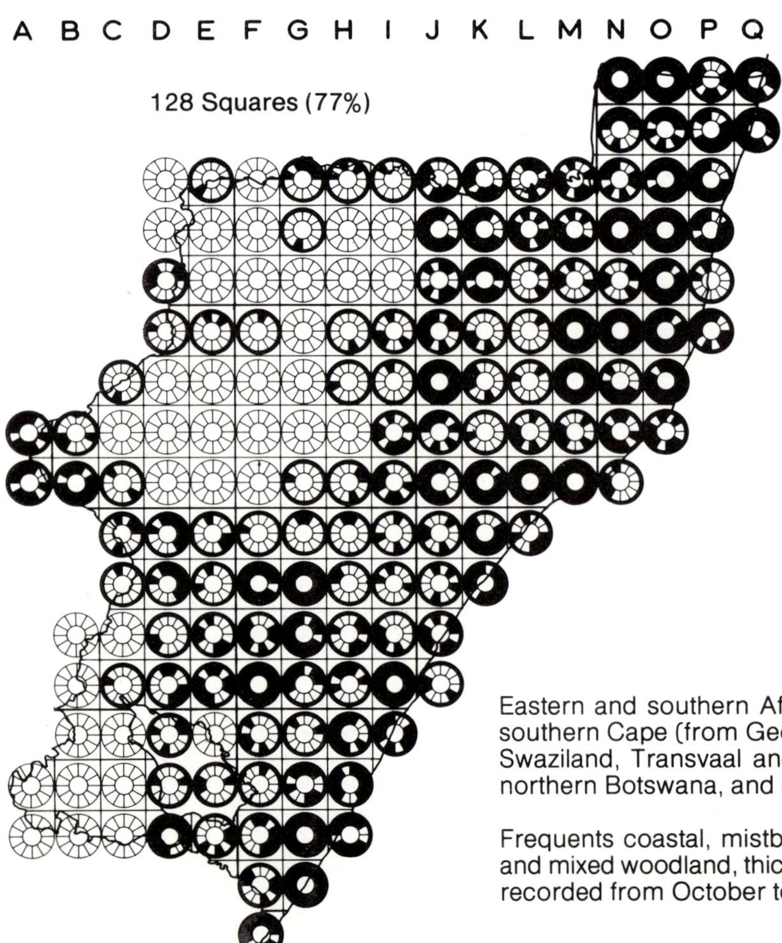

712 Puffback Shrike

Dryoscopus cubla

Eastern and southern Africa, south to the Cape. In Southern Africa, the southern Cape (from George), northeast to Transkei, Natal and Zululand, Swaziland, Transvaal and Moçambique, west to Rhodesia, eastern and northern Botswana, and central and northern S.W.A.

Frequents coastal, mistbelt and Drakensberg evergreen forests, riverine and mixed woodland, thickly-wooded valleys and moist thornveld. Breeding recorded from October to December.

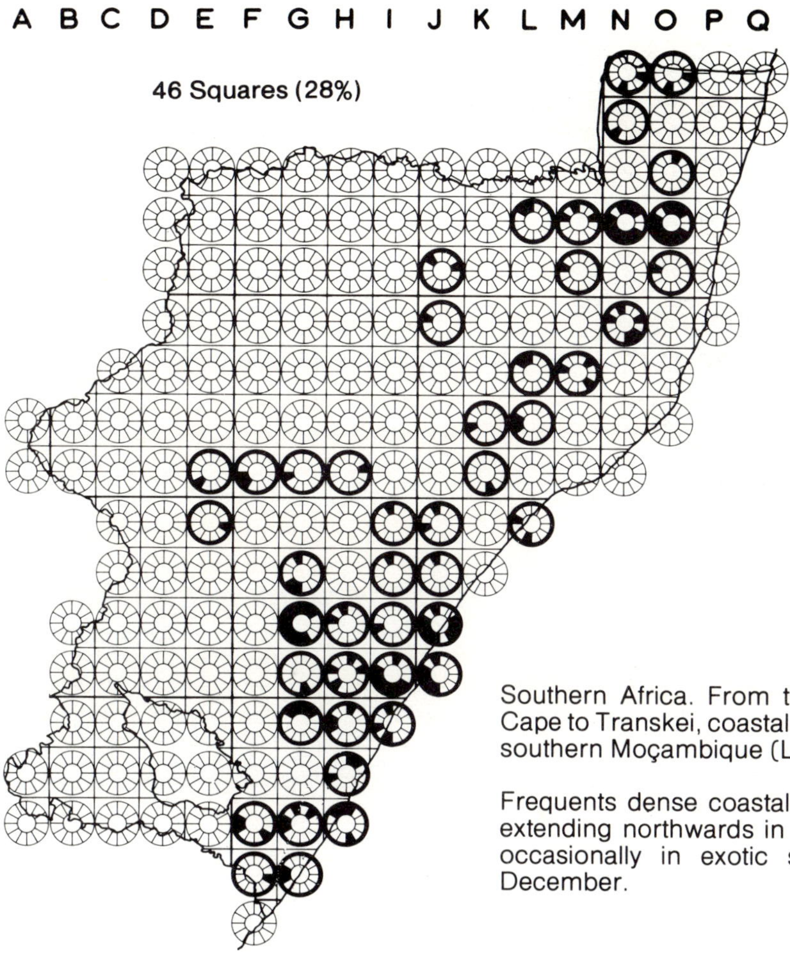

713 Tchagra

Tchagra tchagra

Southern Africa. From the south-western, southern and south-eastern Cape to Transkei, coastal and midland Natal, Zululand, eastern Swaziland, southern Moçambique (Lebombo Range), and eastern Transvaal.

Frequents dense coastal bush and thickly-wooded valleys in the south, extending northwards in thornveld, where generally in the thicker parts; occasionally in exotic scrub. Recorded breeding from October to December.

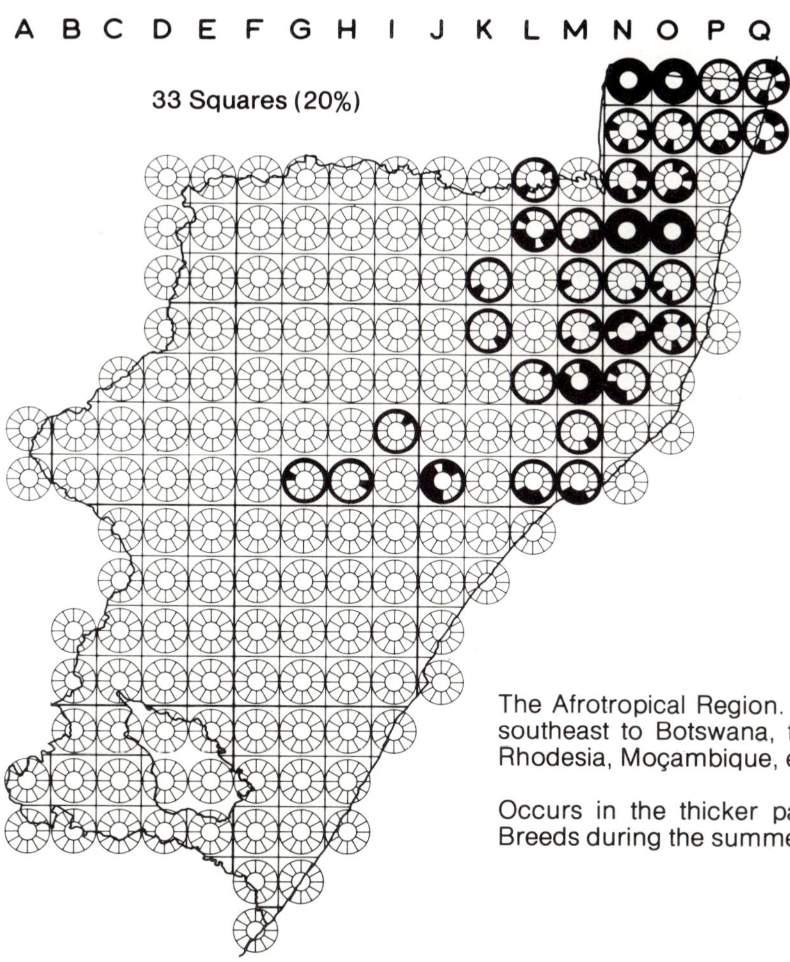

714 Three-streaked Tchagra

Tchagra australis

The Afrotropical Region. In Southern Africa, the northern half of S.W.A. southeast to Botswana, the northern Cape, western O.F.S., Transvaal, Rhodesia, Moçambique, eastern Swaziland and Zululand.

Occurs in the thicker parts of mixed woodland and moist thornveld. Breeds during the summer.

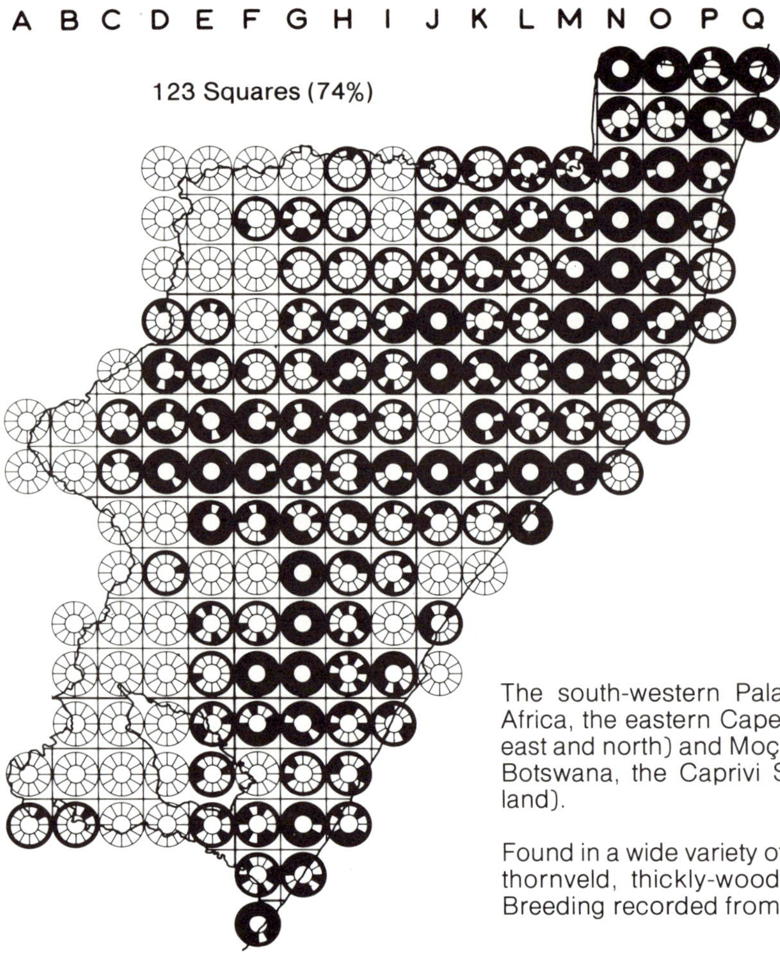

715 Black-crowned Tchagra

Tchagra senegala

The south-western Palaearctic and Afrotropical Regions. In Southern Africa, the eastern Cape, Transkei, Natal and Zululand, Transvaal (in the east and north) and Moçambique, west to Rhodesia, eastern and northern Botswana, the Caprivi Strip and S.W.A. (in Ovamboland and Damaraland).

Found in a wide variety of habitats including coastal bush, mixed woodland, thornveld, thickly-wooded valleys, open bush and exotic plantations. Breeding recorded from October to February.

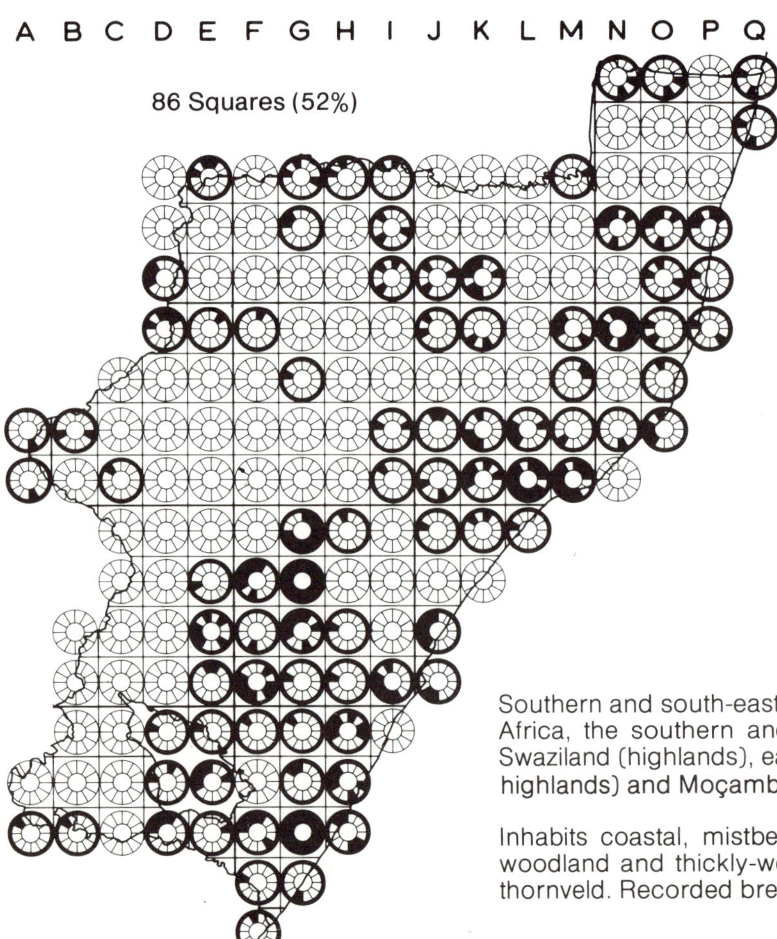

717 Olive Bush Shrike

Malaconotus olivaceus

Southern and south-eastern Africa, from Malawi to the Cape. In Southern Africa, the southern and eastern Cape, Transkei, Natal and Zululand, Swaziland (highlands), eastern and northern Transvaal, Rhodesia (eastern highlands) and Moçambique.

Inhabits coastal, mistbelt and Drakensberg evergreen forests, riverine woodland and thickly-wooded valleys; occasionally in dense patches of thornveld. Recorded breeding from September to December.

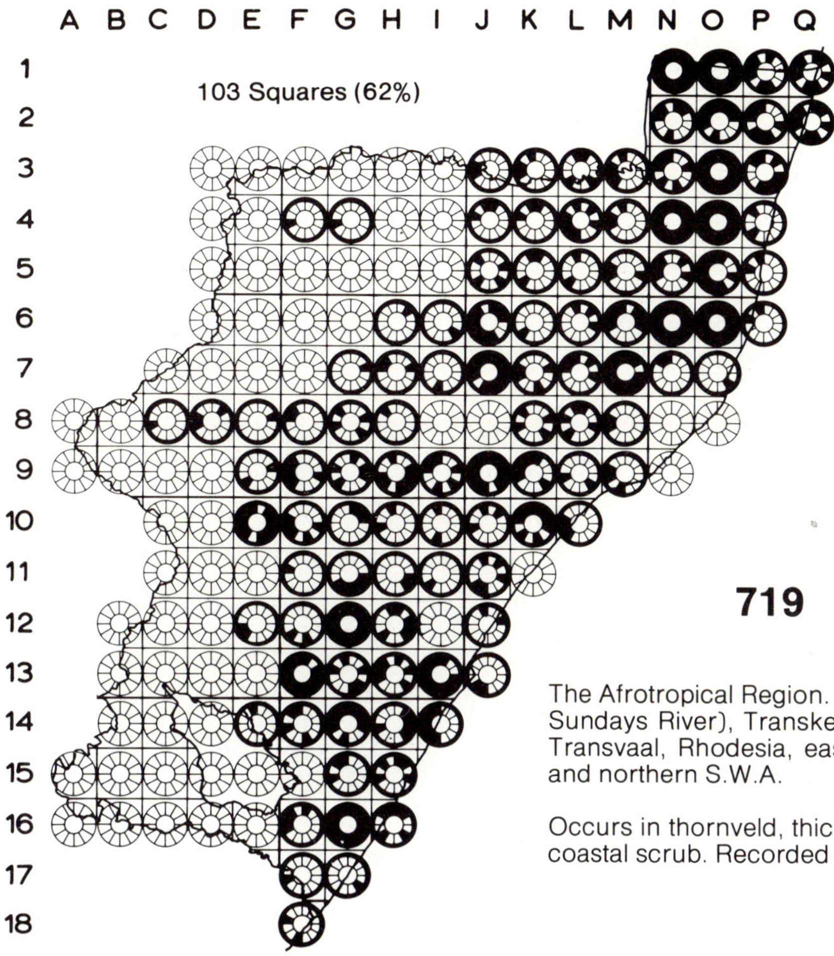

719 Orange-breasted Bush Shrike

Malaconotus sulfureopectus

The Afrotropical Region. In Southern Africa, the eastern Cape (east of the Sundays River), Transkei, Natal and Zululand, Swaziland, Moçambique, Transvaal, Rhodesia, eastern and northern Botswana, the Caprivi Strip and northern S.W.A.

Occurs in thornveld, thickly-wooded valleys and mixed woodland; also in coastal scrub. Recorded breeding from September to January.

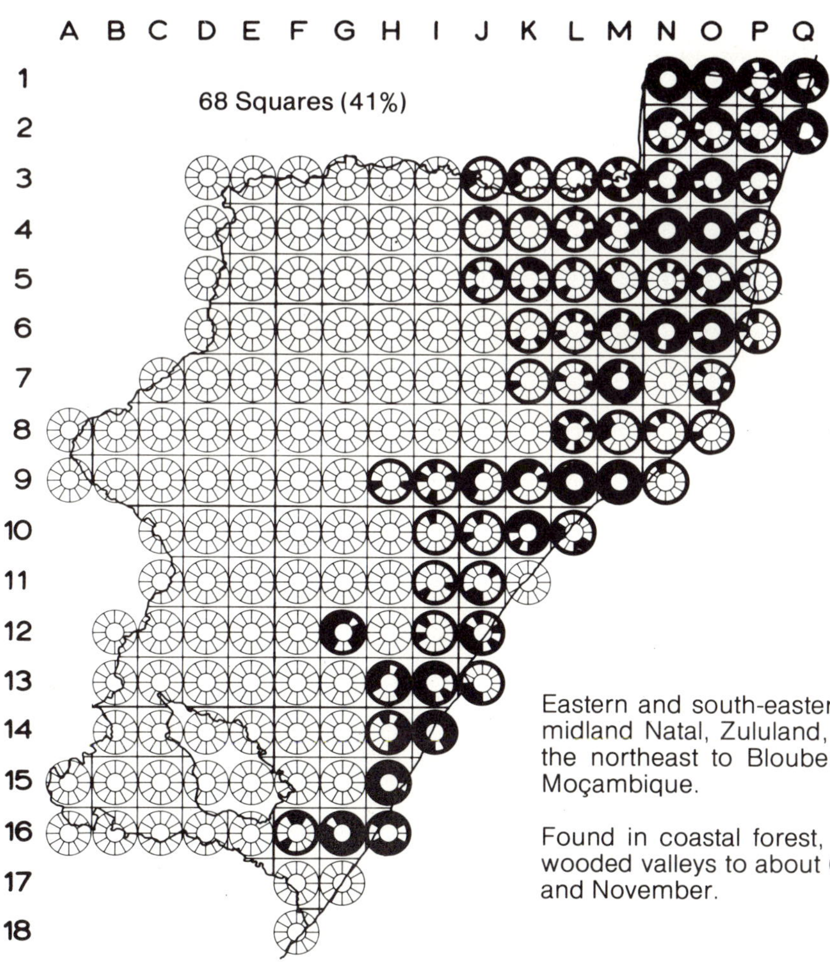

721 Gorgeous Bush Shrike

Telophorus quadricolor

Eastern and south-eastern Africa. In Southern Africa, coastal and lower midland Natal, Zululand, Swaziland, eastern and northern Transvaal (in the northeast to Blouberg), south-eastern and eastern Rhodesia, and Moçambique.

Found in coastal forest, mixed and riverine woodland, and up thickly-wooded valleys to about 600 m a.s.l. Recorded breeding during October and November.

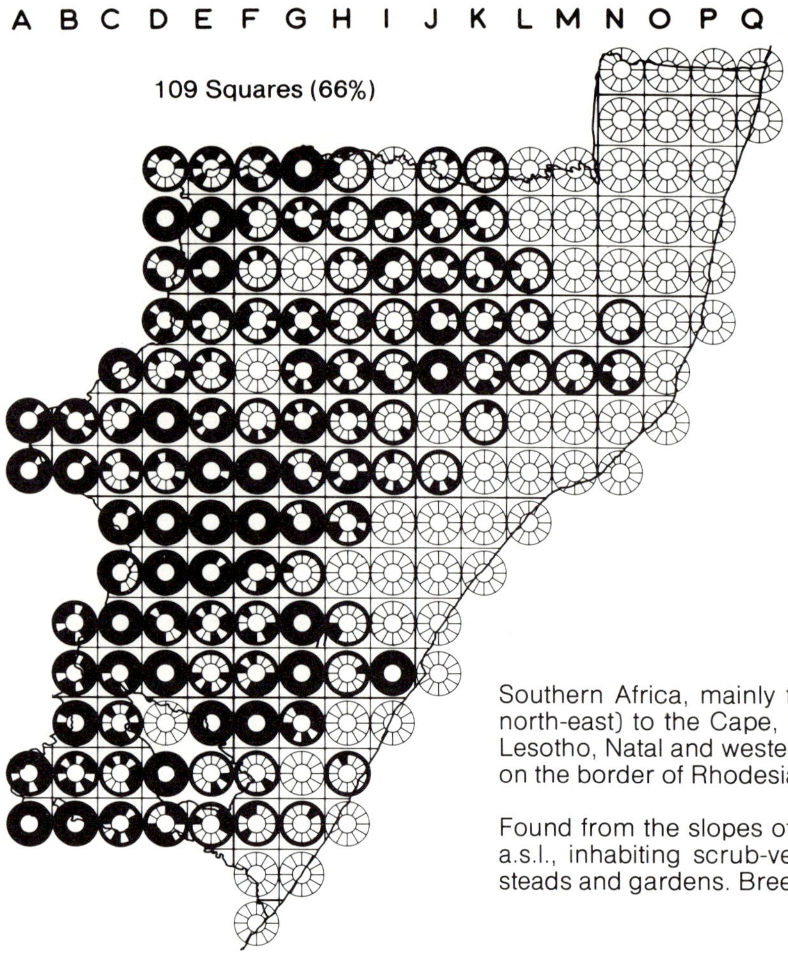

722 Bokmakierie

Telophorus zeylonus

Southern Africa, mainly the interior and west. From S.W.A. (except the north-east) to the Cape, O.F.S., southern Transvaal, western Swaziland, Lesotho, Natal and western Zululand; also in the Chimanimani Mountains on the border of Rhodesia and Moçambique.

Found from the slopes of the Drakensberg Range down to about 600 m a.s.l., inhabiting scrub-vegetation on rocky hillsides; also around farmsteads and gardens. Breeds from September to March.

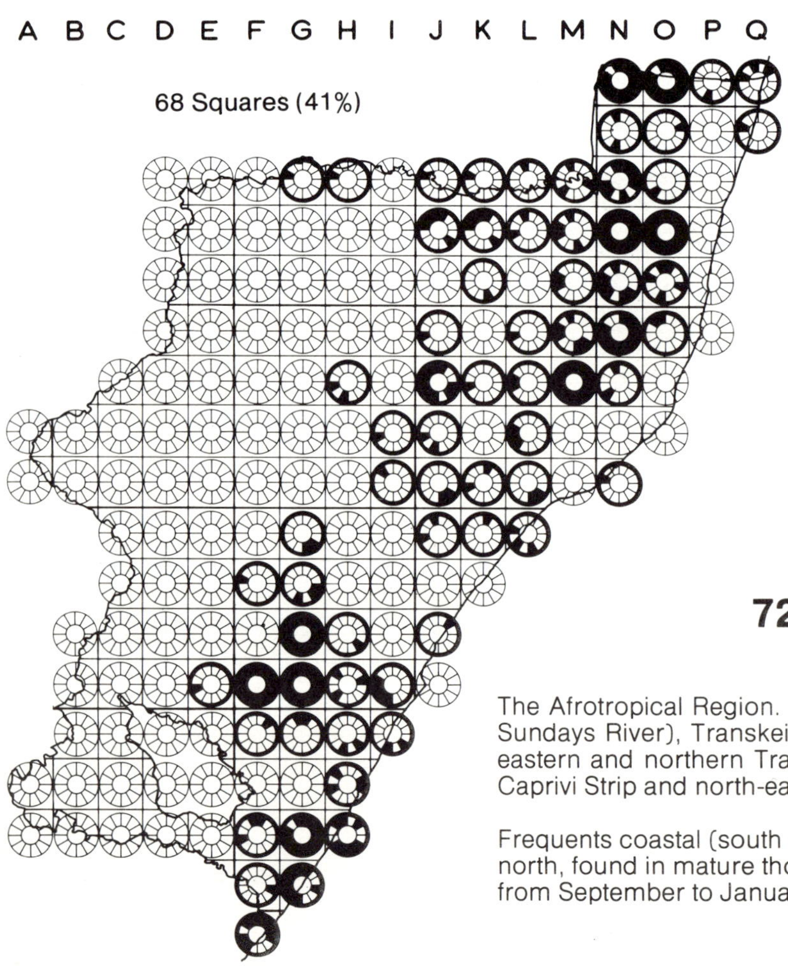

723 Grey-headed Bush Shrike

Malaconotus blanchoti

The Afrotropical Region. In Southern Africa, the eastern Cape (from the Sundays River), Transkei, Natal and Zululand, Swaziland, Moçambique, eastern and northern Transvaal, Rhodesia, north-eastern Botswana, the Caprivi Strip and north-eastern S.W.A.

Frequents coastal (south of Durban) and mistbelt evergreen forest; in the north, found in mature thornveld and woodland, often near water. Breeds from September to January.

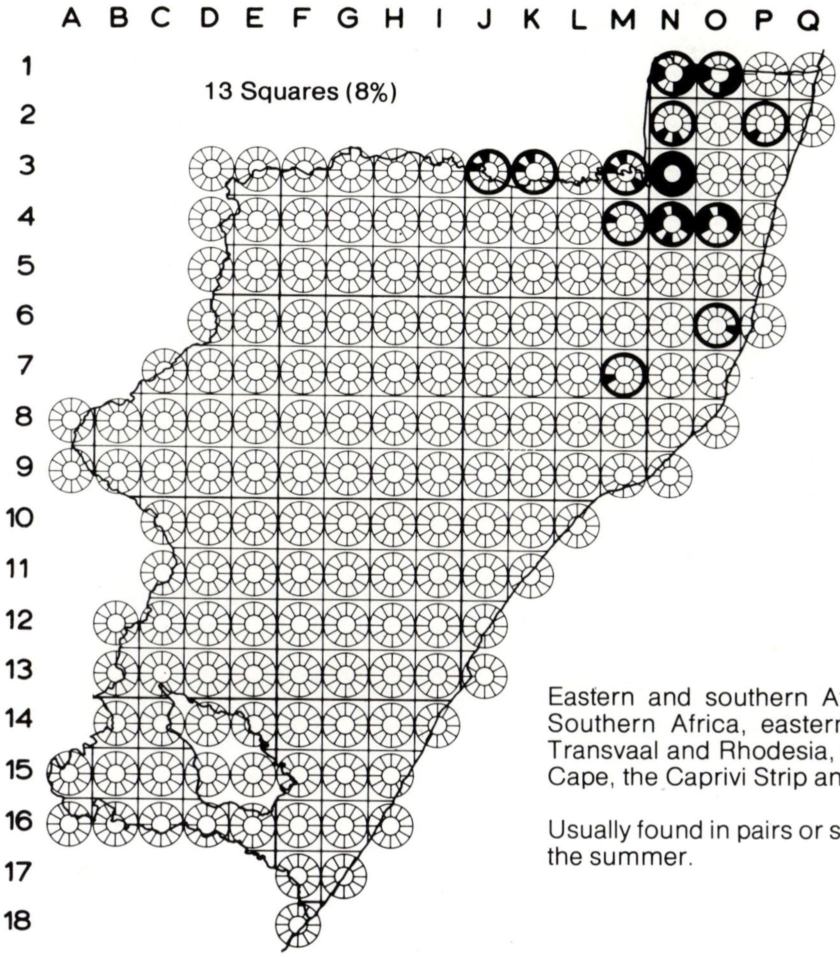

724 Long-tailed Shrike

Corvinella melanoleuca

Eastern and southern Africa, from Angola and Kenya to Zululand. In Southern Africa, eastern Zululand, eastern Swaziland, Moçambique, Transvaal and Rhodesia, west to Botswana, western O.F.S., the northern Cape, the Caprivi Strip and central and northern S.W.A.

Usually found in pairs or small parties in savanna thornveld. Breeds during the summer.

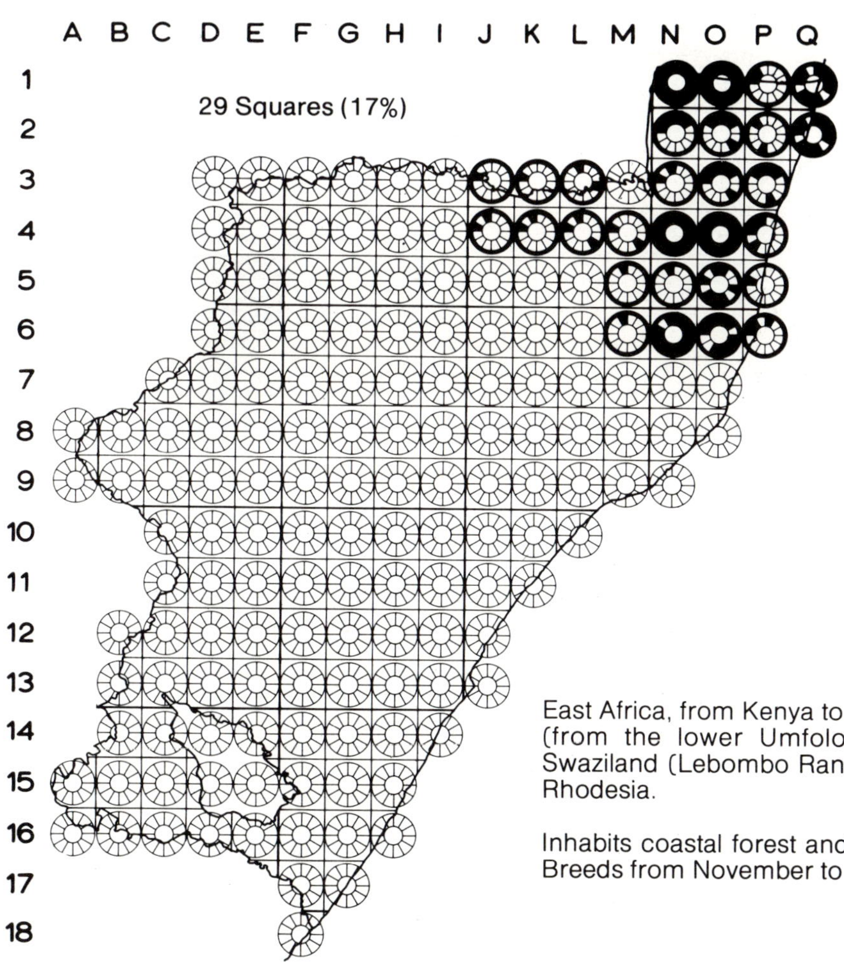

725 Yellow-spotted Nicator

Nicator gularis

East Africa, from Kenya to Zululand. In Southern Africa, eastern Zululand (from the lower Umfolozi River), south-eastern Transvaal, eastern Swaziland (Lebombo Range) and Moçambique to eastern and northern Rhodesia.

Inhabits coastal forest and scrub, mixed woodland and moist thornveld. Breeds from November to January.

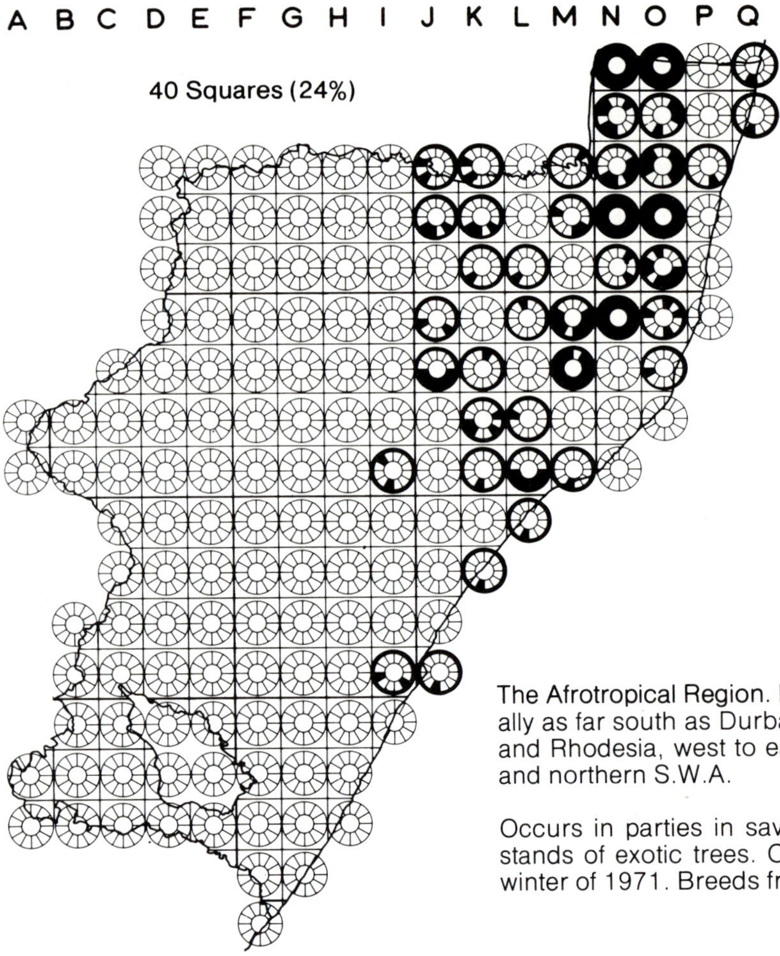

727 White Helmet Shrike

Prionops plumata

The Afrotropical Region. In Southern Africa, Natal and Zululand (occasionally as far south as Durban), western Swaziland, Transvaal, Moçambique and Rhodesia, west to eastern and northern Botswana, the Caprivi Strip and northern S.W.A.

Occurs in parties in savanna woodland and thornveld; exceptionally in stands of exotic trees. Only recorded south of Richards Bay during the winter of 1971. Breeds from September to December.

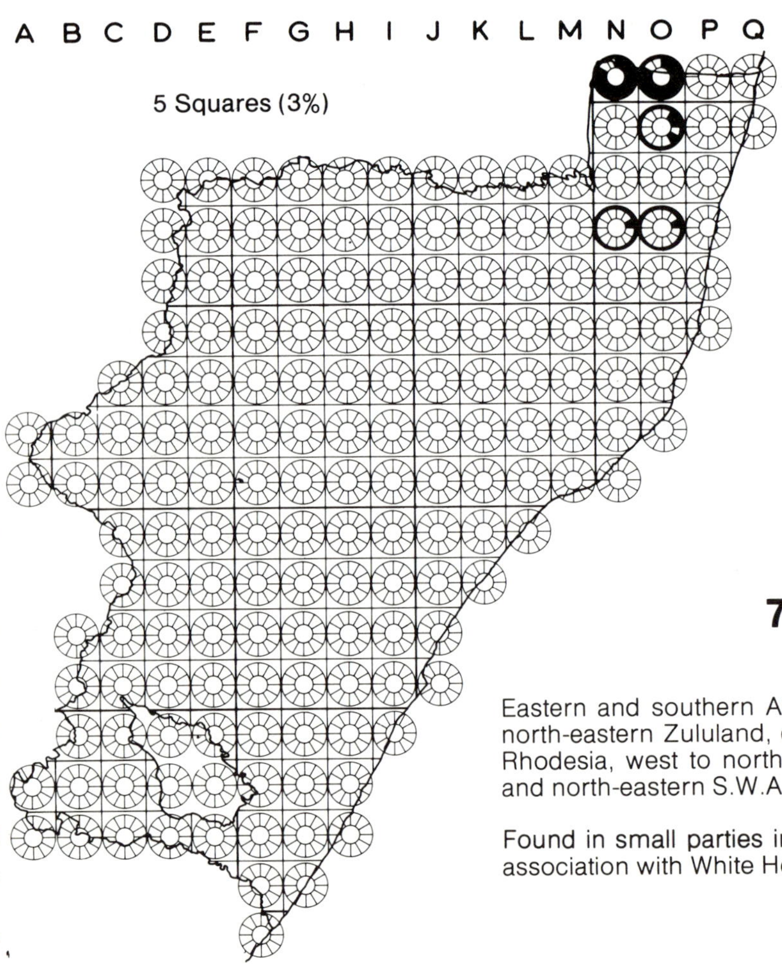

728 Red-billed Helmet Shrike

Prionops retzii

Eastern and southern Africa southeast to Zululand. In Southern Africa, north-eastern Zululand, eastern Swaziland, Transvaal, Moçambique and Rhodesia, west to northern and north-eastern Botswana, and northern and north-eastern S.W.A.

Found in small parties in mature woodland on the littoral plain, often in association with White Helmet Shrike (727). May breed.

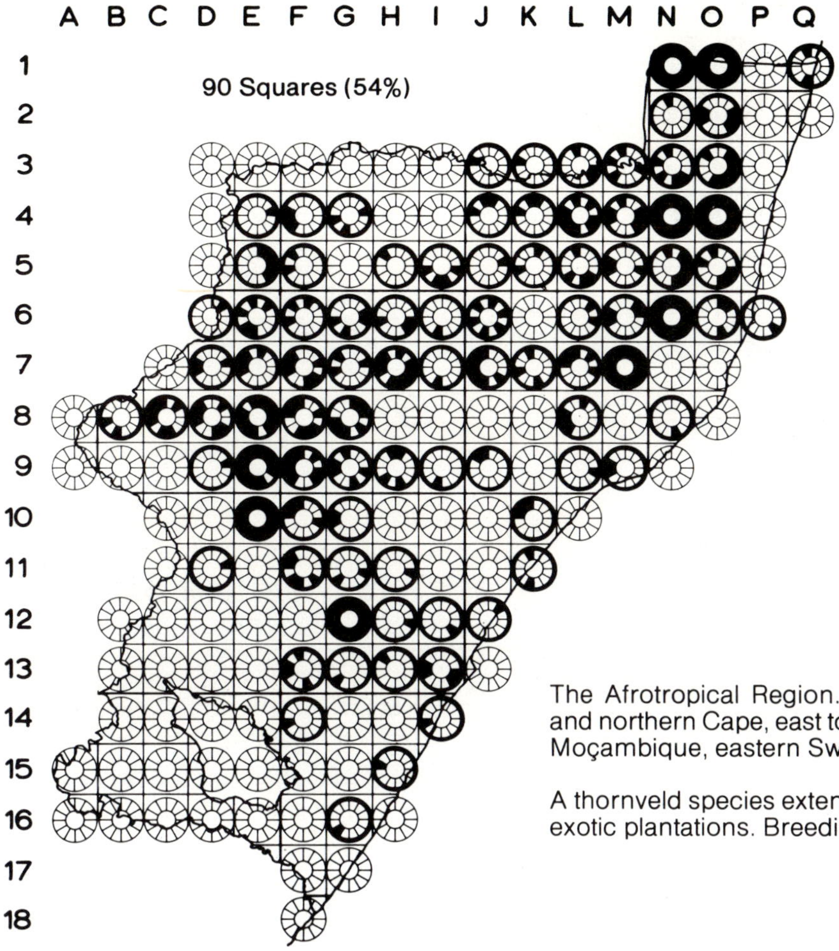

731 Brubru Shrike

Nilaus afer

The Afrotropical Region. In Southern Africa, S.W.A., the north-western and northern Cape, east to Botswana, western O.F.S., Transvaal, Rhodesia, Moçambique, eastern Swaziland, Natal and Zululand.

A thornveld species extending into open bush in the south; occasionally in exotic plantations. Breeding recorded from September to December.

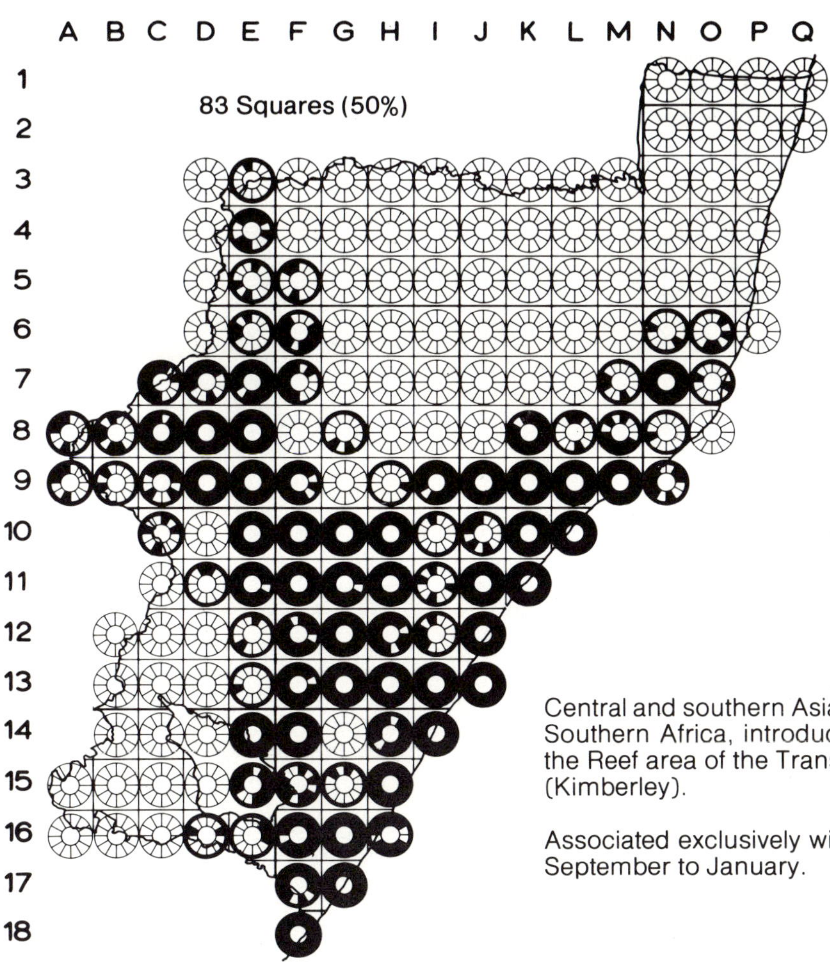

734 Indian Myna

Acridotheres tristis

Central and southern Asia, with introductions in other parts of the world. In Southern Africa, introduced and widespread in Natal and Zululand, and the Reef area of the Transvaal; recently recorded from the northern Cape (Kimberley).

Associated exclusively with human settlements. Recorded breeding from September to January.

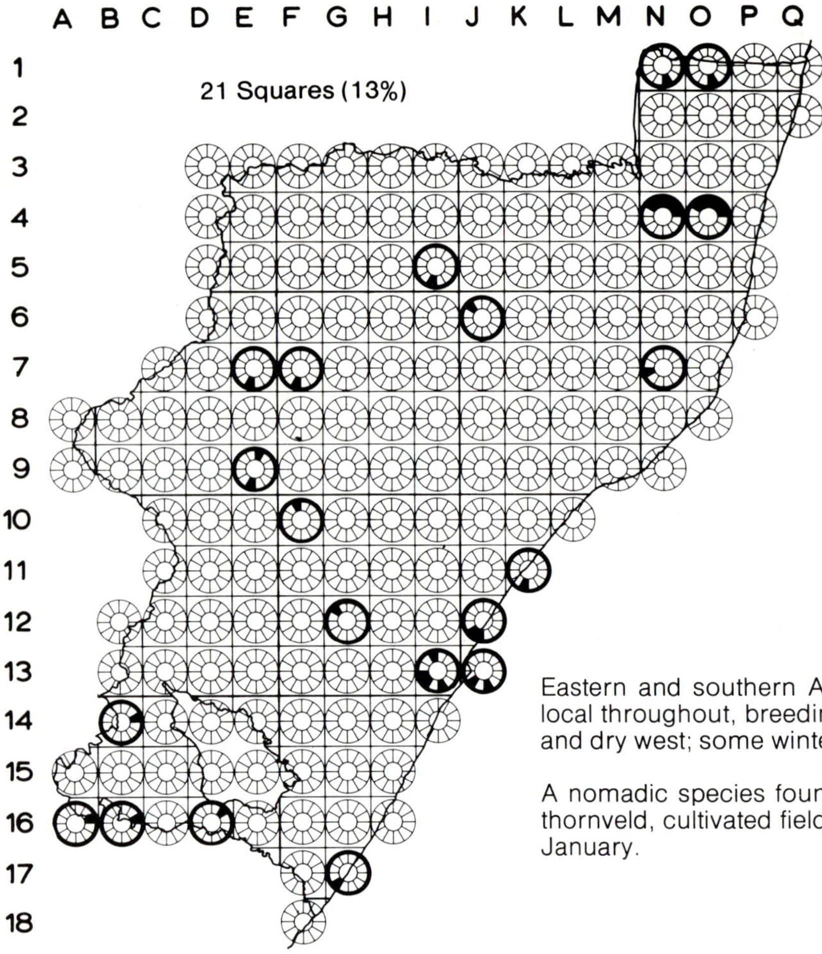

735 Wattled Starling

Creatophora cinerea

Eastern and southern Africa, and southern Arabia. In Southern Africa, local throughout, breeding irruptively mainly in Acacia veld of the interior and dry west; some winter on the south-eastern littoral.

A nomadic species found in flocks, recorded from open woodland and thornveld, cultivated fields and waste ground. Recorded breeding during January.

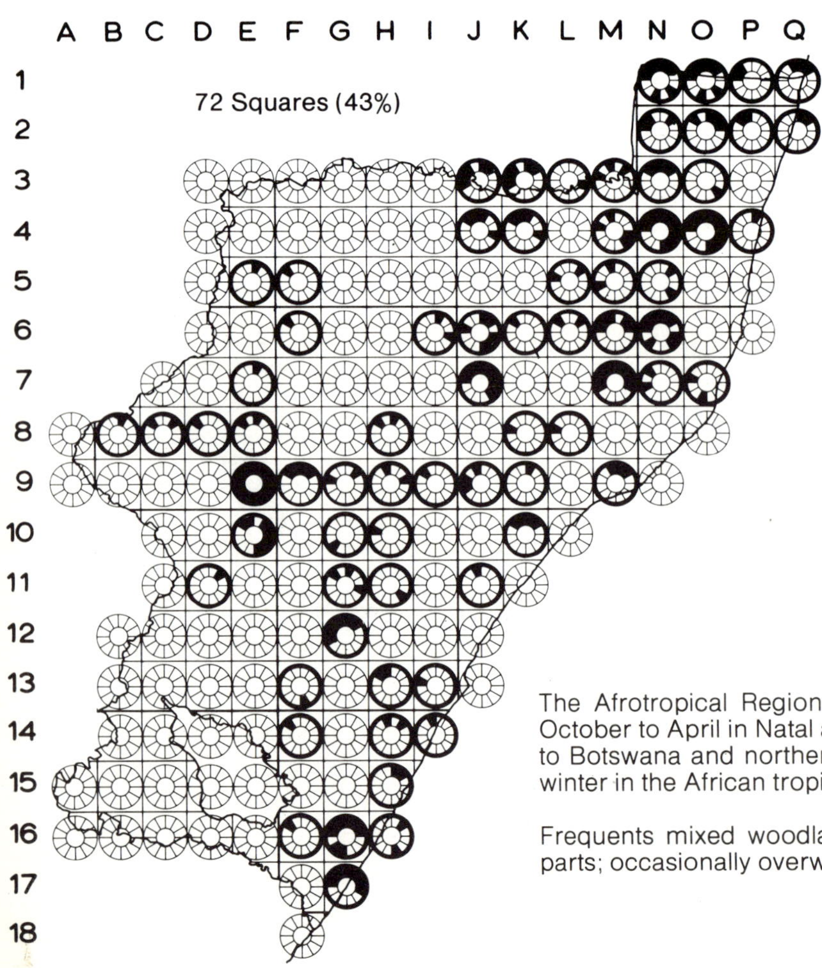

736 Plum-coloured Starling

Cinnyricinclus leucogaster

The Afrotropical Region. In Southern Africa, a summer resident from October to April in Natal and Zululand, Swaziland and Moçambique, west to Botswana and northern S.W.A.; recorded in Transkei. Southern birds winter in the African tropics.

Frequents mixed woodland and thornveld, generally in the more open parts; occasionally overwinters. Breeds from October to February.

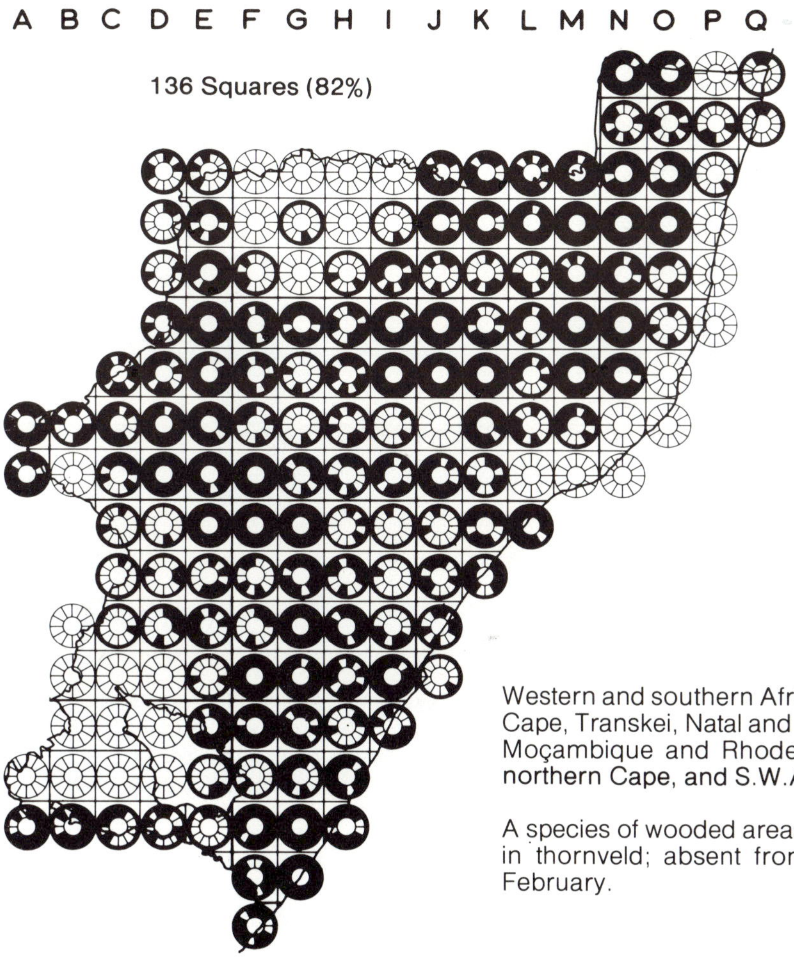

737 Cape Glossy Starling

Lamprotornis nitens

Western and southern Africa. In Southern Africa, the southern and eastern Cape, Transkei, Natal and Zululand, Swaziland, O.F.S., Transvaal, southern Moçambique and Rhodesia, west to Botswana, the north-western and northern Cape, and S.W.A.

A species of wooded areas, including stands of exotic trees, most common in thornveld; absent from evergreen forests. Breeds from October to February.

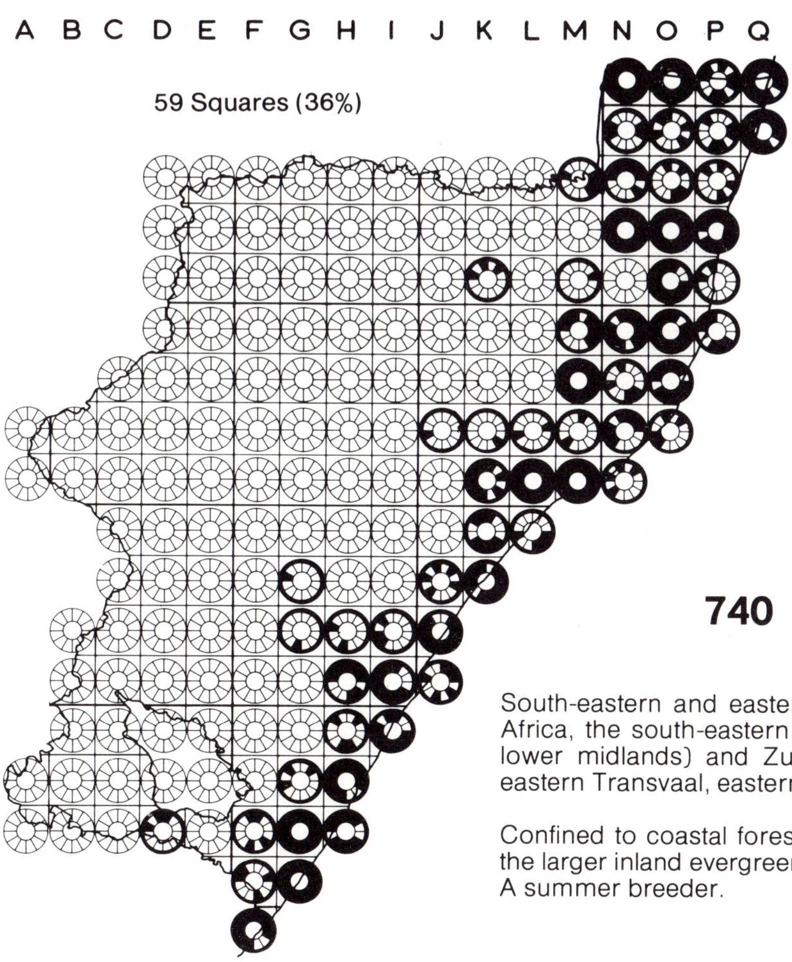

740 Black-bellied Glossy Starling

Lamprotornis corruscus

South-eastern and eastern Africa, mainly along the coast. In Southern Africa, the south-eastern and eastern Cape, Transkei, Natal (coast and lower midlands) and Zululand, eastern Swaziland (Lebombo Range), eastern Transvaal, eastern Rhodesia (lowland) and Moçambique.

Confined to coastal forest and riverine woodland, ranging marginally to the larger inland evergreen forests (Weza, Karkloof, Nkandla and Ngome). A summer breeder.

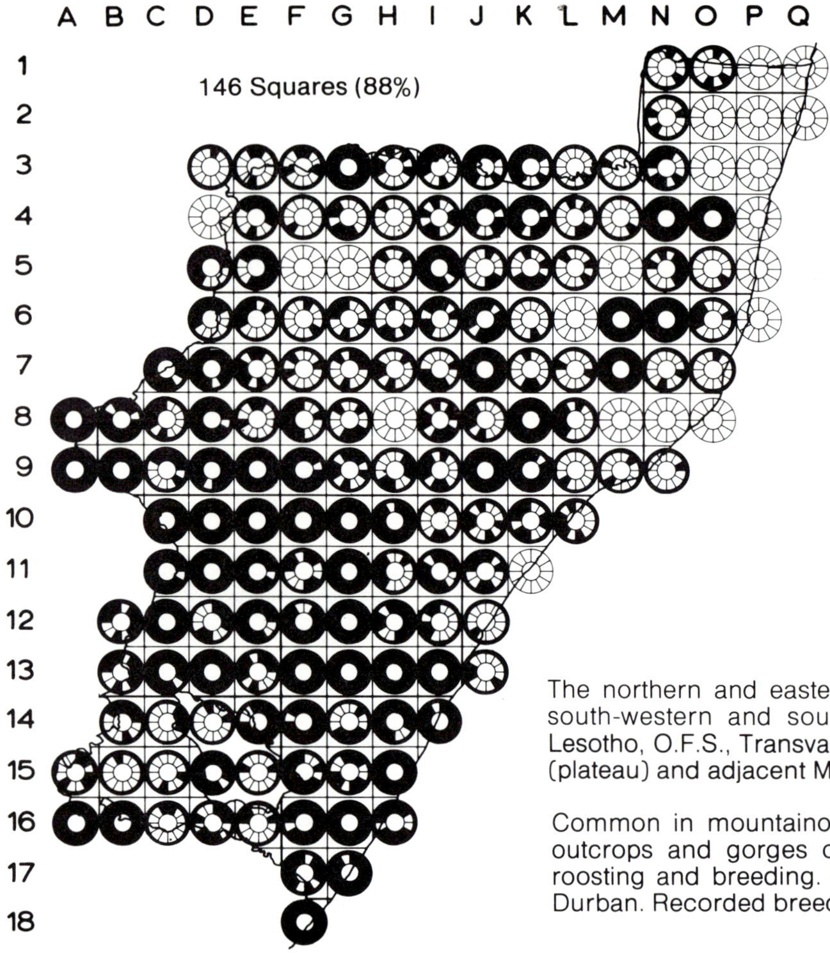

745 Red-winged Starling

Onychognathus morio

The northern and eastern Afrotropical Region. In Southern Africa, the south-western and southern Cape to Transkei, Natal and Zululand, Lesotho, O.F.S., Transvaal, south-eastern Botswana, Swaziland, Rhodesia (plateau) and adjacent Moçambique; also on Mount Gorongosa.

Common in mountainous country and at lower altitudes where rocky outcrops and gorges occur; also takes to bridges and buildings for roosting and breeding. Generally absent from coastal regions north of Durban. Recorded breeding from October to February.

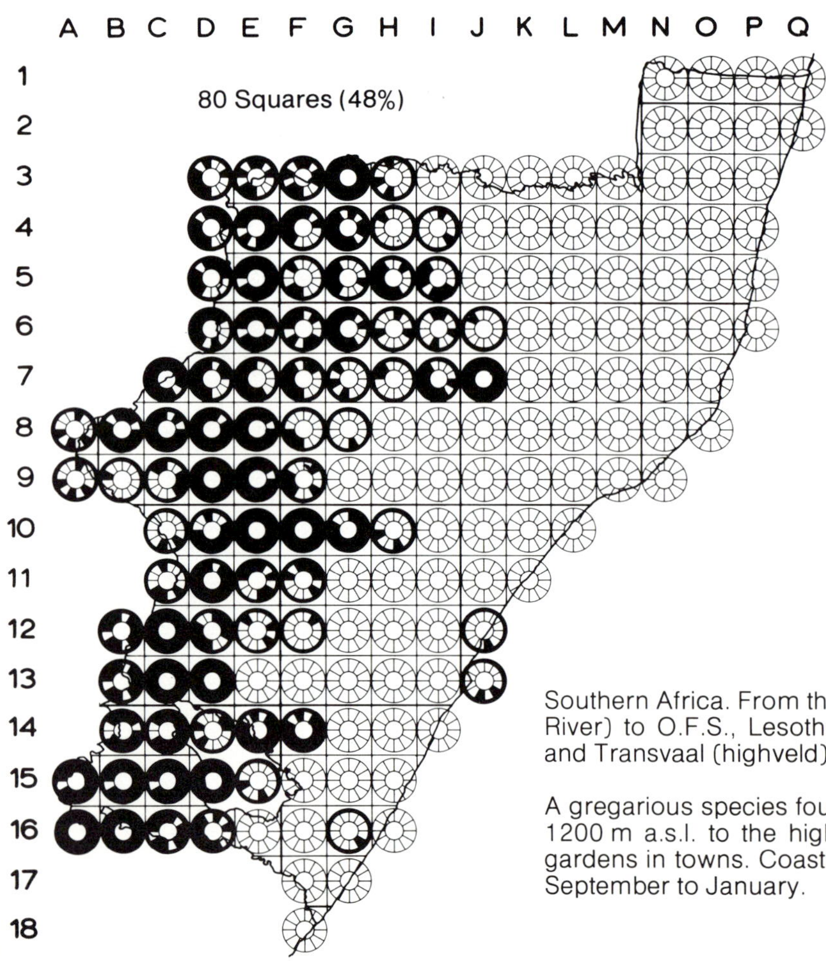

746 Pied Starling

Spreo bicolor

Southern Africa. From the Cape (including the Karoo north of the Orange River) to O.F.S., Lesotho, Natal, western Zululand, western Swaziland, and Transvaal (highveld).

A gregarious species found in open grassland and farmland mainly from 1200 m a.s.l. to the higher reaches of the Drakensberg Range; enters gardens in towns. Coastal records exceptional. Recorded breeding from September to January.

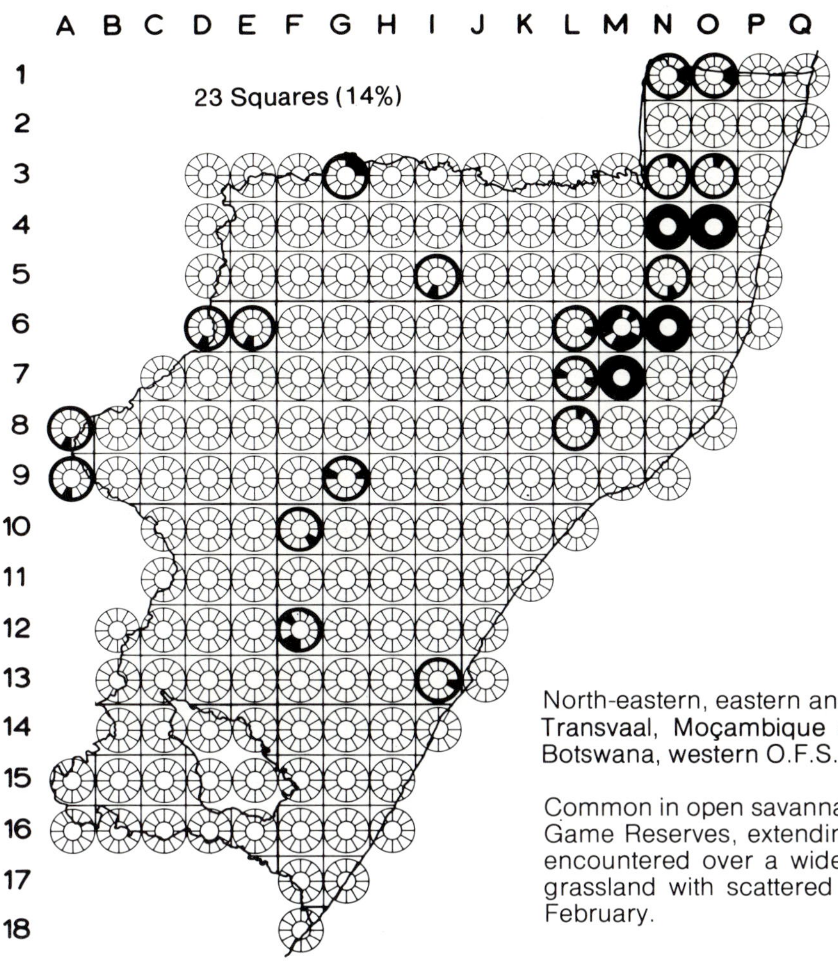

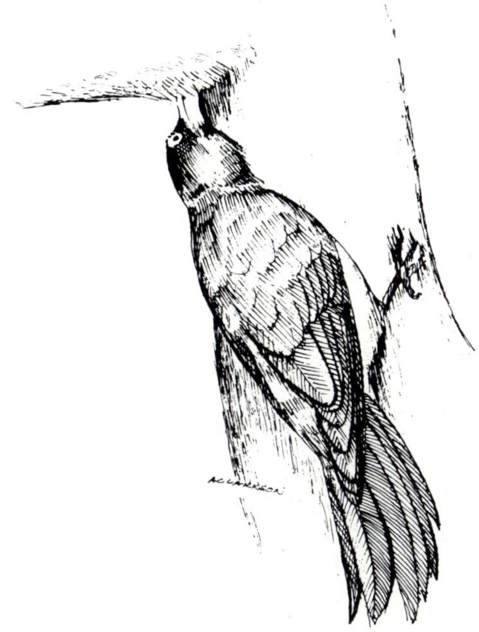

748 Red-billed Oxpecker

Buphagus erythrorhynchus

North-eastern, eastern and southern Africa. In Southern Africa, Zululand, Transvaal, Moçambique and Rhodesia, west to eastern and northern Botswana, western O.F.S., the Caprivi Strip and northern S.W.A.

Common in open savanna woodland of the Mkuzi, Hluhluwe and Umfolozi Game Reserves, extending into adjacent areas. Elsewhere, occasionally encountered over a wide range of localities, showing a preference for grassland with scattered trees. Recorded breeding from September to February.

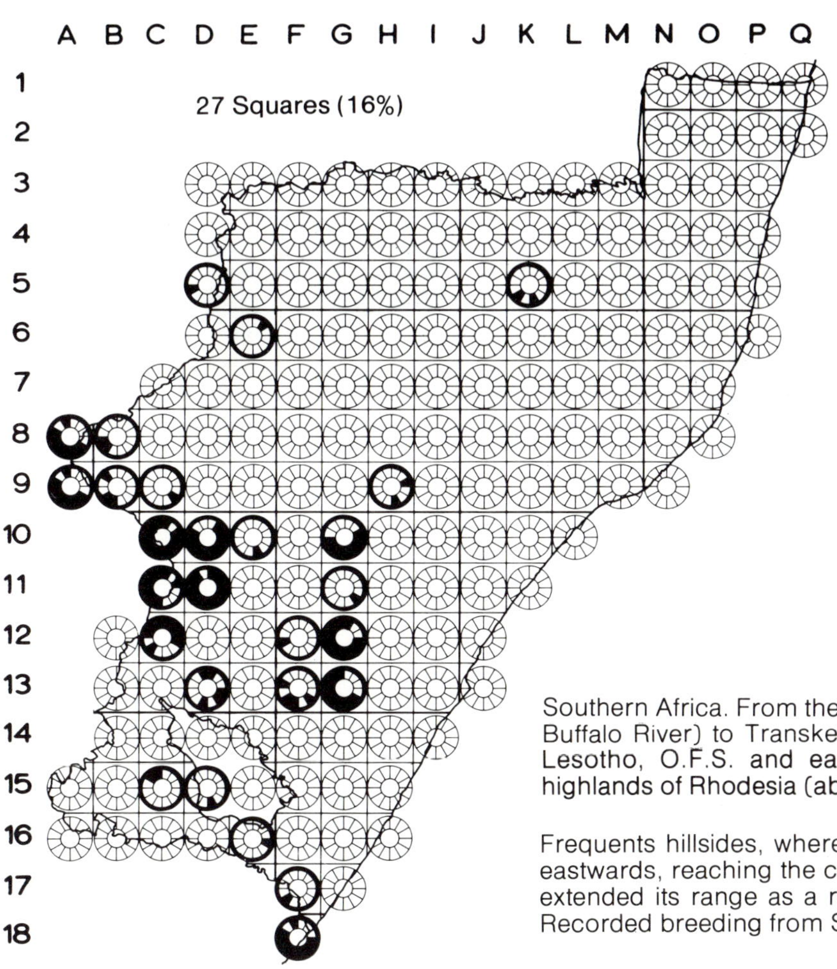

750 Gurney's Sugarbird

Promerops gurneyi

Southern Africa. From the eastern Cape (east of the upper drainage of the Buffalo River) to Transkei, Natal, western Zululand, western Swaziland, Lesotho, O.F.S. and eastern and northern Transvaal; also eastern highlands of Rhodesia (above 1000 m a.s.l.) and adjacent Moçambique.

Frequents hillsides, where proteas occur, from the Drakensberg Range eastwards, reaching the coast in the extreme south. In the midlands, has extended its range as a result of the establishment of protea nurseries. Recorded breeding from September to February.

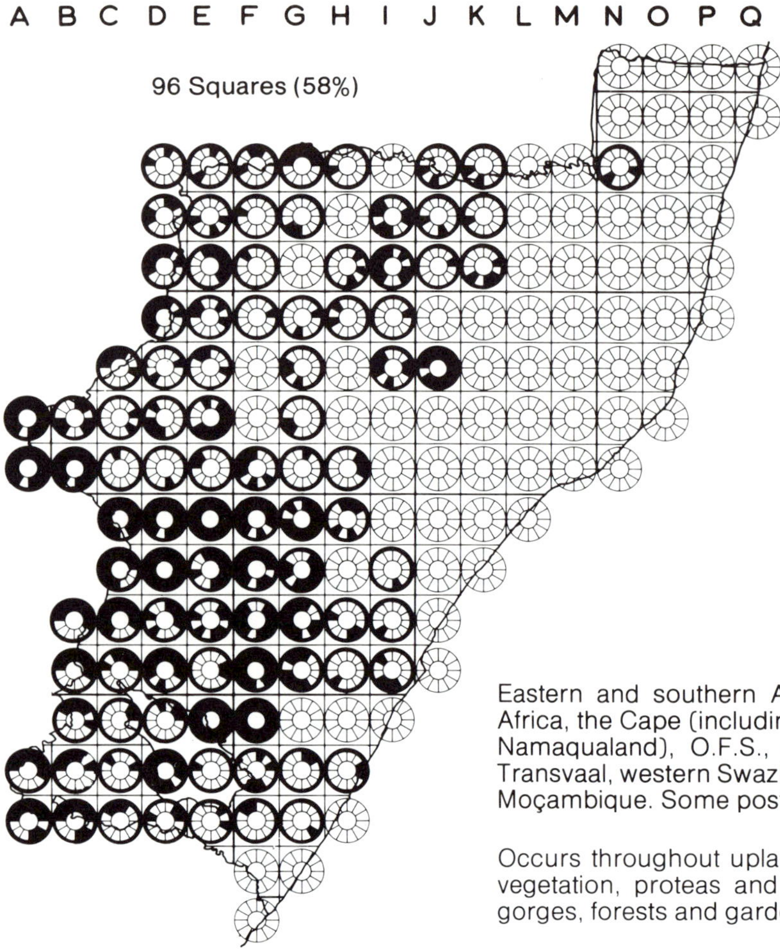

751 Malachite Sunbird

Nectarinia famosa

Eastern and southern Africa, from Ethiopia to the Cape. In Southern Africa, the Cape (including north of the Orange River and southern Great Namaqualand), O.F.S., Lesotho (highlands), Natal, western Zululand, Transvaal, western Swaziland, eastern Rhodesia (highlands) and adjacent Moçambique. Some post-breeding altitudinal movement.

Occurs throughout upland areas where it frequents hillsides with scrub vegetation, proteas and aloes, ranging into adjacent habitats such as gorges, forests and gardens. Breeding recorded from October to April.

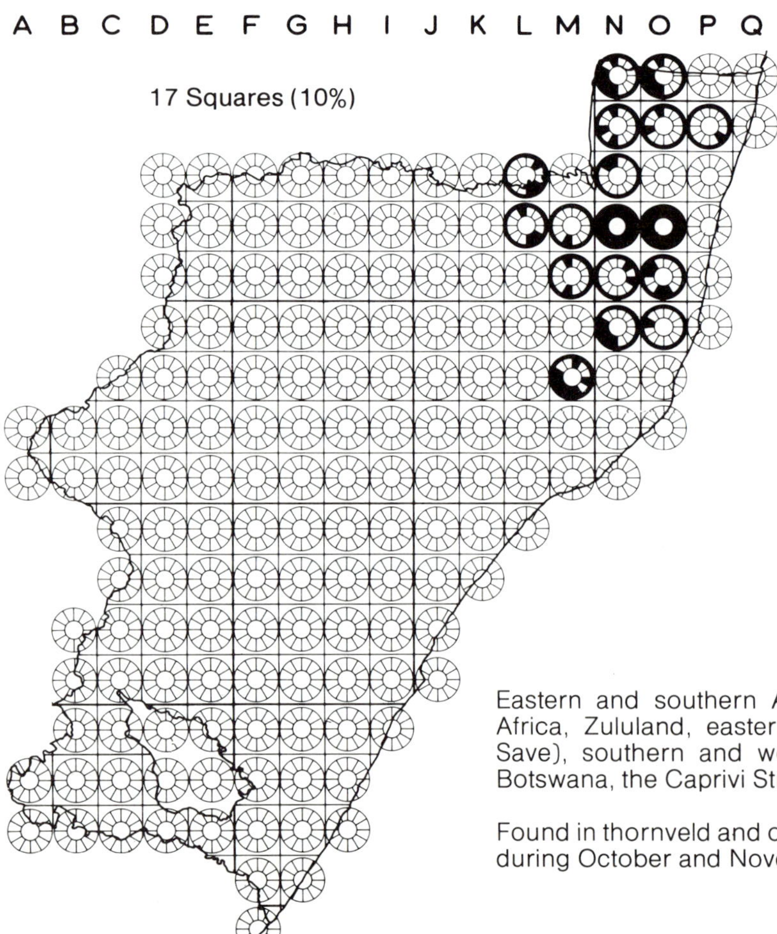

755 Marico Sunbird

Nectarinia mariquensis

Eastern and southern Africa, from Ethiopia to Zululand. In Southern Africa, Zululand, eastern Swaziland, Transvaal, Moçambique (Sul do Save), southern and western Rhodesia, west to the northern half of Botswana, the Caprivi Strip and central and northern S.W.A.

Found in thornveld and occasionally mixed woodland. Breeding recorded during October and November.

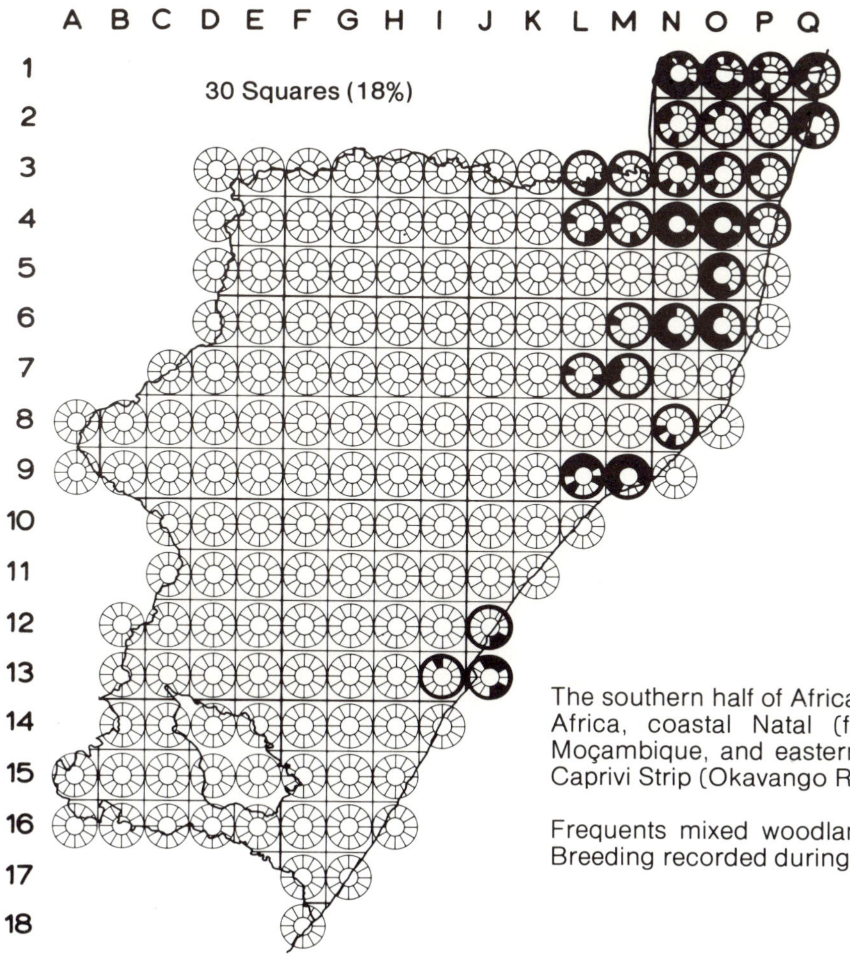

756 Purple-banded Sunbird

Nectarinia bifasciata

The southern half of Africa, from Gabon and Kenya to Natal. In Southern Africa, coastal Natal (from Durban), eastern Zululand, southern Moçambique, and eastern and northern Rhodesia, west to the eastern Caprivi Strip (Okavango River) and northern S.W.A. (Ovamboland).

Frequents mixed woodland and coastal scrub; occasionally thornveld. Breeding recorded during October and November.

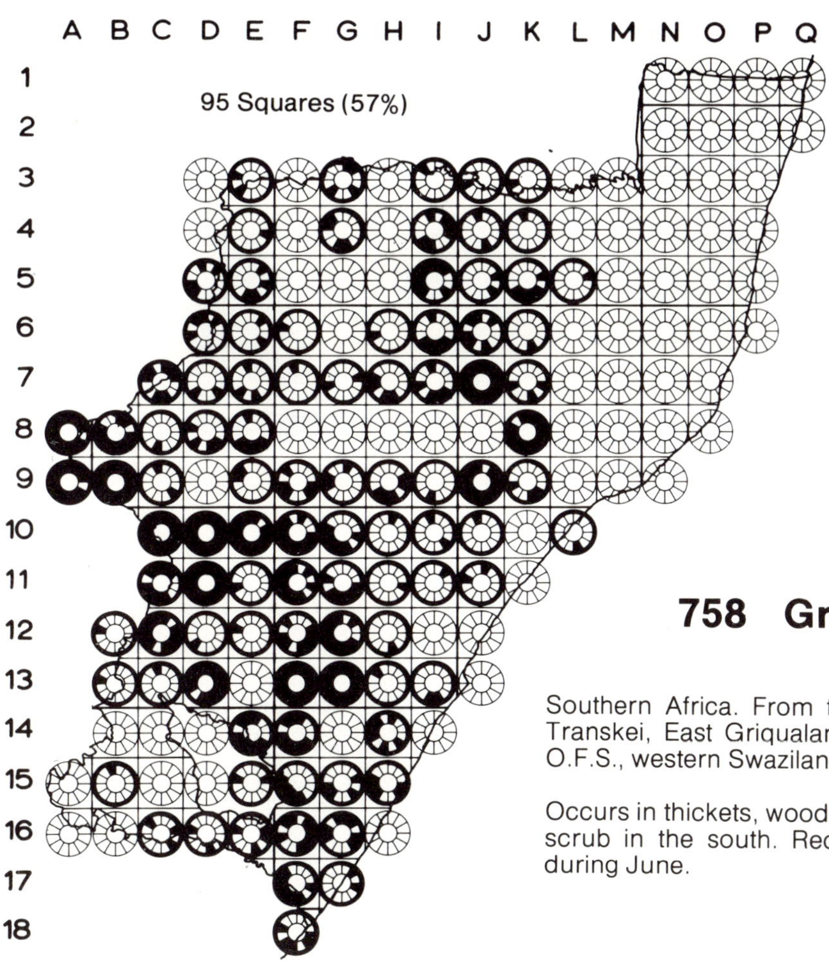

758 Greater Double-collared Sunbird

Nectarinia afra

Southern Africa. From the south-western Cape (from Swellendam) to Transkei, East Griqualand, Natal (interior), western Zululand, Lesotho, O.F.S., western Swaziland, and Transvaal (highveld).

Occurs in thickets, wooded gullies, on the edges of forests, and in coastal scrub in the south. Recorded breeding from August to January and during June.

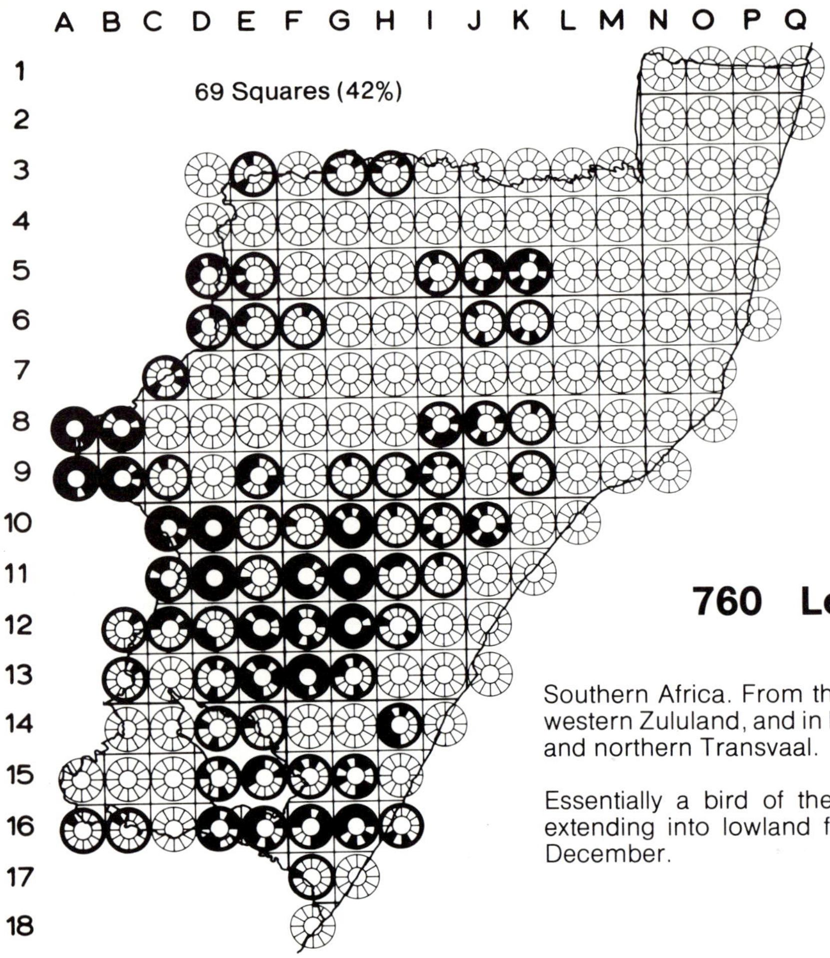

760 Lesser Double-collared Sunbird

Nectarinia chalybea

Southern Africa. From the Cape to Transkei, East Griqualand, Natal and western Zululand, and in highland forest of western Swaziland and eastern and northern Transvaal.

Essentially a bird of the Drakensberg and mistbelt evergreen forests, extending into lowland forest in the south. Breeds from September to December.

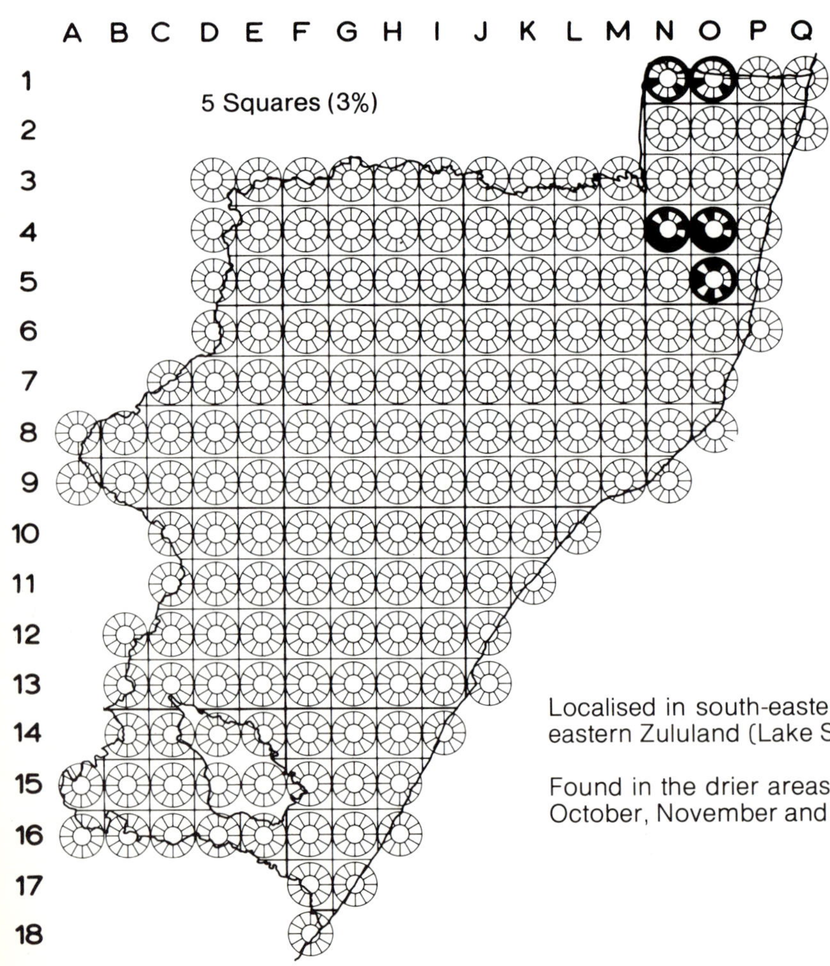

761 Neergaard's Sunbird

Nectarinia neergaardi

Localised in south-eastern coastal Africa. From the littoral flats of north-eastern Zululand (Lake St Lucia) to Moçambique (the Save River).

Found in the drier areas of mixed woodland. Recorded breeding during October, November and July.

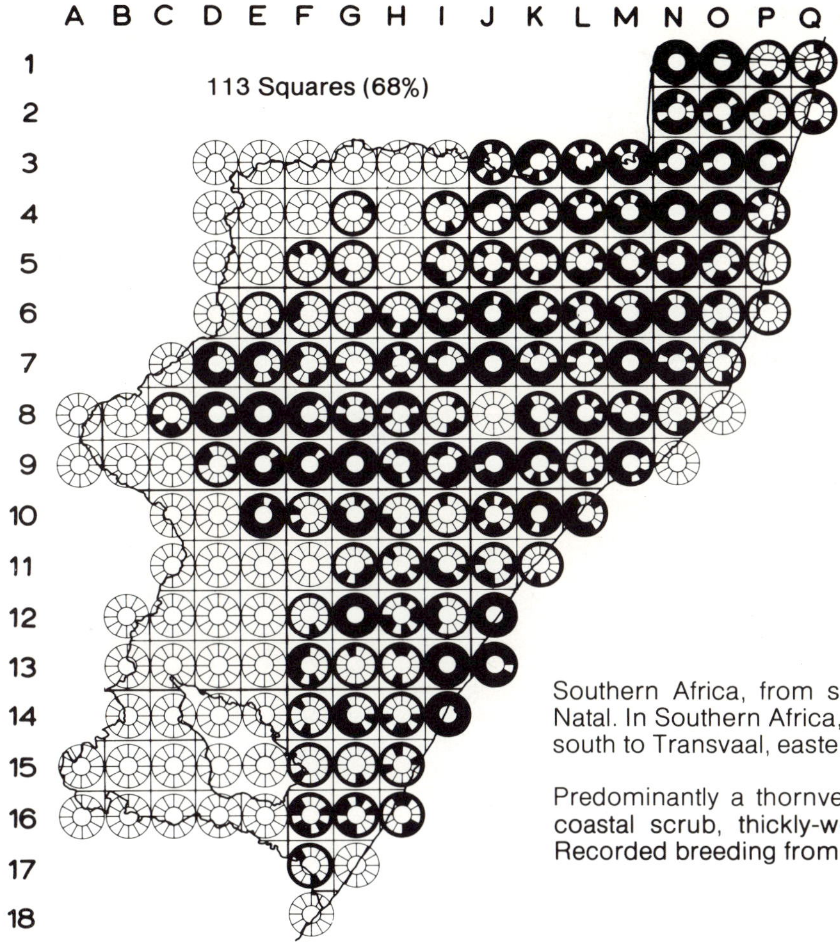

763 White-bellied Sunbird

Nectarinia talatala

Southern Africa, from southern Angola and northern Moçambique to Natal. In Southern Africa, S.W.A., Botswana, Rhodesia and Moçambique, south to Transvaal, eastern Swaziland, Natal and Zululand.

Predominantly a thornveld species, also occurring in mixed woodland, coastal scrub, thickly-wooded valleys and open bush; also gardens. Recorded breeding from August to February.

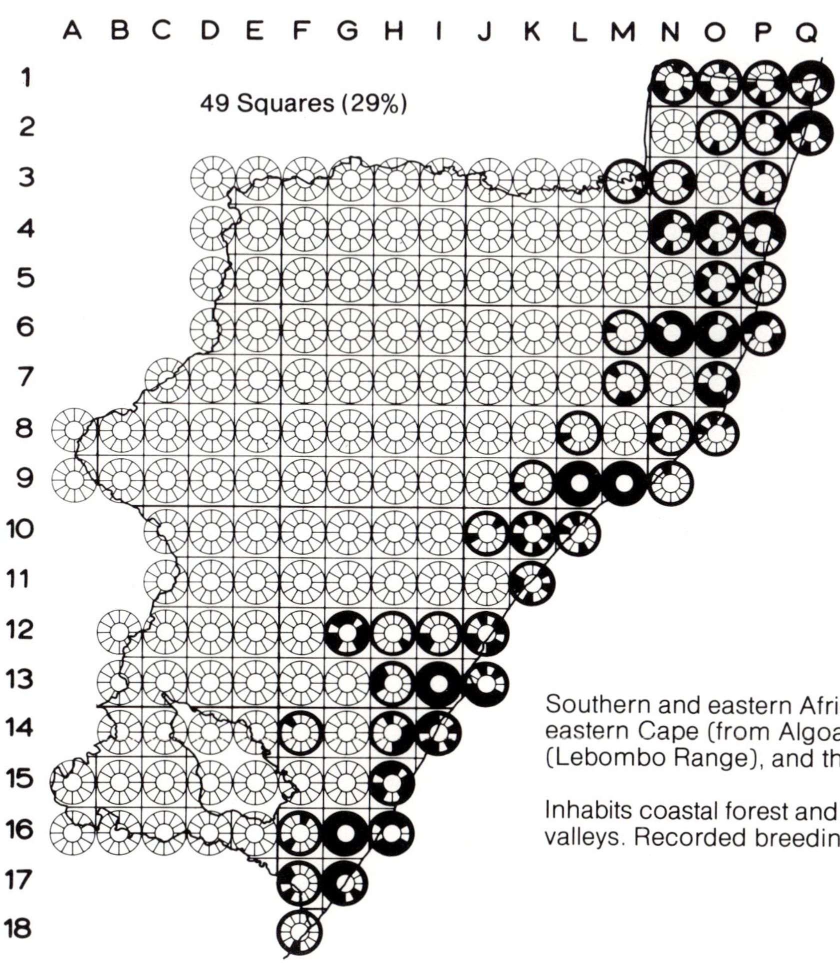

765 Grey Sunbird

Nectarinia veroxii

Southern and eastern Africa in the coastal districts. In Southern Africa, the eastern Cape (from Algoa Bay) to Natal and Zululand, eastern Swaziland (Lebombo Range), and the plain of Moçambique.

Inhabits coastal forest and riverine woodland, extending up thickly-wooded valleys. Recorded breeding from September to January.

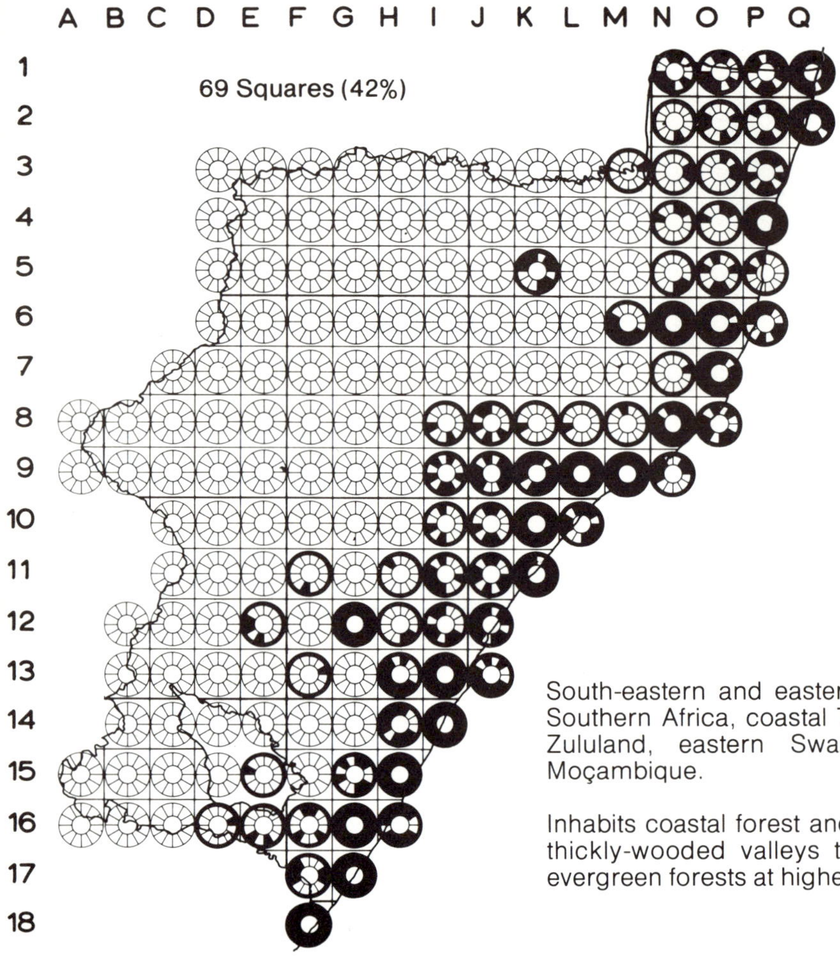

766 Olive Sunbird

Nectarinia olivacea

South-eastern and eastern Africa, west in the north to West Africa. In Southern Africa, coastal Transkei to coastal and midland Natal, eastern Zululand, eastern Swaziland, Rhodesia (eastern highlands) and Moçambique.

Inhabits coastal forest and mixed and riverine woodland, penetrating up thickly-wooded valleys to about 600 m a.s.l.; marginally in mistbelt evergreen forests at higher altitudes. Breeds from September to March.

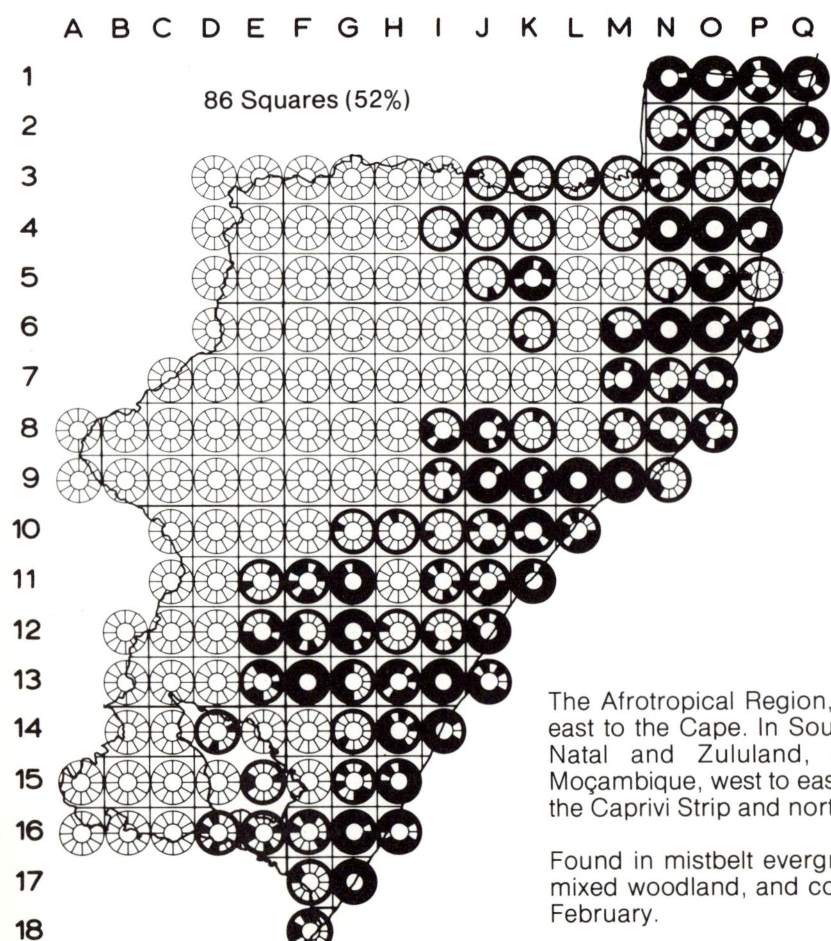

771 Collared Sunbird

Anthreptes collaris

The Afrotropical Region, from Senegal west to Kenya, and south in the east to the Cape. In Southern Africa, the southern and eastern Cape to Natal and Zululand, eastern Swaziland, eastern Transvaal and Moçambique, west to eastern and northern Rhodesia, northern Botswana, the Caprivi Strip and north-eastern S.W.A.

Found in mistbelt evergreen forest, thickly-wooded valleys, riverine and mixed woodland, and coastal forest. Recorded breeding from August to February.

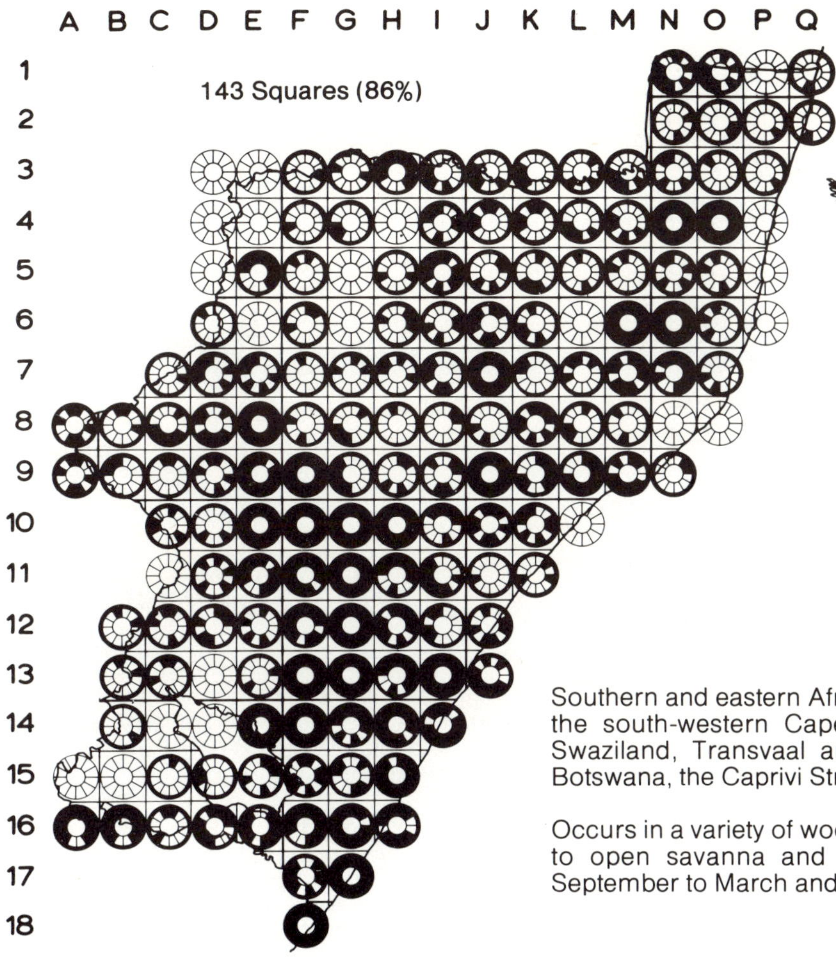

772 Black Sunbird

Nectarinia amethystina

Southern and eastern Africa, from Kenya to the Cape. In Southern Africa, the south-western Cape (from Swellendam) to Natal and Zululand, Swaziland, Transvaal and Moçambique, west to Rhodesia, northern Botswana, the Caprivi Strip and northern S.W.A.

Occurs in a variety of wooded habitats, from the fringe of evergreen forest to open savanna and exotic plantations. Recorded breeding from September to March and during June.

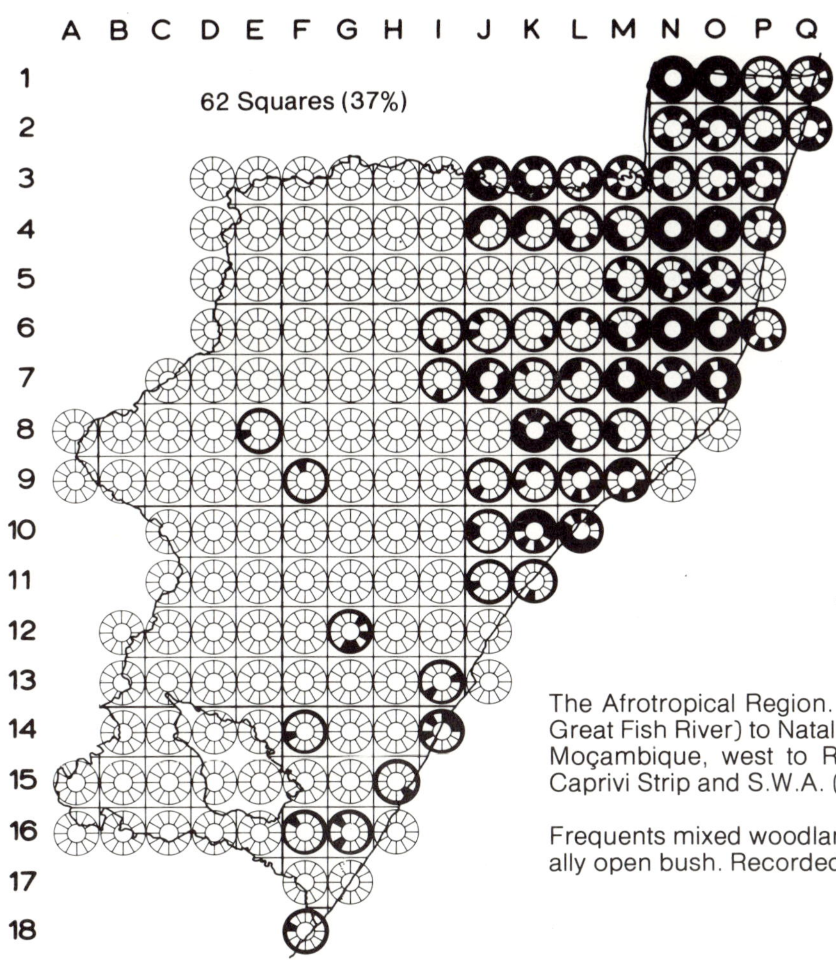

774 Scarlet-chested Sunbird

Nectarinia senegalensis

The Afrotropical Region. In Southern Africa, the eastern Cape (from the Great Fish River) to Natal and Zululand, Swaziland, eastern Transvaal and Moçambique, west to Rhodesia, eastern and northern Botswana, the Caprivi Strip and S.W.A. (from Damaraland northwards).

Frequents mixed woodland, moist thornveld, coastal scrub and occasionally open bush. Recorded breeding from August to January.

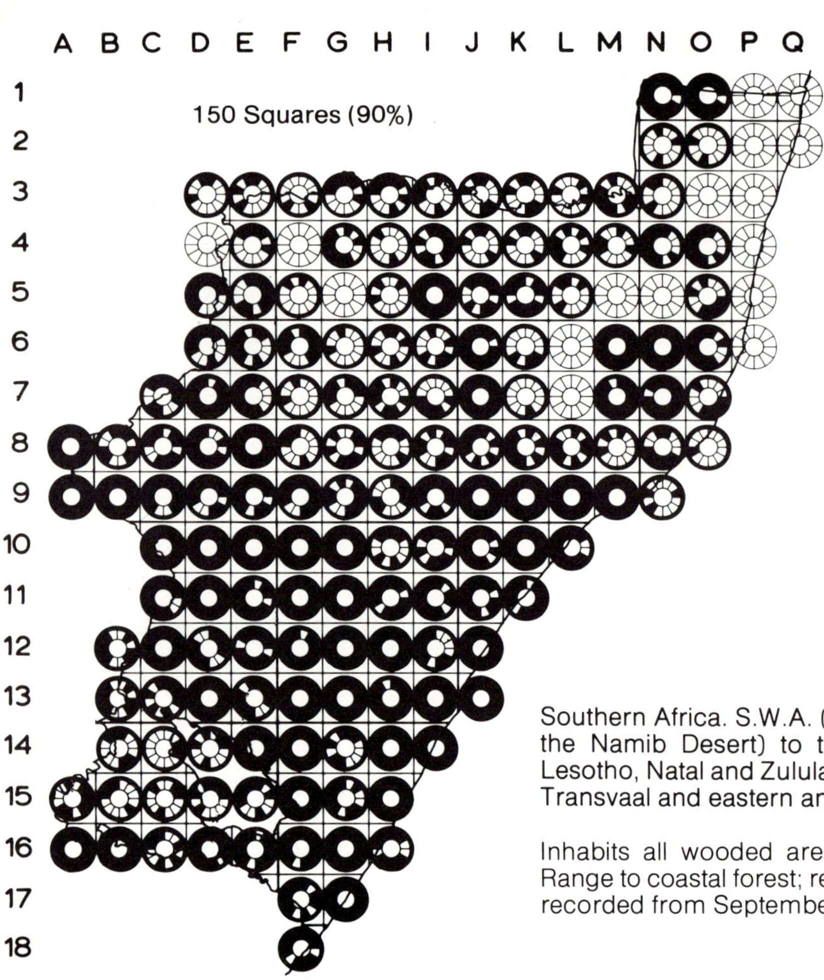

775 Cape White-eye

Zosterops pallidus

Southern Africa. S.W.A. (from the Damaraland highlands and the edge of the Namib Desert) to the Cape, then eastwards to Transkei, O.F.S., Lesotho, Natal and Zululand, Swaziland, Moçambique (Lebombo Range), Transvaal and eastern and south-eastern Botswana.

Inhabits all wooded areas, from scrub vegetation in the Drakensberg Range to coastal forest; restricted to thornveld on the littoral plain. Breeding recorded from September to March and during July.

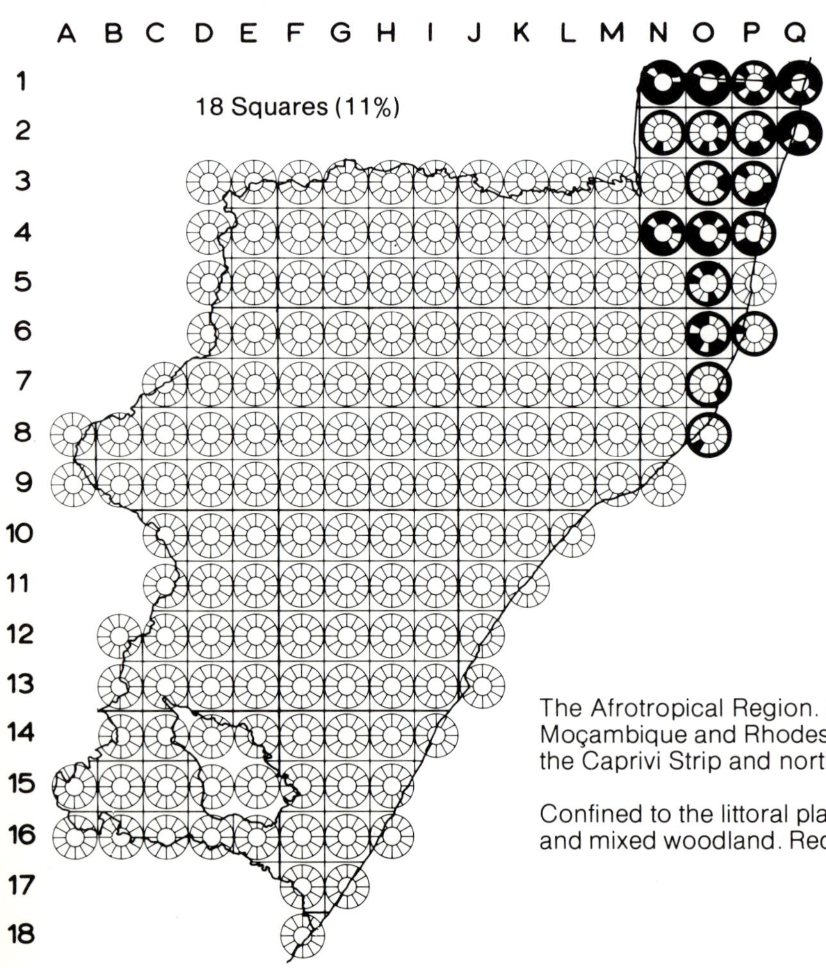

777 Yellow White-eye

Zosterops senegalensis

The Afrotropical Region. In Southern Africa, the eastern flats of Zululand, Moçambique and Rhodesia, west to north-eastern and northern Botswana, the Caprivi Strip and northern S.W.A. (Ovamboland).

Confined to the littoral plain where it is found in coastal forest, and riverine and mixed woodland. Recorded breeding during October and November.

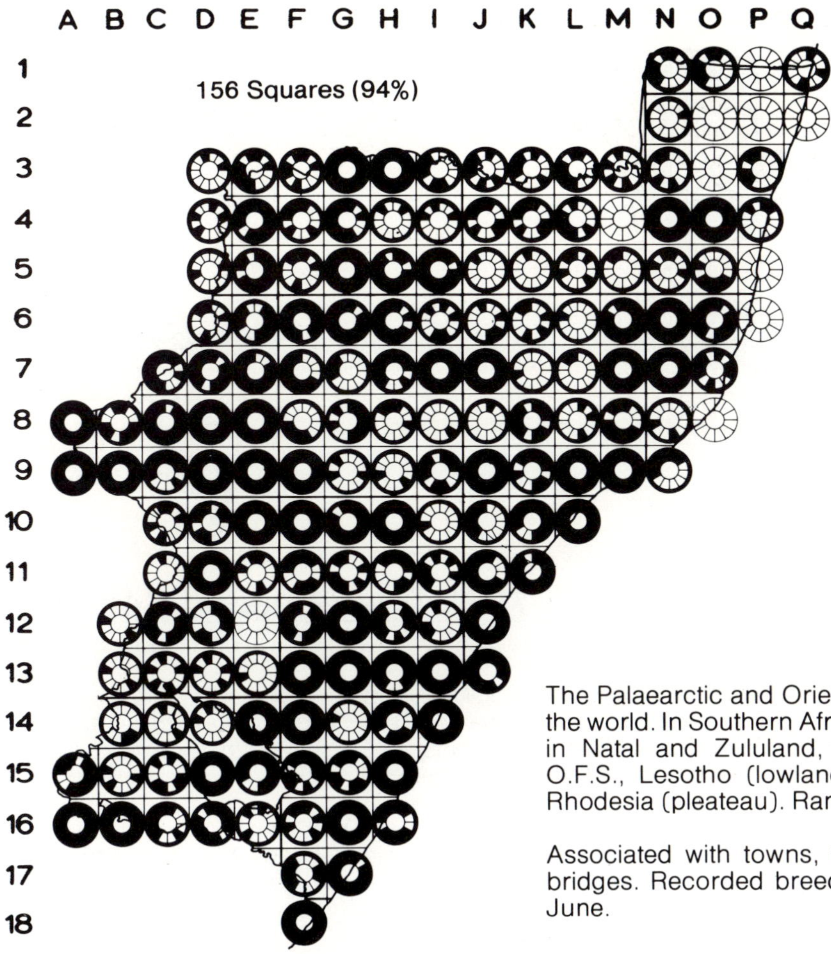

784 House Sparrow

Passer domesticus

The Palaearctic and Oriental Regions, with introductions in many parts of the world. In Southern Africa, introduced originally to Natal, and widespread in Natal and Zululand, Swaziland, southern Moçambique, Transvaal, O.F.S., Lesotho (lowlands), the Cape, S.W.A., eastern Botswana and Rhodesia (pleateau). Range expansion continuing.

Associated with towns, farmsteads and man-made structures such as bridges. Recorded breeding during all months except March, May and June.

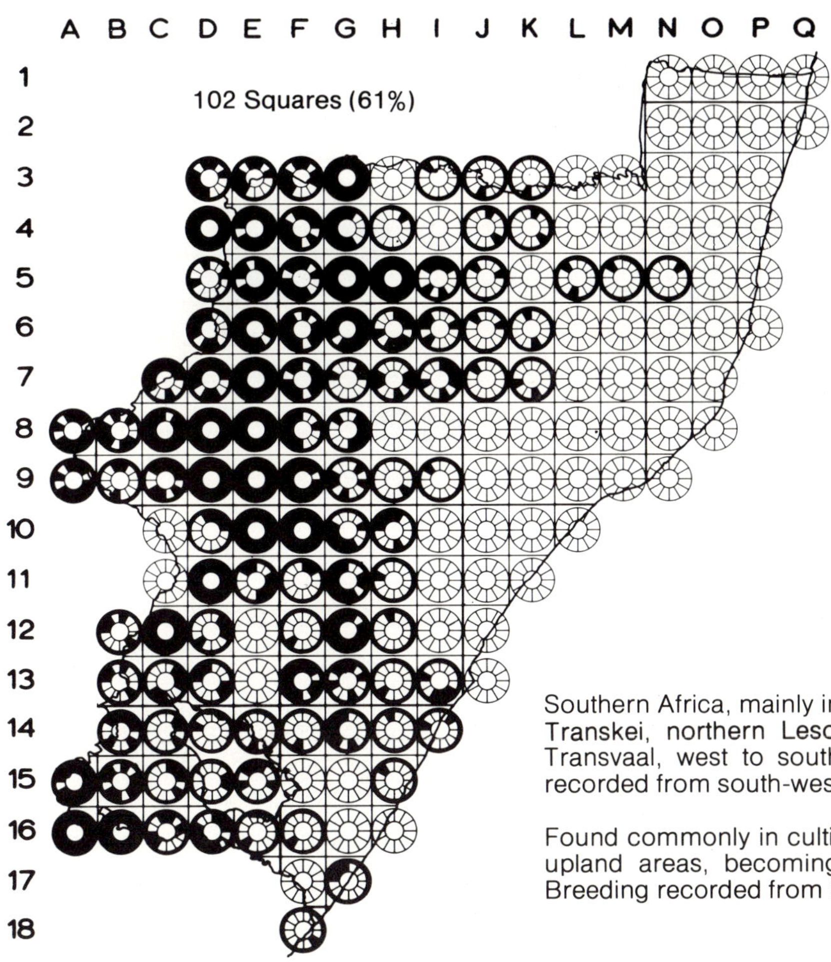

786 Cape Sparrow

Passer melanurus

Southern Africa, mainly in the interior and west. From the Cape to O.F.S., Transkei, northern Lesotho (lowlands), Natal, western Zululand, and Transvaal, west to southern and eastern Botswana, and S.W.A. Also recorded from south-western Rhodesia.

Found commonly in cultivated fields, around farmsteads and in towns in upland areas, becoming uncommon in Zululand and the southeast. Breeding recorded from September to March.

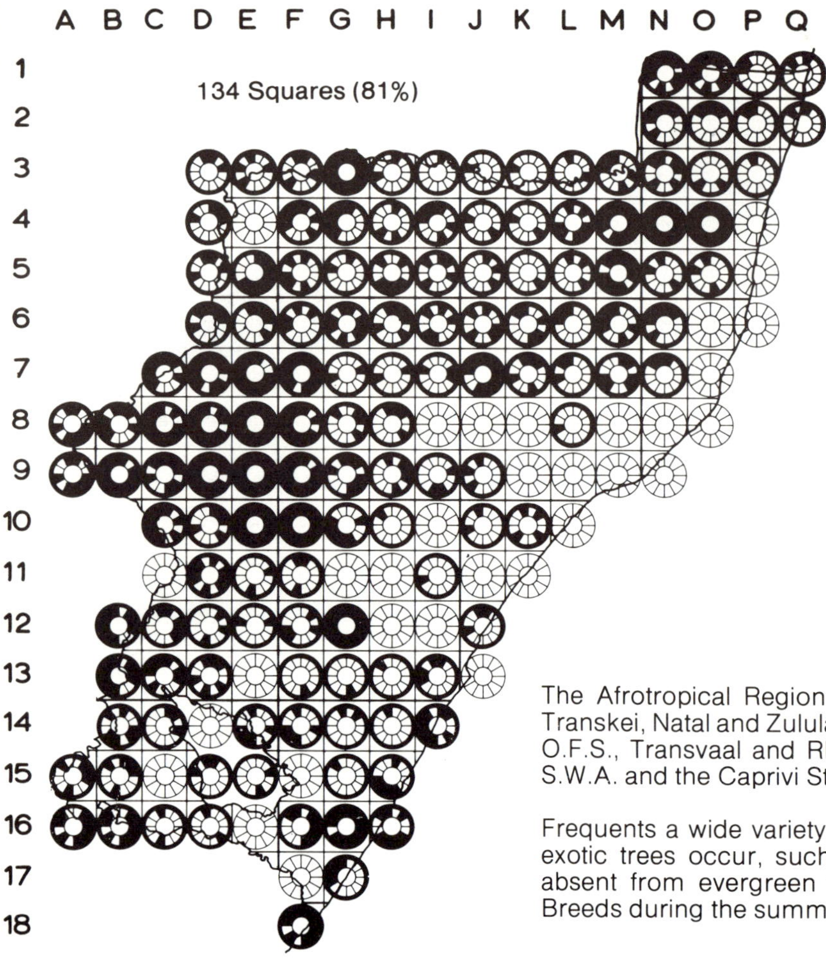

787 Grey-headed Sparrow

Passer griseus

The Afrotropical Region. In Southern Africa, the eastern Cape (local), Transkei, Natal and Zululand, Swaziland, southern Moçambique, Lesotho, O.F.S., Transvaal and Rhodesia, west to Botswana, the northern Cape, S.W.A. and the Caprivi Strip.

Frequents a wide variety of habitats, from open savanna to areas where exotic trees occur, such as farmsteads and edges of cultivated fields; absent from evergreen forests, and from the coast north of Durban. Breeds during the summer.

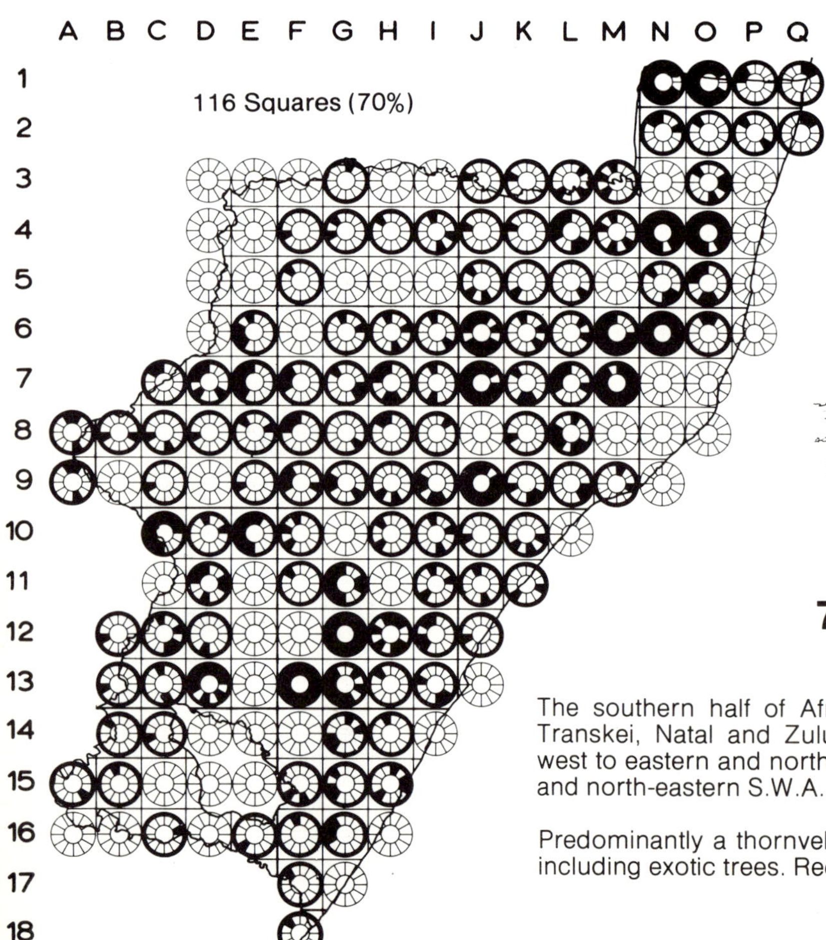

788 Yellow-throated Sparrow

Petronia superciliaris

The southern half of Africa. In Southern Africa, the eastern Cape to Transkei, Natal and Zululand, Moçambique, Transvaal and Rhodesia, west to eastern and northern Botswana, western O.F.S., the Caprivi Strip and north-eastern S.W.A. (Okavango River).

Predominantly a thornveld species, extending into lightly-wooded areas, including exotic trees. Recorded breeding from October to December.

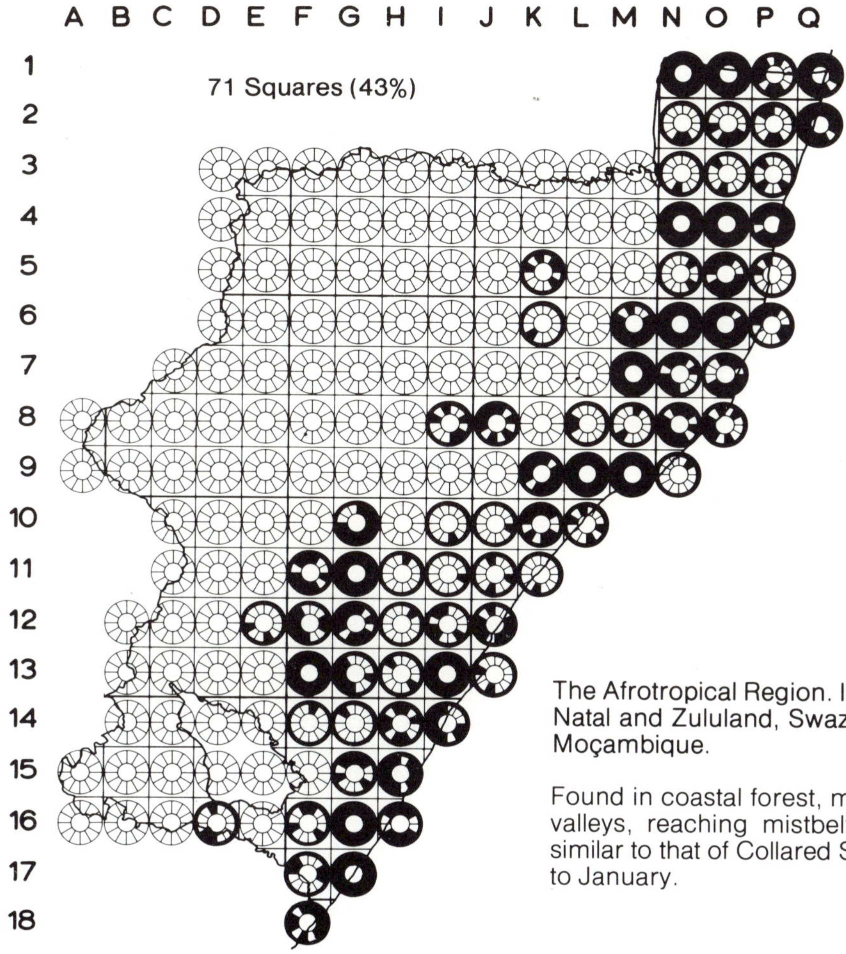

790 Forest Weaver

Ploceus bicolor

The Afrotropical Region. In Southern Africa, the eastern Cape to Transkei, Natal and Zululand, Swaziland, eastern Transvaal, eastern Rhodesia and Moçambique.

Found in coastal forest, mixed and riverine woodland, up thickly-wooded valleys, reaching mistbelt evergreen forest. The distribution pattern is similar to that of Collared Sunbird (771). Recorded breeding from October to January.

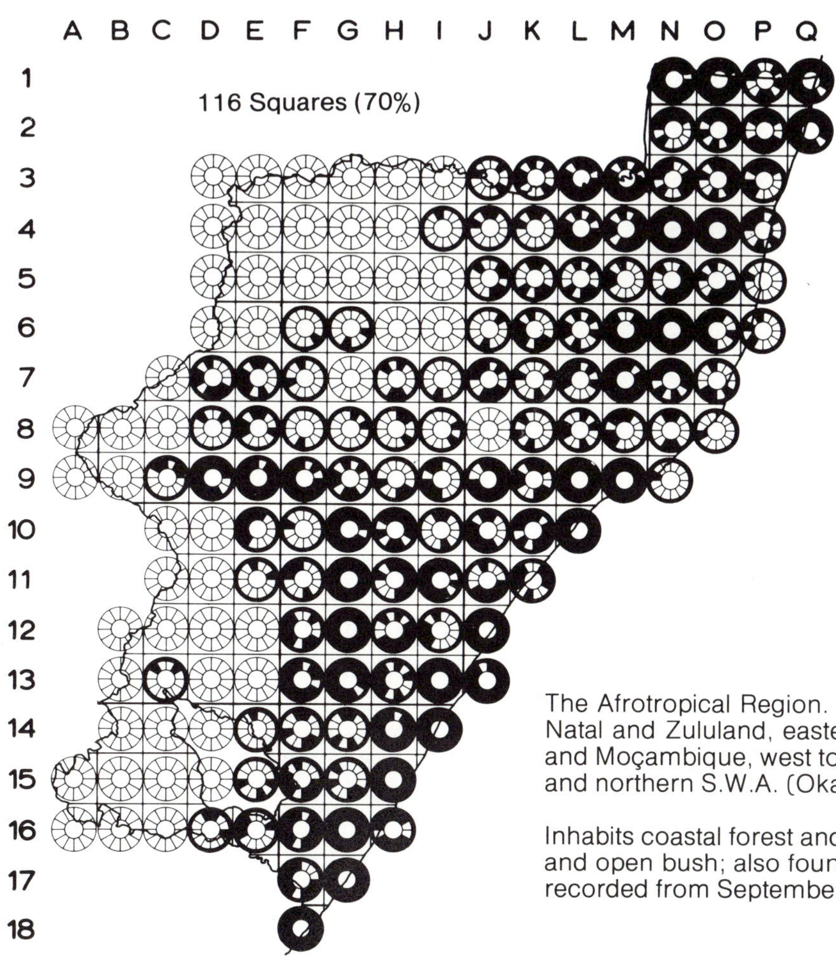

791 Spectacled Weaver

Ploceus ocularis

The Afrotropical Region. In Southern Africa, the eastern Cape, Transkei, Natal and Zululand, eastern Swaziland, eastern and northern Transvaal, and Moçambique, west to Rhodesia, northern Botswana, the Caprivi Strip and northern S.W.A. (Okavango River).

Inhabits coastal forest and scrub, riverine and mixed woodland, thornveld and open bush; also found around farmsteads and in gardens. Breeding recorded from September to February.

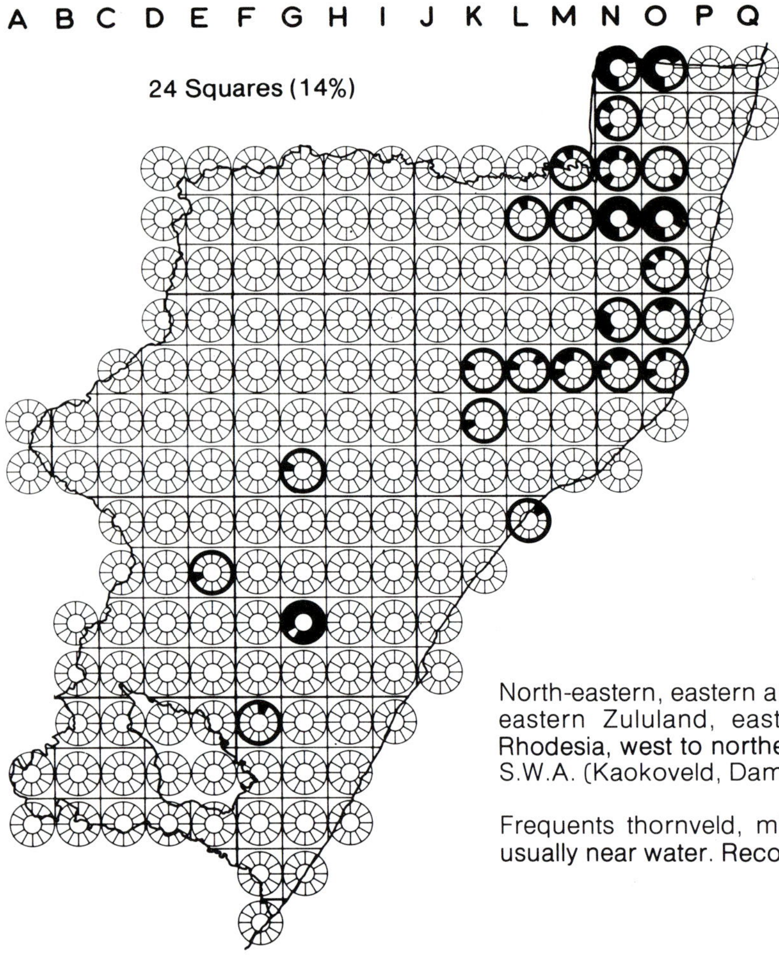

792 Lesser Masked Weaver

Ploceus intermedius

North-eastern, eastern and southern Africa. In Southern Africa, Natal and eastern Zululand, eastern Swaziland, Transvaal, Moçambique and Rhodesia, west to northern Botswana (Ngamiland), the Caprivi Strip and S.W.A. (Kaokoveld, Damaraland and Ovamboland).

Frequents thornveld, mixed woodland and occasionally coastal scrub; usually near water. Recorded breeding from October to February.

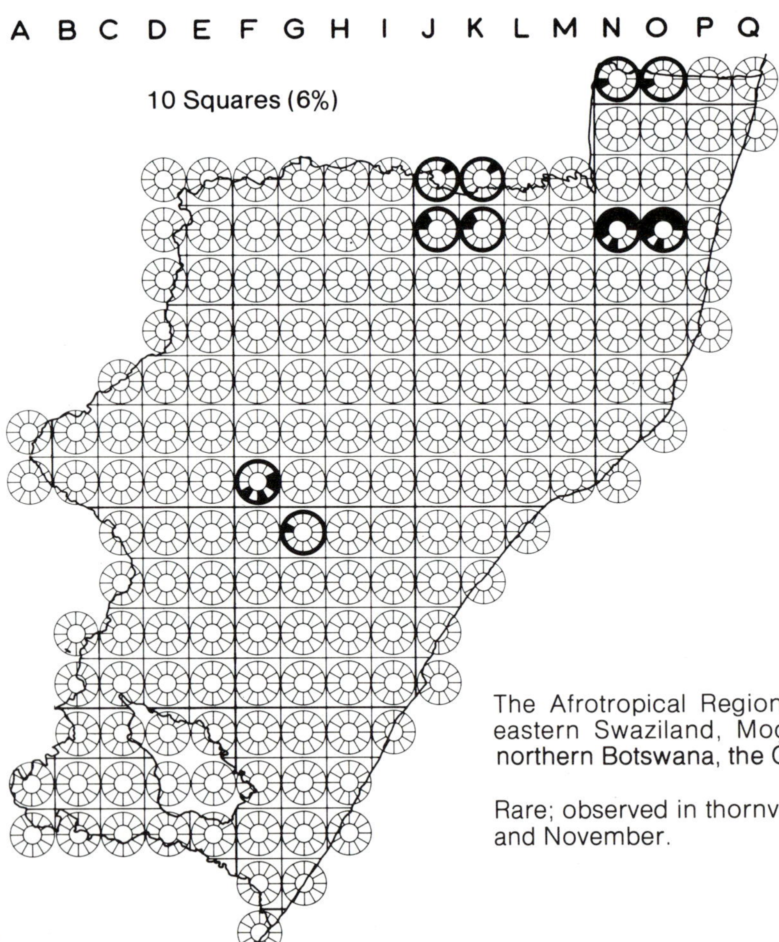

793 Red-headed Weaver

Anaplectes rubriceps

The Afrotropical Region. In Southern Africa, north-eastern Zululand, eastern Swaziland, Moçambique, Transvaal, Rhodesia, eastern and northern Botswana, the Caprivi Strip and northern S.W.A.

Rare; observed in thornveld regions. Recorded breeding during October and November.

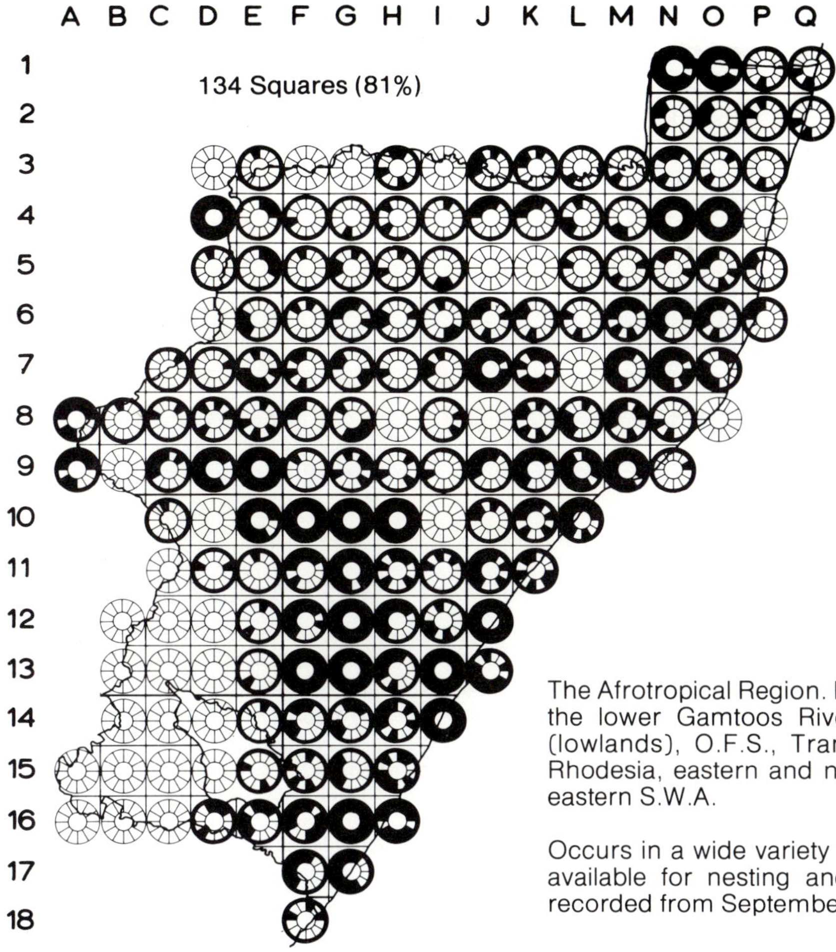

797 Spotted-backed Weaver

Ploceus cucullatus

The Afrotropical Region. In Southern Africa, the south-eastern Cape (from the lower Gamtoos River) to Transkei, Natal and Zululand, Lesotho (lowlands), O.F.S., Transvaal, Swaziland and Moçambique, west to Rhodesia, eastern and northern Botswana, the Caprivi Strip and north-eastern S.W.A.

Occurs in a wide variety of habitats, provided suitable trees or reeds are available for nesting and roosting; often found near water. Breeding recorded from September to February.

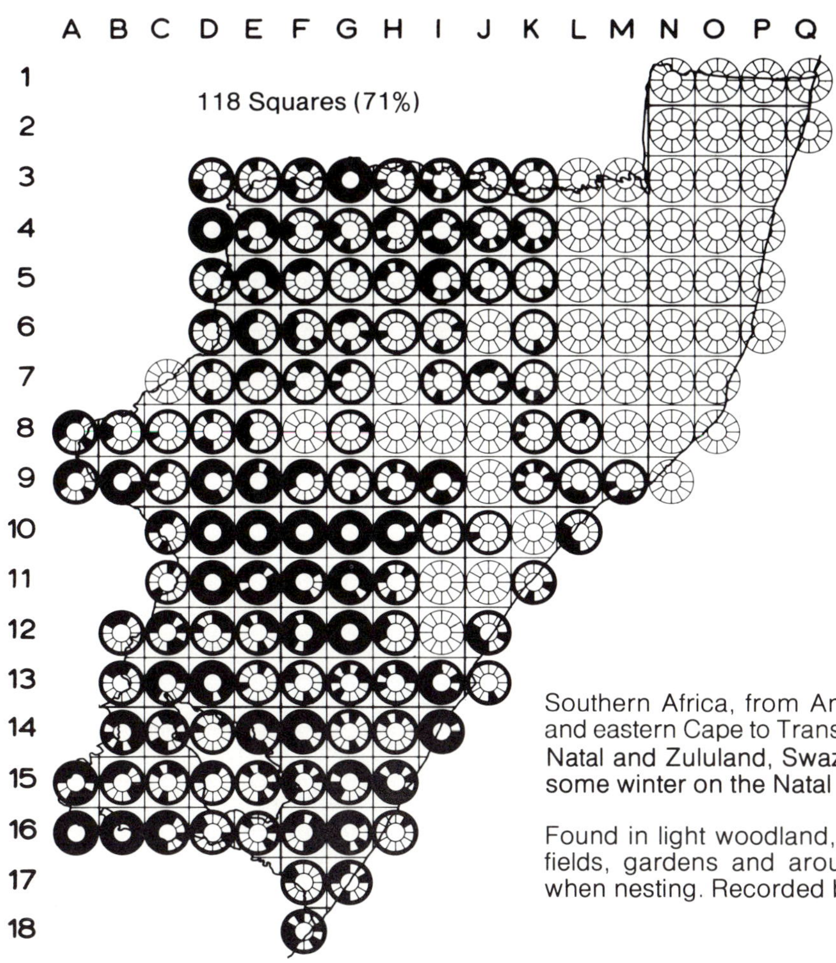

799 Cape Weaver

Ploceus capensis

Southern Africa, from Angola to the Cape. From the western, southern and eastern Cape to Transkei, East Griqualand, O.F.S., Lesotho (lowlands), Natal and Zululand, Swaziland and Transvaal (plateau and escarpment); some winter on the Natal and Zululand coasts.

Found in light woodland, open grassland with scattered trees, cultivated fields, gardens and around farmsteads; usually associated with water when nesting. Recorded breeding from October to February.

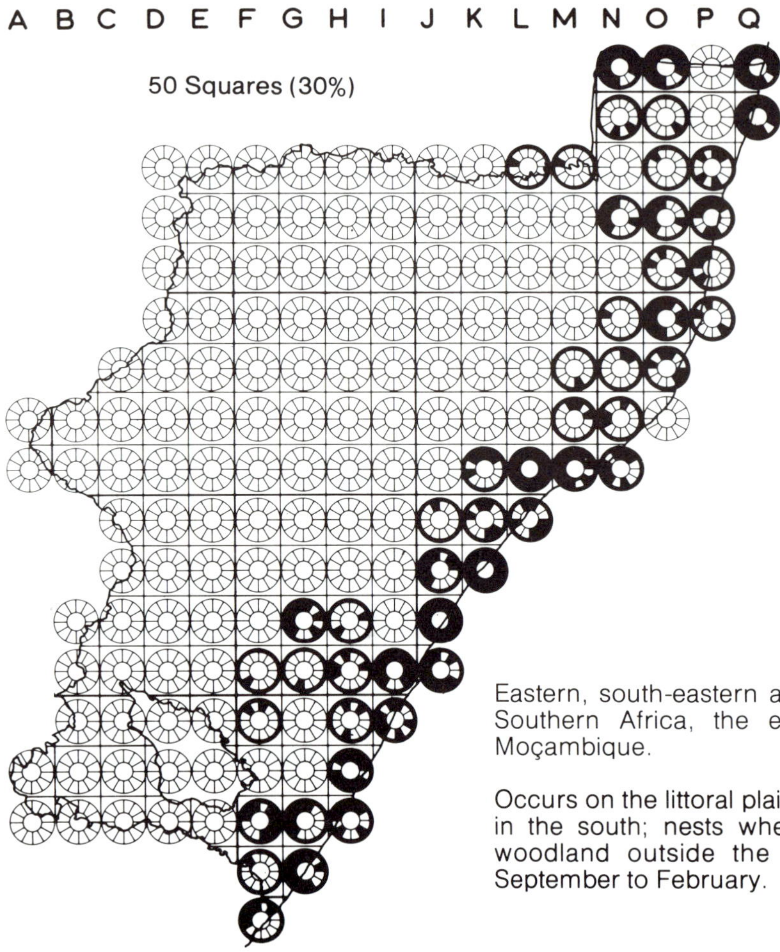

800 Yellow Weaver

Ploceus subaureus

Eastern, south-eastern and southern Africa, mainly along the coast. In Southern Africa, the eastern Cape, Transkei, Natal, Zululand and Moçambique.

Occurs on the littoral plain and along the coast, extending up river valleys in the south; nests wherever reeds are found, ranging into adjacent woodland outside the breeding season. Recorded breeding from September to February.

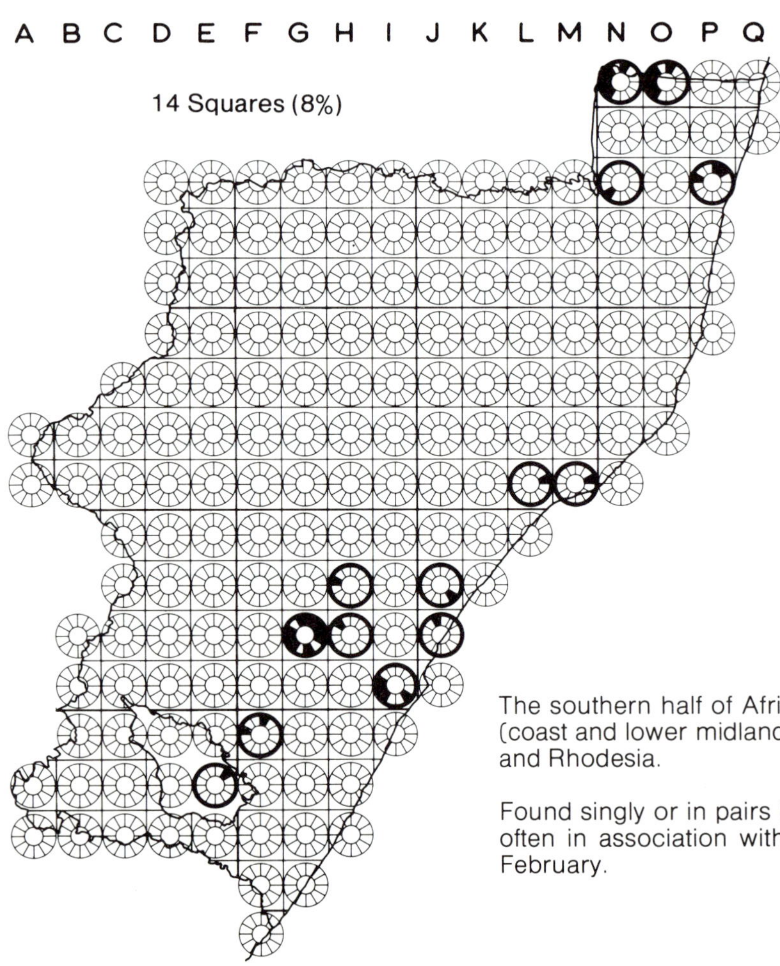

801 Golden Weaver

Ploceus xanthops

The southern half of Africa, southeast to Natal. In Southern Africa, Natal (coast and lower midlands), Zululand, Swaziland, Moçambique, Transvaal and Rhodesia.

Found singly or in pairs in reeds and scattered vegetation close to water; often in association with Cape Weaver (799). Breeds from October to February.

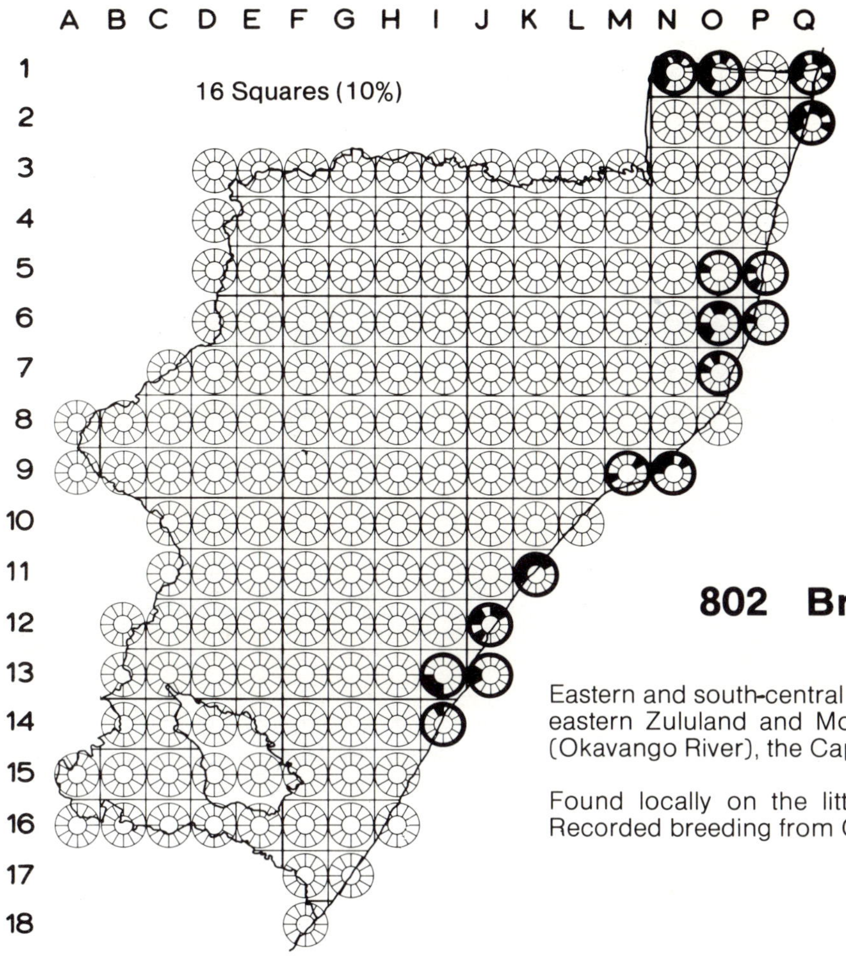

802 Brown-throated Golden Weaver

Ploceus xanthopterus

Eastern and south-central Africa. In Southern Africa, from Natal (coast) to eastern Zululand and Moçambique (littoral); also north-eastern S.W.A. (Okavango River), the Caprivi Strip and northern Botswana.

Found locally on the littoral and coastal plains, in dense reedbeds. Recorded breeding from October to December.

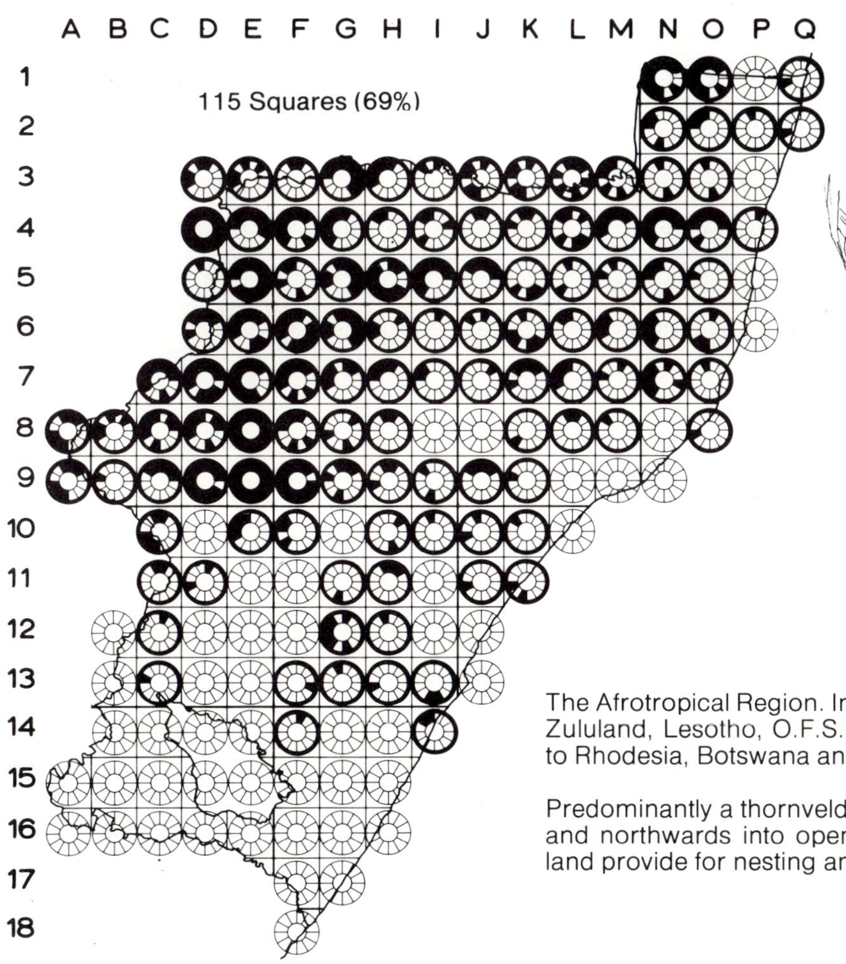

803 Masked Weaver

Ploceus velatus

The Afrotropical Region. In Southern Africa, the Cape, Transkei, Natal and Zululand, Lesotho, O.F.S., Transvaal, Swaziland and Moçambique, west to Rhodesia, Botswana and S.W.A.

Predominantly a thornveld species, extending into open bush in the south and northwards into open grassland, where exotic trees and cultivated land provide for nesting and feeding. Breeds from October to January.

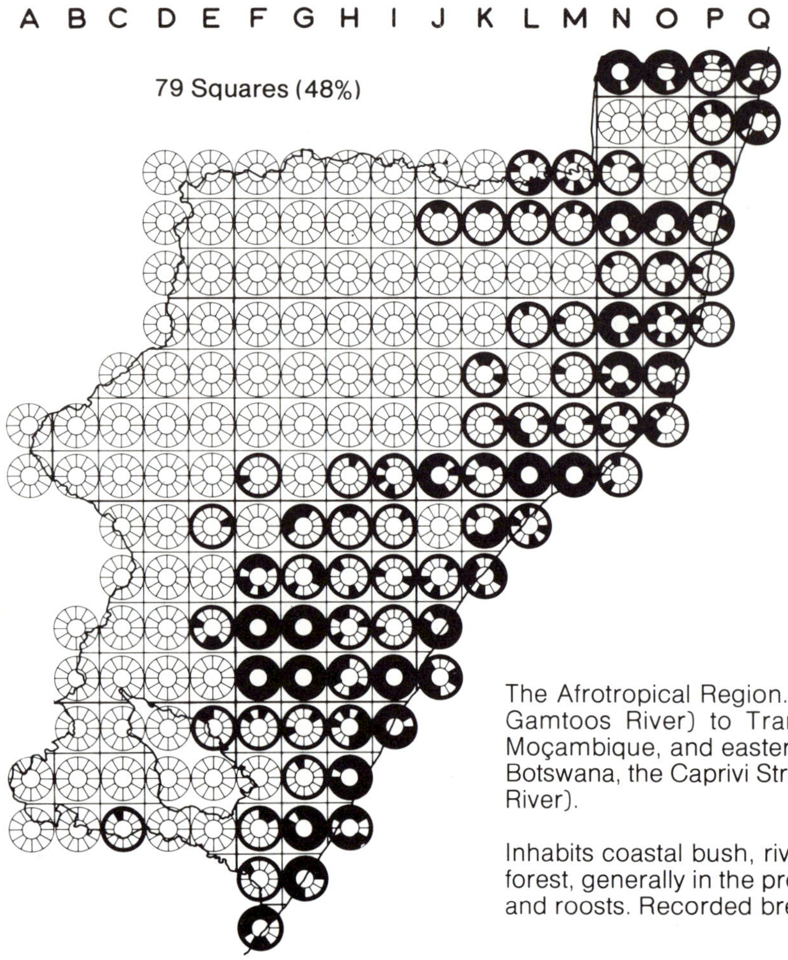

804 Thick-billed Weaver

Amblyospiza albifrons

The Afrotropical Region. In Southern Africa, the eastern Cape (from the Gamtoos River) to Transkei, Natal and Zululand, eastern Transvaal, Moçambique, and eastern and north-western Rhodesia, west to northern Botswana, the Caprivi Strip, and probably north-eastern S.W.A. (Okavango River).

Inhabits coastal bush, riverine woodland, wooded valleys, and evergreen forest, generally in the proximity of vleis, ponds or reedbeds where it nests and roosts. Recorded breeding from September to March.

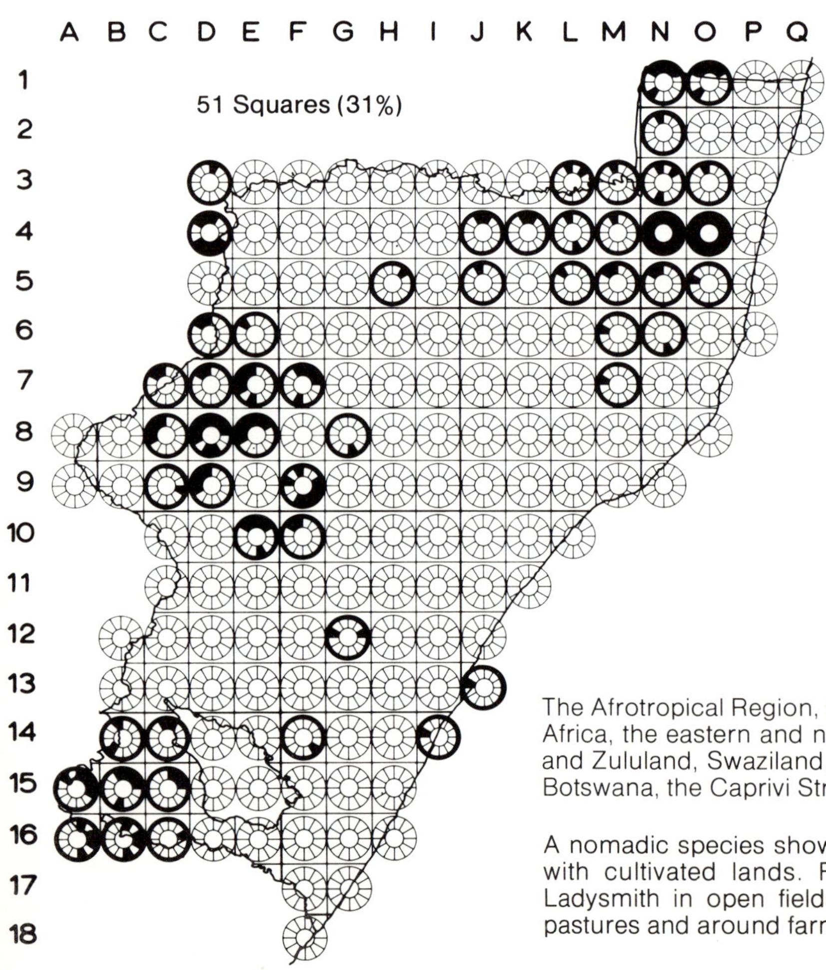

805 Red-billed Quelea

Quelea quelea

The Afrotropical Region, outside the limits of unbroken forest. In Southern Africa, the eastern and northern Cape, O.F.S., Lesotho (lowlands), Natal and Zululand, Swaziland, Transvaal and Moçambique, west to Rhodesia, Botswana, the Caprivi Strip and eastern and northern S.W.A.

A nomadic species showing three distinct concentrations, all associated with cultivated lands. Found in Zululand in open savanna, around Ladysmith in open fields and thorn scrub, and in East Griqualand in pastures and around farmsteads. Breeds during the late summer.

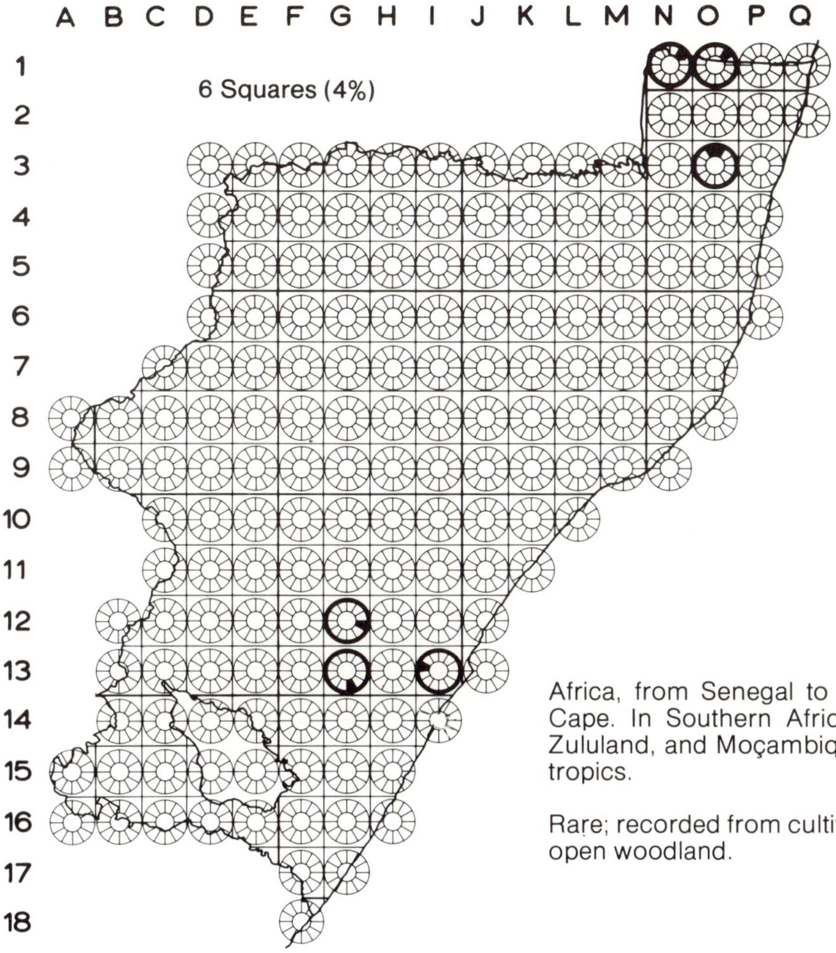

806 Red-headed Quelea

Quelea erythrops

Africa, from Senegal to Ethiopia, and locally south to Angola and the Cape. In Southern Africa, the eastern Cape, Transkei, Natal, eastern Zululand, and Moçambique. A summer resident, wintering in the African tropics.

Rare; recorded from cultivated fields, waste ground and, in Zululand, from open woodland.

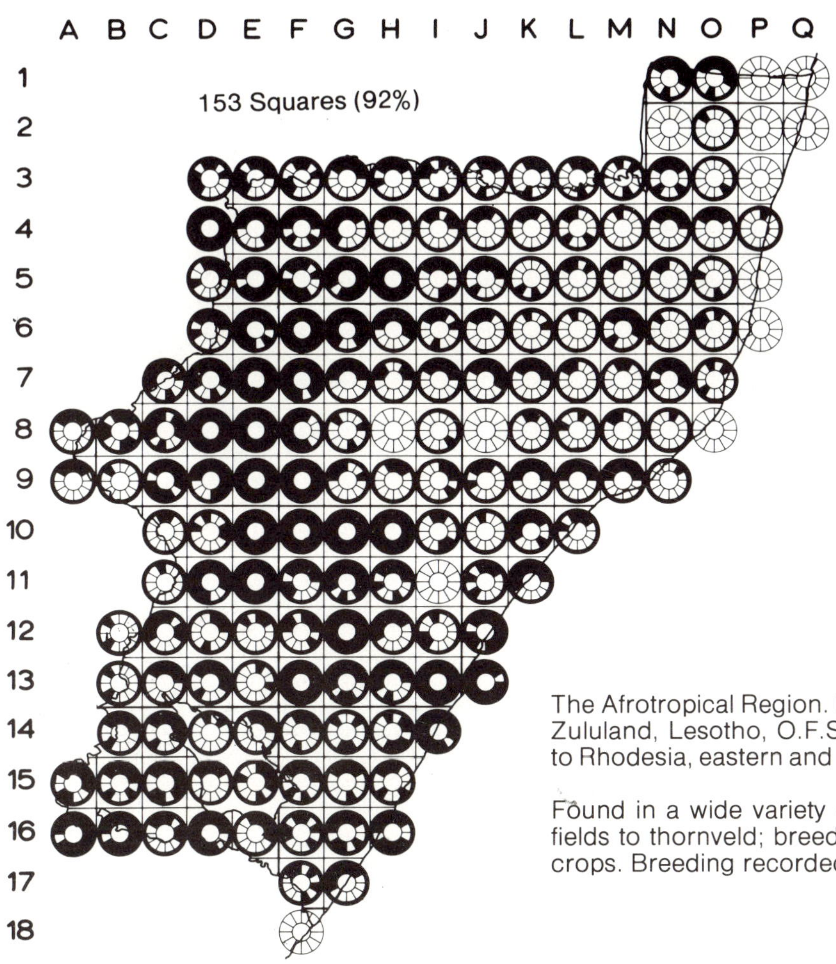

808 Red Bishop

Euplectes orix

The Afrotropical Region. In Southern Africa, the Cape, Transkei, Natal and Zululand, Lesotho, O.F.S., Transvaal, Swaziland and Moçambique, west to Rhodesia, eastern and northern Botswana, the Caprivi Strip and S.W.A.

Found in a wide variety of habitats, from open grassland and cultivated fields to thornveld; breeds in tall grass, reedbeds and occasionally in tall crops. Breeding recorded from October to April.

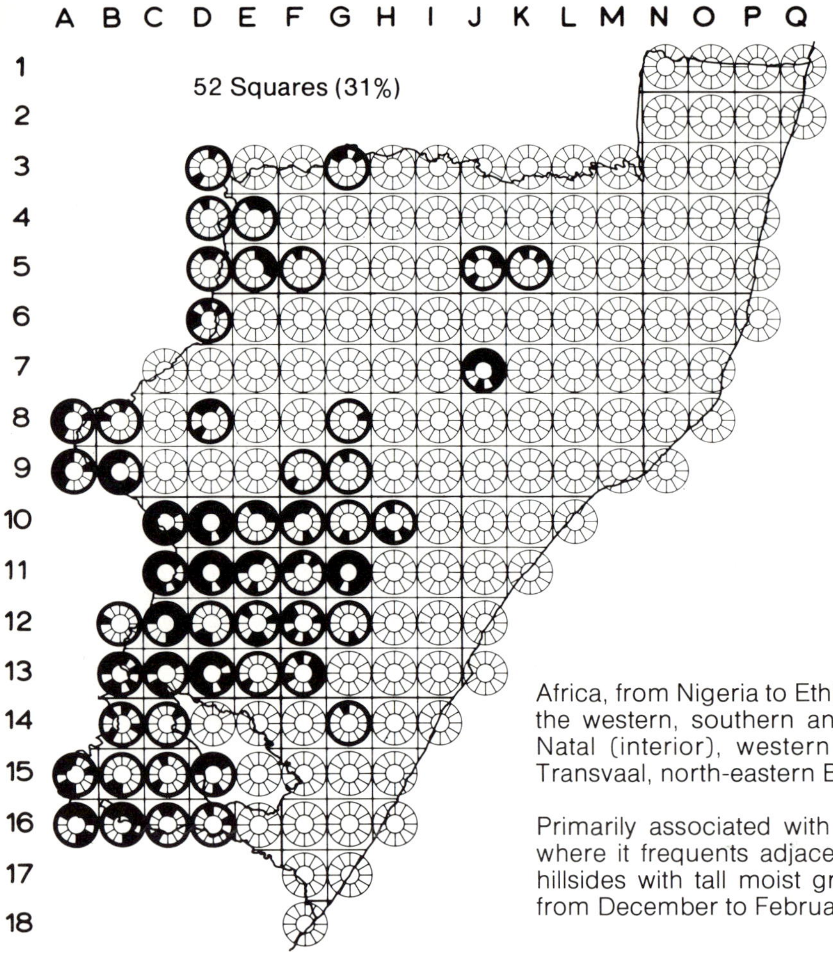

810 Cape Widow

Euplectes capensis

Africa, from Nigeria to Ethiopia and south to the Cape. In Southern Africa, the western, southern and eastern Cape to Transkei, O.F.S., Lesotho, Natal (interior), western Zululand, Swaziland, eastern and northern Transvaal, north-eastern Botswana, Rhodesia and Moçambique.

Primarily associated with Drakensberg and mistbelt evergreen forests, where it frequents adjacent open grassland and scrub; also occurs on hillsides with tall moist grass and scrub vegetation. Breeding recorded from December to February.

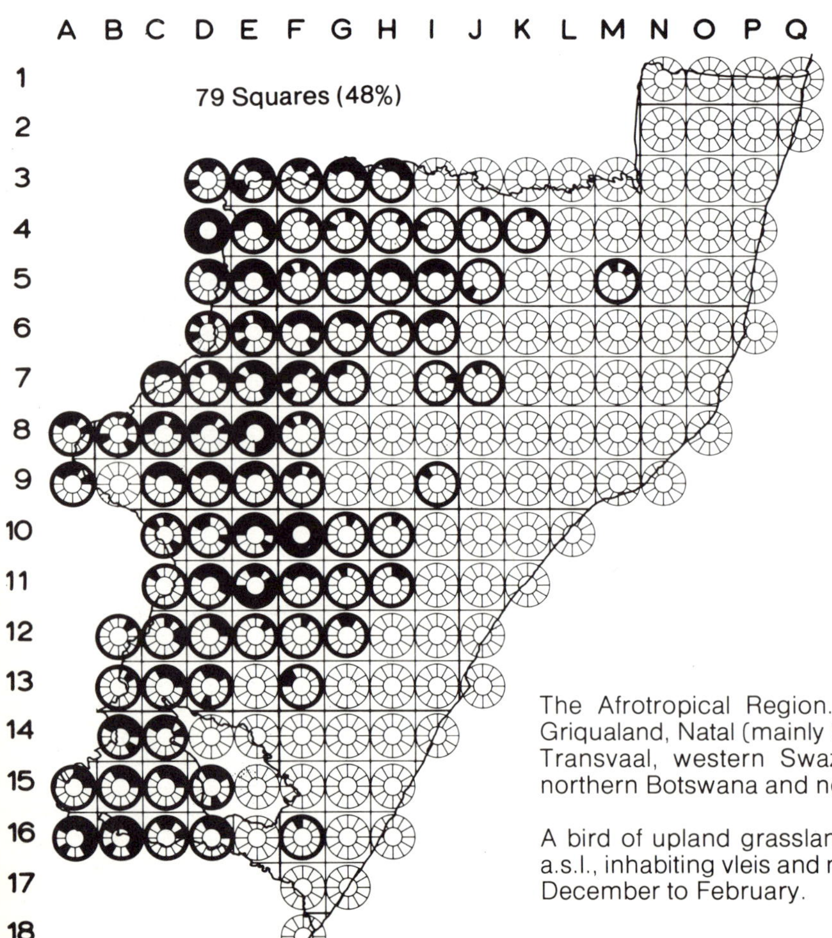

812 Golden Bishop

Euplectes afer

The Afrotropical Region. In Southern Africa, the eastern Cape, East Griqualand, Natal (mainly high interior), western Zululand, O.F.S., Lesotho, Transvaal, western Swaziland, and Moçambique, west to Rhodesia, northern Botswana and northern S.W.A.

A bird of upland grassland and cultivated fields, generally above 900 m a.s.l., inhabiting vleis and marshes when nesting. Recorded breeding from December to February.

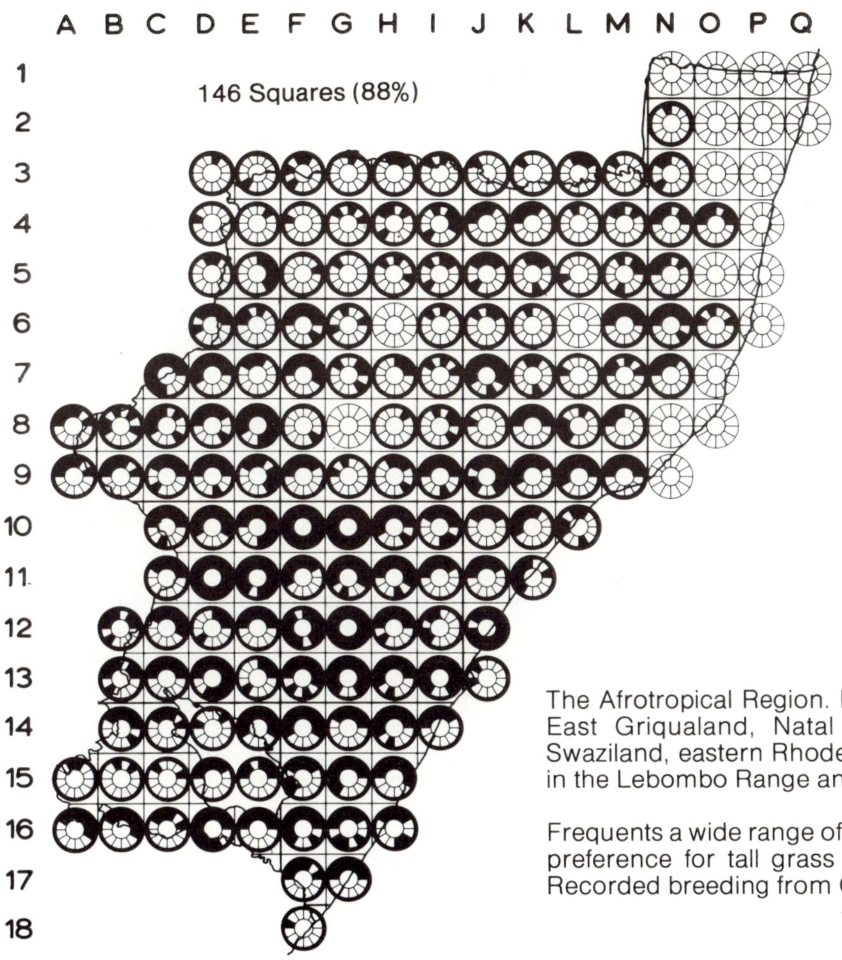

813 Red-collared Widow

Euplectes ardens

The Afrotropical Region. In Southern Africa, the eastern Cape, Transkei, East Griqualand, Natal and Zululand, O.F.S., Lesotho, Transvaal, Swaziland, eastern Rhodesia and adjacent Moçambique (highland); also in the Lebombo Range and on Mount Gorongosa.

Frequents a wide range of habitats from open savanna to grassland, with a preference for tall grass on the edge of cultivated fields and streams. Recorded breeding from October to March.

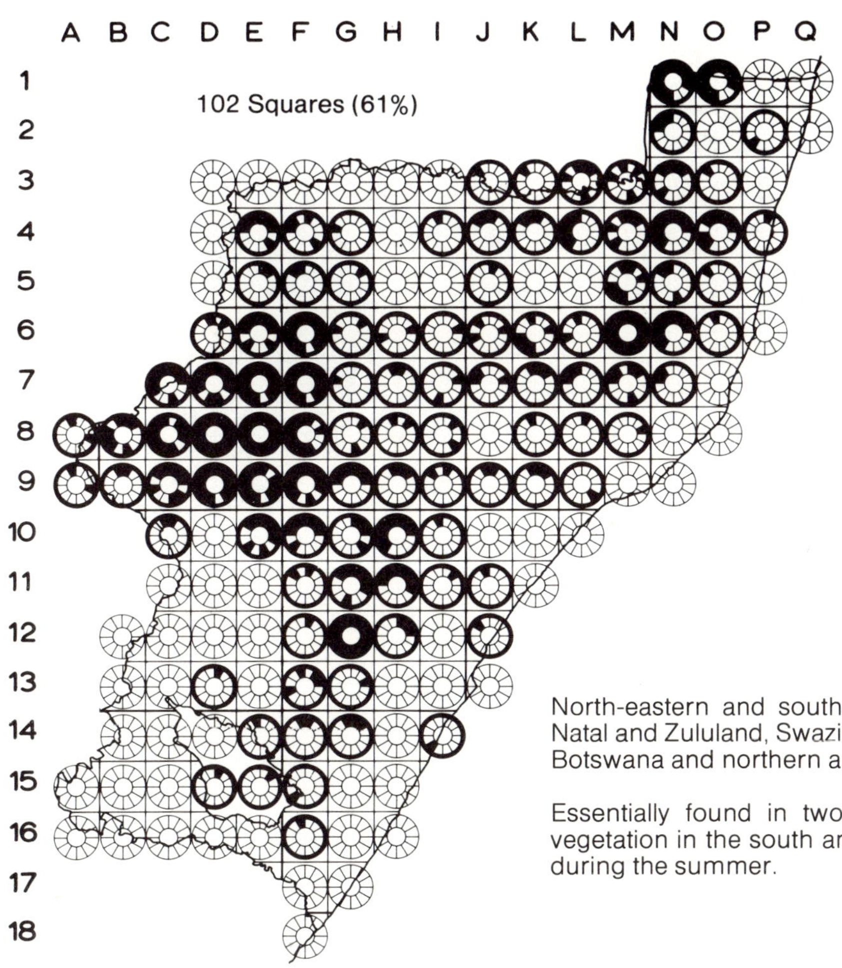

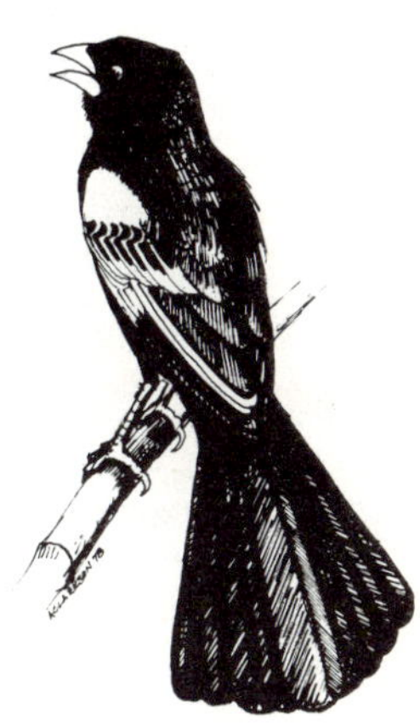

814 White-winged Widow

Euplectes albonotatus

North-eastern and southern Africa south to Natal. In Southern Africa, Natal and Zululand, Swaziland, Transvaal, Moçambique, Rhodesia, eastern Botswana and northern and north-eastern S.W.A.

Essentially found in two habitats; cultivated fields and adjacent rank vegetation in the south and west, and open savanna in Zululand. Breeds during the summer.

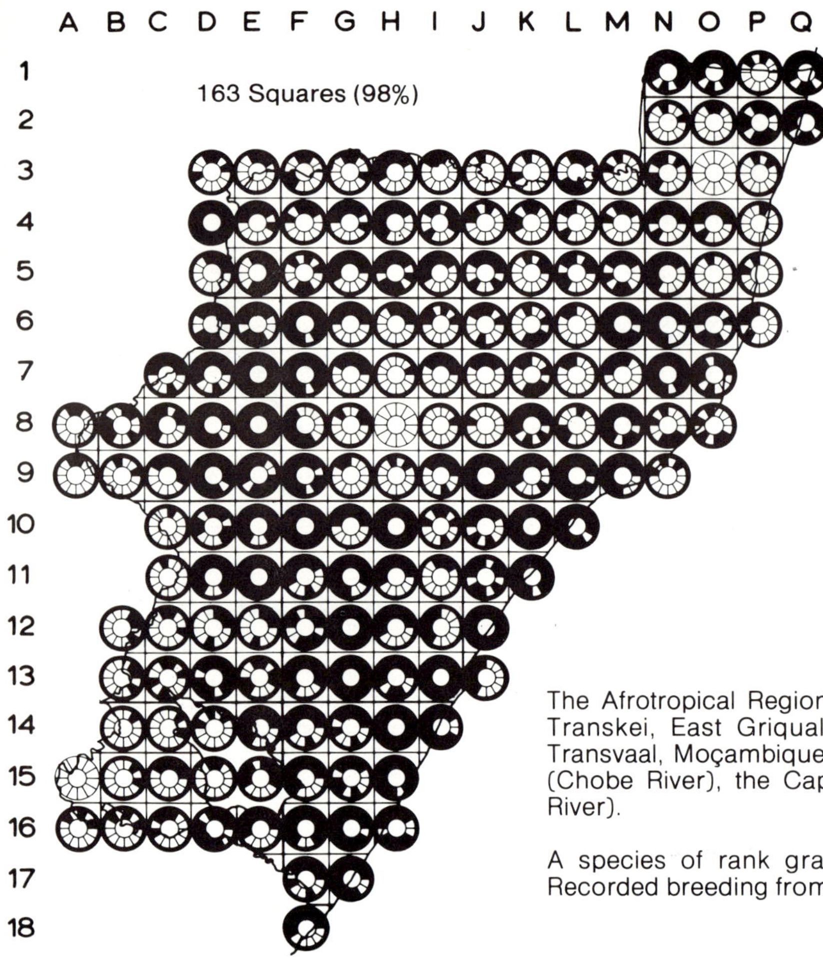

816 Red-shouldered Widow

Euplectes axillaris

The Afrotropical Region. In Southern Africa, from the eastern Cape to Transkei, East Griqualand, Natal and Zululand, Swaziland, eastern Transvaal, Moçambique and eastern Rhodesia; also northern Botswana (Chobe River), the Caprivi Strip and north-eastern S.W.A. (Okavango River).

A species of rank grass and cultivated fields, especially canelands. Recorded breeding from October to March.

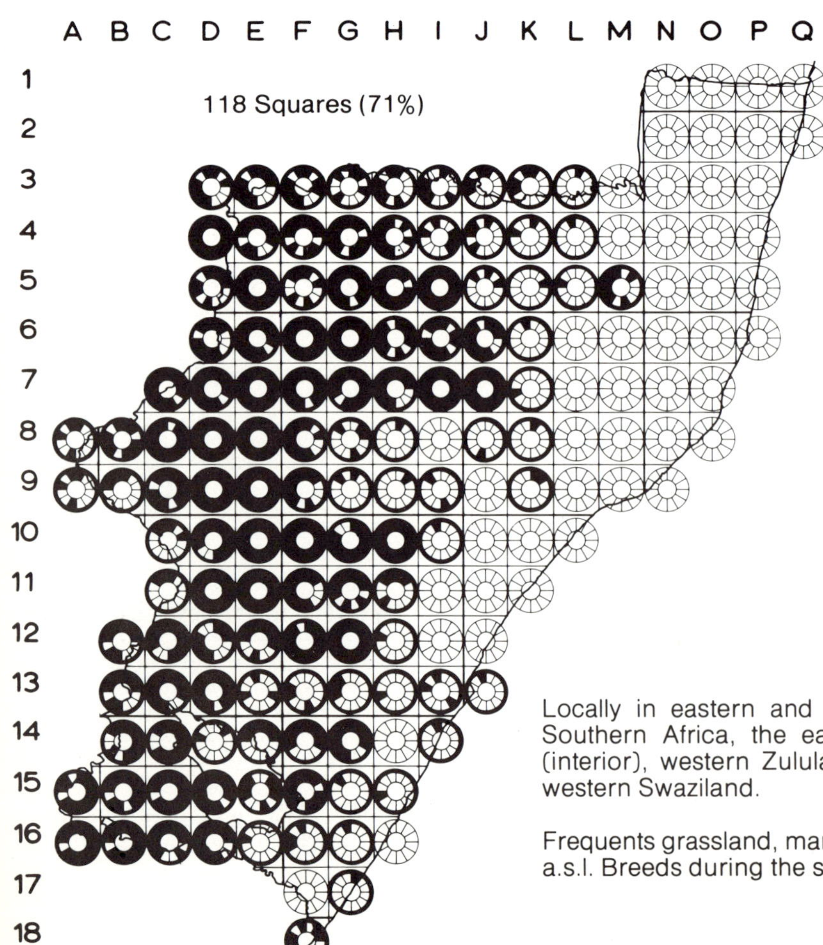

818 Long-tailed Widow

Euplectes progne

Locally in eastern and southern Africa, from Kenya to the Cape. In Southern Africa, the eastern Cape, Transkei, East Griqualand, Natal (interior), western Zululand, O.F.S., Lesotho, Transvaal (highveld) and western Swaziland.

Frequents grassland, marshes and cultivated fields, generally above 600 m a.s.l. Breeds during the summer.

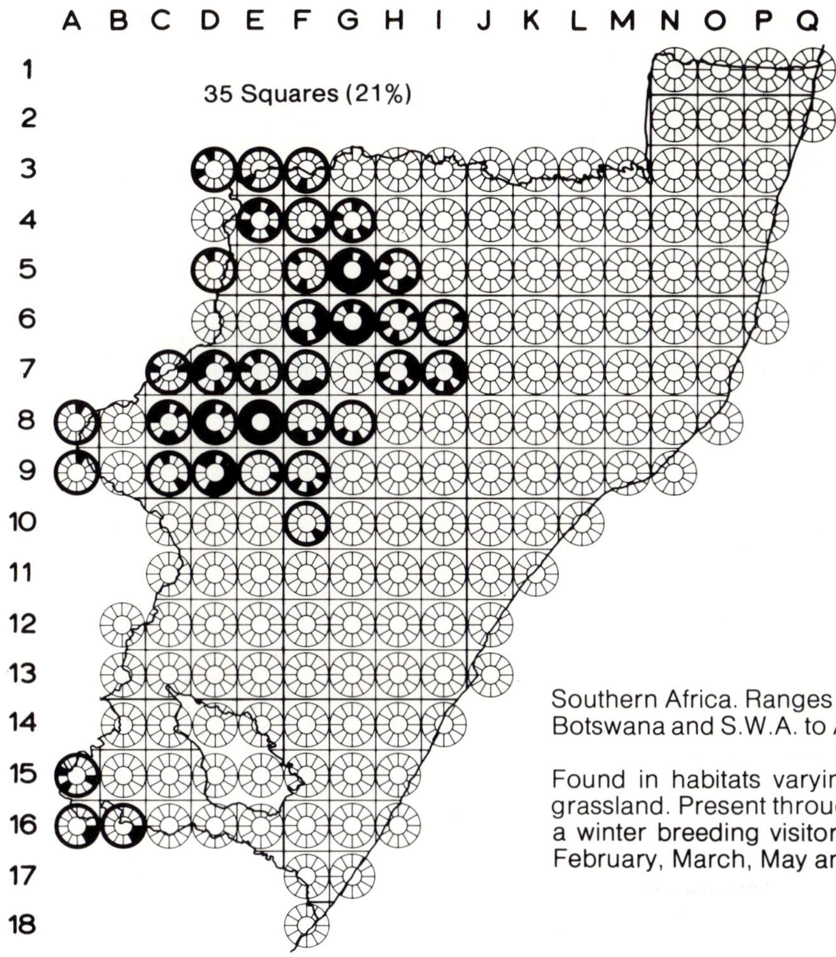

820 Red-headed Finch

Amadina erythrocephala

Southern Africa. Ranges from Natal, Transvaal, O.F.S., the northern Cape, Botswana and S.W.A. to Angola; occasionally in south-western Rhodesia.

Found in habitats varying from thornveld (around Ladysmith) to open grassland. Present throughout the year in the northwest but appears to be a winter breeding visitor to East Griqualand. Recorded breeding during February, March, May and September.

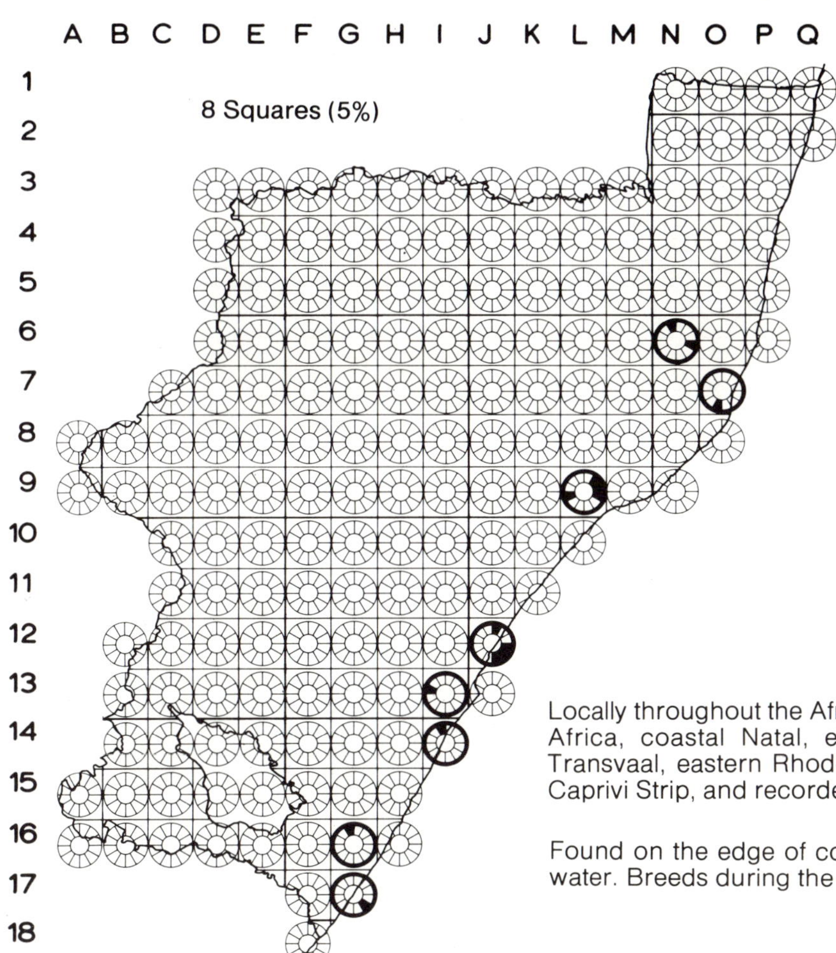

822 Pied Mannikin

Spermestes fringilloides

Locally throughout the Afrotropical Region, southeast to Natal. In Southern Africa, coastal Natal, eastern Zululand, eastern Swaziland, eastern Transvaal, eastern Rhodesia, and Moçambique. Probably occurs in the Caprivi Strip, and recorded from the eastern Cape.

Found on the edge of coastal bush, and in rank vegetation usually near water. Breeds during the summer.

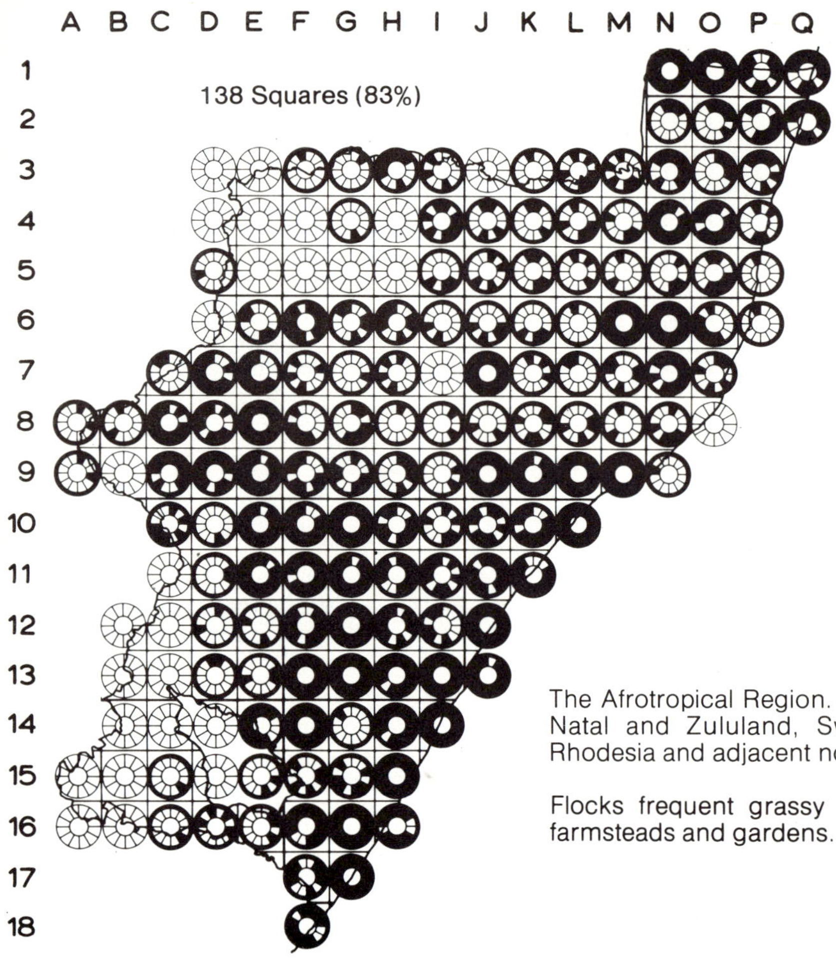

823 Bronze Mannikin

Spermestes cucullatus

The Afrotropical Region. In Southern Africa, the eastern Cape, Transkei, Natal and Zululand, Swaziland, southern Moçambique, Transvaal, Rhodesia and adjacent northern Botswana, and the eastern Caprivi Strip.

Flocks frequent grassy areas in open woodland and thornveld; also farmsteads and gardens. Breeds from September to April.

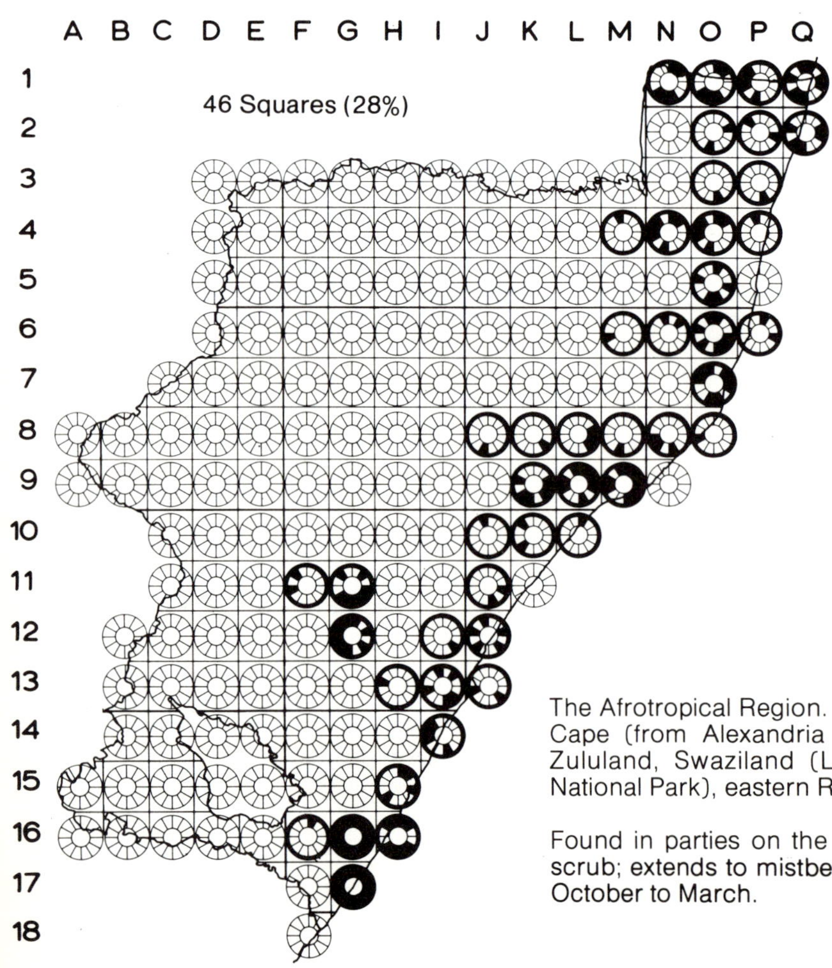

824 Red-backed Mannikin

Spermestes bicolor

The Afrotropical Region. In Southern Africa, coastal forests of the eastern Cape (from Alexandria Forest), Transkei, Natal (mainly coastal) and Zululand, Swaziland (Lebombo Range), eastern Transvaal (Kruger National Park), eastern Rhodesia and Moçambique.

Found in parties on the edges of riverine woodland, coastal forest and scrub; extends to mistbelt evergreen forest in the midlands. Breeds from October to March.

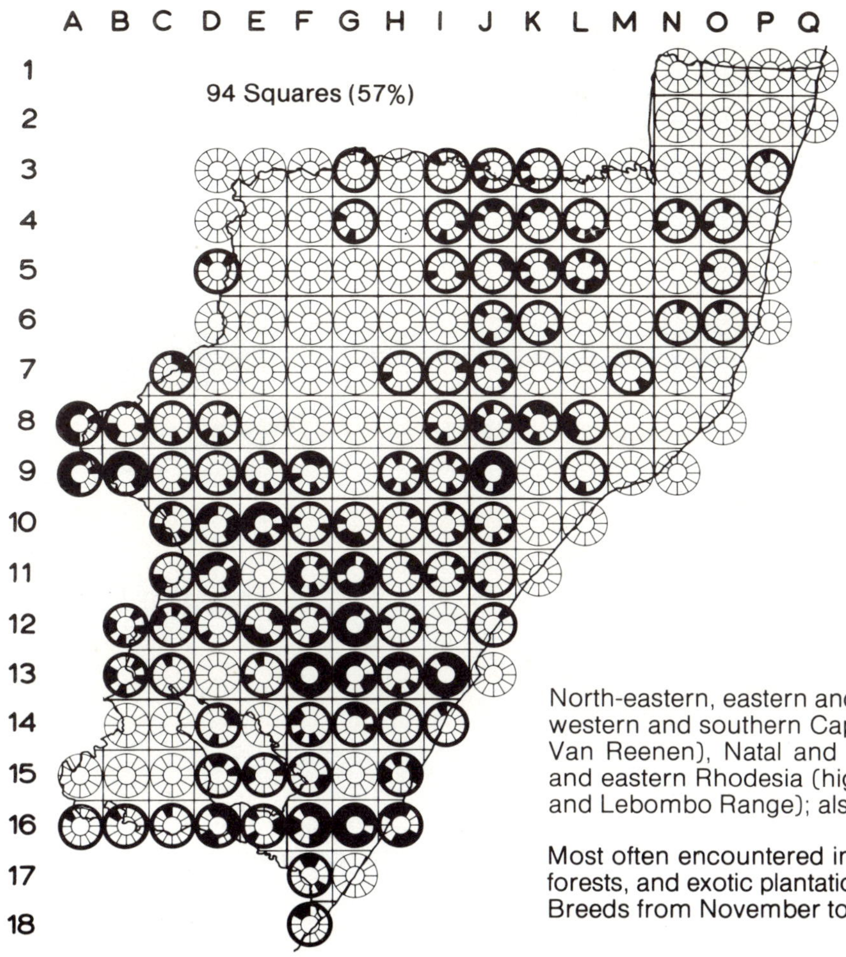

825 Swee Waxbill

Estrilda melanotis

North-eastern, eastern and southern Africa. In Southern Africa, the south-western and southern Cape to Transkei, Lesotho, eastern O.F.S. (west of Van Reenen), Natal and Zululand, Swaziland, Transvaal, south-eastern and eastern Rhodesia (highlands), and Moçambique (western highlands and Lebombo Range); also northern S.W.A. (Ovamboland).

Most often encountered in lowland, mistbelt and Drakensberg evergreen forests, and exotic plantations; also found in scrub vegetation and gardens. Breeds from November to April.

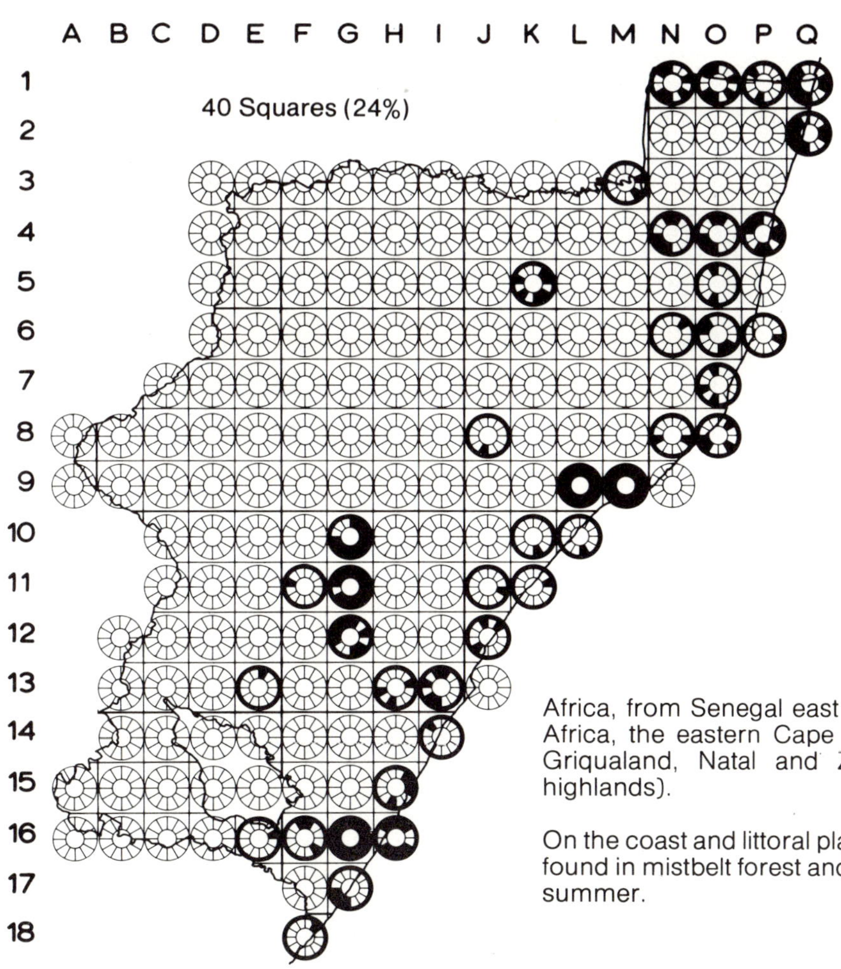

827 Green Twinspot

Mandingoa nitidula

Africa, from Senegal east to Kenya, and south to the Cape. In Southern Africa, the eastern Cape (from the Great Fish River) to Transkei, East Griqualand, Natal and Zululand, Swaziland and Rhodesia (eastern highlands).

On the coast and littoral plain, inhabits forests and adjacent thickets; inland found in mistbelt forest and adjoining exotic plantations. Breeds during the summer.

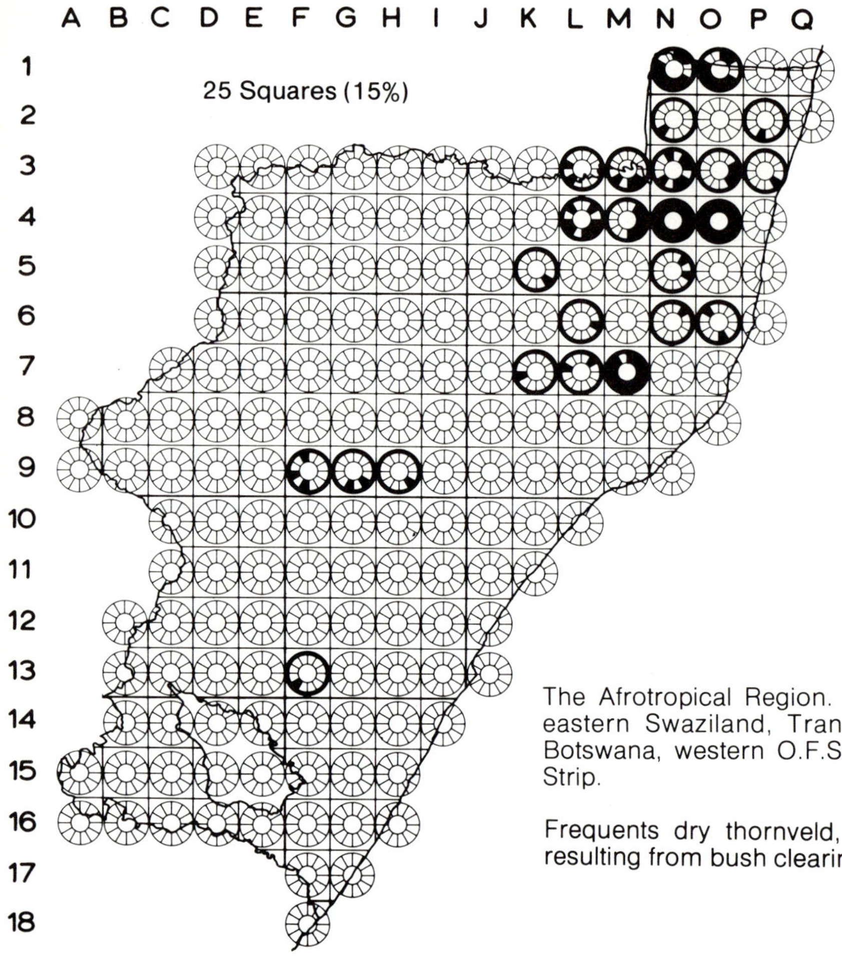

830 Melba Finch

Pytilia melba

The Afrotropical Region. In Southern Africa, Natal (interior), Zululand, eastern Swaziland, Transvaal, Moçambique and Rhodesia, west to Botswana, western O.F.S., the northern Cape, S.W.A. and the Caprivi Strip.

Frequents dry thornveld, inhabiting thickets and piles of dead wood resulting from bush clearing. Breeds mainly during the late summer.

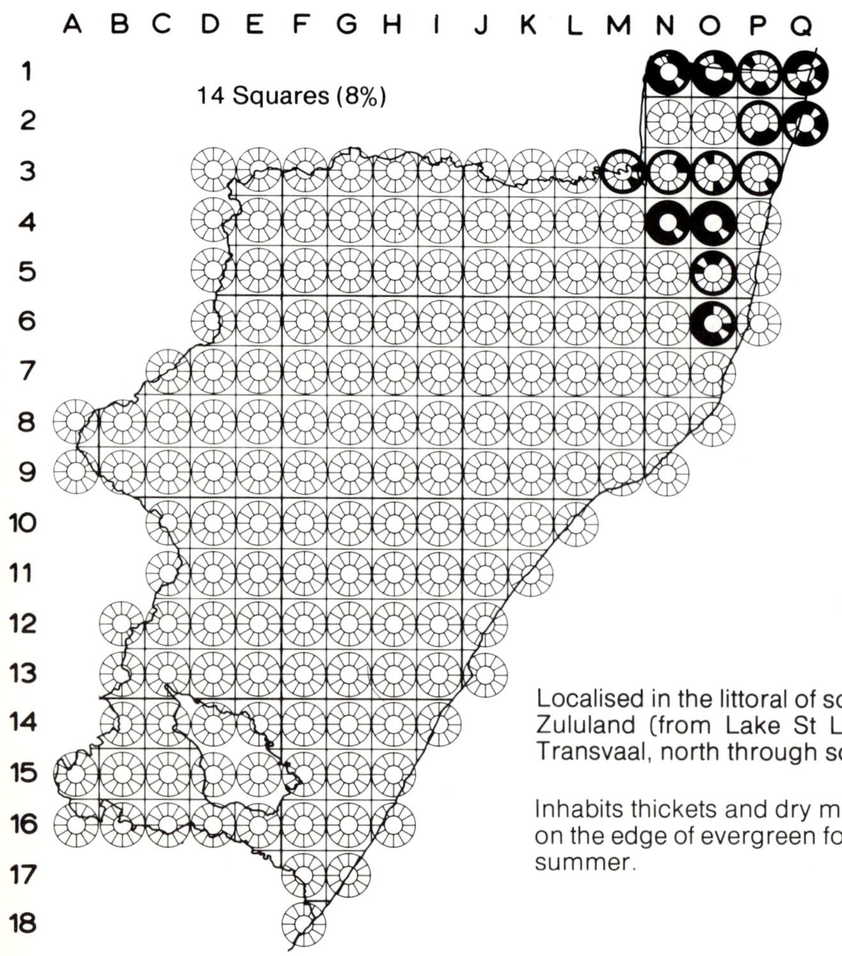

831 Pink-throated Twinspot

Hypargos margaritatus

Localised in the littoral of south-eastern Africa. In Southern Africa, eastern Zululand (from Lake St Lucia), adjacent Swaziland and south-eastern Transvaal, north through southern Moçambique (to the Save River).

Inhabits thickets and dry mixed woodland on the littoral plain; also occurs on the edge of evergreen forest in the Lebombo Range. Breeds during the summer.

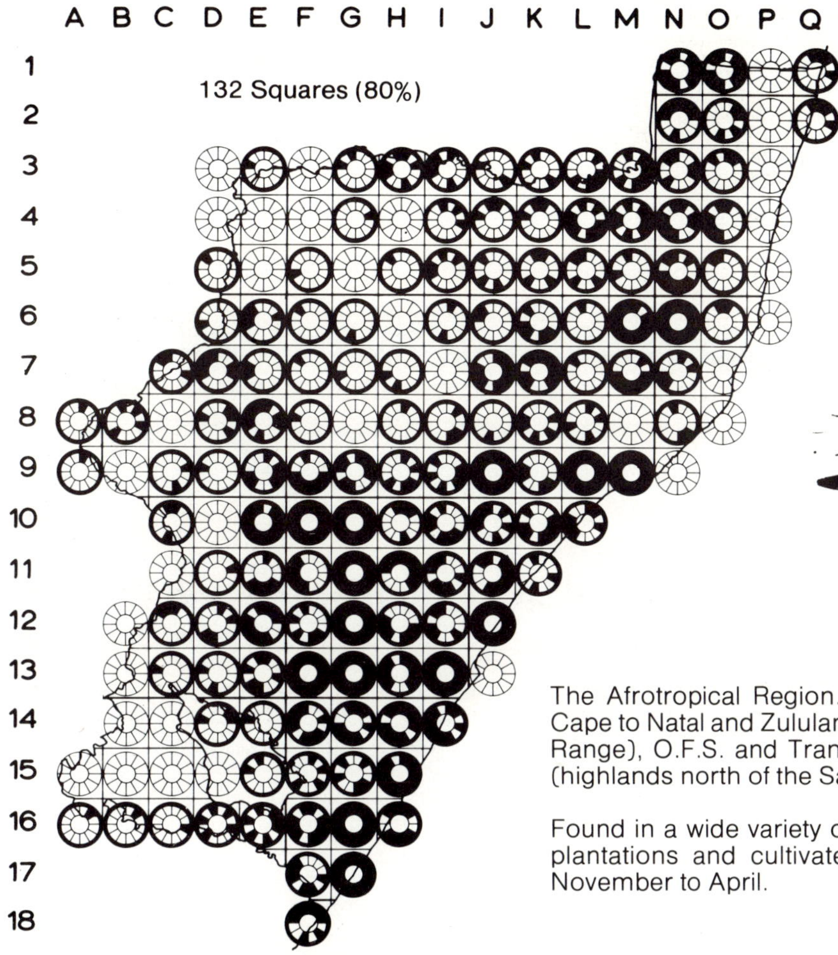

833 Blue-billed Firefinch

Lagonosticta rubicata

The Afrotropical Region. In Southern Africa, the southern and eastern Cape to Natal and Zululand, Swaziland, southern Moçambique (Lebombo Range), O.F.S. and Transvaal; also eastern Rhodesia and Moçambique (highlands north of the Save River).

Found in a wide variety of habitats from thornveld to the edges of forest, plantations and cultivated areas, where thickets occur. Breeds from November to April.

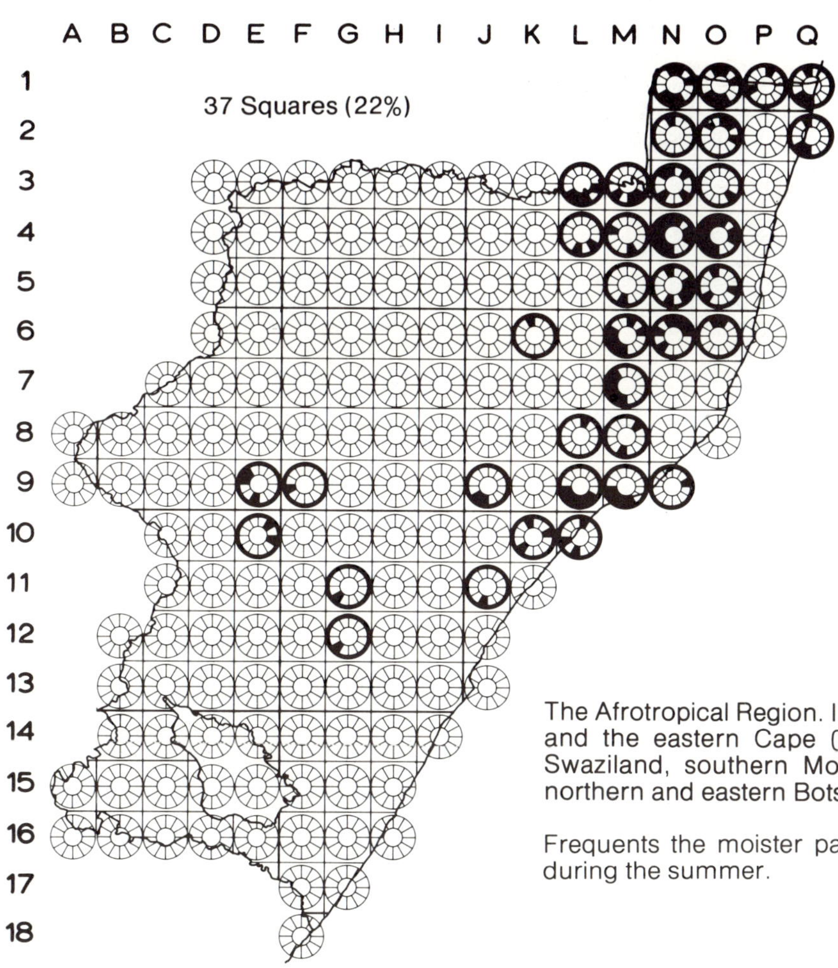

837 Red-billed Firefinch

Lagonosticta senegala

The Afrotropical Region. In Southern Africa, the valley of the Orange River and the eastern Cape (local), O.F.S., Transkei, Natal and Zululand, Swaziland, southern Moçambique, Transvaal and Rhodesia, west to northern and eastern Botswana, the Caprivi Strip and northern S.W.A.

Frequents the moister parts of thornveld and mixed woodland. Breeds during the summer.

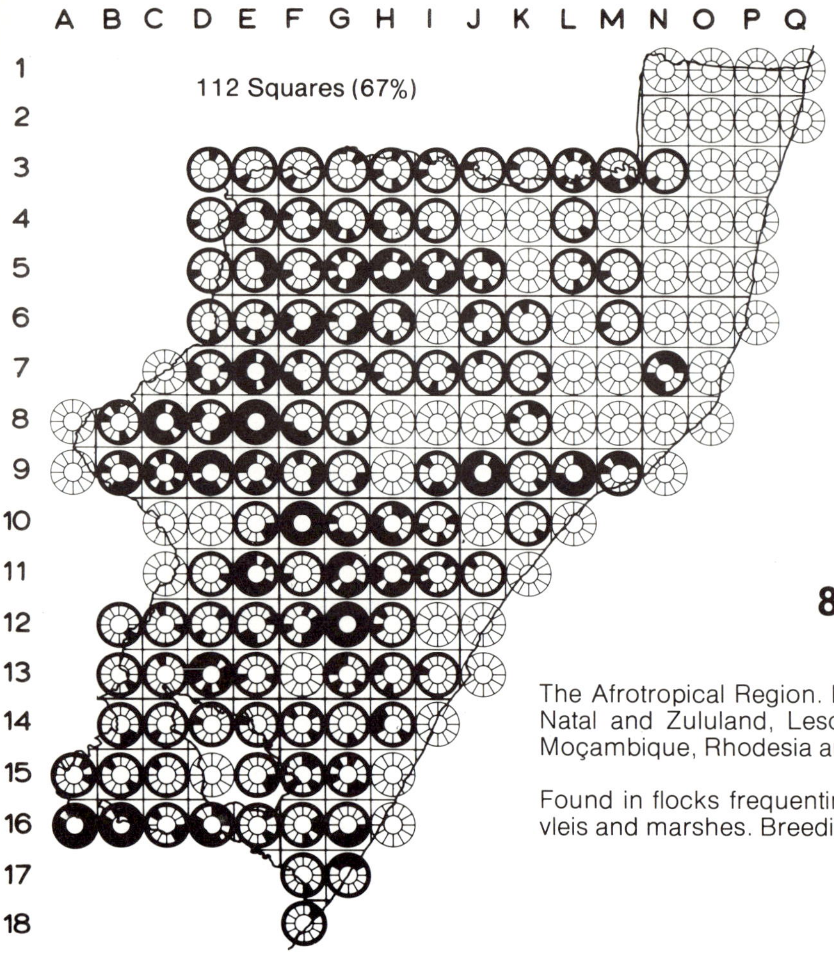

838 Orange-breasted Waxbill

Amandava subflava

The Afrotropical Region. In Southern Africa, the eastern Cape, Transkei, Natal and Zululand, Lesotho (lowlands), O.F.S., Transvaal, Swaziland, Moçambique, Rhodesia and south-eastern Botswana (Kanye).

Found in flocks frequenting moister grassland and cultivated fields near vleis and marshes. Breeding recorded from January to May.

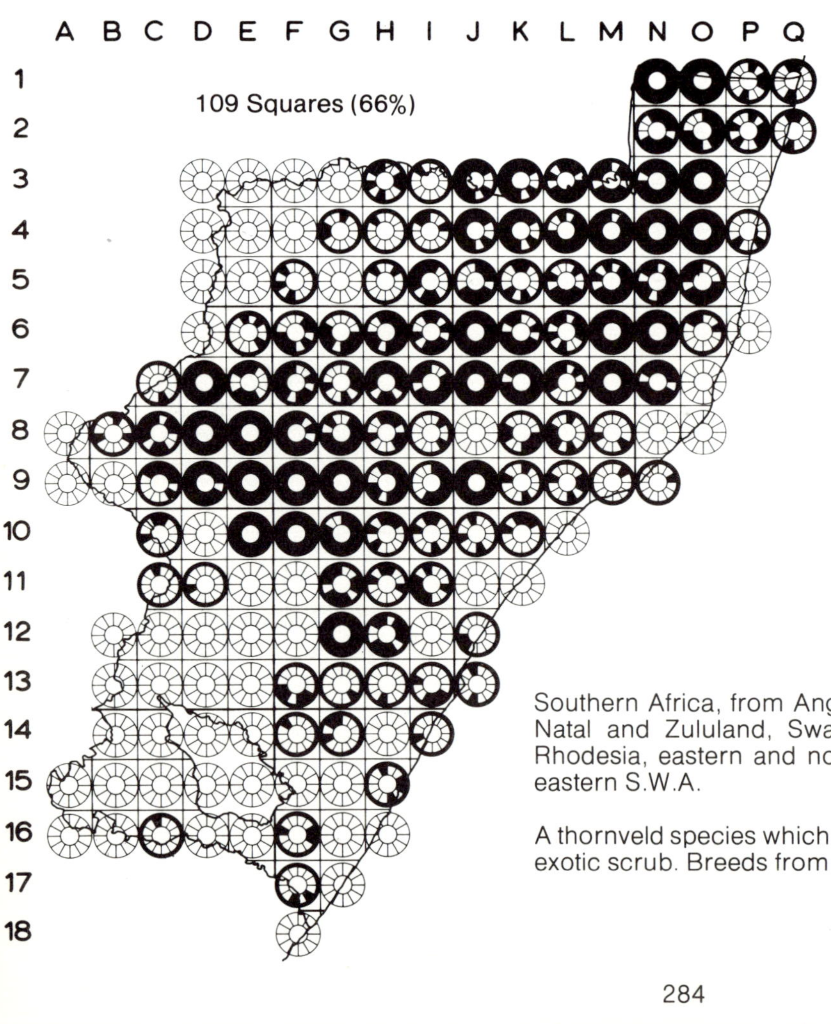

839 Blue Waxbill

Uraeginthus angolensis

Southern Africa, from Angola and Tanzania to Natal. In Southern Africa, Natal and Zululand, Swaziland, Transvaal and Moçambique, west to Rhodesia, eastern and northern Botswana, the Caprivi Strip and north-eastern S.W.A.

A thornveld species which also occurs in mixed woodland, open bush and exotic scrub. Breeds from October to November.

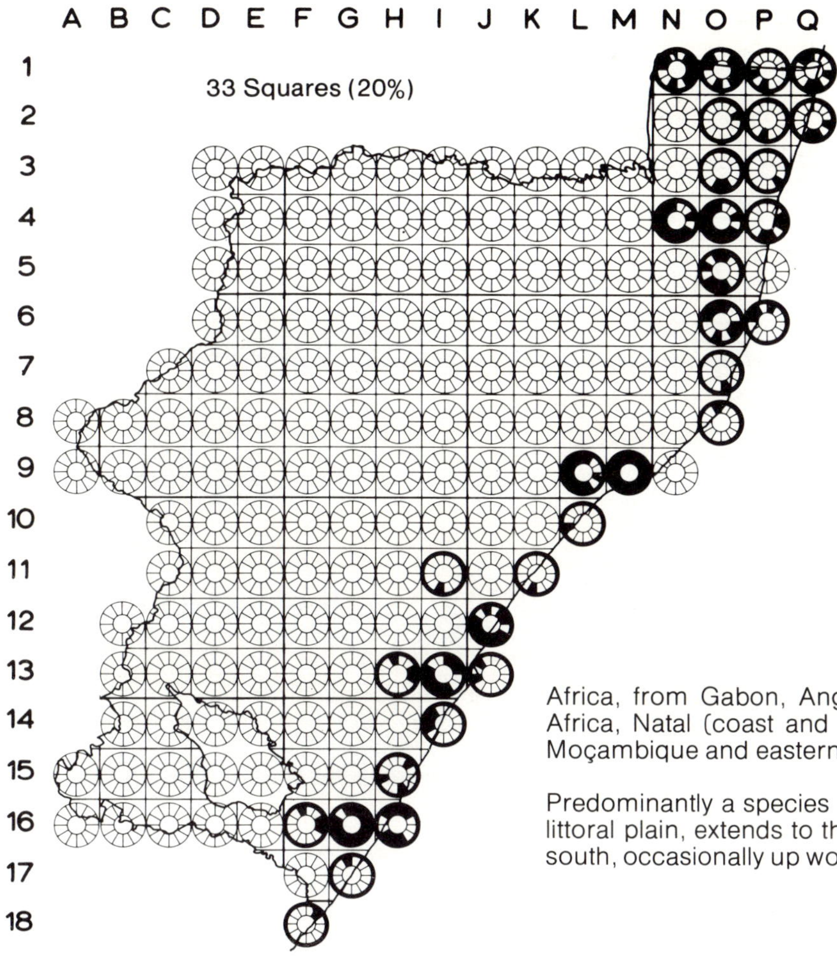

842 Grey Waxbill

Estrilda perreini

Africa, from Gabon, Angola and southern Zaïre to Natal. In Southern Africa, Natal (coast and lower midlands), Zululand, eastern Swaziland, Moçambique and eastern Rhodesia.

Predominantly a species of the more open parts of coastal forest. On the littoral plain, extends to the edges of clearings in mixed woodland; in the south, occasionally up wooded valleys. Breeds from October to February.

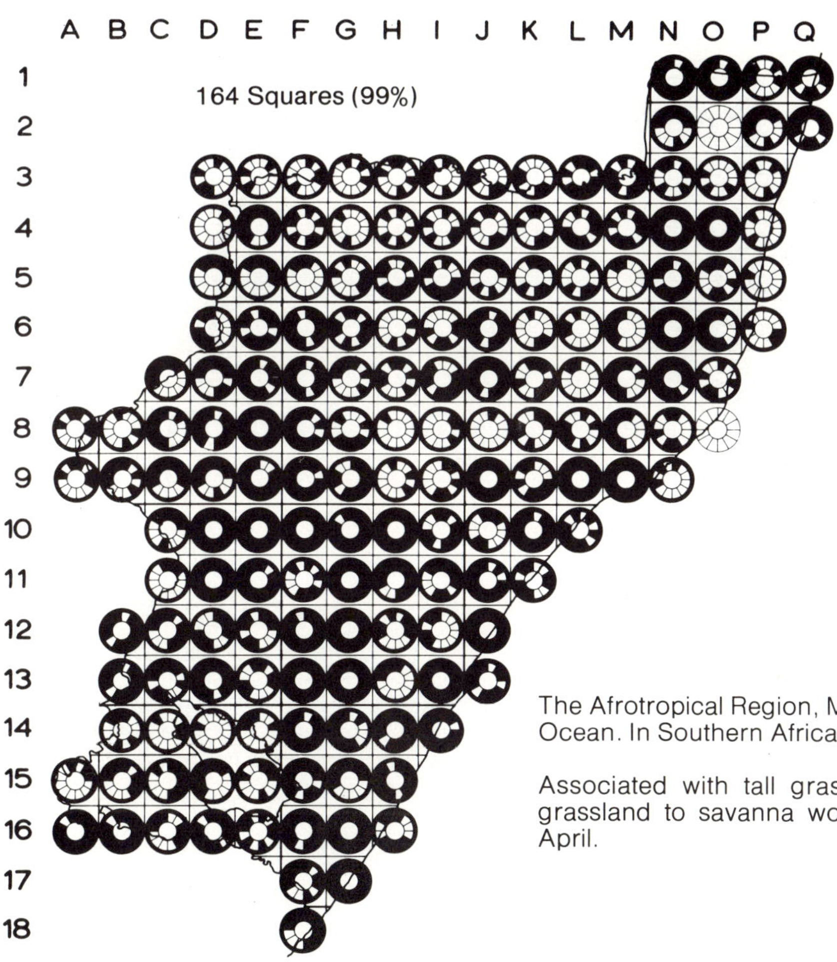

843 Common Waxbill

Estrilda astrild

The Afrotropical Region, Mauritius and other islands in the western Indian Ocean. In Southern Africa, found wherever suitable habitat occurs.

Associated with tall grass, thickets and rank vegetation, from open grassland to savanna woodland. Recorded breeding from October to April.

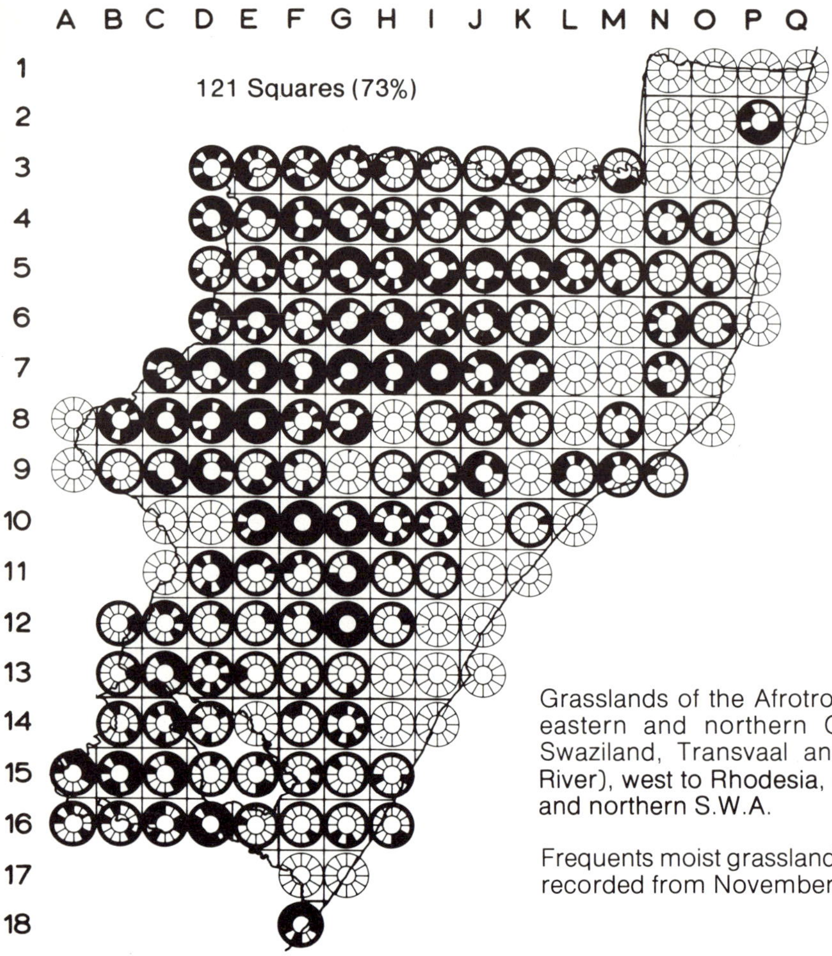

844 Quail Finch

Ortygospiza atricollis

Grasslands of the Afrotropical Region. In Southern Africa, the southern, eastern and northern Cape, Lesotho, O.F.S., Natal and Zululand, Swaziland, Transvaal and southern Moçambique (south of the Save River), west to Rhodesia, most of Botswana, the Caprivi Strip and central and northern S.W.A.

Frequents moist grassland, cultivated fields and the edges of vleis. Breeding recorded from November to April.

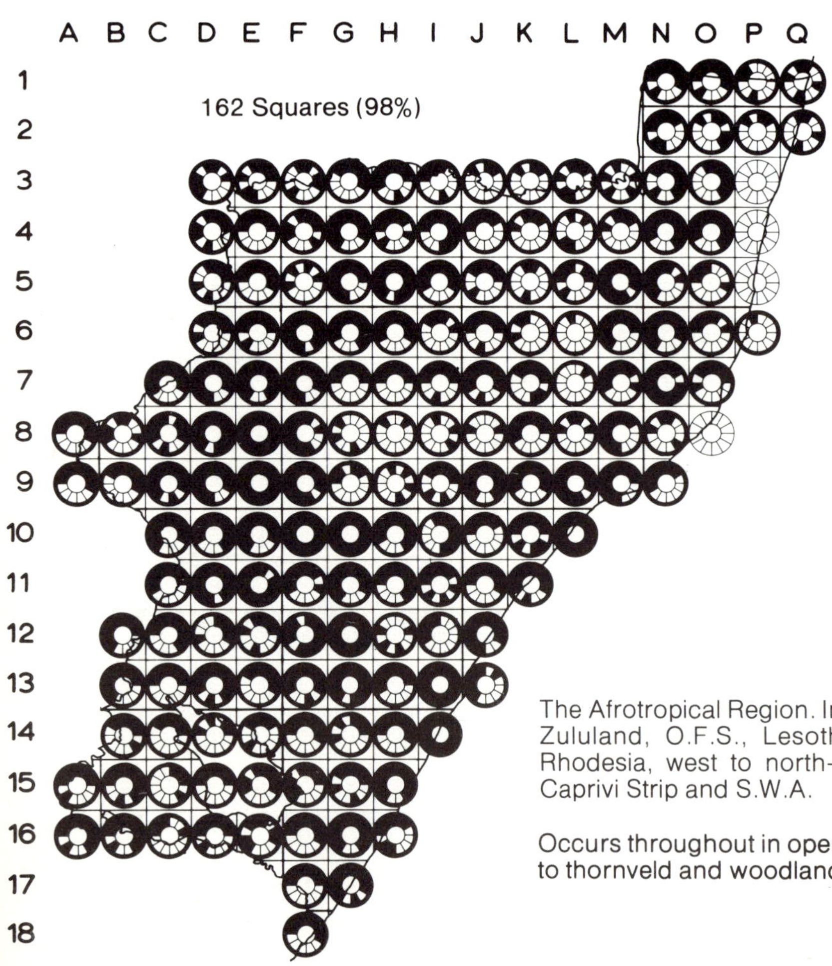

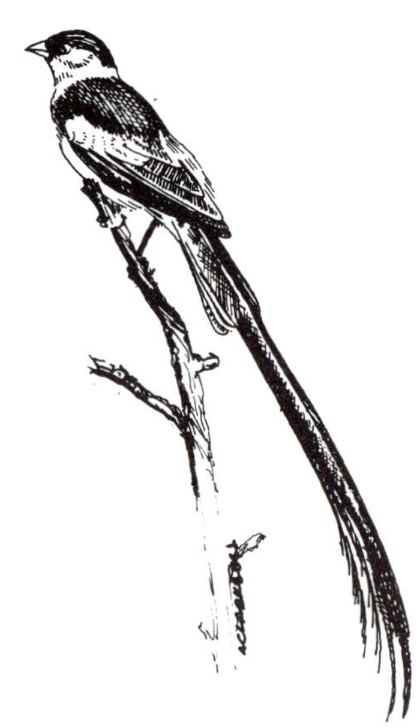

846 Pin-tailed Whydah

Vidua macroura

The Afrotropical Region. In Southern Africa, the Cape, Transkei, Natal and Zululand, O.F.S., Lesotho (lowlands), Transvaal, Moçambique and Rhodesia, west to north-eastern and northern Botswana; possibly the Caprivi Strip and S.W.A.

Occurs throughout in open situations, from grassland and cultivated fields to thornveld and woodland. Breeds from November to April.

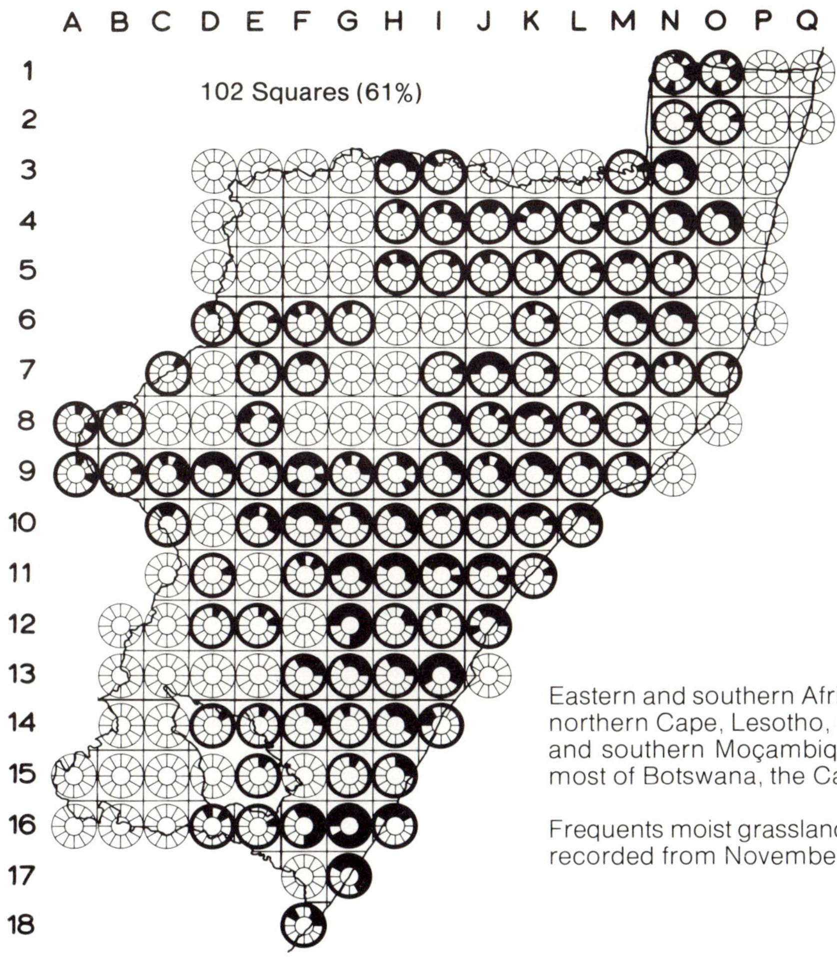

849 Black Widowfinch

Vidua funerea

Eastern and southern Africa. In Southern Africa, the southern, eastern and northern Cape, Lesotho, O.F.S., Natal and Zululand, Swaziland, Transvaal and southern Moçambique (south of the Save River), west to Rhodesia, most of Botswana, the Caprivi Strip, and central and northern S.W.A.

Frequents moist grassland, cultivated fields and the edges of vleis. Breeding recorded from November to April.

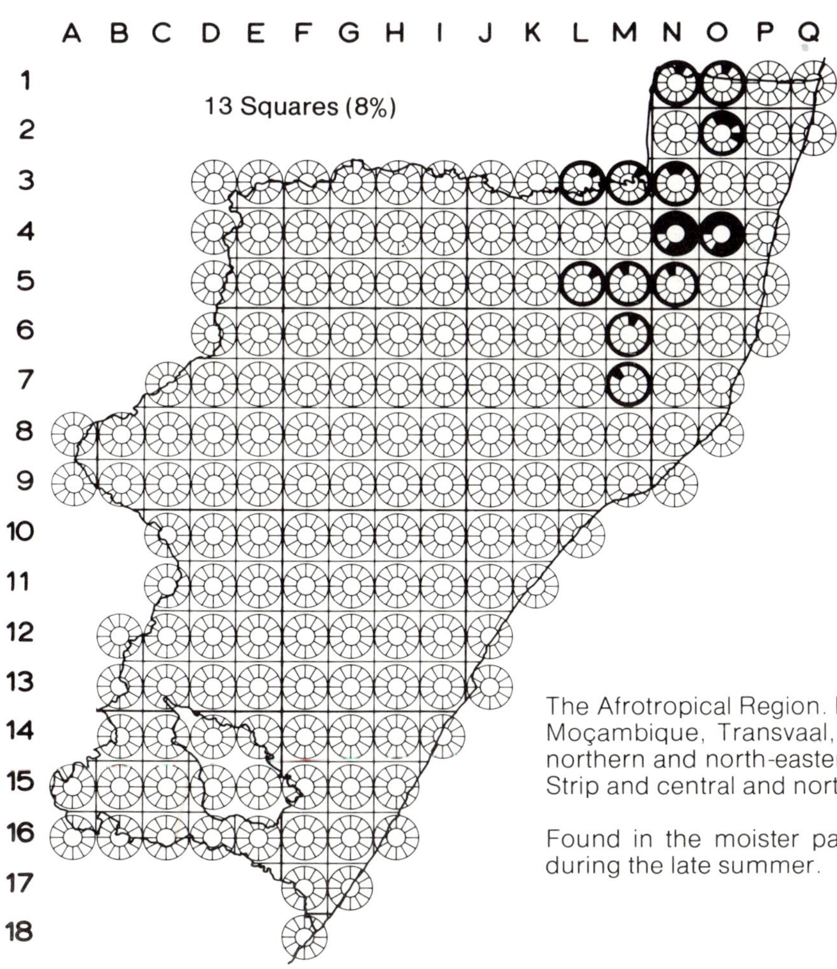

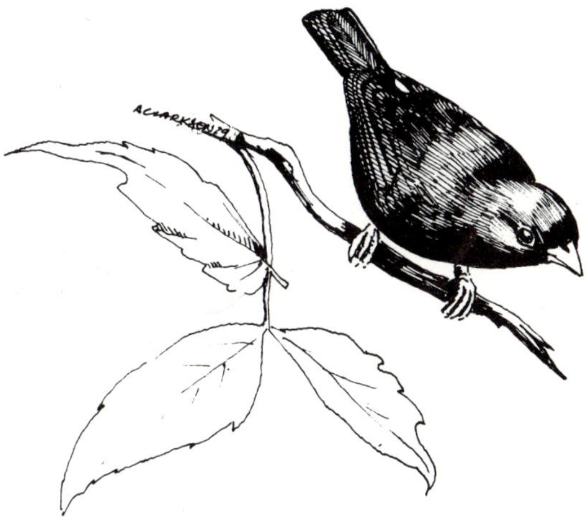

851 Steel-blue Widowfinch

Vidua chalybeata

The Afrotropical Region. In Southern Africa, eastern Zululand, Swaziland, Moçambique, Transvaal, O.F.S. (Modder River) and Rhodesia, west to northern and north-eastern Botswana (Kedia, Dow and Nata), the Caprivi Strip and central and northern S.W.A.

Found in the moister parts of thornveld and mixed woodland. Breeds during the late summer.

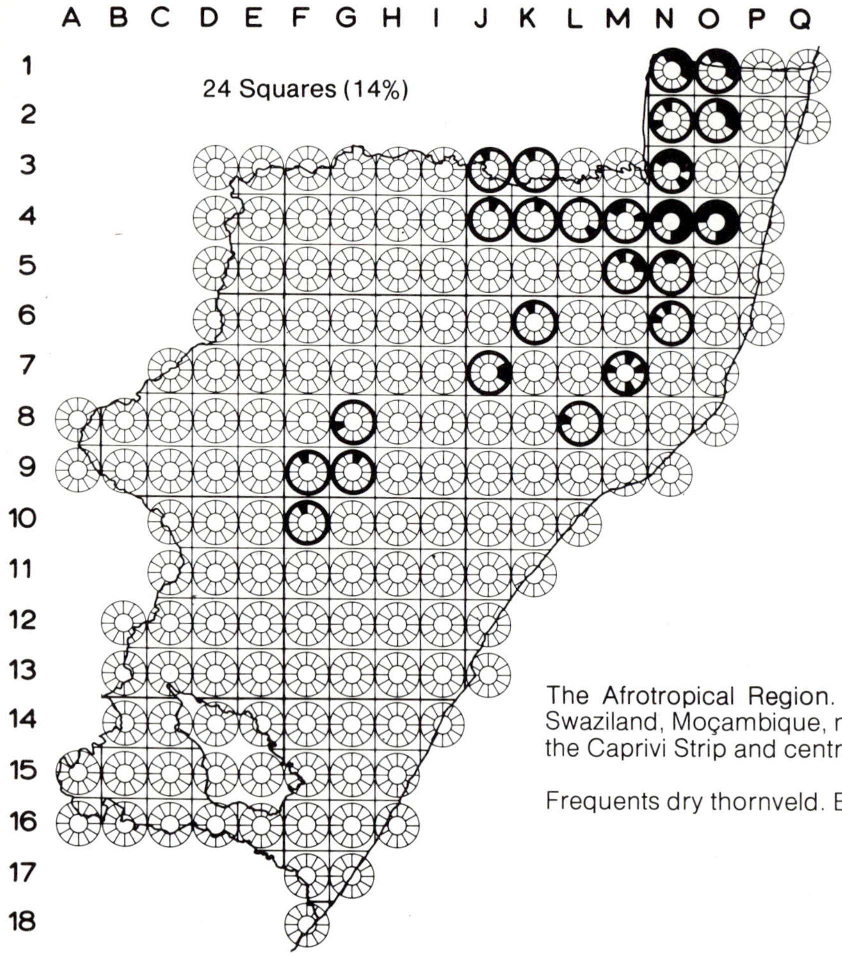

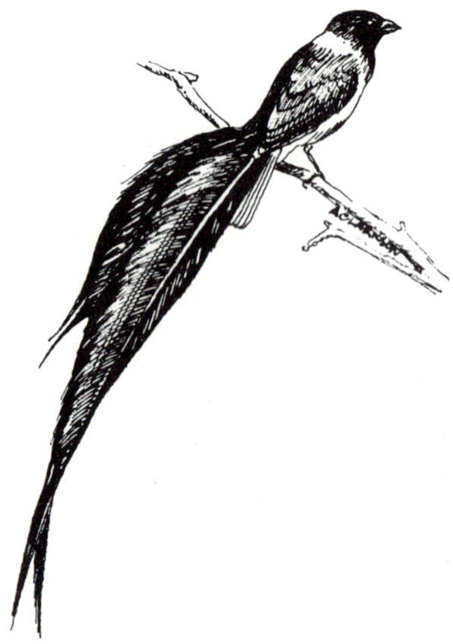

852 Paradise Whydah

Vidua paradisaea

The Afrotropical Region. In Southern Africa, Natal (interior), Zululand, Swaziland, Moçambique, northern O.F.S., Transvaal, Rhodesia, Botswana, the Caprivi Strip and central and northern S.W.A.

Frequents dry thornveld. Breeds from January to July.

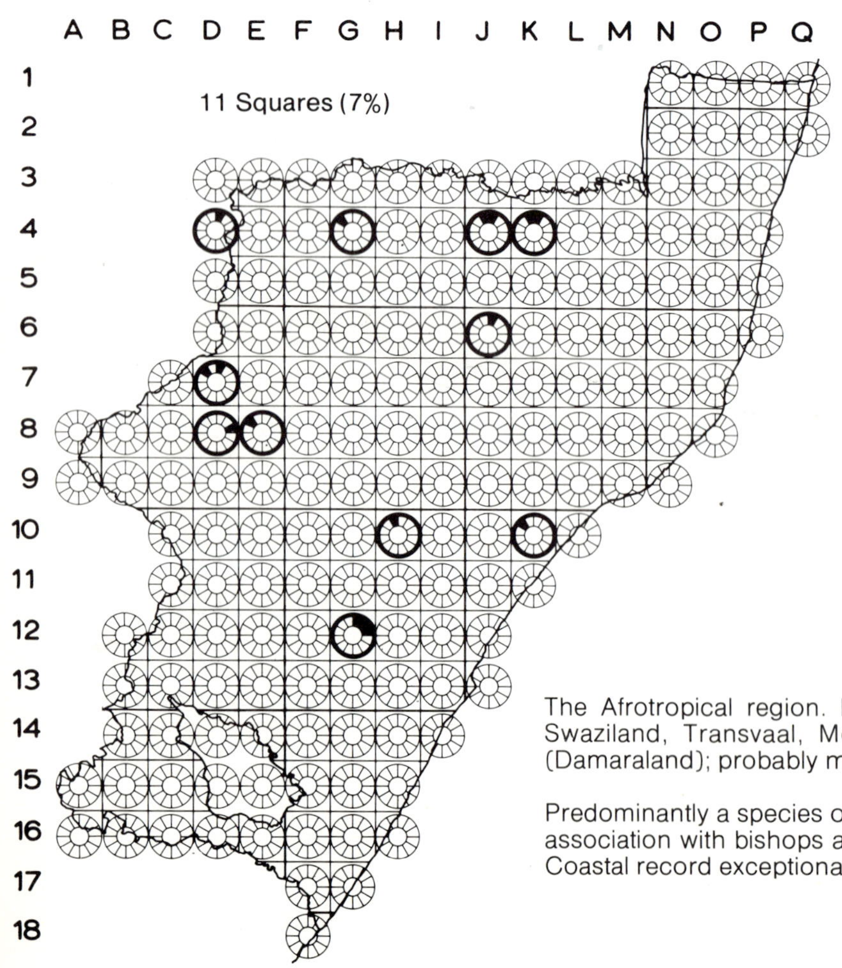

854 Cuckoo Finch

Anomalospiza imberbis

The Afrotropical region. In Southern Africa, Natal (interior), Zululand, Swaziland, Transvaal, Moçambique and Rhodesia, and also S.W.A. (Damaraland); probably more widespread.

Predominantly a species of open grassland in upland areas, often seen in association with bishops and widows; occasionally in thornveld savanna. Coastal record exceptional.

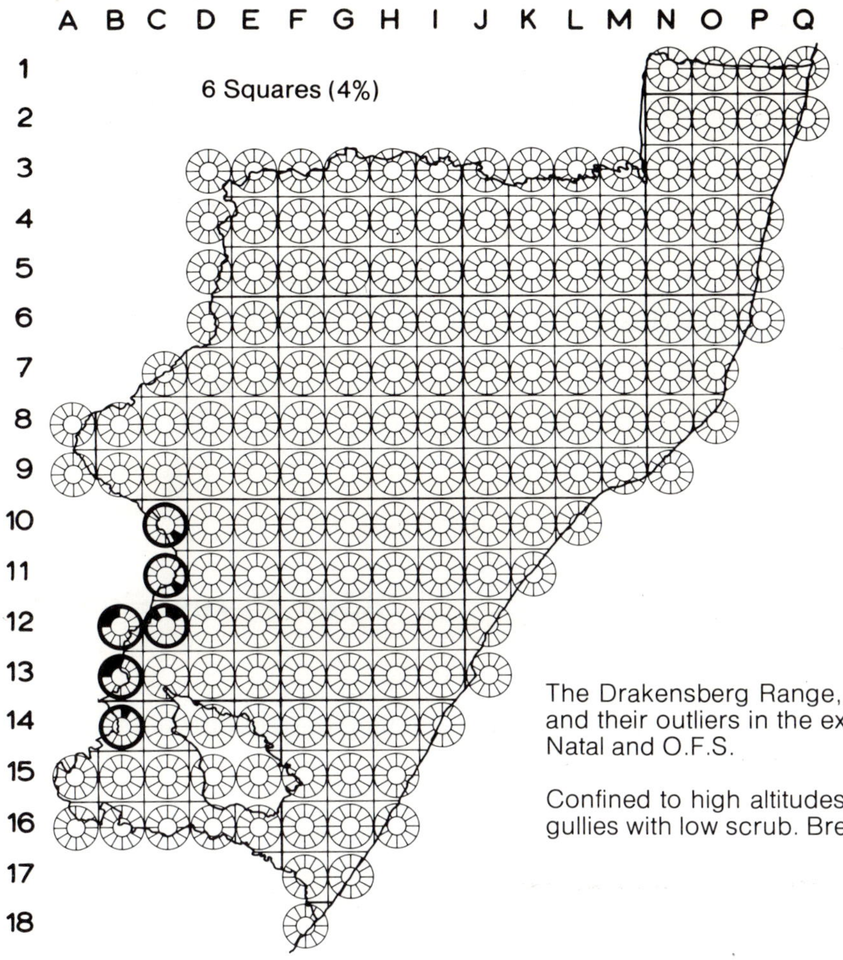

856 Drakensberg Siskin

Serinus symonsi

The Drakensberg Range, Southern Africa. From the Lesotho mountains and their outliers in the extreme north-eastern Cape to Transkei, western Natal and O.F.S.

Confined to high altitudes, where it frequents boulder-strewn slopes and gullies with low scrub. Breeds during the summer.

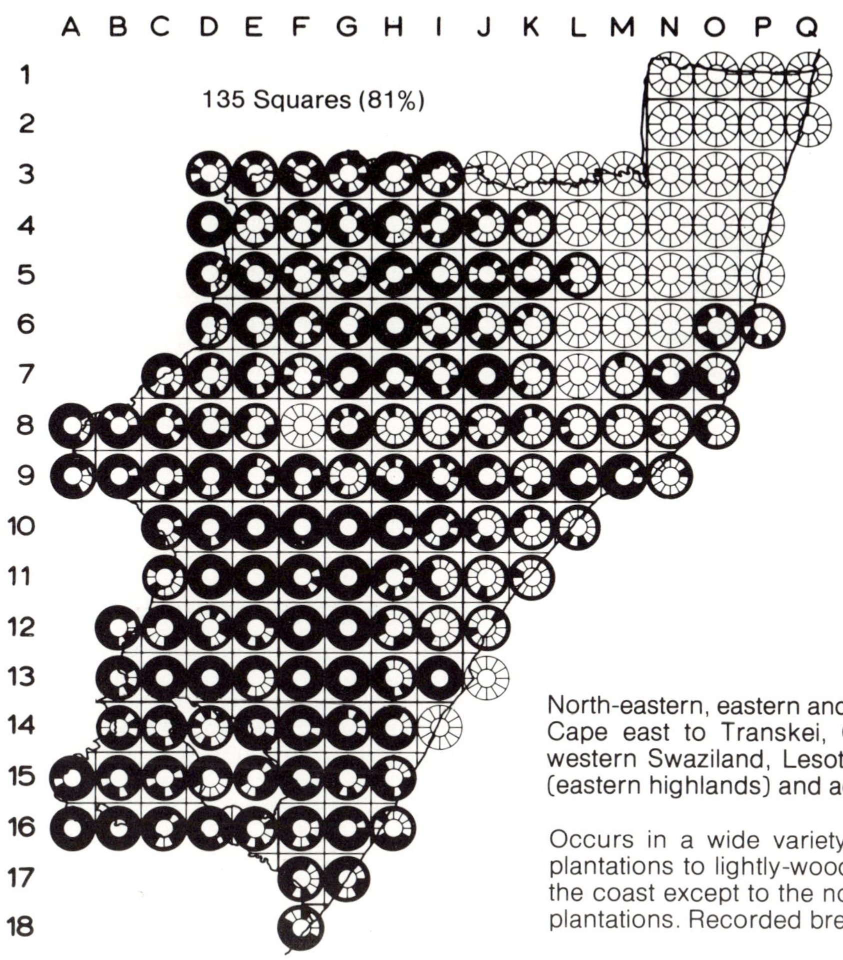

857 Cape Canary

Serinus canicollis

North-eastern, eastern and southern Africa. In Southern Africa, the western Cape east to Transkei, O.F.S., East Griqualand, Natal and Zululand, western Swaziland, Lesotho, eastern and northern Transvaal, Rhodesia (eastern highlands) and adjacent highland Moçambique.

Occurs in a wide variety of habitats, from open grassland and exotic plantations to lightly-wooded areas, gardens and farmsteads. Sparse on the coast except to the north of Richards Bay, where has taken to exotic plantations. Recorded breeding from August to December.

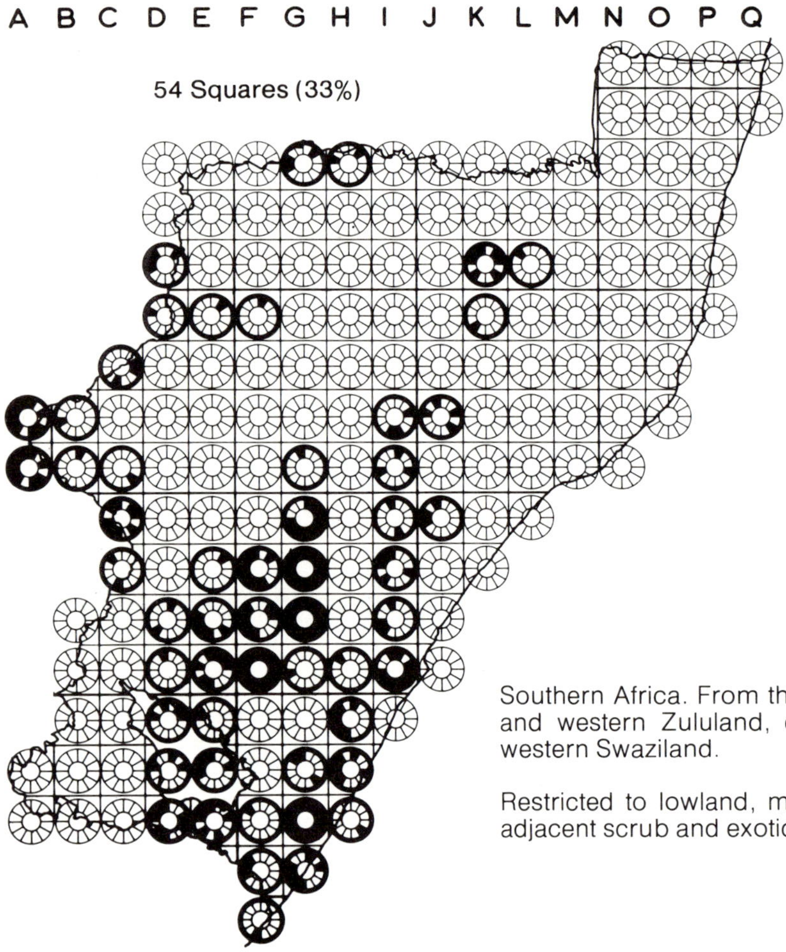

858 Forest Canary

Serinus scotops

Southern Africa. From the southern and eastern Cape to Transkei, Natal and western Zululand, eastern and northern Transvaal and probably western Swaziland.

Restricted to lowland, mistbelt and Drakensberg evergreen forest, and adjacent scrub and exotic vegetation. Breeds from October to January.

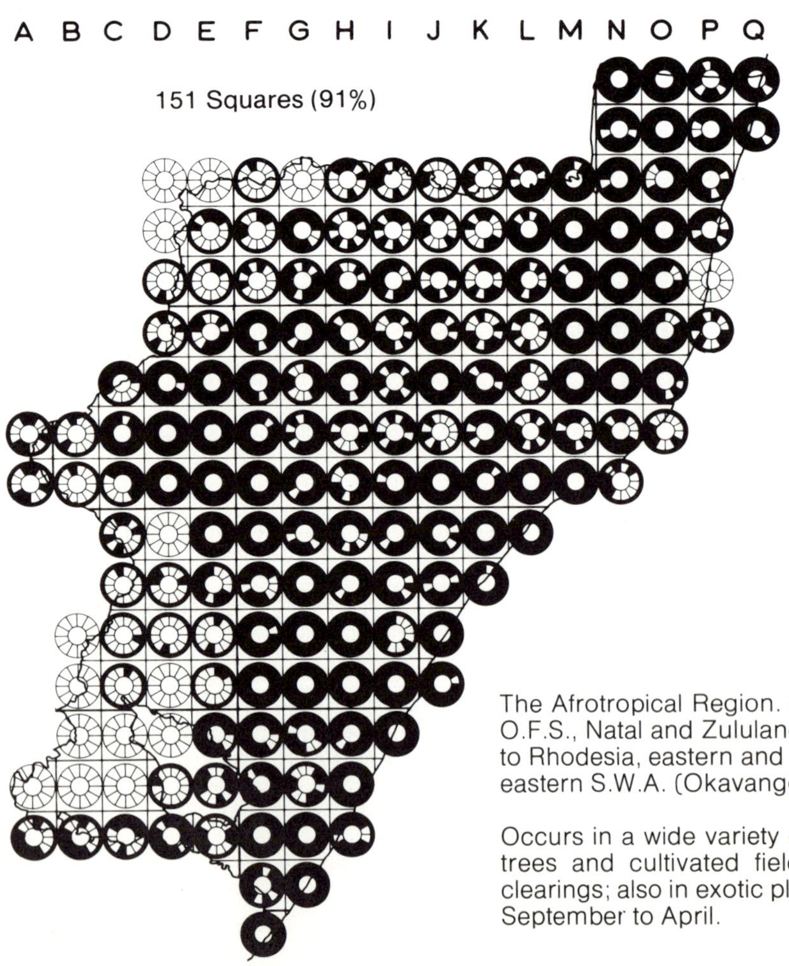

859 Yellow-eye Canary

Serinus mozambicus

The Afrotropical Region. In Southern Africa, the eastern Cape, Transkei, O.F.S., Natal and Zululand, Swaziland, Transvaal and Moçambique, west to Rhodesia, eastern and northern Botswana, the Caprivi Strip and north-eastern S.W.A. (Okavango River).

Occurs in a wide variety of habitats, from open grassland with scattered trees and cultivated fields to thornveld, mixed woodland and forest clearings; also in exotic plantations and gardens. Breeding recorded from September to April.

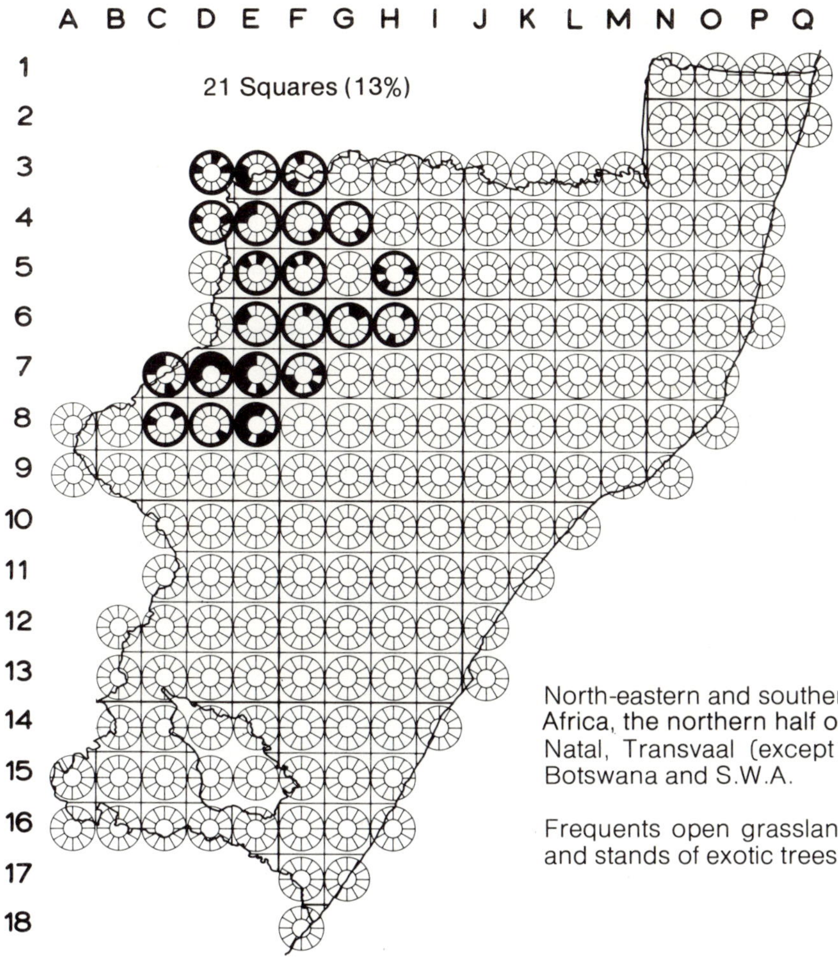

860 Black-throated Canary

Serinus atrogularis

North-eastern and southern Africa, and south-western Arabia. In Southern Africa, the northern half of the Cape, O.F.S., Lesotho, midland and upper Natal, Transvaal (except in the east) and Rhodesia (plateau), west to Botswana and S.W.A.

Frequents open grassland and cultivated fields; also found in thornveld and stands of exotic trees. Breeds during the summer.

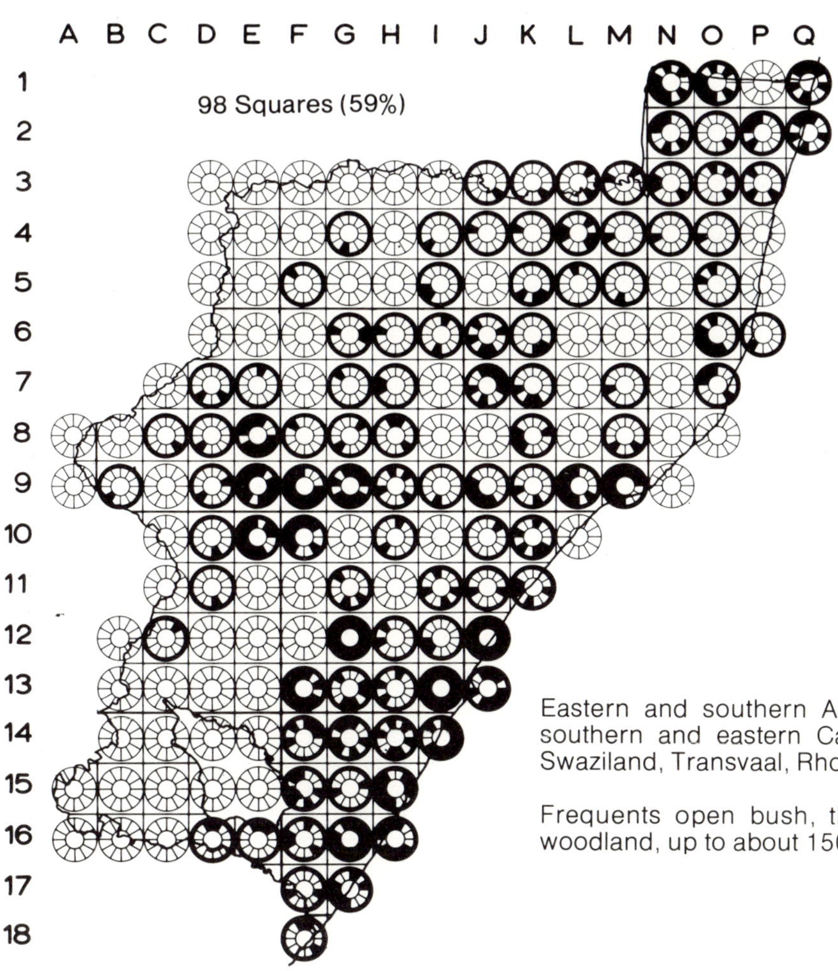

863 Bully Seedeater

Serinus sulphuratus

Eastern and southern Africa. In Southern Africa, the south-western, southern and eastern Cape to Transkei, O.F.S., Natal and Zululand, Swaziland, Transvaal, Rhodesia and Moçambique.

Frequents open bush, thickly-wooded valleys, thornveld and mixed woodland, up to about 1500 m a.s.l. Breeds from October to January.

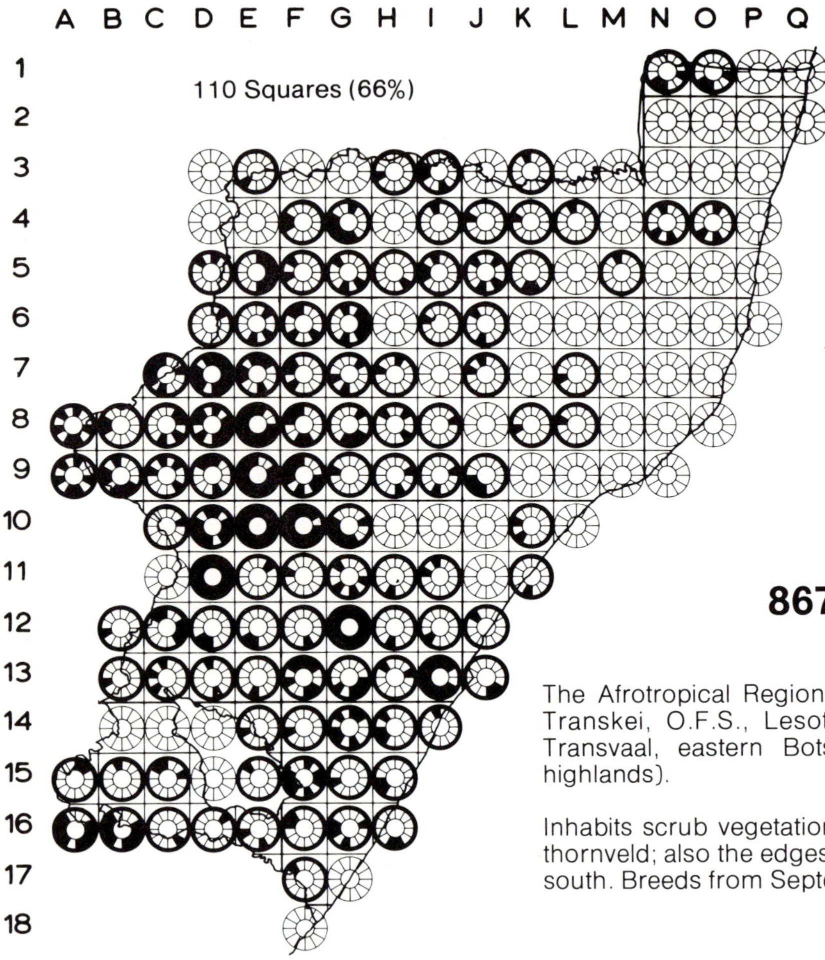

867 Streaky-headed Seedeater

Serinus gularis

The Afrotropical Region. In Southern Africa, most of the Cape east to Transkei, O.F.S., Lesotho, Natal and western Zululand, Swaziland, Transvaal, eastern Botswana, Rhodesia and Moçambique (western highlands).

Inhabits scrub vegetation on the slopes of the Drakensberg Range and thornveld; also the edges of evergreen forest inland and open bush in the south. Breeds from September to February.

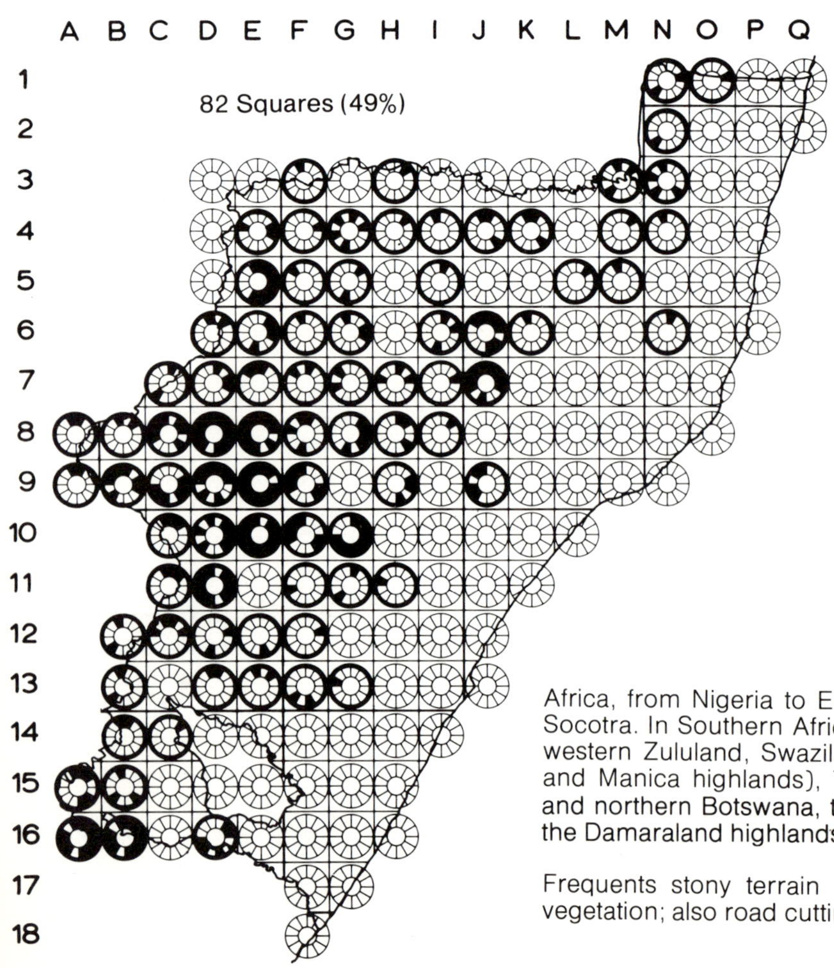

872 Rock Bunting

Emberiza tahapisi

Africa, from Nigeria to Eritrea south to the Cape; southern Arabia and Socotra. In Southern Africa, the eastern Cape, Transkei, Natal (interior), western Zululand, Swaziland, southern Moçambique (Lebombo Range and Manica highlands), Transvaal, O.F.S., Lesotho, Rhodesia, eastern and northern Botswana, the Caprivi Strip and northern S.W.A. (south to the Damaraland highlands).

Frequents stony terrain and hillsides with rocky outcrops and scrub vegetation; also road cuttings. Breeds from November to February.

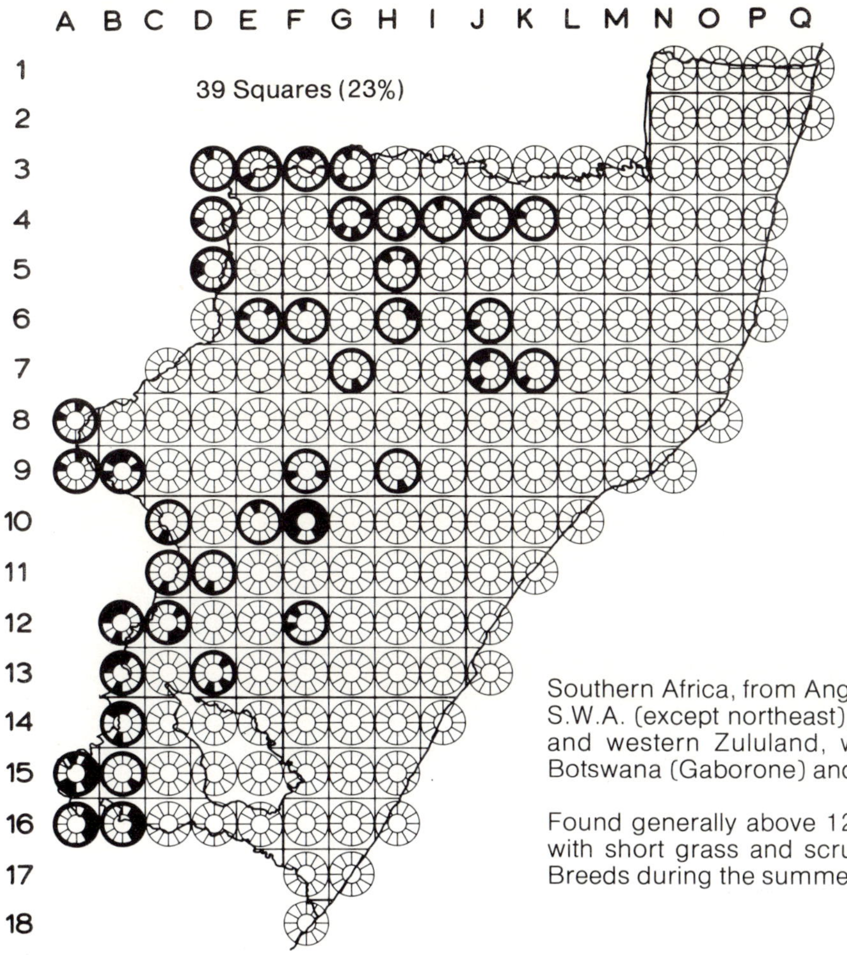

873 Cape Bunting

Emberiza capensis

Southern Africa, from Angola and Malawi to the Cape. In Southern Africa, S.W.A. (except northeast), the Cape (except east), O.F.S., Lesotho, Natal and western Zululand, western Swaziland, Transvaal, south-eastern Botswana (Gaborone) and Rhodesia.

Found generally above 1200 m a.s.l. on rocky or boulder-strewn slopes with short grass and scrub vegetation; also on kranses and in gorges. Breeds during the summer.

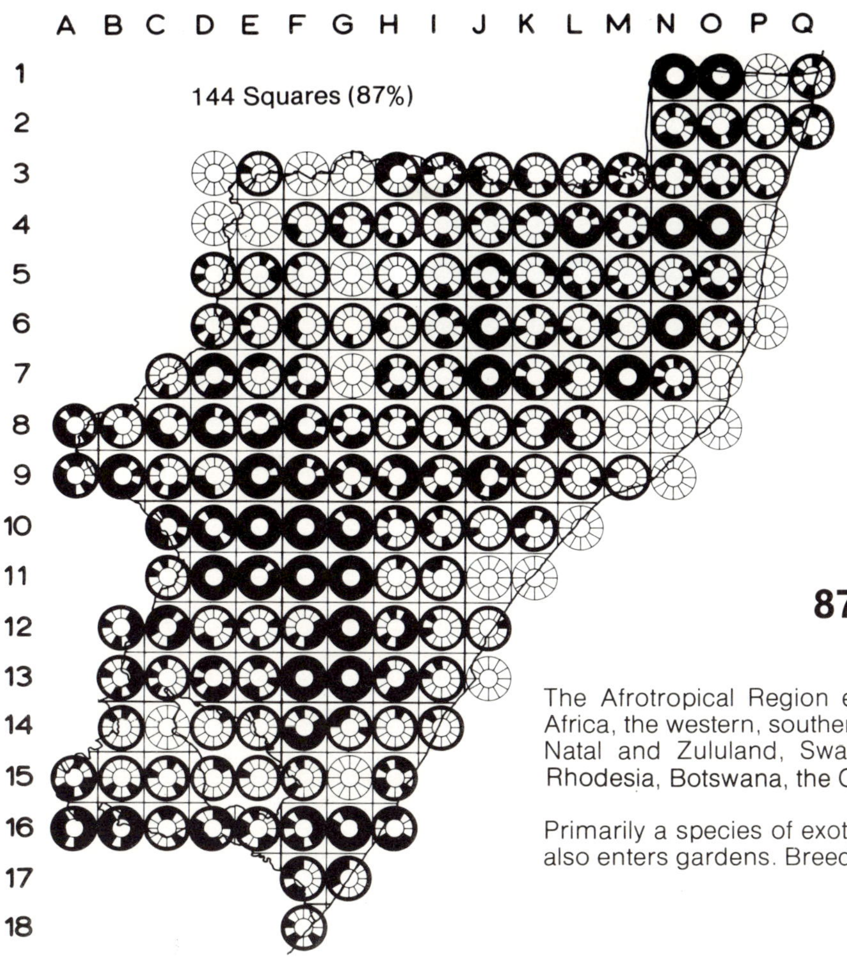

874 Golden-breasted Bunting

Emberiza flaviventris

The Afrotropical Region except for much of West Africa. In Southern Africa, the western, southern and eastern Cape, Transkei, O.F.S., Lesotho, Natal and Zululand, Swaziland, Transvaal and Moçambique, west to Rhodesia, Botswana, the Caprivi Strip and the northern half of S.W.A.

Primarily a species of exotic plantations, thornveld and mixed woodland; also enters gardens. Breeding recorded from October to February.

Great Crested Grebe

Rarer Species and Vagrants

Macaroni Penguin

Eudyptes chrysolophus

Antarctic and Subantarctic Zones of the southern Atlantic and Indian Oceans. In Southern Africa, two records, from Natal and the Cape (Cape Agulhas).

One at Kelso (H 15) in February 1974 (Durban Museum).

4 Great Crested Grebe

Podiceps cristatus

Breeds from western Europe to Siberia, and in eastern and southern Africa, Australia and New Zealand. In Southern Africa, a sparsely distributed breeding resident in the Cape, Transkei, O.F.S., Lesotho (lowlands), Transvaal, Rhodesia and S.W.A. (Walvis Bay, Sandwich Harbour and Etosha); rare in Natal.

Observed at Lake Sibayi (P 3) in all calendar months between 1970-75, usually two birds present (M. N. Bruton). Five birds seen between Hell's Gates, Lake St. Lucia (O 5) in April 1975 (F. Joubert, I. Johnson). A pair near Dannhauser (F 6) in May (breeding) and October 1977 (A. C. V. Clarkson). One at the same locality (F 6) in March 1978 (N. F. Robson). One near Himeville (D 12) in June 1978 (M. G. Taylor). One near Margate (G 17) in February 1979 (D. Mitchell). One near Ladysmith (E 8) in September 1979 (H. E. Wilson).

5 Black-necked Grebe

Podiceps nigricollis

Breeds from western Europe east to Amurland, and in eastern and southern Africa, and North America. In Southern Africa, winters on the coasts of the western Cape and S.W.A., occurring and breeding locally elsewhere in the Cape, Botswana, O.F.S., Lesotho (lowlands), Natal, Transvaal (highveld), and Rhodesia (Bulawayo).

One at Northern Districts S.W., Durban (I 13) from November to December 1978, and in January and April 1979 (J. Hori, H. A. Campbell *et al.*). One at Darvill S.W., Pietermaritzburg (G 12) in March 1979 (H. A. Campbell).

8 Black-browed Albatross

Diomedea melanophris

Southern oceans from about 64°S (occasionally to 69°S), northwards to the Tropic of Capricorn when not breeding. A vagrant to the North Atlantic. In Southern Africa, a visitor to all coasts.

Noted as regular but sparse off Durban (J 13) from July to October. One off St. Lucia Estuary (O 7) in July 1975 (N. F. Robson). Birds found injured at Amanzimtoti (I 14) in July and at Isipingo Beach (I 13) in September 1978 (I. Mellet). One off Umhlanga (J 12) in August 1979 (D. P. Cyrus, N. F. Robson).

9 Grey-headed Albatross

Diomedea chrysostoma

The southern oceans, pelagic when not breeding, ranging north to about 35°S, and farther north in the Australian region. In Southern Africa, an irregular visitor off the coasts of S.W.A. and the Cape; a straggler to Natal.

One found at the Bluff, Durban (J 13) in September 1977 (Durban Museum).

12 Sooty Albatross

Phoebetria fusca

Southern Atlantic and western Indian Oceans; pelagic when not breeding. In Southern Africa, a winter visitor (May to August) to the coasts of the south-western Cape, Natal and Moçambique.

One at St. Lucia Estuary (O 7) in September 1977 (Durban Museum).

Northern Giant Petrel

Macronectes halli

South Atlantic and southern Indian Oceans, east to the New Zealand area. In Southern Africa, occurs seasonally in some numbers in south-western Cape waters, and probably further east.

One at Scottburgh (H 15) May 1977 (per I. Mellet).

15 Silver-grey Fulmar

Fulmarus glacialoides

Antarctic seas, ranging northwards after breeding to about 35°S. In Southern Africa, an uncommon visitor to seas off S.W.A., the Cape and Natal.

One off Durban (J 13) in September 1974 (J. C. Sinclair, N. F. Robson, R. K. Brooke).

18 Schlegel's Petrel

Pterodroma incerta

Southern Atlantic and Indian Oceans. Post-breeding movements uncertain. In Southern Africa, the coastal waters of the western and south-western Cape, east to Natal.

One off Durban (J 13) in August 1973 (J. C. Sinclair).

Kerguelen Petrel

Pterodroma brevirostris

Southern Atlantic and southern Indian Oceans. In Southern Africa, recorded from the Cape (twice, near Cape Town and at Cape Agulhas) and Natal.

One off Amanzimtoti (I 14) in August 1975 (J. C. Sinclair).

19 Soft-plumaged Petrel

Pterodroma mollis

Breeds on oceanic islands in the Atlantic and southern Indian Oceans. Post breeding movements not well documented. In Southern Africa, recorded from the coastal waters of S.W.A., the Cape, Transkei, Natal and southern Moçambique (Inhaca Island, once). Present all months but numbers small.

One at Zinkwazi Beach (K 11) in August 1970 (Durban Museum). One at Durban (J 13) in October 1974 (Durban Museum). Three off Durban (J 13) in August 1975 (G. Bennett, N. F. Robson, J. C. Sinclair). Two off Amanzimtoti (I 14) in August 1975 (J. C. Sinclair). One at Mtunzini (M 9) in April 1976 (Durban Museum). One off Durban (J 13) in June 1978 (Durban Museum). Injured birds found at Salt Rock (J 11) in May 1978, Isipingo Beach (I 13) in September 1978 and Richards Bay (N 9) in August 1979 (per I. Mellet).

21X Medium-billed Prion

Pachyptila salvini

Southern Indian Ocean. In Southern Africa, a visitor to off-shore waters of the Cape, Natal and southern Moçambique (one record).

One found dead at Kosi Estuary (Q 1) in July 1979 (D. P. Cyrus), and another at Umvoti Mouth (K 11) in September 1979 (G. Bennett).

22X Fairy Prion

Pachyptila turtur

Southern oceans, ranging north to at least 30°S. In Southern Africa, rare, recorded off S.W.A., the south-western and eastern Cape, and Natal.

One found dead at Mtunzini (M 9) in June 1970 (Durban Museum).

22Y Slender-billed Prion

Pachyptila belcheri

Southern oceans, particularly the Antarctic Convergence (sea water about 3°C). Pelagic when not breeding, ranging north to about 30°S. In Southern Africa, recorded from the south-western Cape, Natal and Zululand.

Birds found dead at Durban (J 13) in August 1974 (Durban Museum), St. Lucia Estuary (O 7) in July 1978 (Durban Museum), Cape Vidal (P 6) in August 1978 (L. Gillard) and Mpenjati Mouth (G 17) in August 1979 (C. P. M. Rattray).

26 Cory's Shearwater

Calonectris diomedea

The Mediterranean and North Atlantic Ocean, ranging far to the south after breeding. In Southern Africa, the seas off S.W.A. and the Cape (east to Algoa Bay) during the northern winter; also Natal. Recorded as plentiful along the western Cape coast from November to March.

Several noted during February 1972 at sea between Port Edward and Durban (G. J. B. Ross). Off Durban (J 13), one in August 1973 (J. C. Sinclair) and two in March 1979 (D. J. Young).

26X Flesh-footed Shearwater

Puffinus carneipes

Australian and New Zealand seas, ranging northwards in the Indian and Pacific Oceans. In Southern Africa, rare, recorded from the Cape and Natal.

Seven off Uvongo (G 17) in July 1975 (J. C. Sinclair).

26Z Wedge-tailed Shearwater

Puffinus pacificus

The Indian and Pacific Oceans, mainly within the tropics.

One off Umhlanga Rocks (J 12) in August 1975 (J. C. Sinclair).

27 Little Shearwater

Puffinus assimilis

The Atlantic Ocean (except in the north) and seas in the Australian and New Zealand region. In Southern Africa, recorded from the southern and eastern Cape and Natal.

One off Durban (J 13) in June 1975 (J. C. Sinclair).

29 Sooty Shearwater

Puffinus griseus

Southern oceans, migrating as far as the northern Pacific and North Atlantic Oceans. In Southern Africa, a regular visitor to the coasts of S.W.A., the Cape and Natal. Present all year round, but mainly in the winter months.

Recorded off Durban (J 13) annually during August, September and October, with a specimen from December 1976 (Durban Museum). One at Port Edward (F 18) in July 1979 (R. Liversidge, J. Stannard). One at St. Lucia Estuary (O 7) in September 1979 (A. T. Forbes).

30 Storm Petrel

Hydrobates pelagicus

North Atlantic Ocean and the Mediterranean, migrating south after breeding to the seas off western and southern Africa; also the Red Sea. In Southern Africa, a visitor from October to April to all coasts. Seasonally common.

In May 1972, six specimens obtained from Umhlanga Mouth (J 12) (A. Berruti), one from Durban (J 13) (Durban Museum) and one from Hibberdene (H 16) (M. Ward).

33 Wilson's Storm Petrel

Oceanites oceanicus

Antarctic seas and southern oceans, wintering in the North Atlantic, northern Indian, and mid-Pacific Oceans, and the Red Sea. In Southern Africa, a non-breeding visitor to all coasts.

Noted off Durban (J 13) during the months of May, June, August, November and December. Several off Park Rynie (H 15) in June 1974 (G. Bennett) and off Umzumbe (H 16) in July 1979 (N. J. Hori).

36 Black-bellied Storm Petrel

Fregetta tropica

Southern oceans, ranging far to the north of the breeding islands during the winter. In Southern Africa, recorded from the eastern Cape (Algoa Bay and East London) and Natal.

One off Durban (J 13) in September 1974 (N. F. Robson, J. C. Sinclair).

37 White-bellied Storm Petrel

Fregetta grallaria

South Atlantic, Indian and southern Pacific Oceans. In Southern Africa, recorded about 60 km off Cape Town; also seen off Natal and in the Moçambique Channel.

Off Durban (J 13), singletons in September 1974 and August 1979 (D. J. Young).

40 White-tailed Tropicbird

Phaethon lepturus

The Caribbean, mid-Atlantic, Indian and Pacific Oceans. In Southern Africa, a straggler recorded from Natal.

In 1978, one in Durban harbour (J 13) in February (per P. A. Clancey) and March (C. Grieves).

71 Bittern

Botaurus stellaris

The Palaearctic and Afrotropical Regions. In Southern Africa, locally distributed from the western Cape to Transkei, Natal and Zululand, O.F.S., Transvaal and Moçambique; also northern Botswana and S.W.A.

One seen near Harding (E 16) in December 1970 (E. Meyer). One at Otto's Bluff (G 11) in August 1970 (G. L. Maclean).

110 Hooded Vulture

Necrosyrtes monachus

The Afrotropical Region. In Southern Africa, eastern Zululand, eastern Swaziland and southern Moçambique, west to Botswana and S.W.A. Irregularly south to Natal, O.F.S. and the northern Cape. Resident and generally sparse.

At Mkuzi G.R. (N/O 4) recorded in July 1971 (A. Berruti), November 1973 (N. F. Robson, G. Bennett), November 1976 (D. J. Young), January 1977 (D. Schultz), August 1978 (A. C. V. Clarkson), October 1978 (P. G. H. Frost) and December 1979 (N. J. Hori). Two at Hluhluwe G.R. (N 6) in August 1977 (D. H. Day).

116 African Hobby

Falco cuvierii

The Afrotropical Region. In Southern Africa, the eastern Cape, Natal and Zululand, O.F.S., eastern Swaziland, Transvaal, Moçambique and Rhodesia, west to northern Botswana and northern S.W.A. Status uncertain, but probably an uncommon summer breeding resident.

One west of Paulpietersburg (H 3) in May 1979 (R. and S. Hitchner).

120 Western Red-footed Kestrel

Falco vespertinus

Eastern Europe east to the Yenisei, wintering in Africa. In Southern Africa, a visitor from November to March/early April mainly to S.W.A., the Caprivi Strip and Botswana; less numerous elsewhere, but recorded from the eastern Cape, Natal and Zululand, O.F.S., Transvaal and Rhodesia.

Single females north of Jozini (N 3) in February 1975 and near Utrecht (G 4) in February 1977 (D. P. Cyrus). At Estcourt (E 10), a female in March 1978 (D. P. Cyrus, N. F. Robson) and two birds in February 1979 (L. Nutting). Also observed near Franklin (C 15) in January 1979 (J. Mendelsohn).

121 Dickinson's Kestrel

Falco dickinsoni

Africa, from Angola and Tanzania south to southern Moçambique. In Southern Africa, Moçambique (south to Delagoa Bay), eastern Transvaal (south to Hectorspruit) and much of Rhodesia, west to northern Botswana, the Caprivi Strip and north-eastern S.W.A.

Recorded from Ndumu G.R. (N/O 1) during March, July to October and December 1978 (E. Harris).

135 Steppe Eagle

Aquila rapax orientalis

South-eastern Europe east to Siberia, India and Burma; winters to the south. In Southern Africa, a visitor from November to March to S.W.A., Botswana, Rhodesia, Transvaal, Zululand and the central Cape.

Recorded from Mkuzi G.R. (N/O 4) in November 1976 (D. J. Young), December 1976 (G. Bennett, G. H. Nicholls), December 1978 and January 1979 (P. S. Goodman, A. J. Tomkinson) and December 1979 (D. P. Cyrus). One at Hluhluwe G.R. (N 6) in January 1978 (D. H. Day). Observed near Jozini (N 3) in November and December 1979 (C. Hilton-Taylor).

165 Chanting Goshawk

Melierax canorus

South-western Africa, from Angola to the Cape. In Southern Africa, the Karoo regions of the Cape to Griqualand West, western Natal, O.F.S., Lesotho (lowlands), western Transvaal, western Rhodesia, Botswana, the northern Cape, and S.W.A.

Two west of Underberg (C 13) in September 1978 (D. R. Calder).

168 Pallid Harrier

Circus macrourus

Eastern Europe east to central Asia, wintering to the south of the breeding range. In Southern Africa, a visitor from late October/November to April, to the eastern Cape, Natal and Zululand, northwards in the east; in the north, west through Rhodesia and Botswana to S.W.A.

One near Kokstad (C 16) in February 1972 (M. MacDonald). One near Jozini (N 3) in January and March 1977 (M. Sach). One west of Mooi River (E 10) in October and December 1979 (L. Nutting). One east of Mooi River (F 10) in November 1979 (D. E. Manry).

191 Blue Quail

Coturnix adansonii

The Afrotropical Region. In Southern Africa, a sparse summer resident in the eastern Cape, Natal and Zululand,

Swaziland, eastern Transvaal, eastern Rhodesia, and Moçambique.

One found dead in Durban (I 13) in December 1973 (per A. O. Bourquin).

207 Striped Flufftail

Sarothrura affinis

Eastern and southern Africa. In Southern Africa, locally distributed from the south-western Cape to Natal and Zululand, Swaziland and Transvaal (Zuurbron); also the eastern highlands of Rhodesia (Inyanga and the Chimanimani Mountains).

One at Kamberg N.R. (D 11) in November 1974 (D. H. Day). One near Pevensey (D 12/13) in January 1978 (I. F. Rennie). One west of Richmond (F 13) in September 1978 (B. Batchelor). One near Pevensey (D 13) in April and November 1979 (I. F. Rennie).

209 Lesser Gallinule

Porphyrula alleni

The Afrotropical Region. In Southern Africa, the Cape, Natal and Zululand, Lesotho (lowlands), Swaziland and Moçambique to the Zambesi River. In the north, west to the swamp region of Botswana, and northern S.W.A.

One near Portobello (G 17) in February 1973 (J. Stannard). One near Uvongo (G 17) in January, February and July 1975 (D. D. Penn). Singletons at the Kosi Lakes (Q 1/2), in September 1977 (G. O. Jackson), and March and April 1979 (D. P. Cyrus, A. Lawson, C. Walley).

217 Kori Bustard

Ardeotis kori

Southern and eastern Africa. In Southern Africa, locally distributed from the Cape interior to Natal and Zululand (sparse), and Moçambique (Sul do Save, non-breeding visitor), north on the plateau to Rhodesia, Botswana and S.W.A.

One north of Pietermaritzburg (G 12) in February 1976 (A. J. F. Craig).

225 Black Korhaan

Eupodotis afra

Southern Africa in the interior and west. From the Cape and S.W.A. east to Botswana (except the north), western Transvaal, O.F.S. and Lesotho (lowlands).

A pair with young observed northeast of Newcastle (E 4) in December 1979 (S. de Jager).

231 Black Oystercatcher

Haematopus moquini

Coasts of south-western Africa, and the Canary Islands. In Southern Africa, the coasts of S.W.A. and the Cape, irregularly in the east to the Transkei and Natal (seldom).

One at Port Edward (F 18) in July 1972 (D. J. Coghlan). One at Zinkwazi Beach (K 11) in September 1975 (B. Agar). One at Mtunzini, (L 9) in August 1979 (I. F. Garland).

231X European Oystercatcher

Haematopus ostralegus

Europe east to eastern Siberia, Kamchatka and the Commander Islands. In Southern Africa, recorded from Moçambique (specimen), Natal and the coasts of the Cape as far west as Langebaan.

One at Durban Bay (I/J 13) in November 1975 (E. L. Shewell *et. al.*).

234 Mongolian Sandplover

Charadrius mongolus

Central and eastern Eurasia, to eastern Siberia and Kamchatka; also the Commander Islands, wintering to the south of the breeding range. In Southern Africa, regular in small numbers, recorded from S.W.A., (Swakopmund), the eastern Cape (Swartkops River), Natal and Moçambique (north of Beira).

Occurs in small parties each summer on the mudflats in Durban Bay (I/J 13); has overwintered.

236 Chestnut-banded Sandplover

Charadrius pallidus

Southern and eastern Africa. In Southern Africa, estuaries and saline pans in S.W.A., the Cape, Botswana, western Rhodesia, western Transvaal, Natal and Moçambique (north of Beira).

One at Durban Bay (I/J 13) in June 1975 (J. C. Sinclair). Singletons at Umvoti mouth (K 11), in May 1977 (D. J. Young), May and June (S. Jex), October (D. P. Cyrus, N. F. Robson) and November 1979 (D. J. Young). One at Umgeni Mouth (J. 13) in May 1979 (G. H. Nicholls, D. J. Young).

240X Eastern Golden Plover

Pluvialis dominica

Northern Siberia and Arctic North America, wintering to the south of the breeding range, mainly in southern Asia, south to Australia, and on islands in the Pacific; a straggler to Africa. In Southern Africa, recorded from the Cape Peninsula, Zoetendalsvlei (near Cape Agulhas), Port Elizabeth, the Keiskama River, and from Natal.

One at Umhlanga S.W. (J 12) in February 1975 (J. C. Sinclair, D. J. Young). One at Durban Bay (I/J 13) in April 1978 (J. C. Sinclair *et. al.*).

251X Pectoral Sandpiper

Calidris melanotos

North-eastern Asia and Arctic North America, wintering largely in South America. In Southern Africa, recorded from Rhodesia (Bulawayo), the south-western Cape (Rondevlei), Botswana (Botletle River), Transvaal (north of Pretoria) and Natal; evidently regular in small numbers.

Near Karkloof (G 11), one in October and two in November 1973 (G. Bennett, N. F. Robson, J. C. Sinclair *et. al.*).

Buff-breasted Sandpiper

Tryngites subruficollis

Arctic North America, wintering in South America. In Southern Africa, recorded from Natal.

One at Richards Bay (N 9) in December 1977 (N. F. and J. H. Robson).

251Z Broad-billed Sandpiper

Limicola falcinellus

Northern Fenno-Scandia east to north-eastern Siberia, wintering to the south. In Southern Africa, several collected at Swakopmund, S.W.A., between 1965 and 1973. Also recorded from Rhodesia (Lake McIlwaine) and Natal.

Recorded most years in Durban Bay (I/J 13) from September to February. Two observed at Richards Bay (N 9) in January 1977 (G. H. Nicholls).

253X Red-necked Stint

Calidris ruficollis

North-eastern Asia and western Alaska, wintering to the south. In Southern Africa, two obtained in Durban Bay, Natal (November/December 1963), and also recorded from the eastern Cape (Port Elizabeth).

In most years observed in Durban Bay (I/J 13) during September and October. One at Umfolozi Mouth (O 6) in September 1975 (N. F. Robson). Two at Umvoti Mouth (K 11) in September 1976 (G. Bennett, D. P. Cyrus).

259 Green Sandpiper

Tringa ochropus

Northern and eastern Europe to eastern Asia, wintering to the south. In Southern Africa, a sparse visitor from late September to March/April, mainly to the eastern regions, recorded from Rhodesia, eastern Transvaal, Natal and Zululand, and the southern and eastern Cape (seldom).

One at Durban (I 13) in December 1973 (P. A. Clancey). One at Mkuzi G.R. (N/O 4) in January 1977 (G. Nichols). One at Hluhluwe G.R. (N 6) in December 1978 (I. A. W. MacDonald).

261 Redshank

Tringa totanus

Palaearctic, from Europe eastwards to central and eastern Asia. Most populations winter to the south. In Southern Africa, a sparse visitor, recorded from the coasts of S.W.A. (not uncommon at Walvis Bay), the Cape, Natal and Zululand, and Moçambique; also locally inland.

One at Durban Bay (I/J 13) in November 1975 (J. C. Sinclair).

272 Red-necked Phalarope

Phalaropus lobatus

Holarctic, wintering at sea to the south. In Southern Africa, a non-breeding visitor to off-shore waters of S.W.A. and the western Cape. One record from the eastern Cape (Cape St. Francis).

One observed at Umvoti Mouth (K 11) in September 1976 (J. C. Sinclair, N. F. Robson).

Crab Plover

273 Crab Plover

Dromas ardeola

The Red Sea and northern Indian Ocean coasts, wintering farther south. In Southern Africa, a visitor from September to April, to the Moçambique, Zululand and Natal coasts; occasionally to the eastern Cape.

One at Umgeni Mouth, Durban (J 13) from December 1974 to February 1975 (J. C. Sinclair *et al.*). One at Umvoti Mouth (K 11) in December 1974 (J. C. Sinclair).

276 Burchell's Courser

Cursorius rufus

Southern Africa in the drier areas. From the Karoo, eastwards in the Cape to Griqualand West, O.F.S., Transvaal (highveld), upper Natal and western Swaziland, westwards from Botswana (Makgadikgadi Salt Pans) to the northern Cape and S.W.A.

Two north of Newcastle (E 4) in December 1976 (G. L. Maclean). One west of Ladysmith (D 8) in November 1978 (A. Malherbe). Two west of Mooi River (E 10) in December 1979 (L. Nutting).

282 Black-winged Pratincole

Glareola nordmanni

Breeds in south-eastern Europe and south-western Asia, wintering in Africa. In Southern Africa, a visitor from late September/October to March/April (markedly eruptive) mainly to upper Natal, O.F.S., western Swaziland, Transvaal, the northern Cape, Botswana and S.W.A.

One near Volksrust (D 3) in January 1977 (D. H. Day). One near Wakkerstroom (F 3) in March 1978 (N. F. Robson).

284 Arctic Skua

Stercorarius parasiticus

Northern Holarctic, wintering to the south of the breeding grounds. In Southern Africa, a visitor to all coasts, mainly during the northern winter, but sub-adults present at other seasons.

Occurs regularly in small numbers off Durban (J 13) and the Umvoti Mouth (K 11) during the summer.

285 Pomarine Skua

Stercorarius pomarinus

Northern Holarctic, wintering at sea far to the south of the breeding grounds. In Southern Africa, a visitor during the northern winter to the coastal waters of S.W.A. and Natal.

Small numbers noted off Durban (J 13) and Umdloti Mouth (J 12) during the summer months.

287X Lesser Black-backed Gull

Larus fuscus

Western Palaearctic, wintering south as far as South Africa. In Southern Africa, taken at Liambezi Lake in the Caprivi Strip, and regular in Rhodesia (Lake Kariba and other waters); observed in Moçambique (Beira), and on the Natal coast.

Noted at either Durban Bay (I/J 13) or Umgeni Mouth (J 13) during every month. One at Umdloti Mouth (J 12) in May 1977 (D. J. Young).

289 Hartlaub's Gull

Larus hartlaubii

Southern Africa. From the coasts of S.W.A. to those of the western Cape; a vagrant to Natal.

One photographed at Umgeni Mouth (J 13) in April 1976 (G. H. Nicholls). One observed at St. Lucia Estuary (O 7) in May 1979 (D. Schultz).

Franklin's Gull

Larus pipixcan

Breeds in Canada and the United States, wintering on the coasts of Central and South America. In Southern Africa, a straggler recorded from the Cape and Natal.

One at Umgeni Mouth, Durban (J 13) in December 1973 (R. K. Brooke, G. H. Nicholls, J. C. Sinclair) and July 1979 (J. Mendelsohn).

289X Sabine's Gull

Larus sabini

Northern Holarctic, wintering at sea off South America and Africa. In Southern Africa, regular from November to March at sea off S.W.A. and the Cape, east to Natal.

In December 1974, three off Port Durnford (M 9), one at Umvoti Mouth (K 11) and one off Durban (J 13) (J. C. Sinclair). At Umgeni Mouth (J 13) in 1979, one in February (D. J. Young) and three separate individuals during May (G. H. Nicholls, D. J. Young *et. al.*).

290X Gull-billed Tern

Gelochelidon nilotica

From the Palaearctic Region to southern Asia, Australasia, and North and South America. In Southern Africa, a visitor in small numbers to Botswana (taken at Lake Dow and recorded at Lake Ngami), Rhodesia (Salisbury) and Moçambique (Gorongosa National Park).

One at Durban Bay (I/J 13) in July 1975 (J. C. Sinclair).

Black-naped Tern

Sterna sumatrana

Archipelagos and atolls of the Indian and Pacific Oceans. In Southern Africa, recorded from Natal and once from southern Moçambique (Inhaca Island).

At Umvoti Mouth (K 11), four in March 1976 (J. C. Sinclair) and one in October 1978 (S. Jex). One at Umdloti Mouth (J 12) in December 1978 (D. J. Young).

White-cheeked Tern

Sterna repressa

The Red Sea and the Persian Gulf to the west coast of India.

One at Umvoti Mouth (K 11) in November 1975 (J. C. Sinclair, N. F. Robson).

292 Antarctic Tern

Sterna vittata

Island groups in the southern oceans. In Southern Africa, a regular visitor to the coasts of the Cape during the period May to October; recorded from Natal.

An immature found dead at Mandini (K 10) in June 1979 (R. M. F. Berard). An immature seen at Umgeni Mouth (J 13) in June and July 1979 (J. Mendelsohn, D. J. Young).

293 Roseate Tern

Sterna dougallii

North America, West Indies, Europe and North Africa, to southern Africa, southern Asia, islands in the Indian and Pacific Oceans, and Australasia. In Southern Africa, recorded breeding on islets off the Cape coast east to Bird Island, Algoa Bay; also recorded from Natal and Moçambique coasts.

Six at Umvoti Mouth (K 11) in January 1973 (N. F. Robson, J. C. Sinclair). Two at Umgeni Mouth (J 13) in October 1973 (N. F. Robson, J. C. Sinclair). Noted off Port Durnford (M 9) in December 1974 (J. C. Sinclair). At Umgeni Mouth (J 13) in 1979, one in April (D. J. Young), May (D. P. Cyrus, N. F. Robson *et. al.*), June (J. Mendelsohn *et. al.*) and August (H. A. Campbell).

295X Bridled Tern

Sterna anaethetus

The Caribbean, east to West Africa and the Red Sea, islands in the Indian and western Pacific Oceans, and Australia. In Southern Africa, Natal and southern Moçambique (Inhaca Island).

At Umgeni Mouth (J 13), one in March and one in December 1974 (J. C. Sinclair, D. J. Young), one in January, five in February and one in March 1975 (J. C. Sinclair, D. J. Young *et. al.*). Two off Durban (J 13) in June 1975 (N. F. Robson, J. C. Sinclair).

301 White-capped Noddy

Anous minutus

Islands in the South Atlantic and Pacific Oceans, and off Central America. In Southern Africa, a rare straggler recorded from the eastern Cape and Natal.

Two at Umgeni Mouth (J 13) in May 1979 (G. H. Nicholls, D. J. Young *et. al.*). One at St. Lucia Estuary (O 7) in May 1979 (R. Cassidy).

353 Black Coucal

Centropus bengalensis

Africa and southern Asia, and major islands to the south-east. In Southern Africa, coastal Natal (rare), eastern Zululand, Swaziland, Moçambique, eastern Transvaal, Rhodesia, Botswana (Ngamiland) and the Caprivi Strip.

At St. Lucia Game Park (O 7), two in December 1977 (R. Taylor) and one in February 1979 (A. Marchant, P. Ruddle).

377 Pennant-wing Nightjar

Macrodipteryx vexillaria

Africa south of the Equator, wintering farther north. In Southern Africa, a summer resident from August to March in the Caprivi Strip, northern Botswana, Rhodesia and Moçambique; irregularly recorded farther south from Transvaal, Swaziland, Natal, Zululand and East Griqualand.

A male found dead east of Wakkerstroom (G 3) in January 1971 (H. D. Jackson). A male west of Pietermaritzburg (G 12) in October 1974 (G. Bennett, D. P. Cyrus, N. F. Robson). A male west of Himeville (C 12) in December 1978 (A. Davidson).

401 Grey-hooded Kingfisher

Halcyon leucocephala

The Afrotropical Region. In Southern Africa, a summer resident from August/September to April to north-eastern Swaziland, eastern Transvaal, adjacent and coastal Moçambique, and Rhodesia, west to Botswana (south to Lake Ngami, the Botletle and Nata Rivers) and northern S.W.A. Winters in the African tropics.

At Ndumu G.R. (N/O 1) observed in March, April and December 1978 (E. Harris) and January 1979 (P. Thompson). One at Makane's Drift (0 2) in April 1978 (C.A. Bruzas).

407 Carmine Bee-eater

Merops nubicoides

Africa south of the Equator. In Southern Africa, a visitor from August to March to northern Zululand, Swaziland, Moçambique, Transvaal and Rhodesia, west to northern Botswana and S.W.A.; also recorded from Natal and Transkei. Winters north to eastern Zaire and Tanzania.

At Ndumu G.R. (N/O 1), a flock in January 1971 (S. J. M. Blaber), and noted from January to March and in December 1978 (E. Harris).

411 Swallow-tailed Bee-eater

Merops hirundineus

The Afrotropical Region. In Southern Africa, the Cape (from the Orange River) to S.W.A., Botswana, western O.F.S., western and northern Transvaal, Rhodesia and Moçambique; a non-breeding visitor to Natal and Zululand.

Six at Mtunzini (L 9) in August 1970 (I. F. Garland). Observed near Pelindaba (P 2) in January 1976 (M. Sach) and at St. Lucia Narrows (O 7) in December 1977 and February 1978 (R. Taylor). Observed west of Durban (I 13) in June, July and August 1979 with a maximum of six birds (D. C. Grice, J. N. Perryer).

414 Racquet-tailed Roller

Coracias spatulata

Africa south of the Equator. In Southern Africa, northern Zululand (vagrant), Swaziland (occasional), southern Moçambique, eastern Transvaal and Rhodesia.

One at Ndumu G.R. (N/O 1) in May 1972 (K. M. Collins). Collins).

455 Angola Pitta

Pitta angolensis

Tropical Africa. In Southern Africa, a summer resident from late October to April in eastern Rhodesia and in Moçambique (north of the Limpopo River). Individuals recorded farther south in Transvaal (several records), Natal (once) and the eastern Cape (Port Elizabeth).

One found dead in Pietermaritzburg (G 12) in November 1979 (R. Cussons, S. Manson-Smith).

463 Thick-billed Lark

Calandrella magnirostris

Southern Africa. The Cape (except the east coast), Transkei (interior), the highlands of Lesotho and adjacent western Natal, western O.F.S. and south-western Transvaal.

Two at the top of Sani Pass (C 12) in November 1979 (P. G. H. and S. Frost).

464 Dusky Lark

Pinarocorys nigricans

The savannas of Africa south of the Equator. In Southern Africa, a non-breeding visitor from December to May to Zululand, eastern Swaziland, Transvaal and southern Moçambique, west to Rhodesia, Botswana and S.W.A.

One north of Hluhluwe village (N 5) in December 1972 (S. J. M. Blaber). One at Ndumu G.R. (N/O 1) in March 1978 (E. Harris).

498 Pearl-breasted Swallow

Hirundo dimidiata

Africa, from Angola and Zaire to the Cape. In Southern Africa, the Cape (east to the Great Kei River), S.W.A. (north to the Okavango River), and Botswana, to western O.F.S., Lesotho (lowlands), Transvaal and Rhodesia; also recorded from Natal and Swaziland.

One seen north of St. Bernards (B 13) in November 1974 (D. J. Coghlan, D. P. Cyrus). Three west of Underberg (C 12) in November 1976 (M. G. Taylor).

523X House Crow

Corvus splendens

Southern Asia and island groups in the northern Indian Ocean. Introduced to East Africa and the Mascerene Islands. In Southern Africa, self-introduced in the Durban district of Natal where first noted in September 1972; also recorded from the eastern Cape (East London) in November 1975 and reported from southern Moçambique.

Regular in Durban (I/J 13) where first recorded breeding in 1975. Five near Camperdown (H 12) in September 1973 (R. Cowgill).

544 Red-eyed Bulbul

Pycnonotus nigricans

Southern Africa, from Angola to the Cape. In Southern Africa, S.W.A., the Caprivi Strip, Botswana, north-western Rhodesia, north-western and southern Transvaal, O.F.S., the northern and Karoo regions of the Cape, Lesotho and East Griqualand; also recorded from Natal.

One northwest of Swartberg (B 14) in November 1976 (D. P. Cyrus, D. J. Coghlan). Birds present at the same locality from November 1977 to March 1978 (J. B. Shephard) and one in June 1979 (D. J. Coghlan).

572 Sickle-wing Chat

Cercomela sinuata

Southern Africa. Occurs in the western, northern and north-eastern Cape, western O.F.S., and south-western Transvaal; also the highlands of Lesotho, and Natal in winter.

One at the top of Sani Pass (C 12) in April 1973 (D. P. Cyrus, R. T. F. Bernard) and in February 1977 (S. J. M. Blaber). One at Royal Natal National Park (A 8/9) in September 1979 (R. J. Dowsett, F. Dowsett-Lemaire).

577 Whinchat

Saxicola rubetra

Western and central Palaearctic, wintering to the south. In Southern Africa, rare, recorded from S.W.A. (Swakopmund), Rhodesia, Zululand and Natal.

A male at Umhlanga S.W. (J 12) in January 1979 (N. J. Hori).

594 Whitethroat

Sylvia communis

Palaearctic, wintering to the south. In Southern Africa, a visitor from November to April to the Transvaal, Rhodesia, Botswana, the northern Cape and S.W.A.

One observed north of Pietermaritzburg (G 12) in March 1979 (A. J. F. Craig).

597 Olive-tree Warbler

Hippolais olivetorum

South-western Palaearctic, wintering in Africa. In Southern Africa, a visitor mainly during February and March, recorded from Botswana (Tsane), Rhodesia, Transvaal (Blouberg, Naboomspruit and the Vaal River) and Zululand.

One observed at Mkuzi G.R. (O 4) in December 1976 (J. C. Sinclair, G. H. Nicholls).

659 Layard's Titbabbler

Parisoma layardi

Western and interior Southern Africa. From the winter rainfall area of the south-western Cape, northwards in S.W.A. (to the Brandberg) and probably south-western Botswana, to O.F.S., Lesotho (highlands) and south-western Transvaal. Also recorded from Natal along the foothills of the Drakensberg Range.

One near Estcourt (E 10) in June 1974 (G. Symons). One at Cathedral Peak (B 9) in October 1979 (C. Coetser).

Grey Wagtail

Motacilla cinerea

Palaearctic, wintering to the south. In Southern Africa, a vagrant to Natal.

A male photographed at Mkuzi G.R. (N/O 4) in September 1974 (H. Valentine). A female photographed at Mtunzini (M 9) in June 1977 (P. G. H. and S. Frost). A female observed at Karkloof (G 11) in March 1979 (R. Rossiter).

697 Rock Pipit

Anthus crenatus

Southern Africa, from the Cape to south-eastern Transvaal. From the mountains of the south-western and southern Cape, northeast to O.F.S., Lesotho, Transkei, East Griqualand, western and upper Natal, south-eastern Transvaal and western Swaziland.

One near Ingwavuma (N 2) in June 1973 (S. J. M. Blaber). One at Oshoek, east of Wakkerstroom (G 3) in November 1976 (D. H. Day). One northwest of Himeville (C 12) in October 1979 (R. Cowgill). One at Wakkerstroom (F 3) and one east of Volksrust (D 3) in December 1979 (N. F. Robson).

698 Tree Pipit

Anthus trivialis

Western and central Palaearctic, wintering to the south in Africa and southern Asia. In Southern Africa, a visitor from October to April to the northern half of S.W.A., northern and eastern Botswana, Rhodesia, Transvaal and Moçambique; also recorded from Swaziland, Natal and Zululand.

One at Northern Districts S.W., Durban (I 13) in December 1979 (A. Cornell, D. Schultz, P. Scogings).

700 Short-tailed Pipit

Anthus brachyurus

Central and south-eastern Africa. In Southern Africa, coastal and midland districts of Natal and Zululand north to Transvaal (the plateau); also southern Moçambique (in the savanna north of Beira).

One west of Mooi River (F 10) in May 1977 (S. J. M. Blaber). Singletons at Fort Yolland (J 9) in April, June and December 1978 (H. Chittenden).

729 Chestnut-fronted Hemet Shrike

Prionops scopifrons

East and south-eastern Africa. In Southern Africa, north-eastern Zululand (vagrant), Moçambique and south-eastern Rhodesia (Haroni-Lusitu confluence).

Five at Nhlolo Pan, Pongola River (O 2) in May 1974 (S. J. M. Blaber).

730 White-crowned Shrike

Eurocephalus anguitimens

Southern Africa in the southeast, interior and west. From S.W.A. to the northern Cape, Botswana, Rhodesia, Transvaal, southern Moçambique and north-eastern Swaziland.

One near Mkuze village (N 4) in December 1972 (S. J. M. Blaber, A. T. Forbes). One near Jozini (N 3) in February 1975 (M. Sach).

733 European Starling

Sturnus vulgaris

Palaearctic, with introductions in various other parts of the world. In Southern Africa, common and widespread in the south-western and southern Cape, east to the eastern Cape and occasionally Transkei.

One observed at Durban (J 13) in August 1973 (J. C. Sinclair).
Sinclair).

769 Blue-throated Sunbird

Anthreptes reichenowi

East Africa, from coastal Kenya to Zululand. In Southern Africa, north-eastern Zululand, Moçambique and south-eastern Rhodesia (Nuanetsi and Hunyani Rivers).

One male at Ndumu G.R. (N/0 1) in September 1972 (G. Schutte, J. Stannard). Unconfirmed reports received from Mkuzi G.R. (N/0 4) and False Bay Park (0 5).

779 Buffalo Weaver

Bubalornis niger

North-eastern, eastern and southern Africa in the more open savannas. In Southern Africa, north-eastern Zululand, eastern Swaziland, Moçambique (Sul do Save), northern and western Transvaal, south-western and western Rhodesia, Botswana, the Caprivi Strip and central and northern S.W.A.

A small party near Jozini (N 3) in January 1975 (A. C. V. Clarkson).

780 White-browed Sparrow-weaver

Plocepasser mahali

North-eastern, eastern and southern Africa in the drier areas. In Southern Africa, S.W.A., the northern Cape, Botswana, western and northern Rhodesia, Moçambique (Tete district), western Transvaal and O.F.S.; also recorded from Natal.

Up to six birds resident since 1977 south-east of Estcourt (E 10) (G. Reynolds, L. Nutting).

821 Cut-throat Finch

Amadina fasciata

The Afrotropical Region. In Southern Africa, north-western O.F.S., Transvaal, eastern Swaziland, Moçambique (southern Sul do Save) and Rhodesia, west to north-eastern and northern Botswana, the Caprivi Strip and northern S.W.A.

A male found dead near Cedarville (B 15) in June 1972 (K. M. Collins). A pair near Winterton (D 9) in June 1978 (R. and S. Hewett). A male at Underberg (C 13) in March 1979 (E. F. Steineger). A small flock including two males near Jozini (N 3) in December 1979 (C. Hilton-Taylor).

835 Jameson's Firefinch

Lagonosticta rhodopareia

The Afrotropical Region. In Southern Africa, eastern Zululand, eastern Swaziland, Moçambique, Transvaal, Rhodesia, eastern and northern Botswana, the Caprivi Strip and northern S.W.A.

Recorded throughout the year at Mkuzi G.R. (N/O 4). Pairs observed at Jozini (N 3) in December 1978 (D. P. Cyrus, G. Nichols), March 1979 (D. P. Cyrus), and ...y and June 1979 (D. P. Cyrus, N. F. Robson). Col...ted near Weenen (F 9) in September 1979 ...Mendelsohn).

840 Violet-eared Waxbill

Uraeginthus granatinus

The southern half of Africa, mainly in the drier parts. In Southern Africa, the northern Cape, S.W.A., the Caprivi Strip, Botswana, Rhodesia, Transvaal and O.F.S.; taken in Moçambique (Maputo) and observed in Natal.

One near Tugela Ferry (G 8) in February 1979 (L. W. Lawson).

850 Purple Widowfinch

Vidua purpurascens

Central and eastern Africa, south to Zululand in the east. In Southern Africa, from northern Zululand and eastern Swaziland to southern Moçambique, Transvaal (bush-veld regions), north-eastern Botswana (Francistown) and Rhodesia.

One near Pongola (L 3) in January 1977 (D. H. Day). Observed near Jozini (N 3) in March, April and December 1977 (M. Sach).

860X Lemon-breasted Canary

Serinus citrinipectus

The littoral of south-eastern Africa; northern limits unknown. In Southern Africa, north-eastern Zululand, the plain of Moçambique (extending to southern Malawi), south-eastern Rhodesia and possibly eastern Transvaal (lowveld).

First recorded south of Hluhluwe village (O 6) when several small flocks were observed in January, February and March 1971 (J. Feely). Observed in the same area in 1978 in February (S. J. M. Blaber), April (D. P. Cyrus) and December (J. Feely). A small flock west of Mtubatuba (N 7) in May 1971 (R. Wiseman). Near Pelindaba (P 2) in 1979 *c.*30 in May (D. P. Cyrus, N. F. Robson), *c.*50 in June, *c.*30 in July, 100+ in October and 160+ in November (D. P. Cyrus). Recorded from Mkuzi G.R. (N/O 4), Hluhluwe G.R. (N 6) and Umfolozi G.R. (M 7) in October 1979 (G. Bligh).

This species was first recorded in the three major Zululand Game Reserves in October 1979, which suggests a widespread movement in north-eastern Zululand during the latter half of that year. The flock at Pelindaba (P 2) had dispersed by January 1980, when a number of small flocks were noted in squares O 4, O 5 and N 5, on the road between Lower Mkuze and Hluhluwe village (S. J. M. and T. G. Blaber, D. P. Cyrus). On the other hand, this species, which was described as recently as 1960, may have been overlooked in Zululand.

865 White-throated Seedeater

Serinus albogularis

Southern Africa, in the southern interior and dry west. From S.W.A. to the Cape, O.F.S., Lesotho (lowlands) and southern Transvaal; also recorded from Natal.

Observed at Highmoor Forest Reserve (D 11) from April to November 1978 (A. L. and G. S. Ruddock) and at Cathedral Peak (B 9) in June 1978, and in May, August, October and November 1979 (C. Coetser).

866 Yellow Canary

Serinus flaviventris

Southern Africa, mainly in the interior and west. From S.W.A. and the southern half of Botswana to the Cape, O.F.S., Lesotho and southern and western Transvaal, ranging marginally to East Griqualand and western Natal. Also recorded from Rhodesia (Matetsi).

One at Cathedral Peak (B 9) in July 1978 (C. Coetser).

871 Lark-like Bunting

Emberiza impetuani

Southern Africa in the interior and dry west. From S.W.A. to the western and interior Cape, Botswana, western Rhodesia, Transvaal and O.F.S. Also recorded from Lesotho and the Transkei/Natal border.

One southwest of Himeville (C 12) in October 1979 (R. Cowgill).

Checklist

of

Birds Recorded in Natal

.... 2 Jackass Penguin
.... Macaroni Penguin
.... 3 Rockhopper Penguin
.... 4 Great Crested Grebe
.... 5 Black-necked Grebe
.... 6 Little Grebe
.... 7 Wandering Albatross
.... 8 Black-browed Albatross
.... 9 Grey-headed Albatross
.... 10 Yellow-nosed Albatross
.... 11 Shy Albatross
.... 12 Sooty Albatross
.... 13 Southern Giant Petrel
.... Northern Giant Petrel
.... 14 Pintado Petrel
.... 15 Silver-grey Fulmar
.... 16 Great-winged Petrel
.... 18 Schlegel's Petrel
.... Kerguelen Petrel
.... 19 Soft-plumaged Petrel
.... 21X Medium-billed Prion
.... 22 Dove Prion
.... 22X Fairy Prion
.... 22Y Slender-billed Prion
.... 23 White-chinned Petrel
.... 26 Cory's Shearwater
.... 26X Flesh-footed Shearwater
.... 26Z Wedge-tailed Shearwater
.... 27 Little Shearwater
.... 28 Baillon's Shearwater
.... 29 Sooty Shearwater
.... 30 Storm Petrel
.... 33 Wilson's Storm Petrel
.... 36 Black-bellied Storm Petrel
.... 37 White-bellied Storm Petrel
.... 39 Red-tailed Tropicbird
.... 40 White-tailed Tropicbird
.... 41 Pink-backed Pelican
.... 42 White Pelican
.... 44 Cape Gannet
.... 47 White-breasted Cormorant
.... 48 Cape Cormorant
.... 50 Reed Cormorant
.... 52 Darter
.... 53 Greater Frigatebird
.... 54 Grey Heron
.... 55 Black-headed Heron
.... 56 Goliath Heron
.... 57 Purple Heron
.... 58 Great White Egret
.... 59 Little Egret
.... 60 Yellow-billed Egret
.... 61 Cattle Egret
.... 62 Squacco Heron
.... 63 Green-backed Heron
.... 64 Black Egret
.... 65 Rufous-bellied Heron
.... 66 Dwarf Bittern
.... 67 Little Bittern
.... 69 Night Heron
.... 70 White-backed Night Heron
.... 71 Bittern
.... 72 Hamerkop
.... 73 Marabou Stork
.... 74 Openbill Stork
.... 75 Saddlebill Stork
.... 76 Yellow-billed Stork
.... 77 Woolly-necked Stork
.... 78 White-bellied Stork
.... 79 Black Stork
.... 80 White Stork
.... 81 Sacred Ibis
.... 82 Bald Ibis
.... 83 Glossy Ibis
.... 84 Hadeda
.... 85 African Spoonbill
.... 86 Greater Flamingo
.... 87 Lesser Flamingo
.... 88 Spur-winged Goose
.... 89 Egyptian Goose
.... 90 South African Shelduck
.... 91 Knob-billed Duck

.... 92 Pygmy Goose
.... 94 Cape Shoveler
.... 95 Black Duck
.... 96 Yellow-billed Duck
.... 97 Red-billed Teal
.... 98 Cape Teal
.... 99 Hottentot Teal
.... 100 White-faced Whistling Duck
.... 101 Fulvous Whistling Duck
.... 102 Red-eyed Pochard
.... 103 Maccoa Duck
.... 104 White-backed Duck
.... 105 Secretarybird
.... 106 Cape Vulture
.... 107 White-backed Vulture
.... 108 Lappet-faced Vulture
.... 109 White-headed Vulture
.... 110 Hooded Vulture
.... 111 Egyptian Vulture
.... 112 Palm-nut Vulture
.... 113 Peregrine Falcon
.... 114 Lanner Falcon
.... 115 Hobby
.... 116 African Hobby
.... 116Y Sooty Falcon
.... 117 Red-necked Falcon
.... 119 Eastern Red-footed Kestrel
.... 120 Western Red-footed Kestrel
.... 121 Dickinson's Kestrel
.... 122 Greater Kestrel
.... 123 Rock Kestrel
.... 125 Lesser Kestrel
.... 127 Cuckoo Falcon
.... 128 Black Kite
.... 129 Yellow-billed Kite
.... 130 Black-shouldered Kite
.... 131 Bat Hawk
.... 132 Honey Buzzard
.... 133 Black Eagle
.... 134 Tawny Eagle
.... 135 Steppe Eagle
.... 136 Lesser Spotted Eagle
.... 137 Wahlberg's Eagle
.... 138 Long-crested Eagle
.... 139 Booted Eagle
.... 140 Ayres' Hawk Eagle
.... 141 African Hawk Eagle
.... 142 Martial Eagle
.... 143 Crowned Eagle
.... 144 Lizard Buzzard
.... 145 Brown Snake-eagle
.... 146 Black-breasted Snake-eagle
.... 147 Southern Banded Snake-eagle
.... 149 Fish Eagle
.... 150 Bearded Vulture
.... 151 Bateleur
.... 152 Jackal Buzzard
.... 154 Steppe Buzzard
.... 155 Mountain Buzzard
.... Long-legged Buzzard
.... 156 Red-breasted Sparrowhawk
.... 158 Little Sparrowhawk
.... 159 Black Sparrowhawk
.... 160 African Goshawk
.... 161 Little Banded Goshawk
.... 162 Gabar Goshawk
.... 163 Dark Chanting Goshawk
.... 165 Chanting Goshawk
.... 167 African Marsh Harrier
.... 168 Pallid Harrier
.... 169 Black Harrier
.... 170 Montagu's Harrier
.... 171 Gymnogene
.... 172 Osprey
.... 173 Coqui Francolin
.... 174 Crested Francolin
.... 176 Greywing Francolin
.... 177 Shelley's Francolin
.... 178 Redwing Francolin
.... 183 Natal Francolin
.... 185 Swainson's Francolin
.... 188 Red-necked Francolin
.... 189 Common Quail
.... 190 Harlequin Quail
.... 191 Blue Quail
.... 192 Crowned Guineafowl
.... 193 Crested Guineafowl
.... 194 Hottentot Buttonquail
.... 196 Kurrichane Buttonquail
.... 197 Water Rail
.... 198 Corncrake
.... 199 African Crake
.... 200 Striped Crake
.... 201 Spotted Crake
.... 202 Baillon's Crake
.... 203 Black Crake
.... 204 White-winged Flufftail
.... 205 Red-chested Flufftail
.... 206 Buff-spotted Flufftail
.... 207 Striped Flufftail
.... 208 Purple Gallinule
.... 209 Lesser Gallinule
.... 210 Moorhen
.... 211 Lesser Moorhen
.... 212 Red-knobbed Coot
.... 213 African Finfoot
.... 214 Crowned Crane
.... 215 Wattled Crane
.... 216 Blue Crane
.... 217 Kori Bustard
.... 218 Ludwig's Bustard
.... 219 Stanley Bustard
.... 222 White-bellied Korhaan
.... 223 Blue Korhaan
.... 224 Red-crested Korhaan
.... 227 Black-bellied Korhaan
.... 228 African Jacana
.... 229 Lesser Jacana
.... 230 Painted Snipe
.... 231 Black Oystercatcher

.............231X European Oystercatcher
.............232 Turnstone
.............233 Ringed Plover
.............234 Mongolian Sandplover
.............235 White-fronted Sandplover
.............236 Chestnut-banded Sandplover
.............237 Kittlitz's Sandplover
.............238 Three-banded Sandplover
.............239 Great Sandplover
.............240 Caspian Plover
.............240X Eastern Golden Plover
.............241 Grey Plover
.............242 Crowned Plover
.............243 Black-winged Plover
.............244 Lesser Black-winged Plover
.............245 Blacksmith Plover
.............247 Wattled Plover
.............248 Long-toed Plover
.............249 Great Snipe
.............250 Ethiopian Snipe
.............251 Curlew Sandpiper
.............251X Pectoral Sandpiper
.... Buff-breasted Sandpiper
.............251Z Broad-billed Sandpiper
.............253 Little Stint
.............253X Red-necked Stint
.............254 Knot
.............255 Sanderling
.............256 Ruff
.............257 Terek Sandpiper
.............258 Common Sandpiper
.............259 Green Sandpiper
.............261 Redshank
.............262 Marsh Sandpiper
.............263 Greenshank
.............264 Wood Sandpiper
.............265 Black-tailed Godwit
.............266 Bar-tailed Godwit
.............267 Curlew
.............268 Whimbrel
.............269 Avocet
.............270 Black-winged Stilt
.............271 Grey Phalarope
.............272 Red-necked Phalarope
.............273 Crab Plover
.............274 Water Dikkop
.............275 Cape Dikkop
.............276 Burchell's Courser
.............277 Temminck's Courser
.............280 Bronze-wing Courser
.............281 Red-winged Pratincole
.............282 Black-winged Pratincole
.............284 Arctic Skua
.............285 Pomarine Skua
.............286 Antarctic Skua
.............287 Southern Black-backed Gull
.............287X Lesser Black-backed Gull
.............288 Grey-headed Gull
.............289 Hartlaub's Gull
.............289X Sabine's Gull
.... Franklin's Gull
.............290 Caspian Tern
.............290X Gull-billed Tern
.............291 Common Tern
.... Black-naped Tern
.... White-cheeked Tern
.............292 Antarctic Tern
.............293 Roseate Tern
.............294 Arctic Tern
.............295 Sooty Tern
.............295X Bridled Tern
.............296 Sandwich Tern
.............297 Lesser Crested Tern
.............298 Swift Tern
.............299 Little Tern
.............301 White-capped Noddy
.............303 Common Noddy
.............304 White-winged Black Tern
.............305 Whiskered Tern
.............305X Black Tern
.............306 African Skimmer
.............307 Namaqua Sandgrouse
.............311 Rock Pigeon
.............312 Rameron Pigeon
.............313 Bronze-naped Pigeon
.............314 Red-eyed Turtle Dove
.............316 Cape Turtle Dove
.............317 Laughing Dove
.............318 Namaqua Dove
.............319 Tambourine Dove
.............321 Emerald-spotted Wood Dove
.............322 Cinnamon Dove
.............323 Green Pigeon
.............326 Brown-necked Parrot
.............328 Brown-headed Parrot
.............329X Ring-necked Parakeet
.............336 Knysna Loerie
.............337 Purple-crested Loerie
.............339 Grey Loerie
.............340 Cuckoo
.............342 Lesser Cuckoo
.............343 Red-chested Cuckoo
.............344 Black Cuckoo
.............346 Great Spotted Cuckoo
.............347 Striped Cuckoo
.............348 Jacobin Cuckoo
.............350 Emerald Cuckoo
.............351 Klaas's Cuckoo
.............352 Diederik Cuckoo
.............353 Black Coucal
.............356 White-browed Coucal
.............358 Green Coucal
.............359 Barn Owl
.............360 Grass Owl
.............361 Marsh Owl
.............362 Wood Owl
.............363 African Scops Owl
.............364 White-faced Owl
.............365 Pearl-spotted Owl
.............366 Barred Owl

............367 Cape Eagle Owl
............368 Spotted Eagle Owl
............369 Giant Eagle Owl
............370 Fishing Owl
............371 European Nightjar
............373 Fiery-necked Nightjar
............374 Freckled Nightjar
............375 Natal Nightjar
............376 Moçambique Nightjar
............377 Pennant-wing Nightjar
............378 European Swift
............380 Black Swift
............383 White-rumped Swift
............384 Horus Swift
............385 Little Swift
............386 Alpine Swift
............387 Palm Swift
............390 Speckled Mousebird
............392 Red-faced Mousebird
............393 Narina Trogon
............394 Pied Kingfisher
............395 Giant Kingfisher
............396 Half-collared Kingfisher
............397 Malachite Kingfisher
............398 Pygmy Kingfisher
............399 Woodland Kingfisher
............400 Mangrove Kingfisher
............401 Grey-hooded Kingfisher
............402 Brown-hooded Kingfisher
............403 Striped Kingfisher
............404 European Bee-eater
............406 Blue-cheeked Bee-eater
............407 Carmine Bee-eater
............409 White-fronted Bee-eater
............410 Little Bee-eater
............411 Swallow-tailed Bee-eater
............412 European Roller
............413 Lilac-breasted Roller
............414 Racquet-tailed Roller
............415 Purple Roller
............416 Broad-billed Roller
............418 Hoopoe
............419 Red-billed Hoopoe
............421 Scimitar-billed Hoopoe
............422 Trumpeter Hornbill
............424 Grey Hornbill
............425 Red-billed Hornbill
............426 Yellow-billed Hornbill
............427 Crowned Hornbill
............430 Ground Hornbill
............431 Black-collared Barbet
............432 Pied Barbet
............433 White-eared Barbet
............435 Green Barbet
............436 Red-fronted Tinker Barbet
............438 Golden-rumped Tinker Barbet
............439 Crested Barbet
............440 Greater Honeyguide
............441 Scaly-throated Honeyguide
............442 Lesser Honeyguide
............443 Sharp-billed Honeyguide
............445 Ground Woodpecker
............446 Bennett's Woodpecker
............447 Golden-tailed Woodpecker
............448 Knysna Woodpecker
............450 Cardinal Woodpecker
............451 Bearded Woodpecker
............452 Olive Woodpecker
............453 Red-throated Wryneck
............454 Broadbill
............455 Angola Pitta
............456 Singing Bush Lark
............458 Rufous-naped Lark
............460 Sabota Lark
............463 Thick-billed Lark
............464 Dusky Lark
............468 Flappet Lark
............473 Rudd's Lark
............474 Spike-heeled Lark
............475 Long-billed Lark
............484 Chestnut-backed Finchlark
............488 Red-capped Lark
............490 Pink-billed Lark
............493 European Swallow
............495 White-throated Swallow
............496 Wire-tailed Swallow
............497 Blue Swallow
............498 Pearl-breasted Swallow
............499 Grey-rumped Swallow
............501 Red-breasted Swallow
............502 Greater Striped Swallow
............503 Lesser Striped Swallow
............504 Cliff Swallow
............506 Rock Martin
............507 House Martin
............508 European Sand Martin
............509 African Sand Martin
............510 Banded Sand Martin
............511 Black Saw-wing Swallow
............513 Black Cuckooshrike
............516 Grey Cuckooshrike
............517 Fork-tailed Drongo
............518 Square-tailed Drongo
............519 European Golden Oriole
............521 Black-headed Oriole
............522 Pied Crow
............523 Black Crow
............523X House Crow
............524 White-necked Raven
............525 Grey Tit
............527 Southern Black Tit
............530 Grey Penduline Tit
............533 Arrow-marked Babbler
............540B Rockjumper
............542 Bush Blackcap
............544 Red-eyed Bulbul
............545 Black-eyed Bulbul
............546 Terrestrial Bulbul
............547 Yellow-streaked Bulbul
............550 Yellow-bellied Bulbul

. . . . .551 Sombre Bulbul
. . . . .552 Kurrichane Thrush
. . . . .553 Olive Thrush
. . . . .556 Orange Thrush
. . . . .557 Groundscraper Thrush
. . . . .558 Natal Thrush
. . . . .559 Cape Rock Thrush
. . . . .560 Sentinel Rock Thrush
. . . . .564 Mountain Chat
. . . . .568 Capped Wheatear
. . . . .569 Buff-streaked Chat
. . . . .570 Familiar Chat
. . . . .572 Sickle-wing Chat
. . . . .573 Mocking Chat
. . . . .575 Ant-eating Chat
. . . . .576 Stonechat
. . . . .577 Whinchat
. . . . .578 Chorister Robin
. . . . .579 Natal Robin
. . . . .580 Heuglin's Robin
. . . . .581 Cape Robin
. . . . .582 White-throated Robin
. . . . .584 Brown Robin
. . . . .585 Bearded Robin
. . . . .588 White-browed Scrub Robin
. . . . .589 Starred Robin
. . . . .594 Whitethroat
. . . . .595 Garden Warbler
. . . . .596 Icterine Warbler
. . . . .597 Olive-tree Warbler
. . . . .599 Willow Warbler
. . . . .600 Yellow-bellied Eremomela
. . . . .601 Burnt-necked Eremomela
. . . . .602 Greencap Eremomela
. . . . .603 European Great Reed Warbler
. . . . .604 Cape Reed Warbler
. . . . .604X European Reed Warbler
. . . . .606 African Marsh Warbler
. . . . .607 European Marsh Warbler
. . . . .608 European Sedge Warbler
. . . . .609 African Sedge Warbler
. . . . .610 Scrub Warbler
. . . . .611 Knysna Scrub Warbler
. . . . .615 Stierling's Barred Warbler
. . . . .616 Fan-tailed Warbler
. . . . .618 Grassbird
. . . . .621 Crombec
. . . . .622 Bar-throated Apalis
. . . . .624 Rudd's Apalis
. . . . .625 Yellow-breasted Apalis
. . . . .627 Bleating Bush Warbler
. . . . .629 Fantail Cisticola
. . . . .630 Desert Cisticola
. . . . .631 Cloud Cisticola
. . . . .634 Ayres' Cloud Cisticola
. . . . .635 Pale-crowned Cloud Cisticola
. . . . .637 Neddicky
. . . . .639 Wailing Cisticola
. . . . .642 Rattling Cisticola
. . . . .644 Red-faced Cisticola
. . . . .645 Black-backed Cisticola
. . . . .646 Le Vaillant's Cisticola
. . . . .647 Croaking Cisticola
. . . . .648 Lazy Cisticola
. . . . .649 Tawny-flanked Prinia
. . . . .651 Karoo Prinia
. . . . .654 Spotted Flycatcher
. . . . .655 Dusky Flycatcher
. . . . .656 Blue-grey Flycatcher
. . . . .657 Grey Titbabbler
. . . . .658 Titbabbler
. . . . .659 Layard's Titbabbler
. . . . .662 Mouse-coloured Flycatcher
. . . . .664 Black Flycatcher
. . . . .665 Fiscal Flycatcher
. . . . .666 Yellow Warbler
. . . . .671 Yellow-throated Warbler
. . . . .672 Cape Batis
. . . . .673 Chin Spot Batis
. . . . .676 Woodwards' Batis
. . . . .677 Wattle-eyed Flycatcher
. . . . .678 Fairy Flycatcher
. . . . .680 Blue-mantled Flycatcher
. . . . .682 Paradise Flycatcher
. . . . .685 African Pied Wagtail
. . . . .686 Cape Wagtail
. . . . .688 Long-tailed Wagtail
. . . . .689 Yellow Wagtail
. . . . Grey Wagtail
. . . . .692 Richard's Pipit
. . . . .693 Nicholson's Pipit
. . . . .694 Plain-backed Pipit
. . . . .695 Buffy Pipit
. . . . .696 Striped Pipit
. . . . .697 Rock Pipit
. . . . .698 Tree Pipit
. . . . .699 Bushveld Tree Pipit
. . . . .700 Short-tailed Pipit
. . . . .701 Yellow-breasted Pipit
. . . . .703 Orange-throated Longclaw
. . . . .704 Yellow-throated Longclaw
. . . . .705 Pink-throated Longclaw
. . . . .706 Lesser Grey Shrike
. . . . .707 Fiscal Shrike
. . . . .708 Red-backed Shrike
. . . . .709 Boubou Shrike
. . . . .712 Puffback Shrike
. . . . .713 Tchagra
. . . . .714 Three-streaked Tchagra
. . . . .715 Black-crowned Tchagra
. . . . .717 Olive Bush Shrike
. . . . .719 Orange-breasted Bush Shrike
. . . . .721 Gorgeous Bush Shrike
. . . . .722 Bokmakierie
. . . . .723 Grey-headed Bush Shrike
. . . . .724 Long-tailed Shrike
. . . . .725 Yellow-spotted Nicator
. . . . .727 White Helmet Shrike
. . . . .728 Red-billed Helmet Shrike
. . . . .729 Chestnut-fronted Helmet Shrike

. . . . .730 White-crowned Shrike
. . . . .731 Brubru Shrike
. . . . .733 European Starling
. . . . .734 Indian Myna
. . . . .735 Wattled Starling
. . . . .736 Plum-coloured Starling
. . . . .737 Cape Glossy Starling
. . . . .738 Blue-eared Glossy Starling
. . . . .740 Black-bellied Glossy Starling
. . . . .743 Burchell's Glossy Starling
. . . . .745 Red-winged Starling
. . . . .746 Pied Starling
. . . . .747 Yellow-billed Oxpecker
. . . . .748 Red-billed Oxpecker
. . . . .750 Gurney's Sugarbird
. . . . .751 Malachite Sunbird
. . . . .755 Marico Sunbird
. . . . .756 Purple-banded Sunbird
. . . . .758 Greater Double-collared Sunbird
. . . . .760 Lesser Double-collared Sunbird
. . . . .761 Neergaard's Sunbird
. . . . .763 White-bellied Sunbird
. . . . .765 Grey Sunbird
. . . . .766 Olive Sunbird
. . . . .769 Blue-throated Sunbird
. . . . .771 Collared Sunbird
. . . . .772 Black Sunbird
. . . . .774 Scarlet-chested Sunbird
. . . . .775 Cape White-eye
. . . . .777 Yellow White-eye
. . . . .779 Buffalo Weaver
. . . . .780 White-browed Sparrow-weaver
. . . . .784 House Sparrow
. . . . .786 Cape Sparrow
. . . . .787 Grey-headed Sparrow
. . . . .788 Yellow-throated Sparrow
. . . . .790 Forest Weaver
. . . . .791 Spectacled Weaver
. . . . .792 Lesser Masked Weaver
. . . . .793 Red-headed Weaver
. . . . .797 Spotted-backed Weaver
. . . . .799 Cape Weaver
. . . . .800 Yellow Weaver
. . . . .801 Golden Weaver
. . . . .802 Brown-throated Golden Weaver
. . . . .803 Masked Weaver
. . . . .804 Thick-billed Weaver
. . . . .805 Red-billed Quelea
. . . . .806 Red-headed Quelea
. . . . .808 Red Bishop
. . . . .810 Cape Widow
. . . . .812 Golden Bishop
. . . . .813 Red-collared Widow
. . . . .814 White-winged Widow
. . . . .816 Red-shouldered Widow
. . . . .818 Long-tailed Widow
. . . . .820 Red-headed Finch
. . . . .821 Cut-throat Finch
. . . . .822 Pied Mannikin
. . . . .823 Bronze Mannikin
. . . . .824 Red-backed Mannikin
. . . . .825 Swee Waxbill
. . . . .827 Green Twinspot
. . . . .830 Melba Finch
. . . . .831 Pink-throated Twinspot
. . . . .833 Blue-billed Firefinch
. . . . .835 Jameson's Firefinch
. . . . .837 Red-billed Firefinch
. . . . .838 Orange-breasted Waxbill
. . . . .839 Blue Waxbill
. . . . .840 Violet-eared Waxbill
. . . . .842 Grey Waxbill
. . . . .843 Common Waxbill
. . . . .844 Quail Finch
. . . . .846 Pin-tailed Whydah
. . . . .849 Black Widowfinch
. . . . .850 Purple Widowfinch
. . . . .851 Steel-blue Widowfinch
. . . . .852 Paradise Whydah
. . . . .854 Cuckoo Finch
. . . . .856 Drakensberg Siskin
. . . . .857 Cape Canary
. . . . .858 Forest Canary
. . . . .859 Yellow-eye Canary
. . . . .860 Black-throated Canary
. . . . .860X Lemon-breasted Canary
. . . . .863 Bully Seedeater
. . . . .865 White-throated Seedeater
. . . . .866 Yellow Canary
. . . . .867 Streaky-headed Seedeater
. . . . .871 Lark-like Bunting
. . . . .872 Rock Bunting
. . . . .873 Cape Bunting
. . . . .874 Golden-breasted Bunting.

Index of Scientific Names

Index of English Names